NanoScience and Technology

The series NanoScience and Technology is focused on the fascinating nano-world, mesoscopic physics, analysis with atomic resolution, nano and quantum-effect devices, nanomechanics and atomic-scale processes. All the basic aspects and technology-oriented developments in this emerging discipline are covered by comprehensive and timely books. The series constitutes a survey of the relevant special topics, which are presented by leading experts in the field. These books will appeal to researchers, engineers, and advanced students.

Peter Hess

Bonding, Structure, and Performance of Two-Dimensional Materials

An Introduction Based on the Periodic Table

Springer

Peter Hess
Institute of Physical Chemistry
University of Heidelberg
Heidelberg, Germany

ISSN 1434-4904 ISSN 2197-7127 (electronic)
NanoScience and Technology
ISBN 978-3-031-74160-9 ISBN 978-3-031-74161-6 (eBook)
https://doi.org/10.1007/978-3-031-74161-6

This Springer imprint is published by the registered company Springer Nature Switzerland AG
The registered company address is: Gewerbestrasse 11, 6330 Cham, Switzerland

Preface

The book gives a basic introduction into the two-dimensional world of materials and provides an extensive source of bonding types, structural data, diverse physicochemical behavior, and strategies of synthesis. To understand the already studied two-dimensional structures, the nature of chemical bonding is derived from the electronic configuration defining the number and type of valence electrons, which are responsible for covalent bonding and hybridization. The fact that the 'aufbau' principle of the periodic table controls chemical bonding, and thus structure formation and performance of elemental and heteroatomic monolayers and their few-layer sheets, is the basis of the presentation. This 'ansatz' leads to a deeper understanding of questions such as the stability of 2D structures and their physicochemical properties. Dynamic and thermodynamic stability belong to the most critical issues, which can be predicted by theory, as proved by the discovery of several monolayers.

The idea to draft this book originates from a review article that discussed the role played by the main group elements of the periodic table in the formation of two-dimensional materials. This treatment is extended here and includes now binary compounds of the most important transition metals, which also generate two-dimensional structures with fascinating properties. The exploding contributions, specially made by theory, can no longer be covered in a single monography. Therefore, the book includes preferentially recent work, selected with the aim to deliver an overview of the broad range of elements and structures already discovered all over the periodic table to provide an impression of the diversity of the already existing 2D materials.

As expected, the insight into the optimal bonding conditions and the existing restrictions given by the periodic table, helps tremendously to further develop the field of nanoscience and nanotechnology. In reality, the considerable number of elements and the enormous number of possible structural combinations up to now produced only a limited number of promising two-dimensional materials with sufficiently stable structure, favorable physicochemical properties, easy synthesis of monolayers with suitable size and quality, and thus a chance to be used in devices. The already discovered promising candidates involve both light and heavy elements, highly symmetric and low symmetry structures, and cover the entire range of existing electronic configurations and bonding types. Importantly, theoretical predictions are reliable and identified hitherto unknown monolayers based on chemical insight and thus made relevant contributions to the extension of 2D materials.

Each chapter has an own abstract, explains the abbreviations used therein, and therefore can be studied separately. It is recommended, however, to read the introduction first. The concept and selection of the individual chapters was inspired by the periodic table that allowed, grouping together 2D materials with related chemical and physical behavior. This order principle turned out to be remarkably effective to categorize the enormous amount of already gathered information. The last chapter summarizes the relevant information in terms of diverse types of 2D materials, established novel and important properties, and emerging devices. All drawings including the figures adapted from the literature were sketched with the goal, to achieve a consistent nomenclature and style, focusing on the fundamental facts discussed in the text. Primarily, they concentrate on the topics of bonding and the variety of identified structures.

Of course, this first publication contains errors and mistakes, as implied by the observation that it is always possible to find possibilities of improvement of the manuscript even after repeated corrections of a text. Interestingly, the fact that there exists also no perfect 2D crystal in nature, points to the same problem with entropy, but of course, is no excuse for such flaws. I would be thankful to all readers informing me about discovered deficiencies, to avoid them in a following edition of the book.

I am extremely thankful to my wife Bärbel-Jutta. She made this project possible by taking care of nearly everything else, allowing me to fully concentrate on the book project. When starting this enterprise, I was not aware of the enormous consequences, e.g., the time actually needed to complete such an endeavor. Fortunately, owing to her enormous patience, the unforeseen immense extension of the original schedule created only the impression that the book will never be finished.

Now I hope that this book helps many newcomers to the field to find the information needed to get acquainted with the fascinating fields of nanoscience and nanotechnology, and researchers to extend their view by considering the periodic table as a fundamental source of inspiration to new directions.

Heidelberg, Germany Peter Hess, Professor

Contents

Abbreviations

ACSTEM	Aberration-corrected scanning transmission electron microscopy
ADF-STEM	Annular dark-field scanning transmission electron microscopy
AES	Auger electron spectroscopy
AFM	Atomic force microscopy
AHC	Anomalous Hall conductivity
AHE	Anomalous Hall effect
AIMD	Ab initio molecular dynamics
AIREBO	Adaptive intermolecular reactive empirical bond order
ALD	Atomic layer deposition
ALE	Atomic layer etching
APCVD	Ambient pressure chemical vapor deposition
ARPES	Angle-resolved photoemission spectroscopy
BCS	Bardeen–Cooper–Schrieffer
BEC	Bose–Einstein condensation
BLG	Bilayer graphene
BOP	Bond order potential
BTE	Boltzmann transport equation
CA	Ceperley–Alder
C-AFM	Conductive-atomic force microscopy
CALYSO	Crystal structure analysis by particle swarm optimization
CB	Conduction band
CBM	Conduction band minimum
CBO	Conduction band offset
CDW	Charge density wave
CENT	Charge equilibrated neural network technique
CER	Chlorine evolution reaction
CET	Capacitance equivalent oxide thickness
CL	Cathodoluminescence
CMOS	Complementary metal-oxide semiconductor
CO_2RR	Carbon dioxide reduction reaction
CSIF	Critical stress intensity factor

CVD	Chemical vapor deposition
CVT	Chemical vapor transport
DDI	Dipole–dipole interaction
DFPT	Density functional perturbation theory
DFT	Density functional theory
DF-TEM	Dark-field transmission electron microscopy
DMC	Diffusion quantum Monte Carlo
DMFT	Dynamical mean-field theory
DOS	Density of states
DS	David star
DSM	Dirac semimetal
EBL	Electron-beam lithography
EDAX	Energy-dispersive X-ray analysis
EDLT	Electric-double-layer transistor
EDS	Energy-dispersive X-ray spectroscopy
EDX	Energy-dispersive X-ray spectroscopy
EELS	Electron energy loss spectroscopy
EES	Electrochemical energy storage
EI	Exciton insulator
ELF	Electron localization function
EOS	Equation of state
EPC	Electron–phonon coupling
EPI	Electron–phonon interactions
EPMA	Electron probe microanalyzer
EPR	Electron paramagnetic resonance
ESS	Evolutionary structure search
FE	Finite elements
FERE	Fitted elemental phase reference energies
FET	Field-effect transistor
FP-LAPW	Full-potential linearized augmented plane wave
FRUP	Fully relativistic pseudopotential
FUV	Far ultraviolet
GGA	Generalized gradient approximation
GIWAXS	Grazing-incidence wide-angle X-ray scattering
GIXRD	Grazing incidence synchrotron X-ray diffraction
HAADF-STM	High angle annular dark-field scanning transmission electron microscopy
HEDM	High energy density material
HER	Hydrogen evolution reaction
HET	Heterogeneous electron transfer
HOPG	Highly oriented pyrolytic graphite
HRSTS	High-resolution scanning tunneling spectroscopy
HRTEM	High-resolution transmission electron microscopy
HSE	Heyd–Scuseria–Ernzerhof
IC	Integrated circuit

IQE	Internal quantum efficiency
ISB	Inversion-symmetry breaking
ITO	Indium tin oxide
KPFM	Kevin probe force microscopy
LAMMPS	Large-scale atomic/molecular massively parallel simulator
LCAO	Linear combination of atomic orbitals
LDA	Local density approximation
LED	Light-emitting diode
LEED	Low-energy electron diffraction
LHB	Lower Hubbard band
LPCVD	Low-pressure chemical vapor deposition
LPE	Liquid-phase exfoliation
LT-STM	Low-temperature scanning tunneling microscopy
LUMO	Lowest unoccupied molecular orbital
MBE	Molecular beam epitaxy
MD	Molecular dynamics
MFM	Magnetic force microscopy
MGGA	Meta generalized gradient approximation
MHM	Minima hopping method
MIT	Metal-insulator transition
ML	Monolayer
MM	Molecular mechanics
MOCVD	Metal-organic chemical vapor deposition
MPL	Machine-learned potential
NC	Nanocrystalline
NCCDW	Nearly commensurate CDW
NEGF	Nonequilibrium Green´s function
NEMD	Nonequilibrium molecular dynamics
NEP	Neuro-evolution potential
NIR	Near infrared
NMR	Nuclear magnetic resonance
NPR	Negative Poisson ratio
NRR	Nitrogen reduction reaction
NTMDC	Noble-transition-metal dichalcogenide
OER	Oxygen evolution reaction
ONCV	Optimized norm-conserving Vanderbilt
OPE	Oxygen plasma exposure
ORR	Oxygen reduction reaction
OSEP	Orbital selective external potential
PAW	Projector augmented wave
PBE	Perdew–Burke–Ernzerhof
PCE	Power conversion efficiency
PDOS	Partial density of states
PE-ALD	Plasma-enhanced atomic layer deposition
PE-CVD	Plasma-enhanced chemical vapor deposition

PES	Photoelectron spectroscopy
PES	Potential energy surface
PET	Polyethylene terephthalate
PIL	Polymer ionic liquid
PL	Photoluminescence
PLD	Periodic lattice distortion
PLD	Pulsed-laser deposition
PS	Polystyrene sphere
PSO	Particle swarm optimization
PSOA	Particle swarm optimization algorithm
PVD	Physical vapor deposition
QAH	Quantum anomalous Hall
QFM	Quantized fracture mechanics
QP	Quasi-free-standing phase
QP	Quasi-particle
QPI	Quasi-particle interference
QSH	Quantum spin Hall
QSHI	Quantum-spin Hall insulator
ReaxFF	Reactive force-field
REBO	Reactive empirical bond-order
RHEED	Reflection high-energy electron diffraction
ROS	Reactive oxygen species
RPA	Random phase approximation
RSS	Rashba spin splitting
SAED	Selected area electron diffraction
SCAN	Strongly constrained and approximately normed
SCP	Self-consistent phonon
SCRUP	Scalar relativistic pseudopotentials
SE	Spectroscopic ellipsometry
SEM	Scanning electron microscopy
SERS	Surface-enhanced Raman spectroscopy
SHG	Second harmonic generation
SOC	Spin–orbit coupling
SOI	Spin–orbit interaction
SPE	Single-photon emission
SPR	Surface plasmon resonance
SQUID	Superconducting quantum interference device
SSAdNDP	Solid-state adaptive natural density partitioning
SSCHA	Stochastic self-consistent harmonic approximation
STEM	Scanning transmission electron microscopy
STM	Scanning tunneling microscopy
STS	Scanning tunneling spectroscopy
SW	Stillinger–Weber
TAC	Thermally assisted conversion
TB	Tight binding

TCSPC	Time-correlated single-photon counting
TEM	Transmission electron microscopy
TI	Topological insulator
TMD	Transition-metal dichalcogenide
TMDC	Transition-metal dichalcogenide
TMDO	Transition-metal dioxide
TPV	Thermophotovoltaic
TSC	Topological superconductivity
UHB	Upper Hubbard band
UHV	Ultrahigh vacuum
UPS	Ultraviolet photoemission spectroscopy
UV	Ultraviolet
VASP	Vienna ab initio simulation package
VB	Conduction band
VBM	Valence band maximum
VBO	Valence band offset
vdW	Van der Waals
vdWHS	Van der Waals heterostructure
VFF	Valence force field
VMT	Vapor-phase mass transport
VOCs	Volatile organic compounds
VSEPR	Valence-shell electron-pair repulsion
VSS	Valley spin splitting
WS	Weyl semimetal
XAS	X-ray absorption spectroscopy
XMCD	X-ray magnetic circular dichroism
XMR	Extremely large magnetoresistance
XPS	X-ray photoelectron spectroscopy
XRD	X-ray diffraction

Chapter 1
Introduction to Bonding, Structure, and Performance of Two-Dimensional Materials

Abstract A systematic discussion of the plethora of 2D materials follows the '*aufbau*' principle of the periodic table, which explains the existence of various families of chemically related 2D compounds treated in detail in the following chapters of the book. The introduction gives a brief overview of the types of bonding in two-dimensional (2D) space, including the periodicity of bonding behavior based on the number of valence electrons and a definition of covalently bonded monolayers and van der Waals bonded few-layer assemblies of monolayers and their heterostructures. This chapter outlines important theoretical approaches, such as first-principles density functional theory, which often delivers the most accurate results; molecular dynamics simulations, which provide thermal stability at finite temperatures; tight-binding theory that is partially based on chemical intuition; and empirical rules, allowing a deeper understanding of the nature of chemical bonding. The description of the mechanical stiffness and strength of the 2D networks is described by their 2D units because it is possible to measure these quantities and to compare the behavior of different monolayers based on these units. Enormous information is available on the fundamental electronic properties, especially those of semiconductors with direct or indirect band gaps. Of particular interest are unique 2D-based properties and exotic behavior such as topological effects, which derive from specific bonding types and structural arrangements with lower symmetry than in 3D lattices. Moreover, the introduction provides an overview of the various methods used for synthesis and diagnosis of 2D materials and briefly covers the influence of defects on the intrinsic properties of monolayers.

1.1 General: History, Definitions, Connections with Bulk Properties

The investigation of 2D materials has attracted an increasing number of scientists owing to the unique changes in chemical and physical properties that occur when materials are converted from bulk to nanosheets. Notable changes usually occur at thicknesses less than approximately ten monolayers down to single atomic layers. A

P. Hess, *Bonding, Structure, and Performance of Two-Dimensional Materials*, NanoScience and Technology, https://doi.org/10.1007/978-3-031-74161-6_1

reduction in the thickness of multilayers exposes more atoms of the material to the surface, increases the chemical reactivity, e.g., by creating more dangling bonds, and changes the physicochemical properties. Furthermore, quantum-confinement effects appear owing to the restriction to two dimensions, which endows monolayers with exceptional properties. Covalently bonded 2D networks typically contain fewer than five chemically bonded sublayers. Besides single atomic layers such as graphene, three-sublayer assemblies, as realized in MoS_2 monolayers, consisting of a central Mo sublayer sandwiched between two S sublayers, have enormous stability. In principle, these chemically saturated entities are large flat molecules with infinite extension. It is important to discriminate covalently bonded monolayers with sublayers from bilayers, trilayers and few-layer assemblies of these chemically saturated monolayers, ideally without dangling bonds, held together by much weaker van der Waals forces.

Clear experimental evidence for the existence of monolayers namely, the isolation of a one-molecule-thick sheet of MoS_2, synthesized by exfoliation and identified by X-ray diffraction (XRD), was first reported in 1986 [1]. However, this breakthrough has largely been unnoticed because most theoreticians were convinced that monolayers are generally not stable at normal temperatures because they are destroyed by thermal fluctuations [2]. According to elasticity theory, an ultrathin film exhibits instabilities, such as bending, at temperatures > 0 K. After the preparation of graphene by mechanical exfoliation of graphite in 2004, the situation changed completely, and an avalanche of theoretical studies on bonding, structure, and characterization of 2D materials appeared [3]. Today, theoretical predictions include most of the main group elements in the periodic table and a vast number of compounds containing transition metals. Extension to groups not discussed here, for instance, the alkaline earth–halogen monolayers, involves strong ionic bonding due to the significant difference in their electronegativities. Importantly, long-range Coulomb forces are less effective in 2D bonding than in 3D materials because of low localization in the plane [4]. In heteroatomic monolayers, however, different electronegativities induce ionic interactions and polarization of the covalent bonds, which are responsible for additional stability.

Crucial but often not satisfactorily answered questions are the thermodynamic, dynamic, mechanical, and thermal stability of free-standing monolayers [5]. A negative cohesive energy indicates energetic stability at 0 K, as confirmed for most theoretically suggested monolayer structures. A more critical constraint is the dynamic stability especially at higher temperatures that requires the absence of negative (imaginary) phonon frequencies, which lead to unlimited atom displacements due to the absence of restoring forces. The Born–Huang criteria state the conditions to be satisfied by the elastic constants for mechanical stability, and first-principles molecular dynamics (MD) simulations confirm the thermal stability at finite temperatures. Stabilization of a metastable monolayer is possible by additional interactions with a suitable substrate surface. Here, weak physisorption by van der Waals interaction, which maintains the electronic identity of the monolayer, usually is preferred over strong chemisorption that changes the structure and electronic state. On the other hand, such a structural change can lead to novel properties of the chemisorbed monolayer.

1.2 Chemistry and 2D Bonding: Monolayer, Few-Layer Assemblies

Strong covalent and ionic interatomic bonding in 2D materials and weaker van der Waals interlayer interactions between chemically saturated monolayers are the focus of this work. In addition to the quantum–mechanical concepts of electron sharing, electron delocalization, and electron transfer, phenomenological binding models describe the roles of the three fundamental types of bonding interactions, termed covalent, ionic, and metallic bonding [6]. Confining strong chemical bonding to two dimensions creates a large flat molecule ideally without any dangling bonds at the surface. The champion in 2D bonding is carbon. The small carbon atoms allow efficient overlap of the orbitals of planar sp^2 hybrids including delocalized π bonding of $2p_z$ orbitals. This type of hybridization leads to an optimal concentration of bonding power in the plane. Moreover, the four valence electrons of carbon fulfill the 'octet rule' in conventional two-center two-electron bonding, reaching the stable noble gas configuration. This bonding type leads to the largest cohesive energies and the shortest and strongest chemical bonds, which are the basis of the superior mechanical performance not only of purely covalent graphene but also of other planar networks, such as hexagonal boron nitride (h-BN), with covalent-ionic bonding.

In general, conventional two-center, two-electron bonds (2c–2e) represent the most important bonding type of covalent interactions that is defined by the following requirements: (1) A single chemical bond forms when each of two atoms provides an electron to the bond, and both share this electron pair in equal portions. (2) The electron configurations with eight atoms in the outer valence shell possess a lower energy and therefore higher stability than the other configurations due to the stable noble gas configuration ('octet rule').

The increasing size of atoms going down the columns of the periodic table, e.g., from C to Si, reduces delocalized π bonding, and the hybridization of valence electrons changes gradually to mixed sp^2-sp^3 hybrids to further reduce energy and increase stability. Mixing orbitals of planar 2D sp^2 hybrids with 3D orbitals of tetrahedral sp^3 hybrids dramatically extends the possibility of covalent bonding from entirely planar to quasi-two-dimensional corrugated phases. However, the chemical reactivity increases with the introduction of buckled or puckered structures in 2D materials due to partial out-of-plane sp^3 hybridization. Besides the mechanical performance of monolayers, which depends sensitively on the binding energy, chemical stability under environmental conditions is a crucial requirement, which is made important by the extremely large surface area exposed by monolayers to the environment. In addition, corrugation increases the reactivity of the monolayer.

There are two established methods that deliver detailed information on 2D electron distributions and the nature of chemical bonding in extended 2D networks. The interpretation of charge distributions in terms of electron lone pairs, two-center bonds, and delocalized multicenter bonding is usually based on the solid-state adaptive natural density partitioning (SSAdNDP) method [7]. The second method, used for obtaining insight into the electron distribution and type of bonding is the so-called electron localization function (ELF) [8, 9]. The essential ELF values characterizing bonding by the relevant electron densities, are 0, 0.5, and 1, which refer to extremely low electron density, fully delocalized electrons, and extremely high density of electrons, respectively. The latter method is of interest, especially for the detection of delocalized π electrons.

In the periodic table, the position of each element follows the '*aufbau*' principle of the individual electron shells. The resulting intrinsic periodicity of atomic properties determines the overall behavior of elements in 3D but also in 2D bonding, and thus the formation of stable 2D structures. Insight into the type and strength of bonding is the key in the discovery of new innovative 2D materials. The two main factors controlling the type and strength of 2D bonding and the intrinsic structure of the main group IIA–VIA elements result from the number of valence electrons, which vary between two and six, and the size of the involved atoms, which determine both the type of hybridization and the strength of covalent bonding by the extent of orbital overlap [10]. The monoelemental (graphene-like) 2D materials (Xenes), considered in more detail in the following chapters, are among the most promising monolayers for the forthcoming synthesis of high-quality monolayers suitable for practical applications, owing to their chemical and structural simplicity.

The cohesive energy or binding energy per atom (in – eV/atom) describes the energy difference between the free atom and the atom bonded in the individual 2D network. This key property is available for the main group IIA–VIA elements, which are highlighted in Fig. 1.1. The bond length is another sensitive measure of the strength of individual chemical bonds. While accurate values are available for the length of individual bonds and their structural arrangement, accurate information on the bond dissociation energy of single bonds is rare. In most cases, only average cohesive binding energies of the whole structure are available. The most important sources of information are first-principles calculations, which not only predicted the existence of new 2D materials but also help to find possible routes for their synthesis [11–13]. In addition to the class of mono-elemental main-group compounds (Xenes) [14] considered here, most of their binary and isoelectronic compounds are considered in the following chapters [15, 16].

The second well-studied class of monolayers are transition-metal dioxides (TMDOs) [17] and transition-metal dichalcogenides (TMDCs), which normally consist of three chemically bonded sublayers [18, 19]. Figure 1.2 highlights the transition metal (M) elements and chalcogenide (X) elements involved in stable MX_2 monolayers with the three-sublayer arrangement X–M–X. These monolayers have cohesive energies and bond lengths comparable to those of the single-layer structures, however, they exist in a wider variety of stable phases with varying symmetries. As in the case of Xenes, only limited information comes from experiments, namely, the

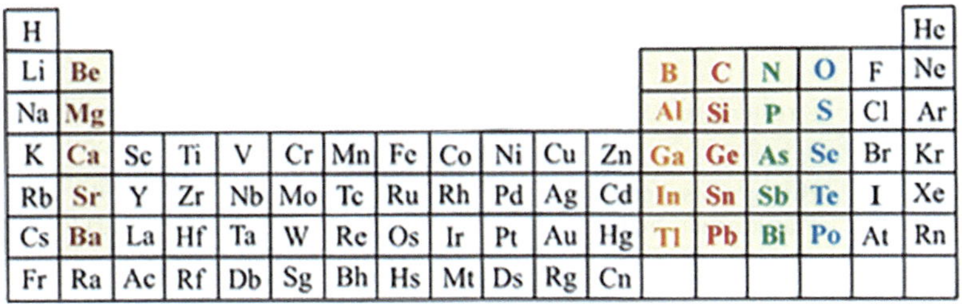

Fig. 1.1 Scheme of the periodic table highlighting the elements of group IIA–VIA elements that form monoelemental monolayers (Xenes) and heteroelemental binary monolayers. Reproduced from ref. [10], Copyright 2022, Wiley–VCH GmbH

few synthesized free-standing monolayers such as MoS_2 and the steadily increasing number of adsorbed van der Waals bonded monolayers grown on stabilizing surfaces with interactions that are weak enough that the original character of the monolayer is not destroyed by substrate interaction.

As mentioned above, mono-elemental Xenes are among the best studied monolayers and are likely to be used first in extended practical applications. The chemical rules governing their formation and structural properties can be derived from the periodic table. Table 1.1 lists the calculated cohesive energies and bond lengths (lattice constants) of the five main groups of elemental monolayers with the group-IIA monolayers (beryllene, magnesene, calcene, strontene, and barene) according to previous nomenclature [20, 21], the elemental group–IIIA monolayer borophene [22–24], aluminene, [25–27] gallenene, [28, 29] indenene [30, 31] thallenene [20, 21], the elemental group–IVA monolayer graphene [32–35], silicene [33–36], germanene [33–35], stanene [33–35], and plumbene [34, 36], the elemental group–VA monolayer nitrogene [37–39], phosphorene, [40, 41] arsenene [41, 42], antimonene [40, 41], and bismuthene[40, 41], and the elemental group–VIA monolayer selenene [42–44], tellurene [43–46], and poloniumene [47, 48].

Figure 1.3 shows the variation in the cohesive energy of the monolayers of the five main group elements IIA–VIA with the period number of the periodic table. The graph reveals the change in binding energies within the diverse columns in the

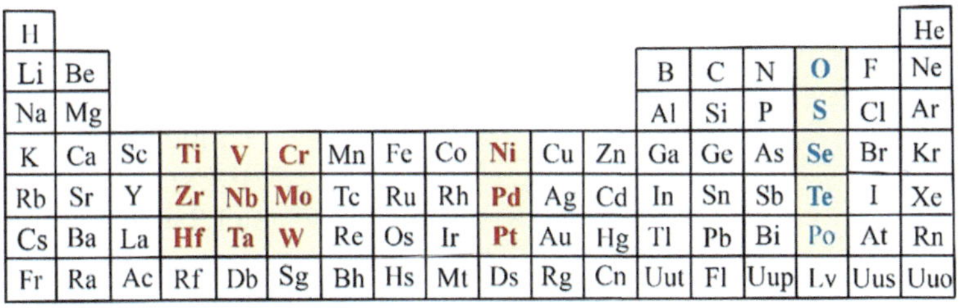

Fig. 1.2 Scheme of the periodic table highlighting the transition metals that form stable MX_2 monolayers with chalcogen atoms (TMDOs and TMDCs)

Table 1.1 A selection of cohesive energies and bond lengths of elemental group IIA–VIA monolayers

Monolayer	Structure, space group	Cohesive energy (eV/atom)	Bond length, d_1, d_2 (lattice const.) (nm)
Beryllene	Hexagonal six neighbors Planar honeycomb	– 2.91 [20] – 3.00 [21]	0.215 [20] (0.213) [21]
Magnesene	Hexagonal six neighbors Planar honeycomb	– 0.91 [20] – 0.95 [21]	0.307 [20] (0.306) [21]
Calcene	Hexagonal six neighbors Planar honeycomb	– 1.09 [20] – 1.18 [21]	0.388 [20] (0.387) [21]
Strontene	Hexagonal six neighbors Planar honeycomb	– 0.85 [20] – 0.97 [21]	0.427 [20] (0.425) [21]
Barene	Hexagonal six neighbors Planar honeycomb	– 1.04 [20] – 1.20 [21]	0.448 [20] (0.447) [21]
Borophene	Planar ß12, *Pmm*2 [22] Planar χ3, *Cmmm* [22]	– 6.15 [22], – 5.71 [23] – 6.16 [22], – 5.72 [23]	0.165 – 0.170 [24] 0.160 – 0.171 [24]
Aluminene	Planar, *P6/mmm* [25] Buckled, *P*-3*m*1 [26] Triangular, *Pmmn* [25]	– 1.96 [25], – 2.32 [26] – 3.27 [26], – 3.47 [27] –2.76 [26]	0.259 [25], 0.257 [26] 0.296 [26], 0.292 [27] 0.268 [26]
Gallenene	Planar, a100-Ga Puckered, b010-Ga	– 2.32 [29]	0.266, 0.267 [28] 0.271, 0.273 [29]
Indenene	Planar, *P6/mmm* [16] Buckled, *P*3*m*1 [16]	– 1.81 [30], – 2.08 [31] – 1.83 [30], – 2.13 [31]	0.286 [30], 0.282 [31] 0.289 [30], 0.302 [31]
Thallenene	Hexagonal six neighbors Planar honeycomb	– 1.67 [20] – 1.65 [21]	0.331 [20] (0.332) [21]
Graphene	Planar, *P6/mmm* [32]	– 7.85 [33], – 7.82 [34]	0.142 [33], 0.142 [35]
Silicene	Low buckled, *P*-3*m*1 [36]	– 3.91 [33], – 3.86 [34]	0.228 [33], 0.227 [35]
Germanene	Low buckled, *P*-3*m*1	– 3.24[33], – 3.19 [34]	0.244 [33], 0.242 [35]
Stanene	High buckled, *P*-3*m*1	– 2.73[33], – 2.65 [34]	0.284 [33], 0.283 [35]
Plumbene	High buckled, *P*-3*m*1	– 2.28 [34]	0.300 [34], 0.302 [36]
Nitrogene	Buckled, hb, *P*-3*m*1	– 6.81 [37], – 3.67 [38]	0.150 [38], 0.150 [39]
Phosphorene	Puckered, sw, *Pmna* [44] Buckled, hb, *P*-3*m*1 [44]	– 3.16 [40], – 3.59 [41] – 3.14 [40], – 3.55 [41]	0.222, 0.226 [41] 0.227 [41]
Arsenene	Puckered, sw, *Pmna* [44] Buckled, hb, *P*-3*m*1 [44]	– 2.85 [40], – 3.13 [41] – 2.89 [40], – 3.14 [41]	0.250, 0.247 [41] 0.250 [41]

(continued)

Table 1.1 (continued)

Monolayer	Structure, space group	Cohesive energy (eV/atom)	Bond length, d_1, d_2 (lattice const.) (nm)
Antimonene	Puckered, aw, *Pmn*21 [44] Buckled, hb, *P*-3*m*1 [44]	– 2.56 [40], – 2.89 [41] – 2.57 [40], – 2.86 [41]	0.285, 0.291 [41] 0.287 [41]
Bismuthene	Puckered, aw, *Pmn*21 [44] Buckled, hb, *P*-3*m*1 [44]	– 2.42 [40], – 2.84 [41] – 2.41 [40], –2.77 [41]	0.308, 0.303 [41] 0.301 [41]
Selenene	Tetragonal Rhombic, *P*-3*m*1 Monoclinic, *Cm* [44]	– 2.71 [42], – 2.77 [43] – 2.56 [42], – 2.72 [43] – 2.65 [44]	0.265 – 0.338 [43] 0.266 [43] 0.240, 0,271 [44]
Tellurene	Tetrag., *P2*/*m*, *P*21/*m* [45] Rhombic, *P*-3*m*1 [45] Monoclinic, *Cm* [43]	– 2.56 [46] – 2.62 [46] – 2.51 [43]	0.275, 0.302 [46] 0.302 [46] 0.287, 0.305 [43]
Poloniumene	Square lattice structure Trigonal, *P*-3*m*1 [48] Hexagonal six neighbors Honeycomb structure	– 2.58 [47] – 3.25 [48] – 2.35 [47] – 2.34 [47]	0.318 [47] 0.320 [48] 0.334 [47] 0.311 [47]

x direction, showing the cohesive energies of the (most) stable structures. Note the comparable cohesive energies of several allotropes in Table 1.1. The characteristic inverse dependence of the cohesive energies of the group members on the period number derives primarily from an increase in atom size within each column with increasing atomic number. Importantly, the atom size controls the efficiency of orbital overlap and thus the strength of the purely covalent bonds in the homonuclear 2D structures that ultimately determine the length of individual chemical bonds. Tight-binding theory predicts an inverse dependence of the bond length 'd' on the orbital overlap 'V' according to the fundamental relation $V \sim 1/d^2$ [49, 50]. Irrespective of the type of orbitals and their principal quantum numbers, the relationship describes the covalent interaction of the valence shell orbitals of all neighboring electrons. Within the errors involved in such calculations, the enormous decrease in the cohesive energy supports the theoretically predicted inverse quadratic dependence predicted by tight-binding theory.

The observed dependences of the cohesive energies in Fig. 1.3 reflect the strong increase in the van der Waals diameter of the atoms between periods two and three compared with the smaller changes between the higher periods. This increase in the atom size induces an alteration of hybridization from perfect planar sp^2 to buckled mixed sp^2-sp^3 hybrids of the higher homologues of graphene such as silicene. Owing to their reduced orbital overlap, mixed sp^2-sp^3 hybrids become energetically more stable, however, with increasing sp^3 content the corrugation of the network increases, while the binding energy decreases for the higher homologues. The tabulated van der Waals diameters yield reasonable estimates of the size of all atoms [51], and measurements of the thickness of planar single-atomic monolayers by atomic force microscopy (AFM) are in good agreement with the tabulated van der Waals diameters [52].

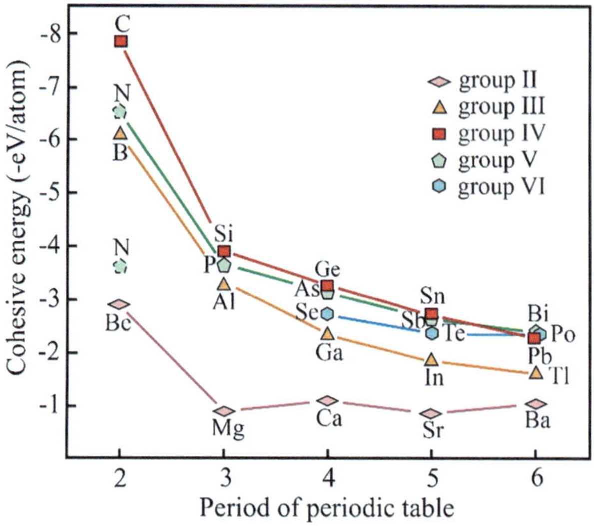

Fig. 1.3 Plot of the cohesive energy of the elements of columns IIA–VIA versus the period number of the periodic table

A plot of the bond lengths of group IIA–VIA monolayers versus period number shows the corresponding significant elongation of bonds between periods two and three, followed by an asymptotic increase in bond length with decreasing binding energy, qualitatively reflecting the smaller increase in the van der Waals diameters of the higher homologues (see Fig. 1.4). In contrast to the mean cohesive energy, the dissociation energy of individual bonds and thus the bond length may vary in complex corrugated structures. For low-symmetry structures with different bond lengths, the plot displays the shortest bond length.

Figure 1.5 shows the calculated cohesive energies of the main group elements versus the column number. The plot again reveals the pronounced change in the binding energies between the second and third periods by the large separation of the binding energies. According to the dependencies from the second to the fifth period, the group-IVA monolayers exhibit the maximum bonding strength. This confirms that four valence electrons form the most stable 2D bonds with their noble gas electron configuration ('octet rule'). Along each period, the dominance of the number of available valence electrons controls bonding rather than minor contributions from the simultaneously varying atom size. The regular pattern of cohesive 2D interaction energies observed for the five periods is clearly a consequence of the 'aufbau' principle of electron shells in the periodic table. The periodic variation of the number of valence electrons reflects the performance of the individual groups with a deficiency or a surplus of valence electrons in comparison to graphene with the preferred four valence electrons and octet stability.

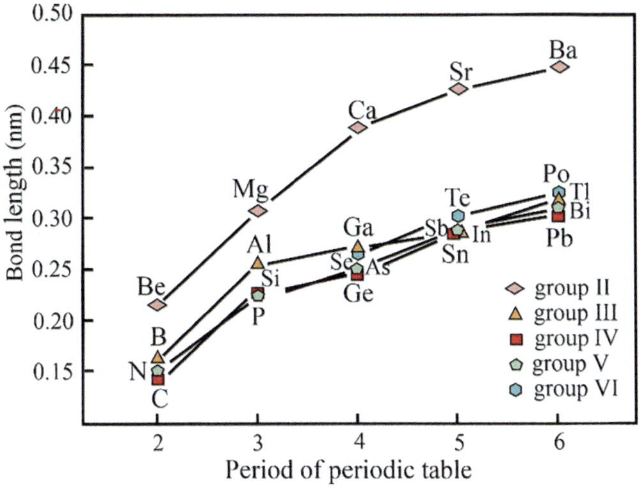

Fig. 1.4 Plot of the bond length of the elements of columns IIA–VIA versus the period number of the periodic table Reproduced from ref. [10], Copyright 2022, Wiley–VCH GmbH

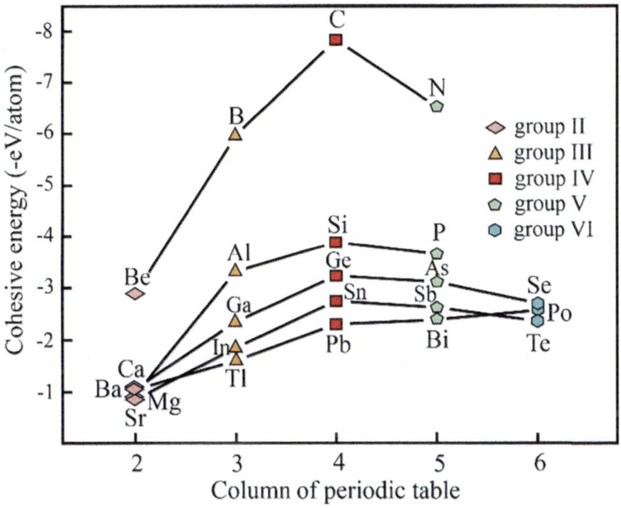

Fig. 1.5 Plot of the cohesive energy of elements of the periods 2–6 versus the column number of IIA–VIA elements of the periodic table. Reproduced from ref. [10], Copyright 2022, Wiley–VCH GmbH

The deficiency of valence electrons in group IIA-IIIA compounds causes a drastic decrease in cohesive energy (see Fig. 1.5). For example, with its three valence electrons, boron can form 2D networks, however, the available bonding orbitals are partially empty. According to tight-binding theory, based on the linear combination of atomic orbitals (LCAO) and the hybridization concept, not all three valence electrons are in bonding orbitals; therefore, the free-standing uniform trivalent network is not stable [53]. Note that the experimentally synthesized planar monolayers of boron, included in Table 1.1, contain a mixture of triangular and hexagonal motifs for stabilization by charge transfer. A change in the type of bonding to weaker metallic interaction leads to an even larger reduction in the cohesive energies in group-IIA monolayers. Metallic bonding usually prefers a large coordination with six nearest neighbors, which, in group-IIA monolayers, has a lower binding energy compared to that of the planar honeycomb lattice (see Table 1.1).

A surplus of valence electrons in group VA–VIA monolayers has a smaller effect on the binding energy. In pnictogens, which have five valence electrons, tetrahedral-like hybrids with a backbone consisting of three σ bonds and a lone-electron pair lead to dynamically stable puckered and buckled structures. Interestingly, the binding energies of the two allotropes are often comparable and may lead to the synthesis of mixtures of the two phases (Table 1.1). While a three-dimensional solid of nitrogen does not seem to be stable, 2D buckled nitrogene is thermodynamically and dynamically stable [37–39]. Owing to the multivalency of chalcogenides with oxidation states of -2 and $+6$, group-VIA elements exist in diverse 2D hybrids, forming low-symmetry structures such as trigonal, tetragonal, rhombic, monoclinic, and square phases.

Figure 1.6 displays the variation in bond length (lattice constants for groupIIA elements) with column number. The minima of the group-IVA bond lengths are consistent with the maxima of the cohesive energies in Fig. 1.5. The diagrams reveal the consequences of electron deficiency or a surplus of valence electrons on bonding. The deficiency of valence electrons leads to only partially filled bonding orbitals with an essential loss of bonding power. Conversely, when the valence shell contains additional electrons, it is possible to fill the bonding orbitals completely, however, the surplus of electrons nevertheless may reduce the binding energy and extend the bond length slightly, e.g., due to the repulsion effect induced by the lone-electron pairs.

Weaker metallic-type interactions appear for group-IIA and group-VIA monolayers, as well as for those of period–VIA monolayers, with a large separation from the graphene position in the periodic table. The graphene position can be considered as the center of perfect covalent bonding. Hence, the position of an element in the periodic table allows general insight into the nature of 2D bonding and the resulting structure, providing a general assessment of physicochemical properties from already published data.

Monoelemental Xenes provide the basis for a general analysis and understanding of the bonding behavior in heteroelemental monolayers of the main group elements IIA–VIA and binary isoelectronic compounds with the same number of valence electrons as the main group elements. Owing to the different electronegativities of the two different atoms, ionic forces change the bonding situation in this group of heteroelemental 2D materials to covalent-ionic interactions with polar bonds and thus novel chemical and physical properties. The main contribution to 2D bonding, however, still comes from covalent interaction.

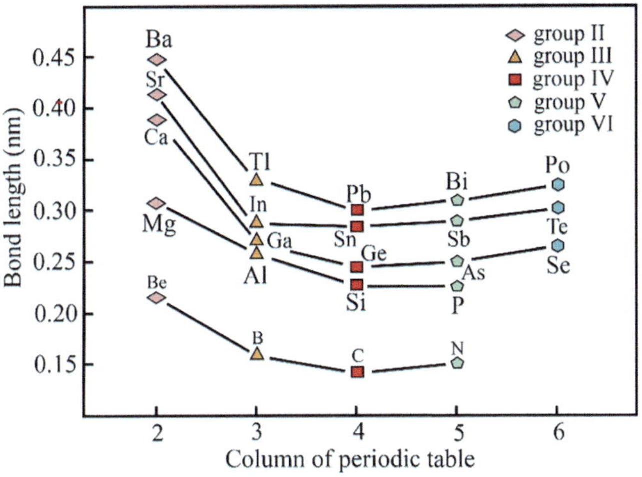

Fig. 1.6 Plot of the bond length of main group elements versus the column number of the group IIA–VIA elements of the periodic table Reproduced from ref. [10], Copyright 2022, Wiley–VCH GmbH

Covalently bonded monolayers form larger multilayer assemblies. If the interlayer interaction forces between the monolayers are weak, e.g., in the case of van der Waals forces, monolayers with different structures combine to weakly bonded compilations with greater interlayer separation. The lack of long-range order in such stacks allows flexible combinations of individual monolayers with distinct properties that can be selected to tailor novel heterostructure devices. Weak van der Waals forces have a minor influence on the bond length and bonding energy of the covalent network; however, they control the cohesion of the whole layered assembly and thus determine key properties of few-layer systems. In this case, various interlayer configurations of an assembly are possible, often with similar interlayer interaction energies. With increasing interlayer interaction, originating, for example, from chemical and electrostatic forces, the structure of multilayers ultimately transforms from a weakly bonded layered structure, such as a few-layer assembly of graphene, as realized in graphite, into a bulk structure with 3D chemical bonding, e.g., pure sp^3 hybridization, as realized in diamond. If a stable layered bulk structure such as graphite is formed, exfoliation is the simplest method to acquire monolayers.

Note that the bottom-up synthesis of layered compounds, using monolayers as building blocks and weak van der Waals interactions for compilation, opens a large new field of artificially assembled 2D materials. Combining monolayers with selected optimized properties provides access to van der Waals superlattices and heterostructures of technological importance [54, 55]. Importantly, the interlayer interaction energy, sometimes called the exfoliation energy, is responsible for the effort needed to separate a monolayer or a specific few-layer assembly from a layered bulk material [10]. If the interlayer interaction energy between the layers is large, only multilayer flakes or nanosheets may be separable from the layered bulk crystal, which means

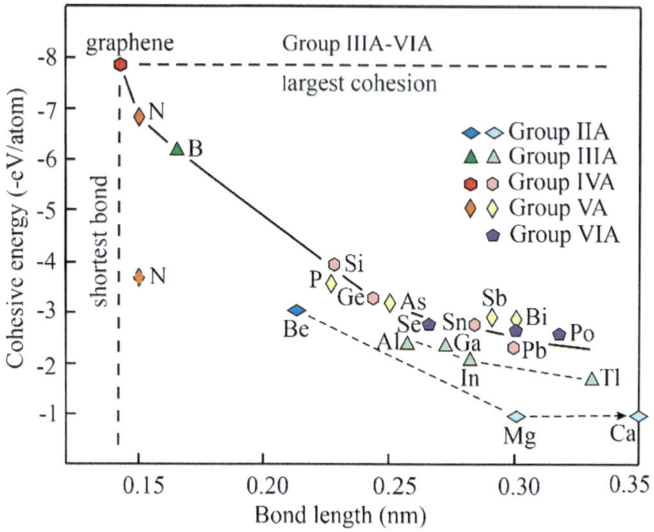

Fig. 1.7 Cohesive or binding energy versus bond length, separating Xenes with strong and short covalent bonds from monolayers with weak and long covalent bonds

that purely synthetic preparation methods must be applied for the isolation of a monolayer.

In a plot of the cohesive energy versus bond length, the monolayers with the largest binding energies and smallest bond lengths are located at the upper left corner, and those with the smallest binding energies and longest bonds are concentrated at the lower right corner in the cohesion–bond length plot (see Fig. 1.7). As expected, the second-period monoelemental Xenes form the strongest monolayers, however, with a large uncertainty for nitrogene. In principle, the monoelemental Xenes of the higher periods can be separated into three groups: those with four or more valence electrons, those with fewer than four valence electrons (group IIIA), and those with metallic bonding such as group—IIA elements.

1.3 2D Structure

The elucidation of new stable 2D structures by theoretical approaches is a topic that has received enormous attention. In addition to the elemental monolayers, heteronuclear and isoelectronic binary compounds, ternary, and quaternary 2D materials of all elements form new combinations and stoichiometries with a plethora of chemical bonding types and structural arrangements. The option of systematically varying the chemical composition represents an additional, not fully explored, chemical degree of freedom. Besides trial and error, 2D structures can be predicted by the method of atomic transmutation, namely, by changing elements in the periodic table into neighboring elements while keeping the total number of valence electrons constant. The theoretical strategies employed to find stable 2D arrangements of atoms are bottom-up approaches, in which the global minimum energy configuration is determined by maintaining the number of valence electrons. One of the most successful methodologies for the discovery of new 2D materials is particle swarm optimization, which is implemented in the CALYPSO (Crystal structure AnaLYsis by Particle Swarm Optimization) code [56]. Clearly, 2D materials with the lowest energy configuration of the global minimum on the potential energy surface (PES) have the highest chance of experimental realization. Quite often, however, additional metastable configurations with a local minimum and a similar cohesive energy exist. If these monolayers turn out to be thermodynamically and dynamically stable, experimental synthesis still may be possible, for example, with additional stabilization by physisorption on a suitable substrate surface that preserves the original geometric and electronic structure.

Interestingly, the hexagonal structure plays a dominant role in 2D phases, as observed in bulk materials. It is noteworthy to mention that the honeycomb structure used by bees is the optimum way to divide a surface into equal areas with the smallest total perimeter. In fact, many networks of the most stable elemental monolayers exhibit hexagonal symmetry in various forms. In graphene, for example, sp^2 hybrids extend in the in-plane direction, with π electrons delocalized parallel to the plane. Figure 1.8 shows the flat hexagonal network of the graphene monolayer. It is important to note that topological mapping on a hexagonal lattice is possible for the often observed three-fold coordinated buckled structure, as realized in silicene (see Fig. 1.9a). Importantly, in addition to hexagonal structures, those with triangles, pentagons, and heptagons play a role in 2D networks, particularly in the stabilization of monolayers of lower symmetry, which may consist of several sublayers. Note that just low-symmetry 2D structures play a key role in the development of structures with exotic properties, e.g., with topological features.

As mentioned above, with increasing size of the elements going down the columns of the periodic table, the structures with the lowest energy contain an increasing portion of sp^3 hybrids because mixed sp^2–sp^3 hybridization allows more efficient orbital overlap with increasing atom size. Mixed hybridization usually leads to a

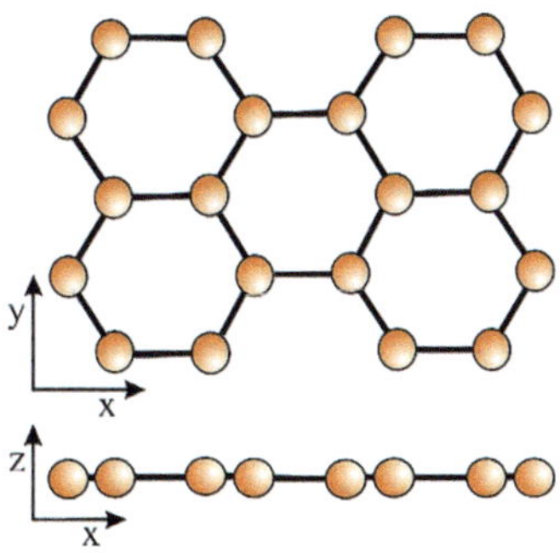

Fig. 1.8 The planar hexagonal or honeycomb lattice of graphene monolayers

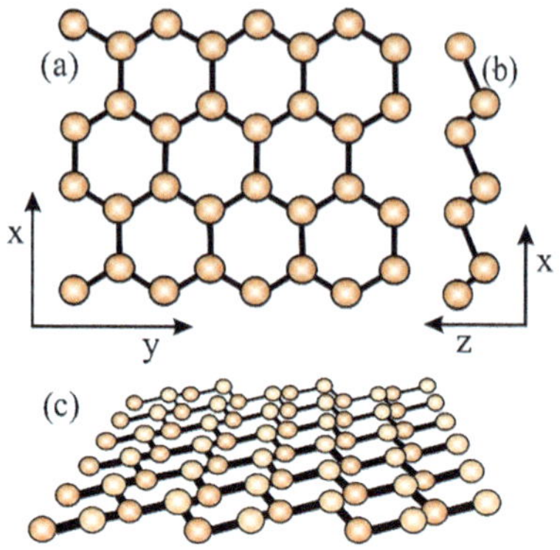

Fig. 1.9 **a** Top view and **b** side view of the 2D buckled hexagonal structure realized in silicene. **c** Perspective view of the hexagonal buckled structure

reduction in the symmetry of the 2D structure and to increased reactivity perpendicular to the layer plane due to corrugation. Figures 1.9a, b show the top and side views, and Fig. 1.9c a perspective view of the hexagonal corrugated or buckled monolayer. As mentioned above, reducing the symmetry may not be detrimental to all monolayer properties.

Monolayers consisting of several chemically bonded sublayers exhibit greater structural variability than single layers. In covalent three-sublayer systems, for example, the most stable monolayers often have hexagonal symmetry and X–M–X stoichiometry, where M is a transition metal that forms the central sublayer sandwiched between two chalcogen sublayers X [18, 19]. Figure 1.10 displays such a three-layer structure as realized in many TMDCs. This structure has a mirror plane in the central layer that disappears by shifting, for example, the atoms of the lower chalcogen layer to the left central metal atom, as illustrated in Fig. 1.11. In most studied TMDCs, the high-symmetry H structure (1H is a single layer of TMDC with trigonal coordination of the transition metal) or the lower-symmetry T structure (1T is a single layer of TMDC with octahedral coordination of the transition metal) have the lowest energy.

An example for a stable structure with much lower symmetry is the pentagonal network realized in some TMDCs, such as PdS_2 monolayers [57]. Figure 1.12 shows such a low-symmetry orthorhombic structure with pentagons. Interestingly, several researchers believe that this structure has a slightly lower energy than the 1T structure of the PdS_2 monolayer. Often, a final decision is not possible owing to the similarity of the relevant properties, such as the involved covalent-ionic bonds and thus the binding energy of the stable and metastable phase, leading to comparable stabilities and the growth of their mixtures.

The symmetry of such a monolayer can also be reduced by replacing the lower chalcogen layer in TMDC monolayers with a different chalcogen element, as shown in Fig. 1.13 (Janus structure) [58]. Interestingly, this chemical modification of the three-sublayer structure to an X–M–Y arrangement leads to stable networks. Importantly, the asymmetric out-of-plane structural configuration breaks the out-of-plane symmetry. This not only changes the chemical but also the physical properties of the

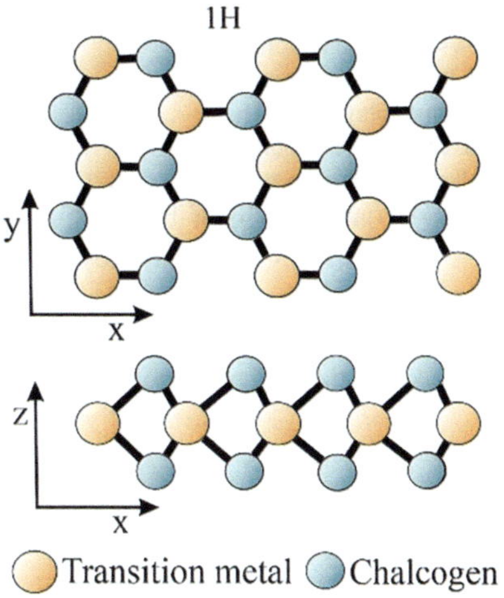

Fig. 1.10 Top and side views of the 1H structure of the MX_2 monolayer of TMDOs and TMDCs. Adapted with permission from ref. [19], Copyright 2021, Elsevier

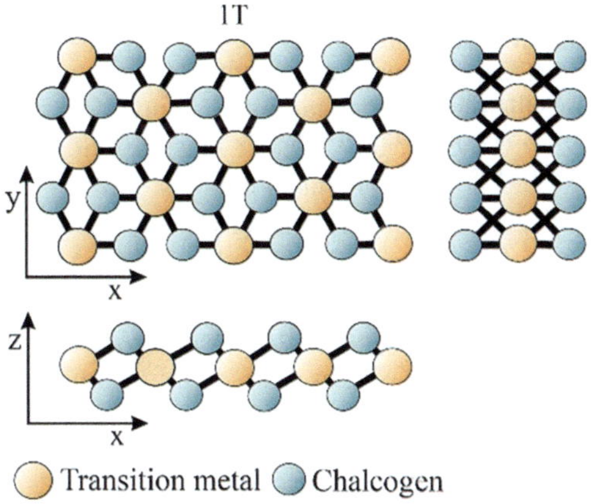

Fig. 1.11 Top and side views of the 1T structure of the MX_2 monolayer of TMDOs and TMDCs. Adapted with permission from ref. [19], Copyright 2021, Elsevier

monolayers, which exhibit, for example, piezoelectric behavior, as discussed below in the relevant chapters on TMDCs.

The explosive growth of theoretical studies on 2D compounds allows the identification of general relationships, e.g., between bonding behavior and structure, and the extraction of basic rules governing the 2D structural world down the periodic table [15]. These relationships result from plots of the cohesive energy per atom versus the bond length of chemically related monolayers. The plots provide fundamental insight into the strong connection between covalent bonding by orbital overlap and the resulting arrangement of atoms to a 2D network and the geometrical lattice parameters of the network. The relations designate 2D-structure formation following the predictions of tight-binding theory, first developed for 3D systems, but also applicable for 2D systems.

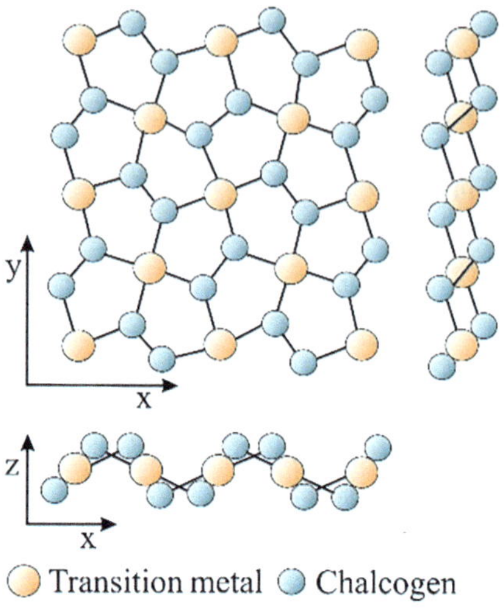

Fig. 1.12 Top and side views of the pentagonal structure of a low-symmetry MX_2 monolayer of TMDOs and TMDCs. Reproduced with permission from ref. [57], Copyright 2015, Royal Society of Chemistry

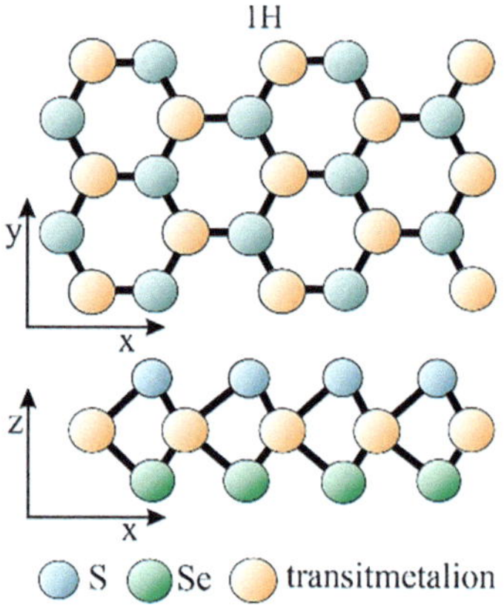

Fig. 1.13 Janus structure of a TMDC monolayer with two different chalcogen layers, such as SMSe, where M is a transition metal

1.4 Mechanics: Definitions, Linear and Nonlinear Behavior

The reliable mechanical performance of 2D materials from manufacture to final use is a key requirement for any application of 2D materials. Comprehensive reviews describe the characterization of nanomaterials and composites [59] and the mechanical properties of 2D materials [60]. While the former study covers fundamental aspects such as linear and nonlinear mechanical properties and failure, the latter work reviews in addition to the presented experimental results the theoretical prediction of elastic properties and especially the role played by inelastic properties, including ultimate strength and intrinsic toughness. The discussion on defects includes point defects such as vacancies and dislocations, topological defects, and line defects such as grain boundaries.

While calculated intrinsic Young's moduli and, in some cases, ultimate strengths are available for various monolayers, reliable toughness values that characterize the fracture strength of large-area defective monolayers seem to be available only for graphene. The following work concentrates primarily on calculated mechanical properties such as intrinsic Young's moduli, Poisson's ratios, ultimate strengths, and ultimate strains of both free-standing and weakly surface-bonded intact monolayers. Most 2D materials exhibit positive Poisson's ratios between 0 and 0.5, where the material contracts in the transverse direction when stretched in the longitudinal direction. For some low-symmetry monolayers, negative Poisson's ratios or auxetic behavior exists for 2D materials. This means, for example, that the material expands perpendicularly to the plane when stretched in the in-plane direction [61].

The fracture toughness describes the ability of a defective material to resist failure and is one of the most important criteria for fracture analysis. Under the same loading conditions, a material with a higher toughness will be less sensitive to preexisting microcracks or lattice distortions, which are inevitably present in any material. Unfortunately, direct measurements of fracture toughness values are rare since, in addition to the real strength, the size and orientation of a preexisting crack must be determined, or the sample must contain a quantifiable precrack to apply one of the fracture models. Since graphene is brittle, its toughness is not an outstanding property, unlike its stiffness and strength, as previously stated in the literature. Analytical molecular mechanics-based models using a Morse potential are available, which provide insight into the relevant factors controlling fracture mechanics [62]. This model gives a toughness of ~ 1.0×10^{-3} N/m$^{1/2}$ for graphene, which is in good agreement with the results of several experiments and an empirical formula based on more realistic interaction potentials. A study of fracture and strength of single-layer hexagonal materials, based on MD simulations, employing Tersoff and Stillinger–Weber potentials, provided the mode-I fracture toughness of 25 hexagonal monolayers at room temperature [63].

A rigorous comparison of the 2D mechanical properties of monolayers requires mechanical properties given in 2D units due to the thickness or volume dependence of 3D mechanical properties. The transformation of 3D units (N/m^2) of stiffness and strength into 2D units (N/m) is possible using the monolayer thickness. Due to the thickness dependence of the 3D values, they are not intrinsic properties of the monolayer and cannot be used to derive other mechanical properties of the monolayer. Note that only 2D mechanical properties can be compared with each other. Moreover, the 2D units employed here are preferable because they can be obtained experimentally.

Unfortunately, reliable measurements of the intrinsic stiffness and ultimate strength of free-standing monolayers, e.g., by nanoindentation using atomic force microscopy (AFM), are available only for graphene, while most measurements for other monolayers, such as graphene oxide, h-BN, black phosphorus, and MoS_2, are still in a preliminary stage [64–68]. Figure 1.14 presents a scheme of an indentation setup that is used to measure the force–displacement curve by optical detection, providing Young's modulus, the ultimate failure strength, and ultimate failure strain [64]. The monolayer thickness often can be estimated for free-standing single layers

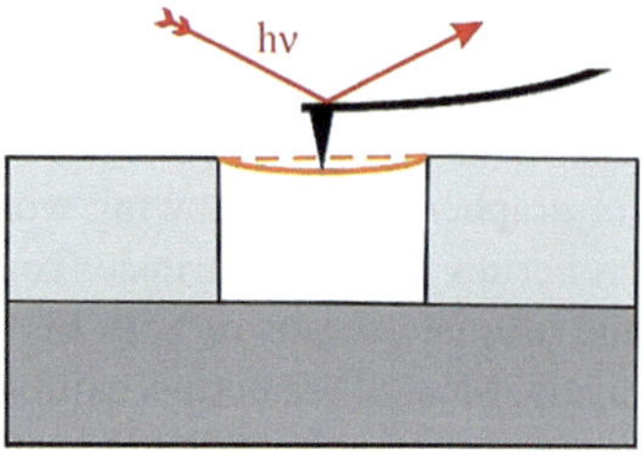

Fig. 1.14 Scheme of a nanoindentation setup using AFM for measuring the force-displacement curve that reflects linear and nonlinear mechanical behavior

using the tabulated van der Waals diameter [52] by including a possible corrugation of the monolayer. Steps in few-layer assemblies, which can be used to measure the thickness by AFM, determine the interlayer spacing. In this case, the extracted formal layer thickness depends on the strength of the interlayer interaction, which controls the distance between monolayers in few-layer assemblies and layered bulk materials.

The lack of high-quality samples, the sophisticated experimental techniques, and critical assumptions made by evaluating the stress–displacement curve limit the accuracy of the experimental in-plane mechanical properties. Alternatively, various numerical approaches, such as first-principles density functional theory (DFT) calculations, force-field-based classical MD simulations, and finite-element (FE) methods, currently provide data for a large number of monolayers [69]. The stress versus strain relationship is a useful tool for calculating or simulating the mechanical behavior of materials under strain. The linear part of the stress–strain curve represents Young's modulus or stiffness, and the nonlinear part represents the ultimate stress or strength and ultimate strain (see Fig. 1.15). The maximum strain a stretched material can withstand before breaking defines the ultimate tensile stress or strength. These stability limits are of practical interest in strain engineering because modification of material properties by strain, such as the band gap, is possible only below the fraction limit.

The mechanical properties of monolayers vary widely because the mechanical stiffness and strength depend sensitively on the binding energy, as demonstrated in Fig. 1.16 for the stiffness. While planar graphene has an outstanding Young's modulus and a large ultimate strength, both the linear and nonlinear mechanical

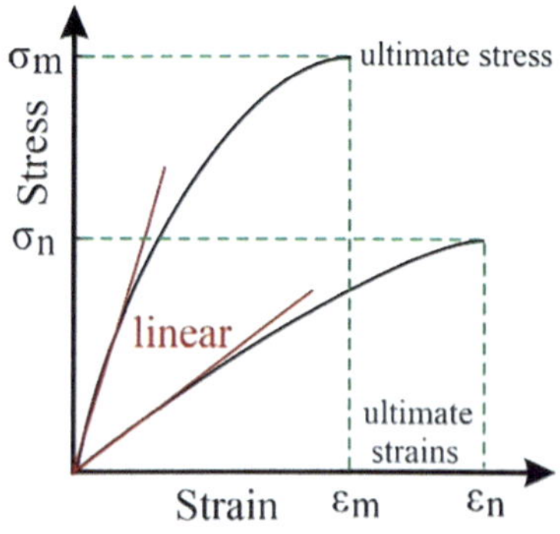

Fig. 1.15 Tensile stress versus strain curves for different directions of a strongly anisotropic monolayer. The red lines illustrate the linear part used to determine Young's moduli. The symbols σ_m, σ_n and ε_m, ε_n indicate the maximum stress (strength) and strain that the structure can tolerate in specified directions, respectively

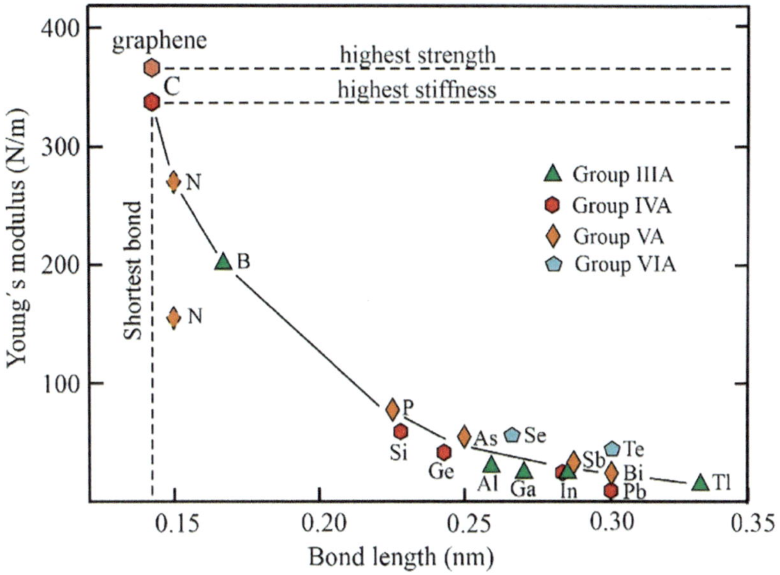

Fig. 1.16 Plot of Young's moduli of monoelemental monolayers versus bond lengths. The pronounced inverse dependence verifies the extreme loss of mechanical stability for group IIA–VIA monolayers going down the columns

properties of the elemental monolayers of columns IIA–VIA deteriorate significantly from the second to higher periods of the periodic table [10, 15]. The drastic deterioration in mechanical performance can be demonstrated by comparison with the cohesive energy, where the stiffness decays from 340 N/m (graphene) to ~ 10 N/m (plumbene), whereas the corresponding cohesive energy changes only from – 7.85 eV/atom (graphene) to approximately – 2.28 eV/atom (plumbene). The ideal ultimate strength is related to the stiffness according to Griffith's rule (stiffness $\approx$ 10 $\times$ strength). This rule or the empirical relationship between cohesive energy and bond length can be used to estimate the upper limit of unknown ultimate strengths. While the planar monolayers of the second period elements with high symmetry partly exhibit outstanding mechanical properties, the mechanical performance of the heaviest compounds with sometimes weakly bonded low-symmetry structures may no longer be sufficient for device applications. Importantly, exactly these low-symmetry structures are often of special interest because they possess desirable exotic chemical, physical, or electronic properties, such as piezoelectricity, ferroelectricity, Dirac cones, and nontrivial topological properties [12].

1.5 Physics and Applications

The best studied classes of monolayers consist of a chemically bonded single layer, two sublayers or three covalently bonded sublayers. As already mentioned, the restriction of chemical bonding to two dimensions alters not only the chemical behavior but also the physical properties at thicknesses less than about 10 layers. The basic types of solid 3D crystals also exist as monolayers, including metals, semiconductors, and insulators, however, the properties of the corresponding nanomaterials are different. Currently, enormous attention is being given to the stability of semiconducting monolayers and their few-layer sheets; their versatile properties, such as the direct or indirect band gap; the thermoelectric properties; the catalytic activity, based on their specific surface chemistry; the efficient energy storage in batteries and supercapacitors; and their applications in optoelectronics.

First-principles DFT calculations combined with the Boltzmann transport equation allow the determination of thermoelectric properties of monolayers. Note that for monolayers with the same structure, a large lattice constant and mass and a low vibration frequency result in a low lattice thermal conductivity. Weaker chemical bonding leads to a larger expansion of the electron cloud, resulting in enhanced phonon scattering and reduced phonon conduction, as demonstrated, for example, for the anisotropic pentagonal monolayers of PtS_2, $PtSe_2$, and $PtTe_2$ [70]. Pentagonal $PtTe_2$ monolayers exhibit the smallest lattice thermal conductivities of 1.77 and 5.17 W m^{-1} K^{-1} at 300 K in the armchair and zigzag directions, respectively. At 300 K, these low thermal conductivities lead to large figures of merit or *ZT* values of 2.60 and 1.14 in the armchair and zigzag directions, respectively. At 600 K, for p-type pentagonal $PtTe_2$ monolayers the larger value increases to $ZT = 5.03$ in the armchair direction. The *ZT* value characterizes the efficiency of thermoelectric conversion. A recent review on the progress made in thermoelectric applications of TMDCs provides insight into the structure–property relationships, with a focus on the electron and phonon transport properties [71].

Due to their widely tunable physicochemical properties, TMDCs are suitable candidates for energy harvesting and energy storage applications. With the goal of further increasing the catalytic activity, the effects of functionalization and vacancy defects on the structural, electronic, and optical response of TiS_2 monolayers were studied within the framework of DFT theory [72]. This work included functionalization at the anion (S) side, substitution with P, N, and C dopants, and incorporation of single sulfur vacancy defects. Besides the density of states (DOS), work-function, and optical absorption spectra, the study provides results on the hydrogen evolution reaction (HER) and oxygen evolution reaction (OER) of functionalized and defective monolayers. New insight into the catalytic activity of these materials is based on the reaction coordinates derived from the adsorption free energies of the intermediates H*, O*, OH*, and OOH*. The promising catalytic performance of functionalized TiS_2 was confirmed by a first-principles study of 33 TMDCs, which revealed that hydrogenated TiS_2 monolayers effectively catalyze the HER over a wide range of hydrogen coverages with a neutral thermal Gibbs free energy and a catalytic ability comparable to that of Pt [73]. A review describes the progress made in the development of TMDCs and their composites as electrocatalysts for acidic and alkaline electrolysis and as photocatalysts [74]. The performance of various TMDC-based materials in the electrochemical splitting of water is discussed in detail, and the application of TMDCs in photoelectrochemical devices is considered.

Layered TMDCs, including VS_2, for example, have attracted great interest as alternative anode materials in ion batteries, which are among the most important rechargeable energy storage devices, and supercapacitors [75, 76]. The adsorption energy and diffusion coefficient of lithium and sodium ions in H-type monolayers and T-type bulk VS_2 electrodes were studied by DFT calculations, which revealed a low adsorption energy and high ion diffusion rates. Moreover, the VS_2 nanosheets show a high reversible capacity. Note that the theoretical Li and Na diffusion coefficients of bulk VS_2 are five and seven orders of magnitude greater than those reported for bulk MoS_2.

The optical absorption coefficients of some graphene-like monolayers and TMDCs are large, with values up to ~ 10^6 cm^{-1}. Furthermore, TMDCs possess an extremely high response rate, a high quantum efficiency, and a broadband photoresponse from IR to UV [77]. Some monolayers have a direct band gap with high absorption efficiency because the top of the valence band and bottom of the conduction band have the same position in reciprocal space. This allows the efficient detection of light by p–n junctions, which convert photons into current. Photodetectors have a wide range of applications, such as light-emitting diodes (LEDs), photomultipliers, photovoltaic devices, displays, and field-effect transistors (FETs), with current controlled by an electric field. MoS_2 was used as the semiconductor in already realized FET devices. The observed light conversion efficiency of an optimized photovoltaic cell composed of three layers, e.g., graphene/MoS_2/n-Si was 11.1% [78].

Notably, exotic properties are found in various monolayers, such as the quantum spin Hall (QSH) effect, which is a topologically insulating (TI) state that also is observed in 2D space [79, 80]. Specifically, topological insulators are characterized by a bulk gap and conducting edge states, which are robust against weak deformations and external perturbations. In 2D materials, this means an insulating gap in the main part of the monolayer and gapless spin-polarized edge states at the edges of the monolayer. The time-reversal symmetry protects the dissipation-less spin currents against backscattering from nonmagnetic impurities, leading to quantized Hall conductance in the absence of a magnetic field. Monoelemental graphene-like Xenes are such potential topological materials [80].

For crystals, the second law of thermodynamics requires the existence of a certain amount of disorder in an otherwise perfect structure [81]. Thus, any preparation process inevitably introduces defects. Most theoretical calculations, however, assume a perfect geometry of the structure. The measurement of intrinsic mechanical properties, for example, is based on the assumption of the existence of essentially defect-free small areas on the free-standing monolayer, sufficient for nanoindentation [67]. In graphene, inherent defects such as Stone–Wales defects, consisting of adjacent pairs of pentagonal and heptagonal rings, single vacancies with dangling bonds, complicated multiple vacancies, line defects introduced by nucleation of different crystallographic orientations during growth, and adatoms can be detected with single-atom resolution by transition electron microscopy (TEM), scanning electron microscopy (SEM), and AFM [82, 83]. Foreign atoms and substitutional impurities introduce additional defects. For example, boron and nitrogen can replace the three bonds of a substituted carbon atom, increasing the reactivity of this region. Note that the doping process may improve cetain properties such as the conductivity or catalytic activity.

1.6 Synthesis: Monolayers and Few-Layer Assemblies

The experimental realization of monolayers lags far behind the predictions provided by theory and found enormous attention with the discovery of mechanical exfoliation of layered materials, such as graphene exfoliated from graphite. For this preparation method, a large reservoir of weakly bonded layered bulk materials exists [84–87]. However, many monolayers are accessible only by entirely synthetic methods, such as those of boron or silicon [88]. Deposition on suitable substrates is possible by bottom-up methods such as chemical vapor deposition (CVD), physical vapor deposition (PVD), and ultrahigh vacuum (UHV) molecular beam epitaxy (MBE). In few cases it is possible to grow a suitable (layered) synthetic compound functioning as a precursor that can be exfoliated in a top-down process.

Mechanical exfoliation of crystals is the simplest method characterized by monolayers or few-layer nanosheets with high crystallinity. Unfortunately, this method lacks large-area uniformity and layer-number controllability [89]. In particular, the CVD method allows the deposition of monolayers and large-area sheets of all kinds of 2D materials without the need for UHV conditions required for MBE [89–91]. As shown in Fig. 1.17, the CVD process involves the evaporation of the precursors and the growth of the monolayer usually occurs in a different temperature zone. Synthesis by MBE combined with in situ diagnosis requires expensive vacuum equipment, however, delivers the best control of the deposition process.

As displayed in Fig. 1.18, molecular/atomic beams are generated in an evaporation cell (Knudsen cell) or an e-beam evaporator. Reviews covering various synthesis

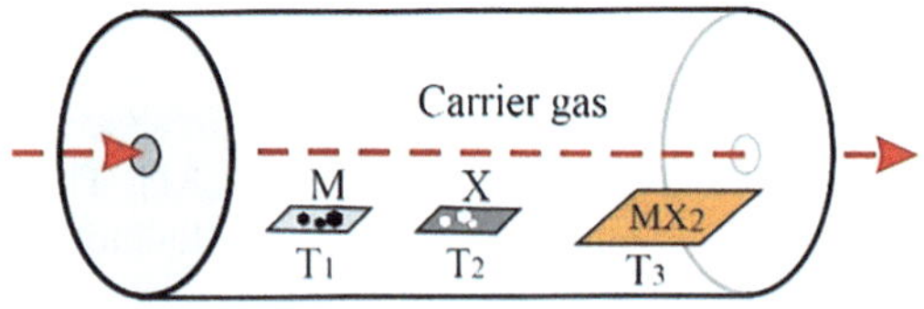

Fig. 1.17 Scheme of a CVD chamber with different temperature zones for evaporation of the precursors and growth of a monolayer on selected substrates under a controlled atmosphere

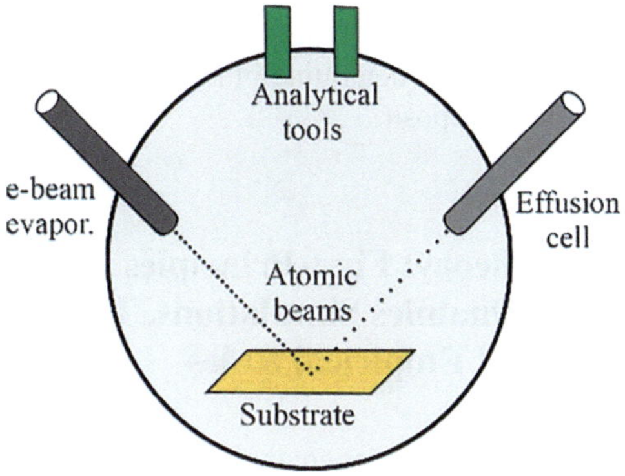

Fig. 1.18 Scheme of a UHV chamber for MBE with an effusion cell (Knudsen cell) to produce chalcogenide beams and an e-beam evaporator for transition metals in the case of TMDC deposition

methods exist for single-element 2D materials [92] and transition metal dichalcogenides [93]. According to a theoretical study, the orientational uniformity of 2D islands can be realized if the symmetry group of the substrate is a subgroup of that of the growing 2D material [94]. A perspective discusses the recent progress made in the growth of TMDCs by MBE employing diverse diagnostic tools and highlights the key challenges and upcoming opportunities [95].

The available UHV-compatible diagnostic strategies employed for structural analysis of single- and few-layer 2D materials include scanning tunneling microscopy (STM), scanning tunneling spectroscopy (STS), in situ microscopy techniques such as SEM, low-energy electron diffraction (LEED), XRD, X-ray photoelectron spectroscopy (XPS), angle-resolved photoemission spectroscopy (ARPES), selected area electron diffraction (SAED), high-resolution transmission electron microscopy (HRTEM), energy dispersive X-ray spectroscopy (EDX), and high-angle annular dark field (HAADF) scanning transmission electron microscopy (STEM). Some of these methods are inevitable for achieving the final goal of controlled growth of high-quality large-area 2D materials with a defined number of layers and an identified structure. Note that various characterization methods allow the quality and uniformity of deposits produced by plasma-enhanced CVD to be controlled [96].

The realization of 2D materials in their free-standing form is quite rare. This approach allows a direct measurement of the intrinsic properties of a sample freely suspended over holes on prepatterned substrates, e.g., by AFM (see Fig. 1.14). As mentioned above, it is also possible to grow quasi-freely-suspended monolayers that basically retain their intrinsic properties on suitable weakly interacting substrates. Another possibility is a weakly interacting buffer layer that may stabilize an otherwise unstable monolayer by van der Waals interactions. However, often stronger covalent and electrostatic interaction forces dominate the interaction of a monolayer with the substrate surface, inducing noticeable changes in the structure and physicochemical properties of the adsorbed monolayer.

A critical issue is the control and measurement of the number of layers during the growth process. Thickness measurements are possible using optical contrast, Raman scattering, and scanning probe microscopy [97, 98]. Scanning AFM allows besides a structural analysis thickness measurements with atomic resolution [97]. This technique determines the number of layers by imaging step edges of the multilayer deposit.

1.7 Theory: First-Principles Calculations, Molecular Dynamics Simulations, Tight Binding Theory, and Empirical Rules

Ab initio calculations are the first choice for computing the pristine structure as well as the chemical and physical properties of 2D materials [11, 99]. First-principles calculations and MD simulations not only predicted the existence of new 2D materials but also helped to find possible routes of synthesis and to improve our understanding of 2D growth processes. In addition, theory helps to evaluate and interpret experimental results. An example is the computationally guided synthesis of 2D boron polymorphs (borophene), which can serve as a prototype for the development of synthetic 2D materials [100]. In addition to the prediction of the existence of individual monolayers, their physical properties, such as the piezoelectricity of molybdenum disulfide, the ferroelectricity of tin telluride, the topological defects in specified transition-metal dichalcogenides, the Dirac cone in borophene, and the high carrier mobility and anisotropy in black phosphorene, were predicted by theory [12].

The most important first-principles approaches are wavefunction-based methods and DFT calculations. The latter approximation treats the many-body system of electrons by considering noninteracting electrons moving in an effective potential. This method plays a leading role in present studies of 2D materials. To take the exchange and correlation effects in the calculation of electron density into account, the local density approximation (LDA) uses Ceperley–Alder (CA) functionals or Troullier–Martins (TM) pseudopotentials. In most DFT calculations, the generalized gradient approximation (GGA) is employed using Perdew–Burke–Ernzerhof (PBE) functionals [99]. State-of-the-art first-principles calculations allow the prediction of structures; supply information on electronic, optical, mechanical, and magnetic properties; and provide information on the influence of defects on the structure and properties of 2D materials, as discussed in a comprehensive review [12].

At present, most results are based on the numerical DFT-GGA-PBE approach, which plays a crucial role in our well-developed understanding of 2D materials. Note that large differences in cohesive energies may be observed between the LDA and GGA approaches, especially for few-layer systems, where in addition to strong covalent in-plane interactions, weak van der Waals interlayer forces or electrostatic interactions play a dominant role (soft layered materials) [13, 55]. This situation is responsible for inaccurate or even contradictory results published on the effective interlayer interaction energies in few-layer systems and the separation of monolayers in multilayer assemblies [99].

Molecular dynamics (MD) simulations complete the investigation of atomic structures and mechanical behavior of monolayers because they can handle systems of thousands or even millions of atoms in comparison with the approximately one hundred atoms usually manageable by DFT calculations [99]. The MD simulation is based on classical Newtonian dynamics, where potential energy functions, which describe the interaction energy between atoms, play a critical role. Typically, the accuracy of such simulations is lower than that of ab initio calculations owing to the lack of accurate potential functions, which can reproduce all relevant properties of a particular 2D system. However, MD simulations not only reveal the intrinsic structure of monolayers and few-layer systems but also provide valuable estimates of the thermal stability and physicochemical properties of 2D materials at temperatures above 0 K [101, 102].

An important potential is the Tersoff potential, which describes the potential energy of the atomic structure as a function of atomic coordinates [99]. The potential considers the effects of all nearest neighbors of an atom by describing the bond strength between two atoms that depends on its environment. Another interaction potential is the Stillinger–Weber potential, which considers bond stretching and bond bending energies. Unfortunately, accurate potential functions and thus accurate MD simulations are available only for a limited number of monolayers. Note that often the results deviate from those of DFT calculations, especially simulations performed at higher temperatures, considering the temperature dependence of specific properties. A potential that describes both weak van der Waals interactions and strong covalent bonding is the four-body adaptive intermolecular reactive empirical bond order (AIREBO) potential [99]. Additionally, this potential considers the torsion effect and changes in bond angles induced by bond rotations.

While the data presented for bonding, structure, and mechanical properties are primarily based on first-principles calculations, we also consider the tight-binding theory of solids, which assumes that the electronic state of a molecule or solid can be described by a linear combination of atomic states of the constituent atoms [50]. This method is closely related to approaches using the linear combination of atomic orbitals (LCAO). The LCAO approach and hybridization concept can deliver analytical expressions for the coupling parameters and the electronic-state energies [50]. This method takes only valence electrons into account, and the interaction between direct neighbors is parametrized based on a one-electron model. The tight-binding approach is the simplest model that can capture the subtleties in bonding that are consequences of the quantum mechanical nature of chemical bonding. Well-known examples of these quantum mechanical features include charge transfer, ionic and covalent bonding, and bond breaking, which are not allowed by MD models. Although this approach is less exact than numerical first-principles computations, simplified bonding models that follow chemical intuition often offer valuable insight and a deeper understanding of the nature of chemical bonding and thus allow the recognition of relationships within chemically related groups of monolayers based on the '*aufbau*' principle of the periodic table.

Besides these theoretical methods, empirical rules help to understand and categorize the general behavior of 2D materials. An example is the rule that most binary monolayers containing at least one element of the second period of the periodic table have a planar structure owing to the small atom size of these constituents. Another important rule originating from the behavior of bulk materials is the Griffith criterion, which postulates a quantitative relationship between Young's modulus and the ultimate strength of brittle materials [103]. According to this rule, the stiffness is approximately a factor of ten greater than the intrinsic strength. This criterion based on the behavior of perfect structures predicts an upper limit of the usually not well-known ultimate strength and thus provides important insight into the mechanical performance of 2D materials. Extensive theoretical data on linear mechanical properties, especially on Young's moduli, allow an estimate of the upper limit of the failure strength of 2D materials. This is important because information on the fracture strength of 2D materials is widely missing [15].

1.8 Concept of the Book

The family of theoretically predicted and experimentally synthesized and characterized 2D materials is growing rapidly by expanding to more elements of the periodic table and by extending to numerous combinations of these elements to binary, ternary, quaternary, and more complex compounds with diverse 2D structures. The strategy of the present treatment is to group chemically related 2D materials together in different chapters of the book, primarily based on their electronic configuration and bonding behavior, as dictated by the '*aufbau*' principle of the periodic table. It is possible to read each chapter of the book separately, preferably together with the introduction, which introduces fundamental knowledge on 2D materials that helps to better understand the following chapters. The review articles cited in this introduction present an in-depth view of the special subject. Owing to the rapidly growing number of publications, it was not possible to include all available publications of the exploding field of nanomaterials. The last chapter summarizes the main results discussed in the previous chapters, identifying 2D solids with distinctive types of chemical bonding, interesting physicochemical properties, and potential applications in devices.

An essential part of the book is the systematic discussion of the main group elements and their binary compounds, including group–IIA elements with preferred metallic bonding, covalently bonded elemental group–IIIA monolayers and their isoelectronic group IIA–IVA compounds, elemental and binary heteroelemental group–IVA monolayers and their isoelectronic group IIIA–VA compounds, elemental and binary heteroelemental group–VA monolayers and their isoelectronic group IVA–VIA compounds, and elemental group–VIA monolayers (see Fig. 1.1). The tables present the quantitative bonding information, structural data, and mechanical properties of these primarily covalently bonded monolayers. Especially monoelemental nanomaterials including one of the most studied monolayers graphene will play a key role in the realization of future applications because they can be easier synthesized with high quality and are more compatible with existing technologies.

The second extensive part of the book covers transition-metal dioxides (TMDOs) and transition-metal dichalcogenides (TMDCs), with the first discovered monolayer of MoS_2 as one of the best studied compounds of this class. Especially the binary compounds with MX_2 stoichiometries IVB–VIA, VB–VIA, VIB–VIA and XB–VIA are discussed in detail in the last four chapters (see Fig. 1.2). The applications expected for these monolayers and nanosheets include electronics, optoelectronics, electrocatalysis, photocatalysis, energy storage and conversion, water splitting, FET devices, and various kinds of 2D sensors. Despite these diverse theoretically predicted potential applications of TMDCs, compared with the progress made in the synthesis of monoelemental Xenes, the growth of 2D TMDCs is still not well developed. In fact, many theoretically well-characterized thermodynamically and dynamically stable TMDC monolayers have not been synthesized.

References

1. P. Joensen, R.F. Frindt, S.R. Morrison, Mater. Res. Bull. **21**, 457 (1986)
2. S. Tongay, Appl. Phys. Rev. **5**, 010401 (2018)
3. K.S. Novoselov, A.K. Geim, S.V. Morozov, D. Jiang, Y. Zhang et al., Science **306**, 666 (2004)
4. S.-H. Lin, J.-L. Kuo, Phys. Chem. Chem. Phys. **16**, 20763 (2014)
5. Y. Wang, Y. Li, Z. Chen, Acc. Chem. Res. **53**, 887 (2020)
6. J.-Y. Raty, M. Schumacher, P. Golub, V.L. Deringer, C. Gatti, et al. Adv. Mater. **1806280** (2018)
7. T.R. Galeev, B.D. Dunnington, J.R. Schmidt, A.I. Boldyrev, Phys. Chem. Chem. Phys. **15**, 5022 (2013)
8. A. Savin, R. Nesper, S. Wengert, T.F. Fässler, Angew. Chem. Int. Ed. **36**, 1808 (1997)
9. K. Koumpouras, J.A. Larsson, J. Phys. Condens. Matter **32**, 315502 (2020)
10. P. Hess, ChemPhysChem, e202100880 (2022)
11. J. Gao, Z. Xu, S. Chen, M.S. Bharathi, Y.-W. Zhang, Adv. Theory Simul. **1**, 1800085 (2018)
12. R. Xu, X. Zou, B. Liu, H.-M. Cheng, Mater. Today **21**, 391 (2018)
13. G. Graziano, J. Klimeš, F. Fernandez-Alonso, A. Michaelides, J. Phys. Condens. Matter **24**, 424216 (2012)
14. W. Tao, N. Kong, X. Ji, Y. Zhang, A. Sharma et al., Chem. Soc. Rev. **48**, 2891 (2019)
15. P. Hess, Nanoscale Horiz. **6**, 856 (2021)
16. S. Guo, Y. Zhang, Y. Ge, S. Zhang, H. Zeng, et al., Adv. Mater. **1902352** (2019)
17. Y. Guo, L. Ma, K. Mao, M. Ju, Y. Bai et al., Nanoscale Horiz. **4**, 592 (2018)
18. V. Sorkin, H. Pan, H. Shi, S.Y. Quek, Y.W. Zhang, Crit. Rev. Solid State Mater. Sci. **39**, 319 (2014)
19. K. Lasek, J. Li, S. Kolekar, P.M. Coelho, L. Guo et al., Surf. Sci. Rep. **76**, 100523 (2021)
20. J. Nevalaita, P. Koskinen, Phys. Rev. B **97**, 035411 (2018)
21. S. Ono, Phys. Rev. B **102**, 165424 (2020)
22. B. Peng, H. Zhang, H. Shao, Z. Ning, Y. Xu et al., Mater. Res. Lett. **5**, 399 (2017)
23. X. Wu, J. Dai, Y. Zhao, Z. Zhuo, J. Yang, X.C. Zeng, ACS Nano **6**, 7443 (2012)
24. B. Mortazavi, O. Rahaman, A. Dianat, T. Rabczuk, Phys. Chem. Chem. Phys. **18**, 27405 (2016)
25. C. Kamal, A. Chakrabarti, M. Ezawa, New J. Phys. **17**, 083014 (2015)
26. I. Lukačević, M.V. Pajtler, M. Mužević, S.K. Gupta, J. Mater. Chem. C **7**, 2666 (2019)
27. B. Anam, N. Gaston, J. Phys. Condens. Matter **33**, 125901 (2021)
28. M. Nakhaee, M. Yagmurcukardes, S.A. Ketabi, F.M. Peeters, Phys. Chem. Chem. Phys. **21**, 15798 (2019)

29. S.V. Badalov, M. Yagmurcukardes, F.M. Peeters, H. Sahin, J. Phys. Chem. C **122**, 28302 (2018)
30. D. Singh, S.K. Gupta, I. Lukačević, Y. Sonvane, RSC Adv. **6**, 8006 (2016)
31. D. Singh, S.K. Gupta, I. Lukačević, M. Muževič, Y. Sonvane, R. Ahuja, Sci. Rep. **9**, 17300 (2019)
32. S.M. Farzaneh, S. Rakheja, Phys. Rev. B **104**, 115205 (2021)
33. J.C. Garcia, D.B. de Lima, L.V.C. Assali, J.F. Justo, J. Phys. Chem. C **115**, 13242 (2011)
34. A. Mahmood, G. Rahman, J. Phys. Condens. Matter **32**, 205501 (2020)
35. R. John, B. Merlin, Cryst. Struct. Theory Appl. **5**, 43 (2016)
36. S. Mahmud, Md.K. Alam, RSC Adv. **9**, 42194 (2019)
37. Y. Zhang, J. Lee, W.-L. Wang, D.-X. Yao, Comput. Mater. Sci. **110**, 109 (2015)
38. F. Ersan, D. Kecik, V.O. Özçelik, Y. Kadioglu, O.Ü. Aktürk et al., Appl. Phys. Rev. **6**, 021308 (2019)
39. J. Lee, W.-C. Tian, W.-L. Wang, D.-X. Yao, Sci. Rep. **5**, 11512 (2015)
40. N. Anugrah, P. Namari, M. Saito, Jpn. J. Appl. Phys. **58**, 061003 (2019)
41. F. Ersan, E. Aktürk, S. Ciraci, Phys. Rev. B **94**, 245417 (2016)
42. D. Wang, L.-M. Tang, X.-X. Jiang, J.-Y. Tan, M.-D. He et al., Adv. Electron. Mater. **5**, 1800475 (2018)
43. L. Ramírez-Montes, W. López-Pérez, R. González-Hernández, C. Pinilla, Int. J. Quantum Chem. **120**, e26267 (2020)
44. L. Xian, A.P. Paz, E. Bianco, P.M. Ajayan, A. Rubio, 2D Mater. **4**, 041003 (2017)
45. Y. Xiang, S. Gao, R.-G. Xu, W. Wu, Y. Leng, Nano Energy **58**, 202 (2019)
46. Z. Zhu, X. Cai, S. Yi, J. Chen, Y. Dai, Phys. Rev. Lett. **119**, 106101 (2017)
47. S. Ono, Sci. Rep. **10**, 11810 (2020)
48. H. Bao, B. Zhao, J. Zhang, Y. Xue, H. Huan, Sci. Rep. **12**, 2129 (2022)
49. W.G. Zeier, A. Zewalkink, Z.M. Gibbs, G. Hautier, M.G. Kanatzidis et al., Angew. Chem. Int. Ed. **55**, 6826 (2016)
50. W.A. Harrison, Pure & Appl. Chem. **61**, 2161 (1989)
51. S.S. Batsanov, Inorg. Mater. **37**, 871 (2001)
52. P. Hess, Nanoscale Horiz. **5**, 385 (2020)
53. H. Tang, S. Ismail-Beigi, Phys. Rev. Lett. **99**, 115501 (2007)
54. M.-Y. Li, C.-H. Chen, Y. Shi, L.-J. Li, Mater. Today **19**, 322 (2016)
55. K.S. Novoselov, A. Mishchenko, A. Carvalho, A.H. Castro Neto, Science **353**, aac9439 (2016)
56. Y. Wang, J. Lv, L. Zhu, S. Lu, K. Yin et al., J. Phys. Condens. Matter **27**, 203203 (2015)
57. Y. Wang, Y. Li, Z. Chen, J. Mater. Chem. C **3**, 9603 (2015)
58. A.-Y. Lu, H. Zhu, J. Xiao, C.-P. Chuu, Y. Han et al., Nat. Nanotechnol. **12**, 744 (2017)
59. R.E. Roman, N.M. Pugno, S.W. Cranford, in *Advanced Computational Nanomechanics*, ed. by N. Silvestre (Wiley & Sons, Chichester, UK, 2016)
60. D. Akinwande, C.J. Brennan, J.S. Bunch, P. Egberts, J.R. Felts et al., Extreme Mech. Lett. **13**, 42 (2017)
61. B. Liu, M. Niu, J. Fu, Z. Xi, M. Lei et al., Phys. Rev. Mater. **3**, 054002 (2019)
62. P. Hess, J. Appl. Phys. **120**, 124303 (2016)
63. M.-Q. Le, Comput. Mater. Sci. **201**, 110854 (2022)
64. C. Lee, X. Wei, J.W. Kysar, J. Hone, Science **321**, 385 (2008)
65. A. Castellanos-Gomez, V. Singh, H.S.J. van der Zant, G.A. Steele, Ann. Phys. (Berlin) **527**, 27 (2015)
66. X. Li, M. Sun, C. Shan, Q. Chen, X. Wei, Adv. Mater. Interfaces **1701246** (2018)
67. K.K. Al-Quraishi, Q. He, W. Kauppila, M. Wang, Y. Yang, Int. J. Smart Nano Mater. **11**, 207 (2020)
68. L. Patra, R. Pandey, Mater. Today Commun. **31**, 103623 (2022)
69. S.I. Yengejeh, S.A. Kazemi, W. Wen, Y. Wang, RSC Adv. **11**, 20232 (2021)
70. W.-L. Tao, Y.-Q. Zhao, Z.-Y, Zeng, X.-R. Chen, H.-Y. Geng, ACS Appl. Mater. & Interfaces **13**, 8700 (2021)
71. W. Zhou, H. Gong, X. Jin, Y. Chen, H. Li et al., Frontiers Phys. **10**, 842789 (2022)

72. T. Das, S. Chakraborty, R. Ahuja, G.P. Das, ChemPhysChem **20**, 608 (2018)
73. Y. Qu, H. Pan, C.T. Kwok, Sci. Rep. **6**, 34186 (2016)
74. R. Sukanya, D.C. da Silva Alves, C.B. Breslin, J. Electrochem. Soc. **169**, 064504 (2022)
75. W. Wang, Z. Sun, W. Zhang, Q. Fan, Q. Sun et al., RSC Adv. **6**, 54874 (2016)
76. L. Cai, Q. Zhang, J.P. Mwizerwa, H. Wan, X. Yang et al., ACS Appl. Mater. Interfaces **10**, 10053 (2018)
77. H. Chen, J. Zhang, D. Kan, J. He, M. Song et al., Crystals **12**, 1381 (2022)
78. Y. Tsuboi, F. Wang, D. Kozawa, K. Funahashi, S. Mouri et al., Nanoscale **7**, 14476 (2015)
79. C. Huang, J. Zhou, H. Wu, K. Deng, P. Jena et al., J. Phys. Chem. Lett. **7**, 1919 (2016)
80. A. Zhao, B. Wang, APL Mater. **8**, 030701 (2020)
81. Z. Lin, B.R. Carvalho, E. Kahn, R. Lv, R. Rao, et al., 2D Mater. **3**, 022002 (2016)
82. W. Tian, W. Li, W. Yu, X. Liu, Micromachines **8**, 163 (2017)
83. G. Yang, L. Li, W.B. Lee, M.C. Ng, Sci. Technol. Adv. Mater. **19**, 613 (2018)
84. P. Miró, M. Audiffred, T. Heine, Chem. Soc. Rev. **43**, 6537 (2014)
85. N. Mounet, M. Gibertini, P. Schwaller, A. Merkys, I.E. Castelli et al., Nat. Technol. **13**, 246 (2018)
86. J. Zhou, L. Shen, M.D. Costa, K.A. Persson, S.P. Ong et al., Scientific Data **6**, 86 (2019)
87. M. Fukuda, J. Zhang, Y.-T. Lee, T. Ozaki, Mater. Adv. **2**, 4392 (2021)
88. A.J. Mannix, B. Kiraly, M.C. Hersam, N.P. Guisinger, Nat. Rev. Chem. **1**, 0014 (2017)
89. Y. Shi, H. Li, L.-J. Li, Chem. Soc. Rev. **44**, 2744 (2015)
90. Y. Wang, J. Pang, Q. Chen, L. Han, Y. Li et al., Nano-Micro Lett. **13**, 143 (2021)
91. J. Wang, T. Li, Q. Wang, W. Wang, R. Shi et al., Mater. Today Adv. **8**, 100098 (2020)
92. K.A. Lozovoy, I.I. Izhnin, A.P. Kokhanenko, V.V. Dirko, V.P. Vinarskiy et al., Nanomaterials **12**, 2221 (2022)
93. T.H. Choudhury, X. Zhang, Z.Y. Al Balushi, M. Chubarov, J.M. Redwing, Ann. Rev. Mater. Sci. **50**, 155 (2020)
94. J. Dong, L. Zhang, X. Dai, F. Ding, Nat. Commun. **11**, 5862 (2020)
95. D.K. Singh, G. Gupta, Mater. Adv. **3**, 6142 (2022)
96. J. Lu, X. Zhang, G. Su, W. Yang, K. Han et al., Mater. Today Phys. **18**, 100376 (2021)
97. C.J. Shearer, A.D. Slattery, A.J. Stapleton, J.G. Shapter, C.T. Gibson, Nanotechnology **27**, 125704 (2016)
98. X. Zhang, A. Beyer, Nanoscale **13**, 1443 (2021)
99. B. Liu, K. Zhou, Prog. Mater. Sci. **100**, 99 (2019)
100. A.J. Mannix, Z, Zhang, N.P., Guisinger, B.I. Jakobson, M.C. Hersam, Nat. Nanotechnol. **13**, 444 (2018)
101. Z.-D. Sha, Q.-X. Pei, K. Zhou, Z. Dong, Y.-W. Zhang, Extreme Mech. Lett. **19**, 39 (2018)
102. Y.-P. Zhou, J.-W. Jiang, Sci. Rep. **7**, 45516 (2017)
103. H. Yin, H.J. Qi, F. Fan, T. Zhu, B. Wang et al., Nano Lett. **15**, 1918 (2015)

Chapter 2
Elemental Group–IIA Monolayers

Abstract In the development of 2D materials, the monolayers of group–IIA elements of the periodic table have played a minor role. Only a few groups are studying monolayers of these elements, which are of interest primarily owing to their metallic behavior and multivalency. Coordination to a larger number of nearest neighbors with a smoother electron distribution than for localized covalent bonding is the main characteristic of these structures with metallic bonding. However, due to the low cohesive energies of single-atomic layers, synthesis is quite difficult in comparison with that of their bulk structures. The monolayers of the alkaline-earth metals Be, Mg, Ca, Sr, and Ba are referred to as beryllene, magnesene, calcene, strontene, and barene, analogous to the nomenclature used for the other main group elements. The known 3D structures of these metals clearly prefer close-packed structures or 3D clusters of atoms rather than extended layered assemblies. This explains why systematic investigations of metallic monolayers for alkaline-earth elements but also other metals are still widely lacking. Since metallic bonding allows atoms to be more mobile than in rigid directional covalent bonding, these 2D networks are of special interest for applications involving surface chemistry, such as catalysis and sensing.

2.1 Chemistry and 2D Bonding

The group–IIA or alkaline-earth elements belong to the main groups of the periodic table, as indicated in Fig. 2.1.

They possess only two valence electrons in their outer valence shell that are available for chemical bonding. Figure 2.2 displays a scheme of the electronic configuration of the outer valence shell, which designates the electronic ground state of the atoms Be (Ne|2s^2), Mg (Ne|3s^2), Ca (Ar|4s^2), Sr (Kr|5s^2), and Ba (Xe|6s^2). To reach the next electronic noble gas configuration six additional valence electrons are needed, however, the alkaline-earth atoms tend to lose the two valence electrons to form Be^{2+}, for example. The reactivity of the metals increases as we go down the column from Be to Ba. The deficiency of valence electrons with respect to the 'octet rule' favors coordination above four, specifically up to six in the plane,

P. Hess, *Bonding, Structure, and Performance of Two-Dimensional Materials*, NanoScience and Technology, https://doi.org/10.1007/978-3-031-74161-6_2

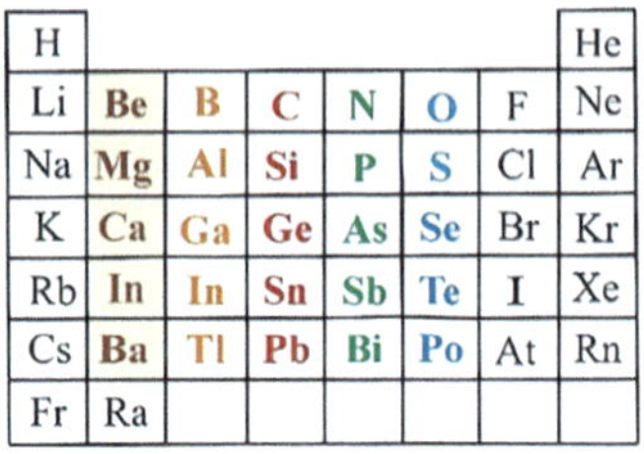

Fig. 2.1 Scheme of main-group elements of the periodic table, highlighting the positions of the alkaline-earth elements (group IIA)

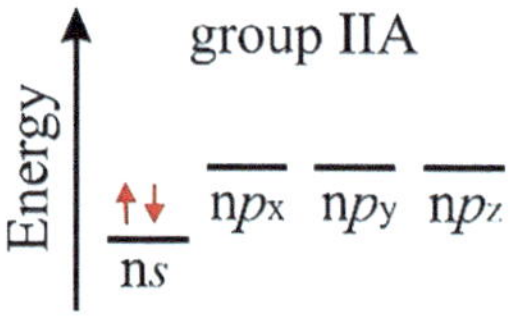

Fig. 2.2 Electronic configuration of the group–IIA elements with two valence electrons in the outer shell involved in chemical bonding

connected with a weakening of individual bonds and elongation of bond lengths. Compared with the thoroughly studied covalently bonded 3D materials, monolayers with metallic bonding, such as the group-IIA monolayers, have found increased interest only recently.

In a systematic study of elemental atomically thin layers density functional theory (DFT) calculations based on the generalized gradient approximation (GGA) using Perdew–Burke–Ernzerhof (PBE) exchange and correlation functionals were employed to derive 2D cohesive energies of the ideal 2D structures of 45 metals from their 3D bulk properties [1]. To validate the method employed, the calculated bulk values were compared with their experimental values. The study included hypothetical triangular-hexagonal, buckled square, and honeycomb structures of the group-IIA elements. The cohesive energy of the triangular-hexagonal structure decreases strongly from $E_c = -2.91$ to -0.91 eV/atom going from Be to Mg and then stays at approximately -1 eV/atom for the higher homologues. It is well-known that *s* metals have lower cohesive energies than *d* metals.

An independent study calculated the cohesive energies of the buckled-hexagonal, honeycomb, and buckled-square structures [2]. A comparison of the cohesive energies of various phases reveals a qualitatively similar behavior. In Table 2.1, the cohesive energies of the two most stable hexagonal structures are displayed [1, 2]. Negative values describe an exothermic reaction of free atoms to a bonded monolayer and indicate energetic stability at zero temperature and pressure.

Metallic bonding prefers close-packed structures with large coordination numbers, e.g., up to twelve nearest neighbors in 3D bulk arrangements [1]. The limitation of the maximum coordination number to six in planar arrangements reduces the binding energy achievable in two dimensions (see Fig. 2.3). Note that the electron density and thus the bonding strengths vary with the coordination number. In fact, the energy of cohesion decreases from the 3D structure with twelve nearest neighbors to a 2D structure with six nearest neighbors to 70% (hexagonal), four nearest neighbors to 65% (square), and three nearest neighbors to 54% (honeycomb) [1]. It

Table 2.1 Symmetry, cohesive energiy, lattice constant, bond length, Young's modulus, and Poisson's ratio of the triangular-hexagonal and buckled-hexagonal structures of group–IIA monolayers

Monolayer	Monolayer symmetry	Cohesive energy, E_c (eV/atom)	Lattice constant a (nm)	Bond length d (nm)	Young's modulus, E_{ac}, E_{zz} (N/m)	Poisson's ratio ν_{ac}, ν_{zz}
Beryllene	Triang., hex. Buckl., hex.	– 2.91 [1] – 3.35 [2]	0.216 [2]	0.215 [1]	73 [1]	0.228 [1]
Magnesene	Triang., hex. Buckl., hex.	– 0.91 [1] – 1.24 [2]	0.309 [2]	0.307 [1]	34 [1]	0.068 [1]
Calcene	Triang., hex. Buckl., hex.	– 1.09 [1] – 1.58 [2]	0.392 [2]	0.388 [1]	16 [1]	0.144 [1]
Strontene	Triang., hex. Buckl., hex.	– 0.85 [1] – 1.33 [2]	0.429 [2]	0.427 [1]	12 [1]	0.147 [1]
Barene	Triang., hex. Buckl., hex.	– 1.04 [1] – 1.59 [2]	0.449 [2]	0.448 [1]	8.0 [1]	0.208 [1]

Fig. 2.3 Schematic representation of the orbitals of a planar structure with six-fold coordination

is important to note that individual bonds may be stronger in 2D networks than in the bulk; however, the greater energy of cohesion in three dimensions leads to a preferred growth of 3D clusters instead of layered assemblies. The correlation of the binding energies of the 2D and 3D structures causes a tradeoff between sufficient cohesion to form a stable 2D structure and low enough 3D cohesion to avoid preferential growth of more stable 3D clusters [1].

2.2 2D Structures

First-principles DFT-PBE calculations were performed revealing the cohesive energies and bond lengths of the hexagonal triangular structure with six-fold coordination for group–IIA monolayers (see Fig. 2.4) [1].

The cohesive energies vary strongly between $E_c = -2.91$ eV/atom (Be) and -1.04 eV/atom (Ba), and the bond lengths between $d = 0.215$ nm (Be) and 0.448 nm

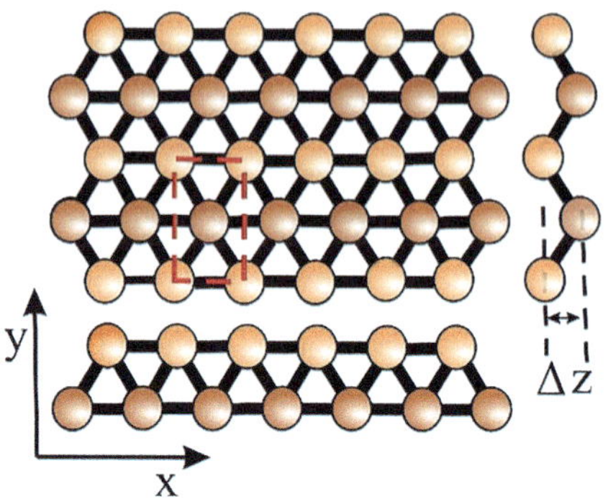

Fig. 2.4 Schematic representation of the triangular-hexagonal structure with a unit cell containing two atoms. Adapted with permission from ref. [1], Copyright 2018, American Physical Society

(Ba), as can be seen in Table 2.1. Only slightly smaller values were obtained for the buckled square structure with four-fold coordination, with the corresponding binding energies of $E_c = -2.57$ eV/atom (Be) and -0.90 eV/atom (Ba) and the bond lengths d = 0.202 nm (Be) and 0.424 nm (Ba). Additional calculations performed for the honeycomb structure with three-fold coordination delivered much smaller cohesive energies but comparable bond lengths. Besides the most stable triangular configuration, the square and honeycomb monolayers with four- and three-fold coordination possess thermodynamic and thermal stability, despite smaller binding energies.

Additional calculations, based on DFT implemented in the quantum ESPRESSO (QE) code determined the cohesive energies, lattice constants, and buckling effect of the buckled hexagonal structure (see Fig. 2.5) of the group-IIA monolayers [2]. Here the cohesive energies varied between $E_c = -3.35$ eV/atom (Be) and -1.59 eV/atom (Ba), the lattice constants between a = 0.216 nm (Be) and a = 0.449 (Ba), and the buckling heights between $\Delta z = 0.096$ nm (Be) and 0.176 nm (Ba). The quite similar values for the buckled square structure are $E_c = -3.18$ eV/atom (Be) and $E_c = -1.55$ eV/ atom (Ba), a = 0.209 nm (Be) and a = 0.434 (Ba), and $\Delta z = 0.080$ nm (Be) and 0.161 nm (Ba) (see Fig. 2.6). For the planar hexagonal structure smaller binding energies and comparable lattice constants were found.

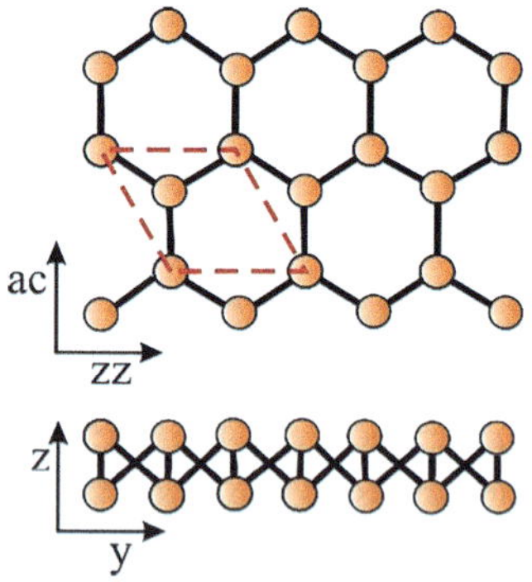

Fig. 2.5 Schematic representation of the buckled-hexagonal structure with a unit cell containing two atoms. Adapted with permission from ref. [1], Copyright 2018, American Physical Society

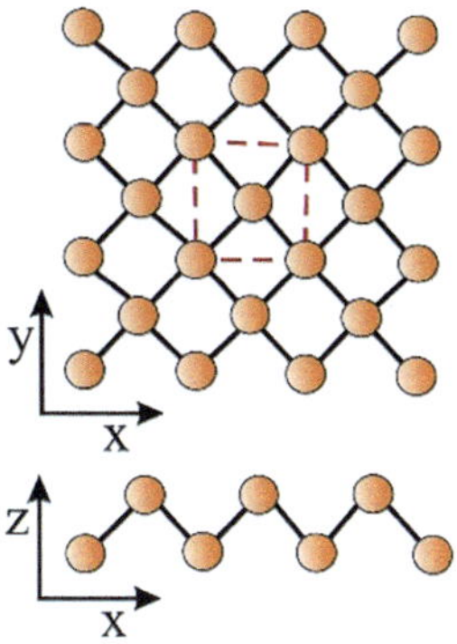

Fig. 2.6 Schematic representation of the buckled-square structure with a unit cell containing two atoms. Adapted with permission from ref. [1], Copyright 2018, American Physical Society

2.3 Mechanics

The presented linear mechanical properties of the triangular-hexagonal structure, specifically the in-plane Young's moduli and Poisson's ratios in the armchair and the zigzag directions, originate from the published elastic constants (see Table 2.1) [1]. The generally low and drastically decreasing stiffnesses of 73 N/m (Be) and 7.8 N/m (Ba) are due to the metallic nature of bonding. Note that these values are a factor of 5 (Be) to 44 (Ba) smaller than those of graphene. While the stiffness decreases drastically, the cohesive energy of the higher homologues is roughly constant. The elastic constants indicate mechanical stability for the monolayers of the triangular-hexagonal and honeycomb structures but not for the square structure [1]. Similarly, the hexagonal and honeycomb structures were found to be dynamically stable, except for Ba, which means that they are stable against out-of-plane vibrations [2]. The reported Poisson ratios exhibit irregular behavior when going to the higher homologues, with a surprisingly low value for magnesene.

2.4 Physics and Applications

The thermodynamic stability of 2D metals is a topic of general interest. Nondirectional metallic bonding makes it difficult to form stable atomically thin layers. Therefore, the term 'two-dimensional' frequently is used for freestanding sheets of ultrathin metallic nanomaterials [3]. The present status of preparation methods and problems encountered in the synthesis of metallic layers, as well as their potential applications, have been reviewed [3, 4]. For example, the investigation of a wide variety of metallic nanomaterials for use in electrocatalysis is an active field of research in which the high surface-to-volume ratio and the existence of dangling bonds and other active sites play important roles [4]. This report not only describes different wet-chemical synthesis techniques but also treats in detail various reactions of current interest in electrocatalysis [4]. Other potential state-of-the-art applications suggested for 2D metals include bioimaging, biomedical engineering, photothermal therapy, sensing, and memory nanodevices [3]. The specific role that alkaline-earth nanomaterials will play in the future development of nanotechnology is not known.

Alkaline-earth metal ions play a significant role in nanotechnology. They are used to decorate monolayers and as doping agents that change the band structure, density of states (DOS), and optical properties of monolayers [5]. A first-principles study showed that in MoS_2 monolayers doped with Be, Mg, Ca, Sr, or Ba the lattice distortion in the doped system suppresses the recombination of photogenerated hole-electron pairs. The simultaneously induced reduction of the band gap enhances the likelihood of electron transfer from the valence to the conduction band. These doped systems with favorable band structures and superior visible absorption qualify for potential application in photocatalytic water splitting.

2.5 Synthesis

The synthesis of free-standing or deposited intact single layers of main-group–IIA elements is still lacking. Various wet chemical synthesis methods using capping agents that reduce the surface energy were employed to synthesize nanomaterials from noble and nonnoble metals [4]. For example, organic ligands such as amines or gas molecules such as CO have been selectively adsorbed to block specific crystal facets and to stabilize the needed 2D structure. Their strong adsorption significantly inhibits the growth of relevant specific facets, resulting in the growth of 2D metals [3]. Consequently, the large surface energy of specific facets is not favorable for 2D growth and is responsible for the intrinsic tendency of metals to form 3D close-packed clusters. Importantly, synthesized metallic networks often consist of nanosheets, which contain an unknown number of monolayers, but they may still exhibit a dominating effect of the exposed surface. On the other hand, the low in-plane stiffness allows flexible adsorption and additional stabilization on suitable surfaces, which helps to realize useful metallic nanomaterials. In addition to surface capping, other strategies for thermodynamic and kinetic control of the growth process include chemical reactions confined to the interlayer space of layered materials (space-confined growth) or template-assisted growth using 2D materials such as graphene oxide or transition-metal dichalcogenides (TMDCs) as templates [3, 4].

References

1. J. Nevalaita, P. Koskinen, Phys. Rev. B **97**, 035411 (2018)
2. S. Ono, Phys. Rev. B **102**, 165424 (2020)
3. T. Wang, M. Park, Q. Yu, J. Zhang, Y. Yang, Mater. Today Adv. **8**, 100092 (2020)
4. Z. Li, L. Zhai, Y. Ge, Z. Huang, Z. Shi et al., Nat. Sci. Rev. **9**, nwab142 (2021)
5. L.-Z. Liu, X.-S. Yu, S.-X. Wang, L.-L. Zhang, X.-C. Zhao et al., Molecules **28**, 6122 (2023)

Chapter 3
Elemental Group–IIIA Monolayers

Abstract The elements in group IIIA of the periodic table have three valence electrons in their outer valence shell and need an additional electron to reach the energetically preferred noble gas configuration. This reactive open-shell configuration results in electron-deficient bonding involving complex structures in two-dimensional space. The diverse polymorphic structures with unique chemical and physical properties have attracted enormous attention, especially of boron monolayers, whereas the higher homologues have not yet received comparable attention. As discussed in the following chapter, all group-IIIA homologues can form extended covalent 2D networks based on sp^2 hybridization. Importantly, electron-deficient multicentered bonding provides access to flexible reactive chemistry. First-principles calculations revealed that the boron homologues have low cohesive energies for the individual bonds due to the missing valence electron. This effect enlarges with increasing atom size. Nevertheless, several energetically and dynamically stable 2D structures and a variety of 2D bonding types exist in borophene, aluminene, gallenene, indenene, and thallenene monolayers formed by B, Al, Ga, In, and Tl. The calculated and measured mechanical properties show strongly decreasing stiffnesses and ultimate strengths for the higher homologues of boron. Potential uses include energy storage, energy conversion, and energy harvesting devices and applications in electronics, optoelectronics, spintronics, various sensors, and biomedicine based on appropriate properties such as high carrier mobility, high thermal conductivity, and superconductivity. Besides a detailed discussion of theory, the chapter outlines recent progress made in the synthesis of the most stable monolayers and few-layer systems. Diagrams elucidate the structure–property relationships of the cohesive energy versus bond length and the linear and nonlinear mechanical properties of group-IIIA monolayers.

P. Hess, *Bonding, Structure, and Performance of Two-Dimensional Materials*,
NanoScience and Technology, https://doi.org/10.1007/978-3-031-74161-6_3

3.1 Boron Monolayers (Borophene)

3.1.1 Chemistry and 2D Bonding

The group–IIIA elements, including boron, are among the main groups of elements in the periodic table, as highlighted in Fig. 3.1.

In its three-dimensional (3D) structures, boron is a metalloid that is neither a metal nor a nonmetal that primarily is of interest owing to its semiconducting properties. According to chemical experience, boron atoms prefer to add an electron to form B^-, which is isoelectronic to carbon, instead of losing an electron to form B^+. This behavior is due to electron deficiency in chemical bonding caused by the missing valence electron, needed to fulfill the octet rule. Since bulk boron has no layered structure, isolation via exfoliation is not possible. However, small boron clusters ($n < 15$) can form covalent quasi-planar molecules with aromatic or antiaromatic binding interactions. Most stable 2D structures grown by chemical synthesis exhibit metallic properties and represent the lightest 2D metals. The reactive open shell configuration allows covalent and ionic bonding, leading to a variety of distinct polymorphic structures.

Corresponding to its position in the periodic table, boron has the electronic configuration ($1s^2 2s^2 p_x{}^1 p_y{}^0 p_z{}^0$) or in a short form ($\mathrm{He}|2s^2p^1$) with the helium electronic configuration of the inner electrons and three valence electrons in its outer valence shell. Therefore, boron cannot form a closed-shell structure by conventional two-center, two-electron bonds. Evidently, the three electrons can form extended covalent networks by sp^2 hybridization, however, only electron-deficient bonding is possible because the $2p_z$ orbitals remain empty. The missing electron reduces the strength of chemical bonding and thus the energy of cohesion of the 2D network. The poor electronic bonding configuration in boron monolayers leads to electron delocalization between three or more atoms. In principle, boron has variable coordination numbers between 3 and 6 and can form delocalized multicenter bonds with up to seven-center, two-electron bond arrangements (7c–2e bonds) [1]. Here, we discuss in more detail the widespread three-center, two-electron bond (3c–2e) involving a motif of three chemically bonded boron atoms that is realized in many 2D networks.

Striped borophene: The following treatment of three-centered bonds is based on tight-binding theory and uses the linear combination of atomic orbitals (LCAO). The LCAO *ansatz* describes the electronic configuration with promoted electrons forming

H							He
Li	Be	B	C	N	O	F	Ne
Na	Mg	Al	Si	P	S	Cl	Ar
K	Ca	Ga	Ge	As	Se	Br	Kr
Rb	Sr	In	Sn	Sb	Te	I	Xc
Cs	Ba	Tl	Pb	Bi	Po	At	Rn
Fr	Ra						

Fig. 3.1 The main groups of the periodic table with the highlighted group–IIIA elements

three equivalent atomic orbitals, where three atoms share the three valence electrons. Figure 3.2 shows the ground state and the promoted electronic configuration leading to an atom arrangement with three-centered bonding. Boron is the lightest atom that forms extended 2D phases by sp^2-type hybrids, forming delocalized (3c–2e) bonds. According to tight-binding theory, overlap of the sp^2-like hybrids yields one bonding orbital and two antibonding orbitals in the triangular boron motif (see Fig. 3.3) [2].

This means that only two electrons are in bonding states, and one is in an antibonding orbital, reducing the bonding energy. With progressing combination reactions, the triangular motifs assemble to form an extended 2D network and the atomic states broaden into bands owing to increasing orbital overlap (see Fig. 3.4) [2].

As mentioned above, only two of the three valence electrons are in an in-plane bonding orbital, but one electron is in an in-plane antibonding orbital and not in the remaining empty $2p_z$-bonding orbital. This is one reason, the planar triangular hybrid is unstable, which in principle should be stable owing to the small atom size of boron. Instead, hybridization with a high-energy electron state and transition into a σ–π bonding combination of in-plane and out-of-plane states with a buckled configuration occurs to gain further stability [2]. As a consequence, in these mixed hybrids a small fraction of electrons occupies bonding π orbitals, and a small fraction of σ states stay unoccupied. Note that the monolayer is metallic due to partial occupancy of the σ and π states. This is one reason for the instability of a planar triangular network and the buckled or stripped monolayer of borophene has a lower energy [2].

Negative cohesive energies define stable chemical bonding and energetic stability at 0 K. The cohesive energy is the amount of energy needed to break the 2D network

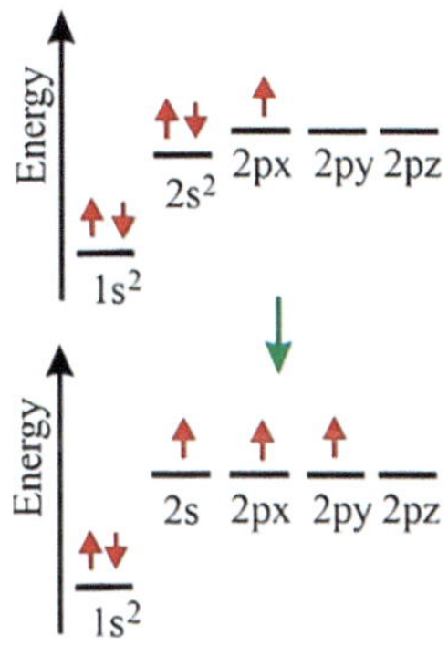

Fig. 3.2 Promotion of the electronic ground state of boron atoms into the triangular bonding state with three equivalent sp^2 orbitals

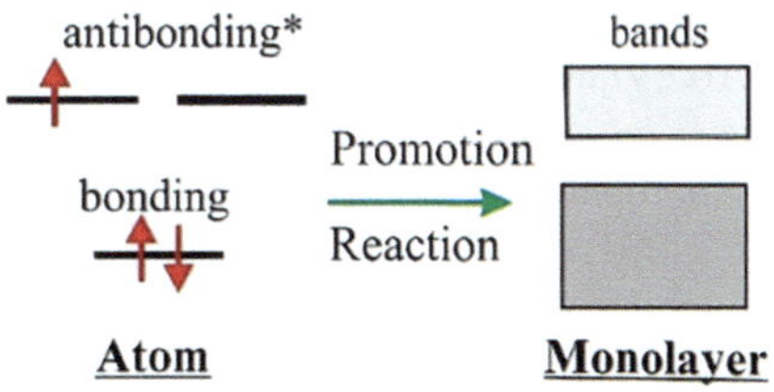

Fig. 3.3 Transition from the ground state of the atom with two bonding states and one antibonding state via promotion to delocalized bonding in monolayers, forming broadened bands

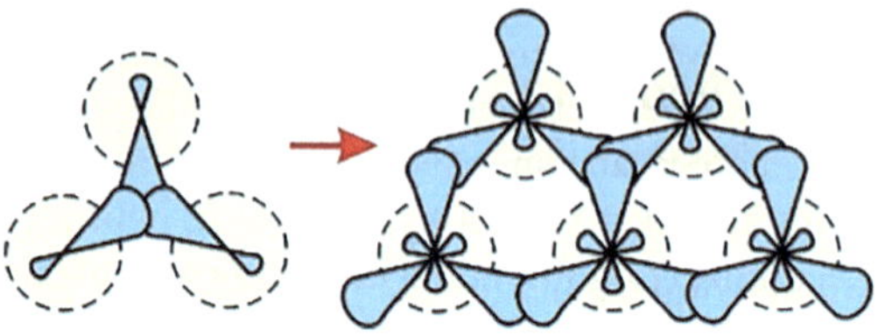

Fig. 3.4 Schematic presentation of monolayer formation via the three equivalent orbitals in boron

completely apart into isolated atoms or the reaction energy released (negative) by combining free atoms to the specific 2D network. Note that the calculated cohesive energies depend sensitively on the theoretical approximation employed and therefore may vary substantially. With the density functional theory (DFT)-local density approach (LDA), the largest binding energy of $E_c = -6.74$ eV/atom was obtained [2]. Application of the DFT approach with the generalized gradient approximation (GGA) and Perdew–Burke–Ernzerhof (PBE) functionals for exchange correlation generally leads to much lower cohesive energies of – 6.10 eV/atom [3], – 6.00 eV/atom [2], and – 5.66 eV/atom [4] for striped borophene. The so-called buckled or striped δ_6 borophene has the space group *Pmmn* [3]. Note that even the latter lower values indicate strong covalent bonding for three-centered binding arrangements. The structure, space group, cohesive energy, lattice constant, bond length, corrugation, and van der Waals diameter of group–IIIA monolayers are presented in Table 3.1.

Planar borophene: Interestingly, in addition to the striped phase, stable planar borophene phases exist. By combining electron donors, as provided by the three-center flat triangular regions with a surplus of electrons in an antibonding state, and two-center hexagonal sites with holes, acting as electron acceptors, in the right proportion, boron can form stable polymorphs consisting of triangular sites and hexagonal sites with holes [2]. A maximum stabilization is achieved when the fraction of occupied nonbonding states fill the fraction of unoccupied bonding σ states in the polymorph. The hexagon hole density is used to characterize these planar phases by the ratio $\eta = m/N$, where m is the number of hexagon holes and N is the number of triangular lattice sites [4, 5]. The resulting highly symmetric perfectly flat free-standing metallic monolayers can achieve high stability and have already been synthesized by deposition on suitable substrates [6]. To explain the high stability of these planar phases, aromatic or antiaromatic bonding has been invoked [1]. From the large number of flat monolayers, two highly symmetric phases called β_{12} and χ_3 borophene with a hole concentration ratio of $\eta = 1/6$ for β_{12} and $\eta = 1/5$ for the χ_3 structure will be discussed in more detail [4].

The reported cohesive energies of striped δ_6 borophene are between $E_c = -6.10$ eV/atom [3] (δ_6) and – 5.66 eV/atom [4] (δ_6), those of planar β_{12} borophenes are $E_c = -6.15$ eV/atom [3] (β_{12}) and – 5.71 eV/atom [4] (β_{12}), and those of planar χ_3 borophenes $E_c = -6.16$ eV/atom [3] (χ_3) and – 5.72 eV/atom [4] (χ_3), based on DFT-PBE calculations. Surprisingly, these cohesive energies of the planar structures are comparable to those reported for striped borophene, despite vacancy-mediated structural motifs creating strongly anisotropic polymorphic structures (see Table 3.1)

Table 3.1 Structure, space group, cohesive energy, lattice constants, bond length, corrugation, van der Waals diameter, and corrugation corrected van der Waals diameter of elemental group–IIIA monolayers

Monolayer	Structure, space group	Cohesive energy, E_c (eV/atom)	Lattice constants a, b (nm)	Bond lengths d_1, d_2 (nm)	Corrugation Δz_1, Δz_2 (nm)	Van der Waals diameter dw [8], $d_{w+\Delta z}$ (nm)
Borophene	Triangular, δ6 *Pmmn* [3]	– 6.00 [2] – 6.10 [3]	0.161, 0.286 [3] 0.322, 0.329 [4]	0.160, 0.186 [2]	0.088 [2] 0.091 [3]	0.36 [8], 0.448
Borophene	Planar, ß12 *Pmm*2 [3]	– 6.15 [3] – 5.71 [4]	0.507, 0.293 [3] 0.292, 0.506 [4]	~ 0.158–0.167 [15]	0 [3]	0.36 [8]
Borophene	Planar, χ3 *Cmmm* [3]	– 6.16 [3] – 5.72 [4]	0.445, 0.445 [3] 0.290, 0.444 [4]	~ 0.158–0.167 [15]	0 [3]	0.36 [8]
Borophene	4-layer slab 8-*Pmmn* [19]	– 6.33 [19]	0.452, 0.326 [19]	~ 0.17–0.19 [19]	0.218 [19]	0.36 [8], 0.578
Borophene	Pucker, acw *Pmn*21 [20]	– 5.73 [20]	0.319, 0.530 [20]	0.157–0.226 [20]	0.126, 0.202 [20]	0.36 [8], 0.562
Aluminene	Planar honeycomb *P6/mmm* [46]	– 1.96 [46] – 2.32 [47]	0.449 [46] 0.446 [47]	0.259 [46] 0.257 [47]	0 [46]	0.42 [8]
Aluminene	Buckled *P*-3*m*1 [49]	– 3.27 [47] – 3.47 [49]	0.272 [47] 0.275 [49]	0.296 [47] 0.275, 0.292 [49]	0.252 [47] 0.245 [49]	0.42 [8], 0.672 0.42 [8], 0.665
Aluminene	Triangular *Pmmn* [50]	– 2.76 [47] – 3.27 [50]	0.268, 0.465 [47] 0.269, 0.272 [50]	0.268 [47]	0.217 [50]	0.42 [8], 0.637
Aluminene	4-layer slab 8-*Pmmn* [49]	– 3.16 [49]	0.674, 0.503 [49]	0.258–0.286 [49]	0.344 [49]	0.42 [8], 0.764
Gallenene	Quasi planar a100-Ga, *Pbam*[58]	– 1.99 [51]	0.787, 0.465 [58]	0.266, 0.267 [58] 0.250, 0.251 [59]	0 [58]	0.42 [8]
Gallenene	Buckled b010-Ga	– 2.32 [60]	0.474, 0.492 [58] 0.465, 0.482 [60]	0.273 [58] 0.269, 0.274 [59]	0.119 [58] 0.132 [59]	0.42 [8], 0.539 0.42 [8], 0.552

(continued)

Table 3.1 (continued)

Monolayer	Structure, space group	Cohesive energy, E_c (eV/atom)	Lattice constants a, b (nm)	Bond lengths d_1, d_2 (nm)	Corrugation Δz_1, Δz_2 (nm)	Van der Waals diameter dw [8], $d_{w+\Delta z}$ (nm)
Gallenene	Triangular Hexagonal buckled	– 2.20 [51] – 2.52 [52]	0.282 [52]	0.277 [51]	0.130 [52]	0.42 [8], 0.55
Indiene	Planar *P6/mmm*[68]	– 1.81 [67] – 2.08 [68]	0.496 [67] 0.489 [68]	0.286 [67] 0.282 [68]	0 [67]	0.44 [8]
Indiene	Buckled *P3m*1[68]	– 1.83 [67] – 2.13 [68]	0.424 [67] 0.424 [68]	0.289 [67] 0.302 [68]	0.154 [67]	0.44 [8], 0.594
Indiene	Pucker, sw	– 1.88 [67]	0.425, 0.568 [67]	0.291, 0.287 [67]		0.44 [8]
Indiene	Triangular Hexagonal buckled	– 1.92 [51] – 2.17 [52]	0.323 [52]	0.316 [51]	0.146 [52]	0.44 [8], 0.586
Thalliene	Triangular Hexagonal buckled	– 1.67 [51] – 1.85 [52]	0.338 [52]	0.331 [51]	0.153 [52]	0.44 [8], 0.593

[2]. In spite of the energetic stability of the planar phases, isolation of their monolayers in free-standing form is still lacking. Note that the name borophene is used for all 2D phases of boron.

Despite the prediction of the basic structure of 2D boron sheets in 2006, experimental synthesis had to wait until 2015 [6, 7]. The reasons for this delay were the nonlayered bulk structure and the special type of complex electron-deficient bonding. Table 3.1, which contains a plethora of quantitative data, shows the progress made in recent years. The table displays the cohesive energies and the structural data of various borophenes, including their space group, lattice constants, bond lengths, buckling, layer thickness, and van der Waals diameter. For corrugated structures, the table provides an estimate of the layer thickness $d_{w+\Delta z}$, obtained by adding the corresponding corrugation effect Δz to the van der Waals diameter d_w [8].

3.1.2 2D Structures

Striped borophene: According to the '*aufbau*' principle, the basic unit of a free-standing boron monolayer consists of regular triangular boron motifs. Theory predicts a surprisingly large cohesive energy of – 6.76 eV/atom for a planar triangular structure [9]. However, the planar configuration occupies a local minimum in the potential energy surface (PES). The pyramid-like B7 complex, with a boron atom in the middle of a hexagon basis is the more stable configuration (see Fig. 3.5) [9]. Thus, a structure buckled vertically along the zigzag direction has a higher stability, where the magnitude of buckling must be determined by theory.

A set of lattice parameters are, for example, the lattice constants a = 0.161 nm, b = 286 nm [3], the bond lengths d_1 = 0.160 nm [2], d_2 = 0.186 nm [2], and the buckling height Δz = 0.088 nm [2]. A schematic representation of the calculated atomic structure of this elemental monolayer is shown in Fig. 3.6. Owing to its highly anisotropic structure, striped borophene shows pronounced anisotropic properties, such as extreme angle-dependent mechanical behavior. Despite a large cohesive

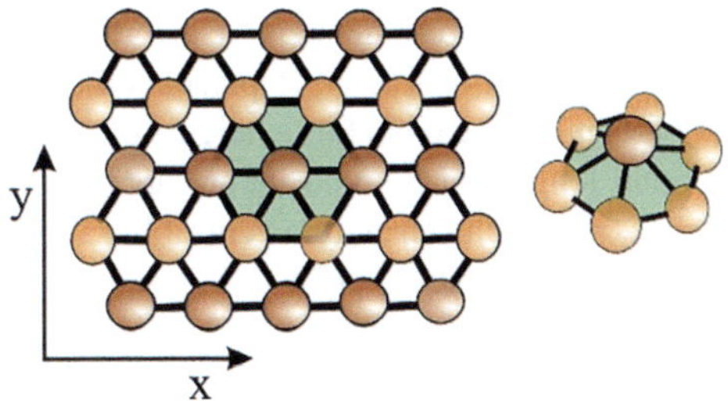

Fig. 3.5 Replacement of the metastable planar arrangement of triangular motifs following the 'aufbau' principle, by the hexagonal pyramidal B7 unit as a more stable network to form 2D structures of boron. Reproduced with permission from ref. [9], Copyright 2006, AIP Publishing

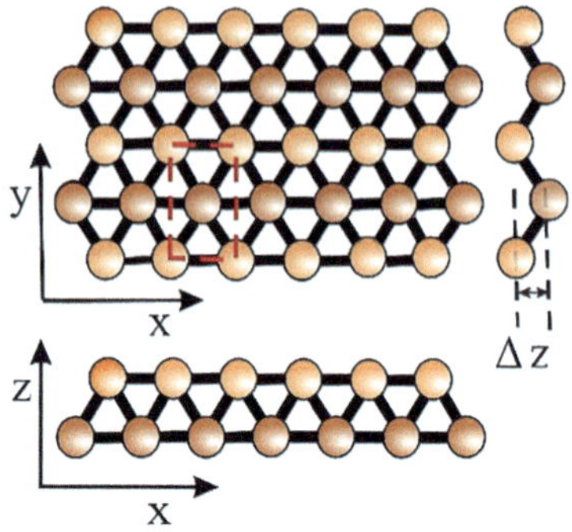

Fig. 3.6 Top, front, and side view structure of the striped borophene monolayer. The red dashed rectangle delineates the unit cell. Partially adapted from ref. [10], Copyright 2018, Springer Nature

energy, the 2D buckled lattice is dynamically unstable at 0 K and thermally unstable at finite temperatures [3, 10].

Different from monolayers derived from layered bulk materials, which form single layers held together by van der Waals forces, triangular striped borophene requires stabilization of the monolayer structure by electron transfer. An interesting question is the stability of the bonding in bilayer borophene and few-layer borophene sheets by covalent-like interlayer interactions to larger chemically bonded entities. Based on first-principles calculations and global minimum search optimization methods, stable bilayer structures with pillars of two atoms between the two layers were constructed [11, 12]. The structural stability and bonding character were analyzed by the electron localization function (ELF) and Bader charge analysis to determine the charge density distribution [12]. In this way, the optimum proportion of pillars and hexagonal holes has been obtained that balances the surplus of electrons. These findings undoubtedly triggered experimental work on the extension of the family of borophenes beyond monolayers to covalently bonded few-layer sheets with novel electronic properties.

Planar borophene: The two planar borophene monolayers shown in Fig. 3.7 exhibit the two highly symmetric phases called β_{12} and χ_3 borophene with the hole concentration ratio $\eta = 1/6$ for ß$_{12}$ and $\eta = 1/5$ for the χ_3 structure [4]. Following first-principles computations probing the synthesis of borophenes [13], epitaxial growth of these monolayers was achieved under ultrahigh vacuum (UHV) conditions. Liquid-phase exfoliation and reduction of borophene oxide afforded free-standing sheets of these metallic monolayers [6, 7, 14]. Table 3.1 presents the structural data such as the lattice constants, bond lengths, and buckling heights of various borophene monolayers. Note that the bond lengths of β_{12} and χ_3 are approximate values estimated from the quasi-planar B_{36} boron cluster with a central hexagonal hole, as predicted theoretically and realized experimentally [15].

The enormous number of possible 2D arrangements of hollow hexagon and triangular sites (variation of boron hexagons with and without a boron atom at the center) makes it possible to tailor lattice properties. In planar borophenes, the introduction of hexagonal holes used as scavengers of extra electrons from filled hexagons solved the problem of excess electrons. In principle, the hole concentration ratio can vary

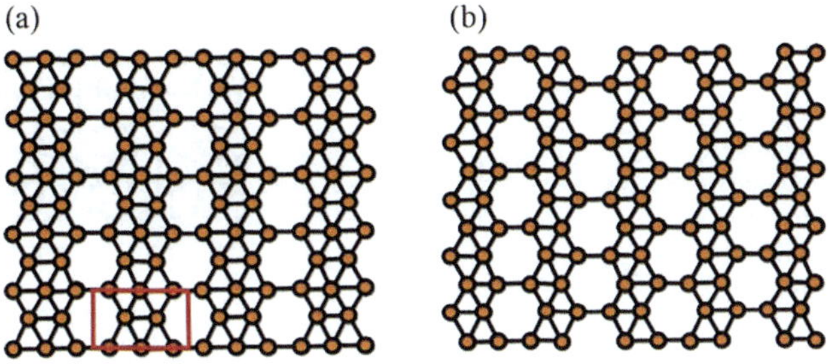

Fig. 3.7 **a** The β_{12} structure and **b** χ_3 structure of two planar borophenes. The red dashed rectangle delineates the unit cell. Adapted from ref. [4], Copyright 2012, American Chemical Society

within the limits of $\eta = 0$ for flat triangular borophene and $\eta = 1/3$ for hexagonal sheets [2].

Several research groups independently demonstrated the existence of strongly bonded bilayers for planar monolayers. Ag(111) substrates allowed the first controlled growth of bilayer borophenes, albeit limited to domains of a few tens of nanometers. Layer-by-layer deposition of two atomic layers on Ru(0001), employing surface segregation of interstitial boron during cooling from elevated temperatures, induced bilayer coverage across larger surface areas [16]. Diffraction measurements confirmed the high crystallinity of the bilayer assembly that grew in a high-order superstructure on the Ru(0001) surface.

Employing MBE on Cu(111) allowed the synthesis of large-size bilayer borophene samples [17]. Remarkably, interlayer bonding by four-coordinated boron atoms leads to better thermal stability than estimated for β_{12} borophene. The calculated energetic stability of $E_c = -6.35$ eV/atom for the free-standing bilayer indicates sufficient stability to detach the bilayer from a weakly interacting substrate [17]. These results were confirmed by analyzing the bilayers using scanning tunneling microscopy (STM) and noncontact atomic force microscopy (AFM), combined with DFT calculations [18]. Field emission resonance spectroscopy revealed distinct interfacial charge transfer doping that is responsible for covalent bilayer stabilization.

8-Pmmn borophene: In addition to the striped and planar phases, a highly anisotropic geometrical structure has been identified that consists of four covalently bonded sublayers with large corrugation (see Fig. 3.8) [19]. The lattice parameters are $a = 0.452$ nm, $b = 0.326$ nm, $d \approx 0.17 - 0.19$, $\Delta z = 0.218$ nm [19]. This hard but flexible material has a large cohesive energy of -6.33 eV/atom, which is in line with the values of the other polymorphs. Five different bond lengths between 0.17 and 0.19 nm characterize the complex structure. In contrast to the failure mechanism of striped borophene, which involves elastic instability in the zigzag direction and phonon instability in the stronger armchair direction, the 8-*Pmmn* structure fails under biaxial and uniaxial tension by elastic instability [19].

Puckered borophene: In the search for other defect-free four-sublayer polymorphs of borophene, an energetically and dynamically stable puckered phase with an

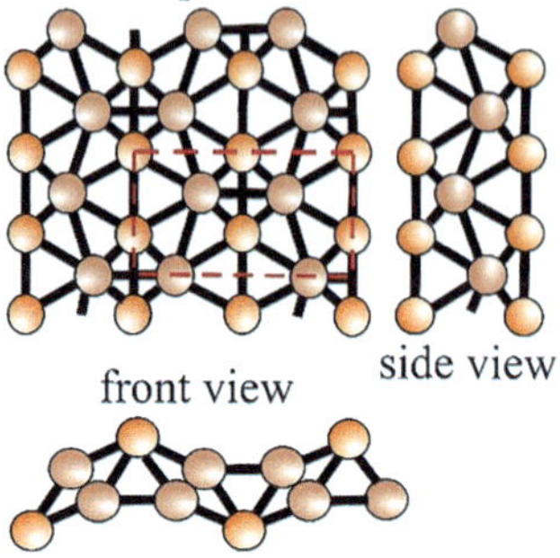

Fig. 3.8 The geometric structure of 8-*Pmmn* borophene with four covalently bonded sublayers. The red dashed rectangle delineates the unit cell. Reproduced from ref. [19], Copyright 2017, Royal Society Chemistry

asymmetric centered-washboard (acw) structure of orthorhombic symmetry has been discovered; unlike other boron sheets, this phase is a semiconductor (centered because there is another atom at the center of the hexagons) [20]. The lattice constants are a = 0.319 nm and b = 0.53 nm, the bond lengths vary widely between 0.157 and 0.226 nm, and the two pucker heights are $\Delta z_1 = 0.126$ nm (distance between inner atoms) and $\Delta z_2 = 0.202$ nm (distance between outer atoms) [20]. According to molecular dynamics (MD) simulations, this semiconducting polymorph is thermally stable up to 800 K. Interestingly, the application of uniaxial or biaxial tensile strain closes the gap and transforms the semiconducting system back to the metallic state, and the positive Poisson ratio becomes negative [21].

Polymorphism is the most unique characteristic of borophene, where multiple 2D structures exist, several of which have comparable stability. Diverse structures, mostly metallic but also semiconducting, were proposed quite early, e.g., by the particle swarm optimization algorithm (PSOA) [21, 22]. Interestingly, for the *Pmmn* structure, which is a non-graphene-like material, theory predicts distorted Dirac cones and novel electrical characteristics typical of a Dirac material [22]. Borophene has been considered as a prototype for the development of synthetic 2D materials [23]. Importantly, when exposed to air, polymorphs such as plane borophenes may partly oxidize and therefore require ultrahigh vacuum conditions for synthesis [6], while others describe boron sheets as quite inert to oxidation [7]. Notably, boron forms the lightest metal and is the lightest atom that can form stable extended 2D networks.

3.1.3 Mechanics

Striped borophene: The DFT-PBE approach has provided the most information on Young's modulus, Poisson's ratio, ultimate strength, and critical strain of important borophene phases (see Table 3.2) [3, 5, 24, 25]. Especially the striped phase has attracted enormous attention due to its extreme structural anisotropy that produces highly anisotropic elastic and metallic behavior. Remarkably, Young's modulus of buckled triangular borophene is approximately E = 380–400 N/m in the armchair direction and is therefore greater than that of graphene (340 N/m), owing to extreme directional bonding. On the other hand, the zigzag direction shows a significantly smaller stiffness of only E = 160–170 N/m [5, 24]. The in-plane Poisson ratio is negative at $\nu = -0.022$ along the [100] direction and $\nu = -0.009$ along the [010] direction [26]. An applied strain along the out-of-plane direction reduces the buckling amplitude of the sheet and induces an increase in the ridge-to-ridge distance due to the lower in-plane modulus across the ridges than along the ridges [5]. The ultimate strength of striped borophene also is strongly anisotropic, with $\sigma_{ac} = 22.8$ N/m and $\sigma_{zz} = 14.0$ N/m [26]. The strong anisotropy of the ultimate strength excludes the application of Griffith's rule. The ultimate tensile strains in the armchair direction of 0.105 and in the zigzag direction of 0.145 confirm the large anisotropy with much smaller critical strain values than those of graphene [24]. The whole structure becomes dynamically unstable, contrary to the announcement of early reports, which described striped borophene as the stiffest material. Both the exceptional stiffness in one direction and the instability of the material along the armchair direction originate from extreme directional covalent bonding that limits the overall mechanical performance [3]. Structural anisotropy also confines the electrical conductivity to the uncorrugated armchair direction and in addition, causes a large optical anisotropy. Foreseen are potential applications as transparent conductor in display technologies and photovoltaics since there is no optical absorption in the visible region [27]. The structure, space group, Young´s modulus, Poisson´s ratio, ultimate strength, and ultimate strain of group–IIIA monolayers are presented in Table 3.2.

β_{12} and χ_3 borophene: The weakly anisotropic planar β_{12} and χ_3 phases of borophene exhibit stiffness values between E ≈ 190 and 210 N/m and ultimate strengths approximately a factor of ten lower, in reasonable agreement with Griffith's rule [3, 5, 24, 25]. The ultimate strengths reported for the β_{12} phase are 22.4 N/m and 20 N/m in the zigzag and armchair directions, respectively [24]. However much lower values can be found in the literature at $\sigma_{zz} = 16.4$ N/m and $\sigma_{ac} = 15.4$ N/m [5]. As expected, the strength values for the χ_3 structure are similar at $\sigma_{zz} = 20.2$ N/m and $\sigma_{ac} = = 19.9$ N/m [24]. Depending on the arrangement and density of the holes, the anisotropy of the ultimate strain varies. For the β_{12} structure, the ultimate strain in the armchair direction of 0.20 is nearly identical to that in the zigzag direction of 0.21, whereas for the χ_3 phase the anisotropic values are $\varepsilon_{ac} = 0.21$ and $\varepsilon_{zz} = 0.16$ [24]. Notably, both structures are mechanically and dynamically stable as free-standing monolayers [3].

In addition to the polymorphic behavior, the out-of-plane bending stiffness of the planar structures is more than fourfold lower than that of graphene, which can be explained by delocalized multicenter bonding with well-ordered holes in the planar phases [5]. Both the bending stiffness and Poisson ratio depend sensitively on the arrangement and the density of holes, and the latter can be used to tune these properties. While 2D materials usually fail at the critical strain by structural failure due to elastic instability, borophenes may undergo phase transitions under tension [5]. For example, the $\eta = 1/6$ sheet is transformed into the $\eta = 1/7$ sheet when stretched across hollow hexagons with a strain $\varepsilon > 0.125$ [5]. The reasons for these different

Table 3.2 Structure, space group, Young's modulus, Poisson's ratio, ultimate strength, and ultimate strain of group–IIIA monolayers

Monolayer	Structure, space group	Young's modulus (N/m) $E_{ac\,(a)}$, $E_{zz\,(b)}$	Poisson's ratio $\nu_{ac\,(a)}$, $\nu_{zz\,(b)}$	Ultimate strength (N/m) $\sigma_{ac\,(a)}$, $\sigma_{zz\,(b)}$	Ultimate strain, $\varepsilon_{ac\,(a)}$, $\varepsilon_{zz\,(b)}$
Borophene	Triangular, δ6 Pmmn [3]	382, 163 [24] 399, 163 [5]	– 0.01 [24] – 0.23, 0 [5]	22.8, 14.0 [24] 20.9, 12.2 [5]	0.11, 0.15 [24] 0.087, 0.14 [5]
Borophene	Planar, ß12 *Pmm*2 [3]	190, 210 [24] 189, 210 [5]	0.18 [24] 0.15, 0.17 [5]	20.0, 22.4 [24] 16.4, 15.4 [5]	0.20, 0.21 [24] 0.12, 0.11 [5]
Borophene	Planar, χ3 *Cmmm* [3]	208, 205 [24] 196, 208 [5]	0.11 [24] 0.11, 0.12 [5]	19.9, 20.2 [24]	0.21, 0.16 [24]
Borophene	4-layer slab 8-*Pmmn* [19]	241, 305 [19] 248, 322 [25]	0.042, 0.062 [19] 0.048, 0.062 [25]	21.6, 27.8 [19]	0.16, 0.17 [19] 0.17, 0.16 [25]
Aluminene	Planar	31 [48]	0.308 [48]		0.26, 0.24 [48]
Aluminene	Triangular	37 [48], 29, 24 [51]	0.489 [48] 0.70, 0.585 [51]		0.24, 0.32 [48]
Gallenene	Triangular	25, 16 [51]	0.566, 0.359 [51]		
Gallenene	Buckled	25 [60]	0.20 [60]		0.30 [60]
Indiene	Planar	28 [48]	0.085 [48]		0.30 [48]
Indiene	Triangular	25 [48]	0.372 [48]		0.28, 0.38 [48]
Thalliene	Planar	15 [51]	0.20 [51]		

behaviors are the low activation energy of bond rotation, which is made possible by variable coordination and delocalized multicenter bonding in borophenes.

Besides DFT calculations, MD simulations have been performed to study the temperature and strain-rate dependence of Young's moduli and failure strengths [28, 29]. Mechanical properties simulated by the Stillinger–Weber potential for striped borophene confirmed the strong anisotropy, and the results are in good agreement with those obtained by first-principles DFT calculations [28]. The MD results reveal a decrease of approximately 50% in the fracture strength and approximately 60% in the ultimate strain for a temperature increase from 10 to 700 K. The main requirement of MD simulations is an accurate empirical interaction potential. The simulations performed with a linear valence force field (VFF) model including bond stretching and bending terms yielded linear mechanical quantities, whereas the empirical Stillinger–Weber (SW) potential that includes linear and nonlinear couplings, also yielded nonlinear physical and mechanical quantities in good agreement with the first-principles DFT calculations [29]. The main advantages of the MD approach are the possibility of treating systems with more than 10^4 atoms and extending the zero-temperature behavior of first-principles calculations to higher realistic temperatures.

8-Pmmn borophene: Using first-principles DFT-PBE calculations the linear and nonlinear mechanical properties of the stable 4-layer phase with symmetry 8-*Pmmn* and superior performance were estimated [19]. The study delivered as linear Young's moduli $E_a = 241$ N/m and $E_b = 305$ N/m, and as Poisson's ratios $\nu_a = 0.042$ and $\nu_b = 0.062$. Furthermore, the nonlinear ultimate strengths $\sigma_a = 21.6$ N/m and $\sigma_b = 27.8$ N/m and ultimate strains $\varepsilon_a = 0.155$ and $\varepsilon_b = 0.165$ were determined. This structure is a hard but flexible 2D material that may find applications as high-strain material [19]. The failure mechanisms under both uniaxial and biaxial tension are elastic instability.

3.1.4 Physics and Applications

The complexity of bonding and structural diversity of the already known borophenes opens new opportunities to discover novel exciting boron nanostructures, e.g., semiconducting ones. Moreover, the unique properties of these materials provide interesting perspectives for a series of potential technological applications. With their flexibility in bonding, mechanical flexibility, in-plane elasticity, metallic character, thermal stability, thermal conductivity, optical transparency, and intrinsic strength, these 2D monolayers and their nanosheets are not only the ideal material for flexible and transparent electronic interconnects, electrodes, and displays [25]. With an optical transparency of up to 100% below 3 eV, buckled borophene surmounts graphene and is an excellent transparent and conductive 2D material for photovoltaics and touch screens [26].

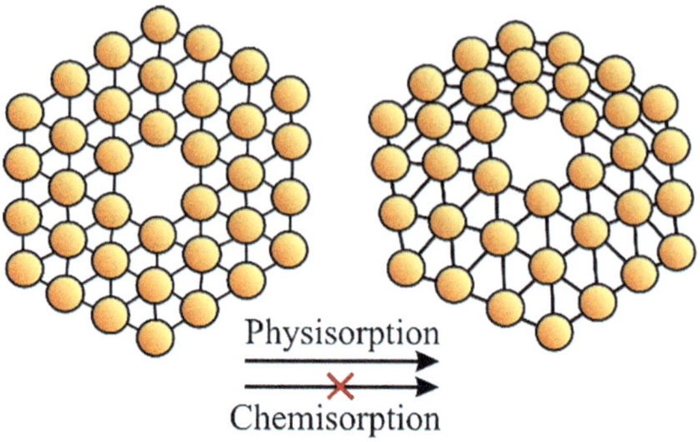

Fig. 3.9 Adsorption behavior of the planar B_{36} cluster. Its stabilization by chemisorption and conversion into the more stable quasi-planar configuration by physisorption of small molecules. Reproduced from ref. [30], Copyright 2015, Indian Academy of Sciences

Other applications proposed for borophenes are based on their distinctive electronic properties, such as high electron density, Dirac states, and control of superconductivity. A comprehensive study reviews potential applications, such as in batteries, hydrogen storage, supercapacitors, sensors, and catalyst for the hydrogen evolution reaction (HER), but also the oxygen reduction and evolution reaction (ORR, OER), and the CO_2 electroreduction reaction [25].

Specifically, triangular borophene could be an innovative component of sensors for use in biomedical devices. In particular, B_{36} borophene is expected to play an important role as a biosensor [15, 30]. The B_{36} cluster, with its central hexagonal hole, has a planar and slightly more stable quasi-planar structure (see Fig. 3.9) [30]. DFT calculations of the adsorption of NO, CO, H_2, H_2O, O_2 and N_2 on the planar cluster indicate (dissociative) chemisorption of NO, O_2, and CO, occurring preferentially at the edge atoms. This causes a stabilization of the planar configuration of the cluster, whereas physisorption induces a conversion to the slightly more stable quasi-planar configuration [30].

Finally, borophene provides a promising allotropic material with bonding options for a variety of objectives owing to its chameleon-like behavior in changing chemical and physical characteristics in various phases [31]. A comprehensive review covers in detail the recent advances in synthesis of boron-based materials and their applications in energy storage, energy conversion, and energy harvesting. Others suggest the use as gas sensor, humidity sensor, and pressure sensor. Additional applications proposed are their use as photodetectors and in information storage devices [32].

3.1.5 Synthesis

The large number of regular publications and progress reports on the synthesis and potential applications of borophenes demonstrate the enormous interest in the development of boron nanomaterials [32–36]. Physical vapor deposition (PVD) under UHV conditions to stabilize Ag(111), provided a well-defined method with an inert surface that allowed the growth of a dynamically unstable quasi-free-standing layer of striped borophene [6]. STM analysis enabled the identification of the typical hexagons with a boron atom in the middle. In situ electronic characterization verified the highly anisotropic electronic properties and metallic character given by first-principles predictions.

Recent progress concerns precise control of the vapor concentration of reactants in CVD deposition processes, which is a prerequisite for the growth of uniform nanosheets with a defined phase and a predetermined number of layers over a large area. Control of the sheet thickness remains a challenge. However, this control is necessary owing to the extreme layer dependence of some properties. To avoid sample transfer to a suitable substrate, a method was developed to grow borophene directly on an insulating quartz substrate via CVD using hydrogen as a carrier gas and $NaBH_4$ as a boron source material [37]. To demonstrate the potential use of the deposited so-called 'borophene glass', the authors present the induced voltage and power density of a borophene hydroelectric generator and compare the results with those of carbon-based generators. The authors suggested the use of borophene glass in energy conversion devices, storage devices, and as sensor material.

Synthesis by MBE has the capability to better control the composition, number of layers, and phase. Due to the enormous effort required for working in a UHV environment, this method qualifies only for fundamental studies, including heterostructures. Following theoretical suggestions, UHV conditions allowed the deposition of van der Waals bonded intact $ß_{12}$ and χ_3 monolayer structures on Ag(111) substrates [38]. Low-energy electron diffraction (LEED), scanning tunneling microscopy (STM), and identification of the types of bonds by X-ray photoelectron spectroscopy (XPS) provide the structural analysis. To specifically visualize the atomic lattice of borophene polymorphs, CO-functionalized AFM probes allow the unambiguous identification of the atomic structure of the polymorphs [39].

Layer growth on Cu(111) substrates delivered large-area single-crystal sheets of up to 100 μm^2 in size. A combination of DFT calculations, STM measurements, and LEED provided the analysis of the deposit [40]. The inert Cu(111) substrate stabilized the interaction without significant covalent contributions, which is helpful for exfoliation. Similarly, a square lattice of Cu(100) served as a substrate for growing micrometer-scale single-crystalline borophene samples [41].

In contrast to the negligible charge transfer from an Ag(111) substrate, strong charge transfer to borophene occurs on a strongly bonded graphene-like honeycomb lattice that is formed because B^- is isoelectronic to carbon. The synthesis of such a monolayer was achieved by MBE on an Al(111) surface under UHV condition [42, 43]. A Bader charge analysis yielded a charge transfer of ~ 1.0 e from the Al(111) surface to each boron atom, which has only a small lattice mismatch with the sheet. The theoretical treatment of this stabilization of honeycomb borophene on Al(111) confirms a large interaction energy with the substrate of approximately $E_{int} = -1.05$ eV/atom and a separation from the surface of only 0.154 nm, suggesting strong covalent bonding by hybridization of the p_z orbitals of borophene with Al(111) [43]. The direct calculation of phonon dispersion validates the kinetic stability of the monolayer.

Enlarging faceted borophene islands to micrometer-sized monocrystals by repeated submersion of boron into copper at ~ 900 K, annealing, and borophene regrowth at 760 K solved the restrictions caused by small and expensive single crystals of silver or copper substrates [44]. A significant step forward in the realization of practical applications was the first isolation of borophene without a substrate under UHV conditions. The progress made in the synthesis of free-standing borophene is based on liquid-phase exfoliation, which yields $ß_{12}$ and χ_3 monolayers and their intermediate phases [45]. X-ray photoelectron spectroscopy (XPS) and STM analyses validated the phase purity and structure of the monolayers, and high-resolution electron transmission microscopy (HRTEM) corroborated the DFT-based band structure calculations. Thickness measurements via AFM nanoindentation delivered the thicknesses of the deposited 1–10-layer sheets.

A problem observed with the realization of an economical synthesis of high-quality large-area crystals is the existence of chemical reactivity and vulnerability of free-standing borophenes, which play a detrimental role. Borophenes are sensitive to environmental contamination and chemical reactions when exposed to air for a longer time because of their reactivity and all-surface nature. A study, using XPS to determine the chemical composition of the material, indicated that three-center-bonded boron atoms at the surface have a certain degree of stability against oxidation, while atoms located at the edges bonded by two-center bonds were preferentially oxidized, as verified by in situ STM analysis [6, 45]. In device applications, the integration of borophenes into van der Waals heterostructures or silicon oxide capping layers avoids direct air exposure of the vulnerable metallic sheets.

As mentioned above, growth by MBE on Cu(111) allowed the synthesis of large-sized bilayer borophenes [40]. A combination of scanning tunneling microscopy (STM) experiments with DFT calculations demonstrated the connection of two $ß_{12}$-like layers with zigzag rows by covalent interlayer B–B bonds, forming an adsorbed stacked bilayer [17]. The large electron transfer to the first layer from the Cu(111) substrate, in comparison to Ag(111), enables the growth of the second layer, as revealed by a Bader charge analysis. Remarkably, interlayer bonding by four-coordinated boron atoms leads to a better thermal stability of the bilayer than that estimated for $ß_{12}$ borophene.

3.2 Aluminum Monolayers (Aluminene)

3.2.1 Chemistry and 2D Bonding

The electronic configuration of aluminum atoms is (Ne|$3s^2p^1$). The three valence electrons, which are available for four bonding orbitals, form electron-deficient delocalized bonds in aluminene. The three electrons are responsible for the similarity of the chemical behavior of aluminene with that of borophene. The resulting specific bonding behavior leads to various polymorphs discussed in the literature, e.g., a planar honeycomb structure of aluminum, a graphene-like planar and buckled structure, a triangular buckled structure, and a four-layer 8-*Pmmn* structure (see Table 3.1). Note that some phases suggested for boron and its homologues still need full characterization and confirmation by other authors and experimental realization. Often, it is unclear, which structure represents the global minimum on the potential energy surface (PES). The latter phases have a greater chance of realization than local minimum structures, which may exhibit quite similar cohesive energies. If the latter metastable monolayers are thermodynamically and dynamically stable, experimental synthesis may still be possible, e.g., on a suitable stabilizing surface. Currently, free-standing aluminene and weakly adsorbed monolayers do not exist. Nonetheless, only experimental confirmation allows a judgement of the accuracy and reliability of available theoretical predictions.

3.2.2 2D Structures

Planar aluminene: The first theoretically studied planar graphene-like aluminene monolayer was a metal with the space group *P6/mmm* [46]. These first calculations performed by the self-consistent DFT-PBE approach and tight-binding computations of the bonding properties of aluminene included the four valence orbitals $3s$, $3p_x$, $3p_y$, and $3p_z$ and considered planar, buckled, puckered, and triangular allotropic configurations. The authors considered aluminene as a hole-doped graphene with a covalently bonded dynamically stable planar structure. The lattice constant of the planar phase is a = 0.449 nm, the bond length d = 0.259 nm, and the cohesive energy $E_c = -1.96$ eV/atom. Because the σ and π orbitals are not filled completely, on the contrary to the situation in graphene with filled bonding orbitals (σ and π) and empty antibonding orbitals (σ^* and π^*), the Dirac point lies 1.618 eV above the Fermi level [46]. The existence of a planar phase of aluminene with a larger cohesion of – 2.32 eV/atom but a similar lattice constant and bond length has been proposed by an independent study [47]. The mechanical properties are E = 31 N/m for the stiffness, $\nu = 0.308$ for Poisson's ratio, and $\varepsilon_{ac} = 0.26$ and $\varepsilon_{zz} = 0.24$ for the ultimate strains [48].

Buckled aluminene: An extensive study investigated the stability of various aluminene allotropes at 0 K, using DFT calculations with PBE parametrization, density functional perturbation theory (DFPT), and ab initio MD simulations at 300 K [47]. The lattice parameters of the buckled structure are a = 0.272 nm, d = 0.296 nm, and $\Delta z = 0.252$ nm. While the free-standing structure of a buckled honeycomb phase with a large buckling height is considered to be dynamically unstable, synthesis on a metallic substrate has been suggested as a way to suppress the unstable flexural phonon mode by transforming the configuration closer to the planar structure [48]. According to DFPT, which is used to resolve the lattice dynamical stability, the buckled monolayer adsorbed on a Cu(111) substrate relaxes to a planar honeycomb structure [47]. Obviously, the higher cohesive energy obtained for this structure of

$E_c = -3.27$ eV/atom is still not sufficient to guarantee thermodynamic stability. Surprisingly, the reported bond length is greater than that of the planar structures despite the much greater bonding energy.

Interestingly, for the buckled honeycomb lattice with space group *P*-3*m*1 a slightly greater cohesive energy of $E_c = -3.47$ eV/atom, a lattice constant at a = 0.275 nm, and a buckling height of $\Delta z = 0.245$ nm were found via ab initio calculations using the DFT-GGA-PBE approach [49]. A phonon dispersion analysis indicated dynamic and thermodynamic stability and thus the possible synthesis of this metallic aluminene structure, which seems to be the allotropic modification with the minimum-cohesive-energy configuration. A charge distribution analysis by ELF revealed that bonding along the two sublayers is typically metallic and bonding between the sublayers is covalent.

For the dynamically and thermodynamically stable four-layer network of 8-*Pmmn* aluminene, covalent sp^3 bonding seems to be stronger than sp^2 bonding according to an ELF analysis [49]. In fact, the analysis indicates that electrons accumulate on top of the upper plane and bottom of the lower plane and that the Al–Al bonds are strongly covalent. The lattice constants are a = 0.674 nm and b = 0.503 nm, and the buckling height $\Delta z = 0.344$ nm. This allotrope with the space group *Pmmn* has a cohesive energy of – 3.16 eV/atom. In the complex structure with four sublayers, the formal monolayer thickness increases to > 0.7 nm owing to the large distance between the outer ridge atoms responsible for the large corrugation effect.

Triangular aluminene: DFT calculations revealed the following structural data for a dynamically stable buckled triangular lattice with the space group 13-*Pmmn* [50]. The lattice constants of this phase are a = 0.269 nm and b = 0.272 nm, and the buckling height is $\Delta z = 0.217$ nm. The energetically stable structure with a cohesive energy of – 3.27 eV/atom showed no phonon mode with a negative frequency. The Al atoms are covalently bonded, and the delocalization of electrons generates metallic behavior. Moreover, the authors predict that superconductivity occurs at a critical temperature of 6.5 K. Tensile strain, e.g., by lattice mismatch with an underlying substrate, may increase the critical temperature. The mechanical properties are E = 37 N/m for the stiffness, $\nu = 0.489$ for the Poisson ratio, and $\varepsilon_{ac} = 0.24$ and $\varepsilon_{zz} = 0.32$ for the ultimate strains [48].

According to two comprehensive DFT studies investigating an enormous number of atomically thin elemental metals, the cohesive energies given for planar honeycomb (– 2.84 eV/atom) and hexagonal triangular aluminene (– 2.74 eV/atom) are in the broad range of those presented by other authors [51, 52]. Unfortunately, only limited structural data are provided, as, for example, the bond length of the honeycomb lattice with d = 0.259 nm and triangular alumina with d = 0.269 nm, which are in good agreement with other calculations (see Table 3.1) [48, 51].

3.2.3 Physics and Applications

Since the excellent thermal properties of aluminum are preserved in 2D sheets, its use as ultrathin thermal insulator or heatsink has been proposed [47]. Bilayers of aluminene have been suggested as suitable material for building nanoscale capacitors since the charge and energy stored on each plate are already significant without an electric field [53]. The stable bilayers of hexagonal aluminene are strongly bonded with an interlayer binding energy of $E_{int} = -0.519$ eV and a layer separation of $d_s = 0.27$ nm. This separation is much smaller than the van der Waals diameter of the Al atom of $d_w = 0.37$ nm and points to strong interlayer interactions between the metallic layers that allow interlayer electron transfer.

A theoretical characterization of different assemblies of bilayer aluminene was based on the DFT-GGA-PBE approach [54]. This study included planar and buckled structures in various stacking orders. Dispersion-corrected DFT calculations, including long-range van der Waals interactions, predicted the already mentioned short interlayer distances, and clearly indicated the minor role of the weak forces. The cohesive energies were in the range reported for single-atom aluminene. AA stacking of planar layers leads to the distribution of a substantial amount of charge in the intralayer region and less in the interlayer region. For AB stacking, where one Al atom is on top of the other Al atom, the originally planar layers buckle without charge accumulation owing to their large interlayer separation. Note that all bilayers are metallic in nature.

The efficient management of storage processes and optimization of the storage capacity of hydrogen are main challenges associated with the upcoming hydrogen economy. A DFT-PBE study investigated the feasibility to use planar hexagonal aluminene (*P6/mmm*) for this purpose [55]. At the most stable bridge site of plane aluminene, hydrogen can be physisorbed without a barrier, it can be physisorbed at the hollow site, and dissociated with an activation energy of $E_A = 1.51$ eV at the top site (see Fig. 3.10). According to the difference in the electronegativities of H (2.2) and Al (1.61), a charge transfer of $\delta q = 0.6$ e occurs from the aluminum atoms to the hydrogen atoms. The desorption energy of hydrogen atoms is 2.37 eV. Due to its physisorption and chemisorption capabilities, aluminene is a suitable candidate for the storage and release of hydrogen.

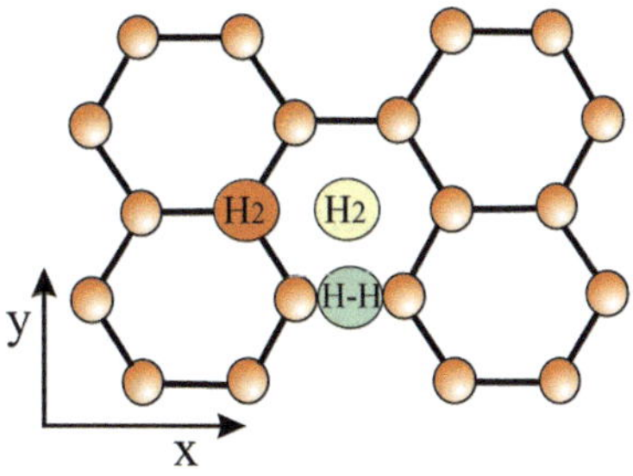

Fig. 3.10 Adsorption (chemisorption) positions of H_2 on the bridge site (physisorption), the top site (chemisorption), and the hollow site (physisorption)

The various examined nanostructures are well suited for gas sensing purposes due to their good chemical and thermal stabilities, large surface-to-volume ratios, and tunable electronic properties. First-principles calculations clarified the adsorption properties of the toxic gases CO and NO on buckled aluminene nanosheets. The adsorption process was studied at the atomic level in terms of the average energy band gap variation between the highest occupied molecular orbital (HOMO) and the lowest unoccupied molecular orbital (LUMO), which is called the HOMO–LUMO gap, the natural bond orbital, and the adsorption energy, which support the suitability of buckled hexagonal aluminene as a chemical nanosensor [56]. The density of states (DOS) provided a clear picture of charge transfer induced by the adsorption of CO and NO on the aluminene nanosheets.

3.2.4 Synthesis

To date, the preparation of free-standing or weakly adsorbed aluminene layers has not been successful. A crucial issue is chemical stability, owing to aluminum's high affinity to oxygen, which is responsible for the formation of highly exothermic oxide compounds in open air. This reactivity requires synthesis and measurements of intrinsic properties under UHV conditions. Suggested solutions for device applications include encapsulation of the reactive sheet or functionalization of dangling bonds.

3.3 Gallium Monolayers (Gallenene)

3.3.1 Chemistry and 2D Bonding

The three valence electrons of the electronic configuration of gallium atoms ($\mathrm{Ar}|4d^{10}4s^2p^1$), which are available for four valence orbitals, form electron-deficient delocalized bonds in gallenene. In its bulk chemical behavior, α Ga is known as the only elemental 'molecular metal' that favors both covalently bonded dimers and weaker metallic bonding in the plane perpendicular to the average alignment of the Ga–Ga dimers [57]. Thus, bulk α Ga combines strong covalent bonds with weaker metallic bonding. The derivation of 2D networks from 3D crystals with nondirectional metallic bonding makes complete reorganization of covalent 2D bonding necessary. Nevertheless, several authors have derived stable monolayer configurations of gallium from the stable surface structures of bulk gallium [57–60]. The cleavage planes for cutting the bulk α-Ga crystal along the [100] and [010] directions and the corresponding a_{100} Ga and b_{010}-Ga monolayers are shown in Fig. 3.11 [61]. The 2D character visible in the conductivity of bulk gallium leads to true metallic gallenene phases [57].

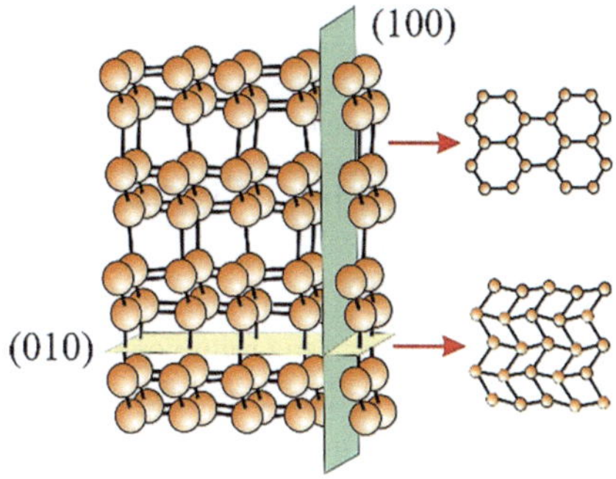

Fig. 3.11 Structure of the bulk α-Ga crystal and its cleavage in the [100] and [010] directions, which results in the formation of a_{100}-Ga and b_{010}-Ga monolayers after structural relaxation. Reproduced with permission from ref. [61], Copyright 2021, IOP Publishing, Ltd.

3.3.2 2D Structures

A_{100} gallenene: A tight-binding model based on the combination of LCAO and ab initio calculations led to the proposal of the existence of a graphene-like planar a_{100} gallenene [58]. This phase can be derived from bulk α Ga by extracting a monolayer of Ga from the (100) surface and allowing relaxation to the stable a_{100}-Ga monolayer (see Fig. 3.12) [58]. The lattice constants of this phase are at a = 0.787 nm and b = 0.465 nm and the bond lengths d_1 = 0.266 nm and d_2 = 0.267 nm. The distorted hexagonal structure consists of three slightly vertically shifted Ga atoms. Accordingly, hybridization does not result in perfect sp^2 hybrids and the interior angles deviate slightly from 120° [58]. The electron charge density reaches a maximum at the center between the covalent Ga–Ga bonds, as demonstrated by an ELF analysis [59]. The honeycomb network, belonging to the space group *Pbam*, needs stabilization, for example, by strain engineering due to instability of its free-standing configuration. In the case of a planar structure, the p_z orbitals remain unoccupied and create Dirac cones lying above the Fermi level [58, 59]. The van der Waals diameter provides a good approximation for the thickness of the planar layer.

b_{010} gallenene: The second phase studied by ab initio calculations and derived by the tight-binding model is the b_{010}-Ga monolayer with a lower symmetry that resembles a zigzag rhombic lattice [58]. The lattice constants are a = 0.474 nm and b = 0.492 nm,

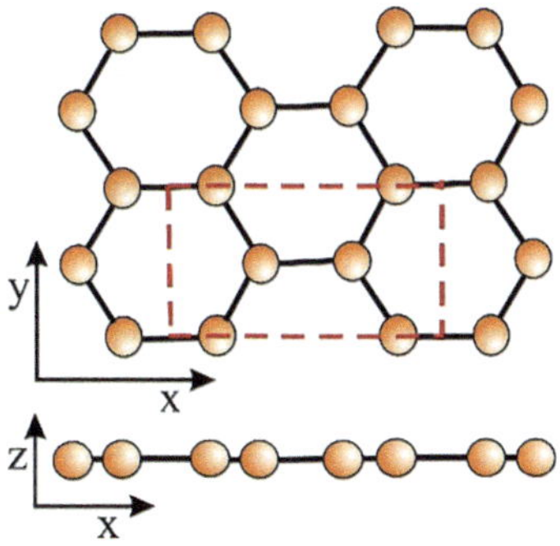

Fig. 3.12 Honeycomb graphene-like a_{100}-Ga lattice of gallenene obtained by cleaving bulk α-Ga along the [100] direction after relaxation. Reproduced with permission from ref. [58], Copyright 2019, Royal Society of Chemistry

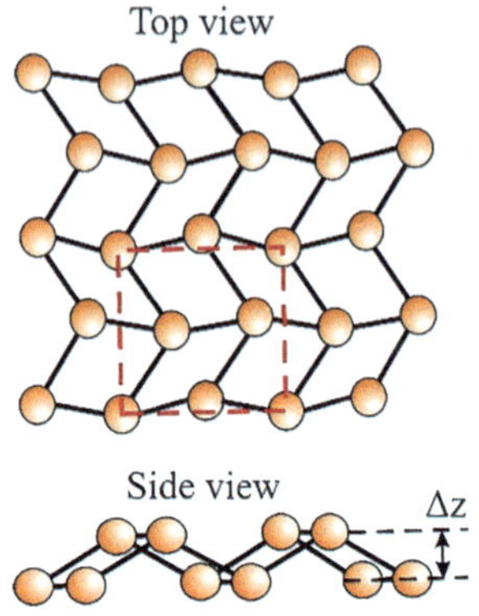

Fig. 3.13 Distorted rhombic lattice of gallenene obtained by cleaving bulk α-Ga along the [010] direction after relaxation. Reproduced with permission from ref. [58], Copyright 2019, Royal Society of Chemistry

the bond length is d = 0.273 nm, and the buckling height $\Delta z = 0.119$ nm [58]. This buckled phase can be derived from bulk α Ga by exfoliation of a sheet of Ga from the most stable (010) surface, allowing relaxation to the stable b_{010}-Ga monolayer (see Fig. 3.13) [58, 59]. The bilayer of gallenene obtained by slicing through covalent Ga–Ga pairs along the conducting plane subsequently stabilizes by significant structural relaxation. In contrast to the situation in real layered materials, the cutting process breaks covalent bonds, and the new bilayer stabilizes by generating new bonds (see Fig. 3.11) [57]. The resulting bond lengths are $d_1 = 0.269$ nm and $d_2 = 0.274$ nm and the separation of Ga dimers of $\Delta z = 0.132$ nm in the two-layer structure in vertical direction defines the buckling effect induced by the mixed hybridization of p_z orbitals with in-plane orbitals [59]. While planar a_{100}-gallenene is purely covalent in nature, in b_{010}-Ga monolayers covalent bonding prevails in the Ga dimers along the zigzag direction, and metallic interactions occur between the Ga pairs in the directions with less directionality. This type of bonding increases the stability of the buckled lattice. The b_{010}-Ga phase is the most stable structure, with a cohesive energy of $E_c = -2.32$ eV/atom [60] and a lower formation energy relative to the bulk α Ga of $E_f = 0.36$ eV/atom compared to $E_f = 0.64$ eV/atom for a_{100}-Ga [59].

The application of biaxial strain or adsorption on a suitable surface stabilizes the b_{010}-gallenene structure, which, however, exhibits imaginary frequencies. Despite the dynamic instability of the monolayer, few-layer assemblies of atomically thin 2D gallenene (~ 4 nm) have been successfully exfoliated from the molten phase of α Ga on a silicon substrate (~ 30 °C), where the substrate stabilizes the dynamically unstable quasi-free-standing layer [59]. The low melting point of Ga (~ 29.7 °C) and the good wettability of oxidized silicon allow solid-melt exfoliation near room temperature. Much higher temperatures lead to oxide formation. The height profile of ~ 4 nm, determined by AFM indentation measurements, points to a gallenene nanosheet of approximately four to six monolayers [59]. Transmission electron microscopy (TEM) measurements confirmed the one-to-one correspondence between the realized structures and theoretically predicted structures.

3.3.3 Physics and Applications

The advent of efficient methods for synthesizing low melting point metals, such as solid-melt exfoliation and the gallium self-propagation technique, has triggered discussions on the potential applications of gallenene. This includes the transmission properties of nanoribbons of monolayers oriented along the armchair and zigzag directions. It is shown that b_{010}-Ga nanoribbons exhibit semiconducting behavior with zero transmission [58]. Other potential applications could be in the field of plasmonics and electrical contacts [59]. In particular, gallenene, with its dynamic stability and robust metallic character, is a promising candidate for nanotechnological applications such as 2D flexible metals and a wide range of interfaces [57]. Recent ab initio calculations indicate that synthesized gallenene sheets of planar a_{100} Ga, which are isostructural with graphene, are the structurally simplest 2D superconductors with topological edge states based on their honeycomb structure [62]. The a_{100}-Ga monolayer exhibits distinct three-gap superconductivity, whereas buckled b_{010} Ga has a single anisotropic superconductivity gap. The critical temperature is in the range of 7–10 K and exceeds that of bulk gallium. Owing to its superconducting and topological properties, gallenene is an interesting partner for the use in heterostructures as building block in the field of quantum information and computing.

3.3.4 Synthesis

A computational search of gallenene using the particle swarm optimization (PSO) algorithm available in CALYPSO revealed the feasibility of the existence of various polymorphs in aluminene, gallenene, and indenene [61]. The low surface energy of bulk Ga facilitates exfoliation. However, it is necessary to stabilize free-standing gallenene nanosheets with a substrate because they tend to assume the bulk form. The influence of the substrate depends on its nature and differs for insulating and metallic substrates.

Extensive DFT-PBE calculations revealed the cohesive energy and thus energetic stability of a bilayer structure of $E_c = -2.88$ eV/atom and a trilayer structure at -3.16 eV/atom [63]. These values are greater than the cohesion of $E_c = -2.32$ eV/atom reported for the b_{010}-Ga monolayer. Nevertheless, the authors conclude that the free-standing bilayer is highly unstable and prefers to add additional layers, whereas the free-standing trilayer is stable up to 100 K, as revealed by DFT-MD simulations for finite temperatures. The thermal stability of 2D Ga increases to 457 K, 350 K, and 433 K for four, five, and six atomic layer sheets, respectively. Sheets with more than 10 layers adopt a bulk-like structure [64].

A metallic gallenene layer with a hexagonal honeycomb structure could be grown epitaxially on a Ga layer with a superstructure $(4 \times \sqrt{13})$ that was deposited on a Si(111) $-$ (7×7) substrate, as demonstrated by low-temperature STM and supported by first-principles calculations [65]. The first Ga layer (buffer), with a thickness of

0.25 nm, was covalently bonded to the dangling bonds of the Si(111) surface, whereas the second gallenene layer, with a hexagonal honeycomb structure, was only weakly bonded to the buffer layer. An STS analysis demonstrated the metallic nature of this polymorph of gallenene.

Large-area fabrication (~ cm^2) of gallenene sheets occurred beneath epitaxial graphene on 6H-SiC under ambient conditions, triggered by liquid gallium intercalation [66]. Optical spectroscopy allowed the observation of stepwise self-propagation and diffusion of Ga atoms to form a gallenene film below the epitaxial graphene surface without further treatment. Subsequent analysis of the graphene-gallenene heterostack by XPS, SEM, and Raman spectroscopy confirmed the fabrication of wafer-scale gallenene.

3.4 Indium Monolayers (Indenene)

3.4.1 Chemistry and 2D Bonding

Investigations of the 2D phases of the heavier In atom are still in their infancy. The electronic configuration of In atoms is (Kr|$4d^{10}5s^2p^1$) with filled $4d$ and $5s$ orbitals and a $5p$ orbital with an unpaired electron. Based on the three $5s^2$ and $5p$ valence electrons, indium is trivalent. However, the heavier group—IIIA elements In and Tl exhibit an increasing tendency toward monovalent bonding behavior owing to the reluctance of the In $5s^2$ (Tl $6s^2$) electron pair to join chemical bonding ('inert pair effect') owing to a decrease in bonding strength. Consequently, the two s electrons behave like a chemically inert closed shell.

3.4.2 2D Structures

First-principles DFT-PBE calculations indicate the existence of two dynamically stable phases of indenene [67, 68]. One is a planar honeycomb phase with space group *P6/mmm* and low cohesive energies of $E_c = -1.81$ eV/atom [67], (PBE) and – 2.08 eV/atom [68], (HSE06), which behaves like a metal. The lattice constant of this planar phase is a = 0.496 nm and the bond length d = 0.286 nm.

The other identified phase is a buckled graphene-like allotrope with the space group $P3m1$ and nearly identical cohesive energies of $E_c = -1.83$ eV/atom [67], (PBE) and –2.13 eV/atom [68], (HSE06). The lattice constant of this phase at a = 0.424 nm is much smaller, and the bond length d = 0.289 nm is comparable to that of the planar phase. The buckled phase is an indirect semiconductor with a band gap of $E_g = \sim 1$ eV that transforms to a metal under compressive and tensile strains. Furthermore, the authors describe a dynamically unstable puckered structure with a comparable cohesive energy [67].

Moreover, in two extensive studies on metallic monolayers, additional phases have been proposed [51, 52]. However, these first-principles studies do not provide the necessary complete characterization of the structure and properties of the indenene phases and unfortunately use different nomenclature. The first paper describes a hexagonal structure with six nearest neighbors and the largest cohesion energy of $E_c = -1.92$ eV/atom, a square structure with four nearest neighbors and a cohesive energy of – 1.85 eV/atom, and a honeycomb phase with three nearest neighbors and the lowest cohesion of – 1.71 eV/atom [51].

The second publication discusses a plane hexagonal structure with a cohesive energy of $E_c = -1.91$ eV/atom, a buckled square phase with – 2.17 eV/atom, and a buckled honeycomb lattice with – 2.14 eV/atom [52]. The allotropes have comparable cohesive energies, and the lattice constants do not show the characteristic inverse correlation with the bonding energy, as observed for covalently bonded systems.

First-principles structural calculations with the software of crystal structure analysis by particle swarm optimization (CALYPSO) provided essential information for aluminene, gallenene, and indenene [61]. This unbiased search included bonding characteristics, phonon properties, and the effect of substrate strain on dynamic stability. The most stable form of the two heavier compounds In and Tl agrees with that of b_{010}-Ga monolayers. Going down the column, the formation energies relative to the bulk increase with similar values for Ga and In. According to phonon dispersion calculations, other allotropes of the three elements stabilize on a suitable surface in the case of matching lattice parameters.

3.4.3 Physics and Applications

The optical properties of planar and buckled indenene allotropes were studied via DFT calculations, which included the complex refractive index, absorption spectrum, electron energy loss spectra (EELS), as well as reflectivity and optical conductivity spectra, which are strongly dependent on the direction of light polarization [68]. For both in-plane and out-of-plane light polarizations, a large optical anisotropy is simulated that can be beneficial in ultrathin optical systems such as light filters for IR–UV light.

Indenene permitted a crucial step forward in the realization of large-gap quantum spin Hall (QSH) insulators based on Dirac fermions suitable for room temperature applications [69]. The investigated monolayer is a triangular indenene lattice epitaxially grown on a SiC(0001) substrate that exhibits nontrivial valley physics driven by an interplay of local spin–orbit coupling (SOC) and inversion-symmetry breaking (ISB). Figure 3.14 illustrates the band ordering effect of the Dirac fermions at the K and K' points, presenting the trivial insulator state and the QSH insulator. Topographic imaging by STM confirmed the (1×1) surface periodicity of the indenene layer on the SiC substrate, as further confirmed by total energy minimization calculations. Owing to the identical triangular lattice, layer and surface are structurally indistinguishable but differ electronically. Notably, electron charges pile up in the

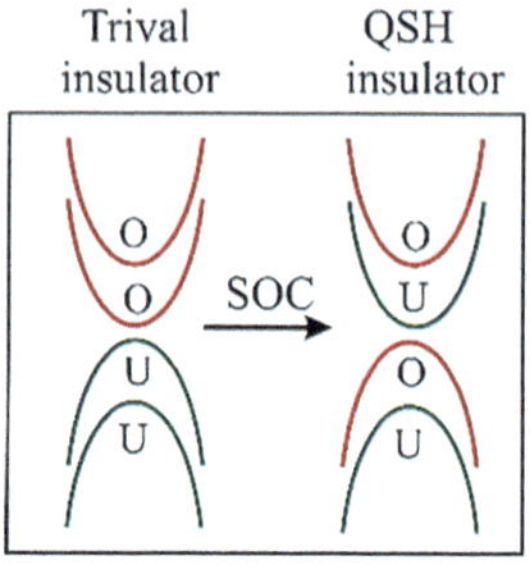

Fig. 3.14 Scheme of band ordering of the Dirac fermions at the K and K' points presenting the trivial insulator state and the QSH insulator. Reproduced from ref. [69], Copyright 2021, Springer Nature

voids of the triangular lattice. The features realized in this work pave the way for room temperature transport applications based on the use of topologically protected edge states. In conventional semiconductors, increasing confinement of electrons enhances scattering at their boundaries and leads to the development of heat. In topological insulating materials, scattering becomes prohibited by symmetry reasons. Therefore, technological applications should use topological materials such as the triangular lattice of indenene, which do not need cooling to low temperatures, as is the case for graphene.

Based on the state-of-the-art applications of 2D metals in bioimaging, photothermal therapy, sensing, memory devices, and catalysis, the thermodynamic stability and synthesis of 2D metals and alloys are topics of primary interest. According to atomistic simulations various top-down and bottom-up methods yield ultrathin 2D metal sheets. The relevant issue is to improve the control of thickness, size, composition, and structural quality. An up-to-date review covers the specific problems encountered in the synthesis of atomically thin monolayers and ultra-thin metallic nanosheets [70]. Specifically, the theoretical work concentrates on the thermodynamic stability of 2D metals and the variety of techniques developed to synthesize 2D metals.

3.4.4 Synthesis

DFT calculations were employed to predict metastable one-atom-thick metallic In overlayers on the Si(111) surface, namely, a striped In/Si(111) $-$ ($\sqrt{7} \times \sqrt{3}$) phase and its low-temperature ($\sqrt{7} \times \sqrt{7}$) phase with coverages of 1.40 and 1.43 monolayers, respectively [71]. The metallic quasi-1D structural features of weakly coupled In chains contrast with those of double-layer In phases with free-electron-like metallic properties, implying an ultimate two-layer limit of metallic behavior. Screening the interaction with the substrate may require one additional buffer layer. The two heaviest group-IIIA elements, indium and thallium, tend to form monovalent bonds in addition to the conventional trivalent state introduced above.

3.5 Thallium Monolayers (Thallene)

The electronic configuration of Tl atoms is (Xe|$5d^{10}6s^2p^1$) with filled $5d$ and $6s$ orbitals and a $6p$ orbital with an unpaired electron. Based on the three $6s^2$ and $6p$ valence electrons, Tl is trivalent. For monolayers of Tl, a topological insulating material, basic information on bonding, structure, and mechanical behavior is still widely missing despite the increasing interest in spintronics [72]. First information can be extracted from the two extensive studies on metallic monolayers mentioned above, were several thallenene phases have been proposed [51, 52]. Interestingly, a nontrivial topology and topological phase transitions have been proposed for 2D hexagonal Tl crystals based on a tight-binding model and DFT calculations [72]. As the heaviest group-IIIA element, thallium has the strongest tendency to form monovalent bonds in addition to forming monomers in the conventional trivalent state owing to the reluctance of the $6s^2$ electron pair to join chemical bonding. This 'inert pair effect' is due to the strongly decreasing bond strength when going down from the light to heavy elements, where the promotion energy needed to activate $6s$ electrons is no longer offset by the surplus of released binding energy [71]. The monovalent character of Tl atoms controls the chemisorption of Tl overlayers on Si(111) − (7 × 7) surfaces, as already reported in 1999 [73]. At room temperature, small 2D clusters are formed with the intact (7 × 7) periodicity of the substrate, as monitored by low-energy electron diffraction (LEED) and STM. At higher temperatures, the monolayer exhibited a (1 × 1) structure with Tl atoms chemisorbed directly above the Si atoms of the unreconstructed Si(111) − (1 × 1) surface, as described for indenene sheets on SiC [71].

3.6 Property–Structure Relationships

Understanding the relationships between individual structures and the resulting properties as a function of chemical composition helps in the design of unknown materials by tremendously improving the search efficiency. Increasing the success rate of experimental trials by theoretical predictions can reduce the cost of experimental tests substantially. As mentioned above, the discovery of borophene started with the prediction of the structural characteristics of nanoclusters and guidelines for their preparation in the laboratory. The '*aufbau*' principle of the periodic table helps enormously to vary the relevant chemical and physical properties in steps by deliberately changing fundamental factors such as the number and nature of valence electrons or the atom size controlling the efficiency of hybridization and thus the strength of covalent bonding.

3.6.1 Bonding–Structure Relationships

Table 3.1 presents the structure, space group, cohesive energy, lattice constants, bond length, corrugation effect, and monolayer thickness of the elemental group-IIIA monolayers for the known planar, buckled, and puckered structures of hexagonal, triangular, square, and 8-*Pmmn* allotropes. A plot, elucidating the change in bonding strength, shows the cohesive energies as a function of the usually well-known bond lengths (see Fig. 3.15) [74]. The cohesive energy describes the interaction of an atom with its surrounding atoms in the 2D network and controls the mechanical stability of the whole 2D network. In the case of nonuniform and corrugated monolayers, we selected the largest bond length in the plot, which is the bond with the lowest critical fracture strength. The inclusion of graphene, with the largest cohesive energy and shortest bond of any uniform single-atom monolayer, illustrates the enormous differences in bonding and physicochemical properties (see Table 3.1). The boron allotropes assemble in a wide range between strong and medium bonds with specific bonding types. Primarily, electron deficiency reduces the overall binding energy in comparison to that of graphene. In comparison with those of borophenes, the monolayer properties of the four heavier group-IIIA elements cluster in a region with much lower cohesive energies and thus with significantly longer and weaker bonds. Importantly, the atomic size, approximated by the van der Waals diameter, of B (0.36 nm), Al (0.42 nm), Ga (0.42 nm), In (0.44 nm), and Tl (0.44 nm) does not increase remarkably for the higher homologues of boron. Only between the second and third period does a substantial increase occur that induces a drastic extension of the bond lengths observed from the borophene (0.16–0.19 nm) to the aluminene phases (0.26–0.29 nm) and its higher homologues.

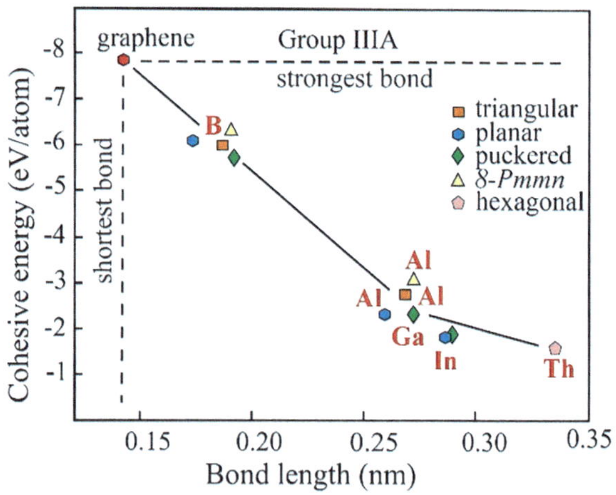

Fig. 3.15 Plot of the cohesive energy versus bond length of group–IIIA monolayers. Adapted from ref. [74], Copyright 2021, Royal Society of Chemistry

3.6.2 Mechanics–Structure Relationships

Selected information on the structure, Young's moduli, Poisson's ratios, ultimate strength, and critical strain of the studied phases of borophene, aluminene, gallenene, indenene, and thallenene is collected in Table 3.2 [75]. The plot of Young's moduli and ultimate strengths as a function of the bond length offers an overview of the mechanical behavior of the elemental group-IIIA monolayers (see Fig. 3.16) [74]. Note the different scales on the left and right ordinates, which differ by a factor of ten and therefore give direct insight into the applicability of Griffith's rule $\sigma = E/10$, where σ is the ultimate strength and E the Young's modulus. This rule connects the elastic stiffness with the approximately ten times smaller ideal critical stress or ultimate strength. The ultimate strength is the maximum stress a perfect crystal can withstand at zero temperature and thus is a measure of the strength of bonding at the ultimate strain the structure can withstand. To judge the mechanical performance of group–IIIA monolayers, we included the outstanding mechanical properties of graphene with the highest Young's modulus of 340 N/m [76], confirmed by experiment and theory, and the theoretical mean strength of σ = 37 N/m for uniaxial tension, in good agreement with 42 N/m measured by biaxial indentation experiments using AFM [77]. Note that the stiffnesses and ultimate strengths of the diverse boron phases vary within a remarkable range, partly connected with an extremely large anisotropy of the nonlinear mechanical properties.

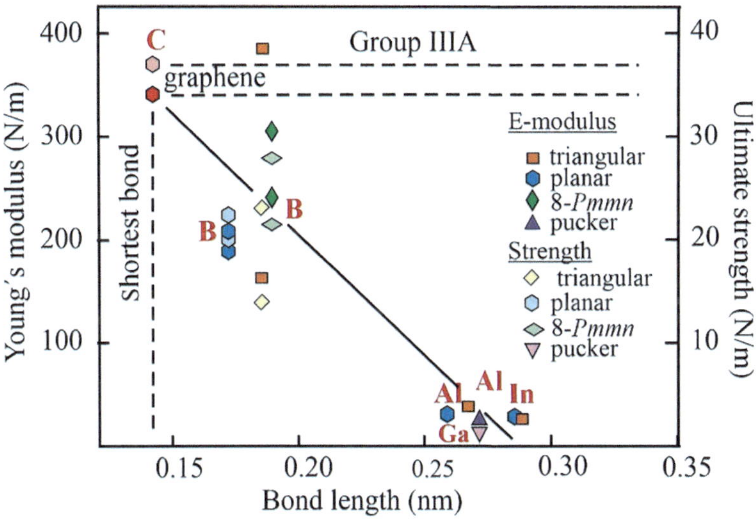

Fig. 3.16 Plot of Young's modulus and the ultimate stress or strength versus bond length of group–IIIA monolayers. Adapted from ref. [74], Copyright 2021, Royal Society of Chemistry

References

1. A.P. Sergeeva, I.A. Popov, Z.A. Piazza, W.L. Li, C. Romanescu et al., Acc. Chem. Res. **47**, 1349 (2014)
2. H. Tang, S. Ismail-Beigi, Phys. Rev. Lett. **99**, 115501 (2007)
3. B. Peng, H. Zhang, H. Shao, Z. Ning, Y. Xu et al., Mater. Res. Lett. **5**, 399 (2017)
4. X. Wu, J. Dai, Y. Zhao, Z. Zhuo, J. Yang et al., ACS Nano **6**, 7443 (2012)
5. Z. Zhang, Y. Yang, E.S. Penev, B.I. Jakobson, Adv. Funct. Mater. **27**, 1605059 (2017)
6. A.J. Mannix, X.-F. Zhou, B. Kiraly, J.D. Wood, D. Alducin et al., Science **350**, 1513 (2015)
7. B. Feng, J. Zhang, Q. Zhong, W. Li, S. Li et al., Nature Chem. **8**, 563 (2016)
8. S.S. Batsanov, Inorg. Mater. **37**, 871 (2001)
9. J. Kunstmann, A. Quandt, Phys. Rev. B **74**, 035413 (2006)
10. L. Kong, K. Wu, L. Chen, Front. Phys. **13**, 138105 (2018)
11. M. Nakhaee, S.A. Ketabi, F.M. Peeters, Phys. Rev. B **98**, 115413 (2018)
12. N. Gao, X. Wu, X. Jiang, Y. Bai, J. Zhao, FlatChem. **7**, 48 (2018)
13. Y. Liu, E.S. Penev, B.I. Jakobson, Angew. Chem. Int. Ed. **52**, 3156 (2013)
14. P. Ranjan, T.K. Sahu, R. Bhushan, S.S.R.K.C. Yamijala, D.J. Late et al., Adv. Mater. **31**, 1900353 (2019)
15. Z.A. Piazza, H.-S. Hu, W.-L. Li, Y.-F. Zhao, J. Li et al., Nat. Commun. **5**, 3113 (2014)
16. P. Sutter, E. Sutter, Chem. Mater. **33**, 8838 (2021)
17. C. Chen, H. Lv, P. Zhang, Z. Zhuo, Y. Wang et al., Nat. Chem. **14**, 25 (2022)
18. X. Liu, Q. Li, Q. Ruan, M.S. Rahn, B.I. Yakobson, Nat. Mater. **21**, 35 (2022)
19. J. Yuan, N. Yu, K. Xue, X. Miao, RSC Adv. **7**, 8654 (2017)
20. S. Ipek, M.E. Kilic, A. Mogulkoc, S. Cahangirov, E. Durgun, Phys. Rev. B **98**, 241408(R) (2018)
21. X. Yu, L. Li, X.-W. Xu, C.-C. Tang, J. Phys. Chem. C **116**, 20075 (2012)
22. X.-F. Zhou, X. Dong, A.R. Oganov, Q. Zhu, Y. Tian et al., Phys. Rev. Lett. **112**, 085502 (2014)
23. A.J. Mannix, Z. Zhang, N.P. Guisinger, B.I. Yakobson, M.C. Hersam, Nat. Nanotech. **13**, 444 (2018)
24. B. Mortazavi, O. Rahaman, A. Dianat, T. Rabczuk, Phys. Chem. Chem. Phys. **18**, 27405 (2016)
25. Z.-Q. Wang, T.-Y. Lü, H.-Q. Wang, Y.P. Feng, J.-C. Zheng, Front. Phys. **14**, 23403 (2019)
26. H. Zhong, K. Huang, G. Yu, S. Yuan, Phys. Rev. B **98**, 054104 (2018)
27. B. Peng, H. Zhang, H. Shao, Y. Xu, R. Zhang et al., J. Mater. Chem. C **4**, 3592 (2016)
28. Z.-D. Sha, Q.-X. Pei, K. Zhou, Z. Dong, Y.-W. Zhang, Extreme Mech. Lett. **19**, 39 (2018)
29. Y.-P. Zhou, J.-W. Jiang, Sci. Rep. **7**, 45516 (2017)
30. Y. Valadbeigi, H. Farrokhpour, M. Tabrizchi, J. Chem. Sci. **127**, 2029 (2015)
31. M. Tatullo, B. Zavan, F. Genovese, B. Codispoti, I. Makeeva et al., Appl. Sci. **9**, 3446 (2019)
32. C. Hou, G. Tai, Y. Liu, Z. Wu, X. Liang et al., Nano Res. Energy **2**, e9120051 (2023)
33. Z. Xie, X. Meng, X. Li, W. Liang, W. Huang et al., Research **2020**, 2624617 (2020)
34. D.J. Joshi, N.I. Malek, S.K. Kailasa, New J. Chem. **46**, 4514 (2022)
35. M. Ou, X. Wang, L. Yu, C. Liu, W. Tao et al., Adv. Sci. **8**, 2001801 (2021)
36. Y.V. Kaneti, D.P. Benu, X. Xu, B. Yuliarto, Y. Yamauchi et al., Chem. Rev. **122**, 1000 (2022)
37. Z. Wu, G. Tai, R. Liu, W. Shao, C. Hou et al., J. Mater. Chem. A **10**, 8218 (2022)
38. G.P. Campbell, A.J. Mannix, J.D. Emery, T.-L. Lee, N.P. Guisinger et al., Nano Lett. **18**, 2816 (2018)
39. X. Liu, L. Wang, S. Li, M.S. Rahn, B.I. Jakobson et al., Nat. Commun. **10**, 1642 (2019)
40. R. Wu, I.K. Drozdov, S. Eltinge, P. Zahl, S. Ismail-Beigi et al., Nat. Nanotech. **14**, 44 (2019)
41. R. Wu, S. Eltinge, I.K. Drozdov, A. Gozar, P. Zahl et al., Nat. Chem. **14**, 377 (2022)
42. W. Li, L. Kong, C. Chen, J. Gou, S. Sheng et al., Sci. Bulletin **63**, 282 (2018)
43. L. Zhu, B. Zhao, T. Zhang, G. Chen, S.A. Yang, J. Phys. Chem. C **123**, 14858 (2019)
44. R. Wu, A. Gozar, I. Božović, NPJ Quant. Mater. **4**, 40 (2019)
45. P. Ranjan, T.K. Sahu, R. Bhushan, S.S.R.K.C. Yamijala, D.J. Late et al., Adv. Mater. **1900353** (2019)

46. C. Kamal, A. Chakrabarti, M. Ezawa, New J. Phys. **17**, 083014 (2015)
47. I. Lukačević, M.V. Pajtler, M. Muževič, S.K. Gupta, J. Mater. Chem. C **7**, 2666 (2019)
48. M. Muževič, Doctoral dissertation, University of Zagreb, Faculty of Science, Zagreb (2019)
49. J. Yuan, N. Yu, K. Xue, X. Miao, Appl. Surf. Sci. **409**, 85 (2017)
50. K.H. Yeoh, T.L. Yoon, Rusi, D.S. Ong, T.L. Lim, Appl. Surf. Sci. **445**, 161 (2018)
51. J. Navalaita, P. Koskinen, Phys. Rev. B **97**, 035411 (2018)
52. S. Ono, Phys. Rev. B **102**, 165424 (2020)
53. S. Kansara, S.K. Gupta, Y. Sonvane, T. Hussain, R. Ahuja, A.C.S. Appl, Energy Mater. **1**, 3428 (2018)
54. D. Pandey, C. Kamal, A. Chakrabarti, J. Chem. Phys. **150**, 194702 (2019)
55. A.R. Villagracia, H.L. Ong, M. David, N. Arboleda Jr, IOP Conf. Ser. Earth Environ. Sci. **268**, 012135 (2019)
56. V. Nagarajan, R. Chandiramouli, Mater. Sci. Eng. B **229**, 193 (2018)
57. D.Z. Metin, L. Hammerschmidt, N. Gaston, Phys. Chem. Chem. Phys. **20**, 27668 (2018)
58. M. Nakhaee, M. Yagmurcukardes, S.A. Ketabi, F.M. Peeters, Phys. Chem. Chem. Phys. **21**, 15798 (2019)
59. V. Kochat, A. Samanta, Y. Zhang, S. Bhowmick, P. Manimunda, Sci. Adv. **4**, e1701373 (2018)
60. S.V. Badalov, M. Yagmurcukardes, F.M. Peeters, H. Sahin, J. Phys. Chem. C **122**, 28302 (2018)
61. B.A. Khalil, N. Gaston, J. Phys. Condens. Matter **33**, 125901 (2021)
62. M. Petrov, J. Bekaert, M.V. Milošević, 2D Mater. **8**, 035056 (2021)
63. K.G. Steenbergen, N. Gaston, Chem. Commun. **55**, 8872 (2019)
64. S. Lambie, K.G. Steenbergen, N. Gaston, Nanoscale Adv. **3**, 499 (2020)
65. M.-L. Tao, Y.-B. Tu, K. Sun, Y.-L. Wang, Z.-B. Xi et al., 2D Mater. **5**, 035009 (2018)
66. S. Wundrack, D.M. Pakdehi, W. Dempwolf, N. Schmidt, K. Pierz et al., Phys. Rev. Mater. **5**, 024006 (2021)
67. D. Singh, S.K. Gupta, I. Lukačević, Y. Sonvane, RSC Adv. **6**, 8006 (2016)
68. D. Singh, S.K. Gupta, I. Lukačević, M. Muževič, Y. Sonvane et al., Sci. Rep. **9**, 17300 (2019)
69. M. Bauernfeind, J. Erhardt, P. Eck, P.K. Thakur, J. Gabel et al., Nat. Commun. **12**, 5396 (2021)
70. T. Wang, M. Park, Q. Yu, J. Zhang, Y. Yang, Mater. Today Adv. **8**, 100092 (2020)
71. J.W. Park, M.H. Kang, Phys. Rev. Lett. **117**, 116102 (2016)
72. J. Zhang, W.-X. Ji, C.-W. Zhang, P. Li, P.-J. Wang, Phys. Chem. Chem. Phys. **20**, 24790 (2018)
73. L. Vitali, F.P. Leisenberger, M.G. Ramsey, F.P. Netzer, J. Vac. Sci. Technol. A **17**, 1676 (1999)
74. P. Hess, Nanoscale Horiz. **6**, 856 (2021)
75. P. Hess, ChemPhysChem **e202100880** (2022)
76. P. Hess, J. Appl. Phys. **120**, 124303 (2016)
77. C. Lee, X. Wei, J.W. Kysar, J. Hone, Science **321**, 385 (2008)

Chapter 4
Group IIA–IVA and IIIA–IVA Monolayers

Abstract Theory revealed the nature of bonding and structure of several representatives of binary IIA–IVA and IIIA–IVA monolayers with novel chemical, physical, and topological properties based on reduced dimensionality. Unfortunately, experimental verification of these 2D materials has not been achieved. The hypercoordinate motifs with multicentered bonding and delocalized π electrons realized in these networks offer an extended platform in 2D materials chemistry, providing access to novel low-dimensional structures with innovative physicochemical properties. Density functional theory calculations and treatment of the exchange correlation energy within the generalized gradient approximation with suitable functionals yielded the binding energy and structural properties. The global minimum structure search with particle swarm optimization using crystal structure analysis via the particle swarm optimization package identified the lowest energy configuration of the potential energy surface for several compounds. Of particular interest in these rule-breaking quasi-planar structures is the electron density distribution. The electron localization function and solid-state adaptive natural density partitioning methods allow notable insight into the type of bonding by elucidating the distribution of electrons. The critical points are the energetic stability given by a negative cohesive energy, the dynamic or kinetic stability requesting the absence of imaginary or negative frequencies, and the thermal stability at elevated temperatures. Despite the enormous relevance of mechanical properties for the preparation and application of 2D materials, only linear properties such as Young's moduli and Poisson's ratios have been determined, with negative Poisson ratios occurring for some compounds. The unique bonding types imply potential applications in various fields, such as electronics, optoelectronics, spintronics, photovoltaics, mechanics, and sensorics. Plots of the cohesive energy versus bond length and of the ultimate stiffness and strength versus bond length allow insight into the fundamental structure–property relationships of these compounds with unconventional bonding behavior and astonishing good stability.

P. Hess, *Bonding, Structure, and Performance of Two-Dimensional Materials*, NanoScience and Technology, https://doi.org/10.1007/978-3-031-74161-6_4

4.1 Be_xC_y. Mg_2C, Ca_2C (IIA–IVA)

4.1.1 Chemistry and 2D Bonding

Theory elucidated the nature of bonding and geometry of the (quasi) planar structures of selected binary IIA–IVA and IIIA–IVA monolayers quite well, but their synthesis is still lacking [1]. Figure 4.1 highlights the elements involved in the so-called hypercoordinated bonding networks elucidated in detail below.

Interestingly, the corresponding heteroelemental monolayers of these main-group elements exhibit the rule-breaking phenomenon of quasi-planar hypercoordinate chemistry. This bonding type is important for the development of novel 2D materials owing to the enforced restriction of covalent bonding to two dimensions. In contrast to conventional 3D tetrahedral sp^3 and planar tricoordinate sp^2 bonds of carbon, binary monolayers with Be, Ca, Mg, B, and Al atoms exhibit quasi-planar tetracoordinate (ptC) moieties as well as quasi-planar pentacoordinate (ppC) and quasi-planar hexacoordinate (phC) moieties [1, 2]. This behavior deviates from that of established tetrahedral bonding behavior of carbon, which has a maximum coordination number of four. The reason for energetic, dynamic, and thermodynamic stabilization of such an unusual coordination in the plane is the delocalization of electrons from the carbon $2p_z$ orbitals by efficient π acceptors and the donation of σ electrons to electron-deficient bonds by suitable charge donors.

The electronic configuration of Be atoms ($1s^22s^2$), Mg atoms (Ne|$3s^2$), Ca atoms (Ar|$4s^2$), B (He|$2s^2p_x{}^1$), Al (Ne|$3s^2p_x{}^1$), and C atoms (($1s^22s^2p_x{}^2$) possess two, three, and four valence electrons. For example, Be is a good π electron acceptor and σ electron donor because of the hybridization of Be $2p_x$ with C $2p_x$ and C $2p_y$ electrons, while the C $2p_z$ electrons delocalize, and the C $2s$ and Be $2s$ electrons form the core states [3]. Moreover, different electronegativities promote electron transfer in these heteroelemental compounds. These effects increase the electron density at the C (Si) moiety and enforce (quasi) planarity of the covalent σ bonds. The stabilization of planar hypercoordination by electronic and steric effects already has been studied intensively for isolated molecular species by theory and experiments, including its mechanical enforcement by steric constraints [4–7]. Although the thermodynamic and kinetic stability of such clusters has been proven, to understand the combination of atoms to realize ptC, for example, is still a serious problem [6]. These peculiar

H							He
Li	Be	B	C	N	O	F	Ne
Na	Mg	Al	Si	P	S	Cl	Ar
K	Ca	Ga	Ge	As	Se	Br	Kr
Rb	Sr	In	Sn	Sb	Te	I	Xe
Cs	Ba	Tl	Pb	Bi	Po	At	Rn
Fr	Ra						

Fig. 4.1 Scheme of the main group elements of the periodic table highlighting the elements involved in the formation of hypercoordinated binary IIA–IVA and IIIA–IVA monolayers

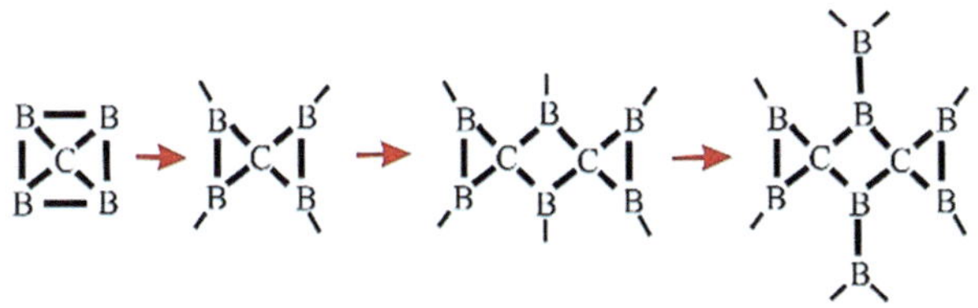

Fig. 4.2 Step-by-step growth of an extended B_2C monolayer containing the hypercoordinated ptC motif, starting from a hypothetical CB_4 molecule

molecular motifs must be extended to infinite layers, starting, for example, from a hypothetical CB_4 molecule (see Fig. 4.2), as schematically illustrated by the step-by-step growth of an extended B_2C monolayer containing the hypercoordinated ptC motif.

The incorporation of planar hypercoordinate motifs in monolayer structures requires a new synthesis strategy that allows the realization of such 2D materials with unconventional bonding. To date this includes, for example, theoretically stable monolayers with hypercoordinate moieties of carbon and silicon with group-IIA and group-IIIA elements. The cohesive bonding energy of these compounds shows the typical inverse dependence on the mean bond length, as predicted by tight-binding theory, also for nonuniform incorporation of these moieties in an anisotropic monolayer structure. The cohesive energies are nearly comparable to those of normally bonded monolayers such as buckled elemental group-IVA monolayers containing mixed sp^2–sp^3 hybrids, such as the monolayer of Si. Similarly, the in-plane stiffnesses seem to be only slightly diminished, despite nonuniform interconnection schemes between the individual hypervalent units. As a rule, hypercoordination of carbon (silicon) with variable coordination is possible in the plane if delocalization of the bonding electrons and π bonding stabilize the structure. However, the observed planarity of the motifs may not be complete with increasing coordination due to an increasing repulsion between the dense hybrids.

Besides electronic effects, mechanical strains can enforce hypercoordinate bonding in 2D space, e.g., by incorporating small rings into the monolayer. Such an additional structural degree of freedom opens the door to novel structures with unforeseen chemical and physical properties, as well as promising applications. It is well known that elements capable of multicenter bonding, such as boron, allow a variety of bonding types and structural allotropes with delocalized charge distributions and satisfactory bonding energies. The specific nature and stability of the interconnects combining the individual hypercoordinate building blocks play a fundamental role in promoting the actual properties of these 2D networks.

There are two methods that deliver the necessary information on charge distribution and thus the nature of bonding in extended 2D systems. The interpretation of the charge distribution in terms of electron lone pairs, two-center bonds, and multi-center bonding is usually based on the solid state adaptive natural density partitioning (SSAdNDP) method [8]. The second method, which provides important insight into the electron distribution and type of bonding, is the electron localization function

(ELF) [9]. The essential ELF values of approximately 0, 0.5, and 1 refer to extremely low electron density, fully delocalized electrons, and extremely high density of electrons, respectively. According to these methods, especially multicenter π bonding is a characteristic feature of hypervalent bonded 2D materials.

The theoretical strategies employed to find such stable hypercoordinated 2D arrangements of atoms include bottom-up approaches, maintaining the number of valence electrons by isoelectronic substitution, and direct global minimum searches. Clearly, 2D materials with the lowest energy configuration of the global minimum on the potential energy surface (PES) have the best chance of experimental realization. Quite often, however, additional metastable configurations with a local minimum and quite similar cohesive energies exist. If these monolayers are thermodynamically and dynamically stable, experimental synthesis may be possible, by providing additional stabilization environments, e.g., by adsorption on a suitable substrate surface.

4.1.2 2D Structure

Theoretical methods supplied structural and bonding information, e.g., the dynamic and thermal stability, the cohesive energy, and the space group presented here for monolayers incorporating hypercoordinated motifs of carbon and silicon. If available, the discussion includes structural data, the film thickness, and assemblies of monolayers to bilayers and few-layer sheets. Table 4.1 lists the cohesive energies, lattice constants, bond lengths, and van der Waals diameters of the compounds with quasi-planar moieties considered in this chapter. The mean van der Waals diameter (d_W) [10, 11], defined by the sum of the tabulated van der Waals radii of the two different elements, allows a rough estimate of the thickness of planar monolayers. Adding the buckling or corrugation effect to the mean van der Waals diameter ($d_{W+\Delta z}$) gives a first estimate of the thickness of corrugated layers. The structure, cohersive energy, lattice constant, bond length, buckling distance, and van der Waals diameter and corrugation corrected diameter of group IIA–IVA and IIIA–IVA monolayers are presented in Table 4.1.

BeC: A global structure search revealed a minimum energy for monolayers of BeC with a complicated structure, containing rings of three, six, and eight atoms (see Fig. 4.3) [12]. The nonuniform complex network contains three types of bonds, namely, covalent bonds, mixed covalent-ionic bonds, and ionic bonds. The Be atoms transfer 2*s* electrons to the C 2*p* orbitals, and the back-donation of $2p_z$ electrons from the C atoms partly fills the empty Be 2*p* orbitals. In this planar low-symmetry structure with ptC moieties, the carbon atoms bind four Be atoms. An explanation for the nearly perfect planarity with an estimated corrugation of only $\Delta z = 0.0085$ nm is the small-ring mechanical strain effect and the efficient π-electron delocalization. The structure has a surprisingly large cohesive energy of $E_{co} = -6.67$ eV/atom with a lattice constant of a = 0.435 nm, a C–C bond length of d = 0.138, which is shorter than that in graphene (0.142 nm), and Be–C bond lengths of 0.160 and 0.166 nm

Table 4.1 Structure, cohesive energy, lattice constants, bond lengths, buckling distance, van der Waals diameter and corrugation corrected diameter of group IIA–IVA and IIIA–IVA monolayers

Mono-layer	Quasi-planar structure	Cohesive energy, E_c (eV/atom)	Lattice constants a, b (nm)	Bond lengths $d_{CX,SiX}$, d_{XX}, d_{CC}, d_{SiSi} (nm)	Buckling distance Δz (nm)	Van der Waals diameter d_w, $d_{w+\Delta z}$ (nm)
BeC	(ptC)	– 6.67 [12]	0.435 [12]	~ 0.163, 0.138 [12]	0.0085 [12]	0.36 [10]
Be_2C	(phC)	– 4.86 [13]	0.299 [13]	0.173, 0.198 [13]	0.092 [13]	0.36 [10], 0.452
Be_5C_2	(ppC)	– 4.58 [17]	0.892, 0.921 [17]	~ 0.17, ~ 0.220 [17]	0.214 [17]	0.36 [10], 0.574
Mg_2C	(phC)	– 3.43 [18]	0.354 [18], 0.355[19]	0.233, 0.271 [18]	0.179 [18]	0.39 [10], 0.569
Ca_2C	(phC)		0.394 [23]	0.256, 0.326 [23]	0.234 [23]	0.41 [10], 0.644
CaSi	(ppSi)	– 3.07 [25]	0.933, 0.686 [25]	~ 0.30, 0.227 [25]	0.076 [25]	0.45 [10], 0.526
B_2C	(ptC)	– 6.47 [29]	0.256, 0.345 [29]	0.156, 0.169 [29]	~ 0.0085 [29]	0.35 [10], 0.359
penta-B_2C	(ptC)	– 6.87 [32]	0.393 [31]	0.158, 0.172 [31]	0.108 [31]	0.36 [10], 0.408
t-B_4C_3	(ptC)	– 6.56 [33]	0.470 [33]	~ 0.156, 0.169 [33]	~ 0.03 [33]	0.35 [10], 0.380
h-B_4C_3	(phC)	– 6.89 [34]	0.269, 0.467 [34]	0.156, 0.176 [34]	0.488 [34]	0.35 [10], 0.838
g-B_4C_3	(ptC)	– 6.63 [35]	0.810, 0.467 [35]		0.075 [35]	0.35 [10], 0.425
B_4C_3-II	Bilayer	– 6.68 [37]	0.805, 0.466 [37]	~ 0.156, ~ 0.170 [37]	0.325 [37]	0.35 [10], 0.675
Al_2C	(ptC)	– 4.49 [38]	0.506, 0.304 [38]	0.196, 0.260 [38]	0 [38]	0.38 [10]

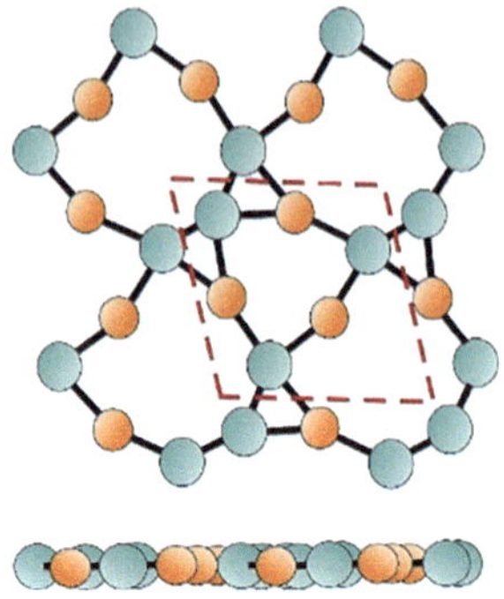

Fig. 4.3 Top and side view of the perfectly planar hypercoordinate BeC monolayer with the quasi-ptC motif. The red dashed rhombus displays the unit cell. Adapted with permission from ref. [12], Copyright 2009, Royal Society of Chemistry

[12]. The cohesive energy suggests that the 2D network formed by the reaction of free atoms has a large formation energy. The negative value of the cohesive energy proves the energetic stability, and the absence of imaginary frequencies suggests a dynamical stability of the lattice. Furthermore, first-principles molecular dynamics (MD) simulations estimated an excellent thermal stability of 2D BeC of up to 2000 K.

Furthermore, BeC is one of the few hypercoordinated monolayers in which bilayer stacking has been studied [12]. For AA stacking, the two monolayers have the same coordinates in the *a* and *b* directions. The AB stacking configuration is found by rotating the top layer by 180°, and AC stacking is obtained by shifting the top layer along the lattice vector *b* by a distance of the unit cell. Total energy calculations revealed that AC stacking is the energetically most stable stacking order, and AA stacking is the least stable configuration. The interlayer separations, which decrease from AA stacking ($d_s = 0.51$ nm) to AB stacking (0.42 nm) and AC stacking (0.37 nm), validate this interpretation [12]. The latter value agrees well with the mean van der Waals diameter of $d_{ew} = 0.35$ nm and thus points to weak interlayer interactions [10, 11].

Be_2C: Based on DFT calculations, PSO, and a global minimum search resulted in the stable incorporation of a quasi-planar phC moiety in a weakly buckled monolayer ($\Delta z = 0.092$ nm) with stoichiometry Be_2C [13]. Each C atom binds to six Be atoms, while each Be atom connects to three C atoms and three Be atoms (see Fig. 4.4) [13–16]. The light green hexagons schematically indicate the electron-rich areas around the carbon atoms. The electron-deficient areas, which are located around the Be atoms, are not highlighted [13].

The transfer of 2*s* electrons from the Be atoms stabilizes the quasi-phC network and induces symmetric buckling of two adjacent Be atoms of $\Delta z = 0.046$ nm from the central C atom plane to reduce repulsion between neighboring cations [13]. This medium bonded structure has a cohesive energy of $E_{co} = -4.86$ eV/atom. The lattice constant is $a = 0.299$ nm and the Be–C and Be–Be bond lengths are $d = 0.173$ nm and 0.198 nm, respectively, and the Be–C–Be bond angle is $\theta = 66.5°$. In addition to the confirmed energetic and dynamic stability, the high melting point at approximately 1500 K confirms excellent thermal stability [13]. Below, a separate section addresses the promising applications of Be_2C monolayers, especially those

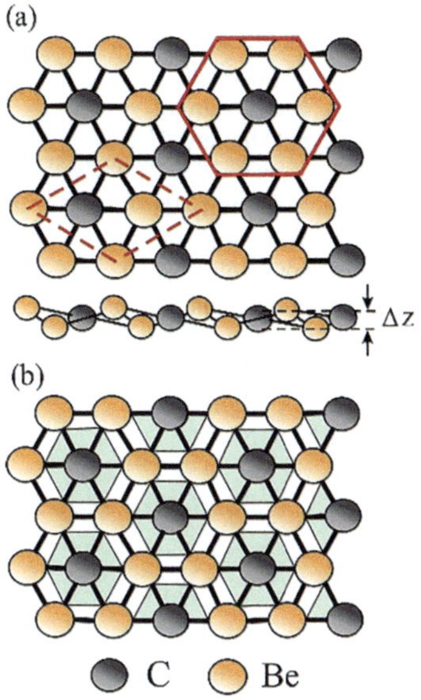

Fig. 4.4 Top and side view of the Be_2C monolayer with the quasi-phC motif. The red line denotes a phC moiety and the red dashed rhombus displays the unit cell. The light green hexagons schematically indicate the electron-rich areas around the carbon atoms. The electron-deficient areas, which are located around the Be atoms, are not highlighted. Partly adapted with permission from ref. [13], Copyright 2014, John Wiley and Sons

with hexagonal crystal structures, three sublayers, and the space group *P*-3*m*1 [14, 15]. These two investigations studied the electronic transport [14], and the use of Be_2C as an anode material in lithium-ion batteries [15].

In addition, DFT calculations, together with the PSO method, predicted three new 2D structures with a Be_2C composition, one with octacoordinate carbon poC moieties and two with phC moieties. All three phases possess a reasonable cohesive energy, positive phonon modes, semiconductive properties, and negative Poisson's ratios [16].

***Be_5C_2*:** The computationally designed Be_5C_2 monolayer contains quasi-planar pentacoordinate carbon motifs (quasi-ppC) in a global minimum structure with stable Be_5C_2 stoichiometry (see Fig. 4.5) [17]. Delocalization of electrons from the Be-2*s* orbitals over the Be–C bonds stabilizes the covalent multicenter bonds of the ppC moieties. A carbon atom binds five Be atoms in quasi the same plane. Remarkably, the Be_5C_2 monolayer contains no classical lone pairs or localized conventional (2c–2e) bonds but consists only of three-, four-, and six-center bonds. The lattice constants are a = 0.892 nm and b = 0.921 nm, the lengths of the two types of C–Be1 bonds are for these shorter bonds d_1 = 0.166 and 0.168 nm. The longer C–Be_2 bonds are d_2 = 0.173 and 0.174 nm, and the much longer Be–Be bonds are in the range of d = 0.202–0.239 nm. Note that while the ppC motifs have good planarity, the large total buckling effect at Δz = 0.214 nm reduces the repulsive interaction between the Be atoms in the monolayer. The SSAdNDP method was employed to improve the understanding of the complex chemical bonding arrangement. The large unit cell contains eight B_5C_2 units with 72 electrons. A negative cohesive energy of E_c = – 4.58 eV/atom reveals energetic stability and positive phonon modes guarantee

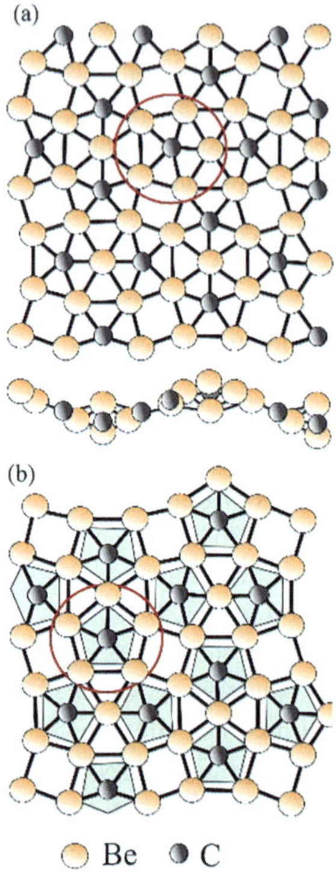

Fig. 4.5 Top and side view of the geometric structure of Be_5C_2 with quasi-ppC moieties. The red circle denotes a ppC moiety. The light green areas schematically signify the electron-rich regions within the quasi-ppC moieties and the electron-deficient regions concentrate around the Be atoms. Partly adapted from ref. [17], Copyright 2016, Springer Nature

dynamic stability. MD simulations suggest thermal stability up to the high melting point of 1500 K.

Mg_2C: First-principles calculations and a global structure search via PSO predicted a quasi-planar hexacoordinate motif (phC) in monolayers of Mg_2C, which represent the global minimum structure [18, 19]. Similar to the previously discussed Be_2C compounds, the s^2-electron configuration of the valence electrons of the metal atom determines the structure. The binding electrons delocalize preferentially around the C atoms and stabilize the hexacoordinate phC motif. Two hexagonal layers of Mg atoms sandwich the basal plane of hexagonal carbon. Each C atom binds six Mg atoms and each Mg atom binds to three Mg and three C atoms. The lattice constant is $a = 0.354$ nm, and the total buckling distance of the monolayer is $\Delta z = 0.179$ nm. The larger Mg–Mg bond length is $d = 0.271$ nm, and the shorter Mg–C bond length is 0.233 nm [18]. These lattice parameters agree well with $a = 0.355$ nm and $d_{Mg\text{-}C} = 0.223$ nm [19]. According to MD simulations, the monolayer maintains its structural integrity up to approximately 900 K, and because no imaginary phonon frequencies exist, the monolayer is dynamically stable. The computed moderate cohesive energy of this monolayer is $E_c = -3.43$ eV/atom [18].

Mg_2C magnesium carbide monolayers may find important applications in optoelectronics and thermoelectrics and provide an intriguing platform for unconventional fermions [19–21]. Analysis of the optical properties revealed sharp absorption peaks in the UV region and a tunable band gap [20]. The thermal conductivity of Mg_2C monolayers is comparable to that of silicene, more than two orders of magnitude lower than that of graphene, and even lower than that of MoS_2 [21]. Reasons for this behavior are strong phonon anharmonicity due to resonant bonding and dual band gaps in the phonon dispersion, limiting the phonon–phonon scattering. Monolayer M_2C is metallic and can be transformed at moderate biaxial strain in a metal-semimetal-semiconductor phase transition, during which several types of unconventional 2D fermions emerge, including the anisotropic Dirac fermions [19].

Ca_2C: DFT calculations, performed using the DFT-GGA-PBE approximation, incorporated the effects of electron–ion and electron–electron exchange–correlation interactions to determine the properties of metallic calcium carbide monolayers, following the first synthesis of Ca_2C [22, 23]. As described for Mg_2C, the C atom is sandwiched between two Ca atoms, forming a similar hexagonal structure. The calculated Ca–C and Ca–Ca bond lengths were d = 0.256 nm and 0.326 nm, respectively. Since the buckling effect is not known, the overall planarity of bonding also is missing. The monolayer is dynamically stable without any negative frequency and is thermally stable up to 700 K [23]. Independent MD simulations predicted a lower thermal stability of up to 400 K, restricting its use as a gas sensor [24].

CaSi: Inspired by the discovery of planar pentacoordinate bonding of Si in the molecular species $Ca_4Si_2^{2-}$, representing a building block following the 18-electron rule, DFT calculations revealed the design of an infinite CaSi monolayer (see Fig. 4.6) [25]. In the top view of this figure the light green areas schematically indicate the electron-rich regions. The electron-deficient areas occur at the Si and Ca atoms. Stabilization of the structure occurs by delocalization of Si $3p_z$ electrons and σ donation of Ca electrons following the donation–backdonation principle. In the CaSi layer, the Si atom binds with five ligands, namely, one Si atom and four Ca atoms (quasi-ppSi). The Ca atoms of this buckled monolayer exhibit a buckling distance of Δz = 0.076 nm and the lattice constants are a = 0.933 nm and b = 0.686 nm. The structure contains variable bonds, e.g., large Si–Ca bonds (0.297 and 0.304 nm) and even longer Ca–Ca bonds (0.361 and 0.398 nm), however, the structure has a short Si–Si bond (0.227 nm). The unit cell contains two two-center Si–Si bonds, eight three-center σ bonds, and two four-center π bonds, accounting for 24 valence electrons. Especially the multicenter delocalized σ and π bonds favor a quasi-planar configuration. The CaSi monolayer is thermodynamically, mechanically, dynamically, and thermally stable up to approximately 1200 K [25]. The cohesive energy of E_c = – 3.07 eV/atom is small, consistent with the large bond lengths, and represents the global minimum structure on the potential energy surface [25].

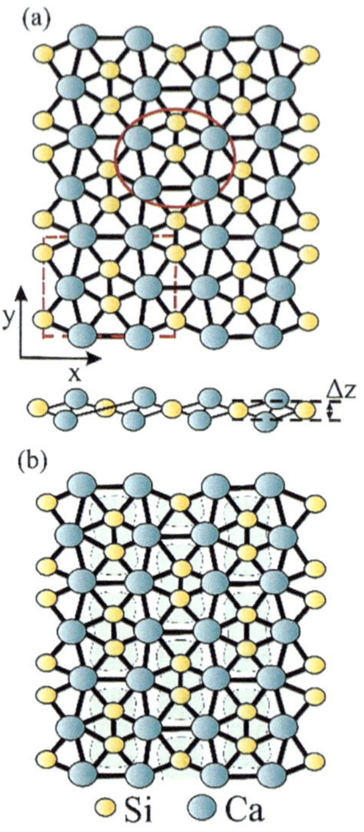

Fig. 4.6 Top and side view of the CaSi monolayer, which contains the quasi-ppSi motif, as indicated by the red circle. The red dashed lines indicate the unit cell. In the top view the light green areas schematically indicate the electron-rich regions. The electron-deficient areas occur at the Si and Ca atoms. Partly adapted with permission from ref. [25], Copyright 2016, Royal Society of Chemistry

4.1.3 Mechanics

Unfortunately, at present, mechanical data are available only on linear mechanical properties such as linear elastic constants, Young's moduli, and Poisson's ratios (see Table 4.2). In addition, the table shows the incorporated quasi-planar hypercoordinated moieties and the space groups of the group IIA–IVA and IIIA–IVA monolayers. All the information comes from calculations and simulations because the synthesis of these compounds is still lacking. In many cases, there is no confirmation of the presented theoretical results by independent studies.

BeC: For the BeC monolayer with perfectly planar tetracoordinate carbon (ptC) motifs theory predicts a complex structure with different ring sizes. In addition to the elastic constants, the authors calculated an in-plane stiffness of E = 145.5 N/m for this beryllium–carbide monolayer [12]. The large cohesive energy of $E_c = -6.67$ eV/atom and the short C–C and Be–C bonds support such a large stiffness.

Be$_2$C: The isotropic stiffness of E = 162 N/m presented for Be_2C monolayers with quasi-planar hexacoordinate carbon (phC) is surprisingly large for the moderate cohesive energy of $E_c = -4.86$ eV/atom and the long Be–C bonds [13]. Furthermore, for the Be_2C monolayer with phC motifs, derived by DFT and a global minimum search, a similar cohesive energy of E_c – 4.87 eV/atom, strongly anisotropic stiffness values of $E_{ac} \approx 175$ N/m and $E_{zz} \approx 46.7$ N/m have been reported, using a layer

Table 4.2 Structure, space group, Poisson's ratio, and Young's modulus of group IIA–IVA and group IIIA–IVA monolayers

Mono-layer	Quasi-planar motifs	Space group	Poisson ratio ν_{ac}, ν_{zz}	Young's modulus E_{ac}, E_{zz} (N/m)
BeC	(ptC)	AMM2 [12]		145.5 [12]
Be_2C	(phC)	*P*-3*m*1 [15]		162 [13], 145, 46.7 [14]
Be_5C_2	(ppC)		– 0.041, – 0.16 [17]	32.7, 130 [17]
Mg_2C	(phC)	*P*-3*m*1 [19]	~ 0.03, – 0.023 [19]	59.6, 56.8 [19]
Ca_2C	(phC)		0.45 [23]	32.6 [23]
CaSi	(ppSi)		– 0.15, – 0.16 [25]	21.8, 24.6 [25]
B_2C	(ptC)	*Pmm*2 [28]	0.148, 0.123 [29]	234, 194 [29]
t-B_4C_3	(ptC), (phC)			35.7 [33]
h-B_4C_3	(phC)	*P*1 [34]	0.18 [34]	431 [34]
B_4C_3-II	Bilayer		0.178, 0.194 [37]	382, 417 [37]
Al_2C	(ptC)	*Pmmm* [39]	0.114, 0.104 [40]	130, 118 [40]

thickness of t = 0.495 nm [16]. Remarkably, in this work, the structure of a novel Be_2C monolayer with octacoordinated carbon (poC) and a similar cohesive energy of – 4.94 eV/atom is presented, which exhibits extremely large slightly anisotropic Young's moduli of $E_{ac} \approx 316$ N/m and $E_{zz} \approx 289$ N/m [16].

***Be*$_5$*C*$_2$**:The semimetallic Be_5C_2 monolayers, equipped by DFT calculations with ppC motifs and a cohesive energy of – 4.58 eV/atom, have strongly anisotropic stiffnesses of $E_{ac} = 32.7$ N/m and $E_{zz} = 130$ N/m [17]. Furthermore, the structure has unusual negative Poisson's ratios of $\nu_{ac} = -0.041$ and $\nu_{zz} = -0.16$.

Mg*$_2$*C: A study on Mg_2C monolayers reported nearly isotropic stiffnesses of $E_{ac} =$ 59.6 N/m and $E_{zz} = 56.8$ N/m and ultimate uniaxial strains of $\varepsilon_{ac} = 0.18$ and $\varepsilon_{zz} =$ 0.22 as linear mechanical properties [19]. The small stiffness values are consistent with the low binding energy of $E_c = -3.43$ eV/atom [18]. The application of tensile strain in the zigzag direction leads to dilatation of the monolayer caused by a negative Poisson ratio of $\nu_{zz} = -0.023$, whereas along the armchair direction, the Poisson ratio is positive at $\nu_{ac} = \sim 0.03$.

Ca*$_2$*C: For Ca_2C monolayers, a smaller isotropic stiffness and a much larger Poisson ratio of E = 32.6 N/m and ν = 0.45, were derived from the elastic constants provided by the authors [23]. The elastic constants satisfy the stability criteria of Born-Huang. The polar diagrams presented for Young's modulus and Poisson's ratio confirm the isotropic mechanical behavior of the Ca_2C monolayer.

CaSi: CaSi monolayers exhibit slightly anisotropic in-plane Young's moduli of E_{ac} = 21.8 N/m and $E_{zz} = 24.6$ N/m, which are smaller than those of Ca_2C owing to the smaller binding energy of only $E_c = -3.07$ eV/atom. Furthermore, 2D CaSi exhibits strongly negative Poisson ratios of $\nu_{ac} = -0.15$ and $\nu_{zz} = -0.16$ [25].

Unfortunately, information on nonlinear mechanical properties such as fracture strength and ultimate strain of monolayers with hypercoordinated motifs is widely missing. This is a fundamental issue since nonuniform networks with bonds of different strengths are prone to bond rupture at the weakest connection between atoms. While the linear mechanical behavior of hypercoordinated compounds seems to broadly follow that of uniformly bonded monolayers, nonlinear properties, such as the ultimate strength or toughness, will respond sensitively to any structural nonuniformity with weak interactions. Since local bond rupture may affect the strength of the whole 2D network, the longest bond with the smallest bonding energy usually will control the overall fracture strength.

4.1.4 Physics and Applications

BeC: This paragraph gives some examples from the widespread discussion of the characteristic properties and potential applications of quasi-planar hypercoordinated monolayers. For example, the exceptionally high carrier mobility is based on the nonbonding character of the valence band maximum (VBM) and the hybrid-bonding character of the conduction band minimum (CBM) [12]. Since the PBE functional tends to underestimate the band gap of the indirect semiconductor BeC (0.49 eV), the Heyd-Scuseria-Ernzerhof (HSE06) functionals are commonly used to calculate the band gap (1.01 eV). The BeC monolayer qualifies for applications in electronic devices and optoelectronics. Several authors suggest synthesis on metal or metal oxide surfaces by chemical vapor deposition (CVD).

Be_2C: The semiconductor Be_2C has a direct medium band gap of $E_g = 2.34$ eV (HSE06) and is of interest for applications in both nanoelectronics and optoelectronics [13]. In addition to this use, the Be_2C monolayer has been suggested as a promising material for thermoelectric applications [14]. By first-principles calculations using Wannier functions the absolute thermoelectric coefficients were determined for Be_2C monolayers. While only moderate figures of merit ($ZT \approx 0.1$) were obtained, Be_2C is a promising material as far as a high-power factor is concerned (~ 3.44 mW m^{-1} K^{-2}). Moreover, Be_2C has been suggested as an anode material in lithium-ion batteries [15]. The metallic character obtained by the adsorption of Li atoms results in a low Li diffusion barrier of 0.11 eV, an open-circuit voltage of 0.33 V, and a maximum theoretical capacity of 1785 mAh/g. Interestingly, three new 2D Be_2C structures were predicted, one with octacoordinated carbon (poC) atoms [16].

Be_5C_2: A further increase in the Be concentration in 2D beryllium carbides leads to a gapless state or semimetallic behavior in the Be_5C_2 monolayer, where the CBM and VBM meet at the Fermi level. An analysis of the density of states (DOS) supports the presence of a Dirac cone like that of graphene [17]. Furthermore, Be_5C_2 could find applications in the field of mechanics as a so-called auxetic material with negative

Poisson ratios. The method suggested for synthesis is like that described above for BeC; however, in this case it is necessary to accurately control the Be/C ratio.

Mg_2C: The Mg_2C monolayer can be strain-tuned from a metal to a semiconductor and it can adopt three different band phases, i.e., metal at its equilibrium state, gapless semimetal, and narrowband semiconductor under mild hydrostatic strains [18]. The Mg_2C monolayer provides a promising platform for studying the properties of unconventional types of fermions beyond the isotropic Dirac fermions in graphene and thus may pave the way for novel nanoscale devices [19]. The monolayer has an excellent mechanical flexibility and an intrinsic negative in-plane Poisson ratio.

Ca_2C: The metallic monolayers of Ca_2C exhibit excellent stability and energetics and retain their metallic character also after adsorption of Li, Na, and K atoms. Thus, these assemblies provide superior charge transfer and superior mobility. The ultralow diffusion barriers allow fast cycling (adsorption and desorption). Therefore, these monolayers are candidates for use as an anode in the next generation of high-performance Li-, Na-, and K-ion batteries [23]. Theoretical studies of the sensitivity of Ca_2C monolayers to the adsorption of toxic pnictogen hydrides such as NH_3, PH_3, AsH_3, and SbH_3 indicate high adsorption energies and a considerable effect on electronic properties to be used for sensitive trace gas detection [24].

CaSi: Promising applications in optoelectronics can be expected for CaSi monolayers with strong optical absorption in the visible and infrared regions of the solar spectrum but also in the ultraviolet range [25]. These monolayers possess good light harvesting capacity and exhibit unique optical anisotropy.

4.1.5 Synthesis

2D materials with multivalent motifs offer a new platform for novel physicochemical properties and applications, e.g., catalytic processes owing to chemically active surfaces or high-speed nanodevices with low dissipation. While the synthesis of isolated but stabilized molecular hypercoordinate moieties was possible in many clusters [6], experimental syntheses of extended monolayers with incorporated hypercoordinated motifs are still lacking, even for monolayers with a confirmed global minimum configuration. A variety of synthesis strategies have been suggested in the literature, which should accelerate the first successful assembly of such a 2D material [1, 12, 17].

4.2 B_7Si, B_2C, B_xC_y, and Al_2C (IIIA–IVA)

4.2.1 Chemistry and 2D Bonding

The electronic configuration of B atoms ($1s^2 2s^2 p_x^{\ 1}$) and Al atoms (Ne|$3s^2 p_x^{\ 1}$) have three valence electrons and that of C atoms four valence electrons ($1s^2 2s^2 p_x^{\ 2}$). These valence electrons control the 2D bonding behavior. The principles of hypercoordinated bonding in binary monolayers of B and Al with C, forming quasi-planar tetracoordinate (ptC), quasi-planar pentacoordinate (ppC), and quasi-planar hexacoordinate (phC) moieties are the same as those already discussed above. The investigation of systems incorporating elements from additional groups such as group-IIIA elements helps to establish bonding–structure relationships, which offer a deeper understanding of the structural possibilities opened by hypercoordination and multivalency for the further development and application of (quasi) planar covalent chemistry. More insight into the general bonding principles and types will help to realize optimized 2D structures with tailored properties. There is an ongoing effort to design novel 2D materials preferentially based on light elements with strong covalent bonding and increasing complexity of lattices with low symmetry.

4.2.2 2D Structures

B_7Si: First-principles calculations, using a global optimization method based on the PSO algorithm revealed that most of the ten studied structures of 2D boron-silicon compounds exhibit a preference for planar sp^2 bonding of silicon [26]. A quasi-planar tetracoordinated silicon atom bonded with four Si atoms (ptSi) exists, for example, in the lowest energy structure of the B_7Si compound. DFT calculations predict that the lowest-energy B–Si monolayer is metallic, and phonon-spectrum calculations confirmed the dynamic stability. The binding energies of 2D B–Si compounds increase with increasing boron content and reach a value of $E_c = -5.686$ eV/atom for the B_7Si composition [26]. Similar to borophenes, boron-rich 2D B–Si compounds such as B_2Si, B_3Si, B_5Si and B_6Si tend to form complex structures with planar sp^2 hybrids, whereas B_4Si exhibits a small buckling effect.

B_2C: For B_2C monolayers, first-principles lattice dynamics and electron–phonon coupling calculations predicted a lattice consisting of a mosaic of hexagons and rhombuses with quasi-ptC moieties (see Fig. 4.7) [27]. The light green areas in this figure signify the electron-rich regions within the quasi-ptC moieties. Electron-deficient regions exist at the B atoms. The ptC atoms form multicenter B–C bonds 0.156 nm in length connected with neighboring motifs by two-center covalent B–B bonds 0.168 nm in length [28–30]. An ELF analysis indicated strong covalent B–B bonds, weaker B–C bonds, and delocalized electrons between B–B bonds. Boron-centered covalent sp^2 hybrids are the backbone of the quasi-planar B_2C network.

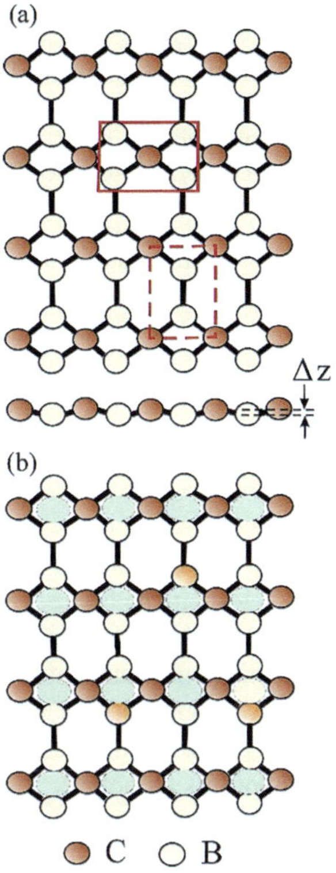

Fig. 4.7 Top and side view of the B_2C monolayer with a quasi-ptC motif. The red rectangular box shows the ptC motif and the red dashed lines the unit cell. In the top view the light green areas signify the main electron-rich regions within the quasi-ptC moieties. Electron-deficient regions exist at the B atoms. Partly adapted with permission from ref. [27], Copyright 2009, Royal Society of Chemistry

The lattice parameters are a = 0.256 nm, b = 0.345 nm and the bond lengths $d_{B–C}$ = 0.156 nm and $d_{B–B}$ = 0.169 nm in excellent agreement with the values given above [29]. Note that the separation between the C layer and B layers is only ~ 0.0085 nm [29] (~ 0.0032 nm [27]). A global minimum search of 2D boron–carbon compounds revealed that this energetically and dynamically stable structure is not the lowest-energy configuration, however, it exhibits a comparable stability [30]. In fact, a possibly metallic metastable network is formed by strong bonds with a cohesive energy of $E_c = -6.47$ eV/atom [29]. The network is dynamically and thermally stable up to approximately 2000 K [28]. The B_2C unit cell contains two B atoms, one C atom, and ten valence electrons [28]. The space group of the low-buckled monolayer is *Pmm*2, which changes to *Pmmm* for the perfectly flat configuration [28].

penta-B₂C: All-electron DFT calculations based on the PBE exchange correlation functional confirmed the stability of the 2D penta-B_2C monolayer [31]. Each pentagon contains three B and two C atoms, each C atom binds to four B atoms (ptC), and the tricoordinate B atom binds two C atoms and one B atom (see Fig. 4.8). The buckled pentagonal network is a trilayer sandwich B–C–B structure with a buckling distance of $\Delta z = 0.108$ nm [31]. The lattice constant of the monolayer is a = 0.393 nm and the bond lengths are 0.158 nm for B–C and 0.172 nm for B–B bonds

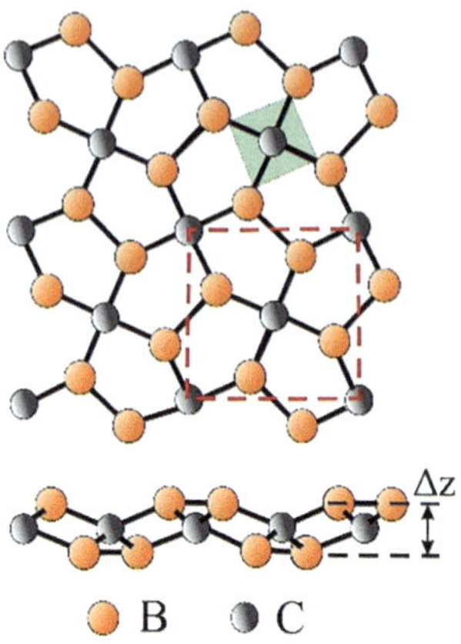

Fig. 4.8 Top and side view of the penta-B_2C monolayer, which contains the quasi-ptC motif. The red dashed lines indicate the unit cell. The light green area shown in one ptC motif schematically indicates the electron-rich region around the C atom. Adapted with permission from ref. [31], Copyright 2015, Royal Society of Chemistry

[31]. In contrast to the planar metallic Be_2C, the penta-Be_2C monolayer is a semiconductor with an indirect band gap of $E_g = 2.28$ eV (HSE). It is dynamically stable and 0.203 eV/atom less stable than the global minimum structure. Ab initio MD simulations indicate that the structure survives at a temperature of 1000 K. The large cohesive energy of $E_c = -6.87$ eV/atom reported for pentagonal Be_2C monolayers makes it interesting for electronic and optoelectronic applications [32]. The space group of this monolayer with low symmetry is $P\text{-}42_1m$ [31].

t-B$_4$C$_3$: Another proposed boron carbide is the B_4C_3 monolayer derived from quasi-planar t-B_4C_3 clusters with tetracoordinate carbon motifs (quasi-ptC) [33]. Theoretical DFT-GGA-PBE computations and structure-searching methods revealed the global minimum structure. The B–C bonds vary between 0.153 and 0.159 nm, where the shorter B–C bond connects different t-B_4C_3 motifs of the slightly buckled monolayer. The bond length of the weaker B–B bonds is 0.169 nm akin to that in borophene. The central B atom of each t-B_4C_3 motif is situated approximately $\Delta z = 0.03$ nm above the hexagonal h-B_3C_3 ring and is responsible for marginal buckling of the monolayer. ELF analysis confirmed σ donation of electrons from B to C atoms and multicenter covalent bonding between C and neighboring B atoms (π acceptance). The strongly bonded energetically stable monolayer has a cohesive energy of $E_c = -6.56$ eV/atom and is not only dynamically, but also thermally stable up to 1600 K. Application of the SSAdNPD method elucidated the roles played by electron lone pairs, classical two-center bonds, and delocalized multicenter bonds. The unit cell of this monolayer contains three conventional two-center two-electron (2c–2e) B–C σ bonds, six (3c–2e) C–B–B bonds, and three (7c–2e) π bonds, accounting for 24 electrons [33].

h-B$_4$C$_3$: An extensive global structure search via the PSO-CALYPSO method offered additional insight into the versatile bonding capabilities of 2D stoichiometric boron carbides. The study reveals the existence of 14 different stoichiometric B–C compositions, including an additional B_4C_3 allotrope [34]. Counterintuitively, a pyramidal tetracoordinate and a hexacoordinate sp^2 carbon moiety were identified for B–C

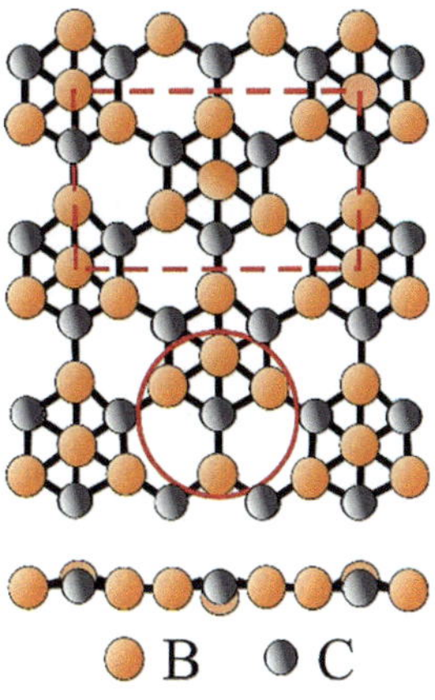

Fig. 4.9 Top and side view of the graphene-like B_4C_3 monolayer, which contains the quasi-ptC motif. The red dashed lines indicate the unit cell and the red circle specifies the ptC motif. Adapted with permission from ref. [35], Copyright 2019, Royal Society of Chemistry

bonding. The energetically favorable h-B_4C_3 allotrope consist of a sandwich arrangement with outer hexagonal B–C lattices and an inner hexacoordinate carbon moiety. The lattice constants are a = 0.269 nm and b = 0.467 nm and the outer B–C bonds 0.156 nm in length are shorter than the inner B–C bonds with 0.176 and 0.177 nm in length. The distance between upper and lower layer was $\Delta z = 0.488$ nm. By ELF analysis, the authors identified a complex structure with two main ELF regions with an isosurface of 0.75. One involves electron accumulation within the inner hexacoordinate carbon moiety, where six B–C bonds surround a C atom. The other one is located around the B–C bonds of the outer hexagonal lattice. For this structure, the authors reported a large cohesive energy of Ec = – 6.89 eV/atom [34]. The dynamically and mechanically stable semiconducting monolayer belongs to the space group *P*1.

g-B_4C_3: First-principles calculations combined with the global minimum search predict a graphene-like g-B_4C_3 monolayer with tetracoordinated C atoms (ptC) akin to B_2C (see Fig. 4.9) [35]. The lattice parameters are a = 0.810 nm and b = 0.467 nm. The boron and carbon atoms alternately appear along the periphery of each hexagon, with B atoms protruding upward and downward from the plane, respectively, resulting in off-plane buckling of $\Delta z = 0.075$ nm. Each carbon atom is tetracoordinated, forming a quasi-ptC motif. The center B atom is hexacoordinate and the periphery B atom is tetracoordinated. The absence of any imaginary frequency in the phonon spectrum confirms the dynamic stability of the material. An extensive PSO search yielded a global minimum for this monolayer. Accordingly, the structure exhibits strong B–C bonds with a cohesive energy of E_c = – 6.63 eV/atom and thermal stability up to 1000 K. Note that the formation energies of all 2D boron carbides are positive owing to the intrinsic phase separation of B and C. However, a positive sign does not mean instability but metastability because strong B–C bonds prevent destabilization. The structure has a rectangular primitive cell and the space group $Pmn2_1$. The semiconducting monolayer has a direct band gap of $E_g = 2.73$ eV. The g-B_4C_3 monolayers exhibit strong absorption of light from blue to ultraviolet light. The section on physics and applications presented below describes results obtained for the sensitive and selective detection of polar gases adsorbed on g-B_4C_3 monolayers [36].

b-B₄C₃: Linking two B–C-bonded monolayers by covalent bonds leads to stable b-B_4C_3 bilayers [37]. DFT calculations using the GGA-PBE functional revealed six different bilayer configurations with metallic and semiconducting properties, which are energetically more stable than the van der Waals bonded bilayers owing to covalent-type interlayer bonding. These b-B_4C_3 bilayers resemble a triple-layer structure with a covalently bonded inner B atom. Owing to their outstanding mechanical properties, we concentrate on two bilayers, namely, b-B_4C_3-II, shown in Fig. 4.10, and b-B_4C_3-V. The former is an indirect semiconductor, and the latter is a direct semiconductor. In b-B_4C_3-V, the inner B atom connecting the bilayers is tetracoordinated by four short B–C bonds ~ 0.17 nm in length, which is typical for sp^3 hybridization. Within the surface layers, B atoms adopt sp^2 hybridization with B–C bonds of 0.154 and 0.158 nm, indicating strong sp^2-like bonds. Tables 4.1 and 4.2 present the structural, bonding, and mechanical properties of the covalent bilayer b-B_4C_3-II. The lattice parameters are a = 0.805 nm, b = 0.466 nm, and the interlayer distance ds = 0.325 nm. Phonon spectrum calculations and MD simulations demonstrated the dynamic and thermal stabilities of the bilayers. The special bonding-based electronic and mechanical properties of these materials are of interest in photovoltaics, nanoelectronics, and nanomechanics.

Al₂C: The structure of the Al_2C monolayer, derived from the planar molecular species $C_2Al_6^{2-}$, follows the previously described structure of B_2C with tetracoordinate carbon (ptC) and tricoordinate Al atoms. A PSO global minimum structure search identified this structure as the lowest energy configuration of all the Al_2C alternatives [38]. Better σ donation due to the larger electronegativity difference between C (2.55) and Al (1.61) than between C and B (2.04) completely avoids buckling by the further influence of the delocalized electron distribution. The effective electronic

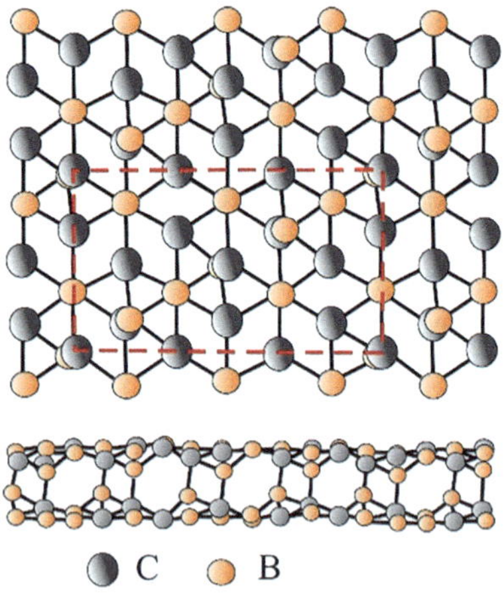

Fig. 4.10 Side view and top view of the geometrical structure of the B_4C_3-II bilayer. The red dashed line denotes the unit cell. Reproduced from ref. [37], Copyright 2021, American Chemical Society

charge configurations of the ptC motif are $2s^{1.61}$, $2p_x{}^{1.65}$, $2p_y{}^{1.68}$ and $2p_z{}^{1.57}$. This is consistent with complete electron delocalization by inductive electron transfer through σ donation by Al atoms and $2p_z$ electron delocalization by C atoms over the four C–Al bonds. The optimized lattice constants are a = 0.506 nm and b = 0.303 nm and the bond lengths $d_{C\text{–}Al} = 0.196$ nm and $d_{Al\text{–}Al} = 0.260$ nm. The semiconducting monolayer with a small indirect band gap of 1.05 eV (HSE06) is energetically and dynamically stable and maintains its structural integrity up to 1500 K according to MD simulations. Note that the larger size of the Al atoms implies reduced orbital overlap, resulting in a moderate cohesive energy of $E_c = -4.49$ eV/atom [38]. The space group of the planar Al_2C monolayer is *Pmmm* [39]. A global minimum search was used to extend the 2D structures to stoichiometric Al_xC (x = 1/3, 1, 2, and 3) monolayers [40]. In this study, the predicted most stable Al–C monolayers had a similar binding energy of $E_c = -4.45$ eV/atom as that of the Al_2C monolayer. While AlC and Al_2C are semiconductors, the Al_3C monolayer is a metal.

4.2.3 Mechanics

B_2C: The 2D B–C allotropes stand out for their strong bonding and excellent mechanical properties. Calculation of the linear mechanical properties of single-layer B_2C with ptC motifs from the published elastic constants yielded $E_{ac} = 234$ N/m and $E_{zz} = 194$ N/m for the in-plane Young's moduli, using a layer thickness of t = 0.371 nm, and $\nu_{ac} = 0.148$ and $\nu_{zz} = 0.123$ for the positive Poisson ratios [29]. This large stiffness is consistent with the large cohesive energy of – 6.47 eV/atom and high thermal stability of this lattice.

t-B_4C_3: Based on DFT-PSO calculations for the t-B_4C_3 monolayer with ptC motifs, the authors obtained an extremely small Young's modulus or in-plane stiffness of only E = 35.7 N/m [33]. A possible reason for the low stiffness is the open structure of the t-B_4C_3 lattice.

h-B_4C_3: The elastic constants calculated for h-B_4C_3 by the DFT-GGA-PBE approach delivered a surprisingly large Young's modulus of E = 431 N/m for a monolayer and a small Poisson ratio of $\nu = 0.18$, indicating extreme stiffness of the 2D lattice.

b-B_4C_3: For the multilayered b-B_4C_3 boron-carbide, similarly an outstanding in-plane Young's modulus of 431 N/m is reported that is greater than that of graphene (340 N/m), and the Poisson ratio is 0.18 [34]. Note that this compound has one of the largest cohesive energies of $E_c = -6.89$ eV/atom observed for the B–C monolayers. Another study on stable b-B_4C_3 bilayers identified the two allotropes b-B_4C_3-II and b-B_4C_3-V as those with the largest Young's moduli among six stable, chemically bonded b-B_4C_3 bilayers [37]. These two bilayers, with the same large cohesive energy of $E_c = -6.68$ eV/atom, have outstanding in-plane Young's moduli of $E_{ac} = 382$ N/m and $E_{zz} = 417$ N/m and small Poisson ratios of $\nu_{ac} = 0.178$ and $\nu_{zz} = 0.194$ for b-B_4C_3-II and $E_{ac} = 401$ N/m and $E_{zz} = 388$ N/m for b-B_4C_3-V, which clearly outperform the stiffness of single-layer graphene. The corresponding elastic constants meet the mechanical stability criteria of Born-Huang.

Al_2C: The published elastic constants of Al_2C allowed an estimation of the stiffness values of $E_{ac} = 130$ N/m and $E_{zz} = 118$ N/m, using a layer thickness of $t = 0.38$ nm for the transformation to 2D mechanical units [40]. These stiffness values are consistent with the medium cohesive energy of $E_c = -4.45$ eV/atom and the lengths of the chemical bonds. The corresponding Poisson ratios are $\nu_{ac} = 0.114$ and $\nu_{zz} = 0.104$.

4.2.4 Physics and Applications

B_2C: The B_2C monolayer seems to be the first intrinsic phonon-mediated 2D superconductor with transition temperatures in the range of 14.3–19.2 K, which are higher than the boiling point of liquid helium [27]. The low-frequency out-of-plane phonon modes contribute most to the coupling. The intrinsic superconductivity in B_2C offers a good possibility to study this effect in 2D systems. Furthermore, topological semimetals, including Dirac semimetals, can coexist in the low-energy electronic state of 2D B_2C monolayers with negligible spin–orbital coupling (SOC) owing to their intrinsic topological nature [28].

t-B_4C_3: The t-B_4C_3 monolayers possess unique physicochemical properties, such as strong visible light absorption, appropriate band edge alignment, and high carrier mobility; thus, they have high potential for use as metal-free catalysts for the CO_2 reduction reaction (CO_2RR) [33]. An additional application of 2D t-B_4C_3 monolayers could be their utilization in gas sensing devices, as indicated by first-principles calculations, which delivered the adsorption energy, recovery time, work function, conductivity, and charge transfer behavior of various gas molecules [36]. While the NO, CO, and SO_2 molecules are charge acceptors, H_2S and NH_3 function as charge donors. The induced variation in the band gap affects both the sensitivity and the selectivity of t-B_4C_3 monolayers.

h-B_4C_3: In addition to exhibiting high efficiency in the use of sunlight, h-B_4C_3 monolayers exhibit fast migration of photoinduced carriers and redox capacity of activated electrons and holes.

Al_2C: The Al_2C monolayer is a semiconductor with a small indirect band gap of $E_g = 1.05$ eV (HSE06) and a quite close direct band gap of 1.28 eV (HSE06), offering interesting applications in electronics and optoelectronics [38]. With its high specific capacity, low diffusion barrier leading to fast lithium-ion diffusion, and safe lithium storage capability, Al_2C monolayers qualify for the design and development of advanced 2D anode materials [39]. Moreover, the semiconducting monolayer with a band gap of 1.05 eV is suitable for photovoltaic applications [40].

A first-principles study on the adsorption and detection behavior of pristine Al_2C monolayers indicated that it might be a superior adsorbent and sensing medium for toxic volatile organic compounds (VOCs), such as acetaldehyde, ethylene oxide, vinyl chloride, and benzene [41]. Upon adsorbate interaction, the band gap changes substantially, and the adsorption energies are greater than those for other substrates. Ab initio molecular dynamics (AIMD) simulations confirmed the stability of the studied complexes under the considered conditions.

4.2.5 Synthesis

There are detailed recommendations in the literature for the synthesis of Al_2C monolayers. This includes CVD and MBE on the (100) or (010) surface of PdO, which matches the Al_2C structure [38, 40]. In fact, the lattice parameters of PdO match those of the Al_2C monolayer quite well, and the bond lengths of Pd–O and O–O at the surface are close to the Al–C and Al–Al bonds of the Al_2C monolayer.

4.3 Hypercoordinate Chemistry of Transition-Metal Borides

The coordination numbers of quasi-planar hypercoordinate motifs of the main group elements considered above are between four and six. The use of transition-metal atoms as central element and small boron atoms as ligands allows the realization of much larger coordination numbers. In fact, first-principles calculations predict the existence of planar motifs with coordination eight and nine. In Fig. 4.11 diverse examples for quasi-planar hypercoordinate samples are presented with low coordination (4–6) dominating in group IIA–IVA and IIIA–IVA monolayers and high coordination (8–10) predicted in transition-metal borides. Such 2D materials may find applications as magnetic material, in ferromagnetic spintronics, and in catalysis.

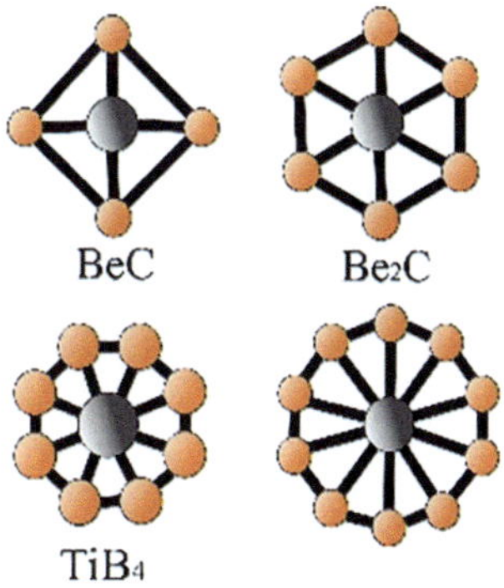

Fig. 4.11 Examples for planar hypercoordinate chemistry. Low coordination (4–6) is observed in group IIA–IVA and IIIA–IVA monolayers and high coordination has been found in transition-metal borides

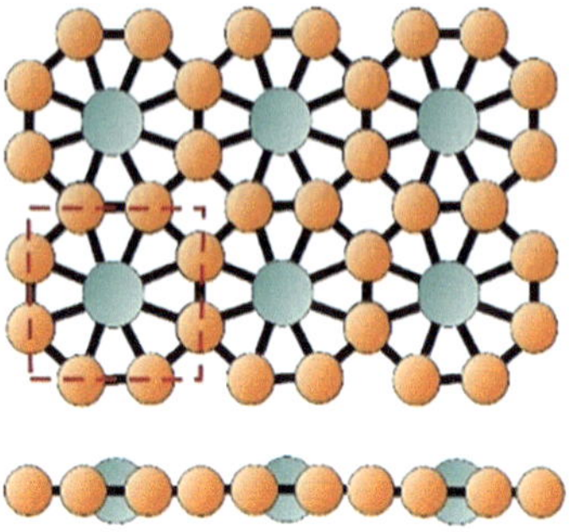

Fig. 4.12 Top and side view of the TiB_4 monolayer, with a planar octacoordinate Ti atom (poTi). Adapted with permission from ref. [42], Copyright 2017, Royal Society of Chemistry

TiB₄: An active field of research is planar hypercoordinate chemistry involving boron and transition metals. For example, in TiB_4 monolayers a planar octacoordinate Ti moiety is realized, where each Ti atom binds eight B atoms in a perfect plane (see Fig. 4.12) [42]. Ab initio calculations confirmed the thermodynamic and dynamic stability of the monolayer that can be attributed to the perfect fit of the atomic diameter of the Ti atom and the size of the 8-membered ring of B atoms. Stabilization of the sp^2 hybridized B network occurs by electron transfer from Ti to B atoms, to compensate the electron deficiency of sp^2 hybrids in B. In addition, quasi-planar octacoordinated chemistry was studied for other transition-metal borides (V, Cr, Mo, W, and Os).

TMB₉: By first principles calculations, the suitability of hypercoordinate planar structures of transition-metal borides with the composition TMB_9 (TM = Mn, Cr, V, and Ti) as tunable 2D magnetic materials was investigated [43]. To study the structural stability phonon dispersions and the elastic constants were determined. As an example, MnB_9 showed a 100% spin polarization ratio at the Fermi level, a high Curie temperature of 240 K, and a robust perpendicular magnetic anisotropy energy of 180 μeV/atom, enabling ferromagnetic spintronics at the nanoscale. Moreover, all investigated TMB_9 structures exhibited excellent catalytic activity for the hydrogen evolution reaction (HER).

TMBₙ: In a recent review regular polygons were used as generally applicable models, to derive relations between the radii of the constitute atoms and the angles of regular polygons such as TMB_n, n = coordination number [44]. Several effective design strategies and principles were suggested. Furthermore, future prospects including diverse potential applications are discussed in this work.

4.4 Bonding–Structure Relationships

4.4.1 Cohesive Energy Versus Bond Length

The plot in Fig. 4.13 displays the cohesive binding energies versus the bond lengths found in the complex hypercoordinated monolayers of BeC, Be_2C, Be_5C_2, Mg_2C, CaSi, B_2C, B_4C_3, and Al_2C. According to the valence-shell electron-pair repulsion (VSEPR) rule, at first glance, hypercoordinated moieties seem to be energetically unfavorable owing to repulsion of the high density of bonding electrons in the plane. However, the plot demonstrates surprisingly efficient bonding of quasi-planar hypercoordinate structures with their unconventional bonding types. In fact, a comparison of some of the discussed quasi-planar tetracoordinated structures with the conventional 3D tetragonal structure of diamond with its cohesive energy of $E_c = -7.71$ eV/atom and bond length of d = 0.154 nm supports this statement [45]. Delocalization of electrons in the bonding region results in efficient stabilization of the quasi-planar moieties. The introduction of both π-accepting and σ-donating groups is essential for the manifestation of the stabilization mechanisms. Thus, π bonding and activation of *p* electrons in quasi-planar multiple bond arrangements are the basis of efficient hypercoordination.

While the cohesive energy gives only a mean bond energy per atom for the formation of a monolayer by the reaction of the free atoms, the different bond lengths

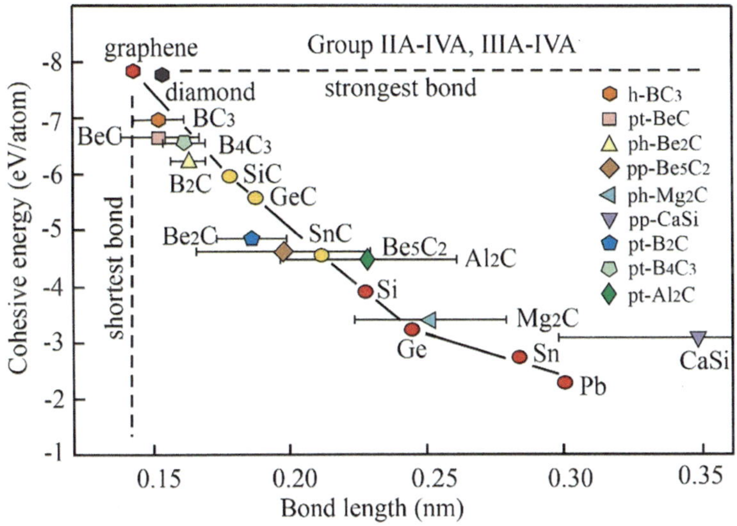

Fig. 4.13 Plot of cohesive energies versus the range of bond lengths realized in group IIA–IVA and IIIA–IVA monolayers. For comparison, the plot displays the corresponding dependence of planar binary and buckled elemental group–IVA monolayers. The binding energies of graphene and the BC_3 monolayer are also included. Adapted from ref. [52], Copyright 2021, Royal Society of Chemistry

of a nonuniform network signify the variation in individual bond energies within the monolayer. Flexible binding schemes usually connect bonded hypercoordinate motifs with fixed coordination sites to neighboring motifs forming an infinite 2D network. Importantly, structural stabilization of nonuniform structures with different ring sizes and varying bonding elements can be achieved by efficiently connecting the hypercoordinated building blocks. Since weak bonds can be compensated by strong bonds, comparable mean binding energies can be achieved, like those observed in uniformly bonded honeycomb 2D networks. However, weak bonds are detrimental to mechanical performance. Among the various possible arrangements, those with global minimum energy configurations have the best chance of experimental realization.

The distinctive inverse dependence of the cohesive energies on the range of existing bond lengths is consistent with the prediction of tight-binding theory (see Fig. 4.13) [46]. In the case of a nonuniform structure, the cohesive energy represents the mean value of the individual bond energies. We plot this mean binding energy versus the resulting spread of bond lengths. Notably, these mean values exhibit the same characteristic inverse behavior as elemental and binary group-IVA monolayers, as discussed in the following chapters. This is a consequence of strong covalent bonding, despite varying ionic contributions. According to the tight-binding model, the strength of covalent orbital coupling, V, which describes the cohesive energy in nonpolar compounds, decreases with increasing bond length, d, as $V \sim 1/d^2$ [47, 48]. Usually, the contribution of electrostatic forces is not dominant and greater for the compounds of C (2.55) with Be (1.57), Al (1.61), and Ca (1.0), whereas the difference in electronegativity and thus the polarity is smaller for bonding of C (2.55) with B (2.04) and Si (1.9).

In the following, the bonding–structure relationships of hypercoordinated compounds are compared with those of the uniform hexagonal monolayer of graphene, characterized by the smallest bond length of d = 0.142 nm and the greatest cohesive energy of $E_c = -7.85$ eV/atom [49]. Furthermore, the BC_3 monolayer can be described as doped graphene with a honeycomb structure, where all the carbon hexagons are connected by boron atoms [50]. This incorporation of boron atoms in 2D BC_3 generates B–C bonds of 0.156 nm in length, whereas the C–C bonds of the carbon hexagons are still ~ 0.142 nm [50]. Accordingly, the cohesive energy decreases to $E_c = -6.86$ eV/atom in 2D BC_3 [51]. The reasons for weaker bonding are the missing valence electrons and the larger atom size of boron. Increasing the boron content, e.g., in hypercoordinated B_2C and B_4C_3 monolayers, surprisingly reduces the cohesive energy only slightly to $E_c = -6.47$ eV/atom and − 6.56 eV/atom, respectively [29, 33]. While 2D B_4C_3 contains two different B–C bonds (0.153 and 0.159 nm) and one B–B bond (0.169 nm), B_2C monolayers have one B–C bond (0.156 nm) and one B–B bond (0.169 nm). Interestingly, the lengths of the B–C bonds are similar, and the weaker B–B bonds have a relatively small influence on the mean binding energy. All these observations imply efficient covalent bonding within the (ptC) moieties. The observed behavior can be correlated with the decrease in the mean number of valence electrons available per atom from 4 (graphene) to 3.75 (BC_3), 3.43 (B_4C_3), and 3.33 (B_2C).

The large cohesive energy of E_c = –6.67 eV/atom for BeC is consistent with the shortest bond of d = 0.138 nm realized for carbon bonds and the two bond lengths of 0.160 and 0.166 nm for the Be–C bonds [12]. The remarkably short C bond originates from the carbon triangles incorporated in this monolayer structure in addition to the hexagonal and octagonal rings. The much longer Be–C bonds of 0.173 nm and even longer Be–Be bonds of 0.198 nm in Be_2C are consistent with the reduced cohesive energy of E_c = – 4.86 eV/atom. The slightly larger size of the Be atoms and the two missing valence electrons per atom strongly reduce the overall bonding power. The bonding in 2D Be_5C_2 fits into the general bonding–structure scheme with E_c = – 4.58 eV/atom and longer Be–C and Be–Be bonds of d = 0.170 and 0.239 nm, respectively. One reason for the specific behavior of the Al_2C and CaSi monolayers with the weakest bonds and the largest alteration in bond lengths is that they contain only one or no small element of the second period of the periodic table. For a comparison with the properties of other monolayers, besides the data for the hexagonal monolayer of graphene, we include the data for the other elemental group-IVA monolayers. Except for the BeC monolayer, all hypercoordinated monolayers follow the inverse dependence of the cohesive energy of the elemental group-IVA monolayers if we correlate with the mean bond lengths.

To benchmark the bonding–structure relationship of hypercoordinated monolayers, we plot the results with those of the planar binary carbides of group-IVA elements SiC, GeC, and SnC and the weakly buckled hexagonal group-IVA elements Si, Ge, Sn, and Pb. This provides deeper insight into the characteristics of hypercoordination. Group-IVA monolayers with four valence electrons follow the energetically preferred noble gas configuration. As is evident in Fig. 4.13, both the planar and buckled group–IVA monolayers show a related inverse behavior, as observed here for the mean bond lengths of the hypercoordinated compounds [52]. Obviously, flexible coordination patterns between the motifs incorporate the quasi-planar molecular moieties effectively into an extended solid monolayer structure. This is surprising since the interconnects must reflect the specific steric requirements of the motifs and their individual shapes. Steric effects, however, may also induce strains that further stabilize a quasi-planar network.

The realization of large mean binding energies, which are only slightly diminished in comparison to those of uniform hexagonal monolayers, is also supported by a comparison of planar hexagonal siligraphene SiC_2 with graphene-like sp^2 hybridization and a binding energy of E_c = – 6.46 eV/atom [53] and SiC_2 with tetracoordinate silicon (ptSi) motifs at – 6.04 eV/atom, containing rings with four and six atoms [54]. Analogous to efficient further stabilization of the σ backbone in graphene by delocalized π electrons, quasi-planar hypercoordinated moieties are stabilized by the delocalization of p_z electrons, leading to novel physicochemical features. As a rule, hypercoordination of carbon or silicon with variable coordination numbers is possible in the plane by efficient delocalization of π electrons. Otherwise, the planarity of the motif may not be complete due to strong repulsion between the spaciously dense hybrids with increasing coordination in the plane.

4.4.2 Mechanical Properties Versus Bond Length

The plot in Fig. 4.14 displays the dependence of the 2D Young's moduli or in-plane stiffnesses on the range of bond lengths for the quasi-planar hypercoordinated monolayers considered in this chapter [52]. It is important to note the enormous sensitivity of the stiffness to the length of bonds. The stiffness decreases drastically by a factor of approximately ten for a change in the bond length of less than a factor of two. One of the main reasons for this extreme decay is the drastic reduction in orbital overlap due to the increasing size of atoms, which ultimately leads to the loss of planarity. However, the bonding of hypercoordinate compounds is more complicated. The existence of longer interconnecting bonds between the motifs in these usually nonuniform 2D networks has an additional influence on the in-plane modulus, as is evident by the comparison with the stiffnesses of elemental group-IVA monolayers. Efficient coordination patterns connecting the quasi-planar hypercoordinate moieties to a planar network permit reasonable stiffness values, as confirmed especially by various B–C compounds.

The peculiar stability of the 2D boron and B–C compounds can be attributed to the optimal filling of the in-plane states because the in-plane σ bonds are stronger than the out-of-plane π bonds [26]. Nevertheless, these interconnects reduce the overall mechanical performance to some extent, where the effect on the stiffness is not as dramatic as that expected for nonlinear properties such as the ultimate strength and ultimate strain. The in-plane stiffness of 145.5 N/m calculated for BeC is not consistent with the behavior of the other compounds. In any case, the correlation

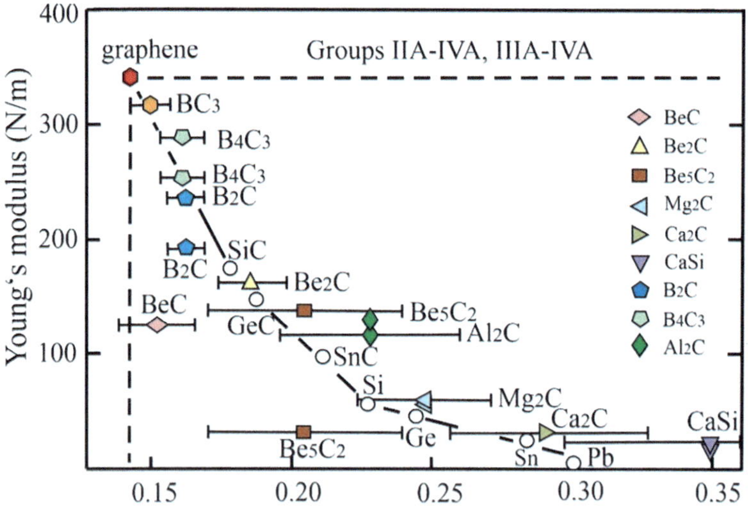

Fig. 4.14 Plot of in-plane stiffnesses of group IIA–IVA and IIIA–IVA monolayers versus bond length. For comparison, the plot displays the corresponding values of planar binary and buckled elemental group–IVA monolayers and those of honeycomb graphene and the BC_3 monolayer. Adapted from ref. [52], Copyright 2021, Royal Society of Chemistry

predicts that the large Young's modulus and the short bonds of the ptC-based BeC monolayer suggest a much greater in-plane stiffness of $\approx$ 300 N/m (see Fig. 4.14). Larger deviations from the correlation are also observed for the Be_5C_2 monolayer.

4.4.3 Young's Modulus Versus Cohesive Energy

The diagram in Fig. 4.15 shows Young's moduli as a function of the cohesive energies of hypercoordinated monolayers. Note that the stiffnesses of the isotropic monolayers, and the stiffnesses along the strong lattice directions in anisotropic lattices, follow mostly a linear increase with increasing binding energy. Only for BeC, Be_5C_2, and Ca_2Si deviates the stiffness in the strong direction significantly from this general behavior. For BeC, the quasi-linear dependence of the stiffness on the binding energy suggests a much larger Young's modulus of $\geq$ 250 N/m. Remarkably, the stiffness decreases by a factor of approximately five in comparison with the cohesive energy, which changes by a factor of approximately two. This correlation allows a reasonable estimation of the Young's moduli of novel structures if the mean binding energy is known and of the ideal ultimate strength if Griffith's rule applies. This linear relationship is surprising considering the complexity of structures estimated for the already characterized hypercoordinated monolayers considered here.

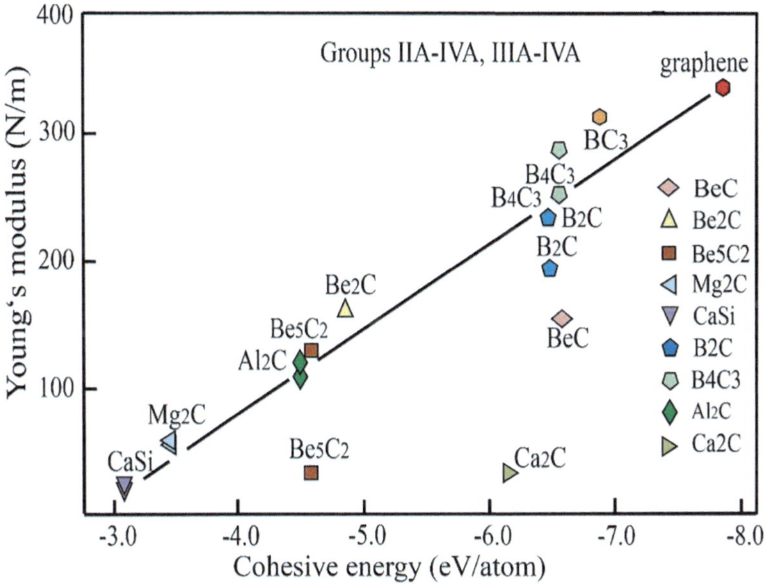

Fig. 4.15 Plots of the Young's moduli of group IIA–IVA and IIIA–IVA monolayers versus their cohesive energy. A straight line indicates linear dependence followed by the stiffest direction of the lattice

References

1. Y. Wang, Y. Li, Z. Chen, Acc. Chem. Res. **53**, 887 (2020)
2. V. Vassilev-Galindo, S. Pan, K.J. Donald, G. Merino, Nat. Rev. Chem. **2**, 0114 (2018)
3. C. Suman, J. Debnarayan, Phys. Sci. Rev. **3**, 162 (2018)
4. R. Hoffmann, R.W. Alder, C.F. Wilcox Jr., J. Am. Chem. Soc. **92**, 4992 (1970)
5. J. Xu, X. Zhang, S. Yu, Y.-H. Ding, K.H. Bowen, J. Phys. Chem. Lett. **8**, 2263 (2017)
6. L.-M. Yang, E. Ganz, Z. Chen, Z.-X. Wang, P.V.R. Schleyer, Angew. Chem. Int. Ed. **54**, 9468 (2015)
7. N. Job, M. Khatun, K. Thirumoorthy, S.S.R. Ch, V. Chandrasekaran et al., Atoms **9**, 24 (2021)
8. T.R. Galeev, B.D. Dunnington, J.R. Schmidt, A.I. Boldyrev, Phys. Chem. Chem. Phys. **15**, 5022 (2013)
9. A. Savin, R. Nesper, S. Wengert, T.F. Fässler, Angew. Chem. Int. Ed.. Chem. Int. Ed. **36**, 1808 (1997)
10. S.S. Batsanov, Inorg. Mater.. Mater. **37**, 871 (2001)
11. M. Mantina, A.C. Chamberlin, R. Valero, C.J. Cramer, D.G. Truhlar, J. Phys. Chem. A **113**, 5806 (2009)
12. C.-S. Liu, H.-H. Zhu, X.-J. Ye, X.-H. Yan, Nanoscale **9**, 5854 (2017)
13. Y. Li, Y. Liao, Z. Chen, Angew. Chem.. Chem. **126**, 7376 (2014)
14. G. Sharma, K.C. Bhamu, https://arxiv.org/abs/2003.07400 (2022)
15. K.H. Yeoh, K.-H. Chew, Y.Z. Chu, T.L. Yoon, D.S. Ong, J. Appl. Phys. **126**, 125302 (2019)
16. S. Qian, X. Sheng, Y. Zhou, X. Yan, Y. Chen et al., J. Phys. Chem. C **122**, 7959 (2018)
17. Y. Wang, F. Li, Y. Li, Z. Chen, Nat. Commun.Commun. **7**, 11488 (2016)
18. L. Meng, S. Ni, M. Zhou, Y. Zhang, Z. Li et al., Phys. Chem. Chem. Phys. **19**, 32086 (2017)
19. S.S. Wang, Y. Liu, Z.-M. Yu, X.-L. Sheng, L. Zhu et al., Phys. Rev. Mater. **2**, 104003 (2018)
20. M. Naseri, Superlattices Microstruct. Microstruct. **102**, 134 (2017)
21. H. Wang, E. Zhou, F. Duan, D. Wei, X. Zheng et al., J. Phys. Chem. Lett. **12**, 10353 (2021)
22. Y.-L. Li, S.-N. Wang, A.R. Oganov, H. Gou, J.S. Smith et al., Nat. Commun.Commun. **6**, 6974 (2015)
23. K. Rajput, V. Kumar, S. Thomas, M.A. Zaeem, D.R. Roy, 2D Mater. **8**, 035015 (2021)
24. Y. Yan, Z. Wei, Mater. Chem. Phys. **281**, 125869 (2022)
25. Y. Wang, M. Qiao, Y. Li, Z. Chen, Nanoscale Horiz. **3**, 327 (2018)
26. J. Dai, Y. Zhao, X. Wu, J. Yang, X.C. Zeng, J. Phys. Chem. Lett. **4**, 561 (2013)
27. J. Dai, Z. Li, J. Yang, J. Hou, Nanoscale **4**, 3032 (2012)
28. P. Zhou, Z.S. Ma, L.Z. Sun, J. Mater. Chem. C **6**, 1206 (2018)
29. X. Wu, Y. Pei, X.C. Zeng, Nano Lett. **9**, 1577 (2009)
30. X. Luo, J. Yang, H. Liu, X. Wu, Y. Wang et al., J. Am. Chem. Soc. **133**, 16285 (2011)
31. F. Li, K. Tu, H. Zhang, Z. Chen, Phys. Chem. Chem. Phys. **17**, 24151 (2015)
32. M. Naseri, J. Jalilian, A.H. Reshak, Int. J. Modern Phys. B **31**, 1750044 (2017)
33. H. Chang, K. Tu, X. Zhang, J. Zhao, X. Zhou et al., J. Phys. Chem. C **123**, 25091 (2019)
34. D. Fan, S. Lu, Y. Guo, X. Hu, J. Mater. Chem. C **6**, 1651 (2018)
35. X. Tian, X. Xuan, M. Yu, Y. Mu, H.-G. Lu et al., Nanoscale **11**, 11099 (2019)
36. S. Ahmed, A. Jalil, S.Z. Ilyas, A. Hassan, A. Ilahi et al., Appl. Surf. Sci. **544**, 148877 (2021)
37. J. Shen, F. Zheng, S. Wang, Z.-Z. Zhu, S. Wu et al., ACS Omega **6**, 13218 (2021)
38. Y. Li, Y. Liao, P.V.R. Schleyer, Z. Chen, Nanoscale **6**, 10784 (2014)
39. N. Lu, K. Wang, J. Jiang, H. Guo, G.Z. Zuo et al., https://doi.org/10.48550/arXiv.2203.06808 (2022)
40. J. Dai, X. Wu, J. Yang, X.C. Zeng, J. Phys. Chem. Lett. **5**, 2058 (2014)
41. R. Rahimi, M. Solimannejad, A. Chaudhari, J. Hazard. Mater. **403**, 123600 (2021)
42. X. Qu, J. Yang, Y. Wang, J. Lv, Z. Chen et al., Nanoscale **9**, 17990 (2017)
43. S. Wang, M. Khazaei, J. Wang, H. Hosono, J. Mater. Chem. C **9**, 9212 (2021)
44. B. Song, L.-M. Yang, WIREs Comp. Molec. Sci. **14**, e1699 (2024)
45. L. Schimka, J. Harl, G. Kresse, J. Chem. Phys. **134**, 024116 (2011)

46. W.A. Harrison, Pure Appl. Chem. **61**, 2161 (1989)
47. B. Peng, H. Zhang, H. Shao, K. Xu, G. Ni et al., J. Mater. Chem. A **6**, 2018 (2018)
48. S. Guo, W. Zhou, B. Cai, K. Zhang, S. Zhang et al., Nanoscale Horiz. **4**, 1145 (2019)
49. J.C. Garcia, D.B. de Lima, L.V.C. Assali, J.F. Justo, J. Phys. Chem. C. **115**, 13242 (2011)
50. Y. Zhang, Z.-F. Wu, P.-F. Gao, D.-Q. Fang, E.-H. Zhang et al., RSC Adv. **8**, 1686 (2018)
51. H. Zhang, Y. Liao, G. Yang, X. Zhou, ACS Omega **3**, 10517 (2018)
52. P. Hess, Nanoscale Horiz. **6**, 856 (2021)
53. L.-J. Zhou, Y.-F. Zhang, L.-M. Wu, Nano Lett. **13**, 5431 (2013)
54. Y. Li, F. Li, Z. Zhou, Z. Chen, J. Am. Chem. Soc. **133**, 900 (2011)

Chapter 5
Elemental Group–IVA Monolayers

Abstract The elemental group-IVA monolayers (Xenes) are among the strongest bonded lattices owing to the four valence electrons forming two-center two-electron bonds, which fulfill the octet rule. Graphene, with the shortest bond and planar honeycomb structure, has the largest cohesive energy and highest mechanical stability of the single-atomic layers. For the homologues silicene, germanene, stanene, and plumbene, the larger atom size reduces delocalized π bonding, and an energetically more stable mixed sp^2–sp^3 hybridization generates monolayers with increasing buckling. The contribution of sp^3 hybrids to mixed bonding and thus buckling increases from Si to Pb. A decrease in the orbital overlap leads to weaker covalent bonding and a drastic decrease in the stiffness and ultimate strength with increasing atom size. While graphene can be prepared as a stable free-standing monolayer, for the other monolayers to date, synthesis was only possible on a suitable stabilizing substrate that prevents the normally preferred tetrahedral sp^3 bonding of these elements. Group–IVA monolayers play key roles in the active search for a new class of nontrivial topological states. Examples are stanene and plumbene with a planar configuration due to strong substrate interaction. Quantum spin Hall (QSH) systems are topological insulators (TIs) that possess edge states of counterpropagating opposite spin states in connection with relativistic spin–orbit coupling and time-reversal symmetry. Such systems are insulating in the bulk owing to an open band gap; however, they have gapless edges or surface states at the boundary, which are immune to imperfections and geometric perturbations. Since dissipationless transport mechanisms are of extreme interest for the semiconductor industry, this is a field of active research. Spin–orbit coupling (SOC), which strongly increases from graphene to plumbene, and time-reversal invariance play essential roles. The realization of large gaps for the heavier homologues of graphene, stanene and plumbene with a graphene-like lattice may allow room-temperature applications of topological devices.

P. Hess, *Bonding, Structure, and Performance of Two-Dimensional Materials*, NanoScience and Technology, https://doi.org/10.1007/978-3-031-74161-6_5

5.1 Carbon Monolayers (Graphene)

5.1.1 Chemistry and 2D Bonding

Carbon atoms with the ground-state electronic configuration ($1s^2 2s^2 p_x{}^1 p_y{}^1 p_z{}^0$) have four electrons in their outer valence shell. All the elements in group IVA, highlighted in Fig. 5.1, have an electronic configuration with four valence electrons following the scheme presented in Fig. 5.2 for graphene with the ground-state electronic configuration and the promoted state with a 2s electron promoted to the empty $2p_z$ state.

The graphene-like elements are called Xenes [1]. The reaction energy, released by the proceeding reaction process forming the hybrids, covers the promotion energy of ~ 4 eV needed for promoting one 2s electron. In the promoted state, carbon atoms can form three planar σ bonds by sp^2 hybridization and an out-of-plane π bond by overlap of the remaining $2p_z$ orbitals with those of neighbors. Figure 5.3 presents a scheme of the C=C double bond formed by planar sp^2 orbitals and delocalized π electrons. Similarly, tetrahedral sp^3 hybrids can be formed in the promoted state, creating four equal bonds with ideal bonding strength in three-dimensional (3D) space (diamond). Importantly, the combination of σ and π bonds generates the strongest bonds in the plane with the largest bond dissociation energy and highest density of binding electrons between the two bonded nuclei.

The planar hexagonal structure formed by carbon atoms based on the overlap of the three sp^2 orbitals and the $2p_z$ orbitals is referred to as graphene. The two hexagons with alternating double bonds (d = 0.135 nm) and single bonds (d = 0.147 nm) shown in Fig. 5.4, which differ only by the position of the three double bonds, transform to the quantum–mechanical superposition of these two conventional configurations. This yields the ground-state bonding structure of graphene with equivalent (partial

Fig. 5.1 Scheme of the main groups of the periodic table highlighting the group-IVA elements considered in this chapter

H							He
Li	Be	B	C	N	O	F	Ne
Na	Mg	Al	Si	P	S	Cl	Ar
K	Ca	Ga	Ge	As	Se	Br	Kr
Rb	Sr	In	Sn	Sb	Te	I	Xe
Cs	Ba	Tl	Pb	Bi	Po	At	Rn
Fr	Ra						

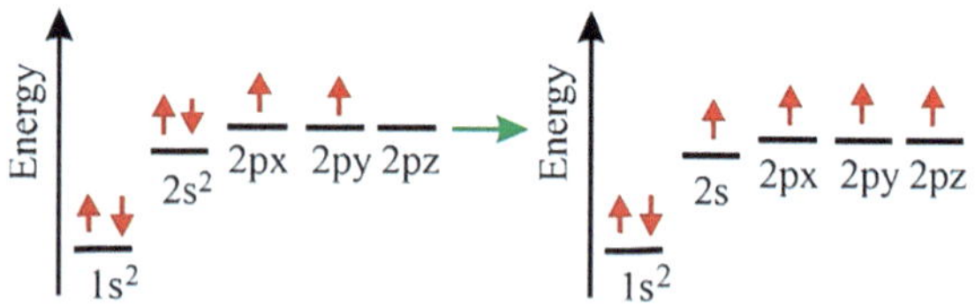

Fig. 5.2 Electronic ground state and promoted electronic state after the promotion of a 2*s* electron to the empty $2p_z$ state, delivering four valence electrons

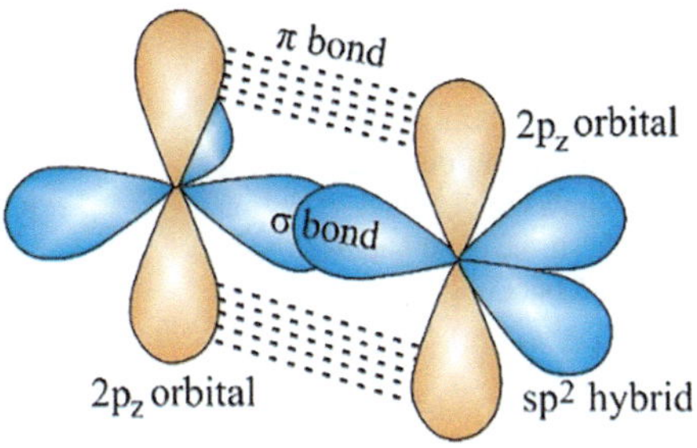

Fig. 5.3 Scheme of the double bond formed by the σ bonds of the sp^2 orbitals in the plane and the delocalized π bonds formed by the p_{pz} orbitals of direct neighbors

double) bonds d = 0.142 nm in length in the monolayer. It is noteworthy that this honeycomb structure used by bees is the optimum way to divide a surface into equal areas with the smallest total perimeter.

Tight-binding theory, based on the linear combination of atomic orbitals (LCAOs), predicts that in graphene all in-plane bonding orbitals are occupied and that the anti-bonding orbitals above the Fermi level are unoccupied. Consequently, graphene is a zero-gap semiconductor (semimetal) because its π-valence bands and π*-conduction bands contact at the Dirac point at the Fermi level [2, 3]. This special bonding configuration is responsible for notable physical properties of graphene, such as its unique electronic properties [1–3]. Since the dispersion relation for the electrons is linear near the Fermi level and since the effective mass depends on the curvature of the energy bands, these electrons behave as massless fermions. Figure 5.5 shows the zero-gap state, the opening of the gap by spin–orbit coupling (SOC) and by corrugation of the lattice, e.g., buckling. Note that the Dirac equation, instead of the Schrödinger equation, describes the propagation of so-called Dirac electrons, which are quasiparticles without an effective mass [2]. The high symmetry of the graphene lattice, where the two carbon atoms in the unit cell see an identical environment, is responsible for the zero-band gap [3]. Opening a band gap in graphene requires the reduction of the high in-plane lattice symmetry by structural or chemical modification. Mechanical deformation is another appealing means to manipulate the electronic structure by strain. However, the strain necessary for opening a sizable band gap in graphene is controversial [3].

Graphene is the building block for forming bilayers and few-layer assemblies with new properties of up to approximately ten layers, where the electronic structure and

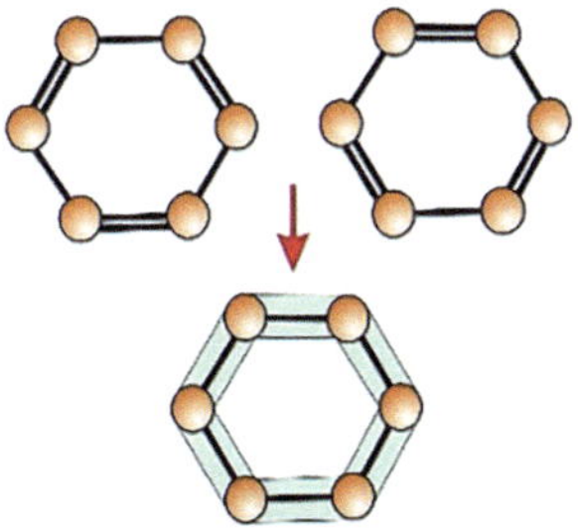

Fig. 5.4 Derivation of quantum mechanical double bonds by the superposition of two conventional bonding configurations with single bonds and isolated double bonds

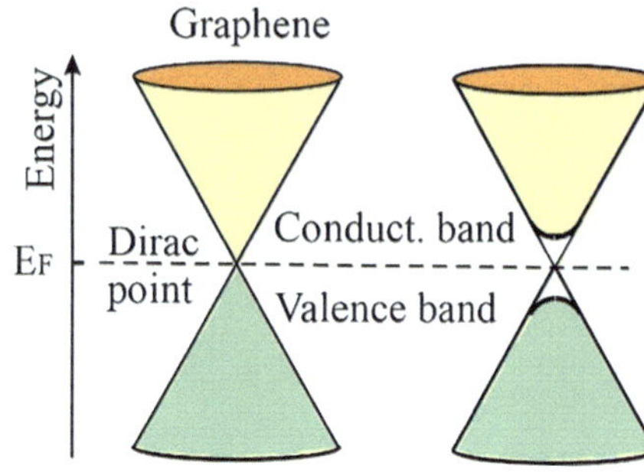

Fig. 5.5 Dispersion forming the so-called Dirac cones touching the Fermi level in graphene. In addition, the figure shows the gradual opening of the band gap by SOC and buckling of the network

properties approache the 3D limit of layered bulk graphite [4]. The interlayer interaction energy (sometimes called the exfoliation energy) and the stacking order are crucial quantities, which control the stability, as well as the chemical and physical properties of the resulting few-layer systems. The published theoretical interlayer binding energies of graphite vary widely between $E_{int} = -22$ and -65 meV/atom [5], and the experimental values range between -31 eV/atom [6] and -52 eV/atom [7]. The enormous differences between the theoretical values are due to the approximations used in the theoretical treatment of soft layered materials with weak interlayer interaction and strong in-plane interaction forces. Several recent publications confirmed the early results of a low interlayer interaction energy of graphite in the range of $E_{int} = -25$ meV/atom [8], e.g., by density functional theory (DFT) and local density approximation (LDA) calculations yielding $E_{int} = -23.7$ eV/atom for AB stacking [9]. In graphite, the most stable layer structure is AB stacking with a bond length of $d = 0.142$ nm, an experimental value of the interlayer spacing of $d_s = 0.342$ nm, and a theoretical value of $d_s = 0.335$ nm (see Fig. 5.6) [10, 11]. In AB or Bernal stacking, half of the atoms lie directly above the center of a hexagon, and half of the atoms lie above an atom, unlike in AA stacking, where the layers are exactly aligned.

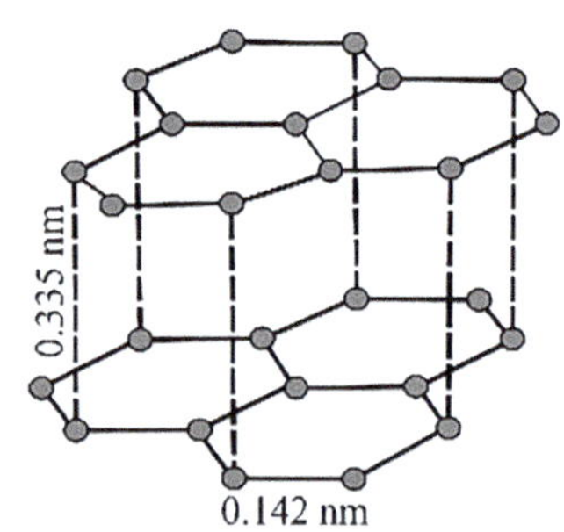

Fig. 5.6 Scheme of the most stable AB stacking order of graphene layers in graphite. Adapted with permission from ref. [10], Copyright 2002, Springer Nature

5.1.2 2D Structures

As mentioned before, graphene is the champion material, with the shortest bond length and the largest cohesive energy (larger than that of diamond) of any single-atomic lattice and the largest dissociation energy of a double bond (see the structural data presented in Table 5.1 [12–16]). The lattice parameter of graphene is a = 0.247 nm and the bond length d = 0.142 nm [12]. A comparison of the bond lengths of graphene (0.142 nm) and diamond (0.154 nm) underlines the high stability of planar sp^2 hybridization. In addition to the bond length, the cohesive energy is of special interest because it describes the energy released by the formation of a 2D network from free atoms and thus the energetic stability. The negative values used here for the binding energy of the network signify that the bonded system lost reaction energy. The graphene values of $E_c = -7.85$ eV/atom [12] (DFT-generalized gradient approximation (GGA)) and – 7.91 eV/atom (Perdew–Burke–Ernzerhof (PBE) level) is slightly greater than that of 3D bonding in diamond, which has a value of – 7.78 eV/atom (PBE level) [15]. The main reason for the outstanding stability of graphene and its homologues is the noble-gas electronic configuration, which is a highly stable shell formed by the bonded electrons ('octet rule'). In sp^2 bonding, the three σ bonds are stronger than the π bond and the four equal bonds of sp^3 hybridization. The tabulated van der Waals diameter of $d_w = 0.34$ nm gives a reasonable value for the thickness of planar graphene [16–18]. The space group of the graphene structure is *P6/mmm* [14]. The space group of graphite is $P6_3/mmc$, which includes the symmetry of few-layer systems with an even number of monolayers of *P*-3*m*1 and an odd number of monolayers with a symmetry of *P*-6*m*2 [14]. The symmetry, space group. cohesive energy, lattice constants, bond angle, bond length, buckling distance, van der Waals diameter, corrected van der Waals diameter, and interlayer spacing of elemental group–IVA monolayers are presented in Table 5.1.

The low value of $E_{int} = -22$ eV/atom calculated for the interlayer interaction energy of graphene bilayers supports the value given above for graphite [19]. Besides the calculation of exact bonding energies, the preparation of defect-free isolated bilayers is extremely difficult owing to self-aggregation. Furthermore, the prepared bilayer is often strongly bound to a surface, where separation of the sheet is difficult. Twisted bilayers are of particular interest because bilayers of graphene with a narrow or magic twist angle transform to strongly correlated 2D electronic systems with properties that are extraordinarily sensitive to carrier density and to experimentally controllable factors such as the twist angle [20]. As a molecular model, cofacially stacked C_{96} cores with stabilizing peripheral substituents were employed to study the properties of graphene bilayers [21]. Nuclear magnetic resonance (NMR) and single-crystal X-ray diffraction (XRD) measurements were used to validate the bilayer structure. These measurements revealed an interlayer separation of $d_s = 0.342$ nm, consistent with the small interlayer interaction energy of the bilayer.

Table 5.1 Symmetry, space group, cohesive energy, lattice constant, bond angle, buckling distance, van der Waals diameter, corrected van der Waals diameter, and layer spacing of elemental group–IVA monolayers

Monolayer	Geometry, space group	Cohesive energy, E_c (eV/atom)	Lattice constant a (nm)	Bond angle θ (°)	Bond length d (nm)	Buckling distance Δz (nm)	Van der Waals diameter d_w [16, 17], $d_{w+\Delta z}$ (nm)	Interlayer spacing d_s (nm)
Graphene	Planar *P6/mmm* [14]	– 7.85 [12] – 7.91 [15]	0.247 [12] 0.246 [13]	120 [12] 120 [13]	0.142 [12] 0.142 [13]	0 [12] 0 [13]	0.340 [16] 0.340 [17]	0.342 [11]
Silicene	Buckled *P3m*1[50]	– 3.91 [12] – 3.90 [53]	0.387 [12] 0.388 [13]	116.2 [52] 116.9 [13]	0.228 [12] 0.227 [13]	0.046 [12] 0.041 [13]	0.420 [16], 0.466 0.420 [17]	0.34 [54]
Germanene	Buckled *P3m*1	– 3.24 [12] – 3.24 [69]	0.406 [12] 0.406 [71]	112.6 [13] 112.3 [71]	0.244 [12] 0.244 [71]	0.069 [12] 0.069 [71]	0.420 [16], 0.489 0.422 [17]	0.254 [69]
Stanene	Buckled *P3m*1[80]	– 2.73[12] – 2.76 [53]	0.464 [12] 0.467 [71]	110.4 [13] 111.3 [71]	0.284 [12] 0.283 [71]	0.095 [12] 0.085 [71]	0.450 [16], 0.545 0.434 [17]	0.33 [87]
Plumbene	Buckled *P-3m*1 [91]	– 2.28 [79]	0.493 [91] 0.493 [92]	110.9 [79]	0.300 [91] 0.299 [92]	0.101 [93] 0.093 [92]	0.460 [16], 0.561 0.404 [17]	

5.1.3 Mechanical Properties

Graphene is the monolayer material with the best studied mechanical properties both theoretically and experimentally. Young's modulus, Poisson's ratio, ultimate strength, and ultimate strain data are collected in Table 5.2 for the considered group–IVA monolayers. It is not surprising that the mean Young's modulus of E = 333 N/m (~ 1 TPa) extracted from DFT calculations using the generalized gradient approximation (GGA) with (PBE) functionals and that of molecular dynamics (MD) simulations [22, 23] agree quite well with the nanoindentation measurements of E = 342 N/m [24]. These experiments were performed on free-standing graphene monolayers of micrometer size, which allowed to probe the intrinsic stiffness. Measurements of Poisson's ratio resulted in $\nu = 0.165$ [24], while DFT calculations yielded a value of $\nu = 0.184$ [13]. In graphene the anisotropy between the armchair and zigzag directions of the honeycomb lattice is negligible for the linear elastic properties, whereas the nonlinear fracture properties show a noticeable angle dependence; thus, it is necessary to consider anisotropy in the latter case. The mechanical properties calculated for free-standing graphene agree well with those of weakly adsorbed graphene, as observed for several metal substrates interacting by van der Waals forces with the monolayer [25]. The structure, Young´s modulus, Poisson's ratio, ultimate strength, and ultimate strain of group–IVA monolayers are presented in Table 5.2.

The less accurate experimental and theoretical values of the anisotropic fracture strength scatter between $\sigma = 30$ N/m (90 GPa) and 42 N/m (125 GPa) [26]. The experimental strength of $\sigma = 42$ N/m [24] obtained by biaxial nanoindentation is in good agreement with DFT calculations of $\sigma_{ac} = 37$ N/m and $\sigma_{zz} = 40$ N/m for uniaxial strain in the armchair and zigzag directions, respectively [26]. By averaging the large number of DFT calculations and MD simulations, a somewhat lower average ultimate strength of $\sigma = 35$–37 N/m is found [27]. Note that usually a higher strength occurs

Table 5.2 Structure, Young's modulus, Poisson's ratio, ultimate strength, ultimate strain of elemental group-IVA monolayers

Monolayer	Structure	Young's modulus E_{ac}, E_{zz} (N/m)	Poisson's ratio ν_{ac}, ν_{zz}	Ultimate strength σ_{ac}, σ_{zz} (N/m)	Ultimate strain ε_{ac}, ε_{zz}
Graphene	Planar	333 [23] 342 [24]	0.184 [13] 0.165 [24]	37, 40 [26] 42 [24]	0.31 [28] 0.25 [24]
Silicene	Low Buckled	61.7, 59 [58] 63 [51]	0.29, 0.33 [58] 0.33 [51]	7.2, 6.0[58] 4.8, 5.9 [57]	0.175, 0.19 [58] 0.089, 0.18 [57]
Germanene	Low buckled	44, 43.4 [58] 42.0, 36.4 [73]	0.29, 0.35 [58]	4.7, 4.1 [58] 5.7, 6.0 [73]	0.20, 0.205 [58] 0.16, 0.18 [73]
Stanene	Low Buckled	25.2, 23.5 [58] 22.4 [53]	0.36, 0.42 [58] 0.40 [53]	2.6, 2.2 [58] 2.59, 2.28 [53]	0.17, 0.18 [58] 0.15, 0.20 [53]
Plumbene	Buckled	8 [94] 12 [95]		3.64 [95]	

in the zigzag direction with a poorly characterized anisotropy of approximately 5–10%. The errors introduced in the measured strengths by defects and inaccuracies in the data evaluation procedure are difficult to estimate. In any case, the available mechanical properties of (nearly) defect-free graphene allows a judgment of the quality of the usually produced defective monolayers.

Nanoindentation experiments resulted in an ultimate equibiaxial strain of $\varepsilon = 0.25$ at room temperature [24]. For biaxial strain, DFT calculations predict a soft-mode phonon instability, which results in a phase transition into an unstable phase at a = 0.213 [28]. In the case of uniaxial strain, graphene fails via the usual elastic instability at a critical strain of $\varepsilon_A = 0.307$ [28]. For single-crystal diamond, the calculated ideal tensile strength and ultimate strain for a stress oriented in the ⟨111⟩ direction yielded $\sigma = 95$ GPa and $\varepsilon = 0.13$, which is in the range of the ultimate strength of graphene [29]. Formally, this corresponds to a 2D strength of 32 N/m. It should be noted that the largest strength measured for natural IIa diamond is only $\sigma = 7.5$ GPa [30], and most of the reported values are $\sigma < 1$ GPa [31].

Several studies have used MD simulations to investigate the temperature dependence of the mechanical properties of graphene. These simulations reveal the increasing C–C bond length and amplitude of intrinsic structural ripples with increasing temperature. The adaptive intermolecular reactive empirical bond order (AIREBO) approach, implemented in the large-scale atomic/molecular massively parallel simulator (LAMMPS) MD simulation package, was used as an empirical potential field in these studies [32, 33]. High-temperature results predict a reduction of Young's modulus of approximately 10%; however, approximately 60% of the ultimate strength is lost between 300 and 2400 K [32]. At lower temperatures up to 800 K, the simulations resulted in 23% and 27% reductions of the ultimate strengths in the armchair and zigzag directions and 54% and 59% reductions in the ultimate strains, respectively [33]. Since all carbon bonds are inclined to the straining direction in the zigzag direction, they carry more strain by changing the bond angle. As mentioned before, the effect on the stiffness is less significant.

In addition to MD simulations, continuum 2D models and quantized fracture mechanics (QFM) have been employed to investigate the effect of defects such as cracks on the fracture strength of graphene. A systematic study including experimental data showed the linear dependence of the critical fracture stress on the square root of the effective defect length from the nanometer to the micrometer scale [27]. Surprisingly, at the subnanometer scale, direct bond breaking models and Griffith's rule seem to provide a consistent description at least for mode I fracture by uniaxial tension [27].

A central but not well studied mechanical quantity with respect to practical applications is the fracture toughness or critical stress intensity factor (CSIF), which in principle describes the ability of a defective material to resist failure. It is important to note that toughness is not an outstanding mechanical property of graphene owing to its brittle nature. For example, currently available information indicates that the toughness of free-standing synthetic polycrystalline diamond plates is nearly a factor of two greater than that of graphene [34]. Specifically, the limitations of chemical vapor deposition (CVD)-grown graphene caused by defects and cracks were studied, including strategies to overcome these problems in real applications [35]. For defective graphene, a mean fracture toughness of $K_{IC} = 1.1 \times 10^{-3}$ N/m$^{1/2}$ (3.3 MPa m$^{1/2}$) can be extracted from available DFT calculations and MD simulations [26, 27]. Direct measurements of fracture toughness are rare since, in addition to the real strength, the size and orientation of preexisting cracks in the sample must be determined. Such measurements performed for large crack sizes between 33 nm and 1.256 μm yielded a toughness of $K_{IC} = 1.3 \times 10^{-3}$ N/m$^{1/2}$ (4.0 MPa m$^{1/2}$) for graphene, in surprisingly good agreement with the theoretical values presented above [36].

Instead of treating graphene as a thermally fluctuating crystalline plate, most studies use the conventional theory of thin plates. At room temperature, thermal fluctuations influence both Young's modulus and the bending rigidity and are currently not considered when evaluating indentation experiments. The bending rigidity of graphene, which depends on temperature and size, is not well known and therefore is often neglected. One reason is the difference between the theoretical predictions at the microscale and the apparent rigidity at the macroscopic scale, which is influenced by thermally induced ripples of the fluctuating crystalline membrane [37]. This effect can induce a dramatic reduction of the stiffness of micrometer-sized samples. In contrast to continuum plate models, which incorporate the thickness of the membrane, the interactions between orbitals, such as bond-angle effects and bonding associated with dihedral angles, are the dominant sources of bending rigidity. The bending rigidity is an important parameter for predicting the performance of nanoelectromechanical devices. The effective bending rigidity of graphene membranes at room temperature is 2.7 eV [38]. The modal analysis applied in this evaluation is based on fitting the resonance frequencies obtained from the Brownian motion of the membrane in MD simulations.

5.1.4 Physics and Applications

Exfoliated graphene samples were sufficient for measuring the outstanding mechanical properties of graphene; however, they are too small for most practical applications. On the other hand, the bottom-up synthesis of useful sample sizes by CVD is connected with the potential formation of a series of defects, such as grain boundaries, topological defects, corrugations, impurities, vacancies, and sp^3 hybridization [39]. For graphene, the most detrimental effects of these disorders have been studied extensively by theory, while a systematic characterization of the mechanical properties of real samples with point and line defects is still missing [39]. To date, it has not been possible to reduce the formation of defects to an extent that the intrinsic stiffness, ultimate strength, and toughness of ideal graphene can be approached in large-scale production. However, for perfect graphene without dangling bonds, environmental stability seems not to be an issue, as this stability may be lost rapidly for defective samples.

The various fields of graphene application based on electronic, thermal, optical, and mechanical properties are covered by several reviews [40–46]. High-quality ultrathin films are of interest in electronics, spintronics, optoelectronics, and photovoltaics, whereas powders are used in batteries, sensors, and catalysts [40]. Vertical van der Waals heterostructures may find use in atomically thin quantum wells, p-n junctions, and Coulomb drag transistors. Based on the most important properties, including high conductivity, strong optical absorption, the quantum spin Hall effect, Dirac fermions, and the thermoelectric effect, the corresponding graphene-based materials for optical devices, electronic and thermal sensors, and energy management systems are described in detail [41].

Biomedical applications are a wide active field, reviewed with emphasis on drug delivery, biosensing, bioimaging, tissue engineering, antibacterial applications, and 3D printing applications [42]. Specifically, biosensing and bioimaging by nanosized graphene and nanosized graphene oxide are reviewed, including application in the field of electrochemistry, where graphene and its derivatives can be modified on the surface to couple with ligands and antigens to detect specific biomolecules such as glucose, deoxyribonucleic acid (DNA), and proteins [43]. Moreover, graphene materials are widely used in photothermal therapy to treat in vivo cancer, and the combination of graphene with anticancer drugs and specific antigens provides efficient drug delivery systems.

A comprehensive discussion of the applications of graphene in mechanical transduction includes sensor and actuator applications, capacitive sensing, optically interrogated displacement, piezoelectricity, triboelectricity, electrostatic actuation, chemomechanical and thermomechanical actuation, and thermoacoustic emission [44]. Progress in the use of graphene as a flexible physical signal sensor has been described, including device structure design and the applications of these devices in wearable materials technology [45]. The relevant properties needed for such applications include flexibility, light weight, conductivity, and transparency. An outlook on diverse topics such as production, additive manufacturing, textile, aerospace, and printing technologies is available that further includes the future horizon of the revolution of emerging materials [46].

The planar honeycomb structure of semimetal graphene with pure sp^2 and delocalized π bonding is a QSH insulator or TI. Such systems are insulating in the bulk owing to a band gap; however, they have gapless edge states at the boundary, which are immune to imperfections and geometric perturbations. These edge states possess counterpropagating opposite spin states in connection with relativistic spin–orbit coupling and time-reversal symmetry (see Fig. 5.7). According to the theory of 2D topological insulators, the opening of the energy gap depends on the size of spin–orbit coupling (SOC). At the atomic level, the electron spin and the magnetic moment resulting from orbital motion couple to the SOC effect. The inclusion of the SOC opens the band gap at the Dirac point. Since carbon is a light element, its SOC is weak, and the opened topological gap is in the 25–50 μeV range with an extremely low implementation temperature [1]. Thus, the development of QSH devices operating at room temperature requires 2D topological materials composed of heavy elements with strong SOCs, e.g., heavy homologues of carbon such as Sn and Pb.

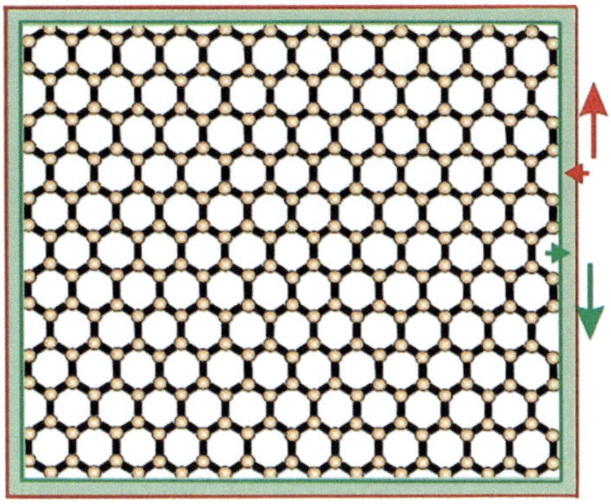

Fig. 5.7 Schematic representation of the SOC-induced conductive 1D helical edge states with opposite spin polarization by the band gap at the Dirac point. Adapted from ref. [1], Copyright 2020, AIP Publishing

5.1.5 Synthesis

In general, the synthesis of 2D materials with the desired perfect structure, domain size, scalability, growth rate, and versatile substrate material is the main challenge in fabricating 2D materials suitable for practical applications. The controlled growth of monolayers and few-layer assemblies of graphene is an active field covered by comprehensive review articles. One overview concentrates on the history and future development of graphene synthesis for research and industry [47], another one treats the progress made in the synthesis of graphene-based materials and covers the practical aspects of these achievements [48]. Other reports reveal the enormous progress made in growing large-area (> cm^2) adlayer-free homogeneous graphene monolayers [40], e.g., by using successfully as-received Cu substrates and common CVD methods to perform controlled layer-by-layer growth [49].

5.2 Silicon Monolayers (Silicene)

5.2.1 Chemistry and 2D Bonding

Since Si, Ge, Sn, and Pb belong to the same column in the periodic table as carbon, they have four valence electrons with an electronic configuration ($ns^2p_x{}^1p_y{}^1p_z{}^0$). However, as is known from chemical experience, these elements prefer sp^3 hybridization with tetrahedrally bonded atoms. Therefore, it is not surprising that DFT calculations predict that only buckled monolayers are dynamically stable because buckling allows a partial inclusion of p_z orbitals in the hybridization process [50, 51]. The main reason for buckling is the increasing atom size from Si to Pb, which destabilizes pure π bonding, owing to the increasing distance between the interacting atoms that decreases orbital overlap. Consequently, besides s, p_x, and p_y orbitals the p_z orbital of Si atoms form variable mixed sp^2–sp^3 hybrid orbitals with three σ bonds, which are no longer restricted to the plane and orbitals perpendicular to the plane that form weaker π bonds with adjacent orbitals. In principle, these mixed sp^2–sp^3 orbitals vary between the limits of planar sp^2 bonding with a bond angle of 120° and ideal tetrahedral sp^3 bonding with a limiting bond angle of 109.5°. Owing to the increasing atom size, in the bonds of the elemental 2D networks of the homologues Ge, Sn, and Pb the sp^3 character of mixed sp^2–sp^3 bonding increases and the bonding angle decreases.

Specifically, silicon atoms have the valence electron configuration (Ne|$3s^2p^2$) and therefore would prefer sp^3 hybridization. With respect to sp^2 hybridization, the standard van der Waals diameter increases substantially from $d_W = 0.34$ nm to $d_W = 0.42$ nm from carbon to silicon [16]. Due to the greater distance between Si atoms, the overlap of the $3p_z$ orbitals is diminished, and their mixing with s orbitals creates a partial sp^3 character, which enhances orbital overlap and stabilizes two corrugated configurations, a weakly and strongly buckled silicene structure. Note that the smaller energetic difference between the 3s and 3p sublevels in comparison to that of graphene favors the mixing process. In the more stable low-buckled silicene, the σ-π orbital mixing is closer to sp^2 than to sp^3 hybridization, with a small buckling effect of $\Delta z = 0.046$ nm [12] (0.045 nm) [13] and a bond angle of 116.9° [13] (116.2°) [52].

5.2.2 2D Structures

Figure 5.8 displays the top and side views of the buckled silicene network. In buckled structures, the bonding power extends parallel and also perpendicularly to the plane. Remarkably, the bond length of the mixed hybridization of silicene at d = 0.228 nm [12] is much greater than that of graphene but still slightly shorter than the bond length in the 3D Si diamond structure at d = 0.235 nm. Similarly, the lattice constant increases substantially from a = 0.247 nm (graphene) to a = 0.387 nm (silicene), the corrugation extents to $\Delta z = 0.046$ nm, and the bonding angle decreases to $\theta = 116.2°$ [12, 52]. A large increase in the bond length from graphene to silicene causes a drastic reduction in the cohesive energy from $E_{co} = -7.85$ to -3.91 eV/atom (projector augmented wave (PAW)-PBE level) [12, 53]. The layer thickness estimated by the van der Waals diameter plus buckling distance agrees reasonably well with atomic force microscopy (AFM) measurements of ~ 0.6 nm [54].

In silicene, partial sp^3-type bonding intensifies the interlayer interaction between bilayers and multilayers and thus reduces interlayer separation. The calculated interlayer separations of the bilayers with AA and AB stacking are $d_s = 0.246$ nm and 0.253 nm, respectively [50]. For free-standing few-layer sheets, interlayer spacings of $d_s = 0.34$ nm were measured by high-resolution transmission electron microscopy (HRTEM) and selected area electron diffraction (SAED) in layered silicene sheets [54]. Bilayers with AA stacking are only slightly more stable than those with AB stacking, and rhombohedral ABC stacking is the minimum energy configuration in

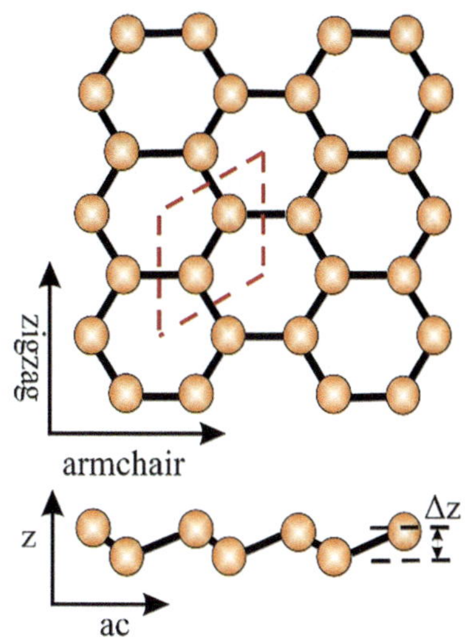

Fig. 5.8 Top view and front view of buckled hexagonal silicene

trilayers [50]. Independent calculations of interlayer separations point to configurations, where the spacing of bilayers decreases because the still existing weak in-plane π bonds are replaced by covalent-type interlayer σ bonds to further minimize the total energy [55, 56]. As a consequence of the stronger covalent-type bonding, the bilayer becomes planar.

5.2.3 *Mechanics*

Young's modulus and ultimate strength values reveal a drastic change in the mechanical performance from graphene to silicene (see Table 5.2) [51, 57, 58]. The enormous loss of binding energy causes a drastic decrease in the intrinsic angle-dependent stiffnesses to $E_{ac} = 61.7$ N/m and $E_{zz} = 59$ N/m that can be associated with the corrugation effect because under tension, the monolayer first undergoes unfolding instead of direct bond stretching. Poisson's ratios of silicene are $\nu_{ac} = 0.18$ and $\nu_{zz} = 0.19$ [58]. The dramatic loss in ultimate strength to $\sigma_{ac} = 7.2$ and $\sigma_{zz} = 6.0$ N/m, which controls failure, is due to structural changes such as a decreased bond density and bond length [58]. A small anisotropy of $\varepsilon_{ac} = 0.18$ and $\varepsilon_{zz} = 0.19$ has been reported for the ultimate strains [58]. Similar values reported for the stiffness and Poisson's ratio are $E = 63$ N/m and $\nu = 0.33$ [51], and the ultimate strength and ultimate strain values are $\sigma_{ac} = 4.8$ N/m and $\sigma_{zz} = 5.9$ N/m, and $\varepsilon_{ac} = 0.089$ and $\varepsilon_{zz} = 0.18$, respectively [57]. While the cohesive energy decreases by a factor of approximately two, stiffness and strength change by a factor of approximately six. While the Poisson ratio doubles, the critical strain decreases, which is consistent with the much lower stiffness and strength.

The bending stiffness of silicene at 38.6 eV is much greater than that of graphene (2.1 eV), resulting from a similar minimization methodology and reactive force field (Raef) potential [57]. This large bending rigidity is due to the buckled structure of silicene. First-principles calculations elucidated the strain and chirality effects under uniaxial tension in different directions [59]. With increasing in-plane tension, a small band gap of up to 0.8 eV opens that disappears before reaching the ultimate strength.

5.2.4 Physics and Applications

In addition to synthesis, characterization, stabilization, and processing several reviews have described a variety of device applications, such as topological bits, quantum sensing, and energy devices [60]. In addition, methods are presented to generate band gaps, different types of field-effect transistors (FETs), and spintronic devices [61]. Furthermore, biomedical applications, such as tumor therapeutics, bioimaging, and antimicrobial materials have been reviewed [62].

Silicene is a novel 2D material with the advantage of compatibility with current industrial processes and possible integration into silicon-based technologies owing to the similarities between the 2D and 3D silicon structures. A problem originating from mixed sp^2–sp^3 hybridization states is the increased reactivity and lower environmental stability. Consequently, in practical applications, silicene and its homologues may require postfabricated passivation by functionalization or sandwich encapsulation.

Like graphene, silicene is a semimetal with linear dispersion around the Dirac point, and its charge carriers behave like massless Dirac fermions. Owing to the broken inversion symmetry, the band gap of silicene can be opened and tuned by an external electric field, contrary to the behavior of graphene [50]. Interestingly, the band structures of planar graphene, buckled silicene, and buckled germanene have a Dirac cone with the p_z orbital, contributing to the K point of graphene, yielding pure π bonds; however, for the higher homologues, p_x and p_y electrons contribute to the top of the filled state in addition to the p_z electrons, yielding mixed σ and π bonds. Inclusion of SOC opens the band gap at the Dirac point from 0 to ~ 1 eV for plumbene, the heaviest homologue. Nontrivial topological properties exist in the free-standing state owing to SOC with time-reversal symmetry and resistance against impurities and geometric perturbations [1, 63]. Figure 5.9 shows a scheme of a QSH system that is insulating in the bulk with topologically protected gapless edge states. Unfortunately, isolated monolayers are not yet accessible for silicene and its higher homologues. The gap size is $E_g \approx 1.5$–2.0 meV for silicene corresponding to 18 K, and the gap of $E_g \approx 24$–30 meV for germanene corresponds to a thermal energy far above the liquid nitrogen temperature (77 K) [1].

The main advantages of silicene over graphene are in addition to the better tunability of the band gap needed for field-effect transistors (FETs) operating at room temperature, the much stronger SOC for realizing the QSH effect at accessible temperatures, easier valley polarization, and better suitability for valleytronics

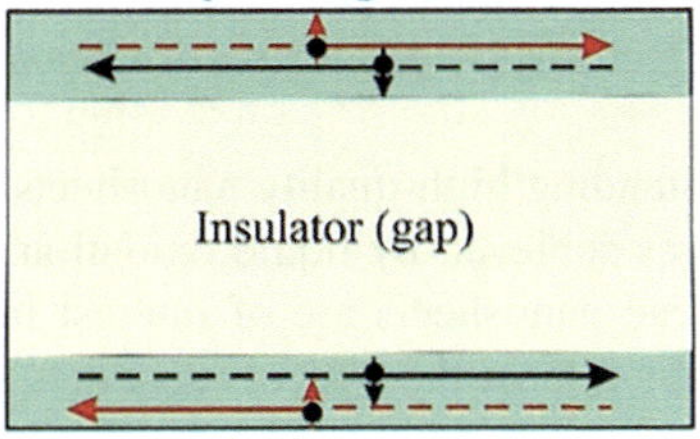

Fig. 5.9 Scheme of a 1D spin system. The upper edge contains two channels one with spin up moving forward and one with spin down moving backward and conversely for the lower edge that is spatially separated by the QSH insulator. Adapted with permission from ref. [63], Copyright 2010, AIP Publishing

studies [51]. Spatial separation is the characteristic feature of a QSH insulator with an energy gap between the valence and conduction bands, however, there is a gapless edge state at the boundary as discussed above. These distinct silicene features stimulated studies on the QSH effect, quantum anomalous Hall effect, quantum valley Hall effect, superconductivity, magnetism, thermoelectric behavior, band engineering, tunneling FETs, spin filters, and spin FETs [51].

5.2.5 Synthesis

Since no layered bulk materials ('silicite') exist in nature for the exfoliation of silicene or its heavier homologues, these compounds must be synthesized either by exfoliation of specially prepared suitable synthetic compounds, by CVD or molecular beam epitaxy (MBE) [60]. Monolayers and few-layer sheets of silicene were first epitaxially grown on Ag(111) substrates, and the atomic structure was analyzed by scanning tunneling microscopy (STM), low-energy electron diffraction (LEED), and atomic force microscopy (AFM) [64, 65]. Various superstructures with respect to the underlying substrate were identified, for example, for the growth of silicene on Ag(111) by MBE the superstructures (4×4), ($\sqrt{7} \times \sqrt{7}$), and ($\sqrt{3} \times \sqrt{3}$) [1].

The structures simulated by theory reproduced the AFM images, which is important because previously reported STM images showed only the top Si atoms [64]. Only, high-resolution AFM imaging, performed at low temperature with functionalized CO tips, was able to determine the atomic buckling of silicene adsorbed on Ag(111) [65]. The larger measured corrugation in the range of $\Delta z = 0.08$–0.11 nm than that found by DFT calculations may be partially due to extended defects. The strongly hybridized Ag-Si states preserved the honeycomb lattice of the adsorbed silicene, whereas the Dirac cones disappeared [65]. Thus, the epitaxial growth of silicene on a metal surface disturbs the π-electron system and changes its fundamental properties. Overall, the electronic structure is sensitive to substrate interactions, surface chemistry, and SOC. Importantly, van der Waals epitaxy on suitable substrates reduces these disturbing interactions. For example, first-principles calculations predicted a passivation effect for the deposition of silicene on a graphene-covered SiC(0001) substrate [66].

The synthesis of free-standing high-quality nanosheets of silicene with one or a few layers of thickness was achieved by liquid oxidation and exfoliation of $CaSi_2$ [54]. The deposited silicene nanosheets are of interest in electrochemical energy storage because they exhibit almost theoretical capacities for double-layer silicene, ultrastable cycling performance, and unique Li adsorption/desorption mechanisms [54].

A recent comprehensive state-of-the-art review considered different epitaxial growth systems [67]. This work covers the strategies developed for the synthesis of low-buckled honeycomb silicene structures and clarifies controversial results, which obviously arise from incomplete characterization. Moreover, the authors present methods to protect reactive layers against oxidation and degradation. A remaining challenge is the realization of free-standing monolayers for the accurate experimental verification of extensive theoretical predictions. The development of an appropriate substrate that preserves all intrinsic properties of the monolayer, including the Dirac cone, seems to be a demanding issue.

5.3 Germanium Monolayers (Germanene)

5.3.1 Chemistry and 2D Bonding

The group—IVA element germanium has four valence electrons in the outer shell, by including the 3*d* electrons in the calculations, the electronic configuration is $(\text{Ar}|3d^{10}4s^2p^2)$. A further minor increase in the bond length from silicene (0.228 nm) to germanene (0.244 nm), which essentially agrees with that of bulk germanium (0.245 nm), causes a slightly greater contribution of sp^3 hybrids to mixed sp^2–sp^3 hybridization. Consequently, the buckling effect increases to $\Delta z = 0.069$ nm, and the bond angle decreases to 112.6° (see Table 5.1). The change in atom size relates to further weakening of bonding. The cohesive energy of the energetically most stable weakly buckled structure with the space group $P3m1$ decreases from $E_c = -3.91$ eV/atom for silicene to – 3.24 eV/atom for germanene [12, 69].

Remarkably, under pressure, the synthesized buckled bilayer of germanene converts into a dynamically stable phase consisting of two covalently bonded flat honeycomb lattices [68]. This new chemically inert bilayer structure is a 2D TI that exhibits the QSH effect with a sizable topologically nontrivial band gap of ~ 0.1 eV [69]. Moreover, for epitaxially grown low-buckled germanene, the first experimental evidence was presented that single-layer germanene is a QSH insulator with a large bulk gap and robust metallic edges [70]. Furthermore, the topological gap can be closed by a perpendicular electric field, which transforms germanene to a Dirac semimetal.

5.3.2 2D Structures

Table 5.1 presents the published structural data for germanene. Interestingly, germanene exists in a planar graphene-like structure and in a low-buckled structure, both of which exhibit topological effects. The greater contribution of sp^3 orbitals to mixed sp^2–sp^3 bonding in germanene causes a reduced binding or cohesive energy in the plane, which implies a larger lattice constant of a = 0.406 nm, a larger bond length d = 0.244 nm, a larger corrugation Δz = 0.069 nm, and a smaller bonding angle of θ = 112.6° [13] compared to those of silicene [71]. Furthermore, hybridization with a larger contribution of $4p_z$ orbitals increases the corrugation of germanene and, most importantly, the reactivity of the monolayer surface [71]. From the tabulated van der Waals radii, nearly identical thicknesses of germanene and silicene can be expected. Note that the 2D honeycomb-like structure exists not only for silicene and germanene but also for stanene and plumbene [72].

As displayed in Fig. 5.10 the high-buckled phase of germanene with the shortest lattice constant has imaginary phonon frequencies and the energy minimum is not a real local minimum of the Born–Oppenheimer surface [51]. For the stable low buckled structure the lattice constant increases substantially and the planar structure of germanene is slightly less stable and exhibits a marginally stretched lattice. Note that for silicene a qualitatively similar behavior has been found.

The greater sp^3 contribution to hybridization in germanene intensifies the inter-layer interaction in bi- and multi-layers in comparison with that in silicene. As already mentioned, in bilayers, the buckled phase converts into a more stable flat bilayer assembly under structural optimization, with stable AA stacking, a larger lattice constant, and essentially the same in-plane bond length and interlayer spacing of d = 0.254 nm and d_s = 0.255 nm, respectively [69]. The practically identical intra-layer and inter-layer bond lengths indicate the same chemical bonding in the plane and in the z direction (see Fig. 5.11). The small interlayer spacing of the bilayer is consistent with the theoretically estimated large interlayer interaction energy of E_{int} = – 300 meV/atom.

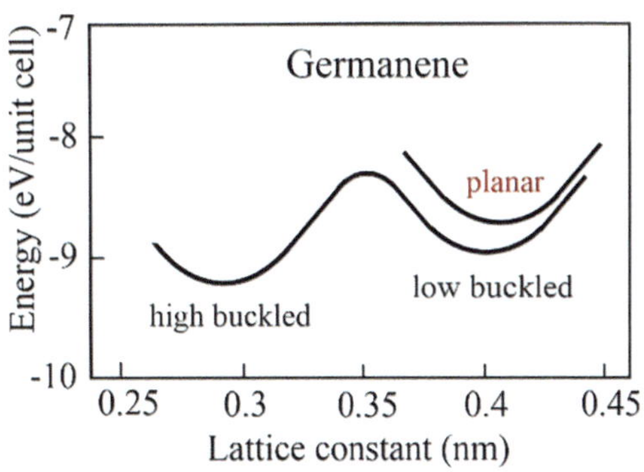

Fig. 5.10 The diagram shows the binding energy of high and low buckled and planar germanene versus the hexagonal lattice constant. Adapted with permission from ref. [51], Copyright 2016, Elsevier

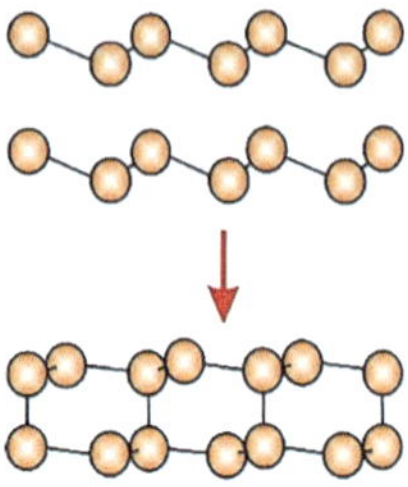

Fig. 5.11 Transition of a buckled monolayer into a planar bilayer with nearly identical intra-layer and inter-layer bond lengths due to strong interlayer interactions

5.3.3 *Mechanics*

The published stiffness and strength values of germanene monolayers and bilayers obtained by different approaches indicate the mechanical behaviors expected for this position in group IVA. The results obtained based on the DFT-PBE approach for the monolayer are $E_{ac} = 44$ N/m and $E_{zz} = 43.4$ N/m with minor anisotropy for Young's moduli, $\nu_{ac} = 0.29$ and $\nu_{zz} = 0.35$ for the Poisson ratios, $\sigma_{ac} = 4.7$ and $\sigma_{zz} = 4.1$ N/m for the ultimate tensile strengths, and $\varepsilon_{ac} = 0.20$ and $\varepsilon_{zz} = 0.205$ for the ultimate strains [58].

MD simulations using an optimized Tersoff potential reveal noticeable anisotropic behavior of the stiffness of bilayer germanene coupled by covalent bonding. The corresponding stiffness, ultimate strength, and ultimate strain values are $E_{ac} = 42.0$ N/m and $E_{zz} = 36.4$ N/m, $\sigma_{ac} = 5.7$ N/m and $\sigma_{zz} = 6.0$ N/m, and $\varepsilon_{ac} = 0.16$ and $\varepsilon_{zz} = 0.18$, respectively [73]. With increasing temperature, both the ultimate strengths and critical strains decrease due to the extension of the bond length. While monolayers and bilayers have similar Young's moduli, the bilayer has a larger ultimate strength and a lower Poisson ratio.

5.3.4 *Physics and Applications*

Weakening of σ bonding by partial hybridization of σ and π orbitals in silicene and germanene affects the electronic properties, optical behavior, and excitonic effects [74]. The corresponding consequences include the geometric structure, the anticrossing phenomenon in the electronic band structure, the corresponding Hove singularities in the density of states (DOS), the prominent optical excitations in the imaginary part of the dielectric functions, and the suppression of the π-like plasmon mode in the energy loss functions.

An application suggested for germanene monolayers is its use as an efficient anchoring material to immobilize lithium polysulfides in Li–S batteries, which can cause capacity losses [75]. Van der Waals corrected DFT calculations indicate that germanene can trap intermediate lithium-polysulfides without disturbing their integrity and enhance the cycling performance of Li–S batteries by preventing dissolution and agglomeration at the electrodes.

Large direct band gaps are predicted by first-principles calculations for chemical functionalization of germanene by halogen atoms (Cl, Br, and F), as confirmed by the Heyd-Scuseria-Ernzerhof (HSE06) hybrid functional [76]. Opening of the gap from $E_g = 0.416$ to 1.596 eV was achieved by the following full and Janus decorations: Cl–Ge–Cl, Br–Ge–Br, F–Gr–Cl, F–Ge–Br, and Cl–Ge–Br. Lower indirect band gaps of $E_g = 0.052$–0.110 eV are found for doping with halogen atoms (Cl, Br, F, and I). Note that the halogen atoms are ionically bonded to host Ge atoms.

The bonding of DNA/deoxyribonucleic acid (RNA) nucleobases with silicene/germanene and Au- and Li-decorated silicene/germanene monolayers as biosensors were studied via quantum simulations [77]. The results indicate marked changes in transmittance due to adsorption of the nucleobases guanine, adenine, and cytosine, suggesting that silicene and germanene are good candidates for fast sequencing of DNA/RNA nucleobases. The large binding capacity and tunable electronic properties of both bare and metal-doped silicene and germanene open the possibility of future use as biosensors of biomolecules.

5.3.5 *Synthesis*

Analysis of the deposition of germanene on Ag(111) under UHV conditions by STM, LEED, and angle-resolved photoemission spectroscopy (ARPES) revealed two distinct phases of low-buckled germanene [72]. The first phase observed at low coverage consisted of a tensile strained honeycomb lattice, partially commensurate with the substrate. At higher coverage, the lattice constant agreed with the theoretical value of free-standing germanene, indicating a quasi-free-standing phase with weak interfacial interactions. To simulate the quasi-free-standing phase (QP) that is incommensurate with the substrate, the authors assumed a lattice constant of $a = 0.39$ nm, which is in good agreement with the theoretical value of free-standing germanene layers [72].

Bernal-stacked germanene bilayers, wherein half of the atoms in the top layer sit on top of the empty centers of hexagons in the bottom layer, were epitaxially grown on a Cu(111) surface and structurally characterized by STM in accordance with DFT calculations [78]. An electron localization function (ELF) analysis provided strong evidence for robust covalent bonding of all Ge–Ge pairs within and between the Ge monolayers, whereas mainly electrostatic interactions with the underlying copper surface occurred. By combining experimental observations and DFT calculations, the authors concluded that the decoupled buckled germanene bilayer carries the Dirac signature of free-standing germanene and suggests its direct use as a functional material.

5.4 Tin Monolayers (Stanene, Tinene)

5.4.1 Chemistry and 2D Bonding

Tin has the electronic configuration (Kr|$4d^{10}5s^2p^2$). Owing to the larger atom size, the π-π bonding between Sn atoms becomes weaker, and the larger contribution of sp^3 hybrids in mixed sp^2-sp^3 bonding further increases the buckling effect in stanene compared to germanene. Conversely, the efficiency of *s-p* mixing decreases with increasing atomic number since the *s* orbital lowers its energy with respect to the *p* orbitals, which reduces corrugation. Independent DFT-GGA-PBE calculations resulted in reduced cohesive energies of $E_c = -2.73$ eV/atom [12] and – 2.65 eV/atom [79]. A somewhat larger binding energy of $E_c = -3.05$ eV/atom (PAW-PBE) comes from a similar approach [53].

5.4.2 2D Structure

The lattice constant a = 464 nm, bond length d = 0.284 nm, bond angle θ = 110.4°, and buckling effect of Δz = 0.095 nm of the buckled hexagonal stanene monolayer, presented in Table 5.1, characterize the extended geometry of the lattice and corrugation [12, 13]. The space group of the buckled stanene structure is *P*3*m*1 [80]. While silicene and germanene prefer the low-buckled configuration, the high-buckled phase becomes energetically more stable for stanene and plumbene [79].

5.4.3 Mechanics

The small in-plane stiffness values of $E_{ac} = 25.2$ N/m and $E_{zz} = 23.5$ N/m and large Poisson ratios of $\nu_{ac} = 0.36$ and $\nu_{zz} = 0.42$ allow extensive rippling or wrinkling of extended 2D structures. The ultimate strength values of $\sigma_{ac} = 2.6$ N/m and $\sigma_{zz} = 2.2$ N/m indicate low mechanical stability against fracture, and the ultimate strains of $\varepsilon_{ac} = 0.17$ and $\varepsilon_{zz} = 0.18$ are surprisingly small (see Table 5.2) [58]. These parameters are in good agreement with other estimations, assuming linear isotropic behavior of E = 22.4 N/m, ν = 0.40, and anisotropic nonlinear behavior of $\sigma_{ac} = 2.59$ N/m, $\sigma_{zz} = 2.28$ N/m, $\varepsilon_{ac} = 0.15$ and $\varepsilon_{zz} = 0.20$ [53].

Notably, Griffith's rule $\sigma \approx E/10$ for fracture strengths, giving an upper limit for the ultimate stress, has been questioned for stanene with a value of $\sigma \approx E/7.4$ [81]. Taking the data presented above, for example, the weaker zigzag direction yields $\sigma \approx 23.5/10 = 2.35$ N/m, which is in good agreement with the strength of $\sigma_{zz} = 2.2$ N/m, taking into account the errors involved in these DFT-PBE calculations. The low in-plane stiffness gives rise to pronounced rippling and/or wrinkling of the monolayer structure that may not be negligible at higher temperatures [78].

5.4.4 Physical Properties

The key features responsible for the expanded interest in stanene include the large spin–orbit gap, the QSH effect, topological superconductivity, quantum anomalous Hall behavior, giant magnetoresistance, tunable band gap, and efficient thermoelectricity near room temperature [82]. These topics open the door to applications in electronics, optoelectronics, spintronics, and as sensors in nanosystems. Moreover, the fundamental properties of the 2D material are tunable because they are susceptible to strain, chemical functionalization, electric fields, doping, and defect engineering.

For stanene theory predicts a topological energy gap at the K point of $E_g = 0.1–0.3$ eV opened by SOC in buckled stanene, hosting the QSH state at room temperature [1]. Furthermore, chemical decoration of stanene induces an inverse band order at the Γ point, where the p_x and p_y bonding states have a higher energy than the antibonding *s* states. This *s-p* band inversion generates a nontrivial QSH phase and enhances SOC of the σ orbitals, opening a large topological gap of ~ 0.3 eV at the Γ point. Theoretical analysis indicated that a Cu(111) substrate not only stabilizes the stretched ultraflat structure of stanene but is also involved in orbital filtering. This filtering effect removes the p_z bands and mimics the chemical decoration effect observed for hydrogenated and halogenated stanene. Consequently, ultraflat stanene adsorbed on Cu(111) has topological properties, as indicated by an energy gap of ~ 0.3 eV in the p_x/p_y band at the Γ point, as confirmed by ARPES.

An enormous step forward toward the realization of 2D topological materials with nontrivial features, including quantum spin/anomalous Hall insulators, has been made recently by the growth of high-quality ultraflat stanene on Cu(111) by low-temperature MBE [83]. Surprisingly, in this system, lattice stretching by strong substrate interactions stabilizes the zero-buckling geometry of stanene. This generates a graphene-like lattice with in-plane *s-p* band inversion induced by SOC with a topological gap of ~ 0.3 eV at the Γ point. Such a substrate stabilization of an ultraflat graphene-like sheet may also be possible for germanene and plumbene and may open the door for exploring 2D topological physics and device applications [82, 83]. Note that previous experiments mostly obtained low-buckled stanene by van Waals epitaxy.

First-principles DFT calculations indicate coherently strained epitaxial binding of stanene on the (0001) surface of alumina [84]. An analysis of the electronic structures of strained stanene revealed a QSH insulator in the stanene/Al_2O_3 system. Furthermore, the effect of in situ fluorine decoration was investigated, including its potential use for mechanical exfoliation.

First-principles calculations served as a tool for developing optimized sensors for the adsorption of toxic gases such as NH_3. The modification of stanene monolayers by embedded transition metals (Ni, Cu, and Au) yields energetically favorable adsorption sites [85]. The favored electron accumulation sites are located at the middle of the metal–molecule bond as shown for Au-embedded stanene, where obviously covalent bonding occurs. Analysis of the band structure indicated that Ni-adsorbed stanene has semiconducting properties, whereas Cu and Au-adsorbed stanene layers have metallic features.

The development of catalytic methods allowing the conversion of CO_2 into fuels or other chemicals has been extended to 2D materials. For example, enhanced catalysis of CO_2 reduction by strain engineering of stanene monolayers weakens the adsorption of CO, which facilitates desorption of the CO product molecules [86]. The adsorption of the intermediate product COOH on stanene is tunable by strain. If the adsorption of the intermediate controls the reduction of COOH, a 10% strain may induce a stronger COOH adsorption and weaker CO adsorption. Thus, catalytic performance can be improved by controlling the adsorption strength of adsorbates using strain engineering. Similarly, the use of strained stanene to improve the selectivity in the hydrogen evolution reaction (HER) was investigated by fist-principles calculations.

5.4.5 *Synthesis*

Based on the possibility of realizing room-temperature topological effects owing to the opening of a spin–orbit gap, stanene has attracted enormous attention. As confirmed by STM and ARPES, the buckled hexagonal lattice of stanene was first fabricated by MBE on a $Bi_2Te_3(111)$ substrate in combination with first-principles calculations, without the signature of a topological state [72]. Using ultrafast laser-materials interactions, few-layer sheets of stanene were deposited and analyzed via HRTEM equipped with energy-dispersive X-ray analysis (EDAX) and a selected area electron diffraction (SAED) detector, allowing the structural and elemental characterization and confirmation of the hexagonal symmetry [87].

Furthermore, high-quality stanene monolayers and nanoribbons were grown by MBE on Sb(111) substrates [88]. The authors concluded that the interaction between stanene and the substrate transforms the large gap 2D TI into a trivial topological system. AFM height profile measurements revealed interlayer separations of $d_s =$ 0.33 nm [87], in agreement with the finding of $d_s = 0.35$ nm for stanene bilayers prepared by MBE [89]. The EDAX measurements showed no traces of oxygen, which was confirmed by the Raman spectra.

As discussed above, MBE allows the growth of graphene-like flat stanene on Cu(111) surfaces [83]. Despite the strong epitaxial effect of the substrate, responsible for the flattening of the monolayer, an in-plane band inversion occurs together with a SOC-induced topological gap. Nontrivial features were observed in ARPES experiments and the topologically derived edge states within the energy gap, observed by STM measurements, support the observation.

5.5 Lead Monolayers (Plumbene)

5.5.1 Chemistry and 2D Bonding

The electronic configuration of lead atoms is (Xe|$4f^{14}5d^{10}6s^2p^2$). Owing to the large atom size, strongly reduced overlap between π orbitals leads to dominant sp^3 hybridization in the mixed sp^2–sp^3 hybrids with the largest buckling effect in this group. For plumbene, DFT-PBE calculations delivered the lowest cohesive energy at $E_c = -2.28$ eV/atom [79]. Owing to the enhanced overlap of σ and π orbitals, the bond angle of plumbene is reduced to $\theta = 110.9°$. Comparison with the tetrahedral angle of 109.47° indicates an almost complete sp^3 hybridization [79].

As revealed by DFT-GGA-PBE calculations and ab initio MD simulations, plumbene forms a bilayer on a MoS_2 substrate consisting of a strongly bonded network with sp^3-like bonding between all nearest neighboring Pb atoms with an interlayer distance of $d_s = 0.364$ nm [90] that is nearly identical with the bond length of d = 0.356 nm in the bulk Pb crystal [79]. Interestingly, the cohesive energy of the bilayer of $E_c = -3.52$ eV/atom is much greater than that of the monolayer, which supports a strongly bonded covalent network forming the bilayer [90].

5.5.2 2D Structures

As a 2D allotrope of lead, plumbene forms a buckled hexagonal structure analogous to that of the other group–IVA members with a lattice constant of a = 0.493 nm, a bond length of d = 0.300 nm, a buckling height of $\Delta z \approx 0.093$ nm (0.101) [93], and a bond angle of $\theta = 110.9°$ [79, 91, 92]. The space group of plumbene is P-3m1 [91]. The estimated thickness of the corrugated monolayer with a buckling height of 0.093 nm [92] is $d_{w+\Delta z} = 0.497$ nm. With increasing sp^3 contribution to mixed bonding of heavier Xenes, a topologically nontrivial electronic structure develops that is very robust in plumbene, the heaviest group member.

The theoretically proposed plumbene bilayer mentioned above is a dynamically and thermally stable structure, as confirmed by the phonon spectrum and ab initio MD simulations [90]. According to theory, on a MoS_2 substrate, plumbene bilayers form a van der Waals heterostructure with a low interaction energy of $E_{int} = -47$ meV, which is considered a new topological phase. With respect to the electronic structure, the bilayer can host a QSH phase under tensile strain, and the QSH gap can be tuned by external strain. The QSH effect was identified by the s-p_{xy} band inversion and helical edge states with a bulk gap [90].

5.5.3 Mechanics

MD simulations predict a stiffness of 16 GPa and an ultimate tensile strength of 12–17 MPa for plumbene [94]. An estimate of the strength, applying Griffith's rule, indicates that this ultimate strength is extremely low. Note that these are the lowest stiffness and strength values of the group-IVA monolayers. Nevertheless, these quantities are greater than those of bulk lead. With the layer thickness of $t = 0.497$ nm estimated above, we obtain a Young's modulus of $E = 8$ N/m and an ultimate strength of $\sigma \approx 0.007$ N/m in 2D units [94]. DFT-GGA-PBE calculations delivered a comparable Young's modulus of 12 N/m but a significantly larger ultimate strength of 3.64 N/m, ascribed to phonon instability under biaxial strain [95]. Consequently, both strength values are not consistent with Griffith's rule.

5.5.4 Physical Properties

Regarding the topological behavior of the homologues, theory predicts the largest energy gap of 0.8–1 eV for plumbene [1]. This increase is due to the stronger SOC effect. Chemical decoration with halogen atoms further increases the band gap [1]. However, since ARPES measurements are still missing, the existence of a band inversion remains an open question for the plumbene monolayer. In addition to having the largest band gap in the group, stable edge states make plumbene an ideal candidate for realizing the QSH effect at room temperature [91].

Contrary to the other four topologically nontrivial 2D compounds of the group, plumbene and its decorated PbX monolayers (X = H, F, Cl, Br, I) are normal insulators and therefore are considered ideal candidates for delivering the QSH effect at higher temperatures [92]. The Pb atoms of the hydrogenated hexagonal network are sp^3 hybridized, the lattice expands, the bond length increases slightly, the buckling decreases, and the band gap more than doubles. As an example, the structure of insulating monolayer PbH and bilayer plumbene decorated by H atoms on both sides is presented in Fig. 5.12 [90, 92]. The topological behavior of PbX monolayers is attributed to s-p_{xy} band inversion due to the lattice symmetry, while SOC only opens a giant gap. The final goal behind the enormous theoretical and experimental activities in this direction is the development of dissipationless transport devices with reduced power consumption and heat production.

New QSH phases with large band gap were found by functionalization of pristine plumbene with amidogen ($-NH_2$), hydroxyl (–OH), and thiol (–SH) groups [96]. Functionalization creates nontrivial topological states with large band gaps in the range of ~ 1.1 eV. The gapless edge states of the chemically decorated monolayers are protected by time-reversal symmetry and they are protected from back-scattering. As suitable substrate to minimize lattice mismatch and to maintain the topological order, which is robust against strain, the authors recommend H-terminated SiC(0001). Consequently, the plumbene QSH insulators provide a promising platform for the

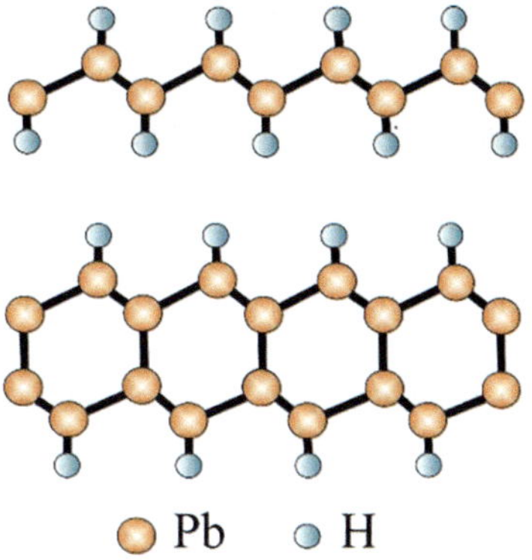

Fig. 5.12 Monolayer and bilayer of insulating plumbene decorated on both sides with hydrogen atoms

future development of low dissipation nanoelectronics and spintronic devices at room temperature.

For the buckled group-IVA monolayers, the effect of electric-field-induced spin splitting was studied by a fully relativistic approach with respect to the spin Hall conductivity [97]. An electric field parallel to the high symmetry axis of an elemental buckled hexagonal structure can tune the topological phase between a QSH insulator and an ordinary band insulator. Calculations based on linear response theory and the Wannier interpolation method indicate that the spin Hall conductivity is not quantized for silicene, germanene, or stanene and decreases significantly as the electric field inverts the band gap and transforms the monolayer into the trivial phase [97]. On the other hand, for plumbene, the electric field-induced band gap does not close because the spin Hall conductivity is robust to an external electric field. In buckled monolayers, SOC plays a vital role in the quantum anomalous Hall effect, and in spintronics applications, such as spin-polarized transistors.

5.5.5 *Synthesis*

Compelling evidence for the epitaxial growth of large-area plumbene sheets by segregation on a $Pd_{1-x}Pb_x(111)$ alloy surface appeared in 2019 [98]. The film showed a planar honeycomb structure, as verified by atomic-scale STM images [72, 99]. Furthermore, DFT calculations rationalized the growth of a flat honeycomb plumbene lattice on Fe monolayers adsorbed on an Ir(111) surface, as detected by STM [100]. According to theory, in free-standing plumbene band inversion leads to the formation of a topologically nontrivial gap. Strong interactions with the Fe layer drive the quantum spin Hall state to a quantum anomalous spin Hall state.

5.6 Property–Structure Relationships

5.6.1 Binding–Structure Relationships

Based on the structural data presented in Table 5.1, Fig. 5.13 shows the cohesive energy plotted versus the bond length of group–IVA monolayers [101]. A drastic decrease in the binding energy appears for the heavier elements. The plot reveals an inverse dependence of the cohesive energy on the bond length with modest changes in the binding energy for the longest bonds. Tight-binding theory describes such an inverse behavior between covalent bonding and the geometric distance between atoms in a network [102, 103]. The tight-binding theory reflects the variation of bonding strength with chemical composition of the system and the geometric arrangement of atoms.

The energy level scheme shown in Fig. 5.14 illustrates the transition of atomic *s* and *p* orbitals to atomic hybrids and bonding orbitals, which are responsible for the formation of bands and eventually a band gap in an extended covalently bonded 2D network [101]. Of interest is the bonding–antibonding splitting 2B that results from the energetic stabilization by the bonding orbitals and determines the binding energy of the 2D network. In nonpolar materials such as elemental monolayers, splitting B is identical to the strength of orbital overlap V, which varies strongly with bond length as $V \sim 1/d^2$ [100, 102]. Consequently, the energy of covalent bonding is inversely

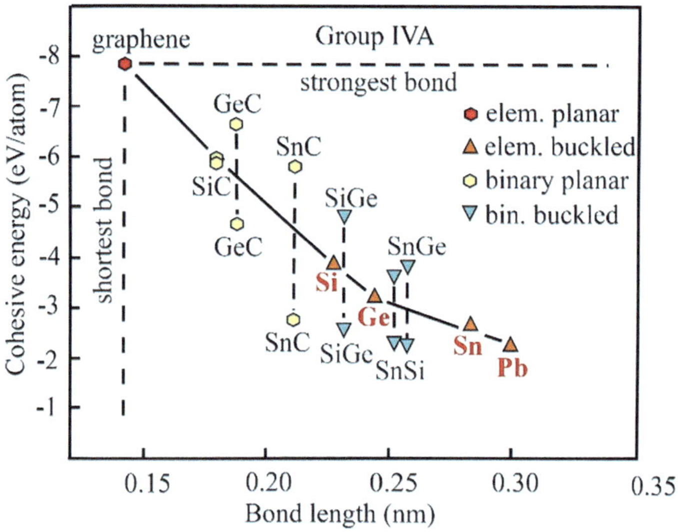

Fig. 5.13 Plot of the cohesive energies versus bond lengths of the elemental IVA monolayers and heteroelemental binary group–IVA monolayers. Adapted from ref. [101], Copyright 2021, Royal Society of Chemistry

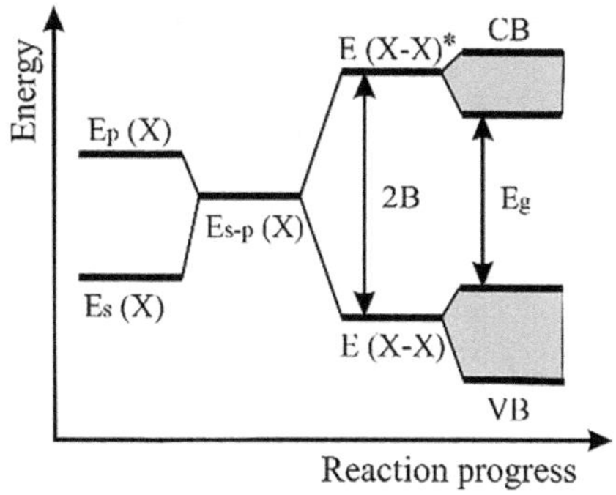

Fig. 5.14 Energy levels of hybridized *s* and *p* orbitals reacting to molecular bonding and antibonding levels with level broadening and band formation. Adapted from ref. [101], Copyright 2021, Royal Society of Chemistry

related to bond length. Figure 5.13 reveals this behavior of the cohesive energy of elemental and in addition, also the binary heteroelemental group–IVA monolayers.

As predicted by tight-binding theory, the binding energy depends sensitively on an increase in the bond length and decreases by a factor of ~ 3.5 on doubling the bond length. Note the enormous loss in the bonding strength within the group and the large separation of graphene in the bonding–structure plot from its buckled homologues, which show only a gradual change in their physicochemical properties. The overall behavior is a consequence of the *aufbau* principle of the periodic table controlling the number of valence electrons and the size of the periods with the closed stable shells of the noble gases [103]. Owing to the availability of exactly four valence electrons with respect to the octet rule, the group-IVA monolayers play a central role in our understanding of bonding behavior and structure formation of elemental monolayers.

5.6.2 *Mechanics–Structure Relationships*

Table 5.2 displays the mechanical properties of group-IVA monolayers termed Xenes. The plot of Young's moduli and ultimate strengths as a function of bond length offers important insight into the linear and nonlinear mechanical behavior of the elemental group-IVA monolayers (see Fig. 5.15) [101]. This reveals a drastic change in the mechanical performance from graphene to silicene. The enormous loss of binding energy from graphene to silicene causes a large decrease in both the intrinsic stiffness and the ultimate strength that controls failure of the 2D network. With their minor increase in atom size, the heavier homologues of Si experience a much smaller weakening of stiffness and strength, mostly following Griffith's rule. Only for plumbene are the published ultimate strength values much smaller or larger than the value ideally expected from the calculated stiffness. While graphene is the stiffest and strongest single-layer 2D material that survives large, applied stress, these properties weaken enormously from graphene to its homologues, for instance, by a factor of approximately five for buckled silicene, however, by a factor of approximately thirty

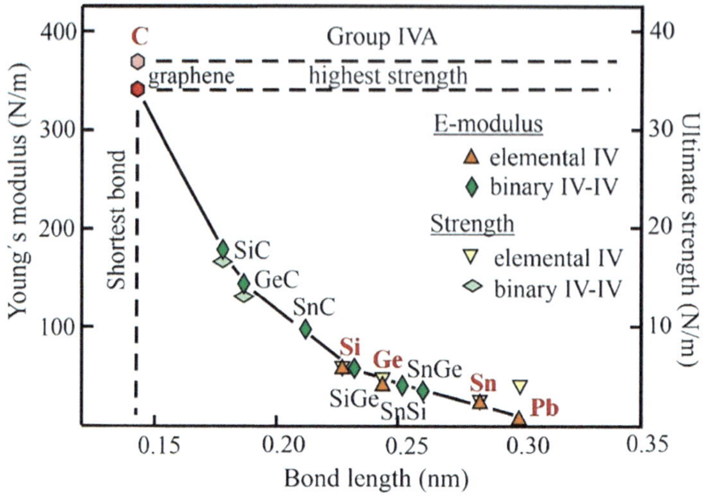

Fig. 5.15 Plots of in-plane Young's moduli and ultimate strengths versus bond lengths of elemental group–IVA monolayers and binary group–IVA monolayers. Adapted from ref. [101], Copyright 2021, Royal Society of Chemistry

to that of plumbene. This drastic degradation of mechanical performance can be a serious problem in practical applications.

The extreme sensitivity of mechanical stability to binding energy has important consequences. Since the mechanical performance decreases so drastically from graphene to plumbene, mechanical stability may become a crucial issue, especially for the monolayers of the heaviest homologues. For instance, low stiffness values give rise to rippling and wrinkling of extended 2D structures, which can have a detrimental influence on the electronic properties. Furthermore, weak mechanical stability permits easier incorporation of impurities and the formation of lattice defects such as dangling bonds. The increasing reactivity may require synthesis under UHV conditions. Moreover, this requires protection by functionalization or encapsulation from the production process to practical applications. On the other hand, as demonstrated above, especially the monolayers of the heaviest elements possess attractive properties, such as nontrivial topological properties.

References

1. A. Zhao, B. Wang, APL Mater. **8**, 030701 (2020)
2. A.K. Geim, Science **324**, 1530 (2009)
3. G. Lu, K. Yu, Z. Wen, J. Chen, Nanoscale **5**, 1353 (2013)
4. G. Yang, L. Li, W.B. Lee, M.C. Ng, Sci. Technol. Adv. Mater. **19**, 613 (2018)
5. G. Graziano, J. Klimeš, F. Fernandez-Alonso, A. Michaelides, J. Phys. Condens. Matter **24**, 424216 (2012)
6. Z. Liu, J.Z. Liu, Y. Cheng, Z. Li, L. Wang et al., Phys. Rev. B **85**, 205418 (2012)

7. R. Zacharia, H. Ulbricht, T. Hertel, Phys. Rev. B **69**, 155406 (2004)
8. M.C. Schabel, J.L. Martins, Phys. Rev. B **46**, 7185 (1992)
9. X. Chen, F. Tian, C. Persson, W. Duan, N.-X. Chen, Sci. Rep. **3**, 3046 (2013)
10. D.D.L. Chung, J. Mater. Sci. **37**, 1475 (2002)
11. J.D. Bernal, Proc. R. Soc. London Ser. A **106**, 749 (1924)
12. J.C. Garcia, D.B. de Lima, L.V.C. Assali, J.F. Justo, J. Phys. Chem. C **115**, 13242 (2011)
13. R. John, B. Merlin, Cryst. Struct. Theory Appl. **5**, 43 (2016)
14. L.M. Malard, D.L. Mafra, M.H.D. Guimarães, M.S.C. Mazzoni, A. Jorio, Phys. Rev. B **79**, 125426 (2009)
15. H. Shin, S. Kang, J. Koo, H. Lee, Y. Kwon, J. Chem. Phys. **140**, 114702 (2014)
16. S.S. Batsnov, Inorg. Mater. **37**, 871 (2001)
17. M. Mantina, A.C. Chamberlin, R. Valero, C.J. Cramer, D.G. Truhlar, J. Phys. Chem. A **113**, 5806 (2009)
18. P. Hess, Nanoscale Horiz. **5**, 385 (2020)
19. T. Gould, S. Lebèque, J.F. Dobson, J. Phys. Condens. Matter **25**, 445010 (2013)
20. E.Y. Andrei, A.H. MacDonald, Nat. Mater. **19**, 1265 (2020)
21. X.-J. Zhao, H. Hou, X.-T. Fan, Y. Wang, Y.-M. Liu et al., Nat. Commun. **10**, 3057 (2019)
22. P. Hess, Mater. Res. Express. **2**, 055601 (2015)
23. G. Cao, Polymers **6**, 2404 (2014)
24. C. Lee, X. Wei, J.W. Kysar, J. Hone, Science **321**, 385 (2008)
25. A. Politano, G. Chiarello, Nano Res. **8**, 1847 (2015)
26. T. Zhang, X. Li, H. Gao, Int. J. Fract. **196**, 1 (2015)
27. P. Hess, J. Appl. Phys. **120**, 124303 (2016)
28. C.A. Marianetti, H.G. Yevick, Phys. Rev. Lett. **105**, 245502 (2010)
29. D. Roundy, M.L. Cohen, Phys. Rev. B **64**, 212103 (2001)
30. J.E. Field, *Properties of Diamond*, ed. by J.E. Field (Academic, London, 1992), pp. 286–310
31. P. Hess, J. Appl. Phys. **111**, 051101 (2012)
32. H. Zhao, N.R. Aluru, J. Appl. Phys. **108**, 064321 (2010)
33. M.A.N. Dewapriya, A.S. Phani, R.K.N.D. Rajapakse, Modelling Simul. Mater. Sci. Eng. **21**, 065017 (2013)
34. M.D. Drory, R.H. Dauskardt, A. Kant, R.O. Ritchie, J. Appl. Phys. **78**, 3083 (1995)
35. S.-M. Lee, J.-H. Kim, J.-H. Ahn, Mater. Today **18**, 336 (2015)
36. P. Zhang, L. Ma, F. Fan, Z. Zeng, C. Peng et al., Nature Commun. **15**, 3782 (2014)
37. J.H. Los, A. Fasolino, M.I. Katsnelson, NPJ 2D Mater Appl. **1**, 9 (2017)
38. B. Sajadi, S. van Hemert, B. Arash, P. Belardinelly, P.G. Steeneken et al., Carbon **139**, 334 (2018)
39. G. Rajasekaran, P. Narayanan, A. Parashar, Crit. Rev. Solid State Mater. Sci. **41**, 47 (2015)
40. S.H. Choi, S.J. Yun, Y.S. Won, C.S. Oh, S.M. Kim et al., Nat. Commun. **13**, 1484 (2022)
41. M. Sang, J. Shin, K. Kim, K.J. Yu, Nanomaterials **9**, 374 (2019)
42. S. Syama, P.V. Mohanan, Nano-Micro Lett. **11**, 6 (2019)
43. W. Wang, H. Su, Y. Wu, T. Zhou, T. Li, J. Electrochem. Soc. **166**, B505 (2019)
44. A.F. Carvalho, B. Kulyk, A.J.S. Fernandes, E. Fortunato, F.M. Costa, Adv. Mater. **34**, 2101326 (2021)
45. X. Li, Y. Chai, 2D Mater. **8**, 022001 (2021)
46. V. Dhinakaran, M. Lavanya, K. Vigneswari, M. Ravichandran, M.D. Vijayakumar, Mater. Today Proc. **27**, 824 (2020)
47. K.A. Madurani, S. Suprapto, N.I. Machrita, S.L. Bahar, W. Illiya et al., ECS J. Solid State Sci. Technol. **9**, 093013 (2020)
48. M. Coroş, F. Pogăcean, L. Măgeruşan, C. Socaci, S. Pruneanu, Front. Mater. Sci. **13**, 23 (2019)
49. C. Shen, X. Yan, F. Qing, X. Niu, R. Stehle et al., J. Materiomics **5**, 463 (2019)
50. C. Kamal, A. Banerjee, A. Chakrabarti, CRC handbook of chemistry and physics, 97th Ed., Chapter 15, *Properties of Two-Dimensional Silicon Versus Carbon Systems* (CRC Press, Boca Raton, FL, 2016), pp. 221–233
51. J. Zhao, H. Liu, Z. Yu, R. Quhe, S. Zhou, Prog. Mater. Sci. **83**, 24 (2016)

52. S. Cahangirov, H. Sahin, G. Le Lay, A. Rubio, *Introduction to the Physics of Silicene and other 2D Materials* (Springer, Heidelberg, 2017), 96 p.
53. H.M. Shodja, F. Ojaghnezhad, A. Etehadieh, M. Tabatabaei, Mech. Mater. **110**, 1 (2017)
54. J. Liu, Y. Yang, P. Lyu, P. Nachtigall, Y. Xu, Adv. Mater. **1800838** (2018)
55. J.E. Padilha, R.B. Pontes, J. Phys. Chem. C **119**, 3818 (2015)
56. P. Xu, Z. Yu, C. Yang, P. Lu, Y. Liu et al., Superlat. Microstruct. **75**, 647 (2014)
57. R.E. Roman, S.W. Cranford, Comp. Mater. Sci. **82**, 50 (2014)
58. B. Mortazavi, O. Rahaman, M. Makaremi, A. Dianat, G. Cuniberti et al., Physica E **87**, 228 (2017)
59. H. Zhao, Phys. Lett. A **376**, 3546 (2012)
60. A. Molle, C. Grazianetti, L. Tao, D. Taneja, Md.H. Alam et al., Chem. Soc. Rev. **47**, 6370 (2018)
61. M.A. Kharadi, G.F.A. Malik, F.A. Khanday, K.A. Sha, S. Mittal et al., ECS J. Solid State Sci. Technol. **9**, 115031 (2020)
62. Y. You, C. Yang, X. Zhang, H. Lin, J. Shi, Mater. Today Nano **16**, 100132 (2021)
63. X.-L. Qi, S.-C. Zhang, Phys. Today **63**, 33 (2010)
64. L. Feng, K. Yabuoshi, Y. Sugimoto, J. Onoda, M. Fukuda et al., Phys. Rev. B **98**, 195311 (2018)
65. R. Pawlak, C. Drechsel, P. D'Astolfo, M. Kisiel, E. Mayer et al., PNAS **117**, 228 (2020)
66. F. Matusalem, D.S. Koda, F. Bechstedt, M. Marques, L.K. Teles, Sci. Rep. **7**, 15700 (2017)
67. L. Masson, G. Prévot, Nanoscale Adv. **5**, 1574 (2023)
68. C. Huang, J. Zhou, H. Wu, K. Deng, P. Jena et al., J. Phys. Chem. Lett. **7**, 1919 (2016)
69. D.K. Sharma, S. Kumar, A. Laref, S. Auluck, Comput. Condens. Matter **16**, e00314 (2018)
70. P. Bampoulis, C. Castenmiller, D.J. Klaassen, J. van Mil, Y. Liu et al., Phys. Rev. Lett. **130**, 196401 (2023)
71. B. Peng, H. Zhang, H. Shao, Y. Xu, G. Ni et al., Phys. Rev. B **94**, 245420 (2016)
72. J. Yuhara, G, Le Lay, Jap. J. Appl. Phys. **89**, SN0801 (2020)
73. M.R. Arshee, S. Adnan, M. Motalab, P. Bose, RSC Adv. **9**, 34437 (2019)
74. V.K. Dien, W.-B. Li, K.-I. Lin, N.T. Han, M.-F. Lin, RSC Adv. **12**, 34851 (2022)
75. P. Panigrahi, Y. Pal, T. Hussain, R. Ahuja, Appl. Surf. Sci. **558**, 149850 (2021)
76. D.M. Hoat, D.K. Nguyen, R. Ponce-Péres, J. Guerrero-Sanches, V. Van On et al., Appl. Surf. Sci. **551**, 149318 (2021)
77. H.H. Gürel, B. Salmankurt, Biosensors **11**, 59 (2021)
78. Z. Qin, J. Pan, S. Lu, Y. Shao, Y. Wang et al., Adv. Mater. **29**, 1606046 (2017)
79. A. Mahmood, G. Rahman, J. Phys. Condens. Matter **32**, 205501 (2020)
80. J.-K. Lyu, S.-F. Zhang, C.-W. Zhang, P.-J. Wang, Ann. Phys. (Berlin) **531**, 1900017 (2019)
81. Z. Shi, C.V. Singh, Nanoscale **9**, 7055 (2017)
82. S.K. Sahoo, K.-H. Wei, Adv. Mater. Interfaces **6**, 1900752 (2019)
83. J. Deng, B. Xia, X. Ma, H. Chen, H. Shan et al., Nat. Mater. **17**, 1081 (2018)
84. S. Eltinge, S. Ismail-Beigl, Phys. Rev. Mater. **6**, 014007 (2022)
85. A. Zhang, H. Yang, Q. Liu, W. Li, Y. Wang, Synthetic Mater. **266**, 116441 (2020)
86. J. Yang, K.E.J. Goh, Z.G. Yu, R.E. Wong, Y.-W. Zhang, Chemosphere **268**, 129317 (2021)
87. S. Saxena, R.P. Chaudhary, S. Shukla, Sci. Rep. **6**, 31073 (2016)
88. J. Gou, L. Kong, H. Li, Q. Zhong, W. Li, P. Cheng et al., Phys. Rev. Mater. **1**, 054004 (2017)
89. F. Zhu, W.-J. Chen, Y. Xu, C.-L. Gao, D.-D. Guan et al., Nat. Mater. **14**, 1020 (2015)
90. L. Zhang, H. Zhao, W.-X. Ji, C.-W. Zhang, P. Li et al., Chem. Phys. Lett. **712**, 78 (2018)
91. X.-L. Yu, L. Huang, J. Wu, Phys. Rev. B **95**, 125113 (2017)
92. H. Zhao, C.-W. Zhang, W.-X. Ji, R.-W. Zhang, S.-S. Li et al., Sci. Rep. **6**, 20152 (2016)
93. S. Mahmud, Md.K. Alam, RSC Adv. **9**, 42194 (2019)
94. D.K. Das, J. Sarkar, S.K. Singh, Comput. Mater. Sci. **151**, 196 (2018)
95. S. Mahdavifar, S.F. Shayesteh, M.B. Tagani, Physica E **134**, 114837 (2021)
96. S.J. Tabassum, T.T. Tanisha, N.T. Hiramony, S. Subrina, Nanoscale Adv. **5**, 3357 (2023)
97. S.M. Farzaneh, Phys. Rev. B **104**, 115205 (2021)
98. J. Yuhara, B. He, N. Matsunami, M. Nakatake, G.L. Lay, Adv. Mater. **31**, 1901017 (2019)

99. G. Bihlmayer, J. Sassmannshausen, A. Kubetzka, S. Blügel, K. von Bergmann et al., Phys. Rev. Lett. **124**, 126401 (2020)
100. W.A. Harrison, Pure & Appl. Chem. **61**, 2161 (1989)
101. P. Hess, Nanoscale Horiz. **6**, 856 (2021)
102. S. Guo, W. Zhou, B. Cai, K. Zhang, S. Zhang et al., Nanoscale Horiz. **4**, 1145 (2019)
103. P. Hess, ChemPhysChem **23**, e202100880 (2022)

Chapter 6
Binary Heteroelemental Group IVA–IVA Monolayers

Abstract The binary heteroelemental monolayers of group-IVA elements have many similarities with the corresponding elemental monolayers. Nevertheless, the monolayers of several combinations of group-IVA elements have novel interesting properties, such as the planar honeycomb carbides SiC, GeC, and SnC, which complement those observed for the elemental compounds and therefore are attracting considerable attention. Of particular interest are the heteroatomic monolayers containing carbon of the second period of the periodic table, due to the small atom size. A well-studied example is graphene-like SiC, which has a series of outstanding properties, such as a direct and wide band gap, high thermal and mechanical stability, suitability for high-power electronics, and hostile-environment electronics. The optical absorption spectra and visible light emission data qualify SiC monolayers for applications in optoelectronics, e.g., for blue light-emitting diodes. The recent success of bottom-up syntheses of 2D SiC is a major step forward to various technical applications. Unfortunately, synthesis of planar GeC and SnC monolayers and low-buckled SiGe, SnSi, and SnGe compounds, which are also promising candidates for various applications in electronics, optoelectronics, spintronics, sensorics, and catalysis, is still lacking. The three planar carbides and the three low-buckled compounds exhibit an inverse relationship between cohesive energy and bond length, as predicted by tight-binding theory. The mechanical properties, which depend strongly on the cohesive energy, exhibit a similar inverse behavior, elucidated for Young's moduli and ultimate strengths with respect to bond lengths. The calculated fracture strengths are only partly in good agreement with Griffith's rule. The functionalization of these monolayers, e.g., by H, F or Cl atoms on one or both sides, transforms the planar monolayers into stable buckled structures. The functionalization introduces topological features of potential interest for applications in spintronics.

P. Hess, *Bonding, Structure, and Performance of Two-Dimensional Materials*, NanoScience and Technology, https://doi.org/10.1007/978-3-031-74161-6_6

6.1 SiC Monolayer (Silagraphene)

6.1.1 Chemistry and 2D Bonding

In Fig. 6.1, the five elements involved in potential binary heteroelemental group IVA–IVA monolayers are highlighted. Note that results on the corresponding compounds containing Pb are still missing.

The currently best studied monolayer of this group is 2D silicon carbide, named 'silagraphene'. The valence electrons controlling chemical behavior are given by the electronic configuration of Si ($3s^2p_x{}^1p_y{}^1p_z{}^0$) and the similar one of C atoms ($2s^2p_x{}^1p_y{}^1p_z{}^0$). In the planar honeycomb SiC monolayer, bonding derives from hybrid sp^2 orbitals forming three backbone σ bonds and strong π bonds formed by the remaining Si $3p_z$ and C $2p_z$ orbitals. For stability reasons, the competition between the usually preferred sp^3 hybridization and the observed sp^2 bonding of Si leads to a hexagonal lattice with planar sp^2 bonding, avoiding the formation of dangling bonds and thus allowing strong covalent bonding. Note that the structure presented in Fig. 6.2a gives a formal correct presentation by conventional single and double bonds; however, this arrangement does not describe the real nature of bonding in SiC monolayers. It is the aromatic character of the bonds that explains the enormous stability of the uniform conjugated system of identical 'double' bonds (Fig. 6.2b). Moreover, the electronegativity decreases from 2.55 to 1.9 from carbon to silicon. Therefore, significant charge transfer of $\delta q = 2.53$ e from the Si atom to the C atom with greater electronegativity leads to a substantial contribution of ionic bonding, which is responsible for an asymmetric distribution of charge between the two nuclei (see Fig. 6.2c) [1]. Consequently, SiC monolayers consist of polarized covalent-ionic bonds. Unlike in elemental monolayers, in addition to different electronegativities, the sizes of the two atoms are different. The van der Waals diameter increases significantly from $d_w = 0.34$ nm for carbon to $d_w = 0.42$ nm for silicon [2]. Moreover, heteroatomic bonds are less rigid and thus more stretchable. The different electronegativities open a wide direct band gap between the π and π* bands of $E_g = 2.56$ eV, calculated by density functional theory (DFT), and $E_g = 2.90$ eV, derived by the Heyd-Scuseria-Ernzerhof (HSE06) approach [3]. The reported cohesive energies of $E_c = -5.97$ eV/atom [1], -5.99 eV/atom [3], (GGA-PW9), and -5.91 eV/atom [4] (PBE) are situated between those of graphene and silicene, however, with

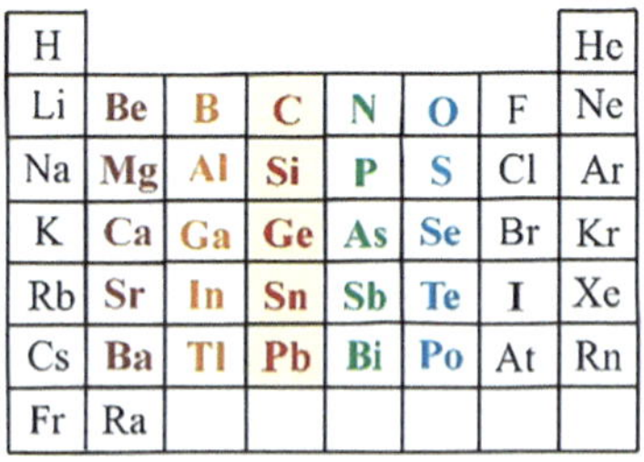

Fig. 6.1 Scheme of the periodic table with the main group elements highlighting the elements involved in binary IVA–IVA monolayers

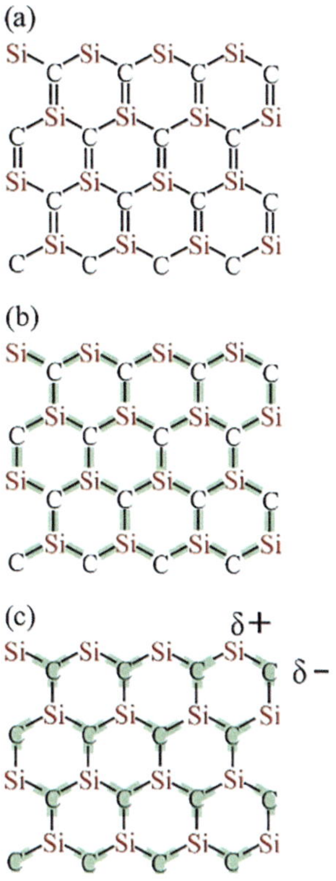

Fig. 6.2 **a** Conventional presentation of the graphene-like structure of binary SiC monolayers with single and double bonds. **b** Schematic representation of conjugated double bonds with aromatic character. **c** Schematic presentation of bond polarization by charge transfer from B to N atoms.

a binding energy near that of graphene. Note that these well agreeing binding energies, based on DFT calculations and the generalized gradient approximation (GGA) with Perdew, Burke, and Ernzerhof (PBE) functionals, are much lower than those obtained by the local density approximation (LDA) of $E_c = -7.63$ eV/atom [5].

6.1.2 2D Structures

The structure of the planar hexagonal SiC monolayer with alternating Si and C atoms is theoretically characterized by the lattice constant a = 0.309 nm, bond length d = 0.179 nm, and bond angle θ = 120°, as presented in Table 6.1 [1]. Importantly, the bond length of the monolayer is shorter than that of the bulk of d = 0.189 nm, consistent with a transition from sp^2 to sp^3 hybridization [1]. The planar geometry is stable because the size of the carbon atom is small enough to allow effective overlap of π orbitals between adjacent atoms. There is no dynamic instability related to phonon modes with imaginary frequency [1]. The monolayer has a large direct band gap of ~ 2.5 eV due to ionicity that is tunable by in-plane strain, whereas multilayers and bulk

SiC possess an indirect band gap [6]. The monolayer thickness can be estimated by the sum of the van der Waals radii of Si and C and is equal to $d_w = 0.38$ nm [2]. Since the unit cell of the SiC monolayer contains two kinds of atoms, its space group is reduced to *P*-6*m*2 in comparison with that of graphene [7]. Among the large variety of monolayers with Si_xC_y stoichiometry, the honeycomb SiC monolayer has the highest stability. The geometry, space group, cohesive energy, lattice constant, bond angle, bond length, buckling distance, van der Waals diameter, and corrugation corrected van der Waals diameter of heteroelemental binary group–IVA monolayers are presented in Table 6.1.

While the effect of stacking order is small for the bilayers of elemental monolayers, for heteroatomic compounds both the interlayer separation and band gap may change considerably. Note that even for fundamental stacking sequences, contradictory theoretical results exist in the literature for the interlayer interaction energy and interlayer distance. A study of several stacking configurations with a van der Waals exchange correlation functional estimated low interlayer binding energies varying between $E_{int} = -34$ and -55 meV/atom for various stacking orders of SiC bilayers [8]. The AA′ stacking order, where Si atoms are located above C atoms and vice versa, showed the lowest energy (see Fig. 6.3). In this configuration, minimization of the forces reduced the interlayer separation to $d_s = 0.232$ nm, and the planar layers started to buckle. Therefore, the authors concluded that no van der Waals bonded layered structure of planar SiC may exist because it is unstable against interlayer linking, as observed in silicene [8]. Obviously, the charge transfer from Si to C and the preference of Si for sp^3 hybridization make the bilayers reactive. Calculations concentrating on the charge transfer properties and stability predicted larger interlayer spacings of $d_s = 0.324$ nm and an extremely large interlayer interaction energy of $E_{int} = -156$ meV/atom for AB stacking, and $d_s = 0.338$ nm and $E_{int} = -154$ meV/atom for AA stacking [9]. Surprisingly, for AA′ stacking the in-plane bond length increases to $d = 0.184$ nm, and the interlayer distance decreases to $d_s = 0.221$ nm [9]. Calculations based on the LDA approach delivered an interlayer spacing of $d_s = 0.212$ nm for Si(C)/C(Si) stacking with buckling of both layers and $d_s = 0.414$ nm for Si(C)/Si(C) stacking, where both layers preserved planar sp^2 bonding [10, 11]. In the ground state, the planar honeycomb monolayer has the highest stability; however, it is not clear at which thickness of a few-layer assembly a phase transition to the bulk structure and the transition to an indirect band gap takes place.

Graphene-like SiC is just one monolayer of a large family of compounds with Si_xC_y stoichiometry. Structures already studied of the large variety of bonding arrangements include SiC_2, SiC_3, SiC_5, SiC_7, and Si_3C [4, 12–15]. Depending on the ratio of Si/C, the individual 2D lattices have different cohesive energies, structures, stabilities, and physicochemical properties. Until recently, most of these monolayers have received less attention than the most stable planar honeycomb SiC monolayer. Nevertheless, some of these compounds could make important contributions to the field if efficient top-down and/or bottom-up methods for high-quality synthesis emerge in the future [16].

Table 6.1 Space group, cohesive energy, lattice constant, bond angle, buckling distance, van der Waals diameter, and corrugation corrected van der Waals diameter of heteroelemental binary group–IV monolayers

Mono-layer	Geometry, space group	Cohesive energy, E_c (eV/atom)	Lattice constant a (nm)	Bond angle θ (°)	Bond length d (nm)	Buckling distance Δz (nm)	Van der Waals diameter d_W [2], $d_{W+\Delta z}$ (nm)
SiC	Planar *P*-6*m*2 [2]	– 5.97 [3] – 5.91 [4]	0.309 [3] 0.3096 [4]	120 [3]	0.179 [3] 0.1786 [4]	0	0.380 [2]
GeC	Planar *P*-6*m*2	– 6.62 [5] – 4.65 [19]	0.322 [5] 0.324 [25]	120 [5]	0.186 [5] 0.187 [25]	0	0.380 [2]
SnC	Planar *P*-6*m*2 [33]	– 5.82 [5] – 2.76 [32]	0.366 [32] 0.349 [33]	120 [33]	0.211 [32] 0.202 [33]	0.001 [32] 0	0.395 [2]
GeSi	Buckled *P*3*m*1	– 2.56 [38] – 2.34 [39]	0.395 [38] 0.389 [5]	114.5 [5]	0.232 [38] 0.231 [5]	0.058 [38] 0.055 [5]	0.420 [2], 0.478
SnSi	Buckled *P*3*m*1 [41]	– 3.61 [41] – 2.28 [32]	0.421 [5] 0.429 [32]	113.3 [5]	0.252 [5] 0.258 [32]	0.067 [5] 0.073 [32]	0.435 [2], 0.502
SnGe	Buckled *P*3*m*1 [41]	– 3.79 [41] – 2.23 [32]	0.427 [5] 0.435 [32]	112.3 [5]	0.257 [5] 0.263 [32]	0.073 [5] 0.080 [32]	0.435 [2], 0.508

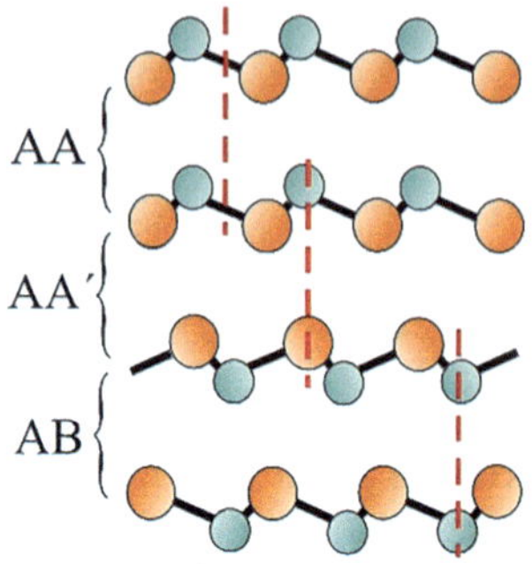

Fig. 6.3 The AA, AA′, and AB stacking configurations of SiC bilayers. Adapted with permission from ref. [11], Copyright 2018, American Chemical Society

6.1.3 Mechanics

The mechanical stability and behavior under strain are crucial issues not only during synthesis and final use but also for understanding the fundamental relationships, e.g., between electronic properties and strain. This is especially true for binary compounds with heterogeneous bonding. Theoretical studies of the linear and nonlinear mechanical behavior of SiC resulted in a Young's modulus of E = 160 N/m, a Poisson ratio of $\nu = 0.314$, ultimate strengths of $\sigma_{ac} = 16.5$ N/m and $\sigma_{zz} = 16.7$ N/m, and ultimate strains of $\varepsilon_{ac} = 0.20$ and $\varepsilon_{zz} = 0.25$ (see Table 6.2) [17]. The stiffness is among the highest values, and the ultimate tensile strength is a respectable value for the maximum strength the monolayer can withstand when stretched to the critical strain. The structure, Young's modulus, Poisson's ratio, ultimate strength, and ultimate strain of heteroelemental binary group–IV monolayers are presented in Table 6.2.

Table 6.2 Structure, Young's modulus, Poisson's ratio, ultimate strength, and ultimate strain of heteroelemental binary group–IV monolayers

Monolayer	Structure	Young's modlus E_{ac}, E_{zz} (N/m)	Poisson's ratio, ν_{ac}, ν_{zz}	Ultimate strength σ_{ac}, σ_{zz} (N/m)	Ultimate strain, ε_{ac}, ε_{zz}
SiC	Planar	160 [17] 163.5 [19]	0.314 [17] 0.300 [19]	16.5, 16.7 [17] 17.6, 20.5 [18]	0.17, 0.21 [17] 0.174, 0.228 [18]
GeC	Planar	144 [28] 140 [19]	0.281 [28] 0.307 [19]	12.7, 12.8 [28] 13.9, 10.5 [19]	0.18, 0.19 [28]
SnC	Planar	98 [5]	0.41 [5]		
GeSi	Buckled	57 [5] 58.6 [39]	0.32 [5] 0.31 [39]	1–1.5 [38]	0.16 [38]
SnSi	Buckled	40 [5] 41.9 [39]	0.37 [5] 0.35 [39]		
SnGe	Buckled	35 [5] 33.3 [39]	0.38 [5] 0.36 [39]		

In an extended theoretical investigation of the mechanical properties of h-SiC_x ($x = 7, 5, 3, 2$, and 1) by the DFT-PBE approach, a Young's modulus of $E = 162$ N/m, Poisson's ratio of $\nu = 0.31$, anisotropic ultimate strengths of $\sigma_{ac} = 19.8$ N/m and $\sigma_{zz} = 20$ N/m, and a critical strain of $\varepsilon = 0.18$ were attained for SiC [4]. The critical strain corresponds to phonon instability and is close to the critical value of elastic instability for tension in the zigzag and armchair directions [4]. Molecular dynamics (MD) simulations with Tersoff and Tersoff-like potentials resulted in the larger anisotropic Young's moduli of $E_{ac} = 180$ N/m and $E_{zz} = 173$ N/m, ultimate strengths of $\sigma_{ac} = 17.6$ N/m and $\sigma_{zz} = 20.5$ N/m, and critical strains of $\varepsilon_{ac} = 0.174$ and $\varepsilon_{zz} = 0.228$ [18]. Calculations using the equation of state (EOS) as an alternative method support an intrinsic stiffness of 163.5 N/m, a high Poisson ratio of $\nu = 0.30$, and lower ultimate strengths of $\sigma_{ac} = 16.4$ N/m and $\sigma_{zz} = 13.3$ N/m, which are in good agreement with Griffith's rule [19]. Currently, single-crystal 2D SiC is one of the toughest and stiffest 2D materials after the planar monolayers of graphene, boronitrene, and borophene. The superior mechanical strength and toughness and satisfactory performance in corrosive and high-temperature environments qualify graphene-like SiC monolayers for special applications.

Considering the enormous problems encountered in producing perfect 2D materials, the mechanical properties of real samples are of foremost importance. MD simulations allow the investigation of the tensile mechanical properties and fracture phenomena of polycrystalline SiC nanostructures. As observed for other nanocrystalline (NC) materials, the NC-SiC sheets exhibit a significantly lower Young's modulus, fracture strength, and fracture strain than single-crystal monolayers [20]. Furthermore, the fracture strength increases with increasing grain size. NC-SiC sheets are more prone to fracture than single-crystal SiC sheets are, and the susceptibility to fracture decreases with decreasing grain size. Crack nucleation starts close to grain boundaries well before reaching the critical strain, but the crack propagation speed is low relative to that of a perfect monolayer. Due to the influence of grain boundaries crack propagation is not perpendicular to the loading direction but complex in NC-SiC materials.

6.1.4 Physics and Applications

Increasing attention has been given to silagraphene because of its potential applications in high-power high-temperature electronics due to its direct wide band gap and thermal robustness. Other applications include hostile-environment electronics, optoelectronics owing to the wide absorption spectrum that covers the entire visible range, blue light emitting diodes (LEDs), and photovoltaic devices. Of interest is the nonmonotonic strain dependence of transport properties, e.g., the low thermal conductivity, which is two orders of magnitude lower than that of graphene owing to the strongly polarized SiC bonds giving rise to large phonon anharmonicity [7]. Quantum confinement and the large surface are sources of distinctive optical and electronic effects. As mentioned above, SiC layers are of special interest for commercial application as LEDs. Owing to their unusual band structure, they are suitable for strain engineering of the band gap and for use as optoelectronic devices [6, 21]. Photoluminescence spectra of ultrathin SiC nanosheets exhibit intense light emission, and a quantum-confinement-induced emission peak at 373 nm [21].

Using spin-polarized first-principles calculations, the change in electronic and magnetic properties of SiC and GeC by the adsorption of H, O, and F atoms were analyzed [22]. Interestingly, significant magnetization by an integer magnetic moment of 1 μ_B is observed by the adsorption of H atoms on SiC, even though monolayer and adatom are non-magnetic. Adsorption of H and F atoms on GeC monolayers yields the same effect. This magnetization effect can be used to generate spin current in spintronic devices by spin filtering. Significant charge transfer from the monolayer to the adatom and deeply negative adsorption energies indicate strong chemisorption of the atoms. Remarkably, chemisorbed O atoms preserve the nonmagnetic nature of SiC and GeC. The results open the door to d^0 2D magnetic materials applicable in spintronic devices.

With the purpose of varying the range of valuable properties of SiC, several studies on the electronic properties of 2D Si_xC_{1-x} monolayers with $0 \leq x \leq 1$ were initiated, which indicate energetic and dynamic stability [11–15]. Depending on the stoichiometry and type of bonding, Si_xC_y monolayers may behave as semiconductors, semimetals, or topological insulators. A universal rule describes the electronic properties of homogeneous Si_xC_{1-x} structures [11]. This family of newly proposed structures offers a variety of useful properties for future applications in electronics, optoelectronics, nanomechanics, bioimaging, sensing, and computing [15].

6.1.5 Synthesis

The synthesis of silicon-carbide monolayers is challenging since bulk SiC is not a layered van der Waals material. The preparation of few-layer assemblies could be achieved by nano-cleavage exfoliation of nanoflakes in polar solutions, as characterized by transmission electron microscopy (TEM), X-ray diffraction (XRD), and X-ray photoelectron spectroscopy (XPS) [21]. The authors assumed that the measured step heights that varied between 0.5 and 1.5 nm for the exfoliated flakes also led to the realization of some monolayer sheets, which could not be characterized in more detail. The first well-documented synthesis of 2D SiC sheets and monolayers were achieved from bulk SiC by a top-down approach via wet exfoliation, using bath sonication in isopropyl alcohol [23]. The planar honeycomb structure was analyzed by scanning electron microscopy (SEM), scanning transmission electron microscopy (STEM), high-resolution transmission electron microscopy (HRTEM), transmission electron microscopy (TEM), atomic force microscopy (AFM), photoluminescence (PL), Raman spectroscopy, and X-ray diffraction (XRD). The generated SiC monolayers were environmentally stable and showed no signs of degradation.

With a novel deposition method, the synthesis of large-area monocrystalline honeycomb SiC monolayers was achieved by bottom-up epitaxial growth [24]. The growth process involved sputter deposition of ~ 10 monolayers of a transition-metal carbide (e.g., TaC) on bulk SiC, followed by annealing at 1700 °C in an argon atmosphere. The tantalum carbide ensured recrystallization upon annealing, which produced surface termination with a monolayer of honeycomb SiC. After relaxation, the SiC layer was weakly buckled, with $\Delta z = 0.0168$ nm. The authors concluded that their results obtained by angle-resolved photoemission spectroscopy (ARPES) can be explained only by the presence of a monolayer of SiC, when additional evidence obtained by low-energy electron diffraction (LEED), scanning tunneling microscopy (STM), core level and particularly DFT band structure calculations are taken into consideration. Band structure calculations together with ARPES measurements indicated hybridization of the π bonds of SiC with underlying TaC surface states to yield a strongly spin–orbit split Dirac-like feature in the electronic band structure.

6.2 GeC Monolayer

6.2.1 Chemistry and 2D Bonding

Like in 2D SiC, Ge and C form a planar graphene-like monolayer, named 'germagraphene' (h-GeC). The electronic configurations of Ge (Ar|$3d^{10}4s^2p^2$) and that of C ($1s^22s^2p^2$) allow the formation of three planar covalent σ bonds by sp^2 hybridization and delocalized π bonds based on the p_z orbitals. The cohesive energy of the planar lattice is only 0.3 eV lower than that of the nonplanar buckled monolayer with mixed sp^2–sp^3 hybridization [25]. Compared with Si atoms, Ge atoms have a slightly larger van der Waals diameter of $d_w = 0.422$ nm versus 0.420 for Si and a smaller electronegativity of 2.01 than carbon atoms with 2.55. Consequently, a minor charge transfer of $\delta q = 0.186$ e occurs from the Ge atom to the C atom, which polarizes the covalent Ge–C bond. Based on the induction of an opening of a band gap by ionic interactions, calculations delivered large direct band gaps of 2.10 eV (DFT-GGA-PBE) and 2.90 eV (HSE06) [25]. Compared with Si, the slightly larger Ge atoms further stretch the Ge–C bond, and the increased bond length reduces orbital overlap and thus strength of bonding. The published cohesive energies deviate remarkably, with a small value of $E_c = -4.65$ eV/atom (DFT-PBE) [19] and an exceptionally large value of – 6.62 eV/atom (DFT-LDA) [5] for the honeycomb GeC monolayer.

6.2.2 2D Structures

Table 6.1 shows the slightly extended lattice constant and bond length of the most stable planar hexagonal h-GeC monolayer in comparison to those of SiC [5, 19, 25, 26]. Confirmed lattice parameters are $a = 0.322$ nm, $d = 0.186$ nm, and $\theta = 120°$ [5]. Phonon dispersion calculations showed no indication of any imaginary or negative frequency, suggesting dynamic stability. Moreover, MD simulations revealed no indication of Ge–C bond breakage below 800 K, verifying good thermal stability of h-GeC [25].

Since electronic and optical properties depend sensitively on interlayer separation, the properties of bi- and few-layer assemblies are receiving increasing attention. AB stacking (Bernal stacking, where the atoms on top of each other are of several types) seems to be the energetically most favorable arrangement of h-GeC bilayers, with an interlayer separation of $d_s = 0.369$ nm [26]. The band gap of multilayers maintains the direct character of the monolayer as the interlayer distance changes. With decreasing interlayer distance, the band gap decreases because the coupling effect increases [26]. Based on DFT calculations for AB and AB′ stacking, where atoms on top of each other are different for AB and are the same for AB′ stacking, bilayer distances of $d_s = 0.350$ nm and $d_s = 0.336$ nm were derived [27]. For comparison, the sum of the tabulated van der Waals radii of the two constituents of $d_w = 0.38$ nm is slightly larger than the calculated interlayer separations as expected [2].

6.2.3 Mechanics

Young's modulus of h-GeC at $E = 144$ N/m indicates a smaller in-plane stiffness than that of SiC because the Si atoms are replaced by the slightly larger Ge atoms, which stretch and thus weaken the bonds [28]. A Poisson ratio of $\nu = 0.28$, nearly identical ultimate strengths of $\sigma_{ac} = 12.7$ N/m and $\sigma_{zz} = 12.8$ N/m, and similar ultimate strains of $\varepsilon_{ac} = 0.18$ and $\varepsilon_{zz} = 0.19$ were calculated by the DFT-PBE approach [28]. The (weak) anisotropy results from the fact that the strain applied in the armchair direction stretches bonds in the same direction, while the strain applied in the zigzag direction first tends to incline bonds. Intrinsic strengths in the same range, with a larger anisotropy of $\sigma_A = 13.9$ N/m and $\sigma_B = 10.5$ N/m, could be derived from the equation of state, confirming an isotropic Young's modulus of 140 N/m and a Poisson ratio of 0.31 [19]. Another study confirmed the Poisson ratio of $\nu = 0.32$ but suggested a smaller stiffness of only 120 N/m [29].

For the fully hydrogenated HGeCH monolayer with a hydrogen atom connected to each Ge and C atom on different sides of the monolayer, dramatically smaller linear mechanical properties of $E = 43$ N/m and $\nu = 0.22$ were obtained [29]. The bonding of hydrogen to monolayers weakens the chemical bonds of the pristine structure and deteriorates its stiffness, ultimate strength, and toughness. As mentioned above, in pristine h-GeC monolayers, a charge of $\delta q = 0.186$ e is transferred from a Ge atom to a C atom, whereas in HGeCH, a net charge of $\delta q = 0.397$ e is transferred from the hydrogen atom to the more electronegative h-GeC monolayer [29]. This charge redistribution leads to an increase in the lattice constant and bond length and thus to a weakening of the overall mechanical stability.

6.2.4 *Physics and Applications*

According to first-principles calculations, h-GeC is a promising highly efficient metal-free oxygen reduction catalyst. The electrocatalytic properties originate from the specific charge distribution owing to the difference in electronegativity of Ge and C. Adsorption and desorption of oxygen occur at room temperature. A study evaluated the activity of h-GeC monolayers as cathode catalyst in fuel cells and in nonaqueous Li-O_2 batteries [30]. The calculations predict low overpotentials comparable to those of Pt-based catalysts.

Another study suggested the use of GeC/SiC van der Waals heterobilayers to boost H_2 fuel production via water splitting [31]. These heterobilayers exhibit direct bandgaps between $E_g = 1.978$ eV and 2.686 eV depending on stacking order, and absorption coefficients > 10^5 cm^{-1}, surpassing the conversion efficiency of most optoelectronic materials. Phonon calculations confirmed the dynamic stability of these heterostructures and the suitability of the stacking-based electronic and optical properties of the proposed structures for various applications. Moreover, h-GeC may qualify as a new anode material for Li/Na-ion batteries [25]. A monolayer of h-GeC can be lithiated or sodiated on both sides, yielding LiGeCLi and NaGeCNa, respectively, with high storage capacities. Fast charge transfer, large Li/Na-ion diffusion, and negligible volume change further enhance the performance of such an anode material.

The optical properties resulting from the capability of direct-to-indirect band gap transitions when h-GeC is under in-plane strain, the wide band gap of $E_g = 2.19$ eV, and a wide energy range of the absorption spectrum make the material suitable for applications in optoelectronics such as light-emitting diodes, photodiodes, and solar cells [26]. Multilayers show a direct gap and new interband transitions between the layers.

6.2.5 *Synthesis*

A breakthrough in synthesis, as discussed for SiC monolayers, allowing the production of high-quality planar h-GeC monolayers and their few-layer assemblies is still missing. There are only reports on the deposition of thin amorphous germanium-carbon films, germanium-carbon coatings, and nanocrystalline germanium-carbide sheets. Nevertheless, various potential applications have already been discussed in the literature.

6.3 SnC Monolayer

6.3.1 Chemistry and 2D Bonding

Like for the other two carbides, it can be assumed that σ and π bonds are formed by the combination of Sn (Kr|$4d^{10}5s^2p^2$) and C ($1s^22s^2p^2$) valence electrons, which constitute a planar sp^2 lattice and a network of delocalized π bonds generated by the overlap of the promoted Sn $5p_z^{\ 1}$ and C $2p_z^{\ 1}$ orbitals. In fact, DFT-GGA-PBE calculations indicate an almost flat structure ($\Delta z = 0.001$ nm) for the indirect semiconductor [32]. The difference in the electronegativity of Sn (1.90) and C (2.55) causes an ionic contribution to bonding, induced by a significant charge transfer of $\delta q = 1.46$ e from a Sn atom to a C atom, leading to an unequal distribution of bonding electrons [33]. Ionicity generates a large indirect band gap of $E_g = 2.37$ eV (HSE06) between the valence band maximum (VBM) at the K point and the conduction band minimum (CBM) at the Γ point. The cohesive energy, based on DFT-PBE calculations, is only $E_c = -2.76$ eV/atom [32], whereas an extremely large binding energy of $E_c = -5.82$ eV/atom was derived with the DFT-LDA approach [5].

6.3.2 2D Structures

The structural data predicted by DFT-PBE calculations presented in Table 6.1 are a = 0.366 nm, d = 0.211 nm, and minimal buckling at $\Delta z = 0.001$ nm [32] or a zero buckling effect for SnC [33]. As monolayer thickness, $d_w = 0.451$ nm is estimated by the sum of the van der Waals radii of Sn and C [2]. The hexagonal structure of SnC has the space group *P*-6*m*2 [33]. The monolayer of graphene-like SnC is dynamically robust because it survives a significant amount of biaxial strain. Pseudopotential DFT calculations revealed buckling and transition to sp^3-type bonding upon chemical functionalization of planar SnC by the halogen atoms Cl and F, which destroy the planarity of bonding [33].

6.3.3 Mechanics

The low Young's modulus of E = 98 N/m and large Poisson ratio of $\nu = 0.41$ agree well with the overall behavior of the other binary group IVA–IVA monolayers [5]. Going down the group–IVA column, varying one constituent of the binary compound, the estimated Young's moduli and ultimate strengths decrease substantially (see Table 6.2).

6.3.4 Physics and Applications

The electronic, vibrational, and optical properties of h-SnC were systematically explored by first-principles calculations in the DFT-LDA framework [34]. For the indirect band gap, this approach delivered a band gap of $E_g = 2.37$ eV using the HSE06 hybrid functional [33]. However, DFT band structure calculations revealed a much smaller gap of ~ 1.25 eV that decreased further to ~ 1.12 eV by introducing the spin–orbit coupling (SOC) effect [34]. The valence band (VB) and conduction band (CB) edges are produced by the bonding and antibonding combinations of the C $2p_z$ and Sn $5p_z$ orbitals, respectively.

The phonon properties of 2D SnC indicate dynamic stability and tolerance to higher compressive and tensile strains. The peaks of the dielectric constant shift to higher photon energies (blueshift) with compressive strain and to lower photon energies (redshift) with increasing tensile strain [34]. The band gap widens with increasing compressive stress but narrows with increasing tensile stress. An indirect-to-direct band gap transition occurs at 6% compressive strain when excluding the SOC effect and at 5% compressive strain a direct band gap is formed, when the SOC is included. Extensive results characterize the honeycomb SnC monolayer as a promising material for strain engineering and applications in optoelectronics and spintronics [34, 35].

Chemical functionalization changes the structural and electronic properties of 2D materials. First-principles calculations based on the pseudopotential DFT approach were employed to study the transformation of sp^2 to sp^3 hybridization in functionalized SnC monolayers, including buckling and changes in the bond angle [33]. Chemical functionalization by half-halogenation to ClSnC and FSnC leads to metallization due to the occupation of surface states. Interestingly, full halogenation, such as ClSnCCl or Janus functionalization ClSnCF, can induce an indirect-to-direct band gap transition (see Fig. 6.4). First-principles calculations based on DFT theory, and the quasiparticle GW approximation showed that tuning of an indirect-to-direct band gap transition by strain engineering is possible for all three carbides SiC, GeC, and SnC [33].

The honeycomb structure of SnC monolayers has attracted attention as photocatalyst owing to its suitable band gap and nontoxicity [36]. A remaining problem is the high energy barrier in the hydrogen evolution reaction (HER) that always required external energy input and/or strongly acidic conditions. Based on first-principles calculations a highly efficient 2D SnC/ZrS_2 van der Waals heterostructure (vdWHS) is proposed for highly efficient photocatalytic water splitting. The intrinsic vdWHS is an S-scheme heterostructure that works for water splitting under acidic conditions owing to the high energy barrier in HER. However, the barrier problem can be solved by doping the SnC monolayer with Si.

First-principles calculations based on DFT theory and many-body perturbation calculations studied the geometric, mechanical, electronic, and optical properties of all three carbides, SiC, GeC, and SnC arranged in honeycomb structures [37]. The type of chemical bonding was investigated through the band-decomposed charge

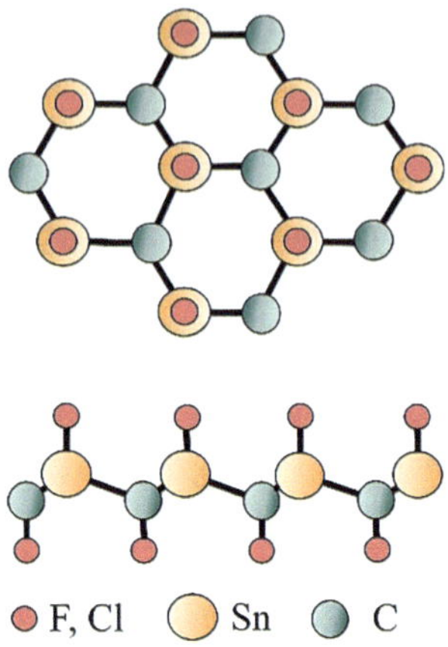

Fig. 6.4 Buckled structure of an on both sides functionalized SnC monolayer, e.g., FSnCF. The halogen atoms (F or Cl) are bonded perpendicularly to the plane

densities, and the appearance of the van Hove singularities in the density of states (DOS). Moreover, the effects of electron–hole interactions in optical responses was analyzed through the imaginary and real parts of the dielectric functions and energy loss functions. For each optical excitation peak it was possible to assign the initial and final electronic states.

6.4 GeSi, SnSi, and SnGe Monolayers

6.4.1 2D Bonding, Structure, Mechanics, Performance, Synthesis

Without any element from the second period of the periodic table, the bonding types of the binary monolayers GeSi, SnSi, and SnGe are similar. Owing to the increased atom size, the s, p_x, and p_y orbitals combine with variable extents with the p_z orbital to form mixed sp^2–sp^3 orbitals. Thus, the characteristic feature of these compounds is a buckled configuration because the corrugated structure has the lowest energy. Three of these orbitals form covalent σ bonds with nearest neighbors, and the fourth orbital, which is directed upward, can form weaker π bonds with neighbors or interact with an adjacent orbital of another layer due to increased reactivity. Corrugation increases with increasing atom size. This finding is not surprising because Si and its heavier homologues prefer sp^3 hybridization. The three compounds have polar bonds since the electronegativity of the group-IVA elements decreases going down the column of the periodic table from Si to Pb. While charge transfer is small in GeSi, it is much greater in SnSi and SnGe. Note that according to theory, the free-standing monolayers of all three compounds are dynamically stable. Synthesis of these compounds either as free-standing monolayers or in an adsorbed form on a substrate as monolayer or few-layer sheet is still lacking.

GeSi: Compared with that of group-IVA carbides, the lattice of GeSi ('siligene') is further extended with a larger lattice constant of a = 0.389 nm, a larger bond length of d = 0.231 nm, and a chair-like buckling effect of $\Delta z = 0.055$ nm according to DFT-LDA calculations [5]. These values have been confirmed by subsequent DFT-PBE calculations reporting similar structural data and an identical buckling distance of $\Delta z = 0.058$ nm [38]. DFT-LDA calculations delivered a cohesive energy of $E_c = -2.56$ eV/atom [38]. This binding energy is in surprisingly good agreement with – 2.34 eV/atom (PBE) [39]. Owing to the small difference in the electronegativities of Si (1.9) and Ge (2.01), charge transfer from Si to Ge and the induced polarity are relatively small.

AB stacking of bilayers is slightly more stable than AA stacking. The strong interlayer bonding energy of $E_{int} = -240$ meV/atom generates a low interlayer distance between the atoms of the nearest adjacent layers of 0.249 nm, whereas the buckling height of the bilayer increases to $\Delta z = 0.072$ nm [38]. A comparison of the mean van der Waals diameter of $d_w = 0.421$ nm and the corrugation corrected value of $d_{w+\Delta z} = 0.493$ nm points to strong interactions by covalent-type interlayer interactions in bilayer GeSi. In a separate study confirming the structural data, an elastic modulus of 58.6 N/m and Poisson ratio of 0.31 were reported [39], which are in good agreement with the findings of an earlier study that reported E = 57 N/m and $\nu = 0.32$ [5]. Transformation of the reported tensile strength of 2–3 GPa into 2D units results in an ultimate strength of 1–1.5 N/m, when using a corrugated layer thickness of $d_{w+\Delta z} = 0.493$ nm [38]. This is an extremely small ultimate strength when compared with the value expected from Griffith's rule. The calculated ultimate tensile strain is $\varepsilon = 0.16$.

GeSi monolayers are excellent bifunctional HER and oxygen evolution reaction (OER) electrocatalysts [40]. As mentioned above, the reduction of π bonding, leading to the low-buckled structure by formation of mixed sp^2–sp^3 orbitals, increases the reactivity of the monolayer surface. Especially the Si sites serve as the most active sites. For both GeSi and SnSi biaxial strain improves HER catalytic activity through optimizing the ΔG_{H*} value and electronic properties in the two Si containing monolayers. In addition, doping by heteroatoms (especially P atoms) further improves the catalytic performance of HER.

SnSi: The values reported for the structural data at a = 0.429 nm, d = 0.258 nm, and the buckling effect $\Delta z = 0.073$ nm of SnSi agree reasonably well, as shown in Table 6.1 [5, 32]. The cohesive energy of $E_c = -3.61$ eV/atom, based on DFT-LDA calculations [41], however, is substantially greater than the – 2.28 eV/atom obtained by the DFT-PBE approach [32]. Phonon calculations indicate that the free-standing monolayer is dynamically stable [41]. Similar to GeSi, the difference in the electronegativities of Sn and Si is very small, with nearly identical values of 1.96 (Sn) and 1.90 (Si). Therefore, the polarity of the Sn–Si bonds and the band gap are small. The elastic modulus of E = 40 N/m and Poisson ratio of $\nu = 0.37$ [5] are in good agreement with the corresponding values of E = 41.9 N/m and $\nu = 0.35$ [39]. By adding the buckling effect of $\Delta z = 0.067$ nm to the sum of the two van der Waals radii of $d_w = 0.427$ nm [1], a monolayer thickness of $d_{w+\Delta z} = 0.50$ nm is attained [39]. The monolayer structure belongs to the space group *P*3*m*1 [32, 39].

In the SnSi monolayer, a direct-to-indirect band gap transition and a semiconductor-to-metal transition occur under strain. The buckled SnSi monolayer is a direct semiconductor with a narrow band gap of $E_g = 0.21$ eV between the valence and conduction band edges at the K point with a gaped Dirac cone and massive fermions [41]. This similarity with graphene is surprising for the noncentrosymmetric buckled structures of SnSi and SnGe. Since the band edges contain only *s* and *p* orbitals, but no *d* orbitals as transition-metal dichalcogenides (TMDCs), the observed valley spin splitting is a nontrivial topological effect due to broken inversion symmetry [42]. Based on the Boltzmann transport equation a high charge carrier mobility of ~ 1000 $cm^2\ V^{-1}\ s^{-1}$ was calculated. The low lattice thermal conductivity at 300 K of < 1.5 $W\ m^{-1}\ K^{-1}$ was attributed to the dominant 'umklapp' scattering, small group velocity, and low Debye temperature. Note that these values surpass those of TMDCs (MoS_2, WS_2), and boron pnictides (BP, BAs, and BSb). Surface decoration, such as hydrogenation and fluorination of SnSi (e.g., FSnSiF), generates nontrivial topological features under strain, leading to large nontrivial SOC band gaps [41].

SnGe: In comparison with SnSi, buckled SnGe has only a marginally larger lattice constant a = 0.435 nm, bond length d = 263 nm, and buckling distance $\Delta z =$ 0.080 nm [32]. Similar to SnSi, the cohesive energies of $E_c = -3.79$ eV/atom [41] and – 2.23 eV/atom [32], calculated by the DFT-LDA and DFT-GGA approaches, respectively, deviate substantially. Since no soft modes exist in the free-standing SnGe monolayer, it is dynamically stable [41]. As expected, for SnGe, we find the lowest Young's moduli of E = 35 N/m [5] and E = 33.3 N/m [39] in the considered group of compounds and the largest Poisson ratios of $\nu = 0.38$ [5] and $\nu = 0.36$ [39]. The buckling effect leads to an estimated monolayer thickness of $d_{w+\Delta z} = 0.509$ nm if $d_w = 0.428$ nm [1] is employed as the mean van der Waals diameter. This monolayer thickness is in good agreement with the published value of t = 0.501 nm [43]. The buckled SnGe monolayer has a small direct band gap of $E_g = 0.23$ eV between the valence and conduction band edges at the K point and belongs to the space group *P*3*m*1, with the polar axis perpendicular to the layer plane [32, 43, 44].

The calculated lattice thermal conductivities and thermoelectric performance of SnSi and SnGe suggest that these materials are promising materials for nanoscale energy storage and transfer [43]. Although SnSi and SnGe nanosheets are trivial band insulators with a gaped Dirac cone, large nontrivial band gaps with *s*-*p* band inversion can be generated by surface decoration, e.g., in strained hydrogenated SnGe monolayers, which are suitable for realizing the room-temperature quantum spin Hall (QSH) effect [41]. As discussed for SnSi, the band edges of the semiconducting SnGe monolayers contain only *s* and *p* orbitals. Therefore, the observed valley spin splitting is a nontrivial topological effect due to broken inversion symmetry [42].

First-principles calculations combined with the Boltzmann transport equation were employed to elucidate the thermal transport properties of the planar and buckled group–IVA (C, Si, Ge, and Sn) binary compounds [45]. The thermal transport properties of the buckled compounds (GeSi, SnSi, and SnGe) are smaller than those of the planar compounds (h-SiC, h-GeC, and h-SnC). Note that factors other than the phonon modes and their amplitude influence this phenomenon, such as the atomic mass and the broken in-plane reflection symmetry. These factors cause an enormous variation in the thermal conductivity, namely, in the range of four orders of magnitude within the heteroelemental binary structures of C, Si, Ge, and Sn [45]. As discussed in this work, the thermal conductivities of planar and buckled monolayers containing Sn atoms are much lower than those of the other compounds.

6.5 Property–Structure Relationships

6.5.1 Binding–Structure Relationships

The diagram in Fig. 6.5 displays deviating values published for the cohesive energies of planar and low-buckled binary group IVA—IVA monolayers versus bond length [46]. The planar group-IVA carbides SiC, GeC, and SnC and the three low-buckled compounds GeSi, SnSi, and SnGe included in the plot, show the typical inverse dependence of cohesive energy on bond length. For all compounds other than SiC, the mean values of the deviating binding energies approximately follow the inverse behavior of the elemental monolayers. One reason for the inverse behavior is the increase in the atomic size from Si to Pb, which gradually reduces efficient π bonding and leads to chair-like buckling with a lower binding energy. The still missing heaviest carbide, namely the PbC monolayer, should follow this general bonding behavior. Considering the quite accurate bond lengths, we suppose that the cohesive energies of the binary compounds are intermediate between the strongly deviating extreme values coinciding with a cohesive energy located between the cohesive energies and bond lengths of the two constituting elemental monolayers as limiting values.

To explain this finding, we apply the tight-binding model to describe the formation of bonding orbitals in the 2D heteroelemental group-IVA compounds. In polar networks, with elements of different electronegativity, the hybrids of the atomic orbitals of atom M and atom X, with an energy difference of 2A, hybridize to form the bonding state of MX and the corresponding antibonding state MX* (see Fig. 6.6) [46]. The bonding-antibonding splitting 2B represents the bonding energy resulting from the stabilization of the bonding orbitals of MX and the destabilization of the antibonding orbitals of MX*. In heteronuclear compounds with different electronegativities, the strength of orbital overlap V also depends on the polar energy A, with $V = (B^2 - A^2)^{1/2}$ [47]. Increasing the polar energy A decreases the importance of covalent energy V or in other words, increasing the electronegativity reduces the contribution of covalent orbital overlap. This is a consequence of the shift of bonding electrons from the center of the two atoms to the atom with the larger electronegativity. Consequently, in polar compounds such as heteroelemental group-IVA

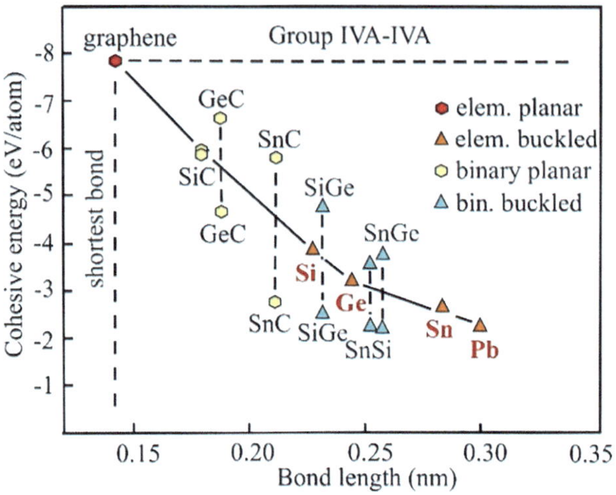

Fig. 6.5 Plot of cohesive energy versus bond length of heteronuclear binary group–IVA monolayers in comparison with the elemental monolayers. Adapted from ref. [46], Copyright 2021, Royal Society of Chemistry

compounds, the bond length depends on both the covalency and ionicity. The orbital overlap determines the covalent energy, which varies strongly with the bond length as $V \sim 1/d^2$ [48]. Accordingly, the energy of covalent bonding, which is the major contributor to chemical bonding in these 2D lattices, decreases strongly with the inversely related length of chemical bonds.

The various physicochemical properties of the binary group-IVA monolayers are tightly connected with the inverse bonding–structure relationship, such as the extremely sensitive behavior of mechanical properties on the binding energy, as discussed in more detail below. The structure and related covalent-ionic bonding determine the size and type of the band gap, which is of special interest in topological materials. The tight-binding model and first-principles DFT calculations were employed to study the atomic structure, electronic structure, and topological

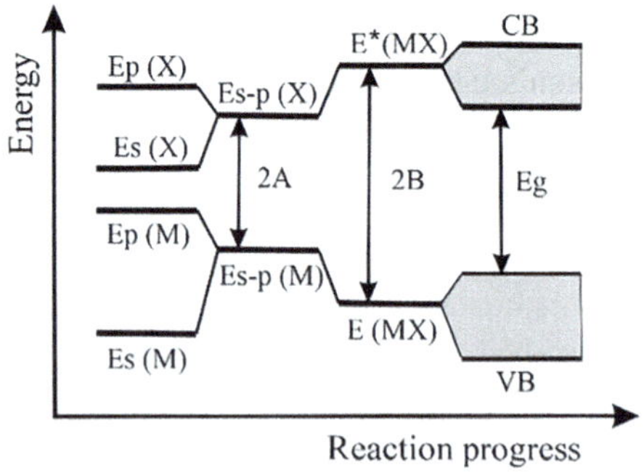

Fig. 6.6 Scheme of the energy levels of heteronuclear binary monolayers according to the tight-binding model, where 2A is the energy difference between the atomic orbitals of X and M and 2B is the bonding-antibonding splitting. Adapted from ref. [46], Copyright 2021, Royal Society of Chemistry

features of the planar and low-buckled binary group-IVA monolayers [48]. This study explored the effects of an external perpendicular electric field and SOC on the two-center overlap integrals of the Slater–Koster model, which is involved in band structures, change of the band gap, and topological phase transitions between the ordinary and QSH regimes. This work opens the door to spintronic applications of 2D topological nanostructures of heteroelemental binary group-IV compounds.

6.5.2 Mechanics–Structure Relationships

The stiffness and ultimate strength of the planar carbides SiC, GeC, and SnC follow the inverse dependence on the bond lengths and those of the elemental group-IVA monolayers quite well (see Fig. 6.7) [46]. The same is true for the low-buckled binary monolayers GeSi, SnSi, and SnGe, which contain no elements from the second row of the periodic table. In fact, their Young's moduli and ultimate strengths are comparable with those of the corresponding elemental compounds. The published strength values of SiC and GeC agree well with Griffith's rule, whereas the calculated ultimate strength of SiGe seems to be too small. The relationships presented in Fig. 6.7 should allow the extraction of reasonable values of missing ultimate strengths of the compounds SnC, SnSi, and SnGe in the case of regular behavior, by applying Griffith's rule. Note the extreme reduction in the stiffness and strength with a moderately decreasing binding energy. As expected, the mechanical properties of the binary compounds are positioned near the intermediate values given by the corresponding properties of the two elemental constituents, as described above for the cohesive energies.

6.5.3 Young's Modulus and Strength Versus Cohesive Binding Energy

The diagram in Fig. 6.8 shows Young's moduli of the elemental and binary group-IVA monolayers as a function of their cohesive energies. Due to the large uncertainty in the cohesive energies of most binary compounds, their mean values are plotted, and the range of published values is displayed. The plot suggests an approximately linear dependence of the mechanical performance on the cohesive energy of both the elemental and binary group-IVA monolayers, except for graphene, which has an extremely large stiffness. Truly remarkable is the enormous decrease in the stiffness by a factor of approximately thirty in comparison to the cohesive energy, which changes by a factor of approximately four. This correlation allows a reasonable estimate of Young's moduli of similar 2D structures if the binding energy is known and vice versa the ultimate strength can be estimated if Griffith's rule applies.

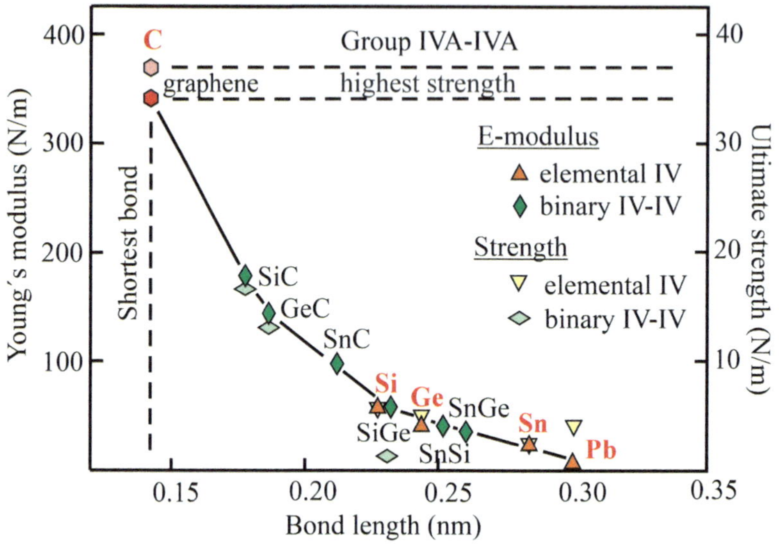

Fig. 6.7 Plot of Young's modulus and ultimate strength values versus bond length of heteronuclear binary group–IVA monolayers in comparison with the elemental group–IVA monolayers. Adapted from ref. [46], Copyright 2021, Royal Society of Chemistry

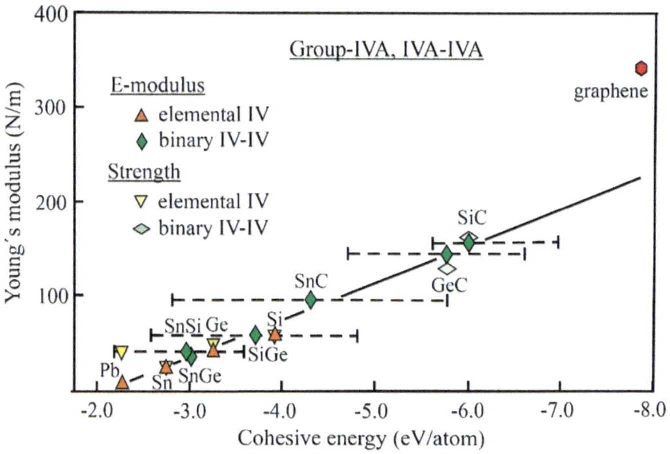

Fig. 6.8 Plot of Young's modulus versus cohesive energy for the elemental and heteronuclear binary group-IVA monolayers. Adapted from ref. [46], Copyright 2021, Royal Society of Chemistry

Taken together, the dependencies between covalent-ionic bonding and structural properties presented above, signify strongly related behaviors of the elemental monolayers and their binary heteroelemental compounds. Changes in the chemical and physical properties of group-IVA monolayers are well documented and qualitatively understood by reliable theoretical studies. Synthesis of the elemental and heteroatomic binary monolayers and final experimental verification of their calculated physicochemical properties are still widely lacking.

References

1. E. Eroglu, M. Tops Akal, S. Cahangirov, S. Circa, Phys. Rev. B **81**, 075433 (2010)
2. S.S. Batsanov, Inorg. Mater. **37**, 871 (2001)
3. H. Dong, L. Zhou, T. Frauenheim, T. Hou, S.-T. Lee et al., Nanoscale **8**, 6994 (2016)
4. X.K. Lu, T.Y. Xin, Q. Zhang, Q. Xu, T.H. Wei et al., Nanotechnology **29**, 315701 (2018)
5. H. Şahin, S. Cahangirov, M. Topsakal, E. Bekaroglu, E. Akturk et al., Phys. Rev. B **80**, 155453 (2009)
6. X. Lin, S. Lin, Y. Xu, A.A. Hakro, T. Hasan et al., J. Mater. Chem. C **1**, 2131 (2013)
7. S.-D. Guo, J.-T. Liu, Phys. Chem. Chem. Phys. **20**, 22038 (2018)
8. T. Susi, V. Skákalová, A. Mittelberger, P. Kotrusz, M. Hulman et al., Sci. Rep. **7**, 4399 (2017)
9. A.V. Kalashnikov, A.V. Tuchin, L.A. Bityutskaya, T.V. Kulikova, J. Phys. Conf. Ser. **1199**, 012009 (2019)
10. M.N. Huda, Y. Yan, M.M. Al-Jassim, Chem. Phys. Lett. **479**, 255 (2009)
11. A. Yaghoubi, K. Masenelli-Varlot, O. Boisron, S. Ramesh, P. Melinon et al., Chem. Mater. **30**, 7234 (2018)
12. Z. Shi, Z. Zhang, A. Kutana, B.I. Yakobson, ACS Nano **9**, 9802 (2015)
13. M. Zhao, R. Zhang, Phys. Rev. B **89**, 195427 (2014)
14. H. Dong, L. Wang, L. Zhou, T. Hou, Y. Li, Carbon **113**, 114 (2017)
15. Q. Wei, Y. Yang, G. Yang, X. Peng, J. Alloys Comp. **868**, 159201 (2021)
16. S. Chabi, K. Kadel, Nanomater. **10**, 2226 (2020)
17. Q. Peng, Mech. Mater. **148**, 103473 (2020)
18. M.-Q. Le, J. Comput. Theor. Nanosci. **11**, 1458 (2014)
19. R.C. Andrew, R.E. Mapasha, A.M. Ukpong, N. Chetty, Phys. Rev. B **85**, 125428 (2012)
20. E.H. Chowdhury, H. Rahman, S. Hong, Comp. Mater. Sci. **197**, 110580 (2021)
21. S.S. Lin, J. Phys. Chem. C **116**, 3951 (2012)
22. C.V. Ha, L.T. Ha, D.T. Hue, D.K. Nguyen, D.T. Anh et al., RSC Adv. **13**, 14879 (2023)
23. S. Chabi, Z. Guler, A.J. Brearley, A.D. Benavidez, T.S. Luk, Nanomaterials **11**, 1799 (2021)
24. C.M. Polley, H. Fedderwitz, T. Balasubramanian, A.A. Zakharov, R. Yakimova et al., Phys. Rev. Lett. **130**, 076203 (2023)
25. N. Khossossi, A. Banerjee, I. Essaoudi, A. Ainane, P. Jena et al., J. Power. Sources **485**, 229318 (2021)
26. Z. Xu, Y. Li, C. Li, Z. Liu, Appl. Surf. Sci. **367**, 19 (2016)
27. L. Pan, H.J. Liu, Y.W. Wen, X.J. Tan, H.Y. Lv et al., Phys. Lett. A **375**, 614 (2011)
28. Q. Peng, C. Liang, W. Ji, S. De, Mech. Mater. **64**, 135 (2013)
29. Z. Sohbatzadeh, H.A. Eivari, D.V. Fakhrabad, Physica B Condens. Matter **547**, 88 (2018)
30. Y. Ji, H. Dong, T. Hou, Y. Li, J. Mater. Chem. A **6**, 2212 (2018)
31. R. Islam, S. Islam, A.F. Mitul, R.H. Mojumder, A.S.M.J. Islam, Sci. Rep. **11**, 17739 (2021)
32. M. Fadaie, N. Shahtahmassebi, M.R. Roknabad, O. Gulseren, Comp. Mater. Sci. **137**, 208 (2017)
33. D.M. Hoat, M. Naseri, R. Ponce-Péreze, N.N. Hieu, J.F. Rivas-Silva et al., Mater. Res. Express **7**, 015013 (2020)
34. R. Islam, Z. Wang, S. Qu, K. Liu, Z. Wang, J. Comput. Electron. **20**, 151 (2021)
35. T.-Y. Lü, X.-X. Liao, H.-Q. Wang, J.-C. Zheng, J. Mater. Chem. **22**, 10062 (2012)
36. D. Yang, X. Zhang, Y. Nie, H. Zhu, G. Xiang, Nanoscale **16**, 4866 (2024)
37. N.T. Han, T.L. Manh, V.K. Dien, Physica E **159**, 115925 (2024)

38. P. Jamdagni, A. Kumar, A. Thakur, R. Pandey, P.K. Ahluwalia, Mater. Res. Express **2**, 016301 (2015)
39. J. Shi, Y. Gao, X.-L. Wang, S.-N. Yun, Chin. Phys. Lett. **34**, 087701 (2017)
40. C. Li, G. Yu, X. Shen, Y. Li, W. Chen, Molecules **27**, 5092 (2022)
41. Y. Wang, Y. Ding, Appl. Surf. Sci. **382**, 1 (2016)
42. M.K. Mohanta, F. Is, A. De Arkar, Phys. Rev. B **107**, 035429 (2023)
43. Y. Ding, Y. Wang, Appl. Surf. Sci. **396**, 1164 (2017)
44. D. Di Sante, A. Stroppa, P. Barone, M.-H. Whangbo, S. Picozzi, Phys. Rev. B **91**, 161401(R) (2015)
45. Q.-Q. Zhang, P.-Z. Pia, X.-K. Chen, W.-X. Zhu, K.-Q. Chen, J. Phys. Condens. Matter **32**, 305301 (2020)
46. P. Hess, Nanoscale Horiz. **6**, 856 (2021)
47. S. Guo, W. Zhou, B. Cai, K. Zhang, S. Zhang et al., Nanoscale Horiz. **4**, 1145 (2019)
48. B. Peng, H. Zhang, H. Shao, K. Xu, G. Ni et al., J. Mater. Chem. A **4**, 3592 (2016)

Chapter 7
Group IIIA–VA Monolayers

Abstract The group of IIIA–VA monolayers of MX (M = B, Al, Ga, In, and Tl; X = N, P, As, Sb, and Bi) with 25 compounds contains several well-studied monolayers, and some of them have already been found to be applicable in electronic and optoelectronic devices. The main available information comes from first-principles calculations, which cover the structural data, cohesive binding energies, electronic properties, and mechanical behavior of these compounds. The electronic structure of group–IIIA atoms is (ns^2np^1) with three valence electrons, whereas that of group–VA atoms is (ns^2np^3) with five valence electrons. These configurations leave an unoccupied p_z orbital in the group–IIIA atom and a lone pair of electrons in the group–VA atom, forming π bonds. The preferred sp^3 bonding in the bulk and the large interlayer interaction energy in few-layer assemblies make the synthesis of monolayers demanding. The different electronegativities cause mixed covalent-ionic bonding of the heteroatomic compounds. Owing to their greater electronegativity, the valence electron distribution of the polar bonds is concentrated on the group-VA atoms. Importantly, planar group-IIIA nitrides exhibit large ionic contributions to bonding, inducing wide bandgaps, which are of interest in lighting and photovoltaics. Owing to the small size of the boron and nitrogen atoms, strong overlap of π orbitals occurs that stabilizes planar sp^2 hybrids, e.g., in h-BN. The other boron–pnictogen monolayers show only minor deviations from planarity for the heaviest compound h-BBi, while the other compounds have a stable planar configuration. Based on their electronic properties, these monolayers provide fundamental building blocks for 2D heterostructures and are promising materials for nanoelectronic and optoelectronic devices. Most of the monolayers possess thermodynamic, thermal, and dynamic stability. The stiffness and ultimate strength decrease enormously from h-BN to the heaviest compound, h-TlBi.

P. Hess, *Bonding, Structure, and Performance of Two-Dimensional Materials*,
NanoScience and Technology, https://doi.org/10.1007/978-3-031-74161-6_7

7.1 BN Monolayer (Boronitrene)

7.1.1 Chemistry and 2D Bonding

The elements involved in the formation of IIIA–VA monolayers are highlighted in Fig. 7.1. All these compounds are well characterized in two comprehensive compilations [1, 2].

Among these, the hexagonal monolayer of boron nitride (h-BN) or boronitrene is one of the best studied monolayers. Owing to the types of 3D bonding of bulk BN, h-BN cannot be obtained by exfoliation from a layered bulk structure of BN that sometimes is referred to as 'white graphene' [3, 4]. Boron and nitrogen are the left and right neighbors of carbon in the periodic table; therefore, carbon and BN are isoelectronic. The electronic configuration of boron ($1s^2 2s^2 2p^1$) has three valence electrons, and that of nitrogen ($1s^2 2s^2 2p^2 p^1$) has five valence electrons. In the planar hexagonal lattice of h-BN, the sp^2 orbitals form a backbone of three σ bonds in a planar structure like graphene. This bonding configuration leaves an unoccupied $2p_z$ orbital at the B atom and a lone pair of electrons at the N atom, which are responsible for additional π bonding. The conjugated B–N bond with a dissociation energy of ~ 4.0 eV is among the strongest chemical double bonds (see Fig. 7.2) [5]. Owing to the different electronegativities of boron and nitrogen, π electrons tend to localize around nitrogen atoms. The resulting mixed ionic-covalent bonding is characterized by a significant Bader charge transfer of $\delta q = 2.06$ e from each B atom to each N atom due to the large difference in the electronegativities of B (2.0) and N (3.0) [6, 7]. According to tight-binding theory, bonding in h-BN is weaker than that in graphene owing to the asymmetric distribution of valence electrons, which shifts charge from the center of the B–N bond to more electronegative N atoms.

As expected, the cohesive energy of $E_c = -8.05$ eV/atom [7] derived by the density functional theory-local density approximation (DFT-LDA) approach is substantially greater than the binding energies of $E_c = -7.10$ eV/atom [7] and -6.67 eV/atom [1] obtained by density functional theory (DFT), using the generalized gradient approximation (GGA) and Perdew-Burke-Ernzerhof (PBE) functionals (see Table 7.1). A comprehensive study explored the interlayer binding of soft-layered materials such as graphite and h-BN multilayers by comparing several theoretical approaches [8].

H							He
Li	Be	B	C	N	O	F	Ne
Na	Mg	Al	Si	P	S	Cl	Ar
K	Ca	Ga	Ge	As	Se	Br	Kr
Rb	Sr	In	Sn	Sb	Te	I	Xe
Cs	Ba	Tl	Pb	Bi	Po	At	Rn
Fr	Ra						

Fig. 7.1 Scheme of the periodic table with the main group of elements, highlighting the elements involved in binary IIIA–VA monolayers

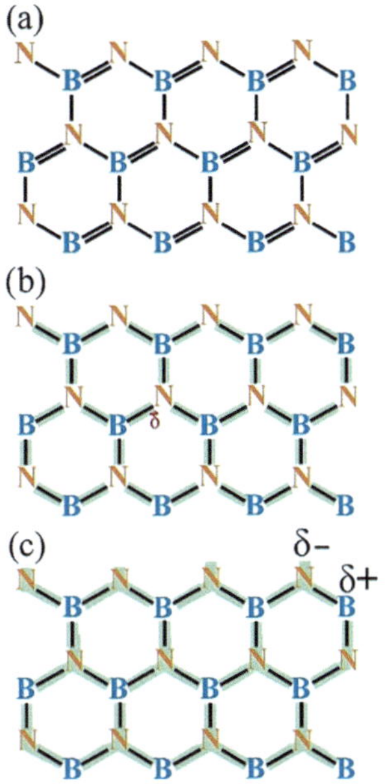

Fig. 7.2 **a** Conventional presentation of the hexagonal structure of binary BN monolayers with single and double bonds. **b** Schematic representation of conjugated double bonds with aromatic character. **c** Schematic presentation of polarization by charge transfer from B to N atoms.

For such calculations, dealing with strong covalent-ionic in-plane bonding and weak van der Waals interlayer interaction, large deviations are found in the literature. The cohesive energy, lattice constant, bond length, buckling distance, bond angle, van der Waals diameter, and corrugarion corrected van der Waals diameter of group IIIA–VA monolayers are presented in Table 7.1.

7.1.2 2D Structures

The planar honeycomb network of h-BN has a lattice constant of a = 0.251 nm and a bond length of d = 0.145 nm (see Fig. 7.3) [1, 2]. In addition to only marginally longer bonds, the widely accepted monolayer thickness of t = 0.333 nm, derived from the interlayer spacing, agrees surprisingly well with that of graphene, despite the occurrence of electrostatic interlayer interactions. This was confirmed by the measured layer separation of $d_s = 0.33$ nm obtained by recent transmission electron microscopy (TEM) and X-ray diffraction (XRD) experiments on 5–50 nm thick h-BN films, synthesized by ambient pressure chemical vapor deposition (CVD) [9]. In excellent agreement with these experimental results, detailed calculations of the interlayer binding energy were performed, and an interlayer spacing of 0.333 nm has been predicted [10]. An estimation of the monolayer thickness by the mean van der Waals diameters using the sum of the van der Waals radii of the boron atom (0.18 nm) and nitrogen atom (0.16 nm) yields 0.34 nm [11]. The good agreement between the experimental and theoretical interlayer separations and estimated thickness clearly

Table 7.1 Cohesive energies, lattice constant, bond length, buckling distance, bond angle, van der Waals diameter, and corrugation corrected van der Waals diameter of group IIIA–VA monolayers

Mono layer	Cohesive energy, E_c (eV/atom)	Lattice constant a (nm)	Bond length d (nm)	Buckling distance Δz (nm)	Bond angle θ (°)	Van der Waals diameter d_w [11], $d_{w+\Delta z}$ (nm)
h-BN	– 6.67 [1], – 7.10 [7]	0.251 [1, 2]	0.145 [1, 2]	0 [1]	120 [1, 2]	0.340 [11]
h-BP	– 4.55 [1], – 5.00 [7]	0.320 [1, 85]	0.185 [1, 85]	0.0005 [1]	120 [1, 85]	0.375 [11]
h-BAs	– 3.86 [1], – 4.33 [7]	0.339 [1]	0.196 [1]	0.0012 [1]	119.99 [1]	0.385 [11], 0.386
h-BSb	– 3.26 [1], – 3.73 [7]	0.374 [1]	0.216 [1]	0.0033 [1]	119.98 [1]	0.400 [11], 0.403
h-BBi	– 2.81 [1], – 3.25 [106]	0.388 [1]	0.229 [1]	0.048 [1]	115.72 [1]	0.410 [11], 0.458
h-AlN	– 5.08 [1], – 5.05 [43]	0.313 [1]	0.181 [1, 44]	0 [1, 44]	120 [1, 44]	0.370 [11]
h-AlP	– 3.36 [1]	0.393 [1]	0.228 [1]	0.018 [1]	119.41 [1]	0.405 [11], 0.423
h-AlAs	– 2.96 [1]	0.406 [1]	0.239 [1]	0.046 [1]	116.26 [1]	0.415 [11], 0.461
h-AlSb	– 2.51 [1]	0.440 [1]	0.262 [1]	0.062 [1]	114.52 [1]	0.430 [11], 0.492
h-AlBi	– 2.28 [1], 2.56 [106]	0.451 [1]	0.272 [1]	0.077 [1]	112.28 [1]	0.440 [11], 0.517
h-GaN	– 3.80 [1], – 4.15 [46]	0.321 [1, 46]	0.185 [1, 46]	0 [1, 46]	120 [1, 46]	0.370 [11]
h-GaP	– 2.77 [1]	0.390 [1]	0.228 [1]	0.038 [1]	117.26 [1]	0.405 [11], 0.443
h-GaAs	– 2.46 [1], – 3.71 [117]	0.405 [1]	0.241 [1]	0.058 [1]	114.45 [1]	0.415 [11], 0.473
h-GaSb	– 2.16 [1]	0.438 [1]	0.263 [1]	0.071 [1]	112.97 [1]	0.430 [11], 0.501
h-GaBi	– 2.01 [1], – 2.31 [106]	0.452 [1]	0.273 [1]	0.082 [1]	111.94 [1]	0.440 [11], 0.522
h-InN	– 3.03 [1], – 3.60 [67]	0.359 [1]	0.207 [1]	0 [1]	120 [1]	0.380 [11]
h-InP	–2.47 [1]	0.422 [1]	0.249 [1]	0.051 [1]	115.97 [1]	0.415 [11], 0.466

(continued)

Table 7.1 (continued)

Mono layer	Cohesive energy, E_c (eV/atom)	Lattice constant a (nm)	Bond length d (nm)	Buckling distance Δz (nm)	Bond angle θ (°)	Van der Waals diameter d_w [11], $d_{w+\Delta z}$ (nm)
h-InAs	– 2.25 [1], – 3.43 [117]	0.436 [1]	0.260 [1]	0.066 [1]	113.89 [1]	0.425 [11], 0.491
h-InSb	– 2.01 [1]	0.468 [1]	0.281 [1]	0.076 [1]	112.97 [1]	0.440 [11], 0.516
h-InBi	– 1.89 [1], – 2.15 [106]	0.481 [1]	0.291 [1]	0.086 [1]	111.65 [1]	0.450 [11], 0.536
h-TlN	– 2.20 [1], – 2.40 [80]	0.373 [1, 78]	0.216 [1, 78]	0.0061 [1]	119.92 [1]	0.380 [11], 0.386
h-TlP	– 1.99 [1], – 2.25 [80]	0.432 [1]	0.256 [1]	0.059 [1]	114.82 [1]	0.415 [11], 0.474
h-TlAs	– 1.86 [1], – 2.14 [80]	0.448 [1]	0.268 [1]	0.069 [1]	113.53 [1]	0.425 [11], 0.494
h-TlSb	– 1.71 [1], – 2.02 [80]	0.479 [1]	0.287 [1]	0.078 [1]	112.88 [1]	0.440 [11], 0.518
h-TlBi	– 1.66 [1]	0.493 [1]	0.297 [1]	0.085 [1]	112.11 [131]	0.450 [11], 0.535

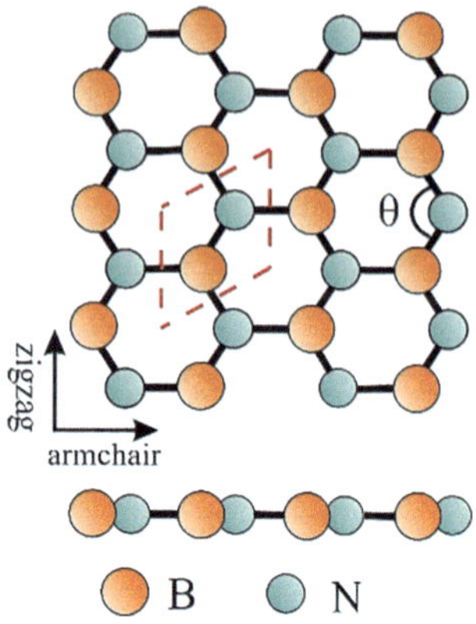

Fig. 7.3 Top and side views of planar honeycomb binary IIIA–VA monolayers. The dashed line indicates the unit cell. Adapted with permission from ref. [2], Copyright 2016, AIP Publishing

supports the dominance of weak van der Waals interlayer forces. The space group of the graphene-like h-BN monolayer, which is the most stable polymorph, is *P*-6*m*2 [6].

Hexagonal boron nitride is an insulating material owing to the buildup of effective charges at the atomic centers with a direct wide band gap of $E_g = 5.67$ eV according to the Heyd-Scuseria-Ernzerhof (HSE06) approach and 4.65 eV according to the PBE approach [1]. The HSE06 gap is in better agreement with the measured bulk gap of 5.97 eV [12, 13]. What is the reason that the boron nitride monolayer is a wide band gap semiconductor and graphene a semimetal with a zero band gap? According to tight-binding theory, the significant difference in electronegativity between B and N leads to a large hybrid polar energy that markedly reduces the hybrid covalent energy. Thus, the ionicity rather than the covalency generates a large electrically insulating gap opening of 5–6 eV [13]. According to tight-binding analysis, the large direct band gap at the K point is related to the difference in the p_z orbital energies of the two constituents [7].

The reason for the striking similarity of the interlayer separation of few-layer systems of h-BN and graphene is that van der Waals forces, rather than electrostatic interactions, dominate and fix the interaction at the equilibrium interlayer distance [14]. Ionic interactions are expected to preferentially control the optimal stacking order and interlayer sliding corrugation [14]. A comparison of the experimentally observed AA′ stacking mode, where opposite charges reside atop each other for optimal electrostatic interaction, with other stacking modes showed that the latter configurations have interaction energies only slightly different from those of the AA′ configuration, which represents the global minimum of the sliding energy landscape. The calculated differences are still within the broad range of reported bilayer interaction energies of $E_{int} = -26.0$ meV/atom and $E_{int} = -38.1$ meV/atom for the PBE approach plus van der Waals interaction energy [10, 14]. The weak interlayer interaction between h-BN monolayers makes multilayers susceptible to staking faults. Interestingly, an essentially free sliding path exists in the layered system. A further comparison of the bulk interlayer interaction energies of graphite and bulk

BN reveals similar interaction energies of up to – 84.7 meV/atom and – 85.9 meV/atom, respectively [10, 14]. Note that the interlayer binding is much greater in the bulk system than in a bilayer because each layer interacts not only with one adjacent layer but also with neighboring layers on both sides and those that are farther away.

7.1.3 Mechanics

It is possible to neglect the angle dependence of the linear Young's modulus and Poisson's ratio, which is not the case for the nonlinear ultimate strength and critical strain of h-BN. Averaging published first-principles DFT calculations, which vary mostly between E = 260–280 N/m, yields a mean stiffness value of E = 270 N/m and a Poisson ratio of $\nu = 0.22$ (see Table 7.2) [15]. Much larger deviations and anisotropies are reported for the nonlinear ultimate strength, varying between $\sigma = 26$ N/m and $\sigma = 36$ N/m, and for the ultimate strains between $\varepsilon_{ac} = 0.23$ and $\varepsilon_{zz} = 0.29$ in the stronger zigzag direction. The space group, Young's modulus, Poisson's ratio, ultimate strength, and ultimate strain of group IIIA–VA monolayers are presented in Table 7.2.

Remarkably, the linear elastic properties of one of the first calculations of E = 267 N/m and $\nu = 0.21$ [16] and the most recent compilation of $E_{ac} = 276.0$ N/m and $E_{zz} = 275.4$ N/m and $\nu_{ac} = \nu_{zz} = 0.225$, based on DFT-GGA-PBE calculations, are in good agreement [1]. Thus, the stiffness of h-BN is much lower than the measured biaxial elastic stiffness of graphene at 340 N/m. An extended DFT study of uniaxial deformation yielded a Young's modulus of 279 N/m, a Poisson ratio of 0.22, and ultimate strength and strain values of $\sigma_{ac} = 23.6$ N/m and $\varepsilon_{ac} = 0.18$ in the armchair and $\sigma_{zz} = 26.3$ N/m and $\varepsilon_{zz} = 0.26$ in the zigzag direction [17]. Note that, based on DFT calculations combined with the PBE functional, much lower Young's moduli of $E_{ac} = 257$ N/m and $E_{zz} = 260$ N/m and larger ultimate strength and ultimate strain values of $\sigma_{ac} = 29$ N/m and $\varepsilon_{ac} = 0.18$ and $\sigma_{zz} = 34$ N/m and $\varepsilon_{zz} = 0.29$ have been reported [18].

Molecular dynamics (MD) simulations based on a Tersoff-like potential and concentrating on the variation in tensile properties with temperature and strain revealed strong softening effects with increasing temperature. Increasing the temperature from 0 to 2000 K decreases Young's modulus of h-BN by 26.5% and the fracture strength and strain decrease by 62.1% and 68.2%, respectively [19]. In addition, this work reviewed previous MD simulations and ab initio studies. Subsequent MD simulations using Tersoff and Tersoff-like interaction potentials obtained anisotropic Young's moduli of $E_{ac} = 253$ N/m and $E_{zz} = 263$ N/m and larger ultimate stresses of $\sigma_{ac} = 29.7$ N/m and $\sigma_{zz} = 36.1$ N/m together with anisotropic ultimate strains of $\varepsilon_{ac} = 0.18$ and $\varepsilon_{zz} = 0.23$ [20]. Furthermore, a stiffness of E = 276 N/m, a Poisson ratio of $\nu = 0.22$, and ultimate strengths of $\sigma_{ac} = 26.0$ N/m and $\sigma_{zz} = 28.0$ N/m have been reported [21]. Recent MD simulations employing both the Tersoff and Stillinger–Weber potentials resulted in substantially lower Young's moduli of $E_{ac} =$

Table 7.2 Space group, Young's modulus, Poisson ratio, ultimate strength, and ultimate strain of group III–V monolayers

Monolayer	Space group	Young's modulus E (N/m)	Poisson ratio, ν	Ultimate strength σ_{ac}, σ_{zz} (N/m)	Ultimate strain ε_{ac}, ε_{zz}
h-BN	*P*-6*m*2 [6]	279 [17] 270 [15], 276 [21],	0.22 [17] 0.22 [15], 0.22 [21]	23.6, 26.3 [17] 26.0, 28.0 [21]	0.18, 0.26 [17] 0.19, 0.20 [22]
h-BP		139 [1], 135 [16]	0.277 [1], 0.28 [16]		
h-BAs		116 [1], 119 [16]	0.282 [1], 0.29 [16]		
h-BSb		~ 84 [1], 91 [16]	~ 0.3 3 [1], 0.34 [16]		
h-BBi		61 [1]	0.30 [1]		
h-AlN	*P*63/mmc [42]	~ 113 [1], 136 [47], 114 [46], 138 [20]	~ 0.46 [1], 0.37 [47] 0.46 [46]	16.2, 15.9 [47] 13.1, 15.0 [20]	0.22, 0.27 [47] 0.17, 0.23 [20]
h-AlP		64 [1], 58 [115]	0.31 [1], 0.47 [115]		
h-AlAs		47 [1], 46 [115]	0.34 [1], 036[115]		
h-AlSb		33 [1], 35 [16]	0.32 [1], 0.37 [16]		
h-AlBi		27 [1]	0.37 [1]		
h-GaN	*P*-6*m*2 [6]	~ 106 [1], 109 [60] 110 [16], 90 [61]	~ 0.45 [1], 0.43 [60] 0.48 [16], 0.47 [61]	11.9, 12.1 [60] 10.1, 10.8 [20]	0.18, 0.22 [60] 0.17, 0.19 [20]
h-GaP		59 [1], 55 [115]	0.31 [1], 0.34 [115]		
h-GaAs		44 [1], 44 [115]	0.33 [1], 0.34 [115]		
h-GaSb		28 [1], 32 [115]	0.39 [1], 0.35 [115]		
h-InN	*P*-6*m*2 [68]	~ 62 [1], 62.0 [70] 87.8, 90.1 [20]	0.58 [1], 0.59 [70] 0.59 [16]	~ 8.0 [70] 8.4, 9.4 [20]	0.18, 0.21 [70] 0.16, 0.20 [20]
h-InP		40 [1], 37 [115]	0.38 [1], 0.42 [115]		
h-InAs		31 [1], 30 [115]	0.38 [1], 0.42 [115]		
h-InSb		25 [1], 22 [115]	0.33 [1], 0.44 [115]		
h-InBi		20 [1]	0.41 [1]		

(continued)

Table 7.2 (continued)

Monolayer	Space group	Young's modulus E (N/m)	Poisson ratio, ν	Ultimate strength σ_{ac}, σ_{zz} (N/m)	Ultimate strain ε_{ac}, ε_{zz}
h-TlN	*P6/mmm* [81]	~55 [1], 34.5 [78]	1.37 [1], 0.689 [78]	5.8, 5.7 [78]	0.17, 0.21 [78]
h-TlP		14 [1]	0.79 [1]		
h-TlAs		17 [1]	0.58 [1]		
h-TlSb		Unstable	0.18 [1]		
h-TlBi		8.1 [1]	0.73 [1]		

221.3 N/m and $E_{zz} = 224.9$ N/m and ultimate strength and strain values of $\sigma_{ac} = 29$ N/m and $\varepsilon_{ac} = 0.193$ together with $\sigma_{zz} = 33.3$ N/m and $\varepsilon_{zz} = 0.198$ [22].

In addition, the uniaxial tensile stress–strain curve of a precracked multilayer sheet was determined up to the fracture point to determine the stiffness and strength of the distorted material. MD simulations with Tersoff and Stillinger–Weber potentials offered insight into crack propagation in precracked boron-nitride monolayers [22, 23]. The fracture toughness or critical stress intensity factor (CSIF) of mode-I fracture at room temperature was determined by creating a central crack perpendicular to the tensile loading direction. An analytical 2D model describes localized bond breaking in perfect monolayers and failure of defective monolayers by the global energy balance concept, in agreement with results estimated by the standard empirical approach using the fracture stress and the initial crack length [24]. The calculated fracture toughness of boronitrene of ~ 2.5 MPa $m^{1/2}$ is lower than that measured via high-resolution transmission electron microscopy (HRTEM) on multilayer boronitrenes of 5.5 MPa $m^{1/2}$ [25]. The measured fracture toughness is greater than the toughness of 3.3 MPa $m^{1/2}$ derived for graphene and thus may suggest a greater toughness of multilayer assemblies [24].

7.1.4 Physics and Applications

With their strong bonding, mechanical stability, large band gap, low dielectric constant, and excellent chemical inertness, planar h-BN monolayers seem to be an ideal insulator for replacing SiO_2 in silicon technology. In addition to being an ultimately thin insulator, h-BN is particularly attractive as a barrier or encapsulant [5, 12, 26, 27]. Based on its chemical inertness, it is resistant to oxidation at elevated temperatures and against corrosion. Owing to its high thermal conductivity, it is a useful material for electronic devices and furthermore, it can be used as an ultraviolet light emitter in optoelectronics. Moreover, h-BN is a useful 2D material for far-ultraviolet (FUV) device applications and it is of interest as a transparent and flexible material in electronics. Of increasing attraction are recently reviewed applications of h-BN-based in-plane and vertically stacked heterostructures in electronic and optoelectronic devices [28]. This review discusses bottom-up and top-down synthesis strategies, including the novel properties of these heterostructures. Finally, the main challenges and future research directions in this field are considered. An important problem is quality control of as-grown monolayers on the initial substrate before transfer that rarely has been achieved [29].

A topical review concentrating on the topics of growth, electrical measurements, and applications in optics and photonics describes primarily experimental work [30]. In addition to the hexagonal structure, first-principles calculations provide the structural, electronic, and optical properties of an orthorhombic phase of 2D BN and the five group IIIA–VA monolayers BP, BAs, AlN, AlP, and GaN [31].

A recent article discussed the use of h-BN in environmental applications [32]. The physico-chemical, mechanical, and biological properties qualify h-BN monolayers for several environmental and energy applications, including photocatalysis, membrane separation, disinfection, environmental sensing, energy conversion, and energy storage. The authors emphasize exceptional environmental capabilities and relevant topics such as toxicity and recyclability. Similarly, a comprehensive article concentrates on the development of green energy systems based on 2D h-BN [33]. This includes physical and chemical modification strategies to convert the insulating material into a conductive material suitable for energy conversion. Moreover, this work reviews the progress made in using h-BN as an electrocatalyst in energy applications such as the hydrogen evolution reaction (HER), oxygen evolution reaction (OER), oxygen reduction reaction (ORR), carbon dioxide reduction reaction (CO_2RR), and nitrogen reduction reaction (NRR).

7.1.5 Synthesis

Current electronic and photonic device concepts still rely primarily on h-BN exfoliated from small bulk crystallites. In an effort to achieve reproducible, tunable, and scalable growth of monolayers and few-layer systems, different approaches, such as CVD, were investigated for precise control of the flow of the BN precursor during growth [34]. An approach focused on obtaining a systematic understanding of Pt-catalyzed crystal growth via an integrated CVD process [35]. Another one, based on high-temperature plasma-assisted molecular beam epitaxy (MBE) on graphite substrates at ~ 1390 °C, achieved controlled monolayer and few-layer growth [36]. Wafer-scale CVD growth of single-crystal monolayers with a lateral size of up to 300 μm could be realized by a high initial nucleation density and low-pressure CVD on polycrystalline copper samples [37].

The possibility of direct probing the electronic properties has been developed for as-grown h-BN flakes, synthesized by low-pressure CVD (LPCVD) (see Fig. 7.4) [29]. Solid NH_3-BH_3 served as precursor in the preferentially surface reaction limited growth process. The thickness of the deposited h-BN sheets was 5–30 nm. A Cu-foil enclosure was used as substrate (~ 1030 °C), to quantitatively analyze the electrical resistivity and inherent Fermi level of the monolayer by employing the AFM PeakForce Tunneling (PF-TUNA) mode and Kevin probe force microscopy (KPFM). Furthermore, direct visualization of the resistance to high-temperature oxidation was achieved. The quantitative analysis suggested the suitability of h-BN as an atomically thin high-quality insulator for 2D electronics, and as a stable high-temperature antioxidation layer.

In a recent breakthrough, epitaxial growth of 100 cm^2 single crystals of h-BN on copper was accomplished, demonstrating unidirectional alignment and seamless stitching on an annealed industrial copper foil [38]. Moreover, single-crystal h-BN, without undesirable grain boundaries, could be grown on a 2-inch Cu(111)/sapphire substrate and successfully detached and transferred to a wafer [39]. Self-aligned

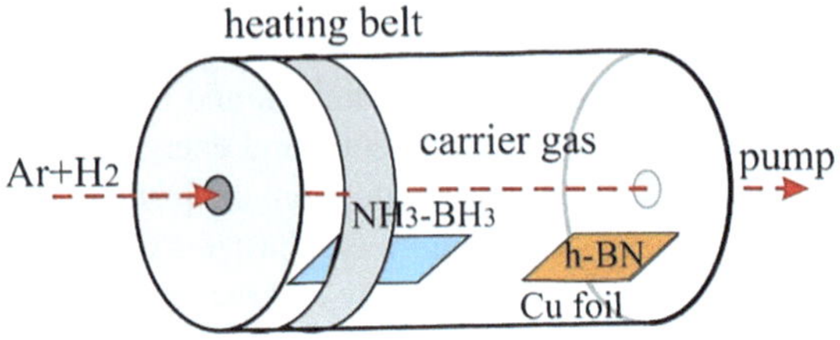

Fig. 7.4 Setup for the growth of 2D h-BN via LPCVD, using solid NH_3–BH_3 as the precursor and a Cu foil as substrate

stitching growth of centimeter-scale quasi-single-crystalline h-BN could be achieved through synergistic CVD growth kinetics and liquid Cu rheological kinetics control [40]. The flowing liquid Cu surface promotes rotation of floating triangular h-BN domains and induces uniform self-alignment upon merging at a critical temperature of 1105 °C.

To perform controlled growth, the h-BN/graphene interface is suggested to enable scalable unidirectional epitaxial growth of h-BN with single-domain crystals on an underlying graphene lattice [41]. Progress in synthesis and applications of h-BN–graphene heterostructures is covered in this work. The discussion demonstrates the ongoing search for practical fabrication methods of high-quality h-BN monolayers.

7.2 AlN Monolayer (Aluminitrene)

7.2.1 Chemical Properties and Bonding

The planar free-standing monolayer of hexagonal AlN is of the same bonding type as that described for the h-BN monolayers. The promoted valence electrons of Al ($3sp^2$) and N ($2sp^4$) form hybrid orbitals with three planar σ bonds with threefold coordination in the ground state. The π bonds of the remaining p_z orbitals of the Al and N atoms, with a lone electron pair in the latter case, are directed out of plane and preserve the plane geometry of the lattice by delocalized 2D bonding. Owing to the large difference in the electronegativities of Al (1.61) and N (3.04), the Al–N bond is highly polar, with a Bader charge donation of $\delta q = 2.26$ e from the Al atom to the N atom [6]. DFT-PBE calculations of the cohesive energy of the 2D network of $E_c = -5.36$ eV/atom and of the 3D crystal of – 6.02 eV/atom point to a local minimum in the Born–Oppenheimer surface for the monolayer [42]. Other theoretical studies based on this approach report slightly smaller binding energies of h-AlN monolayers of – 5.05 eV/atom [43] and – 5.08 eV/atom [1], whereas van der Waals-corrected DFT calculations resulted in an obviously too high cohesion of – 7.40 eV/atom [44].

7.2.2 2D Structures

The lattice constant and bond length of h-AlN are a = 0.313 nm (0.317 nm [43]) and d = 0.181 nm [1, 44]. Note that one study gives a marginal buckling effect of Δz = 0.0001 nm [1]. These structural parameters are in excellent agreement with the comprehensive data collection published for 2D-compound semiconductors [2]. The dynamic stability of the material was confirmed by ab initio phonon calculations at 0 K, and the thermal stability was confirmed by ab initio finite temperature equilibrium and nonequilibrium MD simulations [45]. The strong Al–N bonds are responsible for the estimated high melting point of 3500–4000 K for the h-AlN monolayer. The mean van der Waals diameter of d_w = 0.339 nm suggests a quite similar monolayer thickness as that of boronitrene [11]. Remarkably, the interlayer spacing of d_s = 0.213 nm for bilayers is much smaller than the van der Waals diameter, as discussed in more detail below [42]. DFT-PBE calculations yield an indirect band gap of E_g = 2.9 eV for the monolayer, whereas the DFT-HSE06 functional provides an indirect band gap of 4.06 eV [42]. The space group of h-AlN is $P6_3/mmc$ [42].

An interesting theoretical study covers the structures of 1 L, 2 L, 3 L, 4 L, and multilayer assemblies of h-AlN monolayers [42]. The hexagonal phase is stable for up to ~ 10 layers, and subsequently loses its indirect gap character and transforms to the direct bulk gap. The strongest stacking order of bilayers is AA′ (i.e., hexagons on top of each other with Al above N); that of trilayers is AA′A, and that of 4 L is AA′AA′. The small distances of the bilayer and the bulk interlayer separations of d_s = 0.213 nm and 0.208 nm, respectively, point to a substantial contribution of stronger chemical bonding between the layers, such as anion–cation interactions. This involves, for example, the interaction of the p_z orbital of the N atom of one layer with the p_z orbital of the Al atom of the neighboring layer. This conclusion is supported by the large interlayer interaction energy of E_{int} = – 125 meV/atom based on hybrid DFT-HSE06 functionals with the inclusion of van der Waals interactions [42]. A similar interlayer spacing of d_s = 0.218 nm, however, an even larger interaction energy of E_{int} = – 292 meV/atom was determined for AA′ stacking of bilayers by an independent study [43]. This much larger interlayer interaction energy was obtained with the DFT-PBE approach, including a dispersion correction, by employing the DFT-D2 method of Grimme. Note that few-layer assemblies are of special interest for band gap engineering.

7.2.3 Mechanics

The most extensive DFT-PBE calculations on the mechanical behavior of h-AlN reveal a minor anisotropy in Young's modulus, at E_{ac} = 113 N/m and E_{zz} = 112 N/m, and surprisingly large nearly identical Poisson ratios of ν_{ac} = 0.463 and ν_{zz} = 0.461 [1]. A similar in-plane stiffness of 114 N/m and Poisson ratio of 0.46 have been reported elsewhere [46]. Another study presented a much larger in-plane Young's modulus of 136 N/m, which is roughly half the stiffness of h-BN and a smaller Poisson ratio of 0.366 [47]. In the latter calculations, elastic constants up to the fifth order were taken into account, allowing accurate calculations of nonlinear properties such as strength under strains larger than $\varepsilon = 0.12$.

The reported ultimate strengths in the armchair and zigzag directions are $\sigma_{ac} = 16.2$ N/m and $\sigma_{zz} = 15.9$ N/m, taken from the maximum of the stress–strain curve, and the corresponding ultimate strains are $\varepsilon_{ac} = 0.22$ and $\varepsilon_{zz} = 0.27$ [47]. In a MD study, using a Tersoff and Tersoff-like potential, large Young's moduli of $E_{ac} = 137.9$ N/m and $E_{zz} = 138.1$ N/m, ultimate stresses of $\sigma_{ac} = 13.1$ N/m and $\sigma_{zz} = 15$ N/m, and low ultimate strains of $\varepsilon_{ac} = 0.166$ and $\varepsilon_{zz} = 0.228$ were derived [20]. The agreement of the reported linear and nonlinear mechanical properties is acceptable and fits into the expected overall mechanical behavior of the other binary IIIA–VA monolayers.

7.2.4 Physics and Applications

As a wide band gap material with a direct band gap of ~ 6.45 eV, h-AlN exhibits strong resonances of absorption and second harmonic spectra in the ultraviolet region [48, 49]. Strong peaks of second-order nonlinear susceptibility appear from the blue to deep ultraviolet regions, with a large peak value that is one to two orders of magnitude larger than that of other nonlinear materials in this spectral range. This finding points to promising applications in nonlinear imaging, spectroscopy, and nonlinear photonics in the UV range [48]. In addition, AlN may be used in ultraviolet light emitting diodes (LEDs) and laser diodes [49]. A theoretical analysis revealed that the layered hexagonal phase of AlN is energetically favored up to approximately 12 monolayers [49]. The calculated band gaps vary from $E_g = 2.91$ eV (DFT-PBE) and 4.0 eV (HSE) for the monolayer, to ~ 3.5 eV for 2–4 monolayer assemblies with an indirect gap that ultimately transforms to the direct gap of the bulk at a thickness of approximately 10–12 layers [49]. Monolayer h-AlN exhibits a single absorption peak at ~ 6 eV that is related to important primary and secondary transitions at the Γ, M, and K points in the spectral range of 5.6–7.2 eV.

Single-photon emission (SPE) has attracted enormous amounts of attention in quantum information and communication technologies. It seems that 2D semiconductors such as h-AlN, due to their large band gap, open structure, and low dielectric screening can overcome some shortcomings of traditional bulk systems, such as photon extraction and long radiative lifetimes. Based on a systematic investigation of suitable point defects, the – 1 charged Al vacancy ($V_{Al}-$) and the + 1 charged nitrogen antisite ($N_{Al}+$) are predicted to achieve SPE with zero-phonon lines of 0.77 eV and 1.40 eV, respectively [50]. Naturally occurring and engineered defects make important contributions to material properties and performance. Typical point defects in AlN, GaN, and InN monolayers, including vacancies and antisites, are accessible by first-principles calculations [51]. These defects significantly change the energy band gaps and magnetic properties of group–IIIA nitrides. Interestingly, the semiconducting nitride monolayers are metalized by nitrogen vacancies with the lowest formation energy.

Of general importance are the tuning possibilities of the physical properties of 2D materials by tensile and compressive strain, chemical functionalization, and by changing the number of layers in few-layer arrangements. A DFT study addresses possible methods for tuning electronic and magnetic properties by manipulating the surface and by applying strain [52]. Moreover, potential applications of these materials are discussed, as, for example, gas sensing, with an emphasis on growth strategies involving single-layer and few-layer sheets on different substrates [52]. Detailed DFT calculations on one- to five-layer h-AlN revealed how the electronic and transport properties can be tuned by layer-number modification and strain engineering [44]. In a state-of-the-art first-principles investigation, the much lower thermal conductivity of h-AlN than that of graphene is explained by the existence of isolated electron pairs, which increase the orbit-driven anharmonicity and thus reduce the thermal conductivity [53].

7.2.5 Synthesis

Layered h-AlN assemblies do not exist in nature; therefore, the fabrication of high-quality 2D h-AlN sheets is still far from satisfactory. Single layers of h-AlN are difficult to isolate owing to the strong interlayer interactions. Only when the interaction with the substrate employed for synthesis is weak does the adsorbed h-AlN monolayer preserve the integrity of its free-standing isolated form. Since a detailed description of the various methods currently employed for the development of a feasible chemical synthesis process is available, only a few examples are presented here [54].

The synthesis of the hexagonal form of 2D AlN is possible by plasma-assisted MBE on cleaned Ag(111) single crystals from submonolayer to twelve monolayers [55]. Evidence for the growth of a graphene-like structure came from reflection high-energy electron diffraction (RHEED), scanning tunneling microscopy (STM), X-ray photoelectron spectroscopy (XPS), and ultraviolet photoelectron spectroscopy (UPS).

The development of attractive gate dielectric stacks for well-studied 2D transition-metal dichalcogenide (TMDC) transistors is challenging. While the growth of h-BN layers requires 1000 °C, which is not compatible with complementary metal-oxide semiconductor (CMOS) fabrication, atomic layer deposition (ALD) of high-k 2D h-AlN insulators is an alternative. Successful growth of h-AlN monolayers on TMDC monolayers via van der Waals epitaxy by ALD has been realized [56]. Between 150 and 300 °C growth occurred layer-by-layer in a hexagonal phase, as analyzed by HRTEM and synchrotron-based grazing-incidence wide-angle X-ray scattering (GIWAXS) [56].

7.3 GaN Monolayer (Gallinitrene)

7.3.1 Chemistry and 2D Bonding

Going from the bulk to the 2D structure of GaN, hybridization changes from 3D sp^3 to planar sp^2 hybrids, which increases the strength of covalent bonding in the plane. Bonding of the planar monolayer of h-GaN is primarily based on the promoted valence electrons of Ga ($4sp^2$) and N ($2sp^4$) forming sp^2 hybrid orbitals, which form the already discussed three polar σ bonds arranged in hexagons. The p_z orbitals assemble with the closest neighbor p_z orbitals in delocalized π bonds, and despite their smaller contribution to bonding, they still preserve the planar geometry, as observed for h-AlN monolayers. The large electronegativity difference between 1.81 (Ga) and 3.04 (N) leads to a charge transfer of $\delta q = 1.38$ e from each Ga atom to each N atom and consequently to strongly polarized Ga–N bonds [6]. The negative cohesive energies calculated by the DFT-PBE approach indicate energetic stability, with values of $E_c = -4.15$ eV/atom [46], – 4.07 eV/atom [57], and – 3.80 eV/atom [1]. Note the large reduction in the binding energy compared with that of h-AlN despite a similar atom size and difference in the electronegativities.

7.3.2 2D Structures

The lattice constant and bond length of h-GaN are $a = 0.321$ nm and $d = 0.185$ nm, respectively [1, 46]. These values are in good agreement with the comprehensive data collection for 2D compound semiconductors, which gives $a = 0.325$ nm and $d = 0.188$ nm [2]. Similar to that of h-AlN monolayers, the dynamic stability of the 2D network at $T = 0$ K has been confirmed by ab initio phonon calculations, and the thermal stability was confirmed by ab initio finite temperature MD simulations [46]. Depending on the approximation used, the band gap of the monolayer varies strongly between $E_g = 2.16$ eV (DFT-PBE), 3.42 eV (HSE), and 4.55 eV (G_0W_0) [46]. Surprisingly, the mean van der Waals diameter of $d_w = 0.342$ nm is nearly identical to that of h-AlN [11]. In fact, the lower binding energy and longer bond length of h-GaN cannot be explained by a smaller orbital overlap due to a larger atomic size of Ga, however, charge transfer and thus ionicity is significantly smaller. The space group of h-GaN is *P*-6*m*2 [6].

The optimum stacking of AA′AA′... produces stable geometries with large interlayer interaction energies of $E_{int} = -142$ meV/atom, – 200 meV/atom, and – 327 meV/atom for the plane h-GaN bilayer, trilayer, and multilayer assemblies, respectively [46]. The authors argue that interlayer interactions are dominated by van der Waals forces for the interlayer separations of $d_s = 0.252$ nm and 0.249 nm for the bilayer and trilayer, respectively. Only in multilayers with an interlayer spacing of 0.244 nm do they expect a slight ionic contribution from anion–cation interactions. A previous study based on DFT calculations, included van der Waals interactions evaluated by the vdW-DFT-D2 method of Grimme. These calculations delivered comparable interlayer separations of d_s = 0.247 nm for bilayers and similarly, d_s = 0.249 nm for the trilayer [57]. Contrary to the former conclusion, these authors consider the effective electrostatic interaction of Ga (N) and N (Ga) pairs between the layers as the dominant contribution to the strong interlayer interaction, where the gallium cation attracts the nitrogen anion and repels the other cations. The calculated charge distributions support this assumption. For the most energetically stable AA′ stacking of the bilayer, the cohesive energy of E_c = – 4.34 eV is comparable to the values presented above for the monolayer [57].

In a detailed study of the structure of flat bilayers, the small interlayer separation of d_s = 0.25 nm was confirmed [58, 59]. Interestingly, in addition to the planar configuration, a more stable buckled bilayer could be identified with further reduced interlayer separation from d_s = 0.25 nm to approximately d_s = 0.22–0.23 nm between two buckled layers, each with a buckling height of Δz = 0.03 nm [58]. Under tensile loading, the structures tend to become flat, while under biaxial compression, buckled structures are favored. The buckled bilayer system shows the onset of sp^3 bonding, where the atoms of the two inner sublayers prefer fourfold coordination and the atoms of the two outer sublayers stay threefold. These findings support the gradual transition from a sp^2-bonded planar monolayer, via mixed sp^2–sp^3 hybrid-bonded few-layer assemblies, to the sp^3 structure of bulk wurzite GaN, which is the structure of the global minimum among all allotropes. The calculated cohesive energies increase from E_c = – 4.19 eV/atom for the monolayer to – 4.34 eV/atom for the AA′ stacked bilayer, to – 4.39 eV/atom for the AA′A stacked trilayer, and finally to – 4.87 eV/atom for the wz-GaN bulk crystal [57]. Of interest in few-layer assemblies is the change in the band gap due to the application of uniaxial or biaxial stresses since under compressive stress, an indirect-to-direct band gap transition can be induced [58].

7.3.3 *Mechanics*

DFT-PBE calculations revealed a minor anisotropy in Young's modulus (E_{ac} = 106.6 N/m and E_{zz} = 105.9 N/m) and in the Poisson ratio (ν_{ac} = 0.449 and ν_{zz} = 0.445) for h-GaN [1]. Similar Young's moduli and Poisson's ratios of E = 110 N/m and ν = 0.43 [46], E = 109 N/m and ν = 0.431 [60], as well as E = 110 N/m and ν = 0.48 [16] have been obtained by other researchers. Furthermore, the results of a spring-based molecular mechanics (MM) study using a Stillinger–Weber potential parameterized for h-GaN to describe the interatomic interactions deviate slightly at E = 240 GPa (90 N/m) and ν = 0.47 [61]. The 2D Young's moduli in brackets calculated here are based on a thickness of t = 0.374 nm proposed in the publication, which is marginally greater than the mean van der Waals diameter of d_W = 0.37 nm [61]. The large Poisson ratios of almost ν ≈ 0.43–0.5, predicted by all the authors, characterize an almost incompressible nanomaterial [61].

The ultimate strength and strain, extracted from the maximum of the stress–strain curve obtained by the MM study, are $\sigma = 32.8$ GPa (12.3 N/m) and $\varepsilon = 0.158$, respectively, without considering chirality effects [61]. According to other studies, nonlinear deformation and failure are slightly anisotropic, with ultimate strengths and strains in the armchair direction of $\sigma_{ac} = 11.9$ N/m and $\varepsilon_{ac} = 0.18$ and in the zigzag direction of $\sigma_{zz} = 12.1$ N/m and $\varepsilon_{zz} = 0.22$ [60]. It is important to include the $3d^{10}$ core electrons of Ga in the DFT-PBE calculations. Notably, MD simulations, in which Tersoff and Tersoff-like potentials were used to model interatomic interactions, yielded much smaller stiffnesses, with Young's moduli of only $E_{ac} = 83.3$ N/m and $E_{zz} = 85.1$ N/m, however, similar ultimate stresses of $\sigma_{ac} = 10.1$ N/m and $\sigma_{zz} = 10.8$ N/m, and comparable ultimate strains of $\varepsilon_{ac} = 0.165$ and $\varepsilon_{zz} = 0.188$ [20]. Note that the small ultimate strains that h-GaN can sustain before bond rupture is comparable to that of h-BN.

7.3.4 Physics and Applications

Monolayer and few-layer h-GaN sheets are promising materials for various applications in electronic devices such as transistors, optoelectronic devices such as light emission diodes (LEDs), detectors, and sensors such as gas-sensing devices, piezoelectric devices, energy storage and conversion devices such as batteries, and catalytic materials [48].

The indirect band gap of the monolayer becomes a direct gap in multilayer h-GaN, analogous to h-AlN [57, 58]. The fundamental band gap of vertical structures varies with the stacking sequence, the thickness of the constituent sheets, and the number of sheets in the stack. This allows tuning of diverse properties and thus opens the door to potential applications in flexible nanoelectronics and optoelectronics. With an increasing number of layers, the buckling of the individual layer increases, and the structure approaches the bulk structure. These changes in the intrinsic properties, e.g., lowering of the thermal conductivity, increase of the UV luminescence, and improvement of the mechanical performance, are desirable. Strain engineering provides an additional tool for tuning the atomic and electronic structure of such stacks, as well as that of heterostructures fabricated from different monolayers, such as h-GaN/h-AlN [62–64].

In addition, wide band gap materials are suitable candidates for the fabrication of energy-efficient deep-UV light-emitting diodes, for biomedical applications, sustainable water/air purification, and photovoltaics [60]. Moreover, h-GaN is applicable as multilayer antireflection and protection coating of infrared windows. A remarkable development is the realization of an efficient blue LED, which can provide bright and energy-saving white light sources.

Surface functionalization of a monolayer, for example, by hydrogenation or fluorination, transforms planar sp^2 bonds into stable buckled layers with sp^3-type bonds and generates a direct band gap [64, 65]. This chemical tailoring process blueshifts the optical activity to the ultraviolet region and simultaneously eliminates the existing dynamical instability against compressive strain. Tensile biaxial strain causes a redshift of the optical spectrum to the visible region. Furthermore, surface modifications generate a direct band gap for the monolayers. At the dilution limit of hydrogen coverage, it is possible to systematically tune the band gap. Hydrogen adsorption can induce a magnetic moment due to *s-p* interactions and band structure effects [65, 66].

7.3.5 Synthesis

A recent review describes the enormous effort made in recent years to prepare monolayers of h-GaN and covers in detail the different experimental approaches already applied [54]. For h-GaN, it is necessary to synthesize monolayers and few-layer assemblies by complex chemical reactions on suitable substrates since no layered bulk phase exists in nature. The strong interlayer interactions in few-layer systems mentioned above and the tendency to form energetically preferred buckled layers seem to be the main reasons why the synthesis of single-layer h-GaN is still in its infancy, far away from industrial production standards.

Here, we consider only the successful Ga_2O_3 template method that employs a nitridation reaction. Improvements in the template method allowed the controlled deposition of subnanometer-thick 2D GaN layers by plasma-enhanced CVD using environmentally friendly nitrogen instead of ammonia to nitrogenize the Ga_2O_3 template layer (see Fig. 7.5) [66]. In addition to nitridation, the N_2 plasma etched the ~ 3.0 nm thick Ga_2O_3 surface layer to a thickness of ~ 0.8 nm, corresponding to roughly two layers of GaN. AFM measurements allowed a thickness evaluation of the prepared film, and XPS analysis confirmed the transformation of Ga_2O_3 into GaN. Moreover, HRTEM measurements revealed the atomic structure of the suspended GaN film, and UV–VIS absorption spectra indicated that the deposited layer had a wide band gap of $E_g = \sim 4.9$ eV.

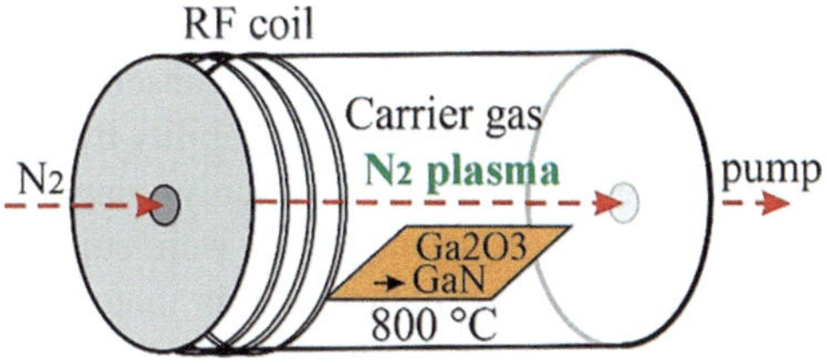

Fig. 7.5 Setup for the growth of 2D GaN via N_2-plasma CVD, using Ga_2O_3 as the precursor. Adapted with permission from ref. [2], Copyright 2016, AIP Publishing

7.4 InN Monolayer (Indanthrene)

7.4.1 Chemistry and 2D Bonding

Like in the other group–IIIA nitrides, the hybrid orbitals resulting from the electronic configuration of In atoms (Kr|$4d^{10}5s^2p_x^{\ 1}$) and N atoms ($1s^22s^2p_x^{\ 2}p_y^{\ 1}$) form a planar hexagonal network with three strong σ bonds. The In p_z and N p_z orbitals form π bonds with their nearest neighbors and stabilize the planar geometry of the monolayer despite the larger In atom. Including the In 4*d* electrons delivers the correct lattice constant of h-InN. The difference in the electronegativities of the In (1.78) and N atoms (3.04) is responsible for a significant ionicity of the In–N bond, which is comparable to that of h-GaN, with a Bader charge transfer of $\delta q = 1.3$ e [6]. The predicted contours of the electronic density distribution indicate strong ionic bonding, clearly identifying indium as a cationic center and nitrogen as an anionic center [67]. Importantly, the authors interpret the nature of this bonding as ionic, with partial contributions from covalent bonding and metallic behavior. The published cohesive energies of $E_{co} = -4.34$ eV/atom (LDA) [68], – 3.60 eV/atom (DFT-PBE) [67], and – 3.03 eV/atom (DFT-PBE) [1] deviate substantially. The more realistic lower values signify further weakening of 2D bonding in h-InN due to the slightly increasing size of In atoms and changes in the electronic structure compared with GaN.

7.4.2 2D Structures

The DFT-based lattice constant a = 0.363 nm and bond length d = 0.210 nm [68] (PBE) are in reasonable agreement with other calculations at a = 0.352 nm and d = 0.203 nm [69] (LDA). Interestingly, the bond length is substantially shorter than that of wurzite InN (0.221 nm), despite a weaker cohesion [68]. The mean van der Waals diameter of $d_w = 0.38$ nm [11] points to only a slight increase in the monolayer thickness within these group of IIIA–VA compounds. The mechanical, dynamical, and thermal stability up to 600 K have been proven by mechanical loading simulations, the kinetic stability by the absence of imaginary frequencies in the phonon band structure, and the thermal stability by using ab initio MD simulations [70]. The stable phase was confirmed in a study comparing six possible monolayer structures of InN [67]. These different phases were studied to better understand the metal–insulator–semiconductor nature of this material. The calculated indirect fundamental band gaps of the semiconducting h-InN monolayer are 0.43 eV (DFT-PBE) [67] and 1.48 eV (DFT-HSE06) [68] and the phase changes under biaxial compressive strain to a direct band gap. Contrary to these results, DFT-LDA calculations revealed that the Γ-Γ direct gap of $E_g = 0.80$ eV almost coincides with the K-Γ indirect gap of 0.81 eV [69]. The space group of the plane hexagonal structure agrees with that of the former group IIIA–VA compounds, namely, *P*-6*m*2 [68]. Note that other authors assigned the space group *P*6/*mmm* to the h-InN monolayer [67].

To determine the optimized structure and interaction energy of bilayers, the LDA approach was employed [69]. For the most stable AA′ stacking sequence, the interlayer separation is d_s = 0.240 nm, and the interlayer interaction energy is E_{in} = 78.8 meV/A^2. The medium interlayer interaction not only occurs due to van der Waals attraction but also originates from the overlap of the p_z orbitals of In and N atoms, located in separate layers, contributing covalent-type interlayer interactions.

For the optimized structures of bilayer, trilayer, and four-layer assemblies, the interlayer distances change slightly from d_s = 0.267 nm to d_s = 0.266/0.278 nm and d_s = 0.256/0.258/0.248 nm, respectively, for different stacking orders [68]. The fundamental band gap decreases steadily from E_g = 1.48 eV, 1.08 eV, 0.95 eV, to 0.40 eV for the 1 L, 2 L, 3 L, and 4 L h-InN assemblies, respectively [68]. These data clearly demonstrate the strong dependence of optoelectronic properties on the number of layers in few-layer assemblies and the decrease in the band gap with increasing thickness of the sheet. Furthermore, as described above, an indirect-to-direct band gap transition takes place by the transition from monolayer to multilayer assemblies.

7.4.3 Mechanics

First-principles calculations yielded the slightly anisotropic Young's moduli E_{ac} = 61.6 N/m and E_{zz} = 62.2 N/m and as Poisson's ratios ν_{ac} = 0.578 and ν_{zz} = 0.583 [1]. Another study reported a consistent isotropic Young's modulus of 62 N/m and Poisson ratio of 0.586 [70]. In addition, this work provided a critical strength of ~ 8 N/m and ultimate strains of ε_{ac} = 0.18 and ε_{zz} = 0.21. Importantly, besides the standard valence electrons of N ($2s^2p^3$) the $4d$ electrons of the In atom with the electronic configuration $4d^{10}5s^2p^1$ were included in these calculations. These results confirmed early calculations of Young's modulus of 67 N/m and a comparable Poisson ratio of 0.59 [16].

MD simulations, using the Tersoff and Tersoff-like potentials, resulted in a greater stiffness and smaller anisotropy between the armchair and the zigzag directions, specifically for Young's moduli E_{ac} = 87.8 N/m and E_{zz} = 90.1 N/m, for the maximal stresses σ_{ac} = 8.4 N/m and σ_{zz} = 9.4 N/m, and the maximal strains ε_{ac} = 0.155 and ε_{zz} = 0.195 [20]. Despite the fact that these values fulfill quite well the $\sigma \approx E/10$ rule, employing the layer thickness of t = 0.335 nm suggested by the authors, a surprisingly large Young's modulus of 269 GPa (90 N/m) in 3D (and 2D) units was obtained. This discussion reveals obvious differences in the calculated stiffnesses of the DFT calculations and MD simulations.

7.4.4 Physics and Applications

A first-principles study on the detection of noxious gases investigated the adsorption structures, adsorption energies, electron transfers, band structures, density of states, and desorption properties of h-InN [71]. Introducing Cu significantly enhances the strong adsorption of SO_2, SF_6, SOF_2, and SO_2F_2, including chemical interactions on pristine h-InN monolayers. A special study of the adsorption conditions and sensing performance of Rh-decorated InN monolayers upon the addition of dissolved H_2 and C_2H_2 species in transformer oil explored the potential of these materials for evaluating the working conditions of transformers [72]. An applied electric field supported the work function analysis, indicating that Rh-InN provides strong potential for field-effect transistor (FET) sensor applications. An investigation of toxic CH_2O revealed physical adsorption on h-BN and h-GaN monolayers but chemisorption on h-AlN and h-InN surfaces, where the latter exhibited excellent adsorption energies and charge transfers [73].

According to DFT and many-body perturbation theory, the optical gap opens dramatically from bulk to 2D InN systems owing to strong quantum confinement and reduced screening [69]. Based on the variation of the band gap with the number of layers, 2D InN sheets are a promising material for optoelectronic devices in the visible and near-infrared spectral ranges. With an increasing number of layers, the direct band gap of multilayers decreases, and thus, modifying the layered structure allows band-gap tuning. For monolayer h-InN the band gap decreases under biaxial tensile strain and remains indirect, while it converts from indirect to direct under biaxial compressive strain [68].

Based on the narrower band gap, smaller effective mass, and greater electron mobility of h-InN compared to those of the lighter group–IIIA nitrides, interest has concentrated on potential applications such as field-effect transistors, solar cells, photodetectors, and thermoelectrics [67].

DFT calculations elucidated the electronic, photocatalytic, and optical properties of fully hydrogenated $InNH_2$ monolayers [74]. Functionalization leads to buckling by rehybridization of plane sp^2 hybrids to sp^3-like hybrids and thus introduces significant changes in the intrinsic physicochemical properties. As shown theoretically, in fully hydrogenated HInNH monolayers, the large indirect band gap of $E_g = 2.59$ eV is tunable, and indirect-to-direct band gap transitions are possible under biaxial strain and external electric fields [74]. The analysis of band edge alignment indicated photocatalytic activity and suitability for water splitting. The monolayers generated by full chemical functionalization of InN with F and Cl atoms are dynamically stable in the buckled configuration, with a buckling effect varying between $\Delta z = 0.460$ nm (FInNF) and 0.516 nm (ClInNCl) [75]. The monolayers are semiconductors with a wide direct band gap, e.g., 1.93 eV (PBE) and 3.27 eV (HSE06) for FInNF, and absorb light in a wide region, especially in the visible spectrum.

7.4.5 *Synthesis*

Bilayers of InN are formed between graphene and SiC during an intercalation process employing metal–organic chemical vapor deposition (MOCVD). These experiments followed the successful synthesis of 2D GaN sheets by such an intercalation technique [76]. Conductive AFM and atomic resolution TEM were used to analyze the geometrical structure. STM measurements operating in scanning tunneling spectroscopy (STS) mode provided an experimental band gap of intercalated InN of $E_g = 2.0$ eV, which agrees with the calculated band gap of 1.99 eV for a buckled InN bilayer.

The first successes in the synthesis of nanoscale InN sheets made their use as gas sensors possible, despite remaining serious problems in depositing high-quality h-InN monolayers. The inevitable defects such as N vacancies incorporated in such sheets due to their low formation energy deteriorate the chemical stability and electronic performance. A first-principles study proposed the use of NO molecules to repair the N vacancies of h-InN via a disproportionation reaction, which is energetically favorable due to the small energy barrier and large energy decrease in the reaction [77]. In particular, decorated h-InN monolayers are promising candidates for sensitive gas detection.

7.5 Tl Monolayer (Thalinitrene)

7.5.1 *Chemistry and 2D Bonding*

In most reports, the bonding configuration proposed for Tl monolayers with the promoted valence electrons of Tl ($5sp^2$) and N ($2sp^4$) in the outer valence shell is a hexagonal structure with the typical sp^2 hybridization and π bonding as discussed for the other group–IIIA nitrides. In fact, there is only a marginal increase in the van der Waals radius from the In atom to the Tl atom [11]. The planar hexagonal structure and its mechanical properties have been elucidated by DFT-based first-principles calculations [78]. According to theory, a perfect or nearly perfect planar conformation is retained in the heaviest group IIIA–VA nitride with ideal bond angles Tl-N-Tl and N-Tl-N of 120° [78, 79] (119.9° [1]). Owing to the substantial difference in the electronegativities of Tl (1.8) and nitrogen (3.04), bonding is partially ionic due to charge transfer. According to the Bader analysis, a medium charge transfer of $\delta q = 0.82$ e occurs from the Tl atom to the N atom [80]. The published cohesive energies of $E_c = -2.40$ eV/atom [80] for the bilayer and – 2.20 eV/atom [1] for the monolayer are substantially smaller than those reported for h-InN monolayers. This reduction in bonding energy is assumed to originate from an increase in the metallic character of bonding with ionic and covalent contributions.

7.5.2 2D Structures

The nearly perfect quasi-planar monolayers possess a lattice constant a = 0.373 nm and a bond length d = 0.215 nm [1, 78]. Some authors postulate a tiny buckling effect of $\Delta z = 0.0061$ nm with a bond angle of $\theta = 119.92°$ [1]. The graphene-like structure has a zero-band gap at its lowest energy state. The space group given for the planar structure is *P*-3*m*1, and for the buckled structure it is *P*6/*mmm* [80, 81]. First-principles DFT-PBE calculations predict large-gap quantum spin Hall (QSH) insulators in the h-TlX group (X = N, P, As, and Sb) and a slightly buckled configuration of the TlN monolayer with a = 0.385 nm, d = 0.224 nm, and $\Delta z = 0.029$ nm (see Fig. 7.6) [81]. In this slightly buckled structure, the monolayer band gap meets at a single point and decreases to zero [81].

The bilayer structure of h-TN is playing an increasing role in the development of novel large band gap 2D topological insulators (TIs) [80]. For hexagonal bilayers, DFT calculations predict a *p-p* band inversion-type QSH effect that is independent of spin–orbit coupling (SOC). The bilayer structure exhibits a larger lattice constant of a = 0.398 nm, and the in-plane bond length of d = 0.247 nm nearly agrees with the out-of-plane interlayer separation of $d_s = 0.251$ nm. This small interlayer spacing is consistent with the calculated very strong interlayer interaction energy of $E_{in} = -370$ meV/atom [80]. This is due to ionic-covalent bilayer bonding, where each Tl (N) atom bonds with the four nearest N (Tl) atoms, where three of them are in the same plane and the other one is in the opposite plane. The bilayer separation of $d_s = 0.262$ nm, obtained by another group, confirms the small distance, which increases in trilayers to 0.284 nm [79]. Both, bilayers and trilayers retain the planar configuration [80]. A comparison of the interlayer spacing of $d_s = 0.251$ nm (0.262 nm) with the van der Waals diameter of $d_w = 0.38$ nm indicates strong ionic-covalent interlayer bonding between bilayers and trilayers.

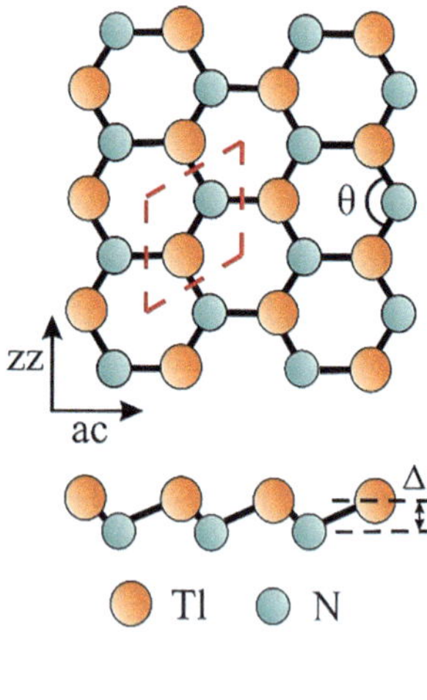

Fig. 7.6 Top and side views of buckled hexagonal IIIA–VA monolayer

7.5.3 Mechanics

For Young's moduli of the h-TlN monolayers, the authors propose a marginal anisotropy of $E_{ac} = 55.1$ N/m and $E_{zz} = 55.0$ N/m [1]. The much lower stiffness of E = 34.5 N/m, is approximately one-tens of the stiffness of graphene, whereas the Poisson ratio of $\nu = 0.689$ is the largest in this group [78]. This Poisson ratio is remarkable and even exceeds the large ratio of h-InN monolayers and the usually given upper limit of $\nu = 0.5$ for stable isotropic materials. The intrinsic strengths of $\sigma_{ac} = 5.8$ N/m and $\sigma_{zz} = 5.7$ N/m are reasonable and the ultimate strains of $\varepsilon_{ac} = 0.17$ and $\varepsilon_{zz} = 0.21$ of h-TlN are surprisingly small [78]. The stiffness of 55 N/m, also extracted by DFT-PBE calculations, is in good agreement with Griffith's rule, e.g., $E/\sigma = 55/5.8 = 9.5$.

7.5.4 Physics and Applications

The investigation of the optical properties of h-TlN monolayers includes the dielectric function, absorption coefficient, refractive index, reflectivity, optical conductivity, and energy loss function [82, 83]. Insight into the intriguing optical behavior of this material is given by a comparison with bulk TlN (wurtzite) [82] and with the monolayers of the other group member h-MN (M = Al, Ga, In) [83].

The thermoelectric properties of nanostructures have recently attracted enormous attention. The thermoelectric figure of merit, known under the name *ZT*, of various graphene-like structures is expected to be high enough for efficient energy conversion [84]. Appropriate nanostructuring and band gap engineering can reduce the lattice thermal conductance and enhance the Seebeck coefficient. A theoretical study compared the thermal conductivities of monolayer, bilayer, trilayer, and bulk h-TlN [79]. Monolayers of TlN exhibit better thermal and electrical conductivities than few-layer sheets, including bulk material, and may be used as heat sink materials, whereas bilayer TlN is a better thermoelectric material, especially at temperatures above 700 K.

7.5.5 Synthesis

The synthesis of high-quality h-TlN monolayers or bilayers seems to be more difficult than that of h-InN. One reason is the strong interlayer interaction. Another reason is the reactivity of the weak Tl–N bonds in comparison with the much stronger Tl–O bonds, leading to rapid degradation in the ambient environment [80]. Therefore, it is necessary to protect the h-TlN surface in a sandwich structure, e.g., between monolayer h-BN, owing to easy oxidation and contamination by atoms and chemical groups.

7.6 Boron–Pnictogen Monolayers (BP, BAs, BSb, and BBi)

7.6.1 BP Monolayer (Borophosphene)

7.6.1.1 Chemistry and 2D Bonding

In the promoted configuration the boron atom has three valence electrons ($2sp^2$) and the phosphorus atom has five valence electrons ($3sp^4$) in the outer shell. As discussed for nitrogen, the typical sp^3 bonding configuration of P atoms may change in the 2D world to planar sp^2 hybridization with a lone pair of electrons in the remaining p_z orbital, which is responsible for π bonding. The off-balanced electronic distribution leads to an unoccupied p_z orbital of the B atom. This bonding behavior results in the formation of a plane honeycomb structure in BP monolayers called borophosphene, where each B atom bonds with three P atoms. The short bond length indicates strong covalent interactions that favor planar structures over buckled or puckered geometries [85]. The ionic contribution is small, with $\delta q = 0.23$ e Bader charge transfer from the B atom to the P atom, consistent with the small difference in the electronegativities of 2.04 (B) and 2.19 (P). The short bond length can be explained by strong covalent forces. In addition to the highly stable planar honeycomb structure, BP forms buckled phases with 4–8 and 3–9 membered rings and sp^3-like bonding of the P atoms. The competition between electrostatic interactions and variable covalent sp^2 and sp^3 hybridizations results in the observed diversity of 2D BP allotropes [85, 86]. The reported cohesive energies of planar h-BP are $E_c = -5.0$ eV/atom [7], – 4.99 eV/atom [85, 86], – 4.93 eV/atom [87] and – 4.55 eV/atom [1]. The PBE-based cohesive energies of approximately – 5.0 eV/atom differ substantially from those of LDA-based calculations of – 5.80 eV/atom [7]. In any case, these large binding energies indicate good energetic stability.

7.6.1.2 2D Structures

The lattice constant a = 0.320 nm and bond length d = 0.185 nm of the h-BP monolayer are considerably smaller than the corresponding values of a = b = 0.356 nm, c = 0.59 nm, and d = 0.217 nm in bulk BP, indicating enormous energetic and mechanical stability [85, 87]. Most studies assume a bonding angle of 120° and an entirely planar network, whereas one work gives a tiny buckling effect of 0.0005 nm [1]. Note that independent calculations performed by the DFT-PBE and DFT-LDA approaches confirmed the short bond length and dynamic stability. The band gaps of the unstrained monolayer obtained by different approaches are $E_g = 0.91$ eV (GGA), 1.37 eV (HSE06), and 1.832 (G_0W_0), where the latter value is regarded as the best approximation [86, 87]. The mean van der Waals diameter of the h-BP monolayer is $d_w = 0.375$ nm [11]. Note that a space groups of *P*-6*m*2 [6, 86] is assigned to h-BP monolayers and *P*6/*mmm* [87] to a semimetal with a similar structure and lower symmetry.

Recent studies have investigated the stacking sequence and interlayer bonding of few-layer systems of h-BP with up to four layers. The most stable AB stacking of bilayers has an interlayer spacing of $d_s = 0.339$ nm and an interlayer interaction energy of $E_{int} = -134$ meV/atom [88]. A charge transfer of $\delta q = 0.05$ e occurs from the bottom layer to the top layer. For AA stacking, the interlayer interaction changes to − 112 eV/atom, and the interlayer spacing increases to 0.393 nm. The trilayers with ABA and ABC stacking have similar structural data as bilayers and an interlayer interaction energy of − 152 meV/atom. Similarly, in the four-monolayer assemblies with ABAB and ABCA stacking, the structures are similar, whereas the interlayer interaction energy increases further to − 161 meV/atom [88]. On the other hand, the band gap decreases from $E_g = 0.955$ eV for the monolayer to 0.467 eV for the ABCA order. Both the monolayer and bilayer with AB stacking possess a direct band gap that is retained in trilayer ABC stacking; however, this band gap changes to an indirect band gap for ABA stacking. Notably, the band gap of $E_g = 1.37$ eV, obtained by using the screened hybrid HSE06 functional, is substantially larger than the band gaps obtained by the other approaches [88].

7.6.1.3 Mechanics

As expected from the short bond length, the h-BP monolayer is a stiff 2D material with published Young's moduli of 163 N/m (DFT-PBE) [7] and 135 N/m (DFT-LDA) [16] and a Poisson ratio of 0.28 [16]. The latter two properties are in good agreement with the slightly anisotropic Young's moduli of $E_{ac} = 139.2$ N/m and $E_{zz} = 139.6$ N/m and Poisson's ratios of $\nu_{ac} = 0.276$ and $\nu_{zz} = 0.277$ [1]. Unfortunately, data on the ultimate strength and ultimate strain are still missing for this stiff compound. A hybrid DFT study revealed the deterioration effects of single B, single P, double BP, and Stone–Wales defects on the Young's modulus [89]. With reference to the isotropic in-plane Young's modulus of 135 N/m for the perfect monolayer, the four defects reduce the stiffness, and the calculated ranges of the four anisotropic stiffness values of these defects are 75–108 N/m (B), 97–104 N/m (P), 94–112 N/m (BP), and 127–132 N/m (Stone–Wales) [89].

7.6.1.4 Physics and Applications

Owing to its moderate direct band gap, as needed for applications in the visible spectral region, the monolayer of h-BP is a promising candidate for optoelectronic applications. Based on its direct band gap, strong optical absorption, ultrahigh carrier mobility, and thermal and mechanical stability, 2D h-BP is a suitable material for optoelectronic applications in photovoltaic devices and as a photocatalyst for water splitting [88, 90–93]. Boron phosphide monolayers satisfy the conditions for water splitting, and the photocatalytic performance can be further improved by biaxial tensile strain [93].

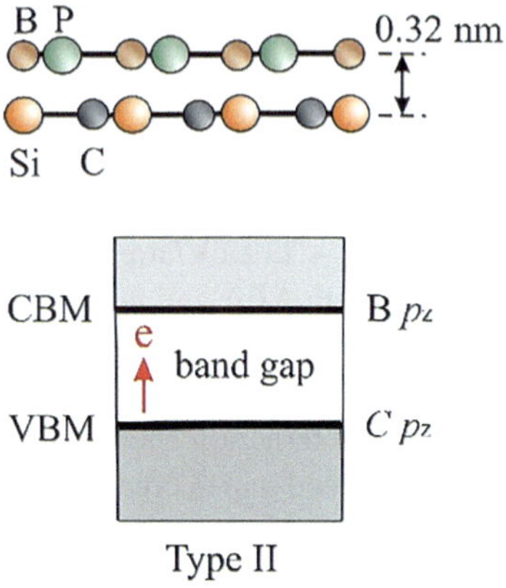

Fig. 7.7 2D boron-phosphide and silicon-carbide van der Waals heterostructure (BP-SiC) with type–II band alignment

First-principles calculations suggest that 2D boron phosphide-silicon carbide (BP/SiC) van der Waals heterostructures with a direct band gap and enhanced optical absorption, compared to those of the constituents, are promising candidates for water splitting at pH = 0–3 [94]. Figure 7.7 presents the optimized geometrical arrangement of the heterostructure with a large interlayer interaction energy of E_{int} = – 240 meV and a separation of the two layers of d_s = 0.319 nm. In the most stable stacking configuration, the B atom is on top of the Si atom. The type-II band alignment, with the valence band maximum (VBM) and conduction band minimum (CBM) belonging to different monolayers, helps to separate the free electrons and holes at the interface (see Fig. 7.8). The introduction of the van der Waals nanocomposite with a gallium nitride-boron phosphide (GaN/BP) structure is based on the concept of constructing heterostructures with desired properties [95]. Biaxial tensile strain changes the type-I band alignment of this direct semiconductor to a type-II alignment, resulting in effective separation of the photogenerated charge carriers. Furthermore, in-plane biaxial strains tune the band gap and band edge positions and enhance optical absorption in the visible and near ultraviolet regions.

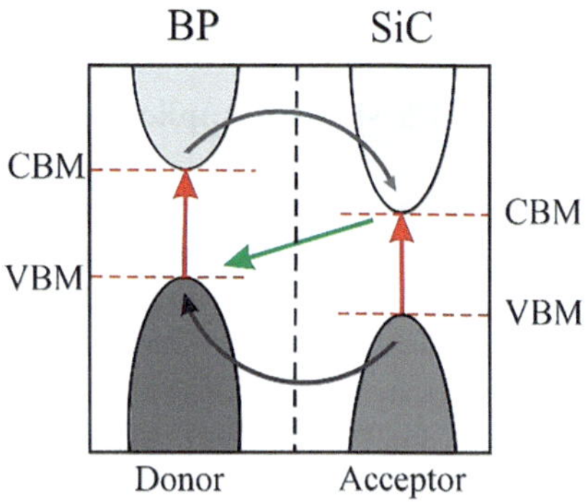

Fig. 7.8 Type-II carrier separation in a heterostructure with a VBM and CBM from a different monolayer

Moreover, BP bilayers, which have a lower lattice thermal conductivity than monolayers and greater carrier mobility, exhibit excellent thermoelectric performance, and good *ZT* values above 1.0 can be achieved at 700 K [96]. The significant reduction in lattice thermal conductivity in comparison to that of monolayers is attributed to the increased crossing of optical and acoustic branches induced by van der Waals interactions. Applying first-principles calculations and the Boltzmann transport theory, a *ZT* value of 0.255 is predicted at 300 K [97].

By functionalization with Cl and Br atoms on one or both sides, the planar h-BP monolayer transforms into a buckled network with a small buckling effect of $\Delta z = 0.06$ nm [98]. The analysis of the phonon spectra and ab initio MD simulations confirmed the stability of the functionalized compounds. All four direct semiconductors possess a larger band gap than the pure h-BP monolayer. The ClBPCl compound had the largest stiffness, with a Young's modulus of 43 N/m.

7.6.1.5 Synthesis

Despite enormous effort, the synthesis of BP monolayers was not successful up to now [87]. The growth of thin films and crystals of BP by the chemical vapor transport (CVT) method has attracted increasing attention because it functions as a precursor for the top-down preparation of phosphorene. The orthorhombic bulk crystals of BP have a structure consisting of puckered layers, which ultimately allows the transformation to phosphorene mono- or few-layer systems by chemical or mechanical forces [90]. One reason why the isolation of boron phosphide monolayers could not be achieved seems to be the reactivity leading to fast degradation of few-layer BP sheets under ambient conditions owing to moisture and oxygen in air. A recent DFT study suggested the induction of an exfoliation process by arsenic incorporation [91].

7.6.2 *BAS Monolayers (Boroarsenene)*

7.6.2.1 Chemistry and 2D Bonding

In the promoted electronic configuration the boron atoms have three valence electrons ($2sp^2$) and the arsenic atoms five valence electrons ($4sp^4$) in the outer valence shell. Among the IIIA–VA compounds, g-BAs has a high degree of covalent interaction [99]. The slightly increasing size of the As atom compared to that of the P atoms yields a mean van der Waals diameter of $d_W = 0.385$ nm and diminishes the overlap of the p_z orbitals. This weakening of π-π bonding reduces the stabilization of the planar structure with a bonding angle of 120°. Nevertheless, values of the electron localization function (ELF) demonstrate a strong localization of electrons in the middle of the B–As bond, indicating high covalent bonding character and thus high thermal, mechanical and chemical stability [99]. In fact, the comparable electronegativities of B (2.04) and As (2.18) lead to a small Bader charge transfer from the B atom to the As atom of $\delta q = 0.26$ e [6]. The reported cohesive energies of $E_c = -5.16$ eV/atom (LDA) [7], – 4.33 eV/atom (PBE) [7], – 4.29 eV/atom [99], and – 3.86 eV/atom (PBE) [1] signify the usually observed differences in binding energy calculations employing diverse approaches. Assuming a comparable sensitivity of the bond length to the binding energy, the longer B–As bond of d = 0.196 nm in comparison to that of B–P d = 0.185 nm seems to exclude a cohesive energy of approximately – 5 eV/atom.

7.6.2.2 2D Structures

The optimized structural data of h-BAs a = 0.339 nm, d = 0.196 nm, $\Delta z = 0.0012$ nm, and $\theta = 119.995°$ [1] are in good agreement with the previously reported data of a = 0.338 nm, d = 0.195 nm, and $\theta = 120°$ [99]. Theory signifies the energetic, dynamic, thermal, and mechanical stability of the h-BAs monolayer. The van der Waals diameter of $d_{\underline{W}} = 0.385$ nm [11] is consistent with the interlayer spacing of $d_s = 0.357$ nm for bilayers and 0.366 nm for trilayers with AB and ABA stacking orders [100]. The direct band gap of $E_g = 0.76$ eV (DFT) changes into an indirect gap by biaxial strain [99]. The space group of 2D h-BAs is *P*-6*m*2 [6].

The estimated Young's modulus of 116 N/m and the Poisson ratio of 0.282 [1] agree well with E = 119 N/m and $\nu = 0.29$. [16]. There are no studies on the ultimate strength and critical strain, however, much larger stiffness values of 136 N/m (PBE) and 146 N/m (LDA) have been reported in the literature [7].

7.6.2.3 Physics and Applications

A case study indicated that by cutting along the (111) cross section of 3D cubic boron arsenide, the thermal conductivity could be decreased by more than one order of magnitude in the 2D system of h-BAs owing to the stereochemically activated lone pair electrons [101]. This specific structure leads to strong phonon anharmonicity. The combination of lone pair electrons and the small electronegativity difference of binary compounds seems to be an efficient way of lowering thermal conductivity. Reducing thermal conductivity is an efficient way to improve thermoelectric performance and boost the direct solid-state conversion of thermal energy to electrical power. Such a design of thermoelectric devices may be transferable to other group IIIA–VA compounds.

Another promising application is the use of 2D h-BAs as potential anode material for alkali-metal (Li/Na/K)-ion batteries [102]. Calculations predict negative adsorption energies for Li, Na, and K ions on the h-BAs surface at the energetically most favorable H-site, above the hollow site of the hexagon that contains three As and three B atoms. An increase in the adsorption energy during the charging process generates no barrier at a crucial distance of ~ 0.8 nm. Thus, the h-BAs surface allows fast diffusion of alkali metals with low energy barriers. The transformation from the semiconducting to the metallic state after alkali adsorption guarantees excellent electronic conductivity as an anode material.

7.6.3 BSb Monolayers (Boroantimonene)

In the promoted configuration the boron atom has three valence electrons ($2sp^2$) and the antimony atoms have five ($4sp^4$) in their outer valence shells. Based on a similar electronic configuration as that of the other group IIIA–VA members, stable hexagonal monolayers of BSb exist. Despite the similar electronegativities of B (2.04) and Sb (2.05), the Bader charge transferred from each B atom to each Sb atom is $\delta q = 0.55$ e [6]. The favored comprehensive theoretical treatment gives a lattice constant of $a = 0.374$ nm, a bond length of $d = 0.216$ nm, a low buckling height of $\Delta z = 0.0033$ nm, and a bond angle of $\theta = 119.98°$ for the h-BSb monolayer [1]. The reported cohesive energies vary widely from $E_c = -3.26$ eV/atom (PBE) [1], and -3.73 eV/atom (PBE) [7], to -4.51 eV/atom (LDA) [7]. The mean van der Waals diameter of h-BSb is $d_w = 0.40$ nm [11] and the space group is *P*-6*m*2 [6].

The low stiffnesses of 109 N/m (PBE) [7], and 91 N/m with a Poisson ratio of 0.34 [16] and the even lower values of $E_{ac} = 83.9$ N/m and $E_{zz} = 84.5$ N/m and Poisson ratios of $\nu_{ac} = 0.328$ and $\nu_{zz} = 0.331$ are in line with the expectations [1]. Using first-principles calculations, two- and quasi-one-dimensional honeycomb structures of BSb monolayers and their defects were studied theoretically [103]. Passivating both the B and Sb edges of a nanoribbon with hydrogen resulted in a direct gap semiconductor with a band gap of $E_g = 0.490$ eV.

A large carrier mobility is of fundamental importance for many future applications of 2D materials. The IIIA–VA monolayers of h-BP, h-BAs, and h-BSb approach the carrier mobility of 3D GaAs but do not reach the high mobility of 3D Si [104]. In particular, h-BSb, which has minimal contributions to electron–phonon coupling, has the highest intrinsic electron and hole mobilities of ~ 9000 $cm^2\ V^{-1}\ s^{-1}$ and 140,000 $cm^2\ V^{-1}\ s^{-1}$, respectively, of all known 2D materials. As discussed above for 2D h-BAs, ab initio calculations also indicate the suitability of h-BSb monolayers as promising anode material for lithium-ion batteries [105].

7.6.4 BBi Monolayers (Borobismuthene)

In the promoted electronic configuration the boron atom has three valence electrons ($2sp^2$) and the bismuth atoms have five electrons ($6sp^4$) in their outer shells. The bonding in h-BBi monolayers shows the largest deviation from the planar geometry in this group. As structural data, DFT-PBE calculations revealed a lattice constant of $a = 0.388$ nm, a bond length of $d = 0.229$ nm, a buckling effect of $\Delta z = 0.048$ nm, and a bonding angle of $\theta = 115.72°$ [1]. The cohesive energy of the network is further decreased to $E_c = -2.81$ eV/atom [1]. An independent study confirmed the lattice

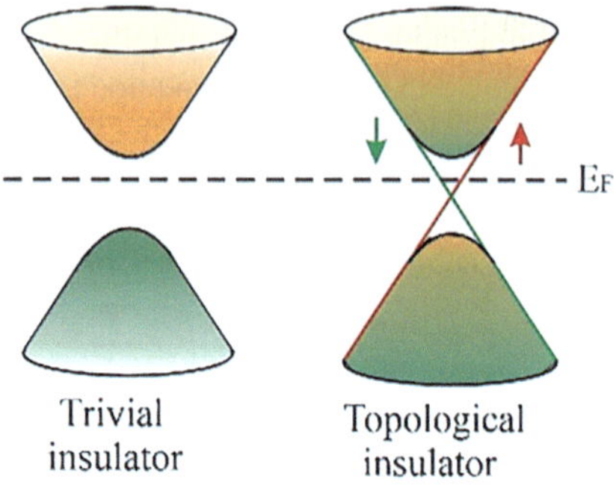

Fig. 7.9 Schematic presentation of the transition from a trivial insulator to a nontrivial insulator. Adapted from ref. [108], Copyright 2024, IOP Publishing Ltd.

constant and buckling effect but reported a larger cohesive energy of $E_c = -3.25$ eV/atom [106]. Adding the small corrugation to the van der Waals diameter of $d_w = 0.410$ nm yields a thickness of $d_{w+\Delta z} = 0.458$ nm [11].

The h-BBi monolayer has the lowest stiffness among the group of boron-pnictogen compounds, with $E_{ac} = 61{,}1$ N/m and $E_{zz} = 60.9$ N/m and nearly identical Poisson ratios of $\nu_{ac} = 0.297$ and $\nu_{zz} = 0.296$ [1]. The binary composition of group–IIIA elements with bismuth in the buckled hexagonal structures were identified as a new class of 2D TIs [107]. Band inversion was identified in pristine GaBi, InBi, and TlBi bilayers. Topological phase transitions by strain engineering were postulated for BBi and AlBi at a tensile strain of ~ 6.6%, which are intrinsic trivial insulators owing to the small SOC of B and Al atoms. The buckled structure may play a role as a versatile platform for hosting nontrivial topological states and spin-polarized Dirac fermions. These states are characterized by an electronic structure with gaped interior and gapless edge states incorporating spin states that counter-propagate at the boundaries with opposite spins. Figure 7.9 elucidates schematically the transition from a trivial insulator to a nontrivial insulator, where the colors indicate the mixed orbital character of the bands. The red and green lines inside the gap symbolize the edge states [108]. Hydrogenation of h-BBi monolayers opens the band gap of this class of 2D TIs and leads to stable crystal structures, which can preserve their topological insulating properties [106].

7.7 Reviews Discussing Specific Topics of Related IIIA–VA Compounds

The most complete collection of structural data and linear mechanical properties of IIIA–VA monolayers has been considered above [1]. In addition to the papers already cited, several comprehensive studies and review-type publications cover selected topics of the considered chemically related compounds in more detail. For the well-studied group–IIIA nitrides, these investigations cover the synthesis and development of devices [109], various properties, deposition methods, and potential applications

[110], effects of strain on the electronic properties [111], and piezoelectric effects in surface engineered group–IIIA nitrides [112]. A first-principles perspective on bilayer hexagonal structures is available for a selected number of monolayers MY (M = Al, Ga, and In; Y = N, P, and As) [113].

Several studies have investigated the piezoelectric and linear mechanical properties of the buckled gallium compounds GaN, GaP, GaAs, and GaSb and the buckled indium compounds InP, InAs, and InSb [114–116]. Since these semiconducting monolayers may be ferroelectric and have nonzero piezoelectric coefficients, they are promising candidates for atomically thin piezoelectric devices [114, 115]. The elastic coefficients needed to calculate the piezoelectric coefficients are available for seventeen different boron, aluminum, gallium, and indium compounds [115].

The application of biaxial mechanical strain or a perpendicular electrical field allows tuning of the energy band gap of the gallium-based compounds GaX (X = N, P, As, and Sb) [117]. The energy band gap increases up to a certain value of compressive strain and then decreases, whereas it always decreases with increasing tensile strain and finally becomes metallic. Under an electric field, the band gap decreases, which induces a semiconductor-to-metal transition.

The DFT-PBE approach provided the structural, mechanical, and electronic properties of the group–IIIA As compounds BAs, GaAs, and InAs [118]. Owing to the weakening of σ and π bonding, h-BAs are planar, and h-GaAs and InAs have buckled configurations. The elastic constants, Young's moduli, bulk moduli, and shear moduli decrease with increasing atom size of the group–IIIA elements. By increasing the number of layers from the bilayer to few-layer assemblies, the width of the band gap decreases until the few-layer system becomes a semimetal or metal. The potential use of the monolayers BAs, GAAs, and AlAs as cost-effective biosensors for detecting COVID-19 biomarkers together with volatile organic compounds (VOCs) has been suggested recently [118]. The study included the investigation of band structure modifications, DOS, work function, difference of electron density, and optical properties such as reflectance and absorbance. Especially GaAs monolayers exhibit a significant level of sensitivity and selectivity towards SARS-CoV-2 biomarkers as chemo-resistive work-function-based and optical sensor.

DFT-PBE calculations yielded Young's moduli and Poisson's ratios for the family of thallium compounds TlN, TlP, TlAs, and TlSb [119]. These monolayers are TIs with room temperature band gaps, topological edge states, and a promising Rashba–Dresselhaus effect. A first-principles study investigated the possible QSH phases of 75 binary compounds of group IIIA–VA (IIIA = B, Al, Ga, In, and Tl; VA = N, P, As, Sb, and Bi) monolayers [120]. Moreover, hydrogenated buckled IIIA–VA compounds with heavy Tl and Bi atoms and robust large gaps form QSH phases at elevated temperatures.

7.8 Property–Structure Relationships

7.8.1 *Binding–Structure Relationships*

As presented in Fig. 7.10, it is possible to collect the planar and buckled group IIIA–VA monolayers into related groups [121]. The groups with the strongest bonding are the group–IIIA nitrides and the boron-pnictogen compounds, which have planar or quasi-planar configurations. The buckled groups consist of the compounds AlX, GaX, TlX, and BiX, with X = P, As, Sb, and Bi. Contrary to the planar phases, the buckled monolayers cluster in a narrow range of low cohesive energies between E_c = – 1.5 eV/atom and – 3.5 eV/atom and long bonds between d = 0.23–0.30 nm. All groups show a correlated inverse behavior in the cohesion–bond length plot. This pronounced inverse behavior shifts systematically to longer bond lengths from the group of nitrides with the strongest bonds to the group of Bi-pnictogen compounds and those with the heaviest constituents. Since, in a uniform hexagonal lattice, the dissociation energy of a single bond is a certain fraction of the cohesive energy, the plot provides new insight into the intrinsic relationships between binding energy and bond length.

The astonishing similarity of the bonding behavior of binary group IIIA–VA monolayers with that of group–IVA monolayers originates from the identical number of valence electrons. This is also apparent from the striking similarity of group–IVA carbides and group–IIIA nitrides. Within the errors involved in such calculations, both

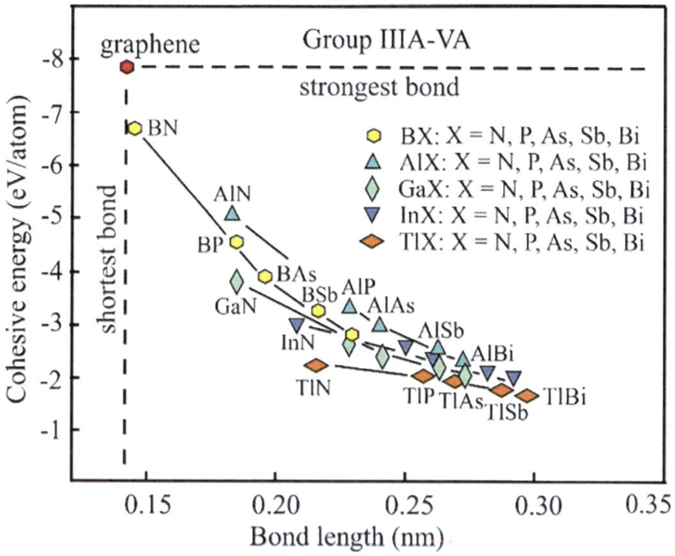

Fig. 7.10 Plot of cohesive energy versus bond length of binary group IIIA–VA monolayers. Adapted from ref. [121], Copyright 2021, Royal Society of Chemistry

groups reveal a similar dependence of the cohesive energy on the length of the corresponding bonds. Owing to the small sizes of the boron, carbon, and nitrogen atoms, these compounds produce planar sp^2 hybrids with minor deviations from planarity only for the heaviest compounds. For example, boron compounds containing the small boron atoms are almost planar, with calculated corrugations of $\Delta z = 0$ nm, 0.001 nm, 0.003 nm, and 0.048 nm for h-BP, h-BAs, h-BSb, and h-BBi, respectively [1]. Plane compounds have shorter bond lengths and larger cohesive energies than the buckled heavier compounds.

7.8.2 *Mechanics–Structure Relationships*

Surprisingly, in the plot of Young's moduli and ultimate strengths versus bond lengths, the Young's moduli of the binary group IIIA–VA monolayers follow the same invers line, which remarkably includes the limiting stiffness of graphene (see Fig. 7.11) [121]. Note that the ultimate strengths of nitrides broadly follow Griffith's rule. The plot allows a first estimate of approximate stiffness and strength values if the monolayer is not characterized otherwise. This behavior is not surprising because all the compounds enclose a small second row atom that stabilizes the planar graphene-like configuration. It is important to note that the line formed by the compounds containing a second-row element defines an upper limit for the mechanical performance of all other IIIA–VA compounds without such an element.

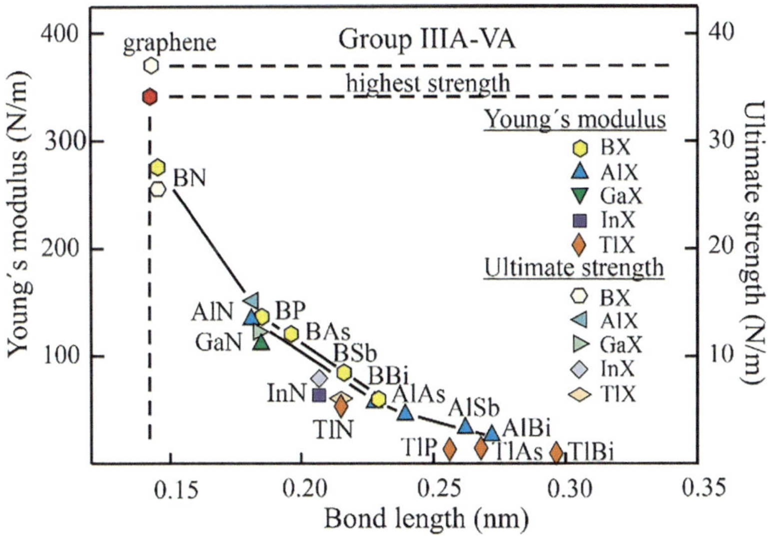

Fig. 7.11 Plot of the Young's modulus and ultimate strength versus bond length of binary group IIIA–VA monolayers. Adapted from ref. [121], Copyright 2021, Royal Society of Chemistry

Inspection of Fig. 7.11 reveals a steady decrease in the mechanical performance for the heavier compounds going down the columns. The reduced stiffness and strength may create serious problems in fabrication of high quality samples because weaker bonds are more susceptible to defect formation and incorporation of impurities when exposed to the environment. Also, during practical use mechanical stability is a critical issue. Nonetheless, one should keep in mind that often monolayers with heavy constituents exhibit specific physical properties of wide-ranging interest. Examples include Tl compounds with QSH phases that must be protected against degradation.

References

1. X.-F. Liu, Z.-J. Luo, X. Zhou, J.-M. Wei, Y. Wang et al., Chin. Phys. B **28**, 086105 (2019)
2. V.O. Özçelik, J.G. Azadani, C. Yang, S.J. Koester, T. Low, Phys. Rev. B **94**, 035125 (2016)
3. J.C. Meyer, A. Chuvilin, G. Algara-Siller, J. Biskupek, U. Kaiser, Nano Lett. **9**, 2683 (2009)
4. A. Pakdel, Y. Bando, D. Golberg, Chem. Soc. Rev. **43**, 934 (2014)
5. L. Chkhartishvili, I. Murusidze, Mater. Sci. Appl. **1**, 223 (2010)
6. C.-J. Tong, H. Zhang, Y.-N. Zhang, H. Liu, L.-M. Liu, J. Mat. Chem. A **2**, 17971 (2014)
7. H.L. Zhuang, R.G. Hennig, Appl. Phys. Lett. **101**, 153109 (2012)
8. G. Graziano, J. Klimeš, F. Fernandez-Alonso, A. Michaelides, J. Phys. Condens. Matter **24**, 424216 (2012)
9. Y. Shi, C. Hamsen, X. Jia, K.K. Kim, A. Reina et al., Nano Lett. **10**, 4134 (2010)
10. O. Hod, J. Chem. Theory Comput. **8**, 1360 (2012)
11. S.S. Batsanov, Inorg. Mater. **37**, 871 (2001)
12. K. Watanabe, T. Taniguchi, H. Kanda, Nature Mater. **3**, 404 (2004)
13. D. Golberg, Y. Bando, Y. Huang, T. Terao, M. Mitome et al., ACS Nano **4**, 2979 (2010)
14. N. Marom, J. Bernstein, J. Garel, A. Tkatchenko, E. Joselevich et al., Phys. Rev. Lett. **105**, 046801 (2010)
15. P. Hess, Nanotechnology **28**, 064002 (2017)
16. H. Şahin, S. Cahangirov, M. Topsakal, E. Bekaroglu, E. Akturk et al., Phys. Rev. B **80**, 155453 (2009)
17. Q. Peng, W. Ji, S. De, Comput. Mater. Sci. **56**, 11 (2012)
18. J. Wu, B. Wang, Y. Wei, R. Yang, M. Dresselhaus, Mater. Res. Lett. **1**, 200 (2013)
19. T. Han, Y. Luo, C. Wang, J. Phys. D Appl. Phys. **47**, 025303 (2014)
20. M.-Q. Le, J. Comput. Theor. Nanosci. **11**, 1458 (2014)
21. R.C. Andrew, R.E. Mapasha, A.M. Ukpong, N. Chetty, Phys. Rev. B **85**, 125428 (2012)
22. M.-Q. Le, Comp. Mater. Sci. **201**, 110854 (2022)
23. M.-Q. Le, R.C. Batra, Comp. Mater. Sci. **203**, 111150 (2022)
24. P. Hess, J. Appl. Phys. **120**, 124303 (2016)
25. X. Wei, S. Xiao, F. Li, D.-M. Tang, Q. Chen et al., Nano Lett. **15**, 689 (2015)
26. R.J.P. Román, F.J.R.C. Costa, A. Zobelli, C. Elias, P. Valvin, et al., 2D Mater. **8**, 044001 (2021)
27. C.R. Dean, A.F. Young, I. Meric, C. Lee, L. Wang et al., Nature Nanotech. **5**, 722 (2010)
28. J. Meng, D. Wang, L. Chen, M. Gao, X. Zhang, Nanotechnology **30**, 074003 (2019)
29. S. Deng, Y. Gu, X. Wan, M. Gao, S. Xu et al., Front. Mater. **8**, 735344 (2021)
30. A.F. Rigosi, A.L. Levy, M.R. Snure, N.R. Glavin, J. Phys. Mater. **4**, 032003 (2021)
31. J. Zhao, H. Zeng, G. Yao, Phys. Chem. Chem. Phys. **23**, 3771 (2021)
32. M. Li, G. Huang, X. Chen, J. Yin, P. Zhang et al., Nano Today **44**, 101486 (2022)
33. M. Rafiq, X. Hu, Z. Ye, A. Qayum, H. Xia et al., Nano Energy **91**, 106661 (2022)
34. A. Azizi, M.A. AlSaud, N. Alem, J. Crystal Growth **496–497**, 51 (2018)

35. R. Wang, D.G. Purdie, Y. Fan, F.C.-P. Massabuau, P. Braeuninger-Weimer et al., ACS Nano **13**, 2114 (2019)
36. T.S. Cheng, A. Summerfield, C.J. Mellor, J. Vac. Sci. Technol. B **36**, 02D103 (2018)
37. Y. Ji, B. Calderon, Y. Han, P. Cueva, N.R. Jungwirth et al., ACS Nano **11**, 12057 (2017)
38. L. Wang, X. Xu, L. Zhang, R. Qiao, M. Wu et al., Nature **570**, 91 (2019)
39. T.-A. Chen, C.-P. Chuu, C.-C, Tseng, C.-K. Wen, H.-S.P. Wong et al., Nature **579**, 219 (2020)
40. Q. Zhang, H. Chen, S. Liu, Y. Yu, C. Wang et al., Nanoscale **14**, 3112 (2022)
41. P. Wang, W. Lee, J.P. Corbett, W.H. Koll, N.M. Vu et al., Adv. Mater., 2201387 (2022)
42. C. Bacaksiz, H. Sahin, H.D. Ozaydin, S. Horzum, R.T. Senger et al., Phys. Rev. B **81**, 085430 (2015)
43. R.B. dos Santos, F. de Brito Mota, R. Rivelino, A. Kakanakova-Georgieva, G.K. Gueorguiev, Nanotechnology **27**, 145601 (2016)
44. Y. Cai, Y. Liu, Y. Xie, Y. Zou, C. Gao et al., APL Mater. **8**, 021107 (2020)
45. L. Zhao, S. Xu, M. Wang, S. Lin, J. Phys. Chem. C **120**, 27675 (2016)
46. D. Kecik, A. Onen, M. Konuk, E. Gürbüz, F. Ersan et al., Appl. Phys. Rev. **5**, 011105 (2018)
47. Q. Peng, X.-J. Chen, S. Liu, S. De, RSC Adv. **3**, 7083 (2013)
48. M.-Il Choe, K.-H. Kim, ChemPhysChem, e202100901 (2022)
49. D. Kecik, C. Bacaksiz, R.T. Senger, E. Durgun, Phys. Rev. B **92**, 165408 (2015)
50. Z. Shi, Z. Qi, H. Zang, K. Jiang, Y. Chen et al., ACS Appl. Mater. Interfaces **13**, 37380 (2021)
51. H. Gao, H. Ye, Z. Yu, Y. Zhang, Y. Liu et al., Superlattices Microstruct. **112**, 136 (2017)
52. M. Beshkova, R. Yakimova, Vacuum **176**, 109231 (2020)
53. H. Wang, L. Yu, J. Xu, D. Wei, G. Qin et al., Int. J. Therm. Sci. **162**, 106772 (2021)
54. Z. Wang, G. Wang, X. Liu, S. Wang, T. Wang et al., J. Mater. Chem. C **9**, 17201 (2021)
55. P. Tsipas, S. Kassavetis, D. Tsoutsou, E. Xenogiannopoulou, E. Golias, Appl. Phys. Lett. **103**, 251605 (2013)
56. S.-J. Chang, S.-Y, Wang, Y.-C. Huang, J.H. Chih, Y.-T. Lai et al., Appl. Phys. Lett. **120**, 162102 (2022)
57. D. Xu, H. He, R. Pandey, S.P. Karna, J. Phys. Condens. Matter **25**, 345302 (2013)
58. A.K.A. Lu, T. Yayama, T. Morishita, M.J.S. Spencer, T. Nakanishi, J. Phys. Chem. C **123**, 1939 (2019)
59. A. Onen, D. Kecik, E. Durgun, S. Ciraci, Nanoscale **10**, 21842 (2018)
60. Q. Peng, C. Liang, W. Ji, S. De, Appl. Phys. A **113**, 483 (2013)
61. G.I. Giannopoulos, S.K. Georgantzinos, Eur. J. Mech. A/Solids **65**, 223 (2017)
62. A.V. Kolobov, P. Fons, Y. Saito, J. Tominaga, B. Hyot et al., Phys. Rev. Mater. **1**, 024003 (2017)
63. A. Onen, D. Kecik, E. Durgun, S. Ciraci, J. Phys. Chem. C **121**, 27098 (2017)
64. H. Shu, X. Niu, X. Ding, Y. Wang, Appl. Surf. Sci. **479**, 475 (2019)
65. R. Gonzáles-Ariza, O. Martines-Castro, M.G. Moreno-Armenta, A. Gonzales-Garcia, W. Lopez-Perez et al., Physica B **569**, 57 (2019)
66. G. Zhang, L. Chen, L. Wang, Z. Lu, H. Dong et al., J. Mater. Chem. A **10**, 4053 (2022)
67. V. Kumar, D.R. Roy, J. Mater. Sci. **53**, 8302 (2018)
68. D. Liang, R. Quhe, Y. Chen, L. Wu, Q. Wang et al., RSC Adv. **7**, 42455 (2017)
69. M.S. Prete, O. Pulci, F. Bechstedt, Phys. Rev. B **98**, 235431 (2018)
70. Q. Peng, X. Sun, H. Wang, Y. Yang, X. Wen et al., Appl. Mater. Today **7**, 169 (2017)
71. D. Chen, X. Zhang, J. Tang, Z. Cui, H. Cui, J. Hazardous Mater. **363**, 346 (2019)
72. H. Wang, H. Wu, H. Cui, Mol. Phys. **120**, e2027535 (2022)
73. C. Feng, H. Qin, D. Yang, G. Zhang, Materials **12**, 676 (2019)
74. B. Pécz, G. Nicotra, F. Giannazzo, R. Yakimova, A. Koos et al., Adv. Mater. **33**, 2006660 (2021)
75. K.D. Pham, T.V. Vu, T.N. Pham, D.D. Vo, P.T. Dang, Chem. Phys. **532**, 110677 (2020)
76. T.V. Vu, K.D. Pham, T.N. Pham, D.D. Vo, P.T. Dang et al., RSC Adv. **10**, 10731 (2020)
77. H. Cui, D. Chen, C. Yan, Y. Zhang, X. Zhang, Nanoscale Adv. **1**, 2003 (2019)
78. Q. Peng. C. Liang, W. Ji, S. De, Model. Num. Simul. Mater. Sci. **2**, 76 (2012)
79. E.V. Sha, D.R. Roy, Appl. Nanosci. **9**, 33 (2019)

80. P. Zhou, L. Xue, L.Z. Sun, J. Mater. Chem. C **5**, 4268 (2017)
81. X. Li, Y. Dai, Y. Ma, W. Wie, L. Yu et al., Nano Res. **8**, 2954 (2015)
82. S.M. Elahi, M. Farzan, H. Salehi, M.R. Abolhasani, Optik **127**, 9367 (2016)
83. N.R. Abdullah, B.J. Abdullah, V. Gudmundsson, Solid State Commun. **346**, 114705 (2022)
84. P. Dollfus, V.H. Nguyen, J. Saint-Martin, J. Phys. Condens. Matter **27**, 133204 (2015)
85. Z. Zhu, X. Cai, C. Niu, C. Wang, Y. Jia, Appl. Phys. Lett. **109**, 153107 (2016)
86. Y. Zhang, J. Kang, F. Zheng, P.-F. Gao, S.-L. Zhang et al., J. Chem. Phys. Lett. **10**, 6656 (2019)
87. J.M. Galicia-Hernandez, J. Guerrero-Sanchez, R. Ponce-Perez, H.N. Fernandez-Escamilla, G.H. Cocoletzi et al., Comp. Mater. Sci. **203**, 111144 (2022)
88. X. Chen, C. Tan, Q. Yang, R. Meng, Q. Liang et al., Phys. Chem. Chem. Phys. **18**, 16229 (2016)
89. F.-Q. Li, Y. Zhang, S.-L. Zhang, Nanomaterials **11**, 1395 (2021)
90. M. Khurram, Z. Sun, Z. Zhang, Q. Yan, Inorg. Chem. Front. **7**, 2867 (2020)
91. O.M. Hernández, J. Guerrero-Sánches, R. Ponce-Pérez, R.G. Diaz, H.N. Fernandez-Escamilla et al., Appl. Surf. Sci. **538**, 148163 (2021)
92. S.-F. Wang, X.-J. Wu, Chin. J. Chem. Phys. **28**, 588 (2015)
93. H. Shu, J. Guo, X. Niu, J. Mater. Sci. **54**, 2278 (2019)
94. T.-N. Do, M. Idrees, B. Amin, N.N, Hieu, H.V. Phuc et al., RSC Adv. **10**, 32027 (2020)
95. H. Shu, M. Zhao, M. Sun, A.C.S. Appl, Nano Mater. **2**, 6482 (2019)
96. Z.Z. Zhou, H.J. Liu, D.D. Fan, G.H. Cao, C.Y. Sheng, Phys. Rev. B **99**, 085410 (2019)
97. M.-S. Li, D.-C. Mo, S.-S. Lyu, Sci. Rep. **11**, 10030 (2021)
98. T.V. Vu, A.I. Kartamyshev, N.V. Hieu, T.D.H. Dang, S.-N. Nguyen et al., RSC Adv. **11**, 8552 (2021)
99. K. Manoharan, V. Subramanian, ACS Omega **3**, 9533 (2018)
100. J. Zhang, H. Liu, Y. Gao, X. Xia, Z. Huang, Phys. Chem. Chem. Phys. **21**, 20981 (2019)
101. G. Qin, Z. Qin, H. Wang, M. Hu, arXiv:1904.00329v1 (2019)
102. N. Khossossi, A. Banerjee, Y. Benhouria, I. Essaoudi, A. Ainane et al., Phys. Chem. Chem. Phys. **21**, 18328 (2019)
103. F. Ersan, G. Gökoğlu, E. Aktürk, J. Phys. Condens. Matter **26**, 325303 (2014)
104. S. Song, S. Liu, Y. Sun, J.-H. Yang, X.-G. Gong, arXiv:2204.11419v1 (2022)
105. A. Elomrani, M. Lamhani, S. Oukahou, K. Sbiaai, S. Lebègue et al., Mater. Chem. Phys. **275**, 125191 (2022)
106. R.R.Q. Freitas, R. Rivelino, F. de Brito Mota, C.M.C. de Castilho, J. Phys. Chem. C **119**, 23599 (2015)
107. F.-C. Chuang, L.-Z. Yao, Z.-Q. Huang, Y.-T. Liu, C.-H. Hsu et al., Nano Lett. **14**, 2505 (2014)
108. A. Tadich, M. Zhao, M.T. Edmonds, J. Phys. Mater. **7**, 022501 (2024)
109. W. Wang, H. Jiang, L. Li, G. Li, Rep. Prog. Phys. **84**, 086501 (2021)
110. J. Ben, X. Liu, C. Wang, Y. Zhang, Z. Shi et al., Adv. Mater. **33**, 2006761 (2021)
111. F. Ghasemzadeh, F. Kanjouri, Sci. Chin. Technol. Sci. **61**, 535 (2018)
112. Y. Guo, H. Zhu, Q. Wang, A.C.S. Appl, Mater. Interfaces **11**, 1033 (2019)
113. M. Faraji, A. Bafekry, M.M. Fadlallah, H.R. Jappor, C.V. Nguyen et al., Appl. Surf. Sci. **590**, 152998 (2022)
114. R. Gao, Y. Gao, Phys. Status Solidi RRL **11**, 1600412 (2017)
115. S. Norouzi, N. Shahtahmassebi, M. Behdani, M.R. Roknabadi, Physica E **102**, 88 (2018)
116. M.N. Blonsky, H.L. Zhuang, A.K. Singh, R.G. Hennig, ACS Nano **9**, 9885 (2015)
117. B.P. Bahuguna, L.K. Saini, R.O. Sharma, B. Tiwari, Physica E **99**, 236 (2018)
118. A. Gonzáles-García, W. López-Pérez, J. Rivera-Julio, F.M. Peteers, V. Mendoza-Estrada et al., Comput. Mater. Sci. **144**, 285 (2018)
119. S. Saha, D.I. Sajib, K. Alam, Phys. Chem. Chem. Phys. **26**, 6242 (2024)
120. C.P. Crisostomo, L.-Z. Yao, Z.-Q. Huang, C.-H. Hsu, F.-C. Chuang et al., Nano Lett. **14**, 6568 (2015)
121. P. Hess, Nanoscale Horiz. **6**, 856 (2021)

Chapter 8
Elemental Group–VA Monolayers

Abstract The search for stable 2D phases of pnictogens revealed two puckered or washboard allotropes as the most stable phases; symmetric washboard (sw) and asymmetric washboard (aw) structures; and a hexagonal buckled (hb) allotrope. The buckled structure has flatter zigzag folds compared with the puckered structures that are characterized by deeper armchair ridges. The stable monolayers are generally named with suffix 'ene', although the structure contains no double bonds, e.g., nitrogene, phosphorene, arsenene, antimonene, and bismuthene. The five valence electrons in the outer shell (ns^2p^3) allow the formation of sp^2-like orbitals, with the remaining orbital in each atom containing a lone electron pair that influences the final structure and physicochemical behavior. The diverse structures have similar cohesive energies and energetic stability. Therefore, the most stable bonding configuration changes with the group element. Essentially, all the monolayers possess energetic, dynamic, and thermal stability as free-standing 2D lattices. The wide-gap semiconductors of this group exhibit both direct and indirect band gaps and cover a broad range of energies. The electronic properties, such as the band gap, are tunable by strain engineering and by increasing the number of layers in few-layer assemblies. The synthesis of the individual monolayers is at dissimilar stages. While nitrogene could not be prepared yet, quasi-free weakly bonded adsorbed monolayers and sheets with few layers exist for the other group members. The high carrier mobility and band gap tunability provide numerous opportunities for practical application. In few-layer systems, some compounds are topological semimetals with the possibility of realizing Dirac cones and topological insulators. Based on their large carrier mobility, wide and tunable band gap, and low thermal conductivity, these 2D materials qualify for applications in electronics, e.g., as field-effect transistors; in optoelectronics, e.g., as photodetectors; and as light emitting devices. Other potential applications include batteries, photovoltaics, spintronics, thermoelectrics, nonlinear optics, catalysis, biosensors, and gas sensors.

P. Hess, *Bonding, Structure, and Performance of Two-Dimensional Materials*,
NanoScience and Technology, https://doi.org/10.1007/978-3-031-74161-6_8

8.1 Nitrogene

8.1.1 Chemistry and 2D Bonding

Independent density functional theory (DFT) calculations confirmed the existence of a variety of stable 2D structures of the group of pnictogens by employing the generalized gradient approximation (GGA) with Perdew–Burke–Ernzerhof (PBE) functionals. The periodic table in Fig. 8.1 highlights the five elements N, P, As, Sb, and Bi of group VA. The N atom with the electronic configuration $(1s^2 2s^2 p_x^2 p_y^1)$ has five valence electrons for covalent bonding.

The most stable phase of nitrogen is the hexagonal lattice with a buckled honeycomb structure named nitrogene [1]. The threefold-coordinated network consists of single-bonded N atoms providing fundamental bonding strength via covalent σ bonds. The remaining two electrons of each N atom form an orbital with relatively high *p* content, which creates π-like bonds between the nearest neighbors. The π-like bonds ensure the planar arrangement of two parallel planes shaped by the pyramidal sp^3-like hybrids of the nitrogene monolayer (see Fig. 8.2) [1]. Most 2D pnictogens are indirect semiconductors with band gaps ranging from 0.43 to 3.7 eV [2]. The cohesive energy of the nitrogene monolayer at $E_c = -3.67$ eV/atom is very low for such a short bond length as d = 0.149 nm [1]. Other first-principles calculations predict much stronger bonding at $E_c = -6.81$ eV/atom for the buckled honeycomb structure and only a slightly lower energetic stability of $E_c = -6.45$ eV/atom for a 2D octa-structure composed of squares and octagons arranged in a buckled 2D lattice [3].

Fig. 8.1 Group–VA elements of the periodic table involved in the formation of elemental pnictogen monolayers

H							He
Li	Be	B	C	N	O	F	Ne
Na	Mg	Al	Si	P	S	Cl	Ar
K	Ca	Ga	Ge	As	Se	Br	Kr
Rb	Sr	In	Sn	Sb	Te	I	Xe
Cs	Ba	Tl	Pb	Bi	Po	At	Rn
Fr	Ra						

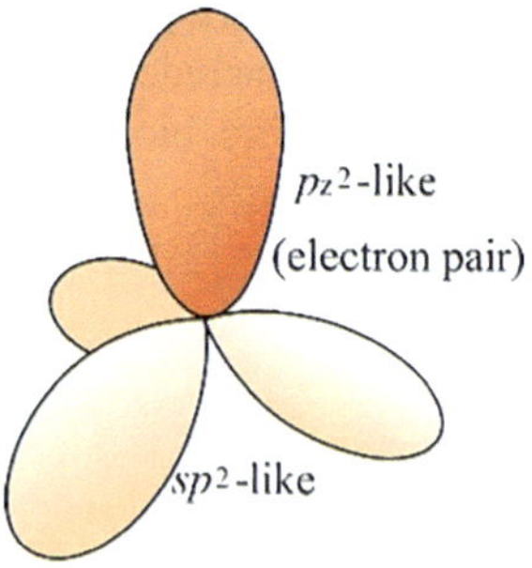

Fig. 8.2 Schematic representation of the geometric structure of the sp^3-like hybrids with a $p_z{}^2$-like hybrid of the lone electron pair, which is not a pure *p* state because it contains *s* state contributions

Note that the nitrogene monolayer has a positive formation energy relative to the triple-bonded N_2 molecule of 1.33 eV [1]. It is important to keep in mind that the bonds in nitrogen molecules belong to the strongest chemical bonds, namely, triple bonds N≡N with a bond energy of 9.89 eV [4]. Consequently, nitrogen exists under ambient conditions in molecular form, and 2D-bonded nitrogene networks possess an enormous energy content. Therefore, these lattices are considered high-energy density materials (HEDMs). 2D phases such as buckled nitrogene are metastable against decomposition into molecular gas. Nevertheless, these metastable structures are at a local minimum on the Born–Oppenheimer surface, which may guarantee dynamic and thermal stability.

8.1.2 2D Structures

The most stable buckled honeycomb structure of nitrogene has a lattice constant of a = 0.227 nm, a bond length of d = 0.149 nm, a bond angle of $\theta = 99°$, and a low buckling distance of $\Delta z = 0.070$ nm [1], in good agreement with the results of other calculations (see Table 8.1) [2, 3]. In the buckled structure of Fig. 8.3, the orbitals extending on both sides perpendicular to the plane with the lone electron pair are indicated schematically by thick bonding symbols. Note that molecular dynamics (MD) simulations indicate that the dynamically stable nitrogene monolayer is thermally stable above room temperature [1]. The space group of this allotrope is *P*-3*m*1 [4]. Nitrogene is an insulator with a large band gap of 3.96 eV (DFT) and 5.96 eV (Heyd-Scuseria-Ernzerhof approach (HSE)), in good agreement with the results of other publications [1]. Adding the buckling effect to the tabulated van der Waals diameter of $d_w = 0.32$ nm yields an estimate of the monolayer thickness of $d_{w+\Delta z} = 0.39$ nm [5]. See Table 8.1 that contains the lattice parameters of the group–VA monolayers).

Nitrogene forms bilayers and multilayers with AB or ABAB stacking, where the N–N bonds in each monolayer become staggered relative to those in the adjacent layers. The bilayer has a minimum energy at an interlayer distance of $d_s = 0.338$ nm and an interlayer interaction energy of $E_{int} = -10$ meV/unit cell. For the 3D multilayer structure ('nitrogenite'), the interlayer distance decreases to 0.326 nm, and the interaction energy increases to – 98 meV/unit cell due to increased interlayer interactions on both sides of the monolayer [1] The theoretically predicted 3D graphitic structure of nitrogenite remains stable above room temperature.

Another predicted allotrope is the buckled octa-structure composed of squares and octagons with geometrical data of a = 0.442 nm for the lattice constant, $d_1 = 0.152$ nm and $d_2 = 0.141$ nm for the bond lengths, and $\Delta z = 0.069$ nm for the buckling distance [3] The unit cell of this structure contains 8 atoms instead of 4 owing to its lower symmetry. The band gap of the indirect semiconductor is $E_g = 2.61$ eV (PBE). As mentioned above, this structure is considered to be energetically and dynamically stable because the cohesive energy is negative and no imaginary vibrational mode could be detected [3] Importantly, other authors have reported an instability of this

Table 8.1 Phase, cohesive energy, lattice constants, bond angles, bond lengths, corrugation effect, van der Waals diameter, corrugation corrected van der Waals diameter, and interlayer separation of elemental group–VA monolayers

Monolayer	Structure	Cohesive energy, E_c (eV/atom)	Lattice constants a, b (nm)	Bond angles $\theta_1, \theta_2, \theta_3; \theta$, (deg.)	Bond lengths d_1, d_2 (nm); d (nm)	Corrugation effect Δz (nm)	Van der Waals diameter d_w [5], $d_{w+\Delta z}$ (nm)	Interlayer spacing d_s (nm)
N	hb	– 3.67 [1], – 3.28 [6]	0.227 [1], 0.229 [6]	99 [1], 94.9 [6]	0.149 [1], 0.156 [6]	0.070 [1]	0.310 [5], 0.390	0.338 [1]
	hb	– 6.81 [3]	0.233 [2]		0.150 [2]	0.070 [2]	0.310 [2], 0.380	0.326 [1]
P	sw	– 3.59 [7], – 2.57 [6]	0.331, 0.455 [7]	96, 104 [7]	0.222, 0.225 [7]	0.212 [7]	0.390 [5], 0.602	0.524 [8]
	hb	– 3.55 [7], – 2.32 [6]	0.328 [6]	93.09 [6]	0.226 [6]	0.123 [6]	0.390 [5], 0.513	0.449 [9]
As	sw	– 3.13 [10], – 1.90 [6]	0.472, 0.367 [10]	94.7, 100.3 [6]	0.251, 0.248 [10]	0.239 [10]	0.410 [5], 0.649	0.566 [10]
	hb	– 3.13 [10], – 2.29 [6]	0.360 [10]	91.4 [11], 91.9 [6]	0.251 [10]	0.140 [10]	0.410 [5], 0.550	0.467 [10]
Sb	aw	– 2.61 [12], – 1.77 [6]	0.474, 0.428 [14]	94.4, 103.5 [7]	0.291, 0.284 [14]	0.282 [7]	0.440 [5], 0.719	0.616 [15]
	hb	– 2.84 [13], – 1.96 [6]	0.408 [11]	90.0 [11]	0.288 [11]	0.166 [11]	0.440 [5], 0.606	0.365 [15]
Bi	aw	– 2.42 [39], – 1.38 [6]	0.487, 0.444 [7]	92.0, 107.4 [7]	0.308, 0.303 [7]	0.272 [7]	0.460 [5], 0.732	
	hb	– 2.41 [39], – 1.44 [6]	0.438 [17, 18]	91.2 [17, 18]	0.307 [17, 18]	0.174 [17, 18]	0.460 [5], 0.634	

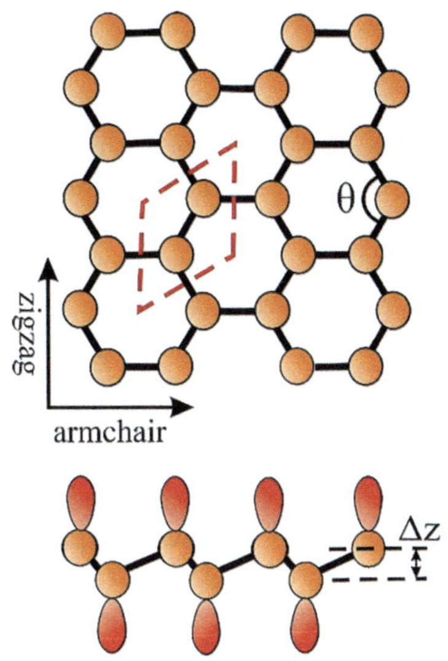

Fig. 8.3 Top view and side view of the geometric structure of the buckled phase with two sublayers and lone electron pairs. The dashed red line indicates the unit cell with two atoms

phase by disintegration into N_2 molecules, observed during structural optimization [1]. Another proposed single-bonded 2D nitrogen allotrope is the zigzag sheet, which has a direct band gap of 1.25 eV [4]. According to the authors, this semiconducting allotrope with the space group *P2mm* is dynamically stable at 0 K.

8.1.3 Mechanics

A detailed study of the mechanical behavior of nitrogene is still missing. MD simulations predict a surprisingly large in-plane stiffness of E = 270 N/m, which is second after that of graphene for a monolayer containing a single element [1]. Such a robust mechanical stability seems not to be consistent with the low cohesive energy of E_c = – 3.67 eV/atom, however in line with the estimated short bond length [1]. The DFT-PBE approach predicts a much smaller stiffness of only E = 154 N/m and strongly differing Poisson ratios of $\nu_{xy} = \nu_{yx} = 0.0752$, $\nu_{xz} = \nu_{yz} = 1.536$, and $\nu_{zx} = \nu_{zy} = 0.0118$, which are derived from elastic constants that satisfy the Born–Huang criteria of mechanical stability [4]. Note that also this much lower stiffness does not correlate well with the calculated low value of the cohesive energy. The linear and nonlinear properties of the five group–VA monolayers are collected in Table 8.2.

Table 8.2 Phase, space group, Young's modulus, Poisson's ratio, ultimate strength, and ultimate strain of elemental group-VA monolayers

Monolayer	Phase	Space group	Young's modulus E_{ac}, E_{zz} (N/m)	Poisson ratio ν_{ac}, $\nu_{zz;}$ ν_{xy}, ν_{yx}	Ultimate strength σ_{ac}, σ_{zz} (N/m)	Ultimate strain ε_{ac}, ε_{zz}
Nitrogene	hb	*P*-3*m1*	270 [1], 154 [4]	0.0752 [4]		
Phosphorene	sw	*Pmna* [12]	25, 93 [20], 22, 56 [21]	0.17, 0.62 [20]	4.5, 10.1 [20]	0.30, 0.27 [20]
	sw	*Pmna* [19]	23, 92.3 [22]	0.175, 0.703 [22]	2.2, 4.5 [23]	0.295, 0.161 [23]
	hb	*P*-3*m*1 [19]	75.5 [7], 79.7 [11]	0.107 [7]	7.06, 6.72 [11]	0.16, 0.21 [11]
Arsenene	sw	*Pmna* [24]	20, 55 [10], 29, 75 [25]	0.33, 0.91 [10]		
	hb	*P*-3*m*1 [24]	58 [10], 52.8 [11]	0.21 [10]	5.15, 4.69 [11]	0.17, 0.20 [11]
Antimonene	aw	*Pmn*2_1 [16]	12, 29 [14]	0.36, 1.20 [14]	5.15, 4.69 [11]	0.32, 0.18 [15]
	hb	*P*-3*m*1 [16]	41 [14], 32.9 [11]	0.21 [14]	3.75, 3.14 [11]	0.20, 0.21 [11]
Bismuthene	aw	*Pmn*21 [16]	10, 25.5 [17]	0.261, 0.648 [17]	4.21 [27] (300K)	0.195, 0.255 [27]
	hb	*P*-3*m*1 [16]	23.9 [17], 26.3 [26]	0.327 [17]		

8.1.4 Physics and Applications

First-principles calculations revealed anisotropic behavior of nitrogene [28]. The discovery of a weak intrinsic and extrinsic spin–orbit coupling (SOC) suggests that doped nitrogene ribbons may be attractive materials for spintronic applications [28]. An in-depth study on the modification of the electronic structure of nitrogene revealed the possibility of engineering the band structure by increasing the number of stacked monolayers, applying biaxial tensile strain, or applying a perpendicular electric field [29]. As the number of layers increases, the band gap decreases linearly due to increasing interlayer coupling. Similarly, the high-energy bands move down with increasing electric field and, at a critical strength of the electric field, the band gap closes. Consequently, all these tuning procedures reduce the band gap and eventually close the gap.

A comprehensive review covers the progress made in the investigation of all main group–VA monolayers (pnictogens), including nitrogene [30]. This work provides an extensive survey of the literature on 2D pnictogens published before 2019, with structural data on the puckered, buckled, and square-octagon phases, the linear elastic properties, and electronic and optical properties. Furthermore, this review discusses stability aspects, thermal properties, chemical functionalization, defects, properties of bilayers and multilayers, and synthesis including stabilization of monolayers by adsorbate–substrate interactions.

8.1.5 Synthesis

Despite the theoretically predicted dynamic and thermal stability of the specified monolayers, bilayers, and multilayers, the synthesis of nitrogene remains a challenge due to the enormous stability of N_2 molecules [1]. The authors showed that for the growth of nitrogene on substrates such as Al(111) or graphene, the pristine properties of nitrogene are maintained owing to weak interactions with the substrate. The open question is whether sufficient stabilization is possible on a suitable surface without disturbing the intrinsic properties of nitrogene. In an early experiment, the authors accomplished the epitaxial growth of 2D atomic sheets of nitrogene embedded in a GaAs host matrix by using the site-controlled N σ-doping technique. However, no structural identification of any of the theoretically predicted phases could be achieved [31]. A unique puckered bulk structure of nitrogene with single-bonded layers could be realized by epitaxial growth at high pressure and high temperature, and a structure also observed for the homologue black phosphorus could be identified [32]. Under normal conditions, however, this structure was not stable.

8.2 Phosphorene

8.2.1 Chemistry and 2D Bonding

Like the other pnictogens, phosphorus has the electronic configuration (Ne|$3s^2 3p^3$) with five valence electrons in the outer shell and needs three extra electrons to reach octet stability. In the phosphorene monolayer, phosphorus forms sp^2-like hybrids, which are intermediates between pure sp^2 and sp^3 hybrids with three σ bonds and a lone pair of valence electrons in a $p_z{}^2$-like orbital that is not a pure *p* state but contains *s* state contributions. The three unpaired electrons of the σ electrons connect with those of the three nearest P atoms, building a pyramid with strong covalent σ bonds. Two neighbors are in the same plane, and one is in the adjacent plane at an angle of 102.4° [33]. The most stable corrugated structure maximizes the distance between repulsive lone electron pairs by positioning alternate electron pairs above and below the phosphorene plane. Of importance is the possibility of mapping such a threefold coordinated structure onto a hexagonal lattice.

The most stable layered bulk structure of orthorhombic black phosphorus (α-phosphorus) is the puckered symmetric washboard (sw) configuration, which is also the most stable structure of phosphorene (see Fig. 8.4) [34]. An initially generated monolayer with an asymmetric washboard (aw) configuration spontaneously transforms into the sw phase. In this stable monolayer, six P atoms of the sw lattice are linked to a hexagonal ring structure, which is puckered (ridged, corrugated) owing to sp^3-like bonding [33]. The reported cohesive energies of the sw phase are E_c = – 3.59 eV/atom [7], – 3.45 eV/atom [12], and – 2.57 eV/atom [6]. It is not explained by the authors why the cohesive energies of all pnictogen monolayers given in the last reference are systematically smaller than those of the other studies, despite the application of the DFT-PBE approach.

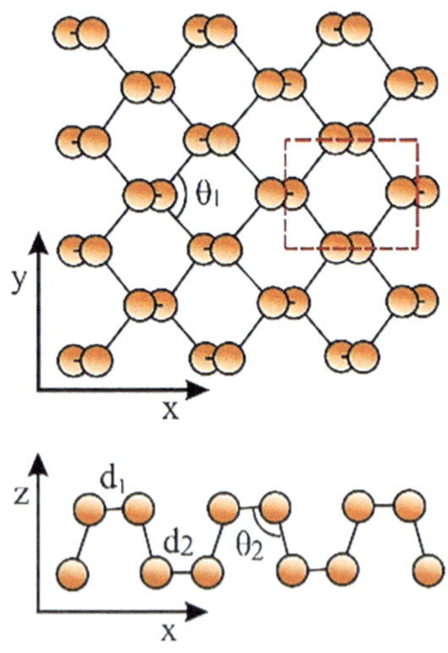

Fig. 8.4 Geometric structure of the puckered symmetric washboard (sw) phase with four sublayers

Another well-studied layered allotrope of bulk phosphorus with similar stability is blue phosphorus (β-phosphorus). The monolayer derived from this structure consists of hexagonal buckled (hb) phosphorene (β-phosphorene) with a cohesive energy quite similar to that of puckered phosphorene of $E_c = -3.55$ eV/atom [7], – 3.48 eV/atom [35], and – 2.32 eV/atom [6]. The similarity of the binding energies of the sw and aw phases is not surprising since both phases comprise a hexagonal structure with only minor differences in the bonding arrangement of the atoms. Note the similarity of the two larger cohesive energies with the lower value reported for nitrogene and the significant difference in the corresponding bond lengths. In addition to the two layered structures already considered, phosphorus has at least ten allotropic phases with similar stabilities but distinct electronic properties [35].

8.2.2 2D Structures

The strongly folded accordion-like structure of sw phosphorene with grooves in the zigzag direction is responsible for the two lattice parameters of a = 0.331 nm perpendicular to the grooves in the armchair direction (a direction, x axis) and b = 0.455 nm in the zigzag direction (b direction, y axis) (see Fig. 8.4) [6, 7]. The pucker effect leads to two different bond lengths: a shorter in-plane bond of $d_1 = 0.222$ nm and a slightly longer bond of $d_2 = 0.225$ nm between the two planes [7] The bonding angle between neighboring P atoms on the same plane is $\theta_1 = 96.1°$ (hinge angle), while the angle between atoms on the two adjacent planes is $\theta_2 = 103.5°$ (dihedral angle) [7]. The published puckering heights or layer distances of the sw phase vary between $\Delta z = 0.212$ and 0.251 nm and thus differ substantially [7, 20].

Determination and control of the monolayer thickness and interlayer separation and thus the layer number in few-layer assemblies are critical issues, for example, during the growth of samples. The methods routinely used for accessing these properties are nanoindentation by atomic force microscopy (AFM) and Raman spectroscopy [36]. Owing to the fast degradation of phosphorene due to the reactive lone electron pairs and large surface area, these methods are not applicable under ambient conditions. Similarly, the use of high-power lasers may lead to damage of phosphorene. A fast and reliable method for determining the layer number during growth is optical interferometry, which is a noninvasive approach [36].

Measurements of the surface corrugation and layer separation of cleaved and exfoliated few-layer phosphorene samples are possible by applying surface-sensitive dynamical microspot low-energy electron diffraction (μLEED) and via high spatial resolution low-energy electron microscopy (LEEM) [8]. These measurements provided an interlayer distance of $d_s = 0.524$ nm. Taking the van der Waals diameter of $d_w = 0.390$ nm for P atoms [5], we estimate a layer thickness of $d_{w+\Delta z} = 0.572$ nm for the corrugated sw monolayer, in good agreement with the suggested effective thickness of t = 0.563 nm [22], and the interlayer spacing of black phosphorus of $d_s = 0.555$ nm [20]. For comparison, nanoindentation measurements of the step height of a single layer of sw phosphorene on a substrate by AFM resulted in a much larger thickness of t = 0.85 nm than the former multilayer-based value [37]. As discussed in more detail below, the pucker configuration causes strongly anisotropic physical properties. The space group of the sw phase is *Pmna* [12].

In the structure of the hexagonal buckled phase of phosphorene the atoms alternatively extend out of the plane. The structural data a = 0.328 nm, d = 0.226 nm, Δz = 0.123 nm, and θ = 93.09° characterize the hb phase [6]. The much greater bond length explains the greater buckling effect in comparison with that of nitrogene. The bond angle between neighboring bonds of θ = 93.09° (92.6° [11]) can be explained by the repulsive interaction between the electron lone pairs and the type of bonding orbitals [6]. Since the size of the band gap is slightly above the photon energy of visible blue light, this material is also referred to as blue phosphorene [38]. Adding the van der Waals diameter and the buckling effect yields a lower thickness of $d_{w+\Delta z}$ = 0.515 nm for the corrugated hb phase than that for the sw phase owing to the smaller corrugation effect. The space group of the hb structure is *P*-3*m*1 [19].

A systematic study based on the DFT-GGA approach compares the cohesive energies of the puckered and buckled structures of group–VA monolayers [39]. According to these calculations, the sw structure is slightly more stable for phosphorene and bismuthene, and the hb structure is more stable for arsenene and antimonene. For example, the calculated cohesive energy for the sw phase of phosphorene is E_c = – 3.59 eV/atom, and that of the hb phase is – 3.55 eV/atom [7]. The sw puckered structures of the group–VA elements have a direct band gap, whereas the hb structures are indirect semiconductors if SOC is included [39]. Generally, hb structures have a larger band gap than sw structures, e.g., E_g = 0.83 eV for the sw structure and 2.10 eV for the hb structure [39].

The electronic and optical properties of phosphorene can be tuned by changing the thickness and stacking order of few-layer assemblies. Systematic studies are available for the monolayers of P (sw), As (sw), and Sb (aw) of the three bilayer stacking orders AA, AB, and AC. A shift of the second layer by ½ of the lattice period along the x or y axis reproduces the AB bulk structure and the most stable stacking configuration (see Fig. 8.5). A shift of 1 lattice period along the x or y axis produces the AC stacking order [30]. For AB stacking, the interlayer distance of d_s = 0.434 nm is smaller than the estimated layer thickness, and the cohesive energy of the bilayer of E_c = – 901 meV/atom is approximately 10 meV/atom smaller than those of the other stacking modes [40]. Surprisingly, this separation is much smaller than that obtained by others for layered sw structures, for which the separation between adjacent AB-stacked layers is d_s = 0.530 nm [33]. For comparison, a very low interlayer interaction energy of E_{int} = – 20 meV/atom has been found for layered black phosphorus, allowing easy mechanical exfoliation [33]. Both the interlayer interaction energy and the required exfoliation energy of 0.339 J/m^2 are extremely close to that of graphene (0.344 J/m^2) [33].

The corrected van der Waals layer thickness of the hb structure at $d_{w+\Delta z}$ = 0.515 nm is in reasonable agreement with d_s = 0.563 nm obtained by DFT-PBE calculations for AB stacking of blue phosphorus [41]. The very low interlayer interaction energy of E_{int} = – 12.5 meV/atom (– 6 meV/atom [41]) for AA and AB stacking points to an even weaker van der Waals interaction than for sw monolayers and the possibility

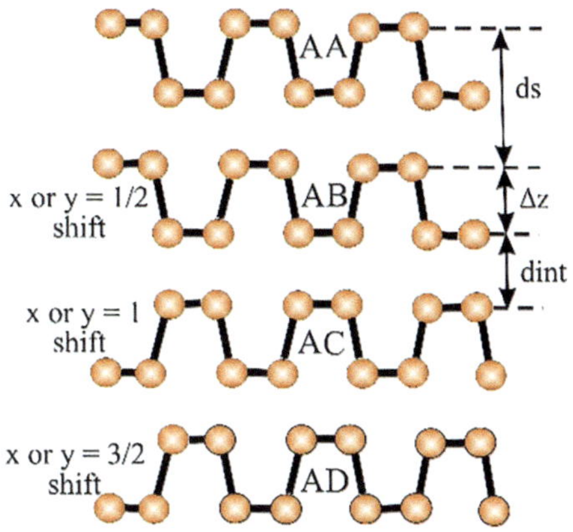

Fig. 8.5 Stacking orders of the AA, AB (x or y = ½ a shifted), and AC (x or y = 1 a shifted) arrangements. Adapted with permission from Ref. [30], Copyright 2019, AIP Publishing

of easily exfoliating single layers [9]. As shown recently, the corrugation of the hb phase leads to additional changes besides the already known high-symmetry AA and AB stacking orders [38]. High-level first-principles computations revealed that the lowest energy configuration of the hb bilayer belongs to a 2D metal phase with a smaller interlayer distance of $d_s = 0.301$ nm and a smaller buckling height of $\Delta z = 0.117$ nm [38].

8.2.3 *Mechanics*

The 2D Young's moduli of the sw phase at $E_{ac} = 44$ GPa (25 N/m) and $E_{zz} = 166$ GPa (93 N/m), calculated by using a layer thickness of $d_s = 0.563$ nm to transform the 3D to 2D units [20], and the 2D stiffnesses $E_{ac} = 21.9$ N/m and $E_{zz} = 56.3$ N/m [21] strongly deviate between the zigzag and the armchair directions. This enormous anisotropy of the stiffness is illustrated in Fig. 8.6, which describes the complete angle dependence of the stiffness of sw phosphorene [20]. The reason is that tension unfolds and flattens the puckered structure predominantly through change of the dihedral angles in the more ductile armchair direction without a significant extension of strong covalent bonds (see Fig. 8.4). Independent calculations support the substantially higher stiffness in the zigzag direction by the strongly deviating values of $E_{ac} = 23.0$ N/m and $E_{zz} = 92.3$ N/m [22]. The reported Poisson ratios of $\nu_{ac} = 0.175$ and $\nu_{zz} = 0.703$ [22], as well as the similar values $\nu_{ac} = 0.17$ and $\nu_{zz} = 0.62$ [20], further confirm the pronounced anisotropy of the sw phase. Interestingly, the puckered structure shows the phenomenon of negative Poisson ratios (NPRs), which makes phosphorene a 2D auxetic material. For example, phosphorene expands in the out-of-plane z direction by stretching in the zigzag direction and contracts in the z direction by compressing in the zigzag direction with a negative value of $\nu_{yz} = -0.027$ [33].

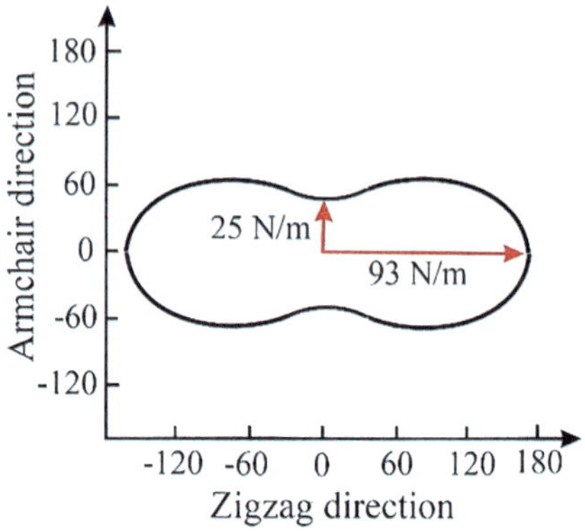

Fig. 8.6 Scheme of the strong angle dependence of the stiffness of phosphorene with limiting values in the armchair and zigzag directions. Adapted with permission from Ref. [20], Copyright 2014, AIP Publishing

In a comprehensive DFT-GGA-PBE study, strain-induced electronic and structural transitions were investigated for monolayers, bilayers, and puckered bulk structures of phosphorus, arsenic and antimony [42]. In addition to the electronic band structures and band gaps, Young's moduli, shear moduli, bulk moduli, and Poisson's ratios are presented as a function of the in-plane orientation and as isotropic averages, elucidating the technologically relevant mechanical and electronic properties. In general, the behavior depends on the underlying anisotropic crystal structure, which changes strongly with the number of layers, with phosphorene typically the stiffest and antimonene the most flexible layered system.

The usually calculated ultimate or fracture strength of chemical bonding describes the maximum stress a perfect crystal can withstand at 0 K. Thus, the intrinsic strength measures the overall mechanical stability of bonding in the 2D network against extensive deformation. The strength of the sw phase shows the expected strong anisotropy with $\sigma_{ac} = 8.0$ GPa (4.5 N/m) in the armchair direction and $\sigma_{zz} = 18$ GPa (10.1 N/m) in the zigzag direction [20]. Perpendicular to the pucker, the ultimate strain is $\varepsilon_{ac} = 0.30$, and parallel to the pucker, it is $\varepsilon_{zz} = 0.27$. These authors assumed that $d_s = 0.555$ nm is the bilayer spacing, which is in excellent agreement with the above-introduced value of 0.563 nm applied here [20].

For completeness, the results of MD simulations are presented as follows: Young's moduli, $E_{ac} = 24.3$ GPa (13.7 N/m), $E_{zz} = 103.8$ GPa (58.5 N/m); ultimate strengths, $\sigma_{ac} = 3.94$ GPa (2.2 N/m), $\sigma_{zz} = 7.95$ GPa (4.5 N/m); and fracture strains, $\varepsilon_{ac} = 0.295$, and $\varepsilon_{zz} = 0.161$ [23]. The much greater fracture strain in the armchair direction, in comparison with that in the zigzag direction, can be explained by additional contributions besides the change due to bond extension in this direction under uniaxial tension. Under these conditions, failure along the armchair direction initially induces changes in the interlayer bond angles, whereas failure in the zigzag direction involves both changes in intralayer angles and breaking of covalent bonds [23]. MD simulations allow the extension of our knowledge of mechanical properties to finite temperatures, which are usually calculated by first principles at 0 K. Such MD simulations, employing a Stillinger–Weber potential, reveal a substantial reduction in fracture strength and strain of more than 65% for a temperature increase from 0 to 450 K [43].

In contrast to the strong anisotropy of the sw phase, the hb phase shows an isotropic stiffness of E = 75.5 N/m and a small Poisson ratio of $\nu = 0.107$ in the plane [7], in good agreement with the finding of E = 79.7 N/m in another study [11]. A slightly larger ultimate strength of $\sigma_{ac} = 7.06$ N/m has been reported for the armchair direction than for the zigzag direction at $\sigma_{zz} = 6.72$ N/m, while the critical strain in the armchair direction of $\varepsilon_{ac} = 0.16$ is smaller than that in the zigzag direction, $\varepsilon_{zz} = 0.21$ [11]. The low uniaxial tensile strains of 0.16 – 0.21 imply that the buckled structure can sustain only a lower mechanical strain and thus has a lower flexibility than the puckered structure with values of approximately $\varepsilon \approx 0.30$. Considering the hb lattice, it is not surprising that the stiffness is isotropic for small strains ($\varepsilon < 0.02$), while a change in the ultimate strength and strain occurs if the hexagonal symmetry is broken by uniaxial strains in the nonlinear part of the stress–strain curve. Note that this is not the case when applying biaxial strain.

8.2.4 Physics and Applications

Phosphorene is a promising candidate for electronic applications as a 2D transistor, light harvesting photocatalyst, and energy storage device, as realized by few-layer assemblies less than 10 nm in thickness. These applications are primarily based on the exceptional anisotropic electronic and optical properties of the sw phase and a wide absorption range of interest in linear and nonlinear optics. Among possible biomedical applications, the most promising may be as biosensors to sensitively detect diseases [44]. The enormous amount of attention gained by the sw phase is based primarily on the direct band gap of the *p*-type semiconducting material with high hole mobility and the possibility of tuning the band gap by in-plane strain or the number of layers in few-layer assemblies. Band gap engineering provides access to gaps between the bandgap of the monolayer $E_g \approx 1.0$ eV and the bulk value of ≈ 0.3 eV [37]. Reviews cover the progress made in synthesis, the electronic, mechanical, thermal, and optical properties, and various potential applications [45, 46].

Similar to various bulk phases, several metastable phosphorene structures, such as γ-P, δ-P, ε-P, ζ-P, η-P, and θ-P, with direct and indirect band gaps, have been characterized [46]. Especially in sensing applications such as gas sensing, humidity sensing, biosensing, ion sensing, and chemical sensing, advances in the development of sw sheets have been predicted [47]. These exceptional sensing properties can be attributed to the high carrier mobility, direct band gap, and lone electron pair of phosphorene.

First-principles DFT calculations and density-functional perturbation theory (DEPT) were used to elucidate the correlation between the vibrational and thermal transport properties of the group–VA monolayers. Lattice thermal transport is correlated with the average atomic mass and is almost inversely proportional to the temperature. Thermal transport properties are of interest for various future applications, such as thermoelectric power generation [12]. The thermal conductivity of phosphorene can be tuned via substitutional As, Sb, and Bi doping, which reduces the thermal conductivity, for example, in the zigzag direction from 109.6 W m^{-1} K^{-1} to 10 W m^{-1} K^{-1}. This should be compared with the ultra-low thermal conductivity of PBi in the armchair direction of 1.5 W m^{-1} K^{-1}.

A serious drawback is the reactivity of phosphorene, which suffers from oxidation and degradation effects under ambient conditions. The degradation is due to the dissolution of oxygen in adsorbed water and accelerated deterioration by light exposure, which leads to the formation of phosphoric acid, and related species [48, 49]. This rapid degradation weakens its function and leads to adverse effects. Oxidative degradation within days requires effective passivation schemes during synthesis, characterization, and processing. Suggested effective strategies to stabilize the material include chemical functionalization, encapsulation, and passivation by noncovalent protection schemes, e.g., aluminum oxide, h-BN, sterically hindered solvent molecules, or other methods, which do not affect the intrinsic properties or functions of the monolayer.

First-principles calculations indicate that all kinds of point defects may be created quite easily with a higher areal density than in graphene. Thus, the deterioration of physicochemical properties will affect the mechanical and electronic behavior of phosphorene [50]. For example, the measured carrier mobility of phosphorene is always lower than theoretically predicted owing to uncontrollable surface and edge defects.

8.2.5 Synthesis

According to theory, both the hb phase and the sw phase form energetically, dynamically, and thermally stable free-standing monolayers and few-layer sheets. Unfortunately, the established mechanical exfoliation of phosphorene monolayers from layered black phosphorus is not suitable for large-scale applications due to the limited flake size and low production rate [48]. In general, top-down methods lack precise control of the lateral dimension and number of layers in the sheet. The deposition of single-layer hb phosphorene was first performed by molecular beam epitaxy (MBE) on Au(111) using black phosphorus as a precursor and employing in situ low-temperature scanning tunneling microscopy (LT-STM) in combination with DFT calculations [51]. The epitaxial growth of single layers of the hb phase on Au(111) was confirmed and the quality was verified by in situ LT-STM and LEED analysis, photoemission spectroscopy (PES), and angle-resolved photoemission spectroscopy (ARPES) [52]. While the sw and hb phases are nearly equally stable, the latter seems to be the preferred phase for growing on a substrate with hexagonal symmetry.

Recent reviews have described in detail the progress made by top-down methods (mechanical, liquid, and electrochemical exfoliation) and bottom-up methods such as chemical vapor deposition (CVD) and MBE [34, 44]. Importantly, by considering the efficiency and scalability of these methods, the authors considered electrochemical exfoliation and shear exfoliation to be the most promising methods [44]. A study on recent progress in synthesis with a special focus on epitaxial growth also addressed the poor chemical stability of phosphorene, leading to deterioration of few-layer systems under collective influence of moisture, water, oxygen, and light [48].

An up-to-date review covers the fundamental properties of group–VA semiconductors including phosphorene and diverse fabrication methods such as the top-down methods of mechanical and ultrasonic exfoliation, plasma etching, hot-pressing, and electrochemical approaches, as well as the bottom-up methods of CVD, van der Waals MBE, and solvothermal techniques. As potential applications of these 2D materials the review discusses implementations as field-effect transistor (FET), photodetector, memristor, sensor, and in batteries [53].

8.3 Arsenene

8.3.1 Chemistry and 2D Bonding

The group–VA element arsenic has five valence electrons in the outer shell with the electronic configuration (Ar|$3d^{10}4s^2p^3$). The planar structure of sp^2 hybrids is not stable owing to the atom size and the resulting weak π-π orbital coupling between the nearest neighbors. In arsenene, sp^3-like hybridization of *s* and *p* orbitals stabilizes a buckled or puckered configuration, as in phosphorene. Three σ back-bonds to the nearest three atoms provide the main rigidity of the corrugated structure, and the remaining two electrons develop π-like states with up- and down-oriented orbitals with the lone electron pairs. In principle, sp^3-type hybridization of *s* and *p* orbitals creates three σ bonds of sp^2-like hybrids and one lone pair of electrons, forming p_z-like π states (see Fig. 8.2). The symmetric washboard structure with a chair-like configuration is energetically more stable than the asymmetric washboard structure at least up to room temperature [10]. The unit cell contains four atoms, and each atom connects with two atoms in the same plane and one atom in the other plane. The selected cohesive energies of the sw phase are $E_c = -3.13$ eV/atom [10], – 2.95 eV/atom [54], and – 1.90 eV/atom [6]. Note the nearly identical binding energies of hb arsenene at $E_c = -3.13$ eV/atom [10], – 2.99 eV/atom [54], and – 2.97 eV/atom [35], excepting the low cohesion of – 2.29 eV/atom [6].

The unit cell of hb arsenene contains two atoms, and each atom has covalent bonds to three atoms of the other plane, with the As atoms arranged in two planes in an alternating manner at the corners of a hexagon. The reported cohesive energies of the hb phase are in the same range as those of the sw phase, with large deviations between the results of different authors. Importantly, in some cases, the deviation between buckled and puckered phase is smaller than the thermal energy at room temperature, which is also true for the two bulk phases. The presence of an arsenic layer adsorbed on an Ag(111) substrate with a lattice constant of a = 0.36 nm clearly points to the existence of buckled arsenene in the adsorbed state [55].

8.3.2 2D Structures

As discussed for phosphorene, the puckered sw phase and buckled hb phase are the two most stable configurations of arsenene and are considered in detail here [10]. The metastable aw structure transforms spontaneously into the more stable orthorhombic sw phase of arsenene with a symmetric puckered configuration upon relaxation. The lattice constants, bond lengths, and pucker effect of the sw phase are a = 0.472 nm, b = 0.367 nm, d_1 = 0.251 nm, d_2 = 0.248 nm, and Δz = 0.239 nm, respectively [10]. With a van der Waals diameter of d_w = 0.410 nm[5] and the pucker distance, a layer thickness of $d_{w+\Delta z}$ = 0.649 nm can be estimated for the sw phase. The calculated bond angles are θ_1 = 94.5° (94.7 [6]) and θ_2 = 100.6° (100.3 [6]) for sw arsenene [12], with the space group *Pmna* [12, 54]. The sw phase is dynamically stable with positive phonon dispersion and is thermally stable above room temperature, as verified by ab initio MD simulations [10]. The indirect band gaps obtained by the standard approaches are E_g = 0.86 eV (PBE-SOC), 1.47 eV (HSE-SOC), and 1.58 (G_0W_0) [10]. Neglecting the SOC effect leads to slightly smaller bandgaps.

Gray arsenic, a bulk form of arsenic, can be considered as the layered precursor of the hb structure [54]. The reported structural data of the buckled phase are a = 0.360 nm and d = 0.251 nm. The buckling effect of the structure is much smaller at Δz = 0.140 nm than the corrugation of the puckered structure [10]. The corresponding bond angles are θ = 91.4° [11] (θ = 91.9° [6]). These structural parameters agree quite well with the results obtained by other studies [6, 11, 54]. Analysis of a quasi-free-standing buckled arsenene on Ag(111) by LEED and scanning force microscopy (STM) delivered a lattice constant of a = 0.36 nm [55]. With a van der Waals diameter of d_w = 0.410 nm and the buckling effect, a thickness of $d_{w+\Delta z}$ = 0.550 nm is estimated for the hb phase [5]. Positive frequencies for all the vibrational modes were found via DFT-PBE calculations, and the thermal stability was predicted via ab initio MD simulations [10]. However, some authors state that the local minima corresponding to these monolayers may not be deep enough to guarantee thermal stability, and others obtained negative frequencies for the transverse acoustic mode with a strong dependence on the computational parameters used for the hb phase [54]. The space group of the hb structure is *P*3*m*1 [54] (*P*-3*m*1) [24]. All approaches yield substantially larger band gaps for the hb phase than for the sw phase, with values of E_g = 1.46 eV (PBE-SOC), 2.10 (HSE-SOC), and 2.64 (G_0W_0) [10]. Since the fundamental gap opens with decreasing bond length, the opening of the gap in the hb phase can be interpreted as a consequence of the confinement effect due to the more compact buckled structure. Recent computations employing the GW approach confirmed a large band gap of E_g = 2.56 eV for the hb phase that is in the range from the near-infrared to the visible region [24].

Bilayer and few-layer materials are important for future technological applications because electronic properties such as the band gap are tunable by the number of layers and their stacking order. Usually, the gap increases with decreasing number of layers due to the confinement effect. Here, we demonstrate the typical differences observed in the theoretical treatment of few-layer systems by different approaches.

According to DFT-GGA-PBE calculations, AA stacking is the most stable configuration for sw bilayers of arsenene, with an interlayer separation of $d_s = 0.566$ nm [10]. Interestingly, the fundamental band gap changes from indirect to direct in the bilayer, where PBE-SOC predicts a semimetal, and the HSE-SOC indicates an open band gap of $E_g = 0.33$ eV. These bilayers remain semiconducting, however, with a substantially reduced band gap.

Normally, the layer separation in multilayers decreases with increasing number of layers, however, it increases from $d_s = 0.566$ nm in the bilayer to $d_s = 0.667$ nm in trilayers of sw arsenene owing to a change in stacking order [10]. As expected, the interlayer spacing decreases from $d_s = 0.467$ nm for bilayers to $d_s = 0.450$ nm for trilayers of hb arsenene [10]. The indirect-to-direct transition and width of the band gap can be tuned by the number of layers and the application of strain. This possibility strongly extends the likelihood of potential application of few-layer systems because strain can be exerted by the substrate.

An alternative tool to study bilayers of arsenene is diffusion quantum Monte Carlo (DMC) analysis combined with DFT calculations [56]. These investigations showed that for sw arsenene AB stacking is more stable, with $E_{int} = -53$ meV/atom interlayer interaction energy, whereas for hb arsenene, AA stacking forms the most stable bilayer, with $E_{int} = -30$ meV/atom. The interlayer distances between the nearest neighbors are $d_{int} = 0.315$ nm and $d_{int} = 0.352$ nm for sw and hb arsenene, respectively [56]. Interestingly, by adding the corrugation effects, we find an interlayer spacing of $d_s = 0.554$ nm for the sw and $d_s = 0.492$ nm for the hb structure owing to the different stacking orders.

Besides the basic washboard and buckled phases of arsenene, As possesses significantly more 2D phases than other elements. For many of these allotropes, predominantly lattice parameters can be found in the literature [57]. These allotropes exist in various poorly studied forms, such as monolayers, nanosheets, nanoribbons, and nanotubes, in comparison with the sw and hb monolayers considered here. The stability of the two latter arsenene phases is based on local minima on the Born–Oppenheimer surface. In fact, the formation energy at $T = 0$ K estimated by subtracting the cohesive energy of the 3D phase from that of the corresponding 2D phase is negative, indicating a greater stability of the 3D phase.

8.3.3 Mechanics

Theoretical studies of the mechanical behavior of sw arsenene up to now cover only linear mechanical properties. As already discussed for phosphorene, the puckered structure of arsenene has the same characteristic anisotropic mechanical behavior. The strongly direction-dependent Young's moduli of the sw phase are $E_{zz} = 55$ N/m, which is substantially greater than the stiffness in the accordion-like armchair direction of $E_{ac} = 20$ N/m [10]. Strongly anisotropic Poisson ratios of $\nu_{xy} = 0.33$ and $\nu_{yx} = 0.91$ have been calculated for sw arsenene [10]. Another study obtained greater anisotropic stiffnesses for the sw phase at $E_{ac} = 29.0$ N/m and $E_{zz} = 74.7$ N/m [25].

The linear and nonlinear mechanical properties of the buckled honeycomb structure are $E = 52.8$ N/m, $\sigma_{ac} = 5.15$ N/m, $\sigma_{zz} = 4.69$ N/m, $\varepsilon_{ac} = 0.17$, and $\varepsilon_{zz} = 0.20$. The calculated stiffness seems to be comparable to that in the zigzag direction of the sw structure, and the ultimate strengths follow Griffith's rule, indicating ideal behavior [11]. A value of $E \approx 55$ N/m for Young's modulus and $\nu \approx 0.20$ for the Poisson ratio of the hb phase are supported by another study, which predicts as linear mechanical properties $E = 58$ N/m and $\nu = 0.21$ [10].

8.3.4 Physics and Applications

A crucial method to tune the properties of 2D materials is the application of compressive, tensile, uniaxial or biaxial strains, which modify the mechanical and electronic properties of monolayers. For example, biaxial strain provides insight into structural stability, such as mechanical stability derived from elastic constants or dynamic stability extracted from phonon spectra [58]. In addition to decreasing the band gap by applying strain, the indirect semiconductor arsenene can be transformed into a direct semiconductor by a compressive strain of – 10% [58]. Under tensile strain, the buckling height decreases linearly, and under sufficient in-plane strain, the buckled monolayer transforms to a planar honeycomb monolayer [58]. The tensile strain-induced reduction of the direct band gap of arsenene closes the gap at 13% strain, followed by band inversion and a transition to a topological insulator (TI) [59]. Since SOC opens a gap of 43 meV, characterization and applications of the TI phase are possible at room temperature. Moreover, arsenene is of interest for electronic devices, which require high electrical conductivity.

The fabrication of thermoelectric devices is a promising application of pnictogens. The thermal conductivity decreases gradually from hb arsenene to hb bismuthene, where the latter exhibits a lower thermal conductivity than most other 2D materials [60]. Using chemical intuition to find monolayers with high thermoelectric performance, sw P, sw As, and aw Sb were identified as potential candidates for high thermoelectric energy conversion efficiency [61]. Based on the chemical concepts of bond length, bond angle, and bond strength, trends in electronic properties, such as band gap, were established, and first-principles calculations confirmed the existence of high carrier mobilities and large Seebeck coefficients. This procedure yields deeper insight into the mechanism through which bonding affects phonon transport.

According to DFT calculations, buckled arsenene is an outstanding anode material for (Li/Na/Mg)-ion rechargeable batteries [62] and magnesium-ion batteries [63]. These calculations indicate important advantages, such as high capacity, low open-circuit voltage, effective adsorption of Mg on arsenene, and low barriers for Mg diffusion. Based on these advantages, hb arsenene may find application in the design and fabrication of anodes for rechargeable batteries.

In a DFT study investigating the suitability of group–VA monolayers as electrocatalysts for producing gaseous chlorine from the anodic chlorine evolution reaction (CER), the best candidate was hb arsenene [64]. This monolayer exhibited an ultralow thermodynamic overpotential of 0.068 V that is close to that of commercially used anodes, and efficiently suppressed parasitical oxygen gas production owing to a high theoretical overpotential of 1.95 V for the oxygen evolution reaction (OER).

Furthermore, due to its excellent optical and electronic properties arsenene has emerged as a promising 2D therapeutic agent for biomedical applications, namely, as a potential therapeutic agent for acute promyelocytic leukaemia cells by acting on nuclear proteins, while showing weak or no toxicity toward normal cells [65]. The relatively week toxicity to normal cells in contrast to cancer cells qualifies arsenene for medical applications [66]. Due to its great toxicity to cancer cells, arsenene can effectively kill these cells with good targeting efficiency. This distinct toxicity to cancer cells is ascribed to the valency increase of arsenic activated by cancer cells, reducing superoxide dismutase activity and producing massive reactive oxygen species (ROS).

Arsenenes have potential as gas-sensitive materials or resistance elements in field sensors [67]. This work reviews the preparation methods, properties, and applications of arsenene sensors developed in recent years. The sensitivity to oxygen allows for the detection of oxidizing or reducing gases. However, due to their reactivity, arsenene sensors are environmentally unstable. Doping arsenene with transition metals such as Ti and Cr results in a sensitive H_2S gas sensor [68]. Moreover, doping introduces magnetic moments, increases the adsorption energy and charge transfer, and narrows the band gap by transforming Ti-doped arsenene to a half-metal. Additional information on tunable electronic structures and potential applications of arsenene-based materials is presented a comprehensive review [69].

Only enormous progress on the experimental side will allow the realization of the theoretically predicted monolayer properties, such as the large band gap, band gap tuning by layered assemblies and strain, nontrivial topological phases, high electron mobility, superconductivity, and the quantum spin Hall (QSH) effect. Transformation to a direct gap semiconductor in few-layer systems or by biaxial strain may pave the way to optoelectronic devices under blue and ultraviolet light [70].

8.3.5 Synthesis

The exfoliation of layered bulk gray arsenic, which exists in nature, allows the isolation of buckled arsenene monolayers. While the production of few-layer systems of the hb structure has been achieved by several methods, such as exfoliation with adhesive tape or ultrasonication, ball milling, CVD, and plasma-assisted synthesis, the realization of monolayers is much more demanding [57, 70, 71]. Based on DFT calculations, the authors predict the epitaxial growth of energetically and thermodynamically more stable sw arsenene on a GeS(001) substrate [72]. Using LEED, STM, and ARPES in combination with DFT calculations for characterization, the formation of a quasi-free-standing buckled honeycomb layer of arsenene on Ag(111) with the correct lattice constant of a = 0.36 nm, a buckling height of $\Delta z = 0.145$ nm, and a semiconducting band structure could be demonstrated [55].

A state-of-the-art review covers the various methods applied in the synthesis of arsenene nanostructures [57]. There is still room to reduce the incorporation of inherent defects and to extend synthesis to large-scale fabrication in the currently fabricated monolayers and bilayer sheets of arsenene.

8.4 Antimonene

8.4.1 Chemistry and 2D Bonding

As discussed for phosphorus and arsenic, the electronic configuration of antimony atoms (Kr|$4d^{10}5s^2p^3$) has besides *d* electrons five valence electrons in the outer shells. Hybridized sp^3-like orbitals created by coupling of the 5*s* states and 5*p* states form puckered and buckled structures, as described above for its analogues. Determination of the cohesive energy of the more stable aw structure (see Fig. 8.7) in comparison with that of the sw structure yielded cohesive energies of $E_c = -2.61$ eV/atom [12] and − 1.77 eV/atom [6] and, for the most stable hb phase, $E_c = -2.84$ eV/atom [13], − 2.62 eV/atom [35], and − 1.96 eV/atom [6]. Note that for antimonene, the asymmetric puckered structure is slightly less stable than the buckled configuration.

As first theoretically discovered monolayers, both antimonene and bismuthene stand out from the other pnictogens because of the formation of a stable planar honeycomb monolayer by adsorption on a SiC(0001) surface [73]. In this unusual

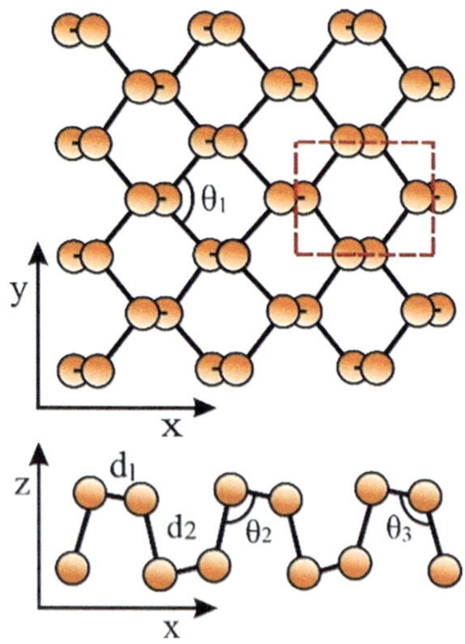

Fig. 8.7 Geometric structure of the puckered asymmetric washboard (aw) phase with four sublayers

bonding type, there is no mutual interaction between the in-plane orbitals, realizing a perfect planar sp^2 configuration, and the out-of-plane orbitals incorporating the lone electron pairs in ${p_z}^2$ orbitals (see Fig. 8.2). This bonding configuration changes the electronic properties drastically from semiconducting to semimetallic behavior, providing access to 2D TIs. The experimental realization of this prediction, would provide a platform for developing room temperature TIs, as further discussed below.

8.4.2 2D Structures

With respect to antimonene, the symmetric washboard (sw) structure transforms into the more stable asymmetric washboard (aw) structure spontaneously. First-principles calculations showed that as free-standing monolayers, both aw and hb monolayers are robust phases and remain stable above room temperature [14]. In the aw lattice, the shorter Sb-Sb bonds are no longer parallel to the 2D plane, as in the sw phase, but are slightly tilted, forming four sublayers (see Fig. 8.7). In the GGA approximation, including van der Waals corrections at the DFT-D_2 level, the lattice constants and bond lengths of the aw phase are a = 0.474 nm, b = 0.428 nm, d_1 = 0.291 nm, and d_2 = 0.284 nm [14]. The bond angles are $\theta_1 = 94.4°$, and $\theta_2 = 103.5°$, and the pucker distance Δz = 0.282 nm [7]. Based on the van der Waals diameter of d_w = 0.440 nm [5], we find a large monolayer thickness of $d_{w+\Delta z}$ = 0.722 nm for the aw lattice by taking the corrugation effect into account. The space group of the aw phase of the Sb monolayer is $Pmn2_1$ [12]

The corresponding structural information of buckled antimonene is a = 0.408 nm, d = 0.288 nm, Δz = 0.166 nm, and the angle between neighboring bonds is $\theta = 90.04°$ [11]. A comparison of different theoretical approaches reveals that the structural parameters are in reasonable agreement [7, 13]. The monolayers of hb and aw antimonene are energetically, dynamically, and thermally stable, and the energetic stability of the buckled phase is slightly greater. Based on the van der Waals diameter of d_w = 0.440 nm [5], we find a monolayer thickness of $d_{w+\Delta z}$ = 0.606 nm for the hb monolayer by taking the corrugation effect into account. The space group of the buckled phase is *P*-3*m*1 [39]. According to first-principles calculations aw Sb has a direct band gap, whereas hb Sb has a larger indirect gap [16]. Another study indicates that both structures are indirect semiconductors, and moderate tensile strains induce the transformation of the indirect band gap into a direct band gap [15].

The structure of antimonene bilayers has recently attracted special interest. The interlayer separation and binding energy of the aw bilayer structure for the preferred AB stacking were determined at d_s = 0.616 nm with an interlayer interaction energy of E_{int} = – 68 meV/atom [15]. Additionally, for AB stacking, the spacing for the more stable hb structure is d_s = 0.365 nm, and the now partially overlapping lone electron pair orbitals increase the interlayer interaction to E_{int} = – 124 meV/atom [15]. These latter results for the buckled phase have been partially confirmed by another group, predicting d_s = 0.381 nm and E_{int} = – 83 meV/atom [14]. For the interlayer distance between the nearest atoms, a much smaller value of d_{int} = 0.257 nm can be found in the literature for AB stacking, which increases to $d_{int} = 0.349$ nm for the less stable AA stacking [74]. While for interlayer binding, a lower interaction energy of $E_{int} = -86.0$ meV/atom has been derived by DFT calculations for buckled antimonene, a value of -89.6 meV/atom has been presented in this study for buckled arsenene [75]. Note that in the case of significant interlayer coupling, bilayers and multilayers may be metallic [14]. Experimentally, a mean thickness of $t \approx 0.40$ nm was found by measuring ~10 layers of a buckled sheet via AFM [76]. Note that a somewhat smaller thickness than the van der Waals diameter is expected due to additional interlayer interactions.

8.4.3 Mechanics

The decrease in the cohesive energy from arsenene to antimonene diminishes the stiffness and ultimate strength of aw and hb antimonene. The chemical and physical properties of the antimonene allotropes resemble those of arsenene. Furthermore, for the aw phase, the expected large anisotropy of the in-plane stiffness, ultimate strength, and ultimate strains is found with stiffness values of $E_{ac} = 12$ N/m and $E_{zz} = 29$ N/m, Poisson ratios of $\nu_{xz} = 0.36$ and $\nu_{yz} = 1.20$ [14], ultimate stresses of $\sigma_{ac} = 2.3$ (~4 GPa) and $\sigma_{zz} = 5.8$ N/m (~10 GPa), and strains of $\varepsilon_{ac} = 0.32$ and $\varepsilon_{zz} = 0.18$ [15].

The deformation behavior of the buckled structure is linear and isotropic up to ~ 2% strain, after which the structure becomes anisotropic. The reported Young's modulus is 32.9 N/m, the uniaxial tensile strength in the armchair direction $\sigma_{ac} = 3.75$ N/m with an ultimate strain of $\varepsilon_{ac} = 0.20$, and in the zigzag direction it is $\sigma_{zz} = 3.14$ N/m with an ultimate strain of $\varepsilon_{zz} = 0.21$ for buckled antimonene [11]. Independent calculations revealed for the hb phase a larger in-plane stiffness of $E = 41$ N/m and a Poisson ratio of $\nu_{xz} = \nu_{yz} = 0.21$ [14], large strength values of $\sigma_{ac} = 6.4$ N/m (~11 GPa) and $\sigma_{zz} = 5.8$ N/m (~10 GPa), and small critical strains of $\varepsilon_{ac} = 0.18$ and $\varepsilon_{zz} = 0.15$ [15]. Another set of mechanical properties of buckled antimonene reported an even greater stiffness, anisotropic Poisson ratios, low strength values, and low strain values of $E = 60$ N/m (104 GPa), $\nu_{xy} = \nu_{yx} = 0.22$, $\nu_{xz} = \nu_{yz} = 0.35$, $\sigma_{ac} = 4.6$ N/m (8 GPa), $\sigma_{zz} = 3.5$ N/m (6 GPa), and $\varepsilon_{ac} = 0.17$, $\varepsilon_{zz} = 0.15$, respectively [13].

8.4.4 Physics and Applications

Different from bilayer AB stacking with zero band gap considered above, AC stacking realized in buckled grey antimony, exhibits a high conversion efficiency in thermophotovoltaic (TPV) cells [74]. The AC stacked bilayer has a narrow band gap of $E_g = 0.62$ eV, under HSE + SOC-based calculations, in the optimum range of the TPV requirements of 0.35–0.75 eV. Furthermore, the van der Waals bonded bilayer has a large interlayer distance of $d_s = 0.499$ nm, resulting from a much smaller interaction energy than described for AB stacking. Remarkably, the TPV device shows a theoretical energy conversion efficiency of $\eta = 31\%$ that is comparable to the efficiencies of good other cells. This concept, based on van der Waals bonded few-layer systems of antimony, seems to be visible experimentally.

Several reviews cover the progress achieved by experimental fabrication of antimonene and discuss chemical and physical properties and potential applications, such as field-effect transistors, optoelectronic devices, electrocatalysis, energy storage devices, and cancer therapy [77]. A recent progress report includes the latest fabrication techniques and potential applications covering all five elements of group VA [78]. Another review provides an extensive treatment of the literature on the approaches employed to produce antimonene monolayers and few-layer sheets by mechanical and liquid phase exfoliation as well as epitaxial growth. Moreover, the work covers the electronic band structure, which is suitable for optoelectronic and thermoelectric applications, the main diagnostic techniques currently employed to characterize monolayers, and the opportunities and challenges of future applications from electronics to biomedicine [79]. An important step forward would be the experimental verification of the intriguing topological properties of antimonene and bismuthene, which are discussed in more detail below.

8.4.5 Synthesis

The deposition of high-quality antimonene sheets with thicknesses down to ~ 4 nm (approximately ten layers) was achieved by van der Waals epitaxy and analyzed by Raman spectroscopy and transmission electron microscopy (TEM) [76]. The analysis confirmed the growth of the theoretically predicted most stable buckled allotrope, which exhibits greater thermodynamic stability in an oxygen atmosphere than phosphorene and arsenene [76]. The growth of buckled monolayers of antimonene could be achieved on a 2D layered $PdTe_2$ substrate by MBE [80]. The deposit was characterized by in situ LEED and STM and by supporting DFT calculations. The analysis confirmed a weak interaction with the substrate, as proposed by the large distance between the nearest atoms of the monolayer and the surface atoms of the substrate ($d_{int} = 0.249$ nm) and a large electron localization function (ELF) value of > 0.8, which indicates strong covalent bonding between Sb–Sb pairs and a much lower value of < 0.3, suggesting a much weaker interaction between monolayer and substrate. Measurements confirmed the high chemical stability of the material in air [79]. The large-area antimonene 2D material deposited by MBE on c-plane sapphire substrates was characterized by Raman spectroscopy and high-resolution X-ray diffraction (XRD). This analysis was employed to evaluate the quality of the antimonene deposit and the competition between 2 and 3D growth processes, providing new insight into the critical parameters that promote 2D growth [81]

Van der Waals heteroepitaxial deposition by MBE allowed the growth of large-scale high-quality monolayers with a puckered honeycomb structure on a T_d-WTe_2

substrate [82]. The lattice constants measured by scanning force microscopy (STM) are in close agreement with those calculated for the free-standing monolayer, and the step-height of Sb islands of ~ 0.78 nm is consistent with the estimate of the layer thickness given above. Importantly, the adsorbed monolayer, which is structurally like sw phosphorene, exhibited chemical inertness under ambient conditions.

The MBE method allowed the deposition of flat antimonene monolayers on an Ag(111) substrate [73]. The analysis of the deposit by LEED and STM measurements confirmed the planar structure in combination with first-principles calculations. The planar configuration completely decouples the in-plane sp^2 hybridized orbitals and the out-of-plane $p_z{}^2$ orbitals, containing the sterically active lone electron pairs (see Fig. 8.2). The transition of a buckled monolayer with lone electron pairs into a planar monolayer with lone electron pairs is illustrated in Fig. 8.8. The monolayer thickness was measured by STM, and ELF analysis revealed weak coupling between the monolayer and the substrate. Note that in the planar structure, there is no mutual hybridization between in-plane and out-of-plane orbitals, and the former leads to a hybridized sp^2 band, while the latter leads to the Dirac cone of π bands. Importantly, these quasi-free-standing atomically planar layers of antimonene with decoupled in-plane sp^2 and out-of-plane $p_z{}^2$ orbitals exhibit nontrivial topological edge states.

A highly efficient preparation method with a yield of micrometer-sized nanosheets above 20% uses a polymer ionic liquid (PIL)-assisted liquid-phase exfoliation method [83]. This development allowed the fabrication of flexible photodetectors based on a hybrid structure of antimonene nanosheets modified by PIL and CdS quantum dots.

DFT calculations combined with particle swarm optimization algorithms (PSOAs) reproduced the experimentally verified growth process of buckled and puckered antimonene on a stabilizing Ag(111) surface [84]. These simulations provide new insight into the growth mechanism, e.g., the stabilization of the antimonene growth process by the immediate formation and reconstruction of Ag_2Sb surface alloys.

A mini-review presents the recent progress achieved in state-of-the-art synthesis, structural stability, electronic properties, and various applications of antimonene and distinct allotropes [85]. Unlike phosphorene 2D antimonene exhibits good stability when exposed to air and water.

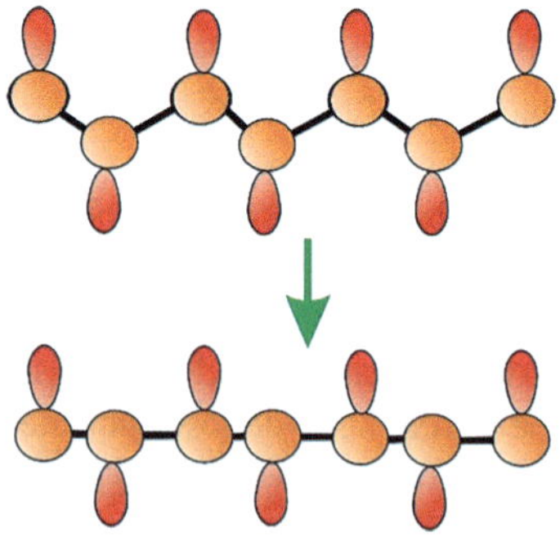

Fig. 8.8 Transition of a buckled monolayer with electron lone pairs into a planar monolayer with electron lone pairs

8.5 Bismuthene

8.5.1 Chemistry and 2D Bonding

As an element of main group VA with an electronic configuration (Xe|$5d^{10}6s^2p^3$), bismuth atoms have five valence electrons in their outer valence shell. Most calculations assume that the *s* and *p* orbitals provide a reasonable description of the bonding arrangement and that the influence of the *d* and *f* orbitals is negligible even for a heavy element such as Bi. This seems to be the case, at least for the calculation of structural data. Like for the other group members, bismuthene exists in a puckered and buckled configuration with three σ-type bonds incorporated in sp^3-like hybridization, including hybridization with the remaining lone electron pairs, located in an orbital that is directed perpendicular to the plane. The calculated cohesive energies of the energetically slightly more stable aw phase are $E_c = -2.42$ eV/atom [39], – 1.97 eV/atom [17], and – 1.38 eV/atom [6], and for the buckled phase they are $E_c = -2.44$ eV/atom [35], – 2.41 eV/atom [39], – 1.95 eV/atom [17], and – 1.44 eV/atom [6]. Note that for the hb structure, a much larger cohesive energy of $E_c = -3.04$ eV/atom [18] and – 3.00 eV/atom [26] was obtained by PBE + SOC calculations, instead of – 2.44 eV/atom (PBE) [18].

8.5.2 2D Structures

As in the case of antimonene, the most stable 2D phase of bismuthene has an aw structure because the sw phase transforms spontaneously into this phase. DFT calculations including SOC using the PBE functional provided the following structural data for the asymmetric washboard structure: a = 0.494 nm and b = 0.455 nm for the lattice constants, and $d_1 = 0.311$ nm and $d_2 = 0.310$ nm for the bond lengths [17] Similar extended lattice parameters at a = 0.487 nm, b = 0.444 nm, $d_1 = 0.308$ nm, $d_2 = 0.303$ nm, $\Delta z = 0.272$ nm, $\theta_1 = 92.0°$, and $\theta_2 = 106.7°$ have been published for the aw phase [13]. The van der Waals diameter of $d_w = 0.460$ nm and the large corrugation yield a thickness of $d_{w+\Delta z} = 0.732$ nm for the aw structure [5]. The space group of the aw phase is *Pmn*21 [39]. In addition, bismuthene exists in a planar structure with perfect sp^2 hybridization separated from the orbitals containing the lone electron pairs that extend perpendicularly to the monolayer plane [73].

The natural layered bulk crystal of bismuth, which has a rhombohedral structure, consists of buckled bismuthene layers. The lattice constant of the hb phase is a = 0.438 nm, the bond length d = 0.307 nm, the buckling effect $\Delta z = 0.174$ nm, and the bond angle $\theta = 91.2°$ [17, 18]. The structural parameters agree well with those of another publication, giving the values a = 0.426 nm, d = 0.303 nm, and $\Delta z = 0.176$ nm [26]. A DFT-GGA-PBE study taking as effective valence electrons the electronic configuration of Bi ($4f^{14}5s^2p^6d^{10}6s^2p^3$) into account, yielded the values: a = 0.434 nm, d = 0.304 nm, $\Delta z = 0.173$ nm and $\theta = 90.96$ [86]. The van der Waals diameter of $d_w = 0.460$ nm [5] yields a monolayer thickness of $d_{w+\Delta z} = 0.634$ nm for the buckled phase, which is in good agreement with the thickness of $t = 0.65$ nm used to calculate the thermal conductivity [27]. The buckled phase is dynamically and thermally stable up to approximately 700 K [17]. The space group of the buckled structure is $P\text{-}3m1$ [39].

The meta-stable planar hexagonal phase of bismuthene (hp-phase) is stabilized through interactions with the surface of the SiC(0001) substrate [73, 87]. The measured lattice constant of $a = 0.501$ nm in the adsorbed state indicates an in-plane binding of the hp structure with a lower buckling effect. This phase is a semiconductor with an indirect band gap of $E_g = 0.51$ eV (PBE) [73]. The chemisorbed bismuthene monolayer has a large binding energy with the substrate due to hybridization via π bonding, inducing a transition from the trivial to the nontrivial topological state, where the substrate controls the relevant orbitals in the 2D QSH insulator [88].

Functionalization of bismuthene (antimonene) by hydrogen on one side breaks the inversion symmetry and therefore has a similar effect on the electronic structure as the presence of a suitable substrate [87]. The hydrogenated planar monolayers possess topologically protected edge states for the zigzag and armchair edges with medium band gaps of $E_g = 1.03$ eV (Bi) and 0.41 eV (Sb). Therefore, these calculations open a new platform for realizing single isolated Dirac cone edge states in 2D TIs with bulk band gaps suitable for room temperature applications.

Different theoretical approaches have been employed to study the structure of bismuthene bilayers. For the most stable AA stacking of the bilayer of buckled monolayers, the interlayer separation is $d_s = 0.464$ nm, and the interlayer interaction energy $E_{int} = -123$ meV/atom [18]. This binding energy between two layers increases to -170 meV/atom by including SOC and van der Waals corrections. The results are in good agreement with an interlayer spacing of $d_s = 0.480$ nm and a binding energy of bilayers of $E_{int} = -136$ meV/atom for the most favorable AA configuration without van der Waals correction [17]. The strong interlayer interaction energy points to a dominant covalent-type interaction between the two bismuthene layers and thus explains the larger thickness values estimated by using the van der Waals diameter.

8.5.3 Mechanics

To date, only linear elastic properties have been estimated for puckered bismuthene. For the aw phase, the stiffness in the armchair direction is $E_{ac} = 10$ N/m, whereas a much larger stiffness of $E_{zz} = 25.5$ N/m was found for the zigzag direction [17]. The corresponding Poisson ratios of $\nu_{xy} = 0.261$ and $\nu_{yx} = 0.648$ reflect the strong anisotropy of the asymmetric washboard structure with a much larger stiffness along the zigzag direction [17].

The buckled phase of bismuthene has a Young's modulus of $E = 23.9$ N/m, which is the lowest stiffness of the group–VA compounds, and Poisson's ratio is $\nu = 0.327$ [17]. The elastic modulus agrees well with that of another publication, which reported a value of $E = 26.3$ N/m and an ultimate strain of $\varepsilon = 0.15$ [26]. Extensive

MD simulations utilizing a Stillinger–Weber potential (SW) were used to study the temperature dependence of the mechanical behavior [27]. For the slightly anisotropic stiffness at 300 K, the authors give $E_{ac} \approx 26.1$ N/m and $E_{zz} \approx 25.5$ N/m. Similarly, at room temperature, the simulated fracture stresses and ultimate strains for the two directions were $\sigma_{ac} \approx 4.21$ N/m and $\varepsilon_{ac} \approx 0.195$ and $\sigma_{zz} \approx 4.22$ N/m and $\varepsilon_{zz} \approx 0.255$. Overall, the ultimate strength decreases from 100 to 500 K by 28.1% (ac) and 20.3% (zz), and the ultimate strains by 36.2% (ac) and 10.2% (zz).

8.5.4 Physics and Applications

Bismuthene is a promising material for high-performance thermoelectric applications. The predicted low thermal conductivity of ~ 1.3 W m^{-1} K^{-1} at 300 K results in a thermoelectric figure of merit of $ZT = 2.4$, which can be further increased to 4.1 at 500 K [27, 89]. These ZT values were obtained by fitting first-principles total energy calculations to construct a modified Morse potential, which was used to predict the lattice thermal conductivity via equilibrium MD simulations [89].

Due to its moderate band gap and tunable broadband optical response, bismuthene qualifies for optoelectronic applications such as photodetectors, light-emitting devices, and photovoltaic cells [86]. Biaxial compressive and tensile stresses are suitable tools for tuning the band gap. The study revealed strong optical absorption, low light loss, and 8% metallic properties. Furthermore, the influence of adsorption on the elastic properties of bismuthene was evaluated via DFT calculations [90]. This approach allowed the calculation of the elastic and plastic properties of the pristine nanosheets and those with adsorbed H, F, Cl, and Br atoms under uniaxial and biaxial loading [90]. Under uniaxial loading, the yield stress decreases by adsorption, whereas for the yield strain, no unique behavior could be observed under biaxial loading.

Owing to the strong SOC, buckled bismuthene is the topological insulator with the largest nontrivial band gap of > 0.60 eV in its natural state and therefore shows great promise for applications in spintronics at room temperature [27]. A too small bulk band gap is one of the limiting factors that prohibits QSH applications at higher temperatures. A promising approach is functionalization of bismuthene or bismuthene compounds such as BiX (X = Ga, In, and Tl) on one or both sides (see Fig. 8.9) [91]. This approach yields a nontrivial topological phase by band inversion from the band order *p*-*p*-*s* for halogens on both sides to *p*-*s*-*p* order for halogen on one side and hydrogen on the other side. The band order transition leads to a *p*-*s* bulk gap, originating from the strong hybridization of the *s* orbitals with the $p_{x,y}$ and p_z orbitals orbitals of F-GaBi-H with an energetics sensitive to the structural deformation. Due to the sensivity of the *p*-*s* band gap on an external electric field, it may be possible to engineer the quantum spin Hall edge transport.

The electronic structure of few-layers assemblies depends on the effective SOC strength, the number of layers including stacking order, and epitaxial strain. In bismuthene bilayers first-principles calculations identified TIs, topological

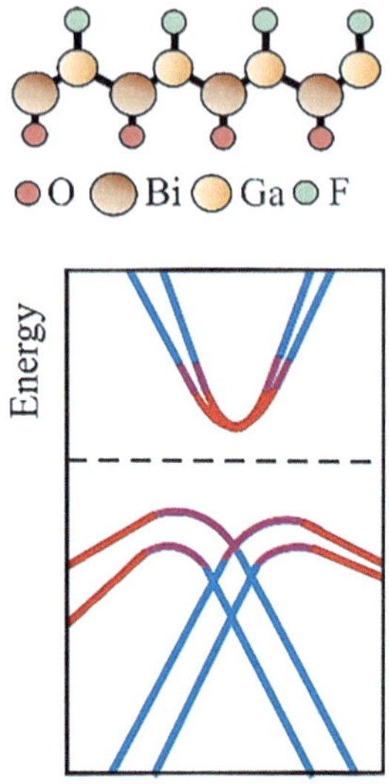

Fig. 8.9 GaBi layer functionalized on the top side by F atoms and the bottom side by hydrogen atoms. The band inversion of the topological insulator is schematically indicated by colors. Adapted from Ref. [91], Copyright 2016, Springer Nature

semimetals, trivial semimetals, and nontrivial insulators of bismuthene (antimonene) [92]. For example, double-layer bismuthene in AB stacking order with strong atomic SOC has clearly been identified as TI phase through band inversion by first-principles calculations.

As revealed by first-principles computations, even weak interactions between buckled bismuthene and hexagonal SiC(0001), Si(111), and SiO_2(0001) substrates can stabilize topologically nontrivial phases by breaking inversion symmetry and changing the electronic structure [93]. Note that the size of the band gap of the hb phase varies drastically with the nature of the substrate. Hydroxylated SiO_2 is a substrate that opens the topological gap. Time-reversal symmetry prevents helical edge states from mixing and is responsible for a protected crossing point in the edge spectrum and for quantized longitudinal conduction. According to theory, hexagonal armchair QSH edge states remain gapless under an in-plane magnetic field in the direction of the edge, whereas an out-of-plane magnetic field opens a gap [94]. This behavior is a hallmark of topological crystalline protection.

Environmental stability is a principal issue for all group–VA monolayers, as already discussed in more detail for phosphorene. In an extensive report on the oxidation mechanisms and structural stability of 2D pnictogens under oxygen and water exposure, the authors concluded that the higher stability of antimonene and bismuthene originates from the acceptor role of water, which impedes the formation of acids by the interaction of water molecules and oxygen species [95]. Conversely, on the phosphorene surface, water molecules act as donors, and the formation of acids leads to rapid degradation of the ideal structure.

8.5.5 *Synthesis*

Owing to the increasing interest in QSH materials, which hold promise for dissipationless spin currents, the growth of bismuthene is an active field of research. Atomic Bi atoms are heavy and therefore have a strong SOC and an inverted gap. The deposition of supported bismuthene using a SiC(0001) substrate yielded a honeycomb superstructure of Bi atoms with a lattice constant of a = 0.501 nm, which is considerably larger than that of the buckled monolayer and belongs to an energetically more favorable planar configuration [73]. Bonding with the substrate has an important influence on the kind of hybridization. The band structure consists of σ bond contributions of Bi $6s$, $6p_x$, and $6p_y$ orbitals and π bond contributions formed by $6p_z$ orbitals [88]. Hybridization of the π bonds with the substrate breaks inversion symmetry and changes the contribution of $6p_z$ orbitals at low energies. As a result, σ-type contributions of the $6p_x$ and $6p_y$ orbitals predominantly control this energy region around the Fermi level, pushing out the $6p_z$ orbital of the low-energy region. This orbital filtering gives rise to a large SOC. Remarkably, the large band gaps of hexagonal bismuthene and antimonene may lead to room-temperature QHS applications.

8.6 Property–Structure Relationships

8.6.1 *Binding–Structure Relationships*

Figure 8.10 shows a plot of the presented cohesive energies versus the bond lengths of the corresponding stable sw, aw, and hb phases of the elemental group–VA monolayers. This cohesive energy–bond length plot reveals the expected inverse behavior of the cohesive energy on the bond length of both the puckered and buckled phases, which exhibit comparable binding energies. Importantly, most of the included studies were based on DFT calculations and involved a treatment of the exchange–correlation functional with the PBE form of GGA.

Note that Fig. 8.10 also includes systematically smaller cohesive energies, including the SOC effect [6]. It is not clear whether this is the reason for the much smaller binding energies reported in this work. An even larger difference of nearly a factor of two appears for the reported cohesive energies of nitrogene. The large value fits into the correlation with the other group–VA elements, where graphene is the leading monolayer with the largest cohesive energy and shortest bond length. Interestingly, the calculated large cohesive energy of nitrogene at $E_c = -6.81$ eV/atom is consistent with the short bond length of d = 0.149 nm given by all the authors. Such an agreement is not found for the binding energies with much lower values of $E_c = -3.67$ eV/atom and -3.28 eV/atom suggested by other authors [1, 6].

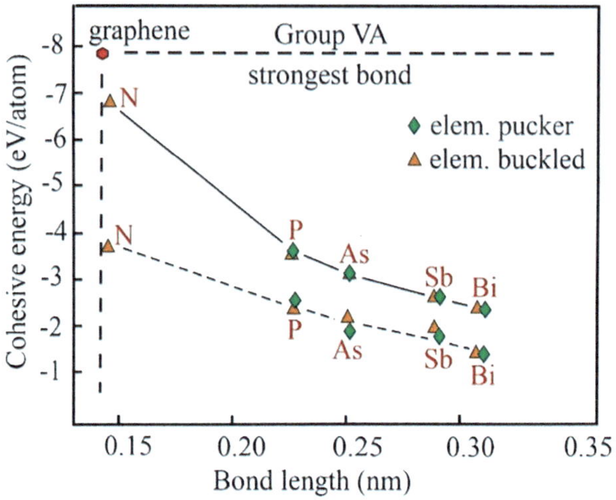

Fig. 8.10 Cohesion energies of the monolayers of group–VA elements as a function of bond length

8.6.2 *Mechanics–Structure Relationships*

Figure 8.11 shows the presently available Young's moduli and ultimate strengths of the group–VA elements versus bond lengths (see Table 8.2). Notably, the large anisotropy of Young's moduli between the armchair and the zigzag directions of the puckered phases decreases abruptly for the heavier elements. Unfortunately, calculations of several ultimate strengths and strains are still missing. Since the already available strength values agree quite well with Griffith's rule, this correlation allows a reasonable first prediction of the ultimate stress values of sw arsenene and aw bismuthene.

Fundamental information on the nonlinear mechanical properties of nitrogene is still missing. The available Young's moduli deviate by nearly a factor of two. The second largest stiffness of 270 N/m after graphene for an elemental monolayer is consistent with the second shortest bond in 2D space and the large value of the reported cohesive energy [1]. This would imply an enormous ultimate strength of the dynamically and thermally stable nitrogene. Alternatively, the much lower stiffness of 154 N/m correlates well with the other buckled monolayers of group–VA elements and their low cohesive energy [4].

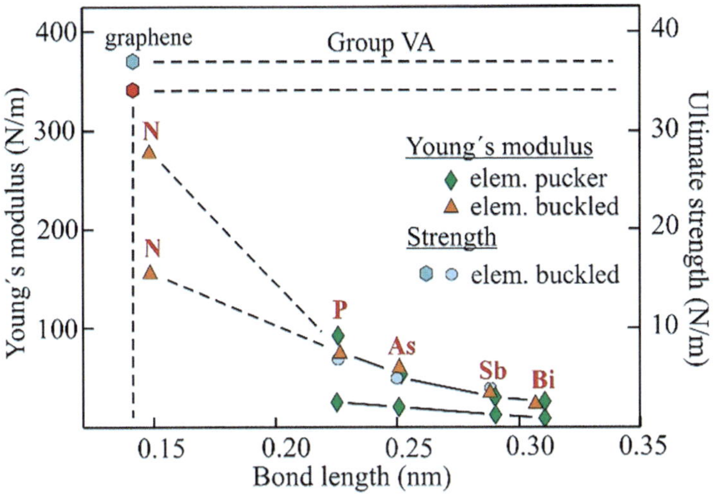

Fig. 8.11 Young's moduli and ultimate strengths of the monolayers of group–VA elements versus bond length

References

1. V.O. Özçelik, O.Ü. Aktürk, E. Durgun, S. Ciraci, Phys. Rev. B **92**, 125420 (2015)
2. J. Lee, W.-C. Tian, W.-L. Wang, D.-X. Yao, Sci. Rep. **5**, 11512 (2015)
3. Y. Zhang, J. Lee, W.-L. Wang, D.-X. Yao, Comput. Mater. Sci. **110**, 109 (2015)
4. S.V. Bondarchuk, B.F. Minaev, Comput. Mater. Sci. **133**, 122 (2017)
5. S.S. Batsanov, Inorg. Mater. **37**, 871 (2001)
6. S. Guo, W. Zhou, B. Cai, K. Zhang, S. Zhang et al., Nanoscale Horiz. **4**, 1145 (2019)
7. F. Ersan, E. Aktürk, S. Ciraci, Phys. Rev. B **96**, 205434 (2017)
8. Z. Dai, W. Jin, J.-X. Yu, M. Grady, J.T. Sadowski et al., Phys. Rev. Mater. **1**, 074003 (2017)
9. Y. Mogulkoc, M. Modarresi, A. Mogulkoc, Y.O. Ciftci, Comp. Mater. Sci. **124**, 23 (2016)
10. D. Kecik, E. Durgun, S. Ciraci, Phys. Rev. B **94**, 205409 (2016)
11. G. Liu, Z. Gao, J. Zhou, Phys. E **112**, 59 (2019)
12. T. Kocabas, D. ҫakir, O. Gülseren, F. Ay, N.K. Perkgöz et al., Nanoscale **10**, 7803 (2018)
13. D.R. Kripalani, A.A. Kistanov, Y. Cai, M. Xue, K. Zhou, Phys. Rev. B **98**, 085410 (2018)
14. O.Ü. Aktürk, V.O. Özçelik, S. Ciraci, Phys. Rev. B **91**, 235446 (2015)
15. G. Wang, R. Pandey, S.P. Karna, A.C.S. Appl, Mater. Interf. **7**, 11490 (2015)
16. Y. Xu, B. Peng, H. Zhang, H. Shao, R. Zhang et al., Ann. Phys. **529**, 1600152 (2017)
17. E. Aktürk, O.Ü. Aktürk, S. Ciraci, Phys. Rev. B **94**, 014115 (2016)
18. M.-Y. Liu, Y. Huang, Q.-Y. Chen, Z.-Y. Li, C. Cao et al., RSC Adv. **7**, 39546 (2017)
19. J. Ribeiro-Soares, R.M. Almeida, L.G. Cançado, M.S. Dresselhaus, A. Jorio, Phys. Rev. B **91**, 205421 (2015)
20. Q. Wei, X. Peng, Appl. Phys. Lett. **104**, 251915 (2014)
21. J.-W. Jiang, H.S. Park, J. Phys. D Appl. Phys. **47**, 385304 (2014)
22. L. Wang, A. Kutana, X. Zou, B.I. Yakobson, Nanoscale **7**, 9746 (2015)
23. N. Liu, J. Hong, R. Pidaparti, X. Wang, Nanoscale **8**, 5728 (2016)
24. J.M.G. Hernandez, H.N. Fernandez-Escamilla, J.G. Sanchez, N. Takeuchi, Sci. Rep. **12**, 20979 (2022)
25. Z. Zhang, J. Xie, D. Yang, Y. Wang, M. Si et al., Appl. Phys. Exp. **8**, 055201 (2015)
26. S.M. Mozvashi, S.I. Vishkayi, M.B. Tagani, Phys. E **118**, 113914 (2020)

27. E.H. Chowdhury, Md.H. Rahman, P. Bose, R. Jayan, Md.M. Islam, Phys. Chem. Chem. Phys. **22**, 28238 (2020)
28. P. Jureczko, M. Kurpas, Sci. Rep. **12**, 3201 (2022)
29. J.-S. Li, W.-L. Wang, D.-X. Yao, Sci. Rep. **6**, 34177 (2016)
30. F. Ersan, D. Kecik, V.O. Özçelik, Y. Kadioglu, O.Ü. Aktürk et al., Appl. Phys. Rev. **6**, 021308 (2019)
31. Y. Harada, M. Yamamoto, T. Baba, T. Kita, Appl. Phys. Lett. **104**, 041907 (2014)
32. C. Ji, A.A. Adeleke, L. Yang, B. Wan, H. Guo, et al., Sci. Adv. **6**, eaba9206 (2020)
33. V. Sorkin, Y. Cai, Z. Ong, G. Zhang, Y.W. Zhang, Crit. Rev. Solid State Mater. Sci. **42**, 1 (2017)
34. S. Zhang, S. Guo, Z. Chen, Y. Wang, H. Gao et al., Chem. Soc. Rev. **47**, 982 (2018)
35. Y. Nie, M. Rahman, D. Wang, C. Wang, G. Guo, Sci. Rep. **5**, 17980 (2015)
36. M. Akhtar, G. Anderson, R. Zhao, A. Alruqi, J.E. Mroczkowska, et al., NPJ 2D Mater. Appl. **1**, 5 (2017)
37. H. Liu, A.T. Neal, Z. Zhu, Z. Luo, X. Xu et al., ACS Nano **8**, 4033 (2014)
38. J. Arcudia, R. Kempt, M.E. Cifuentes-Quintal, G. Merino, T. Heine, Phys. Rev. Lett. **125**, 196401 (2020)
39. N.A.P. Namari, M. Saito, Jpn. J. Appl. Phys. **58**, 061003 (2019)
40. M.Y.H. Widianto, A. Zaharo, N.A.P. Namari, M. Saito, Jap. J. Appl. Phys. **60**, 061001 (2021)
41. Z. Zhu, D. Tománek, Phys. Rev. Lett. **112**, 176802 (2014)
42. G. Moynihan, S. Sanvito, D.D. O'Regan, 2D Mater. **4**, 045018 (2017)
43. Z.-D. Sha, Q.-X. Pei, Z. Ding, J.-W, Jiang, Y.-W. Zhang, J. Phys. D Appl. Phys. **48**, 395303 (2015)
44. M. Pica, R. D'Amato, Inorganics **8**, 29 (2020)
45. V. Chaudhary, P. Neugebauer, O. Mounkachi, S. Lahbabi, A. El Fatimy, 2D Mater. **9**, 032001 (2022)
46. Y. Jing, X. Zhang, Z. Zhou, WIREs Comput. Mol. Sci. **6**, 5 (2016)
47. A.K. Tareen, K. Khan, S. Rehman, M. Iqbal, J. Yu et al., Prog. Solid State Chem. **65**, 100336 (2022)
48. Y. Wang, S. Sun, J. Zhang, Y.L. Huang, W. Chen, SmartMat **2**, 286 (2021)
49. A.J. Mannix, B. Kiraly, M.C. Hersam, N.P. Guisinger, Nat. Rev. Chem. **1**, 0014 (2017)
50. W. Hu, J. Yang, J. Phys. Chem. C **119**, 20474 (2015)
51. J.L. Zhang, S. Zhao, C. Han, Z. Wang, S. Zhong et al., Nano Lett. **16**, 4903 (2016)
52. W. Zhang, H. Enriquez, A.J. Mayne, G. Dujardin, H. Oughaddou, Small **14**, 1870249 (2018)
53. S. Batool, M. Idrees, S.-T. Han, Y. Zhou, Adv. Sci. **10**, 2203956 (2023)
54. C. Kamal, M. Ezawa, Phys. Rev. B **91**, 085423 (2015)
55. J. Sha, W. Wang, H.M. Sohail, R.I.G. Uhrberg, 2D Mater. **7**, 025013 (2020)
56. Y. Kadioglu, J.A. Santana, H.D. Özaydin, F. Ersan, O.Ü. Aktürk et al., J. Chem. Phys. **148**, 214706 (2018)
57. B. Bhuvaneswari, V. Nagarajan, R. Chandiramouli, Surf. Interf. **28**, 101610 (2022)
58. H. Cao, Z. Yu, P. Lu, Superlatt. Microstruct. **86**, 501 (2015)
59. G. Rahman, A. Mahmood, V.M. Garcia-Suárez, Sci. Rep. **9**, 7966 (2019)
60. D.-C. Zhang, A.-X. Zhang, S.-D. Guo, Y.-F. Duan, RSC Adv. **7**, 24537 (2017)
61. B. Peng, H. Zhang, H. Shao, K. Xu, G. Ni et al., J. Mater. Chem. A **6**, 2018 (2018)
62. H. Benzidi, M. Lakhal, M. Garara, M. Abdellaoui, A. Benyoussef, Phys. Chem. Chem. Phys. **21**, 19951 (2019)
63. X.-J. Ye, G.-L. Zhu, J. Liu, C.-S. Liu, X.-H. Yan, J. Phys. Chem. C **123**, 15777 (2019)
64. J. Liu, J.J. Hinsch, H. Yin, P. Liu, H. Zhao et al., Catalysts **12**, 296 (2022)
65. X. Wang, Y. Hu, J. Mo, J. Zhang, Z. Wang et al., Angew. Chem. Int. Ed. **59**, 5151 (2020)
66. Y. Hu, J. Liang, Y. Xia, C. Zhao, M. Jiang, et al., Small 2104556 (2021)
67. R. Zhou, N. Xu, R. Guo, G. Ling, P. Zhang, J. Phys. D Appl. Phys. **55**, 163002 (2022)
68. Y.-P. Tian, C.-B. Wang, W.-J. Gong, RSC Adv. **13**, 2234 (2023)
69. J. Zhao, Z.-H. Qi, Y. Xu, J. Dai, X.C. Zeng et al., WIREs Comput. Molec. Sci. **9**, e1387 (2019)
70. M. Pumera, Z. Sofer, Adv. Mater. **29**, 1605299 (2017)

71. D.S. Sumanth, R. Shwetharani, H.R. Chandan, C. Lavanya, B. Geetha, Adv. Mater. Interf. **9**, 2200442 (2022)
72. B. Shi, Y. Song, W. Zhang, J. Phys. Condens. Matter **35**, 24 (2023)
73. Y. Shao, Z.-L. Liu, C. Cheng, X. Wu, H. Liu et al., Nano Lett. **18**, 2133 (2018)
74. M. Xie, S. Zhang, B. Cai, Y. Gu, X. Liu et al., Nano Energy **38**, 561 (2017)
75. S. Zhang, Z. Yan, Y. Li, Z. Chen, H. Zeng, Angew. Chem. Int. Ed. **54**, 3112 (2015)
76. J. Ji, X. Song, J. Liu, Z. Yan, C. Huo et al., Nat. Commun. **7**, 13352 (2016)
77. X. Wang, J. Song, J. Qu, Angew. Chem. Int. Ed. **58**, 1574 (2018)
78. P. Yuan, T. Zhang, J. Sun, L. Liu, Y. Yao et al., J. Semicond. **41**, 081003 (2020)
79. P. Ares, J.J. Palacios, G. Abellán, J. Gómez-Herrero, F. Zamora, Adv. Mater. **30**, 1703771 (2018)
80. X. Wu, Y. Shao, H. Liu, Z. Feng, Y.-L. Wang, et al., Adv. Mater. 1605407 (2017)
81. M. Gu, C. Li, Y. Ding, K. Zhang, S. Xia et al., Appl. Sci. **10**, 639 (2020)
82. Z.-Q. Shi, H. Li, Q.-Q. Yuan, Y.-H. Song, Y.-Y. Lv et al., Adv. Mater. **31**, 1806130 (2019)
83. Q. Xiao, C.-X. Hu, H.-R. Wu, Y.-Y. Ren, X.-Y. Li et al., Nanoscale Horiz. **5**, 124 (2020)
84. K. Liu, K. Bai, J. Wang, J. Song, Y. Liu, Front. Phys. **10**, 856526 (2022)
85. R. Bhuvaneswari, V. Nagarajan, R. Chandiramouli, J. Molec, Graphics Model. **122**, 108473 (2023)
86. F. Ghasemzadeh, M. Esmaeilzadeh, M.E. Shayan, Int. J. Smart Grid **4**, 190 (2020)
87. C.-H. Hsu, Z.-Q. Huang, F.-C, Chuang, C.-C. Kuo, Y.-T. Liu, et al., New J. Phys. **17**, 025005 (2015)
88. F. Reis, G. Li, L. Dudy, M. Bauernfeind, S. Glass et al., Science **357**, 287 (2017)
89. L. Cheng, H. Liu, X. Tan, J. Zhang, J. Wei et al., J. Phys. Chem. C **118**, 904 (2014)
90. P. Aghdasi, R. Ansari, S. Rouhi, M. Goli, Superlatt. Microstruct. **135**, 106242 (2019)
91. Y. Kim, W.S. Yun, J.D. Lee, Sci. Rep. **6**, 33395 (2016)
92. X. Wang, G. Bian, C. Xu, P. Wang, H. Hu et al., Nanotechnology **28**, 395706 (2017)
93. N. Wittemeier, P. Ordejón, Z. Zanolli, J. Phys. Mater. **5**, 035002 (2022)
94. F. Dominguez, B. Scharf, G. Li, J. Schäfer, R. Claessen et al., Phys. Rev. B **98**, 161407(R) (2018)
95. A.A. Kistanov, S.K. Khadiullin, K. Zhou, S.V. Dimitriev, E.A. Korznikova, J. Mater. Chem. C **7**, 9195 (2019)

Chapter 9
Binary Heteroelemental Group VA–VA Monolayers

Abstract The members of the group of heteroatomic binary VA–VA compounds consist of multivalent atoms with nonmetallic (N and P), semimetallic (As), and metallic (Sb and Bi) behavior; versatile sp^2-type versus sp^3-type hybridization; and several types of bonding, such as covalent, ionic, and metallic bonding. The most widely investigated phases are buckled and asymmetric washboard structures owing to their better energetic stability. The different atom sizes in mixed compounds reduce the symmetry of the networks, which are no longer centrosymmetric. Furthermore, ionic contributions to bonding occur by charge transfer due to the different electronegativities of the elements. Atom size and electronegativity are important elemental properties, controlling the physicochemical behavior of these binary compounds. Combining elements with selected atom size and electronegativity allows band gap engineering of the heteroatomic monolayers. The compounds discussed in the following chapter are energetically stable, with negative cohesive energies, and dynamically stable as free-standing monolayers or as adsorbate on a suitable substrate. Eventually, molecular dynamics (MD) simulations confirmed the thermal stability at finite temperatures. There are no reports on the synthesis of these compounds, even in stabilized adsorbed form. The published structural data are in good agreement; however, large deviations of more than a factor of two exist for the pertinent cohesive energies, despite the use of standard density functional theory (DFT) with the generalized gradient approximation (GGA) and Perdew–Burke–Ernzerhof (PBE) approach. Here, the medium cohesive energies are preferred, which allow a consistent straightforward description of all available information. The heavier compounds of these monolayers, with an anisotropic electronic structure and lone electron pairs, are 2D ferroelectric materials with spontaneous electric polarization that can be switched by an electric field. For example, aw SbN and aw BiN are excellent candidates for integrated ferroelectric devices such as nonvolatile memory devices and field-effect transistors. Furthermore, various compounds are topological insulators (TIs) and exhibit the quantum spin Hall (QSH) effect, endowing these 2D materials with potential applications in dissipationless electronics and topical quantum computing.

P. Hess, *Bonding, Structure, and Performance of Two-Dimensional Materials*, NanoScience and Technology, https://doi.org/10.1007/978-3-031-74161-6_9

9.1 Group VA–Nitride Monolayers (PN, AsN, SbN, and BiN)

9.1.1 PN Monolayer

9.1.1.1 Chemistry and 2D Bonding

The type of atomic bonding in binary group–VA compounds follows the bonding behavior of group–VA elements with five valence electrons (ns^2np^3). Specifically, chemical behavior and 2D bonding of phosphorus nitride depend on the valence electrons of P ($3s^2{p_x}^2{p_y}^1$) and N ($2s^2{p_x}^2{p_y}^1$) in the outer shell. First-principles calculations of the electron-electron interactions of binary compounds of the five main group elements (see Fig. 9.1) via the DFT-GGA-PBE approach made the main contributions to the presently available knowledge.

The interaction between the valence and core electrons determined by the projector-augmented wave (PAW) method revealed that P atoms favor the formation of sp^3 hybrids, while N atoms prefer sp^2 hybrids [1, 2]. Similar to group–VA elements, the specific bonding configuration of tetrahedrally coordinated phosphorus atoms with three nitrogen atoms results in the formation of three nonplanar σ bonds, involving either buckling, forming a hexagonal buckled (hb) structure, or puckering, forming an asymmetric washboard (aw) structure (see Figs. 9.2 and 9.3). Since long P–N (pnictogen-nitrogen) bonds weaken the overlap of π orbitals, buckling or puckering lowers the total energy of the system. The sp^2 hybrids combine with the p_z orbital, forming sp^3-like orbitals with the ${p_z}^2$-like orbital that extends perpendicular to the plane and contains the two remaining electrons.

The partially ionic bonds of the puckered phase are characterized by a large Bader charge transfer of $\delta q = 3.26$ e that occurs from the P atom, with an electronegativity of 2.19, to the N atom, with an electronegativity of 3.04 [1]. A nearly identical Bader charge transfer of $\delta q = 3.11$ e occurs in the hexagonal buckled (hb) structure. Importantly, the stability of the structure increases with increasing charge transfer to the N atom [3]. This outcome follows from the decrease in the formation energy with increasing ionicity, which describes the energy required to synthesize the monolayer from the corresponding bulk material. Notably, the formation energy of the puckered phase of $E_f = -0.062$ eV is only marginally smaller than that of the buckled phase at $E_f = -0.02$ eV. Remarkably, low cohesive energies of $E_c = -6.76$ eV/atom (aw)

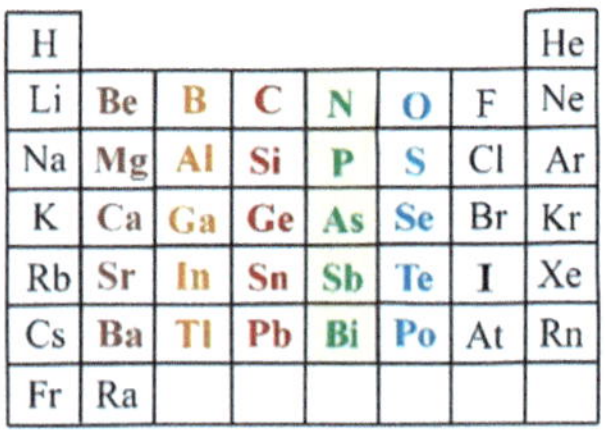

Fig. 9.1 Group–VA elements involved in the formation of heteroatomic binary VA–VA monolayers

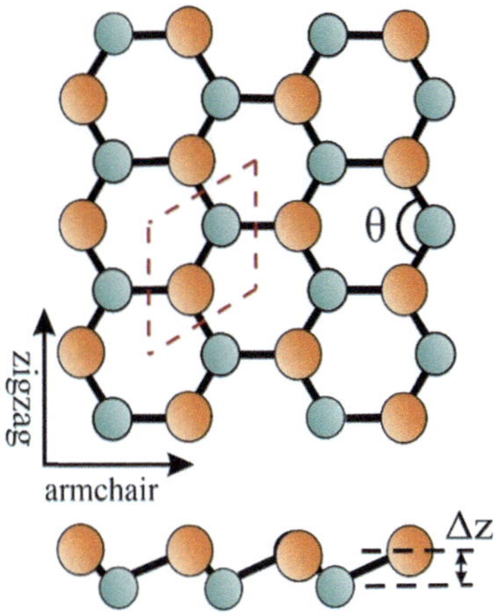

Fig. 9.2 Top and side views of the buckled structure of binary VA–VA monolayers

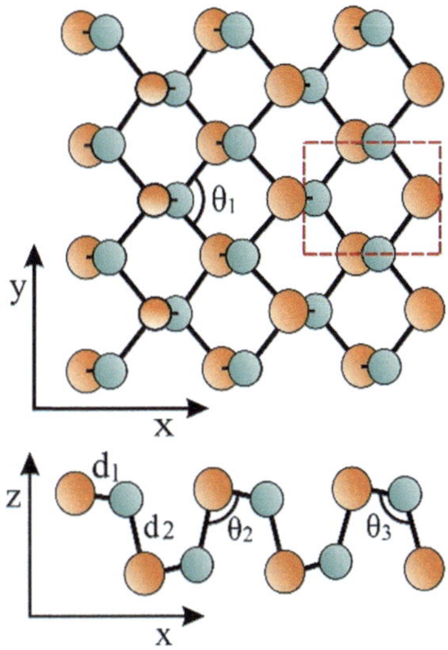

Fig. 9.3 Top and side views of the aw puckered structure of binary VA–VA monolayers

and – 6.23 eV/atom (hb) [2] have been reported in comparison to much lower binding energies of – 4.41 eV/atom (aw) and – 4.09 eV/atom (hb) [1] or even – 3.78 eV/atom (aw) and – 3.45 eV/atom (hb) [4] reported by other groups. This points to a not well understood chemical bonding behavior and stability of the two structures. Per definition, a negative binding energy designates energetic stability of the structure.

9.1.1.2 2D Structures

2D bonding and the resulting chemical behavior of phosphorus nitride depend on the available valence electrons provided by the electronic configuration of P (Ne|3s^2p^3)

and N (He|$2s^2p^3$). To characterize the 2D phases of PN monolayers we follow the nomenclature introduced for phosphorene, with two stable structures considered in more detail, the hexagonal hb phase and the puckered aw phase (note the different but not consistent classifications used in various publications). The DFT-PBE-calculated lattice constants of aw PN are a = 0.417 nm and b = 0.270 nm, the bond lengths are d_1 = 0.172 nm and d_2 = 0.182 nm, the pucker effect is Δz = 0.190 nm, and the bond angles are θ_1 = 103.2° and θ_2 = 97.65° and θ_3 = 123.7° [4, 5]. Based on the mean van der Waals diameter of PN of d_w = 0.355 nm [6], the addition of the corrugation effect provides an estimate of the monolayer thickness of $d_{w+\Delta z}$ = 0.545 nm for the aw phase. Since no imaginary frequencies could be found, the free-standing network is dynamically stable. The thermal stability was verified by ab initio molecular dynamics (AIMD) simulations [1]. The first reported indirect band gaps of E_g = 1.67 eV (PBE) and 2.82 eV (Heyd-Scuseria-Ernzerhof (HSE06)) are only slightly smaller than those given in subsequent publications [3]. The space group of the puckered PN monolayer is *Pmn*21 [1]. In bilayers, the shortest interlayer distances between nearest neighbors are d_{int} = 0.4464 nm and d_{int} = 0.4445 nm for the AA and the AB stackings, respectively, and taking the pucker effect into account, comparable layer spacings of d_s = 0.636 nm and d_s = 0.635 nm are obtained, respectively [1]. The structural properties of the heteroatomic binary VA–VA compounds are collected in Table 9.1.

For hb PN, the most complete compilation of structural data [4] gives a = 0.272 nm, d = 0.180 nm, Δz = 0.086 nm, and θ = 98.76° in good agreement with the findings of other studies [1, 3]. Based on the mean van der Waals diameter of PN of d_w = 0.355 nm [6], the addition of the corrugation effect amounts to a smaller layer thickness of $d_{w+\Delta z}$ = 0.441 nm for the buckled phase. The indirect band gaps at E_g = 1.82 eV (PBE) and 2.92 eV (HSE06) deviate little from those of the puckered phase [3]. The space group of buckled PN is *P*3*m*1 [1]. In hb PN bilayers, a distance between the nearest neighbors of d_{int} = 0.490 nm for AA and d_{int} = 0.541 nm for AB stacking was obtained, yielding interlayer spacings or layer thicknesses of d_s = 0.576 nm and d_s = 0.627 nm, respectively, if buckling is included [1]. Accordingly, the thickness based on the nearest interlayer distance is much greater for AB stacking than for AA stacking, which agrees well with the van der Waals-based thickness estimate for AA stacking. Such data on the interlayer spacing in few-layer and multilayer systems are especially important because both the band gap and band type vary with the order and number of stacked monolayers in few-layer assemblies [1].

9.1.1.3 Mechanics

As expected, the stiffness of aw-PN monolayers is strongly anisotropic and much greater than that of phosphorene due to the influence of the N atoms. Young's moduli and Poisson's ratios, calculated from the elastic stiffness coefficients, are available for the aw and hb phases of all group–VA nitrides [5]. The strongly anisotropic stiffnesses of the puckered phase of aw PN are E_{ac} = 39.0 N/m and E_{zz} = 198.6 N/m with the Poisson ratios ν_{xy} = 0.105 and ν_{yx} = 0.536 [3]. These values agree

Table 9.1 Phase, cohesive energy, lattice constants, bond angles, bond lengths, corrugation effect, van der Waals diameter, and corrugation corrected van der Waals diameter of binary group VA–VA monolayers

Monolayer	Phase	Cohesive energy, E_c (eV/atom)	Lattice constants a, b (nm)	Bond angles $\theta_1, \theta_2, \theta_3$ (deg.), θ (deg.)	Bond lengths d_1, d_2 (nm), d (nm)	Corrugation effect Δz (nm)	Van der Waals Diameter d_w [6], $d_{w+\Delta z}$ (nm)
PN	aw	− 4.41 [1], − 3.78 [4]	0.417, 0.270 [4]	103.2, 97.7, 123.7 [4]	0.172, 0.182 [4]	0.190 [5]	0.355 [6], 0.545
	hb	− 4.09 [1], − 3.45 [4]	0.272 [1]	98.76 [4]	0.180 [4]	0.086 [4, 5]	0.355 [6], 0.441
AsN	aw	− 3.46 [9], − 3.21 [4]	0.420, 0.296 [4]	102.3, 95.5, 119.7 [4]	0.190, 0.196 [4]	0.213 [5]	0.365 [6], 0.578
	hb	− 3.44 [10], − 3.38 [4]	0.297 [4]	98.26 [4]	0.197 [4]	0.096 [4, 5]	0.365 [6], 0.461
SbN	aw	− 3.90 [11], − 3.04 [4]	0.417, 0.330 [4]	103.5, 91.2, 117.3 [4]	0.211, 0.211 [4]	0.242 [5]	0.380 [6], 0.622
	hb	− 3.59 [11], − 2.76 [4]	0.327 [4]	99.45 [4]	0.214 [4]	0.101 [4, 5]	0.380 [6], 0.481
BiN	aw	− 2.71 [4]	0.376, 0.352 [4]	100.5, 86.8, 108.7 [4]	0.229, 0.221 [4]	0.261 [14]	0.390 [6], 0.651
	hb	− 2.38 [4]	0.347 [4]	100.33 [4]	0.226 [4]	0.104 [4]	0.390 [6], 0.494
AsP	aw	− 3.15 [15], − 2.32 [4]	0.462, 0.351 [4]	95.4, 99.6, 103.8 [4]	0.237, 0.238 [4]	0.235 [5]	0.400 [6], 0.635
	hb	− 3.18 [16], − 2.34 [4]	0.344 [4]	91.95 [4]	0.239 [4]	0.133 [4]	0.400 [6], 0.533
SbP	aw	− 2.97 [15], − 2.14 [4]	0.428, 0.391 [4]	96.9, 88.3, 105.0 [4]	0.261, 0.255 [4]	0.285 [5]	0.415 [6], 0.700
	hb	− 2.96 [16], − 2.13 [4]	0.372 [4]	91.61 [4]	0.259 [4]	0.145 [4]	0.415 [6], 0.560
BiP	aw	− 2.85 [15], − 1.98 [4]	0.445, 0.410 [4]	96.7, 90.9, 101.1 [4]	0.274, 0.264 [4]	0.293 [17]	0.425 [6], 0.718
	hb	− 2.79 [16], − 1.91 [4]	0.387 [4]	92.21 [4]	0.269 [4]	0.149 [4]	0.425 [6], 0.574

(continued)

Table 9.1 (continued)

Monolayer	Phase	Cohesive energy, E_c (eV/atom)	Lattice constants a, b (nm)	Bond angles $\theta_1, \theta_2, \theta_3$ (deg.), θ (deg.)	Bond lengths d_1, d_2 (nm), d (nm)	Corrugation effect Δz (nm)	Van der Waals Diameter d_W [6], $d_{W+\Delta z}$ (nm)
SbAs	aw hb	− 2.77 [15], − 2,05 [4] − 2.79 [16], − 2.07 [4]	0.455, 0.402 [4] 0.386 [4]	95.6, 90.0, 103.9 [4] 91.44 [4]	0.271, 0.265 [4] 0.270 [4]	0.262 [19] 0.152 [4, 19]	0.425 [6], 0.687 0.425 [6], 0.577
BiAs	aw hb	− 2.69 [15], − 1.93 [4] − 2.66 [16], − 1.90 [4]	0.460, 0.423 [4] 0.400 [4]	96.1, 91.6, 99.7 [4] 91.72 [4]	0.285, 0.277 [4] 0.279 [4]	0.295 [17] 0.155 [17]	0.435 [6], 0.730 0.435 [6], 0.591
BiSb	aw hb	− 2.53 [15], − 1.78 [4] − 2.52 [16], − 1.77 [4]	0.478, 0.448 [4] 0.423 [4]	94.7, 87.0, 102.55 [4] 90.85 [4]	0.304, 0.296 [4] 0.297 [4]	0.335 [17] 0.169 [4, 17]	0.450 [6], 0.785 0.450 [6], 0.619

Table 9.2 Phase, space group, Young's modulus, Poisson's ratio of binary group VA–VA monolayers

Monolayer	Phase	Space group	Young's modulus E_{ac}, E_{zz} (N/m)	Poisson's ratio ν_{ac}, ν_{zz}; ν_{xy}, ν_{yx}
PN	aw	*Pmn*2$_1$	39.0, 198.6 [3]	0.105, 0.536 [3]
	hb	*P*3*m*1	149.2 [3], 147 [5]	0.138 [3], 0.191 [5]
AsN	aw	*Pmn*2$_1$	17.6 132.8 [3]	0.121, 0.914 [3]
	hb	*P*3*m*1	101.7 [3], 100.8 [5]	0.217 [3], 0.217 [5]
SbN	aw	*Pmn*2$_1$	7.2, 97.2 [3]	0.06, 0.815 [3]
	hb	*P*3*m*1	77.9 [3], 75.1 [5]	0.205 [3], 0.261 [5]
BiN	aw	*Pmn*2$_1$	20, 66.5 [3]	0.108, 0.359 [3]
	hb	*P*3*m*1	53.4 [3]	0.275 [3]
AsP	aw	*Pmn*2$_1$	14.5, 60.6 [5]	0.234, 0.979 [5]
	hb	*P*3*m*1	61.5 [5], 52 [20]	0.151 [5], 0.24 [20]
SbP	aw	*Pmn*2$_1$	7.2, 44.5 [5]	0.223, 1.33 [5]
	hb	*P*3*m*1	45.2 [5]	0.186 [5]
SbAs	aw	*Pmn*2$_1$	4.5, 25.7 [5]	0.293, 1.66 [5]
	hb	*P*3*m*1	39.2 [5]	0.196 [5]

well with the elastic moduli of $E_{ac} = 40$ N/m and $E_{zz} = 190$ N/m and the Poisson ratios of $\nu_{xy} = 0.120$ and $\nu_{yx} = 0.564$, which were derived from the elastic stiffness coefficients [5]. The reason for the remarkable anisotropy of the mechanical behavior is that strain in the armchair direction primarily changes the bond angles rather than the bond length during the first stage of strain application. The data presented for the aw structure satisfy the mechanical Born–Huang stability criteria. The mechanical properties of the heteroatomic binary VA–VA monolayers are collected in Table 9.2.

The corresponding values for the quasi-isotropic hb-PN monolayer are somewhere between those of the anisotropic aw phase at $E = 149.2$ N/m and $\nu = 0.138$ [3]. These values are also in reasonable agreement with the stiffness of 147 N/m and the Poisson ratio of 0.191 derived from the elastic stiffness coefficients [5]. The values of the buckled structure also satisfy the mechanical stability criteria.

9.1.1.4 Physics and Applications

A comparison of the cohesive energies indicates that the P–N bond is stronger than the P–P bond of phosphorene. Therefore, the higher dynamic stability and oxidation resistance of PN monolayers are not surprising [1]. The PN monolayer is an indirect semiconductor with a wide band gap suitable for the application as a high-speed transistor, power electronic device, or efficient light emission diode (LED).

While elemental group–VA monolayers have no intrinsic piezoelectricity owing to their centrosymmetric structure, heteroatomic binary group–VA monolayers possess remarkable anisotropic piezoelectric properties that are comparable to those of the group–IVA monochalcogenides [5]. Compared with those of the buckled phase, the piezoelectric coefficients have surprisingly high values for the aw phase. This may be partly due to their flexible structure and, of course, their noncentrosymmetric symmetry [5]. The relevant piezoelectric coefficients controlling the efficiency of mechanical-to-electrical energy conversion correlate with the atomic polarizability, the transferred Bader charge, and the lattice parameters.

The iterative solution of the Boltzmann transport equation, together with first-principles calculations, supplied information on the phonon transport and thermal conductivity of aw PN and symmetric washboard (sw) phosphorene [7]. Surprisingly, aw-PN monolayers exhibit a higher thermal conductivity in the armchair direction (37.7 W m^{-1} K^{-1}) than in the zigzag direction (18.7 W m^{-1} K^{-1}), which is opposite to the behavior of sw phosphorene. The authors conclude that in addition to the directional dependence of the phonon group velocity, the avoided crossing of the acoustic and optical phonon branches plays a key role in determining the phonon relaxation time and thermal conductivity. Notably, these results provide important guidance for future studies on the anisotropy of thermal phonon transport in low-dimensional materials.

A first-principles study investigated the possibility of enhancing the electrochemical performance of lithium-sulfur batteries with hb PN. Due to the chemical interactions between electrodes and intermediates produced during the charging and discharging process, 2D electrode materials can improve electrochemical performance by overcoming the shuttle effect and by increasing the diffusion rates [8]. During the discharge process sulfur (S_8) and its intermediate lithium polysulfide products (Li_2S_n, n = 1, 2, 4, 6, and 8) with low conductivity and long-chains dissipate in the electrolyte (shuttle effect). The hb-PN monolayer can suppress the shuttle effect and the extremely small diffusion barriers for Li ions and, for example, Li_2S_8 allow a higher battery charge and discharge rate. The chemisorption of the Li atoms in lithium polysulfides at the N atoms of the hb-PN surface effectively suppresses the shuttle effect.

9.1.1.5 Synthesis

The realization of PN monolayers should be possible either as free-standing few-layer sheet or as monolayer stabilized on a suitable substrate surface. Since there is theoretical evidence that the two configurations considered here are both dynamically and thermally stable at room temperature, chemical vapor deposition (CVD) was suggested for the possible growth of the aw-PN and hb-PN phases on Ag substrates. As precursors, cyclic phosphazenes with a 1:1 mol fraction of P and N atoms were proposed to overcome difficulties associated with the strong nitrogen triple bonds [1]. A large separation between the adsorbed PN monolayer and the Ag substrate is expected, and theoretical scanning tunneling microscopy (STM) images suggest the presumed nonplanar periodic structure of the adsorbed PN monolayer. Notably, the synthesis of all group–VA nitrides is still in its infancy. To date, neither monolayers nor few-layer samples have been prepared for any of the four 2D group–VA nitrides.

9.1.2 AsN Monolayer

9.1.2.1 Chemistry and 2D Bonding

Chemical behavior and 2D bonding of arsenic nitride depend on the available valence electrons provided by the electronic configuration of As (Ar|$3d^{10}4s^2p^3$) and N atoms (He|$2s^2p^3$). Like group–VA elements, the specific bonding configuration of tetrahedrally coordinated arsenic atoms leads to bonding with three nitrogen atoms, which results in the formation of three nonplanar σ bonds. Note that the sp^2 hybrids in combination with the p_z^2 orbital, which contains the two remaining electrons, form sp^3-like orbitals with a p_z^2-like orbital that extends perpendicular to the plane. The sizable difference between the electronegativities of 2.20 (As) and 3.07 (N) introduces polar bonds with partial ionic character. There are small cohesive energies at $E_c = -6.02$ eV/atom [2] (aw) and $E_c = -5.58$ eV/atom [2] (hb) for both phases, and medium values indicating much weaker binding for the puckered phase of – 3.46 eV/atom [9] (aw) and – 3.21 eV/atom [4] (aw), and the buckled phase of – 3.44 eV/atom [10] (hb) and – 3.38 eV/atom [4] (hb).

9.1.2.2 2D Structures

For the aw phase of the AsN monolayer, the lattice constants are a = 0.420 and b = 0.296 nm, the bond lengths are $d_1 = 0.190$ and $d_2 = 0.195$ nm, and the bond angles are $\theta_1 = 102.3°$, $\theta_2 = 95.52°$, and $\theta_3 = 119.7°$ [4]. These values are in good agreement with those of another study, in which a pucker height of $\Delta z = 0.213$ nm has been identified [5]. The mean van der Waals diameter of $d_w = 0.365$ nm [6] and the corrugation yield an estimate of the monolayer thickness of $d_{w+\Delta z} = 0.578$ nm for the puckered structure. Compared to aw PN, the increases in the lattice constants and bond lengths are a consequence of a decrease in the binding energy. The space group of the puckered phase is *Pmn*21. The calculated band gaps of the indirect semiconductor are $E_g = 2.12$ eV (PBE) and 2.84 eV (HSE06) [3].

The structural parameters of hb-AsN monolayers are a = 0.297 nm, d = 0.197 nm, $\Delta z = 0.096$ nm, and $\theta = 98.26°$ [4, 5]. The mean van der Waals diameter of $d_w = 0.365$ nm [6] and the corrugation effect yield an estimate of the monolayer thickness of $d_{w+\Delta z} = 0.461$ nm. The space group of the buckled monolayer is *P*3*m*1. As mentioned above, substantially deviating cohesive energies of $E_c = -5.58$ eV/atom [2] and those with much lower binding strength of – 3.44 eV′atom [10] and – 3.38 eV/atom [4] exist for buckled AsN. The calculated indirect band gaps of the semiconductor are $E_g = 1.97$ eV (PBE) and 2.57 eV (HSE06) [3] The results published on the dynamic stability are contradictory. While one study found softening phonons, another one postulates dynamic stability [3, 9].

Manipulation of the indirect band gap of puckered AsN monolayers is possible by the application of strain [9]. In the armchair direction, a considerable compressive strain (– 12%) can trigger an indirect-to-direct band gap transition, and an even larger tensile strain (18%) can transform aw AsN into a stable metal. In the zigzag direction, compressive strains of – 12% and a smaller 8% tensile stretching process induce an indirect band gap-to-metal transition.

9.1.2.3 Mechanics

The puckered structure produces strongly anisotropic Young's moduli of $E_{ac} = 17.6$ N/m and $E_{zz} = 132.8$ N/m and Poisson's ratios of $\nu_{xy} = 0.121$ and $\nu_{yx} = 0.914$ [3]. The strong anisotropy is confirmed by $E_{ac} = 18.4$ N/m and $E_{zz} = 129$ N/m for the stiffness and $\nu_{xy} = 0.123$ and $\nu_{yx} = 0.864$ for the Poisson ratios, which were derived from the published elastic constants [5].

The linear elastic properties for hb AsN are $E = 101.7$ N/m and $\nu = 0.217$ [3]. These values are within the errors involved in such calculations and are practically identical to the elastic modulus of 100.8 N/m and Poisson ratio of 0.217 estimated from the published elastic constants [5].

9.1.2.4 Physics and Applications

While the structural, electronic, and mechanical properties have been studied by several groups, additional studies are widely missing for this compound. This can be explained be the fact that the properties of AsN monolayers follow well the expected group behavior without any presently known distinctive features of interest.

9.1.3 SbN Monolayer

9.1.3.1 Chemistry and 2D Bonding

Chemical behavior and 2D bonding of antimony nitride depend on the available five valence electrons in the outer shell of the electronic configuration of Sb (Kr|$4d^{10}5s^2p^3$) and N (He|$2s^2p^3$). As described for PN and AsN, in SbN the specific bonding configuration of tetrahedrally coordinated antimony atoms leads to bonding with three nitrogen atoms, which results in the formation of three nonplanar σ bonds by sp^3-like orbitals with the $p_z{}^2$-like orbital extending perpendicular to the plane. These hybrids possess the lowest energy. The bonds are strongly polar owing to the difference in electronegativities of 3.04 (N) and 2.05 (Sb). The lowest cohesive energies are $E_c = -5.73$ eV/atom [2] (aw) and – 5.31 eV/atom [2] (hb), mean values are – 3.90 eV/atom [11] (aw) and – 3.59 eV/atom [11] (hb), and the values with the weakest binding found in the literature are – 3.04 eV/atom [4] (aw) and – 2.76 eV/atom [4] (hb) for the two most stable aw and hb phases of SbN monolayers.

9.1.3.2 2D Structures

For the aw-SbN monolayers the lattice parameters are a = 0.417, b = 0.330 nm, $d_1 = d_2 = 0.211$ nm, $\Delta z = 0.242$ nm, $\theta_1 = 103.45°$, $\theta_2 = 91.17°$, and $\theta_3 = 117.28°$ [4, 5]. The thickness was calculated by adding the mean van der Waals diameter of $d_w = 0.380$ nm [6] to the pucker distance, which yields $d_{w+\Delta z} = 0.622$ nm. The calculated band gaps of the indirect semiconductor are $E_g = 1.78$ eV (PBE) and 2.58 eV (HSE06) [3].

In addition, the structural and electronic properties of the binary compounds SbN and SbP with *R*-3*m* symmetry were studied via first-principles calculations [12]. The monolayers exhibit indirect band gaps and undergo semiconductor-to-metal transitions as the number of layers increases.

The corresponding lattice parameters of hb-SbN monolayers are a = 0.327 nm, d = 0.214 nm, $\Delta z = 0.101$ nm, and $\theta = 99.45°$ [4, 5]. The thickness was calculated from the mean van der Waals diameter and the corrugation effect, yielding $d_{w+\Delta z} = 0.481$ nm. Note the extreme deviation between the published cohesive energies of $E_c = -5.31$ eV/atom [2], and – 2.76 eV/atom [4]. The band gaps of the indirect semiconductor, calculated by different approaches, are $E_g = 1.68$ eV (PBE) and 2.45 eV (HSE06) [3].

9.1.3.3 Mechanics

The values calculated for the strongly anisotropic Young's moduli and Poisson's ratios of puckered SbN are $E_{ac} = 7.2$ N/m and $E_{zz} = 97.2$ N/m and $\nu_{xy} = 0.06$ and $\nu_{yx} = 0.\ 815$ [3], in good agreement with $E_{ac} = 6.2$ N/m and $E_{zz} = 85.6$ N/m, however, much greater Poisson ratios of $\nu_{xy} = 0.121$ and $\nu_{yx} = 1.67$ derived from the elastic constants [5].

The buckled monolayer has a somewhat lower stiffness of E = 77.9 N/m and a larger Poisson ratio of $\nu = 0.205$ [3]. Qualitatively, this agrees well with the stiffness at E = 75.1 N/m and the Poisson ratio $\nu = 0.261$, derived from the published elastic constants [5].

9.1.3.4 Physics and Applications

The observation of negative Poisson ratios (NPRs) in two dimensions, where a transverse expansion/contradiction occurs when a tensile/compressive strain is exerted is of great interest.

Fully oxidized graphene is an interesting example of a 2D auxetic material with a great NPR value of – 0.567. Moreover, the aw and δ phases of the 2D group–VA nitrides are auxetic materials with large in-plane NPRs, except for δ BiN [3]. The δ phase has a ridge-like puckered structure with the space group *Pca*21. Of special interest is the δ SbN monolayer with an extreme NPR of – 0.628 in the zizag direction and – 0.296 in the armchair direction. This structure can sustain strains up to $\varepsilon = 0.22$

and $\varepsilon = 0.35$ at maximum stresses of 12.1 and 9.8 GPa in the zigzag and armchair directions, respectively. A strain of $\varepsilon = 0.18$ along the armchair direction introduces a maximal lateral expansion of 6.6%. The observed superior mechanical flexibility is of interest in designing 2D electromechanical and optoelectronic devices.

Another active field is the theoretical characterization of 2D ferroelectric materials showing a quantum order effect, occurring by spontaneous alignment of electronic dipoles. This spontaneous electric polarization is switchable by an electric field [13]. The sizable difference in the electronegativities and the large displacement of atoms in the polarization direction lower the symmetry of aw-SbN monolayers and simultaneously lower the energy of the system. In fact, aw SbN has one of the largest spontaneous polarizations of 7.8×10^{-10} C m^{-1}. The spontaneous polarization and switching barrier can be tuned by strain and the material can retain the ferroelectricity until melting at 1700 K [11]. The AA and AC stacking orders of aw-SbN sheets realize parallel polarization for each monolayer with strong ferroelectricity in the multilayer, contrary to the behavior of the elemental group–VA monolayers, which arrange the two adjacent layers in an antiferroelectric manner. The calculated effective layer spacing of the multilayers is $d_s = 0.590$ nm [11]. This agrees reasonably well with the mean van der Waals diameter plus pucker effect of $d_{w+\Delta z} = 0.622$ nm. Consequently, aw SbN and also BiP are excellent candidates for integrated ferroelectric devices such as nonvolatile memory devices, field-effect transistors, and solar cells.

The majority of the noncentrosymmetric aw and hb phases of binary group–VA monolayers possess piezoelectric properties. To describe the efficiency of mechanical-to-electrical energy conversion the piezoelectric coefficient d_{11} is used. Of particular interest are the aw phases of SbN, SbP, and SbAs with their theoretically predicted extremely high piezoelectric coefficients at 118.3, 142.4, and 243.5 pm/V, respectively [5]. These so-called giant piezoelectric effects are ascribed to the special symmetry and flexible structure of the aw lattice. To understand the origin of the anisotropic giant piezoelectric effect more parameters of the mechanical and electronic behavior are needed than the atomic polarizability that is sufficient to explain the 1–2 orders of magnitude smaller piezoelectric effect of the monolayers with hb structure.

The piezoelectric coefficients estimated by the DFT-GGA-PBE approach are comparable to those calculated for group–IVA monochalcogenides such as SnSe, SnS, GeSn, and GeS (d_{11} = 75–250 pm/V). The values suggest potential applications of these 2D materials as nanosized flexible sensors, actuators, and energy conversion devices.

9.1.4 BiN Monolayer

9.1.4.1 Chemistry and 2D Bonding

The 2D bonding and chemical behavior of bismuth nitride monolayers depend on the available five valence electrons provided by the electronic configuration of Bi

(Xe|$5d^{10}6s^2p^3$) and N (He|$2s^2p^3$). Like in the other binary group–VA compounds, the specific bonding configuration of tetrahedrally coordinated bismuth atoms leads to bonding with three nitrogen atoms, which results in the formation of three nonplanar σ bonds based with sp^3-like orbitals and $p_z{}^2$-like orbitals containing the lone electron pairs. Bonding is polar due to strongly deviating electronegativities of 1.9 (Bi) and 3.04 (N). The currently available cohesive energies for BiN monolayers are the low binding energies $E_c = -2.71$ eV/atom (aw) and – 2.38 eV/atom (hb) [4].

9.1.4.2 2D Structures

For aw-BiN monolayers, the lattice constants are a = 0.376 nm and b = 0.352 nm; the bond lengths are $d_1 = 0.229$ nm and $d_2 = 0.221$ nm; and the bond angles are $\theta_1 = 100.5°$, $\theta_2 = 86.83°$, and $\theta_3 = 108.7°$ [4]. The pucker distance is $\Delta z = 0.261$ nm in the network of four atom planes with the sequence Bi–N–N–Bi [14]. DFT-PBE calculations, which adopted the pseudo-potentials of Bi and N, suggest thermal and dynamical stability. For the indirect band gap, values of $E_g = 1.46$ eV (PBE) and 2.23 eV (HSE06) have been published [3]. The addition of the mean van der Waals diameter of BiN, $d_w = 0.390$ nm [6] and the buckling effect result in a monolayer thickness of $d_{w+\Delta z} = 0.651$ nm. The space group of the puckered structure is *Pmn*21. Both, the elastic moduli and Poisson ratios of $E_{ac} = 20$ N/m and $E_{zz} = 66.5$ N/m as well as $\nu_{xy} = 0.108$ and $\nu_{yx} = 0.359$ confirm the large anisotropy of the puckered structure [3].

The structural data of hb-BiN monolayers calculated for the lattice constant, bond length, buckling effect, and bond angle are a = 0.345 nm, d = 0.225 nm, $\Delta z = 0.104$ nm, and $\theta = 100.3°$, respectively [3, 4]. The addition of the mean van der Waals diameter of $d_w = 0.390$ nm [6] to the buckling effect provides an estimate for the thickness of $d_{w+\Delta z} = 0.495$ nm. The buckled configuration has the lowest binding strength of $E_c = -2.38$ eV/atom found for group–VA nitrides, which relates to the longest bonds observed for hb-BiN monolayers [4]. The calculated band gaps of the indirect semiconductor are $E_g = 0.86$ eV (PBE) and 1.65 eV (HSE06) [3]. The space group of the buckled phase is *P*3*m*1.

9.1.4.3 Mechanics

Calculations based on the DFT-PBE approach revealed the anisotropic stiffnesses of $E_{ac} = 20.0$ N/m and $E_{zz} = 66.5$ N/m and Poisson ratios $\nu_{ac} = 0.108$ and $\nu_{zz} = 0.359$, which demonstrate a superior mechanical flexibility [3]. The corresponding values of E = 53.4 N/m and ν = 0.275 for the buckled phase confirm the expectation that this phase has the lowest stiffness in this group of nitride compounds [3].

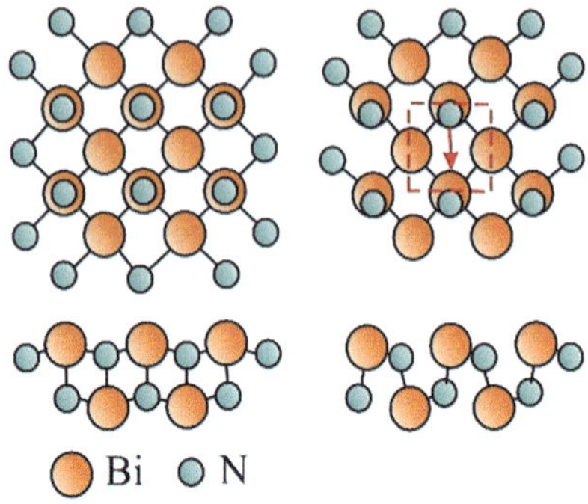

Fig. 9.4 **a** Top view and side view of the reference phase of BiN monolayers and **b** top and side view of the induced ferroelectric phase. The red arrow indicates ferroelectric distortions. Adapted with permission from Ref. [4], Copyright 2019, American Chemical Society

9.1.4.4 Physics and Applications

As previously mentioned for the behavior of the aw-SbN monolayers, the ridged structure with the $6s^2$ lone electron pairs of Bi induces 2D ferroelectricity in aw-BiN monolayers (see Fig. 9.4) [14]. Phonon spectra and MD simulations indicate that in the monolayer the ferroelectric polarization can reach a value of 580 pC/m. It is possible to switch the spontaneous electric polarization between the (100) and the (010) direction by applying tensile (010) stress or compressive (100) stress, corresponding to a low strain of approximately 2%. The reason for this easy switching of the polarization direction is the superior mechanical flexibility of BiN monolayers. Note that the phase transition involves a change of the lattice constants. The application of uniaxial tensile stress parallel to the polarization direction can fix the polarization and change the energy gap from direct to indirect.

9.2 Group VA–Phosphide Monolayers (AsP, SbP, and BiP)

9.2.1 AsP Monolayer

9.2.1.1 Chemistry and 2D Bonding

In the arsenic phosphide monolayer, AsP, the As atom has the valence electrons ($4s^2{p_x}^2{p_y}^1$) and the P atom ($3s^2{p_x}^2{p_y}^1$) in the outer valence shell. As described for group–VA elements, the specific bonding configuration of tetrahedrally coordinated arsenic atoms leads to bonding with three P atoms, which results in the formation of three nonplanar σ bonds realized by sp^3-like orbitals with a ${p_z}^2$-like orbital that contains the lone electron pair and extends perpendicular to the plane. Owing to the nearly identical electronegativities of As (2.18) and P (2.19), the As–P bonds are only slightly ionic, with a minimal charge transfer from the As atom to the P atom. In addition to the cohesive energies of $E_c = -4.31$ eV/atom [2] (aw) and -4.58 eV/atom [2] (hb) with strong bonding, medium values of $E_c = -3.15$ eV/atom [15] (aw) and – 3.18 eV/atom [16] (hb), and values indicating very low binding strength of E_c = – 2.32 eV/atom [4] (aw) and – 2.34 eV/atom [4] (hb) have been published for AsP monolayers.

9.2.1.2 2D Structures

The structural data of the AsP, SbP, and BiP monolayers were calculated within the DFT framework using the GGA approximation and the PBE formalism [4, 17]. The best studied and most stable bonding configurations of AsP monolayers are the aw and hb structures. For aw AsP, the lattice constants are a = 0.462 and b = 0.351 nm, the bond lengths are d_1 = 0.237 and d_2 = 0.238 nm, the pucker height is Δz = 0.235 nm [5], and the three bond angles are θ_1 = 95.37°, θ_2 = 99.64°, and θ_3 = 103.81° [4, 5]. The mean van der Waals diameter of d_w = 0.400 nm [6] yields a thickness of $d_{w+\Delta z}$ = 0.636 nm for the puckered monolayer. The space group of puckered AsP monolayers is *Pmn*21 [17].

A recent publication describes a novel structure, derived for a binary group–VA compound, which the authors call the β phase (e.g., β AsP) [18]. Note that the hb or β phases are normally assigned to hexagonal buckled structures. The novel monolayer structure with a direct band gap introduced by these authors seems to be a modified asymmetric puckered structure and therefore should be better labeled an aw-like phase.

Several studies have focused on the properties of the hb-AsP monolayer, which has desirable electronic properties. The structural parameters are a = 0.344 nm, d = 0.239 nm, Δz = 0.133 nm, and θ = 91.95° [4]. The mean van der Waals diameter of d_w = 0.400 nm [6] yields a thickness of $d_{w+\Delta z}$ = 0.533 nm for the buckled monolayer. The intermediate cohesive energy of E_c = – 3.18 eV/atom [16] and the much lower binding strength of – 2.34 eV/atom [4] are presented in Table 9.1. The space group of buckled AsP is *P*3*m*1 [17] The latter study presents the structural data of seven different 2D AsP allotropes [17].

9.2.1.3 Mechanics

The determination of Young's moduli and Poisson's ratios of puckered AsP monolayers from the published elastic stiffness coefficients resulted in the strongly anisotropic stiffnesses E_{ac} = 14.5 N/m and E_{zz} = 60.6 N/m and the strongly anisotropic Poisson ratios ν_{ac} = 0.234 and ν_{zz} = 0.979 [5].

Determination of the elastic modulus and Poisson ratio of buckled AsP from the published elastic stiffness coefficients resulted in E = 61.5 N/m and ν = 0.151 [5]. Independent calculations delivered a somewhat smaller stiffness and a substantially larger Poisson ratio of E = 52 N/m and ν = 0.24 for hb AsP monolayers [20].

9.2.1.4 Physics and Applications

First-principles calculations provided information on the possibility of tuning the band gap of aw-AsP monolayers [21]. This monolayer has a desirable direct band gap that can be easily transformed between direct–indirect-direct phases with increasing tensile and compressive strain, within a narrow strain range. Simultaneously, significant tuning of the effective mass of electrons and holes and the carrier mobility is possible by applying strain. Since the effective masses of both electrons and holes are an order of magnitude smaller in the armchair direction than in the zigzag direction, electron and hole transport occur preferentially in the armchair direction.

A comparative study of strain-dependent phonon transport in hb-AsP monolayers provided information on thermal transport by solving the linearized phonon Boltzmann equation within the single-mode relaxation time approximation [20]. The lattice thermal conductivity of buckled AsP first increases with increasing mechanical strain and then decreases. Moreover, first-principles molecular dynamics (MD) simulations provided information on the lattice thermal conductivity [22]. These results support the idea of engineering thermal transport properties by simple mechanical strain for applications in thermoelectrics and nanoelectronics. Lattice strain may also induce topological phase transitions in direct and indirect semiconductors [16]. Analysis of the band structure evolution indicates that the band inversion is due to the evolution of bonding and antibonding states under lattice strain. This evolution process is different from the mechanisms previously described for 2D TIs.

The modification of the electronic and magnetic properties by adsorbed molecules was investigated for both buckled and puckered AsP [23]. The presence of adatoms such as N, P, As, or Bi induces resonance states in the band continua, which may be attributed to permanent magnetic moments and may induce half-metallic character in multilayer AsP sheets. Another study investigated the sensing properties of aw AsP for the adsorption of SF_6, H_2O, and O_2 and the decomposition of SF_6, SO_2, and H_2S [24]. AsP chemisorbs only SO_2 with a robust adsorption energy and charge transfer, allowing sensitive and selective detection of SO_2 molecules.

The discovery of a series of novel energetically favored hb-AsP compounds was based on ab initio swarm intelligence, which is a global minimum structure-searching method [25]. The compound with the lowest energy is called I-AsP and exhibits a tunable quasidirect band gap, an ultrahigh electronic mobility, and a high optical absorption coefficient. Since the band gap can be tuned to an optimal value for photovoltaic applications by in-plane strain, this structure provides a promising material for highly efficient solar cells.

9.2.2 *SbP Monolayer*

9.2.2.1 Chemistry and 2D Bonding

In antimony phosphide monolayers, the Sb atom has the valence electrons ($5s^2p_x{}^2p_y{}^1$) and the P atom ($3s^2p_x{}^2p_y{}^1$) in the outer valence shell. As in the other group VA–VA compounds, the specific bonding configuration of tetrahedrally coordinated Sb atoms leads to bonding with three P atoms, which results in the formation of three nonplanar σ bonds by sp^3-like orbitals with the $p_z{}^2$-like orbital extending perpendicular to the plane. Owing to the small difference in the electronegativities of Sb (2.05) and P (2.19), the Sb–P bonds are slightly ionic, with charge transfer from the Sb atom to the P atom. In addition to the cohesive energies of Sb monolayers with strong cohesion $E_c = -4.51$ eV/atom [2] (aw) and -4.33 eV/atom [2] (hb), medium values of $E_c = -2.97$ eV/atom [15] (aw) and -2.96 eV/atom [16] (hb), and values indicating low binding strengths of $E_c = -2.14$ eV/atom [4] (aw) and -2.13 eV/atom [4] (hb) have been reported in the literature.

9.2.2.2 2D Structures

The lattice constants, bond lengths, pucker heights, and bond angles of aw-SbP monolayers are a = 0.428 nm, b = 0.391 nm, $d_1 = 0.261$ nm, $d_2 = 0.255$ nm, $\Delta z = 0.285$ nm [5], $\theta_1 = 96.94°$, $\theta_2 = 88.30°$, and $\theta_3 = 105.01°$ [4, 5]. Note that lattice constants and corrugation effects have been reported for five different monolayer structures of SbP [17]. The estimate of the thickness based on the van der Waals diameter of $d_w = 0.415$ nm [6] delivers $d_{w+\Delta z} = 0.700$ nm for the puckered configuration. The space group corresponds with that given for aw AsP, namely, *Pmn*21 [17]. The indirect band gap of puckered SbP is Eg = 0.85 eV [26]. Calculations of Young's moduli and Poisson's ratios of puckered SbP from the published elastic stiffness coefficients resulted in $E_{ac} = 7.2$ N/m and $E_{zz} = 44.5$ N/m and $\nu_{ac} = 0.223$ and $\nu_{zz} = 1.33$, respectively [5].

The corresponding lattice parameters for hb SbN monolayers are a = 0.372 nm, d = 0.259 nm, $\Delta z = 0.145$ nm, and $\theta = 91.61$ [4, 5]. The thickness based on the van der Waals diameter of $d_w = 0.415$ nm [6] extends to $d_{w+\Delta z} = 0.560$ nm for the buckled configuration. The space group of *P*3*m*1 corresponds with that given for buckled AsP [17]. The calculated interlayer spacing of buckled SbP multilayers in the ABC rhombohedral stacking order is $d_s = 0.334$ nm and thus much smaller than the layer thickness [12]. Determination of the elastic modulus and Poisson ratio of buckled SbP monolayers, using the elastic stiffness coefficients, resulted in E = 45.2 N/m and $\nu = 0.186$ [5].

9.2.2.3 Physics and Applications

Both aw-SbP and aw-BiP monolayers are suitable thermoelectric materials with large Seebeck coefficients and high electrical conductivity for building thermoelectric devices, capable of generating electrical power from waste heat [26]. The puckered BiP monolayer with a smaller thermal conductivity showed a large *ZT* value of 4.59, whereas it was only 1.34 for aw-SbP monolayers. Using first-principles calculations and an iterative solution of the phonon Boltzmann transport equation, an extensive number of binary group VA–VA monolayers were investigated [15]. The lattice thermal conductivities of all these compounds could be correlated with their average atomic masses and were almost inversely proportional to the temperature for $\geq$100 K. Similar calculations were performed for AsP, SbP, and SbAs monolayers, with the conclusion that the thermal conductivities of these compounds are between those of their constituting elements [27].

As already discussed for aw-AsP monolayers, a first-principles study also investigated the decomposition of SF_6 by a SbP monolayer, which confirmed the preferred detection of SO_2. This process is characterized by an adsorption energy of – 0.541 eV and a charge transfer of $\delta q = 0.232$ e. These findings suggest chemisorption by orbital hybridization [28].

The ferroelectricity of the allotropes γ-SbAs and γ-SbP, which have excellent ferroelectric properties and in-plane spontaneous polarizations of ~ 3.80×10^{-10} C/m (γ SbAs) and ~ 3.47×10^{-10} C/m (γ SbP), has attracted considerable attention [29]. Notably, the ferroelectric state has been maintained above 700 K (γ SbAs) and 600 K (γ SbP).

9.2.3 *BiP Monolayer*

9.2.3.1 Chemistry and 2D Bonding

In the bismuth phosphide monolayer BiP, the Bi atom has the valence electrons ($6s^2p_x{}^2p_y{}^1$) and the P atom ($3s^2p_x{}^2p_y{}^1$) in their outer valence shell. Like in the other heteroatomic group VA–VA compounds, the specific bonding configuration of tetrahedrally coordinated Bi atoms leads to bonding with three P atoms, which results in the formation of three nonplanar σ bonds, based on sp^3-like orbitals with the $p_z{}^2$-like orbital extending perpendicular to the plane. Owing to the small difference in the electronegativities of Bi (1.9) and P (2.19), the Bi–P bonds are slightly ionic, with charge transfer from the Bi to the P atom. Ab initio DFT-PBE calculations produced a wide range of cohesive energies with strong bonding at $E_c = -3.19$ eV/atom [11] (aw) and – 3.09 eV/atom [11] (hb), medium bonding strengths at – 2.85 eV/atom [15] (aw) and – 2.79 eV/atom [16] (hb), and very weak bonding at – 1.98 eV/atom [4] (aw) and – 1.91 eV/atom [4] (hb).

9.2.3.2 2D Structures

From the monolayer structures studied theoretically, the most stable one is the aw-BiP monolayer, with the lattice constants a = 0.445 and b = 0.410 nm, the bond lengths d_1 = 0.274 nm and d_2 = 0.264 nm, a pucker distance of Δz = 0.293 nm [17], and the three bond angles $\theta_1 = 96.69°$, $\theta_2 = 90.93°$, and $\theta_3 = 101.09°$ [4, 17]. An estimate of the monolayer thickness of $d_{w+\Delta z}$ = 0.718 nm can be obtained by adding the average van der Waals diameter of d_w = 0.425 nm [6] and the pucker effect. The space group of the puckered phase is *Pmn*21 [17]. The interlayer spacing for AA stacking is d_s = 0.606 nm, which is the ground state of bilayers [11]. This separation of layers is consistent with the slightly larger estimated thickness of the monolayer due to missing interlayer interaction. Notably, aw BiP has a polarization that is close to that of other strong 2D ferroelectric materials, such as aw SbN, and therefore qualifies for integrated ferroelectric devices.

The structural parameters of hb-BiP monolayers are a = 0.387 nm, d = 0.269 nm, Δz = 0.149 nm, and $\theta = 92.21°$ [4, 30]. Adding the mean van der Waals diameter of d_w = 0.425 nm [6] and the buckling effect delivers an upper limit of the monolayer thickness of $d_{w+\Delta z}$ = 0.574 nm. The space group of hb BiP is *P*3*m*1 [17]. Contrary to the large interlayer separation observed for puckered BiP sheets, the interlayer separation in ABC stacking of buckled BiP is only d_s = 0.342 nm [12]. Because of its calculated direct band gaps at E_g = 1.45 eV (PBE) and 2.16 eV (GAU) and enhanced carrier mobility, the hb-BiP monolayer holds greater promise for practical applications in nanoelectronics and optoelectronics than phosphorene [30]. Note that the monolayers of hb AsP and hb SbP are indirect band gap semiconductors with larger gaps [30].

9.3 Various Binary Group–VA Monolayers (SbAs, BiAs, and BiSb)

9.3.1 SbAs Monolayer

9.3.1.1 Chemistry and 2D Bonding

In monolayers of SbAs, the Sb atom has the five valence electrons ($5s^2p_x{}^2p_y{}^1$) and the As atom ($4s^2p_x{}^2p_y{}^1$) in the outer valence shell. As discussed above, the specific bonding configuration of tetrahedrally coordinated Sb atoms leads to bonding with three As atoms, which results in the formation of three nonplanar covalent σ bonds based on sp^3-like orbitals with the $p_z{}^2$-like orbital extending perpendicular to the plane. Owing to the small difference in the electronegativities of Sb (2.05) and As (2.18), the Sb–As bonds are only slightly ionic, with marginal charge transfer from the Sb atom to the As atom. Besides the quite large cohesion of SbAs monolayers of E_c = – 4.26 eV/atom [2] (aw) and – 4.92 eV/atom [2] (hb), a medium cohesion of

E_c = – 2.77 eV/atom [15] (aw) and – 2.79 eV/atom [16] (hb), and results indicating even lower cohesion at E_c = – 2.05 eV/atom [4] (aw) and – 2.07 eV/atom [4] (hb) have been reported in the literature.

9.3.1.2 2D Structures

The lattice constants, bond lengths, corrugation, and bond angles of aw-AsSb monolayers are a = 0.455, b = 0.402 nm, d_1 = 0.271, d_2 = 0.265 nm, Δz = 0.262 nm [17, 19], θ_1 = 95.64, θ_2 = 90.03°, and θ_3 = 103.85 [4, 17, 19]. In addition to the aw and hb structures considered here, additional 2D SbAs polymorphs have been characterized [17, 19]. The mean van der Waals diameter is d_w = 0.425 nm [6], and adding the pucker height yields a layer thickness $d_{w+\Delta z}$ = 0.687 nm. The space group of *Pmn*21 agrees with that of puckered AsP monolayers. Determination of the elastic modulus and Poisson ratio of puckered SbAs from reported elastic stiffness coefficients resulted in the strongly anisotropic stiffnesses E_{ac} = 4.5 N/m and E_{zz} = 25.7 N/m and Poisson ratios ν_{ac} = 0.293 and ν_{zz} = 1.66 [5].

The corresponding data for the hb-SbAs monolayer are a = 0.386 nm, d = 0.270 nm, 0.152 nm, and θ = 91.44° [4]. Furthermore, the lattice parameters of additional allotropes are available [17, 19]. The van der Waals diameter is d_w = 0.425 nm [6], and by taking the buckling effect of 0.152 nm into account, an estimate of monolayer thickness yields $d_{w+\Delta z}$ = 0.577 nm. For the lowest-energy configuration of hb SbAs, spin–orbit coupling (SOC) is significant and reduces the gap by 200 meV [19]. Closing and reopening of the band gap of 1.47 eV (PBE) occurs under biaxial tensile strain, concomitant with a change in the band shape. This property and the gapless edge states are associated with TI characteristics. Therefore, this polymorph could be a potential candidate for the QSH effect [19]. Determination of the elastic modulus and Poisson's ratio of buckled SbAs from the published elastic stiffness coefficients resulted in a small mechanical stiffness of E = 39.2 N/m and ν = 0.196 [5].

9.3.2 *BiAs Monolayer*

9.3.2.1 Chemistry and 2D Bonding

In bismuth arsenide monolayers, the Bi atom has the five valence electrons ($6s^2p_x{}^2p_y{}^1$) and the As atom ($4s^2p_x{}^2p_y{}^1$) in the outer shell. As discussed above, the specific bonding configuration of tetrahedrally coordinated Bi atoms leads to bonding with three As atoms, which results in the formation of three nonplanar covalent σ bonds based on sp^3-like orbitals with the $p_z{}^2$-like orbital extending perpendicular to the plane. Owing to the small difference in the electronegativities of Bi (1.9) and As (2.18), the Bi–As bonds are slightly ionic, with charge transfer from the Bi atom to the As atom. In addition to very strong cohesive interaction in BiAs monolayers at E_c = – 4.17 eV/atom [31] (aw) and – 4.14 eV/atom [31] (hb), values suggesting medium interactions at $E_c = -2.69$ eV/atom [15] (aw) and – 2.66 eV/atom [16] (hb), and very low binding interactions at $E_c = -1.93$ eV/atom [4] (aw) and – 1.90 eV/atom [4] (hb) have been reported.

9.3.2.2 2D Structures

From the five polymorph structures characterized by first-principles calculations in combination with a tight-binding model, the results for the lattice constants, bond lengths, pucker, and bond angles of the aw-BiAs monolayer are a = 0.460, b = 0.423 nm, $d_1 = 0.285$ nm, $d_2 = 0.277$ nm, $\Delta z = 0.295$ nm [17], $\theta_1 = 96.10°$, $\theta_2 = 91.56°$, and $\theta_3 = 99.7°$ [4, 17]. For the puckered phase, we estimate a monolayer thickness of $d_{w+\Delta z} = 0.730$ nm with a mean van der Waals diameter of $d_w = 0.435$ nm [6]. The space group of puckered BiAs monolayers is *Pmn*21 [17].

The corresponding structural properties for hb-BiAs monolayers are a = 0.400 nm, d = 0.279 nm, $\Delta z = 0.156$ nm, and $\theta = 91.72$ [4]. For the more important buckled phase, we obtain a monolayer thickness of $d_{w+\Delta z} = 0.591$ nm with the average van der Waals diameter of $d_w = 0.435$ nm [6]. The space group of buckled BiAs monolayers is *P*3*m*1 [17].

9.3.2.3 Physics and Applications

First-principles calculations and the Boltzmann transport theory were used to determine the band structures, phonon spectra, lattice thermal conductivities, electronic transport properties, Seebeck coefficients, and the figures of merit for hb-BiAs and hb-BiSb monolayers [32]. The results show that the Seebeck coefficients and figures of merit are greater than those of the elemental hb-Bi monolayer.

High-throughput first-principles calculations were employed to determine the ground-state structure of group–VA binary monolayers [33]. The authors propose two stable structures, β_1-BiAs and β_2-BiAs, with a direct band gap and extremely high hole mobilities of interest for nanoelectronic and optoelectronic applications. The new structural arrangement reduces the total energy of β_1-BiAs monolayers by 0.359 eV compared with that of aw-BiAs monolayers and by 0.866 eV in comparison with that of the hb-BiAs phase.

Stretching the bonds of monolayers by biaxial tensile strain is a powerful method for tuning the electronic and topological properties of 2D materials containing the heavy elements As, Sb, and Bi with a strong SOC, such as observed for hb-BiAs monolayers [34]. With increasing tensile strain, the conduction band shifts downward, and the valence band shifts upward, forming a semimetal at $\varepsilon \leq 0.12$ (see Fig. 9.5). Further increasing tensile strain and including the SOC opens a band gap of $E_g = 0.28$ eV and induces band inversion (see Fig. 9.6). Thus, first-principles calculations combined with a tight-binding model indicate the transition of 2D hb BiAs to a nontrivial TI with the prospect of dissipationless transport at room temperature. Theory indicates dynamical stability up to a tensile strain of $\varepsilon = 0.17$.

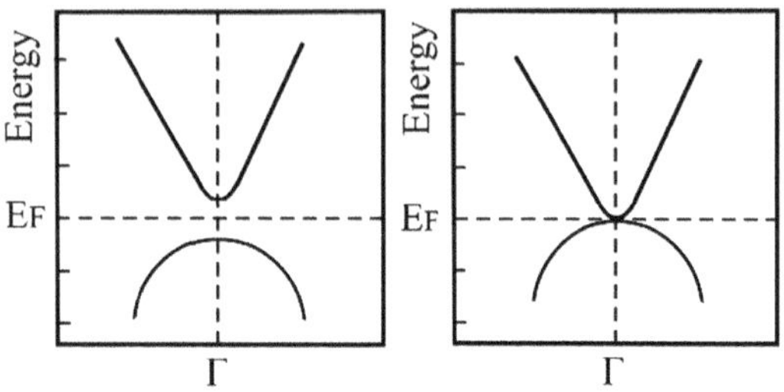

Fig. 9.5 Transformation of the semiconducting hb-BiAs monolayer to a semimetal by a strain of $\varepsilon \approx 0.12$ without SOC. Adapted with permission from Ref. [34], Copyright 2018, American Chemical Society

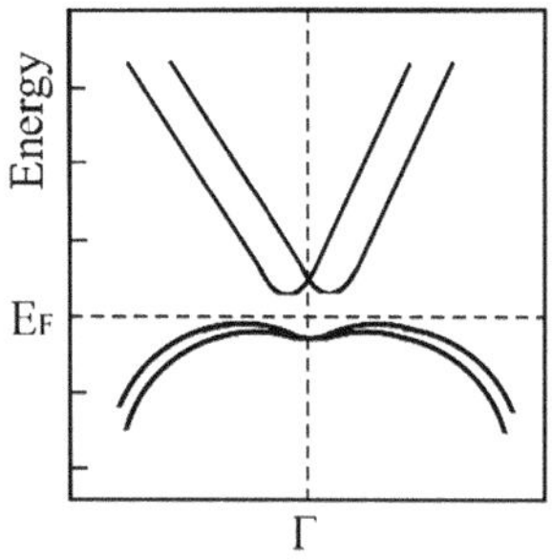

Fig. 9.6 Increasing strain beyond $\varepsilon = 0.12$ and the inclusion of SOC open a gap in the hb-BiAs monolayer and lead to band inversion at the Γ point and generates a TI. Adapted with permission from Ref. [34], Copyright 2018, American Chemical Society

9.3.3 BiSb Monolayer

9.3.3.1 Chemistry and 2D Bonding

In the BiSb monolayer, the Bi atom has the valence electrons ($6s^2p_x{}^2p_y{}^1$) and the Sb atom ($5s^2p_x{}^2p_y{}^1$) in the outer valence shell. As discussed above, the specific bonding configuration of tetrahedrally coordinated Bi atoms leads to bonding with three Sb atoms, which results in the formation of three nonplanar covalent σ bonds based on sp^3-type orbitals with the $p_z{}^2$-type orbital extending perpendicular to the plane. Owing to the small difference in the electronegativities of Bi (1.9) and Sb (2.05), the Bi–Sb bonds are slightly ionic, with charge transfer from the Bi atom to the Sb atom. As mentioned before, calculations predict in addition to strong cohesion in BiAs monolayers at $E_c = -3.85$ eV/atom [31] (aw) and – 3.86 eV/atom [31] (hb), medium cohesive interaction at $E_c = -2.53$ eV/atom [15] (aw) and – 2.52 eV/atom [16] (hb), and weak cohesive interaction at $E_c = -1.78$ eV/atom [4] (aw) and – 1.77 eV/atom [4] (hb).

9.3.3.2 2D Structures

The lattice constants, bond lengths, pucker, and bond angles of aw-BiSb monolayers are a = 0.478, b = 0.448 nm, $d_1 = 0.304$ nm, $d_2 = 0.296$ nm, $\Delta z = 0.335$ nm, $\theta_1 = 94.72$, $\theta_2 = 87.03°$, and $\theta_3 = 102.55°$ [4, 17]. An estimate of the thickness with a mean van der Waals diameter of $d_w = 0.450$ nm [6], including corrugation, results in $d_{w+\Delta z} = 0.785$ nm.

The calculated structural parameters of hb-BiSb monolayers are a = 0.423 nm, d = 0.297 nm, $\Delta z = 0.169$ nm, $\theta = 90.85°$ [4, 17, 35]. An estimate of the thickness with the mean van der Waals diameter of $d_w = 0.450$ nm [6] results in $d_{w+\Delta z} = 0.619$ nm when buckling is included. The space group of the buckled structure is $P3m1$ [16].

9.3.3.3 Physics and Applications

The hb-BiSb monolayer is a potential candidate for achieving the QSH effect. The most stable phase of buckled BiSb transforms from a normal insulator to a nontrivial topological phase with significantly larger band gap under strain engineering along with the SOC effect [35].

First-principles calculations combined with Boltzmann transport theory elucidated the effect of Rashba SOC on the thermoelectric performance of BiSb monolayers [36]. The results reveal that Rashba SOC promotes the thermoelectric performance enormously due to the reduced dimensionality of the density of states (DOS) that makes BiSb monolayers to function as an effective 1D thermoelectric material. For example, the *ZT* value is 0.32 at room temperature with Rashba SOC and only $ZT = 0.18$ in the spin degenerate case. The induction of one-dimensional-like DOS and a prolonged carrier lifetime enhance the response. The internal electric field in the Rashba system increases the anharmonicity of phonons. This further improves the thermoelectric performance owing to the reduced thermal conductivity.

A detailed study revealed the structural dependence of the calculated optoelectronic properties of hb- and aw-Sb, aw-Bi, and aw-BiSb monolayers on the applied approaches PBE, PBE-SOC, HSE, HSE-SOC, and G_0W_0 [37]. Their calculated optical activities extend from the IR to the UV region, either partially or entirely, and exhibit isotropic or anisotropic optical responses. Specifically, BiSb qualifies for applications as a solar cell, light emitting device, photodetector, or light modulator. As discussed in more detail for BiAs, elemental and binary BiSb monolayers possess high SOC and potential TI properties upon engineering their structural and chemical properties.

The status and challenges of binary VA–VA compounds are reviewed in a comprehensive work covering theoretical predictions and experimental advances. This includes the crystal structure, stability, and electronic structure; tunability by composition, defects, doping, strain, and substrate; synthesis by vapor transport, molecular beam epitaxy (MBE), and melt growth; applications in field-effect transistors and as topological spintronic device, photodetector, energy conversion and energy storage device, catalysis, rechargeable batteries, and thermoelectric materials [38].

9.4 Property–Structure Relationships

9.4.1 Binding–Structure Relationships

As mentioned above, the calculated cohesive energies of binary group–VA monolayers vary by more than a factor of two depending on the approximation applied. Figure 9.7 shows a plot of the published small, medium, and large cohesive energies versus bond lengths of the stable aw, and hb phases of elemental and binary group–VA compounds [39]. The cohesive energy–bond length plot reveals the typical inverse behavior and comparable binding energies of the puckered and buckled phases of the elemental and heteroatomic monolayers. Unfortunately, a more detailed description of the theoretical approximations used by the authors in these calculations is often missing, as well as a detailed discussion of the deviations of their results from previous work. In any case, the cohesion presented by one research group for PN, AsN, SbN, and AsP seem to be too large. For example, the cohesive energy of AsP should be located between the limiting values of elemental phosphorene and arsenene because the data from several sources follow the principle that binary compounds have cohesive energies comparable or intermediate to those of the two corresponding elemental constituents. Importantly, nitrides possess the strongest bonding and thus deviate from the other groups, because in the nitrides in addition to the efficient overlap of orbitals owing to the small size of nitrogen atoms, strong ionic interactions contribute to in-plane interactions.

Certain cohesive energies show inverse and almost parallel dependencies, one with medium cohesion and the other one with low cohesion. The cohesive energies of both groups reveal a consistent description of the postulated relationship between

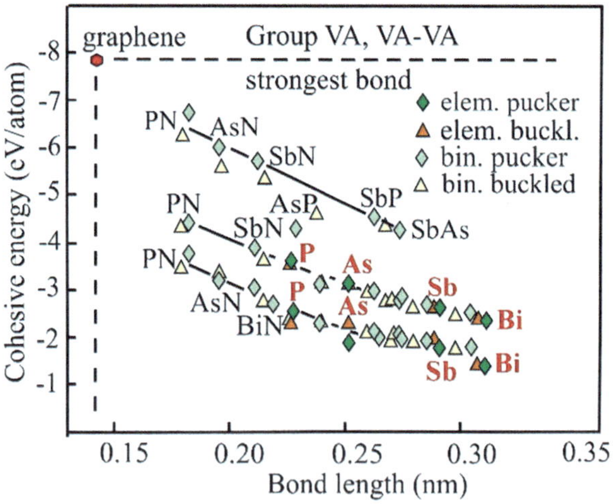

Fig. 9.7 Cohesive energies of elemental and binary group–VA monolayers as a function of bond lengths. Adapted from Ref. [39], Copyright 2021, Royal Society of Chemistry

the cohesive energies of elemental and binary compounds, which requires an intermediate bonding strength of the mixed networks. Importantly, most of the included studies were based on DFT calculations and involved a treatment of the exchange-correlation functional by the GGA-PBE approach. A difference recognized is that the calculations of the weakest cohesive interactions considered the SOC effect [4]. To which extent this explains the observed deviations in the binding energies is not clear yet.

9.4.2 Mechanics–Structure Relationships

Figure 9.8 shows Young's moduli and available ultimate strengths versus bond lengths of the group–VA elements and their binary VA–VA compounds [39]. Notably, the large anisotropy of the elastic moduli between the armchair and the zigzag directions of the puckered phase decreases abruptly with the inclusion of heavier elements in the composition. The stiffness of the buckled structures widely follows that of the stronger zigzag direction of the puckered phases. A comparison of the results obtained for elemental and binary compounds supports the close relationship between their bonding and their structure due to the similarity of the electronic configurations.

As shown in Fig. 9.8, the strong mechanical performance of binary VA–VA monolayers in the zigzag direction follows the periodic trend of their elements, where the nitrides PN, AsN, SbN, and BiN stand out owing to the greater rigidity of these monolayers. The observed correlations clearly point to comparable bond energies in homonuclear and heteronuclear monolayers despite additional ionic interactions in

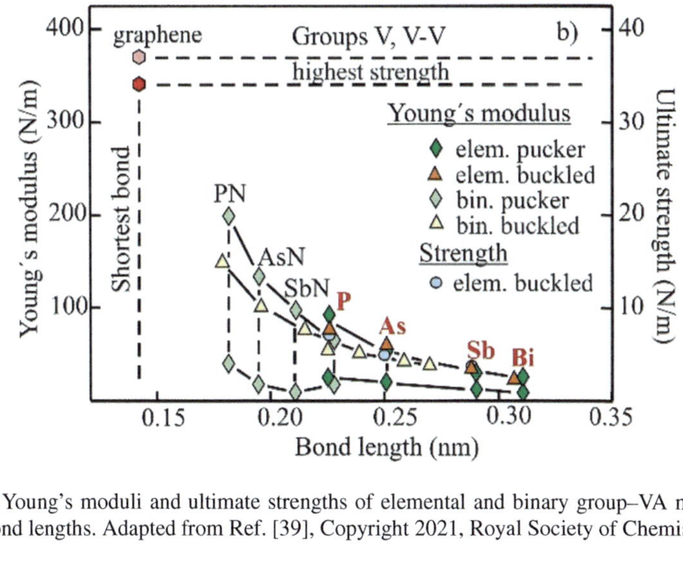

Fig. 9.8 Young's moduli and ultimate strengths of elemental and binary group–VA monolayers versus bond lengths. Adapted from Ref. [39], Copyright 2021, Royal Society of Chemistry

the heteroatomic compounds. Unfortunately, not only experiments but also calculations of the ultimate strengths and strains of binary VA–VA compounds are still widely lacking. The fact that the few currently available strength values agree well with Griffith's rule allows first estimates of the strengths of binary monolayers not yet characterized by theory based on this correlation.

References

1. S. Ma, C. He, L.Z. Sun, H. Lin, Y. Li et al., Phys. Chem. Chem. Phys. **17**, 32009 (2015)
2. W. Yu, C.-Y. Niu, Z. Zhu, X. Wang, W.-B. Zhang, J. Mater. Chem. C **4**, 6581 (2016)
3. W.-Z. Xiao, G. Xiao, Q.-Y. Rong, L.-L. Wang, Phys. Chem. Chem. Phys. **20**, 22027 (2018)
4. S. Guo, W. Zhou, B. Cai, K. Zhang, S. Zhang et al., Nanoscale Horiz. **4**, 1145 (2019)
5. H. Yin, J. Gao, G.-P. Zheng, Y. Wang, Y. Ma, J. Phys. Chem. C **121**, 25576 (2017)
6. S.S. Batsanov, Inorg. Mater. **37**, 871 (2001)
7. A. Taheri, C.V. Singh, Phys. Rev. Mater. **5**, 034009 (2021)
8. X. Zheng, S. Lin, D. Kong, Y. Wei, K. Pang et al., Adv. Theory Simul. **5**, 2100305 (2021)
9. P. Liu, Y.-Z. Nie, Q.-L. Xia, G.-H. Guo, Phys. Lett. A **381**, 1102 (2017)
10. M.-Y. Liu, Y. Huang, Q.-Y. Chen, C. Cao, Y. He, Sci. Rep. **6**, 29114 (2016)
11. C. Liu, W. Wan, J. Ma, W. Guo, Y. Yao, Nanoscale **10**, 7984 (2018)
12. Z. Yang, D. Han, G. Chen, S. Chen, Mater. Res. Express **5**, 035904 (2018)
13. R.M. Ur, H. Chenqiang, L. Yunhao, Chin. Phys. B **29**, 057304 (2020)
14. P. Chen, X.-J. Zhang, B.-G. Liu, A.C.S. Appl, Nano Mater. **2**, 58 (2019)
15. T. Kocabas, D. ςakir, O. Gülseren, F. Ay, N.K. Perkgöz et al., Nanoscale **10**, 7803 (2018)
16. Y. Nie, M. Rahman, D. Wang, C. Wand, G. Guo, Sci. Rep. **5**, 17980 (2015)
17. W.-Z. Xiao, G. Xiao, Q.-Y. Rong, L.-L. Wang, Mater. Res. Express **5**, 035903 (2018)
18. F. Chen, H. Xu, J. Wang, Z. Wang, X. Liu et al., J. Appl. Phys. **125**, 214303 (2019)
19. S. Zhang, M. Xie, B. Cai, H. Zhang, Y. Ma et al., Phys. Rev. B **93**, 245303 (2016)
20. S.-D. Guo, J. Phys. D Appl. Phys. **51**, 265307 (2018)
21. Y. Mao, Z. Yao, R. Wu, X. Zhou, Y. Du, Mater. Res. Express **9**, 025009 (2022)
22. Y. Sun, Z. Shuai, D. Wang, Phys. Chem. Chem. Phys. **20**, 14024 (2018)
23. I. Ozdemir, H.D. Ozaydin, H. Arkin, E. Aktürk, Mater. Res. Express **6**, 065032 (2019)
24. W. Jin, Y. Guofeng, X. Junjun, L. Jianming, C. Qing et al., Sci. Rep. **8**, 12011 (2018)
25. X. Cai, Y. Chen, B. Sun, J. Chen, H. Wang et al., Nanoscale **11**, 8260 (2019)
26. X. Liu, D. Zhang, Y. Chen, H. Wang, H. Wang et al., Phys. Chem. Chem. Phys. **23**, 24598 (2021)
27. L. Liu, C. Peng, S. Jiang, Z. Feng, G. Zhang, et al., Mater. Res. Express **8**, 075007 (2021)
28. H. Guo, K. Zheng, H. Cui, F. Zhang, J. Yu et al., Chem. Phys. Lett. **756**, 137868 (2020)
29. S. Shen, C. Liu, Y. Ma. B. Huang, Y. Cai, Nanoscale **11**, 11864 (2019)
30. L. Zhu, T. Zhang, J. Nanopart. Res. **20**, 237 (2018)
31. H. Zhang, M. Chen, J. Mater. Chem. C **6**, 11694 (2018)
32. C.Y. Wu, L. Sun, J.C. Han, H.R. Gong, J. Mater. Chem. C **8**, 581 (2020)
33. W. Xu, Y. Jin, B. Zheng, H. Xu, J. Phys. Chem. C **122**, 27590 (2018)
34. T. Teshome, A. Datta, J. Phys. Chem. C **122**, 15047 (2018)
35. W. Yu, C.-Y. Niu, Z. Zhu, X. Cai, L. Zhang et al., RSC Adv. **7**, 27816 (2017)
36. J. Yuan, Y. Cai, L. Shen, Y. Xiao, J.-C. Ren et al., Nano Energy **52**, 163 (2018)
37. D. Kecik, V.O. Özçelik, E. Durgun, S. Ciraci, Phys. Chem. Chem. Phys. **21**, 7907 (2019)
38. S. Guo, Y. Zhang, Y. Ge, S. Zhang, H. Zeng et al., Adv. Mater. **31**, 1902352 (2019)
39. P. Hess, Nanoscale Horiz. **6**, 856 (2021)

Chapter 10
Group IVA–VA Monolayers

Abstract Most of the information presently available on binary group IVA–VA monolayers with the stoichiometry AB (A = C, Si, Ge, Sn, and Pb; B = N, P, As, Sb, and Bi) is based on first-principles calculations. Monolayers of these compounds with the chemical formula A_2B_2 exist as hexagonal four-layer assemblies with the configurations B–A–A–B, named the α and β phases, with the space groups *P*-6*m*2 and *P*-3*m*1, respectively. Cutting the covalent A–A bonds yields two buckled monolayers. Their unit cell consists of two A and two B atoms. Furthermore, a less symmetric monoclinic phase consisting of hexagons and pentagons has been synthesized. These and other allotropes often possess similar binding energies due to the similarity of the bonding arrangements, and therefore the most stable phase often is not known. Notably, enormous differences in the calculated cohesive or binding energies exist in the literature. Some structures possess a high anisotropy owing to differences in chemical bonding in the armchair and the zigzag directions. In particular, nonlinear mechanical properties such as the ultimate strength and strain depend sensitively on anisotropic bonding. The monolayers or few-layer sheets are direct or indirect semiconductors with band gaps of 0.3–5.0 eV; in a few cases, they have metallic character. Usually, these 2D materials are excellent absorbers of visible and ultraviolet light. With their strain-tunable bandgaps and band edges, they are of interest for optoelectronic applications and as photocatalysts, e.g., for photocatalytic decomposition of water. The controlled mechanical or liquid-phase exfoliation of high-quality monolayers and few-layer assemblies from synthesized layered bulk crystals is difficult and has been successful in only a few cases. The synthesis of these compounds by bottom-up methods such as chemical vapor deposition (CVD) and molecular beam epitaxy (MBE) is still lacking.

P. Hess, *Bonding, Structure, and Performance of Two-Dimensional Materials*, NanoScience and Technology, https://doi.org/10.1007/978-3-031-74161-6_10

10.1 C–VA Monolayers (CN, CP, CAs, CSb, and CBi)

10.1.1 CN Monolayer

10.1.1.1 Chemistry and 2D Bonding

The elements involved in binary group IVA–VA monolayers are displayed in Fig. 10.1. The group of C–VA monolayers includes covalent-ionic bonding of C atoms with the pnictogens N, P, As, Sb, and Bi. The main sources of information on C–VA monolayers are first-principles calculations based on density functional theory (DFT) using the generalized gradient approximation (GGA) and Perdew–Burke–Ernzerhof (PBE) functionals. In carbon–nitrogen (CN) monolayers, the carbon atom with the valence electrons $(2s^2p_x^{\ 2})$ and the nitrogen atom with the valence electrons $(2s^2p_x^{\ 2}p_y^{\ 1})$ have one more electron than needed for energetically preferred octet bonding. This additional electron strongly increases the structural diversity of this family of 2D materials. The valence electrons form various stable, primarily covalently bonded 2D networks. The different electronegativities, e.g., of C (2.55) and N (3.04) introduce polarity with an ionic contribution to bonding. The polarity decreases going down the column of the periodic table from N to Bi.

The best-studied bonding type is the hexagonal four-layer network of α CN, which is the most stable phase with the space group *P*-6*m*2 and consists of two inner sublayers of C atoms and two outer layers of N atoms, where the C atom is bonded to four neighbors and the N atom is bonded to three neighbors [1]. Note that in this structure the *p*-type valence electrons of N play a dominant role. For α-CN monolayers, a large charge transfer of $\delta q = 1.85$ e from the inner C atoms to the outer N atoms generates negatively charged outer sublayers with a fractional ionic character of 5.8% [1]. The cohesive energy of the α phase is $E_c = -6.16$ eV/atom, which is the strongest binding energy in the family of IVA–VA monolayers owing to the small size of the C and N atom [1]. Much stronger covalent-ionic bonding in α-CN monolayers is suggested in a recent review predicting a cohesive energy of – 8.45 eV/ atom [2].

Besides α CN the other group–IVA nitrides, except for the α-CBi monolayer, are dynamically and energetically stable owing to strong covalent bonding and ionicity [2]. Dynamic instability is caused by negative (imaginary) frequencies that lead to unlimited atom displacements due to the absence of restoring forces. In addition to

Fig. 10.1 Periodic table of the main group elements, highlighting the elements involved in binary group IVA–VA monolayers

H							He
Li	Be	B	C	N	O	F	Ne
Na	Mg	Al	Si	P	S	Cl	Ar
K	Ca	Ga	Ge	As	Se	Br	Kr
Rb	Sr	In	Sn	Sb	Te	I	Xe
Cs	Ba	Tl	Pb	Bi	Po	At	Rn
Fr	Ra						

hexagonal bonding, first-principles calculations combined with ab initio molecular dynamics (MD) and tight-binding theory predict an all-pentagon tetragonal structure of penta CN_2 [3]. This dynamically stable phase is isostructural to pentagraphene with a puckered three-layer structure consisting of a C atom sublayer located in the middle between two N sublayers on the top and bottom of the monolayer. In this network, each C atom transfers a Bader charge of $\delta q = 1.2$ e to each N atom, generating polar C–N bonds. Since the N_2 units form single bonds, these monolayers qualify as high-energy density materials (HEDMs). Note that the metastable single N–N bond has a bond energy of 160 kJ/mol and the bond energy of the stable triple N≡N bond is 954 kJ/mol (HEDM material).

Another important 2D material is the graphitic g-C_3N_4 monolayer, which can be exfoliated from bulk graphitic carbon nitride (g-C_3N_4) [4]. Moreover, many 2D C_xN_y compounds can be fabricated from C-rich and N-rich precursors [5]. Besides g-C_3N_4, other well-studied and already synthesized monolayers have C_2N, C_3N, and C_5N_2 stoichiometries. Chemical bonding in these compounds and other less studied C_xN_y derivatives, such as g-CN, g-C_2N_2, g-C_2N_3, and g-C_5N, is based on the planar sp^2-hybridized structural motif arranged in various basic benzene ring assemblies.

The enormous flexibility of covalent bonding in this group of 2D IVA–VA compounds originates from potential planar sp^2 hybrids competing with nonplanar sp^3 hybridization of C atoms and nonplanar sp^3-type bonding of VA atoms with a lone electron pair. Examples for such structural variations, obtained by changing the geometric arrangement of the elements, can be seen in Figs. 10.2 and 10.3. The first structure consists of hexagons and the second one is formed by pentagons and heptagons, where the C atoms are bonded to four neighbors and the N atoms are bonded to three neighbors and each N atom incorporates a lone electron pair. Note that in the hexagonal structure, each N atom has one neighbor of the same kind, whereas in the pentagon-heptagon structure, there are two neighbors of the same kind.

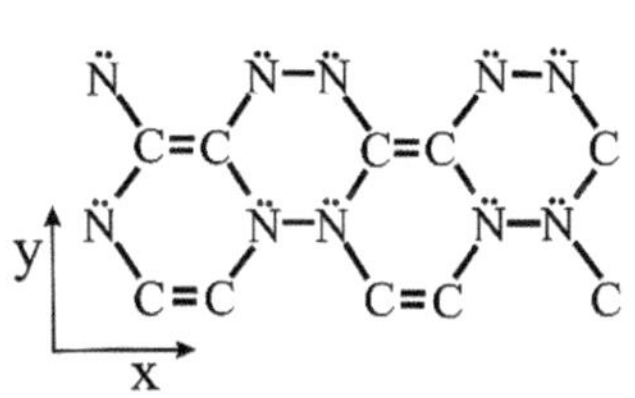

Fig. 10.2 Schematic presentation of bonding configuration of CN monolayers consisting of hexagons. The carbon atoms (group–IVA atoms) are fourfold coordinated, and the nitrogen atoms (group–VA atoms) are threefold coordinated, with the lone pair of electrons indicated in the figure by two points

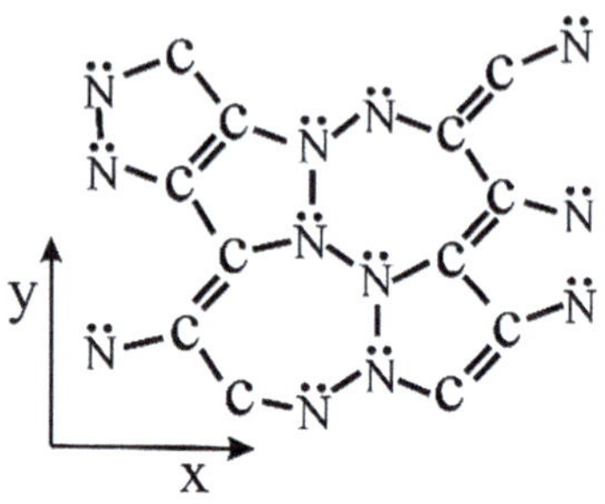

Fig. 10.3 Bonding configuration of the CN monolayer consisting of pentagon-heptagon pairs. The carbon atoms (group–IV atoms) are fourfold coordinated and the nitrogen atoms (group–V atoms) are threefold coordinated

10.1.1.2 2D Structures

The best studied structures of the family of group IVA–VA monolayers are the hexagonal α phase with symmetry *P*-6*m*2 (see Fig. 10.4) and the structurally similar β phase with symmetry *P*-3*m*1 (see Fig. 10.5), both of which have the four-layer configuration N–C–C–N of C_2N_2 monolayers. Note that depending on the size of the atoms and the type of bonding, other configurations, such as IVA–VA–VA–IVA or IVA–VA–IVA–VA, are possible in a four-layer arrangement [6]. The structure of the latter mixed four-layer configuration shown in Fig. 10.6 is a Janus structure [6]. These different arrangements of chemical elements in various allotropes lead to changes in their chemical and physical properties. Unfortunately, the nomenclature presently used in the literature for the different structures is not consistent. Here, the VA–IVA–IVA–VA arrangement with hexagonal symmetry *P*-6*m*2, we name the α phase, and the VA–IVA–IVA–VA arrangement with symmetry *P*-3*m*1, we name the β phase. The Janus structure with the space group *P*3*m*1, we name the γ phase. The symmetry or space group of the monolayer is presented whenever necessary to avoid confusion.

The stable four-layer α-CN phase has a lattice constant of a = 0.238 nm, a bond length between C atoms of $d_{C\text{–}C} = 0.284$ nm, a distance between N atoms of $h_{NN} = 0.265$ nm, a bond length between C and N atoms of $d_{C\text{–}N} = 0.146$ nm, and the bond angles $\theta_{CCN} = 110.35°$ and $\theta_{NCN} = 108.58°$ [1, 6]. Adding the van der Waals diameter

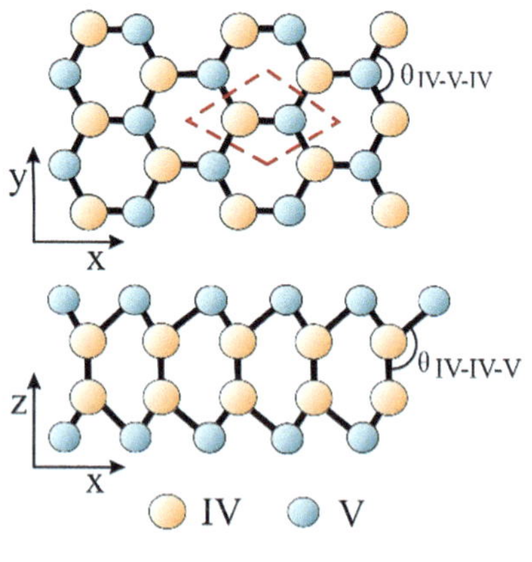

Fig. 10.4 Top and side views of the four-layer α-CN structure with symmetry *P*-6*m*2. The group–IVA atoms form the two inner sublayers, and the group–VA atoms form the two outer sublayers. The unit cell is indicated by the red dashed line

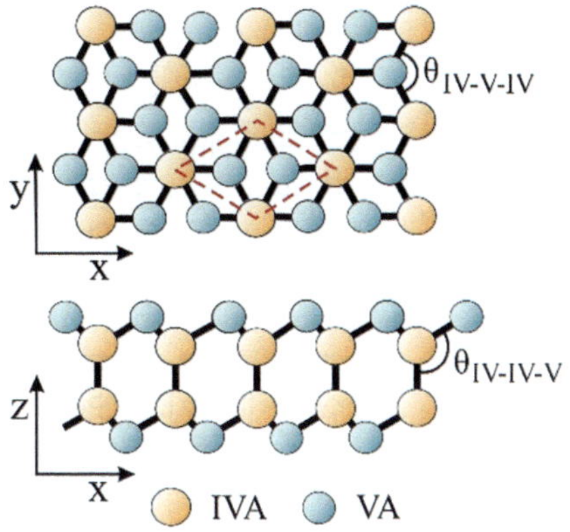

Fig. 10.5 Top and side views of the four-layer β-CN structure with symmetry *P*-3*m*1. The group–IV atoms form the two inner sublayers, and the group V atoms form the two outer sublayers. The unit cell is indicated by the red dashed line. Adapted from Ref. [6], Copyright 2021, IOP Publishing Ltd

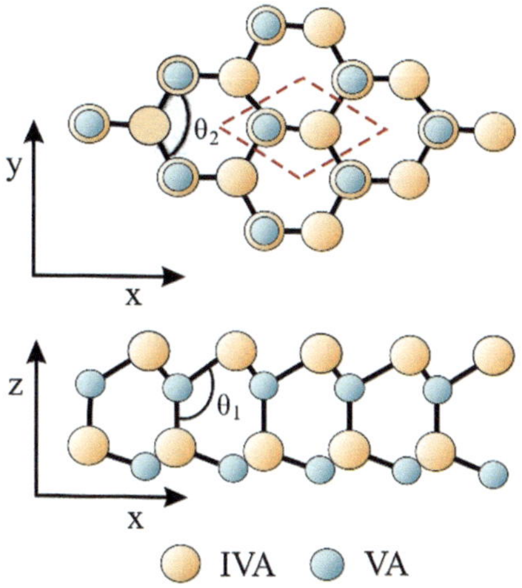

Fig. 10.6 Four-layer configuration with mixed arrangement of the four-layer γ-CN structure with symmetry *P*3*m*1 (Janus structure: IVA–VA–IVA–VA). Adapted from Ref. [6], Copyright 2021, IOP Publishing Ltd

of N atoms $d_w = 0.32$ nm and distance between N atoms $h_{NN} = 0.265$ nm yields an estimate of the monolayer thickness of $d_{w+h} = 0.585$ nm. Independent theoretical approaches have been used to determine the band gap of the indirect semiconductor, which has a large gap of $E_g = 3.71$ eV (PBE) and 5.14 eV (Heyd-Scuseria-Ernzerhof (HSE06)) [1, 6]. The space group, cohesive energy, bond lengths, bond angles, and thickness of the most stable IVA–VA monolayers are presented in Table 10.1.

The lattice parameters of the tetragonal penta-CN_2 structure are a = 0.331 nm, $d_{C\text{-}N} = 0.147$ nm, and $d_{N\text{-}N} = 0.145$ nm (see Fig. 10.7) [3]. Note that the realization of single nitrogen bonds is possible by a pucker effect of $\Delta z = 0.152$ nm (HEDM material). This metastable structure is dynamically, mechanically, and thermally stable. Penta CN_2 is an insulating material with a large band gap of $E_g = 4.83$ eV (PBE) and 6.53 eV (HSE06) [3]. The space group of this tetragonal structure is $P42_1m$.

To date, various structural parameters of g-C_3N_4, g-C_2N, and g-C_3N monolayers have been calculated [5]. The thickness of t = 0.33 nm calculated for the atomic

Table 10.1 Space group, cohesive energy, lattice constant, bond length, V–V distance, bond angle, and corrugation corrected van der Waals diameter of group IVA–VA monolayers

Mono-layer	Space group	Cohesive energy, E_c (eV/atom)	Lattice constant a (nm)	Bond length d_{IV-IV} (nm)	Distance h_{VV} (nm)	Bond length d_{IV-V} (nm)	Bond angle $d_{IV-IV-V}$ (°)	Bond angle d_{V-IV-V} (°)	Thickness d_{w+h} (nm)
CN	*P*-6*m*2	– 6.16 [1]	0.238 [1]	0.163 [1]	0.265 [1]	0.146 [1]	110.35 [1]	108.58 [1]	0.585
CP	*P*-3*m*1	– 5.37 [1]	0.290 [1]	0.153 [1]	0.367 [1]	0.189 [1]	117.35 [1]	100.56 [1]	0.757
CAs	*P*-3*m*1	– 4.82 [1]	0.312 [1]	0.150 [1]	0.386 [1]	0.204 [1]	118.02 [1]	99.73 [1]	0.796
CSb	*P*-3*m*1	– 4.45 [1]	0.342 [1]	0.149 [1]	0.416 [1]	0.225 [1]	118.77 [1]	98.77 [1]	0.858
CBi	*P*-3*m*1	– 4.18 [1]	0.360 [1]	0.144 [1]	0.433 [1]	0.239 [1]	119.63 [1]	97.67 [1]	0.893
SiN	*P*-6*m*2	– 5.59 [1]	0.290 [1]	0.243 [1]	0.354 [1]	0.176 [1]	108.40 [1]	110.52 [1]	0.674
SiP	*P*-6*m*2	– 4.19 [1]	0.353 [1]	0.237 [1]	0.441 [1]	0.228 [1]	116.50 [1]	101.62 [1]	0.831
SiAs	*P*-6*m*2	– 3.85 [1]	0.370 [1]	0.236 [1]	0.457 [1]	0.240 [1]	117.38 [1]	100.54 [1]	0.867
SiSb	*P*-6*m*2	– 3.50 [1]	0.402 [1]	0.236 [1]	0.482 [1]	0.262 [1]	117.92 [1]	99.85 [1]	0.922
SiBi	*P*-6*m*2	– 3.31 [1]	0.417 [1]	0.235 [1]	0.492 [1]	0.273 [1]	118.07 [1]	99.66 [1]	0.952
GeN	*P*-6*m*2	– 4.28 [1]	0.310 [1]	0.257 [1]	0.390 [1]	0.191 [1]	110.40 [1]	108.53 [1]	0.710
GeP	*P*-6*m*2	– 3.60 [1]	0.366 [1]	0.251 [1]	0.465 [1]	0.237 [1]	116.91 [1]	101.12 [1]	0.855
GeAs	*P*-6*m*2	– 3.36 [1]	0.382 [1]	0.250 [1]	0.480 [1]	0.249 [1]	117.54 [1]	100.33 [1]	0.890
GeSb	*P*-6*m*2	– 3.12 [1]	0.412 [1]	0.250 [1]	0.501 [1]	0.269 [1]	117.84 [1]	99.95 [1]	0.941
GeBi	*P*-6*m*2	– 2.98 [1]	0.426 [1]	0.249 [1]	0.509 [1]	0.278 [1]	117.85 [1]	99.94 [1]	0.969
SnN	*P*-6*m*2	– 3.78 [1]	0.342 [1]	0.297 [1]	0.444 [1]	0.211 [1]	110.30 [1]	108.63 [1]	0.764
SnP	*P*-6*m*2	– 3.28 [1]	0.395 [1]	0.289 [1]	0.522 [1]	0.256 [1]	117.05 [1]	100.94 [1]	0.812
SnAs	*P*-6*m*2	– 3.10 [1]	0.409 [1]	0.288 [1]	0.537 [1]	0.267 [1]	117.74 [1]	100.04 [1]	0.947
SnSb	*P*-6*m*2	– 2.89 [1]	0.439 [1]	0.287 [1]	0.558 [1]	0.287 [1]	118.13 [1]	99.59 [1]	0.998
SnBi	*P*-6*m*2	– 2.78 [1]	0.451 [1]	0.286 [1]	0.566 [1]	0.296 [1]	118.26 [1]	99.43 [1]	1.026
PbN	*P*-6*m*2	– 3.04 [1]	0.363 [1]	0.317 [1]	0.478 [1]	0.224 [1]	110.92 [1]	107.98 [1]	0.798

(continued)

Table 10.1 (continued)

Mono-layer	Space group	Cohesive energy, E_c (eV/atom)	Lattice constant a (nm)	Bond length d_{IV-IV} (nm)	Distance h_{VV} (nm)	Bond length d_{IV-V} (nm)	Bond angle $d_{IV-IV-V}$ (°)	Bond angle d_{V-IV-V} (°)	Thickness d_{w+h} (nm)
PbP	*P*-6*m*2	– 2.83 [1]	0.412 [1]	0.306 [1]	0.549 [1]	0.267 [1]	117.08 [1]	100.91 [1]	0.939
PbAs	*P*-6*m*2	– 2.72 [1]	0.425 [1]	0.305 [1]	0.562 [1]	0.277 [1]	117.66 [1]	100.18 [1]	0.972
PbSb	*P*-6*m*2	– 2.58 [1]	0.453 [1]	0.303 [1]	0.582 [1]	0.296 [1]	118.01 [1]	99.75 [1]	1.022
PbBi	*P*-6*m*2	– 2.50 [1]	0.463 [1]	0.302 [1]	0.588 [1]	0.303 [1]	118.10 [1]	99.62 [1]	1.048

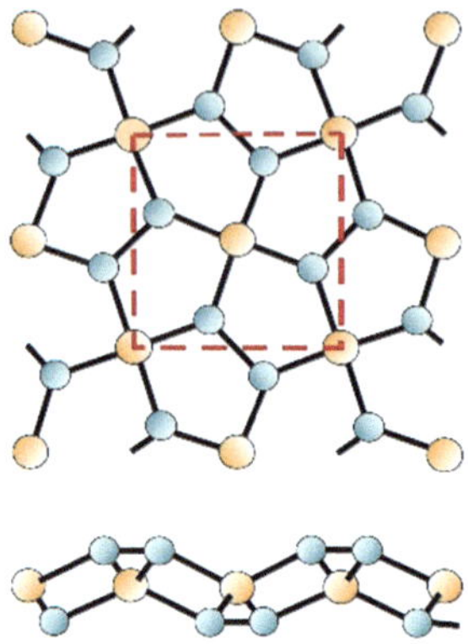

Fig. 10.7 Top and side views of penta-monolayer AB_2, which consists of three sublayers. In this arrangement of pentagons, the group–IVA atoms are fourfold coordinated, and the group–VA atoms are threefold coordinated. The unit cell is indicated by the dashed line. Examples include the pentagonal SiN_2 and $SiSb_2$ monolayers. Adapted with permission from Ref. [3], Copyright 2016, American Chemical Society

layer of g-C_3N_4 is in reasonable agreement with a topographical height of ~ 0.45 nm measured by atomic force microscopy (AFM) [4]. These values confirm the planar graphitic structure of this monolayer with good environmental stability. Spherical aberration-corrected high-resolution transmission electron microscopy (HRTEM) allowed direct observation of the g-C_3N_4 unit cell, in agreement with the simulated HRTEM image.

10.1.1.3 Mechanics

The bond lengths obtained for α-CN monolayers indicate strong bonding of the four-layer network. In fact, calculations predict an enormous Young's modulus of 1013 GPa (576 N/m) [8], however, other authors find a much smaller stiffness at 288.0 GPa (164 N/m) [2]. The Poisson ratio is exceedingly small at $\nu = 0.12$ [2]. While the 3D stiffness is comparable to that of graphene, the value in 2D units, estimated by assuming a layer thickness of $t = 0.569$ nm, is much greater than the stiffness of graphene (340 N/m). The thickness used here is supported by the thickness $d_{w+\Delta z} = 0.585$ nm, obtained by adding the mean van der Waals diameter of N atoms $d_w = 0.32$ nm and the distance between N atoms. It should be noted that two authors obtained extremely large cohesive energies of $E_c = -8.45$ eV/ atom [2] and $E_c = -8.36$ eV/atom [8], which exceed the binding energy of graphene. The strong C–N bonds and the four-layer assembly with a large volume (thickness) may explain a 2D Young's modulus that is much larger than that of graphene. The symmetry, Young's modulus, Poisson's ratio, ultimate strength, and ultimate strain of stable IVA–VA monolayers are presented in Table 10.2.

Table 10.2 Symmetry, space group, Young's modulus, Poisson's ratio, ultimate strength, and ultimate strain of binary group IVA–VA monolayers

Mono-layer	Symmetry, space group	Young's modulus E_{ac}, E_{zz} (N/m)	Poisson's ratio ν_{ac}, ν_{zz}	Intrinsic strength σ_{ac}, σ_{zz} (N/m)	Ultimate strain ε_{ac}, ε_{zz}
CN	α phase	576 [8]		36, 51 [8]	0.11, 0.21 [8]
CP	α_1 phase	20.4, 197.2 [12]		3.2, 21.0 [12]	0.54, 0.32 [12]
C_3As_2	*P6/mmm*	88.6 [21]	0.296 [21]		
CSb	α phase	129 [23]	0.208 [23]		
CBi	β phase	50 [27]	0.30 [27]		
SiN	α phase	263 [30]	0.244 [30]		
SiP	γ phase	77.2, 100.4 [36]	0.14 [36]	8.1, 13.7 [36]	0.15, 0.31 [36]
SiAs	γ phase	64.7, 86.3 [36]	0.11, 0.16 [36]	7.5, 11.4 [36]	0.16, 031 [36]
SiSb	α phase	(72) [23]	(0.224) [23]		
SiBi	α phase	77 [46]	0.279 [46]		
GeN	γ phase	86.5, 126.4 [51]	0.18, 0.26 [51]	4.66, 10.2 [51]	0.19, 0.26 [51]
GeP	γ phase	65.6, 82.1 [36]	0.15, 0.16 [36]	6.0, 10.1 [36]	0.15, 0.31 [36]
GeAs	γ phase	60, 72.2 [36]	0.16 [36]	5.6, 8.7 [36]	0.15, 0.31 [36]
GeSb	γ phase	38.9, 51.2 [51]	0.15, 0.20 [51]	3.42, 4.03 [51]	0.19, 0.29 [51]
GeBi	γ phase	33.2, 43.0 [51]	0.18, 0.23 [51]	3.21, 3.67 [51]	0.20, 0.28 [51]
SnP	α phase	80.8 [65]	0.218 [65]		
	β phase	83.3 [65]	0.229 [65]		
SnAs	α phase	68.8 [65]	0.236 [65]		
	β phase	77.7 [65]	0.244 [65]		
SnSb	α phase	53.8 [65]	0.265 [65]		
	β phase	52.2 [65	0.254 [65]		
SnBi	α phase	48.8 [65]	0.268 [65]		
	β phase	48.9 [65]	0.277 [65]		
PbN		67.4 [72]	0.15 [72]		

The large ultimate strengths of $\sigma_{ac} = 63$ GPa (36 N/m) and $\sigma_{zz} = 89$ GPa (51 N/m) seem to be consistent with the extreme Young's modulus [8]. Under tensile loading, the in-plane interactions compete with the out-of-plane bonding, causing complex behavior in the four-layer system. The layer height may change from decreasing to increasing with increasing biaxial tension, resulting in a negative Poisson ratio. An explanation of the auxetic behavior is the existence of additional degrees of freedom in the out-of-plane relaxation process. For uniaxial tension, the ultimate strains at the maximum stress are $\varepsilon_{ac} = 0.11$ and $\varepsilon_{zz} = 0.21$, indicating strong anisotropy [8]. Under biaxial tension, the failure mechanism is a combination of phonon instability and elastic instability.

10.1.1.4 Physics and Applications

The g-C_xN_y-type monolayers with graphene-like structures exhibit an open band gap, a large surface area, good chemical stability, and variable surface chemistry. These properties enable applications in electronics, photocatalysis, electrocatalysis, energy storage devices, and gas adsorption/separation, which are discussed from an experimental and theoretical point of view [4, 5]. 2D materials possess the prerequisites needed for nonmetal heterogeneous catalysis. In fact, the abundant distribution of Lewis basic functions and Brönsted basic functions, originating from the different electronegativity of the C and N atoms, serve as catalytic sites for diverse reactions [5]. In addition to the intensely studied photocatalytic water splitting reaction, electrocatalytic reactions, such as the oxygen evolution reaction (OER), the hydrogen evolution reaction (HER), the CO_2 reduction reaction (CO_2RR), and the nitrogen reduction reaction (NRR) have been studied.

Visible light irradiation allowed photocatalytic disinfection of *Escherichia coli* on the surface of single-layer g-C_3N_4 with a thickness of 0.5 nm [9]. Transmission electron microscopy (TEM) indicated a damage of the cell wall after 8 h irradiation and a complete destruction of the bacterial cell after 12 h irradiation. The low charge transfer resistance and efficient separation of photogenerated charges explain the enhancement observed for single layers. Electrochemical impedance spectroscopy and photocurrent measurements confirmed this explanation.

The recently synthesized g-C_3N_4, g-C_2N, g-C_3N, and g-C_5N_2 monolayers opened a viable route to the realization of heterostructures with widely tunable properties [5]. Van der Waals heterostructures provide additional degrees of freedom to modify and improve the properties of 2D materials. A study investigated the generation of metal-free monolayer catalysts with outstanding photocatalytic properties, electrocatalytic properties, and reliable photocatalytic disinfection under visible light irradiation [9]. The enhancement of photocatalytic efficiency was attributed to efficient photogenerated charge separation in the single-layer structure.

10.1.1.5 Synthesis

Reports on the realization of the theoretically well-studied four-layer-type CN monolayers are rare. Atomic layers of graphitic carbon nitride g-C_3N_4 were successfully prepared by a bottom-up process, where a cyanamide precursor was intercalated into a 2D mica template and subjected to thermal polymerization reactions [4]. In the next step, the g-C_3N_4 monolayer was separated by selective etching of the cyanamide-mica hybrid, as confirmed by spectroscopic and microscopic analysis. Single layers with a thickness of ~ 0.5 nm could be separated by ultrasonic exfoliation [9]. As mentioned previously, the g-C_3N_4 monolayer showed outstanding photocatalytic activity, e.g., for H_2O_2 generation, and excellent electrocatalytic activity for the oxygen reduction reaction (ORR) [4].

10.1.2 CP Monolayer

10.1.2.1 Chemistry and 2D Bonding

In the carbon phosphorus monolayer CP, the carbon atom has the electronic configuration ($2s^2p_x^2$), and the phosphorus atom ($3s^2p_x^2p_y^1$). According to chemical experience, C atoms bind with three or four partners, and the same is possible for P atoms in 2D and 3D dimensions. This competition between sp^2 and sp^3 hybridization leads to a variety of dynamically and energetically stable 2D CP structures. There are stable threefold-coordinated allotropes of CP that form a single layer with a puckered or buckled structure. In other stable monolayers, the C atoms are fourfold, and the P atoms are threefold coordinated, forming the previously discussed monolayers consisting of four covalently bonded sublayers.

In the best studied β-CP monolayers, the valence electrons form a monolayer consisting of four sublayers with the symmetry $P\text{-}3m1$, where the two inner sublayers are C atoms and the two outer sublayers are P atoms [1]. Here, each C atom binds four atoms, one C atom and three P atoms, and each P atom forms three bonds with C atoms. A charge transfer of $\delta q = 0.27$ e from the P atom to the C atom due to the slightly different electronegativities of P (2.19) and C (2.55) leads to negatively charged inner C sublayers and a low fractional ionic character of bonding of 3.2% [1]. Note that the cohesive energy of $E_c = -5.37$ eV/atom of the possibly most stable β phase of CP monolayers differs only marginally from that of the α phase at – 5.35 eV/atom.

A stable CP polymorph formed by hexagons consists of planar sp^2-hybridized double-bonded C atoms and nonplanar sp^3-hybridized P atoms, which share three of their valence electrons with neighboring C and P atoms, forming either a puckered α_1 phase (see Fig. 10.8) or a buckled β_1 phase with isolated dimers (see Fig. 10.9) [7, 10]. The identical cohesive energies of the α_1 and β_1 phases of $E_c = -5.32$ eV/atom (α_1) and – 5.33 eV/atom (β_1) [7] indicate a lower energetic stability than that of the α and β phases [10].

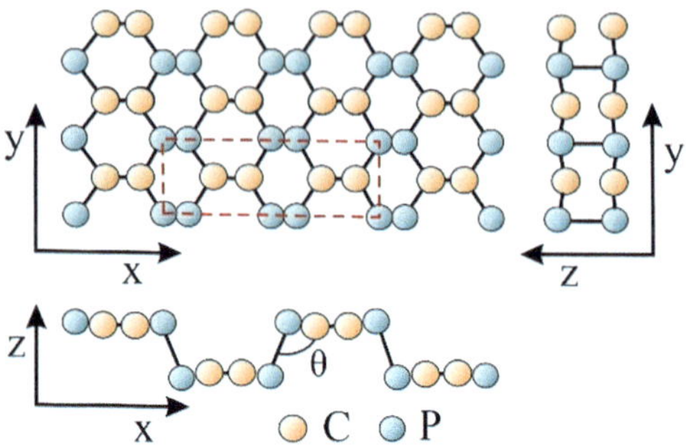

Fig. 10.8 Top, front, and side views of the puckered α_1 phase of the CP monolayers. The unit cell contains 8 atoms. Adapted with permission from Ref. [7], Copyright 2016, Royal Society of Chemistry

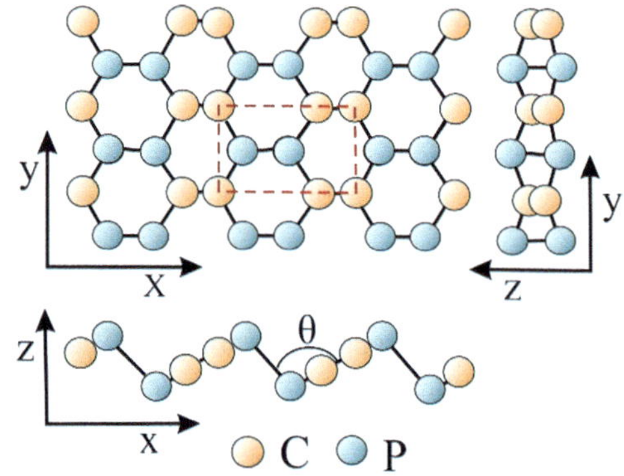

Fig. 10.9 Top, front, and side views of the buckled β_1 phase of CP monolayers. The unit cell contains 4 atoms. Adapted with permission from Ref. [7], Copyright 2016, Royal Society of Chemistry

In the two energetically stable phases of the puckered α_2-CP and the buckled β_2-CP monolayer with threefold coordination of C and P atoms, each atom has two like neighbors and one unlike neighbor [10]. Note that these structures consist not of hexagons but of pentagon–heptagon pairs. Again, the puckered α_2 phase and buckled β_2 phase of CP have the same cohesive energy of – 5.20 eV/atom.

10.1.2.2 2D Structures

The energetically most stable phase with a hexagonal four-layer structure is β CP, for which the lattice parameters are a = 0.290 nm, $d_{C–C}$ = 0.153 nm, $d_{C–P}$ = 0.189 nm, h_{PP} = 0.367 nm, and θ_{CCP} = 117.35° and θ_{PCP} = 100.56° [1]. Adding the van der Waals diameter of P atoms d_w = 0.390 nm and distance between P atoms h_{PP} = 0.367 nm yields an estimate of the monolayer thickness of d_{w+h} = 0.757 nm. The calculated band gaps of the β phase are E_g = 1.91 eV (PBE), 1.89 eV (PBE + SOC), and 2.77 eV (HSE06) [1].

In the α phase, which has nearly identical bond lengths and binding energies, the distance between the outer P atoms of h_{PP} = 0.329 nm is smaller [1]. The calculated band gaps of the α phase at E_g = 1.82 eV (PBE), 1.81 eV (PBE + SOC), and 2.74 eV (HSE06) are nearly identical with those of the β phase [1]. Note that there is no final agreement concerning the most stable phase of CP monolayers due to the similarity of the cohesive energies and lattice parameters [1].

The completely threefold coordinated CP structures with different lattice constants but practically identical bond lengths and cohesive energies of the puckered α_1 phase and buckled β_1 phase have the hexagonal structural parameters a = 0.873 nm (α_1) a = 0.476 nm (β_1), b = 0.295 nm, d_{C-C} = 0.138 nm, d_{P-P} = 0.236 nm, and d_{C-P} = 0.184 nm [10]. The puckered α_1 phase is a semiconductor with an indirect band gap of 1.1 eV (HSE06) [10]. The buckled β_1 phase has a band gap of 0.80 eV (HSE06).

The corresponding lattice parameters of the puckered α_2 phase of CP monolayers with pentagon-heptagon pairs are a = 0.984 nm, b = 0.511 nm, d_{C-C} = 0.144 nm, d_{P-P} = 0.229 nm, and d_{C-P} = 0.185 nm [10]. For the buckled β_2 phase, only the lattice constant of a = 1.06 nm deviates from the geometrical parameters of the puckered α_2 phase. Remarkably, these two structures incorporate short C–C bonds (0.144 nm) and P–P bonds (0.229 nm). Importantly, the C and P sites of these structures satisfy the octet rule of highly stable bonding arrangements. The large rectangular unit cell of these phases contains sixteen atoms.

The metastable γ phase of the four-layer CP monolayer is a direct band gap semiconductor with a gap of ~ 2.60 eV, superior structural stability and resistance to deformation, ultrahigh conductivity, and the ability to recover upon interaction with environmental molecules [11]. For this allotrope, a systematic investigation revealed the geometry, energetic stability, and electronic structure of typical point defects, such as antisites, single C or P vacancies, and double C and P vacancies (see Fig. 10.10). Since these defects influence the structure, physicochemical properties, and stability of the network, identification of these defects is necessary for evaluating the performance of real 2D materials.

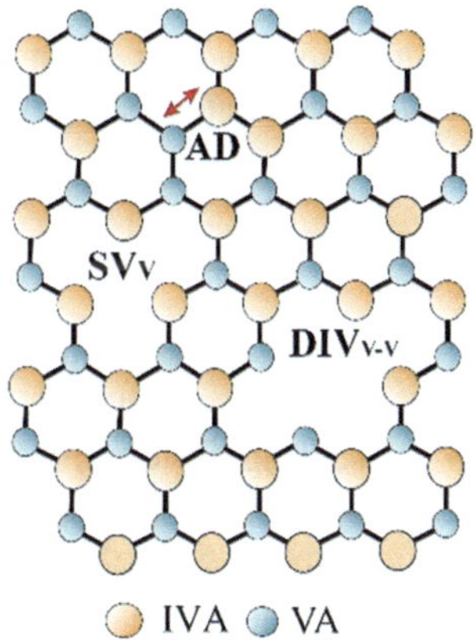

Fig. 10.10 Schematic presentation of antisite defect AD, single vacancy SV of V atom, and double vacancy of IV and V atoms. Adapted from Ref. [11], Copyright 2021, American Chemical Society

10.1.2.3 Mechanics

The elastic properties of CP monolayers, such as Young's moduli, Poisson's ratios, and ultimate strengths and strains of the α_1 CP and β_1 CP phases, were studied by density functional tight-binding theory (DFTB) [12]. The elastic modulus was calculated by subjecting the monolayer to small uniform tensile strains along the armchair and the zigzag directions ($\varepsilon < 0.3$). For the puckered hexagonal α_1 CP monolayer, the strongly anisotropic stiffnesses are $E_{ac} = 35.3$ GPa (20.3 N/m) and $E_{zz} = 343.0$ GPa (197.2 N/m), and for the buckled β_1 CP monolayer, they are $E_{ac} = 81.3$ GPa (51.6 N/m) and $E_{zz} = 288.9$ GPa (183.5 N/m). The values in brackets in 2D units are based on the thickness values $t = 0.575$ nm and $t = 0.635$ nm for the α_1-CP and β_1-CP monolayers, respectively [12]. These results are in reasonable agreement with $E_{ac} = 18.8$ N/m and $E_{zz} = 171.5$ N/m for the α_1 phase and $E_{ac} = 46.6$ N/m and $E_{zz} = 158.3$ N/m for the β_1 phase [7]. The reason for the massive anisotropy is the easy flattening of the puckered and buckled structure via bond rotation in the armchair direction, however, direct stretching of the covalent C–P bonds by deformation in the zigzag direction.

At large strains, strongly anisotropic failure of the pristine structures occurs. The ultimate strengths of α_1 CP are $\sigma_{ac} = 5.49$ GPa (3.2 N/m) and $\sigma_{zz} = 36.5$ GPa (21 N/m), and those of β_1 CP are $\sigma_{ac} = 7.15$ GPa (4.5 N/m) and $\sigma_{zz} = 27.5$ GPa (17.5 N/m). Along the soft armchair direction, the critical strains of $\varepsilon_{ac} = 0.54$ for α_1 CP and $\varepsilon_{ac} = 0.52$ for β_1 CP are nearly identical, whereas much smaller values are found for the stiffer zigzag direction of $\varepsilon_{zz} = 0.32$ for α_1 CP and $\varepsilon_{zz} = 0.34$ for β_1 CP owing to the strong C–P bonds [12]. As discussed for the stiffness, the failure mechanism depends on the direction of tensile strain. In the armchair direction, the P–P bonds first rotate and flatten the structure. Only after a certain extension at the critical strain do the weak P–P bonds break by bond elongation. In the zigzag direction, tensile deformation involves direct stretching of the stronger C–P bonds until bond rupture occurs at the critical strain.

10.1.2.4 Physics and Applications

Carbon phosphorus monolayers with semimetallic behavior in the armchair direction and semiconducting behavior in the zigzag direction are of interest for use in gas sensing devices. First-principles calculations allowed the elucidation of the adsorption mechanisms of CO, CO_2, NH_3, NO, and NO_2 on CP monolayers [13]. The adsorption strength of the gas molecules was much stronger than that observed for other monolayers, such as graphene or phosphorene. A charge transfer analysis explains this behavior by efficient charge sharing between the CP monolayer and the adsorbate. Moreover, the authors present an extensive study of the transport properties of sensor devices using CP monolayers. A similar theoretical study investigated the adsorption of NH_3, NO, NO_2, H_2O, and O_2 on α_1-CP and β_1-CP monolayers [14]. While NH_3, H_2O, and O_2 physisorbed, NO and NO_2 chemisorbed on the surface. The two latter molecules are acceptors with a charge transfer of $\delta q \approx 1$ e per molecule.

The layered semiconductor black phosphorus carbide (b-PC) is predicted to have the lightest electrons and holes among the 2D semiconductors, with a p-type mobility of ~ 10^5 cm^2 V^{-1} s^{-1}, which is approximately five times greater than that of black phosphorus [15]. A field-effect transistor (FET), consisting of a composite fabricated by a novel carbon doping technique with few-layer b-PC sheets, achieved a high hole mobility of 1995 cm^2 V^{-1} s^{-1} at room temperature. Moreover, the absorption spectrum of the material covers a region of the infrared spectrum of interest for range-finding applications, because the atmosphere is widely transparent in this spectral region.

First-principles DFT calculations combined with a nonequilibrium Green's function model provided information on various CP nanoribbons [16]. For example, edge configurations of α_1 nanoribbons with hydrogen edge termination mimic black phosphorus nanoribbons with a field-effect-tunable band gap. A nanoribbon with a bare edge terminated by phosphorus atoms becomes graphene-like and exhibits large spin splitting in the ferromagnetic state. In the antiferromagnetic state, tuning into a half metal is possible under an external transverse electric field. Moreover, the authors investigated the spin-resolved transport properties of zigzag α_1 CP nanoribbons.

As mentioned above, a comprehensive DFT-based study revealed the formation energetics, stabilities, and surface densities of seven kinds of point defects, which are stable in γ-CP monolayers [11]. These defects included antisites, single C or P vacancies, and double C and C vacancies, which are stable in γ CP up to 400 K. The energy input needed to form single and double vacancies is comparable to that found for graphene, however, about two-times higher than that for phosphorene. The Stone–Wales defect, which is unlikely to form in γ CP monolayers owing to its transition-metal dichalcogenide (TMDC)-like structure, is not included. These special defects may significantly change the electronic structure by inducing hole/electron doping.

Hexagonal γ-CP monolayers are promising anode materials for lithium-ion batteries [17]. Each adsorbed Li ion strongly binds with three neighboring phosphorus atoms with an average adsorption energy of – 1.2 eV, which is beneficial for the charging and discharging process. The diffusion barrier of Li can be as low as 77 meV, and the theoretical specific capacity of the monolayer may reach 623.3 mA h g^{-1}. For the γ phase of CP monolayers an enormous mechanical stability with the elastic moduli $E_{ac} = 445.7$ GPa (337 N/m) and $E_{zz} = 383.1$ GPa (290 N/m) and the Poisson ratios $\nu_{ac} = 0.075$ and $\nu_{zz} = 0.064$ have been reported [17]. The rough estimate of the 2D-stiffness values given in brackets is based on the layer thickness of the α phase of $t = 0.757$ nm and demonstrates an enormous in-plane stability of the γ-CP monolayer.

Single-atom transition-metal-doped CP monolayers are candidates for HER catalysis [18]. DFT calculations indicate that Ir-α-CP monolayers with high physical and chemical stability have the optimal Gibbs free adsorption energy for H atoms. The relationship between the *d*-band center and HER activity shows a volcano-like curve, and the calculation of the reaction energy barrier indicates that the Volmer–Heyrovsky step is more favorable in this case than the Volmer-Tafel step.

10.1.2.5 Synthesis

Up to now, realization of the theoretically predicted CP allotropes has not been achieved. However, mechanical exfoliation of few-layer black phosphorus samples from a bulk black phosphorus crystal in an argon-filled glove box was achieved on a doped Si substrate covered by a pregrown thermal oxide [15]. The synthesis of black phosphorus carbide (b-PC) requires doping with hydrocarbons immediately after preparation of the black phosphorus sample inside an atomic layer deposition (ALD) chamber using trimethylaluminum (TMA) as a precursor. The optimum number of purging cycles with TMA plus N_2 can be determined by measuring the carrier mobility of the growing b-PC sheet. The authors describe the synthesis of a few-layer b-CP sheet-based high-performance composite FET employing the carbon doping method.

10.1.3 CAs Monolayer

10.1.3.1 Chemistry and 2D Bonding

In carbon-arsenic monolayers CAs, the carbon atom has the valence electrons ($2s^2p_x^{\ 2}$), and the arsenic atom ($4s^2p_x^{\ 2}p_y^{\ 1}$). The additional 4*p* electron, which is not needed to fulfill the octet rule, has an enormous influence on the bonding types, variety of structures, and performance of various CAs allotropes. Owing to the different electronegativities of C (2.55) and As (2.18), the C–As bonds are slightly ionic, with charge transfer from the As atom to the C atom.

As discussed for β-CP monolayers, the valence electrons can form a covalently bonded monolayer consisting of four covalently bonded sublayers, where the two inner sublayers are formed by C atoms and the two outer layers are As atoms (see Fig. 10.5). For the β phase of CAs, only a small charge transfer of $\delta q = 0.38$ e occurs from the outer As atoms to the negatively charged inner C sublayers, with a fractional ionic character of 3.4% [1]. The cohesive energy of $E_c = -4.82$ eV/atom for the energetically most stable β-CAs phase is only slightly smaller than that of the metastable α phase of CAs of − 4.79 eV/atom [1]. A comparison of the binding energy with that of β-CP monolayers shows weaker bonding for the heavier compound, primarily due to increasing atom size.

First-principles calculations predict the existence of a monolayer involving pentagons with the stoichiometry CAs_2, which is a semiconductor with an indirect band gap of 1.62 eV [19]. In this three-layer lattice, each C atom is bonded to four As atoms, and each As atom is bonded to one As and two C atoms (see Fig. 10.7). This thermodynamically and dynamically stable monolayer exhibits a surprisingly large cohesive energy of – 5.69 eV/atom. Remarkably, the monolayer shows strong absorption for UV light; however, almost no absorption occurs in the visible region of the electromagnetic spectrum.

According to theory, a stable silicene-like buckled monolayer should exist for the CAs_3 stoichiometry with the space group $P\text{-}3m1$ [20]. In CAs_3 monolayers, C atoms prefer sp^2-like hybridization, where one s and two p orbitals form three covalent bonds with the nearest As atoms, while the As atoms prefer sp^3-like hybrids. The two electrons in the $2p_z$ orbital will partly hybridize with sp^2 orbitals, as indicated by the deviation of the carbon bond angle at $\theta_{AsCAs} = 116.1°$ from the normal value of 120°. The binding energy of the semi-metallic buckled monolayer is $E_c = -3.58$ eV/atom, which is lower than the binding energy of the corresponding planar structure of -3.26 eV/atom owing to energetically favorable mixing to the sp^3-like hybrids of As. Notably, the application of a tensile strain of $\varepsilon = 0.2$ leads to the realization of the planar phase with the symmetry *P6/mmm*.

Finally, the stable stoichiometric C_3M_2, (M = As, Sb, and Bi) hexagonal monolayers were predicted, in which the C atoms adopt the sp^2 hybridization of graphene coordinated with one C and two M atoms (e.g., As) and the M atoms assume sp^3-type hybridization with threefold coordination and a lone electron pair [21]. The calculated cohesive energy of C_3As_2 is $E_c = -5.16$ eV/atom, which surprisingly is lower than that of related allotropes; however, C_3As_2 has much weaker bonding than graphene and much stronger bonding than arsenene. These compounds exhibit desirable band gaps and intrinsic acoustic-phonon-limited carrier mobilities.

Interestingly, based on the first-principles framework and the particle swarm optimization (PSO) approach, two CAs allotropes were added to the ever-increasing number of allotropes, with the prediction of puckered and buckled honeycomb structures [22]. The predicted characteristic properties are excellent stability, plateaus for hole conduction, optical and photocatalytic applications, and tunable band gaps with large absorption coefficients.

10.1.3.2 2D Structures

The most stable β phase of the four-layer hexagonal CAs monolayers has lattice parameters of a = 0.312 nm, $d_{C\text{–}C} = 0.150$ nm, $d_{C\text{–}As} = 0.204$ nm, $h_{AsAs} = 0.386$ nm, and as bond angles $\theta_{CCAs} = 118.02°$ and $\theta_{AsCAs} = 99.73°$ [1]. Adding the van der Waals diameter of As atoms $d_w = 0.410$ nm and distance between As atoms h_{AsAs} yields an estimate of the monolayer thickness of $d_{w+h} = 0.796$ nm. The calculated band gaps of the indirect semiconductor are 1.14 eV (PBE) and 1.84 eV (HSE06). Note that others consider the α phase to be the energetically most favorable structure of the CAs polymorphs owing to the quite similar lattice parameters, bond angles, and binding energies [6].

Additionally, theory predicts thermodynamic, dynamic, and thermal stability for the puckered and buckled honeycomb CAs structures at 300 K [22]. The intrinsic band gaps of the puckered and buckled phases are $E_g = 1.27$ eV and 1.78 eV, respectively. The band gaps, appropriate edge positions, and strong absorption in the visible and ultraviolet regions make both monolayers promising candidates for photocatalytic and optoelectronic applications. The large hole carrier mobility at room temperature renders these monolayers for p-type FETs.

The lattice constant, bond lengths, and bond angles of the buckled CAs_3 monolayer are a = 0.678 nm, $d_{C\text{–}As}$ = 0.192 nm, $d_{As\text{-}As}$ = 0.257 nm, θ_{AsCAs} = 116.09° and θ_{CAsAs} = 105.41° [20]. The buckling distances of the four-layer arrangement of the silicene-like CAs_3 monolayer with space group *P*-3*m*1 are $\Delta z_{C\text{–}As}$ = 0.0385 nm and $\Delta z_{As\text{-}As}$ = 0.157 nm. As predicted by electronic structure calculations, the monolayer is a semimetal with linear dispersion and a Dirac cone around the Fermi level. Thus, at low excitation energies, the charge carriers behave as massless Dirac fermions. In addition, this work presents the lattice parameters of the planar CAs_3 monolayer with the space group *P*6/*mmm*. In this less stable structure, the As atom shows sp^2 hybridization instead of sp^3 hybridization [20].

The structural parameters of the planar C_3As_2 monolayer are a = 0.586 nm, $d_{C\text{–}C}$ = 0.133 nm, and $d_{C\text{–}As}$ = 0.200 nm [21]. The band gaps calculated for the intrinsic semiconductor are 1.42 eV (PBE) and 2.27 eV (HSE06), and the space group *P*6/*mmm*. The theoretical electron mobility of C_3As_2 at 300 K of 4.45×10^5 cm^2 V^{-1} s^{-1} is greater than that of graphene (3×10^5 cm^2 V^{-1} s^{-1}). Moreover, the heterojunction C_3As_2/C_3Sb_2 delivers solar cells with an impressive power conversion efficiency of ~ 23%.

10.1.3.3 Mechanics

Linear mechanical properties are available for the C_3As_2 allotrope, with a Young's modulus of 88.6 N/m and a Poisson's ratio of 0.296 [21]. These values were determined from the calculated elastic constants, which meet the requirements of the Born–Huang mechanical stability criteria for 2D hexagonal structures. Moreover, this work reports the decreasing stiffness values of C_3Sb_2 (65.7 N/m) and C_3Bi_2 (51.8 N/m) and the increasing Poisson ratios $\nu = 0.327$ and $\nu = 0.340$, respectively [21].

10.1.3.4 Physics and Applications

The abovementioned potential applications of various allotropes of C_xAs_y monolayers are based on widely tunable band gaps and band edges, sufficient mechanical stability, and promising optoelectronic properties owing to strong absorption in broad visible and ultraviolet spectral regions. Some monolayers exhibit metallic character with superconductivity, semimetallic properties with an anisotropic Dirac cone, and direct/indirect semiconducting properties with superior carrier mobility.

10.1.4 CSb Monolayer

10.1.4.1 Chemistry and 2D Bonding

In the carbon antimony monolayer CSb, the carbon atom has the electronic configuration ($2s^2p_x^2$), and the Sb atom ($5s^2p_x^2p_y^1$). According to theory, the four-layer structure of the β-CSb monolayer is slightly more stable than that of the α-CSb monolayer. Owing to the different electronegativities of C (2.55) and Sb (2.05), the dominant covalent bonding in the CSb polymorphs has a partial ionic character. The polarity of the C–Sb bonds is induced by a charge transfer of $\delta q = 1.11$ e from the Sb atom to the C atom, which induces two negatively charged inner C sublayers and a fractional ionic character of 6.1% [1]. The cohesive energy of $E_g = -4.45$ eV/atom for the stable β-CSb phase is only marginally smaller than that of the α-CSb phase at -4.42 eV/atom. Both phases have weaker bonding than their lighter homologues. Notably, the DFT-GGA-PBE approach also delivered a much lower cohesive energy with stronger bonding of $E_c = -5.55$ eV/atom for the α-CSb monolayer [23].

In addition, CSb_2 monolayers exist as buckled pentagonal structures with three sublayers and a buckling distance of $\Delta z = 0.281$ nm [24]. In this stable structure, each carbon atom bonds with its four neighboring antimony atoms, and each antimony atom bonds with two carbon atoms and an antimony atom. Surprisingly, the binding energy of this phase is $E_c = -4.45$ eV/atom, which is comparable to that of the high-symmetry four-layer structures.

10.1.4.2 2D Structures

The slightly more stable β phase of the four-layer CSb monolayer has lattice parameters of $a = 0.342$ nm, $d_{C\text{–}C} = 0.149$ nm, $d_{C\text{–}Sb} = 0.225$ nm, $h_{SbSb} = 0.418$ nm, $\theta_{CCSb} = 118.77°$ and $\theta_{SbCSb} = 98.77°$ [1]. Adding the van der Waals diameter of Sb atoms of $d_w = 0.440$ nm and the distance between Sb atoms h_{SbSb}, yields an estimate of the monolayer thickness of $d_{w+h} = 0.858$ nm. While both the PBE and the PBE + SOC approaches predict metallic behavior of β-CSb monolayers, an indirect semiconductor with a band gap of $E_g = 0.43$ eV is predicted by the HSE06 approximation [1]. The study suggesting stronger bonding of α CSb (– 5.55 eV/atom) reported a substantially shorter distance between the Sb atoms of $h_{SbSb} = 0.371$ nm and thus a smaller thickness, whereas the lattice constant and C–C bond length have similar values [23].

The structural parameters of the penta-CSb_2 monolayer are $a = 0.483$ nm, $d_{C\text{–}C} = 0.149$ nm, $d_{Sb\text{–}Sb} = 0.297$ nm, and $d_{C\text{–}Sb} = 0.221$ nm, $\theta_{SbCSb} = 101.39°$ and $\theta_{CSbC} = 100.28°$ [24]. According to DFT calculations and first-principles MD simulations, this local minimum structure is energetically, dynamically, and thermally stable up to 500 K. The monolayer is an indirect semiconductor with a band gap of about 0.90 eV.

10.1.4.3 Mechanics

The elastic modulus of the quasi-isotropic four-layer α-CSb monolayer is $E_{ac} \approx E_{zz} = 150$ GPa, and the Poisson ratio is $\nu_{ac} \approx \nu_{zz} = 0.208$, assuming that the mechanical properties in this study are given in GPa because the units are missing [23]. A layer thickness of $d_{w+\Delta z} = 0.858$ nm is used to estimate the stiffness in 2D units. This yields a reasonable stiffness of 129 N/m in 2D units. Such a value of ≈129 N/m correlates well with the stiffnesses of the other group members. MD simulations with an optimal empirical potential delivered extremely large values of the elastic and shear moduli of E = 383.2 N/m and G = 171.9 N/m, respectively, for the pentagonal CSb_2 monolayer [25]. The presented Poisson ratio of $\nu = 0.394$ also is surprisingly large.

10.1.4.4 Physics and Applications

An interesting theoretical work discusses the use of Janus monolayers as a novel platform for the integration of topological physics, i.e., spintronics and valleytronics. The Janus monolayer $CSb_{1.5}Bi_{1.5}$ represents a spin-valley-coupled quantum spin Hall (QSH) insulator with topological Rashba-splitting edge states due to the missing horizontal mirror symmetry [26]. The investigation of four-layer AB_3-type atomic sheets was inspired by recent studies on CP_3 monolayers to be used as an anode in sodium-ion batteries and CAs_3 monolayers associated with massless Dirac-fermions. The Janus $CSb_{1.5}Bi_{1.5}$ monolayer is a direct band gap semiconductor with and without SOC, and the valence band maximum (VBM) and conduction band minimum (CBM) are at the valley points. The inequivalent valleys have opposite Berry curvatures and spin moments and therefore can generate a nontrivial spin-valley-coupled quantum spin-valley Hall insulator (svcQSHI) that is robust against biaxial strain. It is worth noting that the energy band gap and spin splitting at the valley point are greater than the thermal energy at room temperature (25 meV).

Based on its band edges and visible light absorption C_3Sb_2 is a promising opto-electronic material for photocatalytic water splitting [21]. As mentioned above, a power conversion efficiency of $\eta = 23\%$ has been estimated for C_3Sb_2/C_3As_2 excitonic solar cells and C_3Sb_2/C_3Bi_2 heterojunctions possess features of 2D node-line semimetals in the absence of SOC.

10.1.5 CBi Monolayer

10.1.5.1 Chemistry and 2D Bonding

In the carbon bismuth monolayer CBi, the carbon atom has the electronic configuration $(2s^2p_x^{\,2})$, and the Bi atom $(6s^2p_x^{\,2}p_y^{\,1})$, with an additional valence electron with respect to optimal bonding according to the energetically favored octet configuration. The stable hexagonal four-layer structure of the β-CBi monolayer has been well studied. Here the valence electrons form two inner sublayers of C atoms and two outer sublayers of Bi atoms (see Fig. 10.5). Owing to the different electronegativities of C (2.55) and Bi (1.90) atoms, the covalent C–Bi bonds have an ionic contribution to covalent bonding. The different electronegativities induce a charge transfer of $\delta q = 0.62$ e in β CBi from the less electronegative Bi atoms to the inner C sublayers, generating negatively charged C sublayers with a fractional ionic character of bonding of 6.8% [1]. The β-CBi monolayer exhibits metallic behavior. The cohesive energy of $E_c = -4.18$ eV/atom for the energetically most stable β-CBi phase is only marginally smaller than that of the α-CBi phase at − 4.13 eV/atom.

In addition, a buckled honeycomb CBi monolayer exists that has a meta-stable structure, in which C and Bi atoms are sp^2 hybridized as in β-phase bismuthenes [27]. The different electronegativities induce covalent-ionic bonding via a charge transfer of $\delta q = 0.80$ e from the Bi atom to the C atom. The unit cell of this phase contains four atoms, and the cohesive energy is $E_c = -4.30$ eV/atom. By including the SOC effect, theory predicts metallic behavior for this buckled phase of CBi monolayers [27].

Furthermore, a stable three-layer monolayer with CBi_2 stoichiometry exists that is composed of 5-membered rings with single C–Bi bonds and double Bi–Bi bonds [27]. Both atoms are sp^3 hybridized involving fourfold bonding of C atoms and threefold bonding of Bi atoms. Here, covalent- ionic bonding is based on the transfer of a charge of $\delta q = \sim 0.51$ e from the Bi atoms to the C atoms. The unit cell of this phase contains six atoms, and the cohesive energy is $E_c = -4.29$ eV/atom. This phase is a semiconductor with a direct band gap of $E_g = 0.403$ eV according to the HSE06 functional and an indirect band gap of $E_g = 0.126$ eV based on GGA-PBE calculations.

Besides the already mentioned phases, a theoretically studied hexagonal C_3Bi_2 monolayer has been described that consists of a four-layer Bi–C–C–Bi configuration, consisting of large 12-membered rings [21]. In this monolayer, C atoms undergo sp^2 hybridization, and Bi atoms possess threefold coordination to ambient C atoms via sp^3-like hybridization, including lone electron pairs. This phase has the lowest cohesive energy of the studied C_xBi_y allotropes at $E_c = -4.66$ eV/atom. First-principles calculations indicate that C_3Sb_2/C_3Bi_2 heterojunctions are 2D node-line semimetals in the absence of SOC.

10.1.5.2 2D Structures

The most stable four-layer structure is the hexagonal β-CBi monolayer, with the lattice parameters of $a = 0.360$ nm, $d_{C\text{–}Bi} = 0.239$ nm $d_{C\text{–}C} = 0.144$ nm, $h_{BiBi} = 0.433$ nm, $\theta_{CCBi} = 119.63°$ and $\theta_{BiCBi} = 97.67°$ [1]. Adding the van der Waals diameter of Bi atoms $d_w = 0.460$ nm and distance between Bi atoms h_{BiBi}, yields an estimate of the monolayer thickness of $d_{w+h} = 0.893$ nm. Note that for both β CBi and α CBi, the PBE, PBE + SOC, and HSE06 approaches predict metallic behavior.

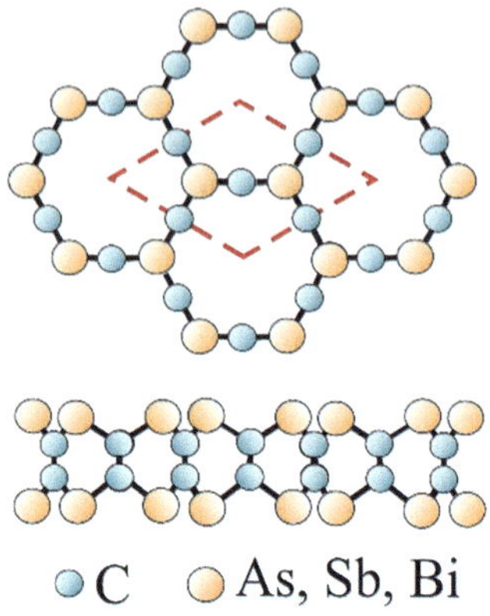

Fig. 10.11 Top and side views of the M_2C_3 structure M = As, Sb, and Bi. The dashed line indicates the unit cell. Reproduced with permission from Ref. [21], Copyright 2018, Royal Society of Chemistry

The lattice parameters of the C_3Bi_2 monolayer are a = 0.670 nm, $d_{C–Bi}$ = 0.231 nm, and $d_{C–C}$ = 0.133 nm [21]. This structure has large periodic pores with a diameter of d_p = 0.670 nm that agree with the lattice constant (see Fig. 10.11). In the hexagonal lattice C atoms adopt sp^2 hybridization, whereas the pnictogen atoms accept three-fold coordination with lone electron pairs. The distinctive structural features may be of interest in gas separation and storage devices. The semiconducting monolayer has an intrinsic band gap of E_g = 0.81 eV (PBE) and 1.23 eV (HSE06). Similarly, C_3As_2 and C_3Sb_2 monolayers form such open structures, which are semiconductors with band gaps of 2.27 eV (HSE06) and 1.58 eV (HSE06), respectively.

10.1.5.3 Mechanics

For the essentially isotropic Young's moduli of the hexagonal buckled CBi monolayers, the authors give the similar values E_{ac} = 49.8 N/m and E_{zz} = 49.6 N/m [27]. The Poisson ratio determined from the published elastic constants is $\nu = 0.30$. The strongly anisotropic stiffness values obtained from the elastic constants for penta CBi_2 are E_{ac} = 12.4 N/m and E_{zz} = 31.6 N/m, and for the Poisson ratios the authors estimated the anisotropic values $\nu_{ac} = 0.151$ and $\nu_{zz} = 0.383$ [27]. The stiffness and Poisson ratio of C_3Bi_2 monolayers are E = 51.8 N/m and $\nu = 0.340$ [21]. The various CBi monolayers possess the smallest stiffness values and largest Poisson ratios in this group of carbon compounds. The calculated elastic constants meet the mechanical stability criteria of Born–Huang for 2D structures.

10.1.5.4 Physics and Applications

The metallic buckled CBi monolayer and the direct semiconductor consisting of penta-CBi_2 monolayers are promising candidates for the adsorption and capture of CO_2 molecules and therefore they are of interest for mitigating greenhouse effects

[27]. Moreover, hexagonal buckled CBi is a suitable battery material and therefore may find interest in nanoelectronics. The penta-CBi_2 monolayer, with its narrow direct band gap and good carrier mobility, could find a wide scope of practical applications in optoelectronics and photocatalysis. Unfortunately, experimental verification of these theoretically predicted stable monolayers is still lacking.

10.2 Si–VA Monolayers (SiN, SiP, SiAs, SiSb, and SiBi)

10.2.1 SiN Monolayer

10.2.1.1 Chemistry and 2D Bonding

In the silicon nitrogen monolayer SiN, the electronic configuration of Si with four valence electrons ($3s^2p_x{}^2$) and that of the N atom with five valence electrons ($2s^2p_x{}^2p_y{}^1$) determine 2D bonding. These valence electrons allow the formation of various dynamically stable covalently bonded allotropes. In the perhaps most stable α-SiN monolayer with a hexagonal four-layer structure, each Si atom binds four atoms, one Si atom and three N atoms, and each N atom forms three bonds with Si atoms. The significant difference in the electronegativities of Si (1.90) and N (3.04) induces a considerable ionic contribution to covalent bonding. As a result, in the monolayer with its four sublayers, an enormous charge transfer of $\delta q = 4.04$ e occurs from the Si atoms of the inner sublayers to the negatively charged outer N sublayers. This transfer induces a substantial fractional ionic character of 27.7% [1]. The cohesive energy of $E_c = -5.59$ eV/atom presents the strongest binding energy in this group of 2D Si–VA compounds [1].

In addition to the hexagonal α phase, a structure consisting of pentagons named penta SiN_2 has been proposed [28]. This monolayer can be formally derived by replacing the two sp^3-hybridized carbon atoms in the unit cell of pentagraphene by silicon atoms and by substituting the four remaining three-coordinated carbons by sp^2 hybridized nitrogen atoms [28, 29].

10.2.1.2 2D Structures

The stable four-layer α phase of SiN has a lattice constant of a = 0.290 nm, a bond length between Si atoms of $d_{Si–Si} = 0.243$ nm, a Si–N bond length of $d_{Si–N} = 0.176$ nm, a distance between the N atoms of $h_{NN} = 0.354$ nm, and the bond angles $\theta_{SiSiN} = 108.40°$ and $\theta_{NSiN} = 110.52°$ [1]. Adding the van der Waals diameter of N atoms of $d_w = 0.320$ nm and the distance between N atoms h_{NN}, yields an estimate of the monolayer thickness of $d_{w+h} = 0.674$ nm. The band gaps calculated by various approaches for the indirect semiconductor are $E_g = 1.74$ eV (PBE), 1.74 eV (PBE + SOC), and 2.73 eV (HSE06) [1]. It is of interest to compare these properties with those of the bilayer with a stacking order AA1 of a = 0.291 nm, $d_{Si–Si}$ = 0.240 nm, $d_{Si–N}$ = 0.178 nm, and h_{NN} = 0.357 nm [30]. For this stacking order with a large interlayer separation of d_s = 0.322 nm and low van der Waals interaction forces, essentially undisturbed monolayer properties are expected. The smaller band gap of the bilayer (1.40 eV), which is an indirect semiconductor, increases by the application of out-of-plane tensile strain, whereas compressive strain induces a semiconductor-to-metal transition [30].

The lattice parameters of penta-SiN_2 monolayers are a = 0.405 nm, d_{N-N} = 0.144 nm, and $d_{Si–N}$ = 0.173 nm, with a buckling effect of Δz = 0.062 nm [28]. Interestingly, the pentagonal phases of pentagraphene, penta-SiC_2, and penta-SiN_2 monolayers have comparable structures and corrugation effects. Nevertheless, the strain dependence of the thermal conductivity is different, with a one order of magnitude enhancement in penta-SiN_2 monolayers due to the strain-induced transition from the buckled to the planar configuration, as discussed in more detail below.

10.2.1.3 Mechanics

Linear mechanical properties are available only for the bilayer of the α phase of four-layer SiN monolayers. DFT calculations, taking into account the van der Waals interaction between the layers by Grimme's correction to the PBE functional, yielded elastic constants, which gave an extremely large Young's modulus for the bilayer of E = 525 N/m, supporting a very large stiffness of the monolayer [30]. The medium Poisson ratio of the bilayer is ν = 0.244 [30]. This study implies an isotropic behavior of the indirect semiconductor with an astonishingly large elastic modulus.

10.2.1.4 Physics and Applications

The structural, electronic, and thermoelectric properties of hexagonal Si–VA monolayers were determined by DFT calculations and the semiclassical Boltzmann transport theory [31]. The calculations revealed a moderate indirect band gap for the semiconductors. Tuning the band gap by strain initially leads to an increase and then to a decrease in the band gap that finally becomes metallic at higher strains. The deformation potential theory yields remarkable carrier mobilities, which indicate p-type semiconducting behavior for SiN, SiP, and SiBi and n-type semiconducting behavior for SiAs and SiSb. For p-type SiN, a maximum figure of merit of ZT = 0.99 was observed at 300 K. These binary compounds exhibit high Seebeck coefficients, high power factors, and low electronic and thermal conductivities.

The behavior of thermal conductivity of penta SiN_2 is unique under tensile strain [28]. The thermal conductivity of the unstrained penta-SiN_2 monolayer is 185 W/mK. As the tensile strain increases to approximately ε = 0.08, the thermal conductivity decreases by approximately 10%, while the structure remains buckled. If the tensile strain increases further to approximately ε = 0.10, the structure becomes perfectly planar, and the thermal conductivity jumps to 1822 W/mK and then decreases to 1129

W/mK as the tensile strain further increases from $\varepsilon = 0.10$ to 0.15. The substantial increase in phonon lifetime induced by the structural change can explain the dramatic increase in thermal conductivity. Accordingly, interatomic bonding and the induced structural deformation under tensile strain are the key factors in understanding this unexpected enhancement effect.

Since the band gap of single-layer group IVA–VA monolayers exhibit an almost linear response to biaxial strain, this material may be applicable in mechanical sensor devices [32]. Because of the strong absorption from the visible to the far-ultraviolet region and suitable band gaps and band edges, these monolayers are suitable candidates for applications in optoelectronics, solar energy conversion, and photocatalytic water splitting.

10.2.2 SiP Monolayer

10.2.2.1 Chemistry and 2D Bonding

In the silicon phosphorus monolayer SiP, the Si atom with the valence electrons ($3s^2p_x^2$) and the P atom with the valence electrons ($3s^2p_x^2p_y^1$) form a variety of covalently bonded structures with comparable binding energies. In the well-studied hexagonal four-layer α-SiP monolayer, each Si atom binds four atoms, one Si atom and three P atoms, and each P atom forms three bonds with Si atoms. The small difference in the electronegativities of Si (1.90) and P (2.19) induces a charge transfer that generates polar S–P bonds. Explicitly, the two outer P layers are negatively charged by an enormous charge transfer of $\delta q = 2.84$ e from the inner Si sublayers to the outer P sublayers, producing, however, a surprisingly small fractional ionic character of only 2.1% [1]. The cohesive energy of hexagonal α-SiP monolayer of $E_c = 4.19$ eV/atom designates a much smaller binding energy than that of the α-SiN monolayer.

Besides the highly symmetric four-layer structure, SiP monolayers form a disturbed low- symmetry phase, which consists of the motif of a hexagon surrounded by a pair of pentagon rings (see Fig. 10.12) [33]. Interestingly, this less symmetric monoclinic phase with the space group *Cm* has the same coordination as the symmetric hexagonal phase; only the bond lengths vary slightly, as revealed below [33–35]. Within the errors involved in such calculations of cohesive energies, the value of $E_c = -4.21$ eV/atom essentially agrees with that of the symmetric α phase [34]. Note that the large rectangular primitive cell of this complex structure contains 12 Si and 12 P atoms [35].

10.2.2.2 2D Structures

The stable α phase of SiP has the following lattice parameters: $a = 0.353$ nm, $d_{Si\text{–}Si} = 0.237$ nm, $d_{Si\text{–}P} = 0.228$ nm, $h_{PP} = 0.441$ nm, $\theta_{SiSiP} = 116.50°$ and $\theta_{PSiP} = 101.62°$

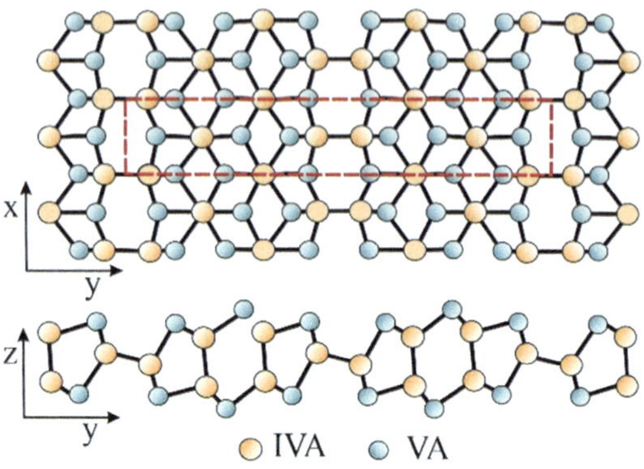

Fig. 10.12 Top and side views of the γ structure with a lower symmetry of *C*2/*m*, incorporating hexagons surrounded by two pentagons. The large unit cell is indicated by the red dashed line. Well-studied examples are SiP, SiAs, GeP, and GeAs monolayers. Adapted with permission from Ref. [33], Copyright 2016, AIP Publishing

[1]. Adding the van der Waals diameter of P atoms $d_w = 0.390$ nm and distance between P atoms h_{PP}, yields an estimate of the monolayer thickness of $d_{w+h} = 0.831$ nm. The calculated band gaps of the indirect semiconductor are $E_g = 1.52$ eV (PBE), 1.51 eV (PBE + SOC), and 2.22 eV (HSE06) [1]. The space group of this symmetric phase is *P*-6*m*2. This phase seems to be the most stable monolayer structure of SiP [33].

Remarkably, the lattice parameters of the low-symmetry monoclinic structure of SiP with hexagons combined with pentagons of a = 0.352 nm, b = 2.048 nm, $d_{Si\text{–}Si} = 0.235$ nm, and $d_{Si\text{–}P} = 0.226 – 0.230$ nm are nearly identical to those of the hexagonal α-SiP phase, despite the lower symmetry of the space group *C*2/*m* (*Cm*), as mentioned above [33–35]. However, the structure has an extreme lattice constant in the elongated zigzag direction of b = 2.05 nm.

Bulk orthorhombic SiP with the symmetry $Cmc2_1$ undergoes a transition from an indirect to a direct gap when the thickness decreases to the monolayer thickness [34, 35]. The calculated direct band gaps of the monolayer are 1.86 eV (PBE) and 2.59 eV (HSE06). Note that the formation energies of the symmetric phase and the disturbed phase differ by less than 10 meV/atom [33]. Therefore, polymorphic behavior occurs during bulk exfoliation, and synthesis via deposition on different substrates, often is yielding mixed phases. For the energetically and dynamically stable monoclinic SiP monolayer MD simulations indicate thermal stability up to 900 K [34].

10.2.2.3 Mechanics

Strongly anisotropic mechanical behavior was demonstrated for the monolayer of the low-symmetry monoclinic phase of SiP by the DFT-GGA-PBE approach [36]. For Young's modulus, the values of $E_{ac} = 77.2$ N/m and $E_{zz} = 100.4$ N/m indicate a strongly varying in-plane stiffness, while an isotropic Poisson ratio of $\nu = 0.14$ is proposed. The reason for the lower stiffness for loading in the armchair direction is the orientation of the weak Si–Si bonds exactly in this direction. As expected, the anisotropy further increases for the nonlinear properties such as the ultimate strengths of $\sigma_{ac} = 8.1$ N/m and $\sigma_{zz} = 13.7$ N/m and even more for the ultimate strains of $\varepsilon_{ac} = 0.15$ and $\varepsilon_{zz} = 0.31$ [36]. The armchair values show good agreement, and the zigzag values qualitatively agree with Griffith's rule of $E/\sigma \approx 10$. The band gap of $E_g = 2.58$ eV (HSE 06) derived by these authors is in excellent agreement with the gap given above.

10.2.2.4 Physical Properties and Applications.

An important topic in sensing applications is band gap engineering by tensile or compressive strain or an external electric field [36, 37]. The electronic band gap of α-SiP monolayers can be tuned between 1.91 and 0.7 eV by applying a biaxial compressive strain of 0–10% and between 1.91 and 1.0 eV by changing the tensile strain between 0 and 12% [37]. A transition to metallic behavior occurs above 12% compressive strain and beyond 14% tensile strain. Importantly, the effect of an external electric field on the electronic band structure is negligible, which may be useful for device applications.

Low-symmetry monoclinic SiP nanosheets with anisotropic optical and transport properties were suggested for solar water splitting by first-principles DFT calculations [38]. The band gap of stress-free monolayers is within a suitable range, and the band positions required for the H^+/H_2 and O_2/H_2O reactions satisfy a wide range of environmental pH values (see Fig. 10.13). In addition to suitable band gaps and band edges, this 2D material also provides the necessary optical absorption in the visible region of the spectrum for use as a photocatalyst. According to theory, besides SiP also the monolayers of SiAs, GeP, and GeAs satisfy the requirements for photocatalytic water splitting.

First-principles calculations combined with semiclassical Boltzmann transport theory provided information on the thermoelectric properties of SiP monolayers [39]. Highly anisotropic phonon and electron transport occurs, where both the thermal and electrical conductivity along the armchair direction are lower than in the zigzag direction. The lattice thermal conductivity in the armchair direction, for example, is 11.05 W/mK at 300 K.

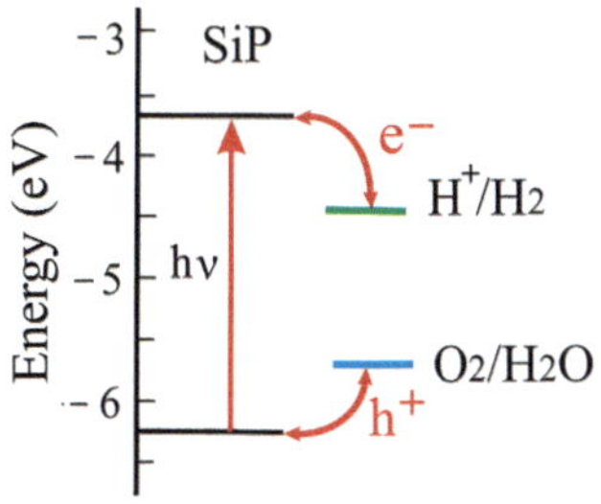

Fig. 10.13 Band gaps of the stress-free monolayer SiP with the required band positions to permit the H^+/H_2 and O_2/H_2O reactions in solar water splitting/

10.2.2.5 Synthesis

For a layered bulk precursor, the formation energy is the difference in energy between the 2D phase (e.g., monolayer or sheet) and the most stable bulk phase. This is the energy required, for example, for exfoliation of a single monolayer. The estimated exfoliation energy is 0.26 J/m^2 for SiP, which is lower than that of graphite (0.32 J/m^2). Accordingly, the synthesis of SiP monolayers and few-layer assemblies is energetically feasible. In fact, the successful synthesis of SiP monolayers on a mica substrate has been achieved by employing the chemical vapor transport (CVT) method [40]. The temperature in the reaction zone determines whether the transport agent I_2 can react with the two reactants Si powder and P powder, and the temperature in the growth region (850 °C) controls nucleation and growth of the SiP layers. The knowledge of the accurate growth conditions may trigger the deposition of other semiconductors of the IVA–VA family.

10.2.3 SiAs Monolayer

10.2.3.1 Chemistry and Bonding

In the silicon arsenic monolayer SiAs, the valence electrons of the Si atom ($3s^2p_x{}^2$) and the As atom ($4s^2p_x{}^2p_y{}^1$) form a variety of covalently bonded structures with comparable covalent bonding strength. In the four-layer hexagonal α phase of the SiAs monolayer, each Si atom binds four atoms, and each As atom forms three bonds, as discussed for α SiP. The bonds are polar owing to the deviating electronegativities of Si (1.90) and As (2.18), which induce ionic interactions, leading to polar Si–As bonds. From the two inner Si sublayers, a surprisingly large charge transfer of $\delta q = 2.52$ e occurs to the negatively charged outer As sublayers, causing a surprisingly low fractional ionic character of 1.9% [1]. The weaker bonding indicated by the cohesive energy of the α-SiAs monolayer of $E_c = -3.85$ eV/atom in comparison to α SiP is predominantly caused by the larger atom size of As atoms in comparison to P atoms.

In addition to the hexagonal α phase, the monoclinic phase of SiAs comprises two pentagonal rings and one hexagonal ring with a large primitive cell of 24 atoms [41]. The cohesive energy of this structure is $E_g = -3.49$ eV/atom and suggest a lower stability than that of the hexagonal α phase. Nevertheless, this phase has been considered the energetically most stable phase of monolayer SiAs [42]. The corresponding much larger cohesive energy of $E_c = -5.13$ eV/atom presented for the monoclinic SiAs monolayers seems not to be consistent with the value of $E_c = -3.85$ eV/atom presented above for the hexagonal α phase [42].

Moreover, first-principles DFT-PBE calculations predict a thermodynamically and dynamically stable monolayer with $SiAs_2$ stoichiometry and another surprisingly large cohesive energy of $E_c = -4.98$ eV/atom for this structure consisting of pentagons [42].

10.2.3.2 2D Structures

The hexagonal four-layer α phase with the chemical formula Si_2As_2 has the following lattice parameters: a = 0.370 nm, d_{Si-Si} = 0.236 nm, d_{Si-As} = 0.240 nm, h_{AsAs} = 0.457 nm, θ_{SiSiAs} = 117.38° and θ_{AsSiAs} = 100.54° [1]. Adding the van der Waals diameter of As atoms d_w = 0.410 nm and distance between As atoms h_{AsAs}, yields an estimate of the monolayer thickness of d_{w+h} = 0.867 nm. The calculated band gap of the indirect semiconductor, based on several approaches, is E_g = 1.63 eV (PBE), 1.54 eV (PBE + SOC), and 2.27 eV (HSE06) [1].

The lattice parameters for the monoclinic SiAs monolayer, extracted from the bilayer structure of a = 0.369 nm, b = 1.083 nm, d_{Si-Si} = 0.235 nm, d_{Si-As} = 0.239–0.243 nm, h_{AsAs} = 0.486 nm, and θ_{SiAsSi} = 99.81°, are nearly identical to those of the hexagonal α phase [42]. Each Si atom forms four bonds with its nearest neighbors, and each As atom forms only three bonds, while the lone electron pair of the p_z orbital augments the three bonding orbitals and influences the structure and corrugation. The indirect band gap of this phase of SiAs is 1.72 eV (PBE) and 2.39 eV (HSE06) [42]. Note that another study suggested a direct band gap, however, with similar values of E_g = 1.69 eV (GGA) and 2.35 eV (HSE) [35].

The prime cell of the low-symmetry penta-$SiAs_2$ structure with the lattice parameters a = 0.368 nm and b = 1.057 nm contains 4 Si atoms and 8 As atoms [42]. The bond lengths are d_{As-As} = 0.250 nm, d_{Si-As} = 0.241–0.245 nm, and the large buckling height is Δz = 0.509 nm. Each Si atom has four nearest As atoms, and each As atom has three nearest neighboring Si atoms and forms one covalent bond with a neighboring Si atom and two covalent bonds with itself. The estimates of the almost direct band gaps of the semiconducting monolayer are 1.42 eV (PBE) and 2.07 eV (HSE06).

10.2.3.3 Mechanical Properties

The DFT-GGA-PBE approach predicts strongly anisotropic mechanical properties for monolayers of the low-symmetry monoclinic phase of SiAs [36]. The linear mechanical behavior was determined by fitting the linear part of the stress–strain curve with a strain below ε = 0.02. For the elastic modulus, the calculated values at E_{ac} = 64.7 N/m and E_{zz} = 86.3 N/m indicate a strongly varying in-plane stiffness, and the anisotropic Poisson ratios are ν_{ac} = 0.11 and ν_{zz} = 0.16. As expected, for the nonlinear properties, the anisotropy increases, and the ultimate strengths are σ_{ac} = 7.5 N/m and σ_{zz} = 11.4 N/m, and the ultimate strains ε_{ac} = 0.16 and ε_{zz} = 0.31 [36]. The large anisotropy correlates with the 2D symmetry of the lattice. That direct load is stretching the homonuclear Si–Si bonds oriented along the armchair direction, explains the lower tensile strength or stretchability in this direction. Uniaxial loading induces the greatest elongation and easiest bond breaking in this direction.

10.2.3.4 Physics and Applications

First-principles calculations of the band structure of the monoclinic SiAs monolayer revealed an indirect-to-direct band gap and even a metal transition by the application of isotropic strain along two in-plane directions [42]. Furthermore, theory elucidated a carrier mobility that is higher than that of MoS_2, and anisotropic carrier transport. To gain more insight into the electronic structure, the acoustic phonon-limited carrier mobilities of electrons and holes in the armchair and zigzag directions were determined based on deformation potential theory at 300 K. In the low-energy regime, electron-acoustic-phonon scattering dominates carrier transport.

From first-principles calculations it was concluded that the monoclinic phase, not the hexagonal α phase, of SiAs is slightly favored by a negative formation energy, and its energetic and dynamic stability was confirmed [41]. This SiAs monolayer is a direct semiconductor with a calculated band gap of 1.84 eV (PBE) and 2.50 eV (HSE06). Note that this implies a lower binding energy than given above for the hexagonal α phase. The authors noted the suitability of the band gap and band edge positions of monoclinic SiAs for application as photocatalyst for hydrogen production (water splitting).

First-principles calculations revealed the properties and potential applications of three thermally and dynamically stable novel hexagonal SiAs structures, named α-, β-, and γ-SiAs phases by the authors [43]. The suggested 2D crystal structures are like those of the GaSe monolayer. The study includes electron transport properties, wide-band gap optoelectronic applications, and photoelectronic devices used for catalytic water splitting (α and β SiAs), as well as strong absorption in the visible-light region and high carrier mobility (γ SiAs).

10.2.3.5 Synthesis

Based on the synthesis of 3D layered SiAs single crystals with weak van der Waals interactions by vapor growth and CVD methods, it should be possible to produce monolayers and few-layer sheets of SiAs by mechanical exfoliation, especially by keeping in mind the low estimated exfoliation energy of 0.27 J/m^2. This indicates good prospects for the experimental realization of high-quality SiAs monolayers and nanosheets in the near future [36].

10.2.4 SiSb Monolayer

10.2.4.1 Chemistry and 2D Bonding

In the silicon antimony monolayer SiSb, the Si atom has the valence electrons ($3s^2p_x{}^2$), and the Sb atom ($5s^2p_x{}^2p_y{}^1$). The electronic configuration of Sb with an additional valence electron opens the door to various allotropes with comparable binding energies. One of the best studied monolayers is the four-layer hexagonal α-SiSb phase, where each Si atom binds four atoms and each Sb atom forms three bonds with Si atoms and carries a lone electron pair. The covalent Si–Sb bonds are slightly polar owing to the minor difference in the electronegativities of Si (1.90) and Sb (2.05). A charge transfer of $\delta q = 0.52$ e to the outer Sb sublayers induces a small fractional ionic character of 0.6% [1]. Compared to its lighter homologues, hexagonal α SiSb has a smaller binding energy of $E_c = -3.50$ eV/atom [1]. However, independent DFT-PBE calculations indicate that the cohesive energy of α SiSb monolayers may be much greater at $E_c = -4.42$ eV/atom [23].

Furthermore, theory predicts a thermodynamically, kinetically, and thermally stable penta-$SiSb_2$ monolayer, as confirmed by the negative cohesive energy, lack of negative modes in the phonon dispersion spectra, and MD simulations indicating kinetic stability up to 500 K [44]. In the three-layer structure, the Si plane is sandwiched between two antimony layers with a distance between the Si and Sb atoms of $d_{Si\text{–}Sb} = 0.166$ nm. The unit cell of this pentagonal structure contains six atoms, two four-coordinated Si atoms and four three-coordinated Sb atoms bonded to an antimony atom and two silicon atoms. The calculated cohesive energy of $E_c = -4.58$ eV/atom indicates a comparable binding energy as given above for the lower cohesive energy of the hexagonal α-SiSb structure.

10.2.4.2 2D Structures

The stable hexagonal α phase of SiSb has the lattice parameters $a = 0.402$ nm, $d_{Si\text{–}Si} = 0.236$ nm, $d_{Si\text{–}Sb} = 0.262$ nm, $h_{SbSb} = 0.482$ nm, $\theta_{SiSiSb} = 117.92°$ and $\theta_{SbSiSb} = 99.85°$ [1]. Adding the van der Waals diameter of Sb atoms $d_w = 0.440$ nm and distance between Sb atoms h_{PP}, yields an estimate of the monolayer thickness of $d_{w+h} = 0.922$ nm. The band gaps of the indirect semiconductor obtained by various approaches are $E_g = 1.18$ eV (PBE), 0.99 eV (PBE + SOC), and 1.76 eV (HSE06). Interestingly, the study that reported the strong binding presented quite similar lattice parameters for α SiSb [23].

The lattice parameters of the pentagonal structure are $a = b = 0.553$ nm, $d_{Sb\text{–}Sb} = 0.289$ nm, $d_{Si\text{–}Sb} = 0.261$ nm, $\theta_{SbSiSb} = 113.76°$ and $\theta_{SbSbSi} = 96.78°$ [44]. The buckling effect occurs at $\Delta z = 0.332$ nm. For the indirect band gap of the semiconductor theory delivered $E_g = 1.53$ eV (PBE) and 2.1 eV (PBE).

10.2.4.3 Mechanics

The isotropic elastic modulus and Poisson ratio of the hexagonal α-SiSb monolayer derived from the elastic constants C_{11}, C_{22}, and C_{12}, published without units, are $E = 78.1$ GPa and $\nu = 0.224$, assuming that the mechanical properties are presented in GPa units in this publication [23]. The monolayer thickness of $d_{w+\Delta z} = 0.922$ nm, determined by adding the van der Waals diameter of Sb and distance between Sb atoms, yields a similar stiffness of 72 N/m in 2D units.

10.2.4.4 Physics and Applications

The electrical and thermal transport properties of the indirect band gap semiconductor α SiSb were studied via first-principles calculations [45]. The lattice thermal conductivity of the monolayer was 4.75 W m^{-1} K^{-1}, which leads to a high figure of merit value of $ZT \approx 2$, when combined with the corresponding electrical properties. The authors suggest the use of α SiSb as a model system to understand the effects of dimensional reduction on the thermoelectric properties.

For the pentagonal $SiSb_2$ phase, the dielectric function was determined to describe the optical absorption and polarization characteristics [44]. The real part of the complex dielectric function defines the ability of the 2D material to store energy from an outside field. The penta-$SiSb_2$ monolayer exhibits almost zero absorption and extremely low reflection (12–15%) in the visible region; however, the considerable absorption and reflection in the ultraviolet range of the spectrum have potential for applications in nano-optoelectronic devices.

10.2.5 SiBi Monolayer

10.2.5.1 Chemistry and 2D Bonding

In the silicon bismuth monolayer SiBi, the Si atom has the valence electrons ($3s^2p_x^2$), and the Bi atom ($6s^2p_x^2p_y^1$). These valence electrons form slightly polar bonds in covalent-ionic 2D networks as already discussed. Like in the other hexagonal four-layer structures, in hexagonal α-SiBi monolayers each Si atom binds to one Si and three Bi atoms, and each Bi atom forms three bonds with Si atoms. Owing to the similar electronegativities of Si (1.90) and Bi (2.02), the ionic contribution to bonding is very small [46]. A charge transfer of $\delta q = 0.66$ e occurs from the Si atoms of the inner Si sublayers to the negatively charged Bi atoms of the outer Bi sublayer, inducing a small fractional ionic character of 0.4% [1]. Note that according to other calculations, the Bi atom gains only a charge of $\delta q = 0.04$ e from a Si atom [46, 47]. The cohesive energy of α SiBi at $E_c = -3.31$ eV/atom is the smallest in this group of Si–VA compounds [1]. Additionally, a substantially lower cohesive energy of – 4.65 eV/atom, indicating much stronger bonding, was derived employing the DFT-GGA-PBE approach [46, 47].

10.2.5.2 2D Structures

The stable α phase of the four-layer α-SiBi structure has the following lattice parameters: a = 0.417 nm, $d_{Si–Si} = 0.235$ nm, $d_{Si–Bi} = 0.273$ nm, $h_{BiBi} = 0.492$ nm, $\theta_{SiSiBi} = 118.07°$ and $\theta_{BiSiBi} = 99.66°$ [1]. The divers band gaps calculated for the indirect semiconductor are $E_g = 0.64$ eV (PBE), 0.27 eV (PBE + SOC), and 1.15 eV (HSE06). Adding the van der Waals diameter of Bi atoms of $d_w = 0.460$ nm and distance between Bi atoms $h_{Bi\text{-}Bi}$, yields an estimate of the monolayer thickness of $d_{w+h} = 0.952$ nm. Note that the structural parameters given in the work suggesting a much stronger binding of α SiBi agree with the lattice parameters presented above within the errors involved in such calculations [46, 47].

10.2.5.3 Mechanics

To calculate the linear mechanical properties of α-SiBi monolayers, uniaxial strains were applied to a rectangular cell with a maximum strain of $\varepsilon = 0.02$ to extract the relevant linear lattice parameters. For the isotropic structure of hexagonal α-SiBi monolayers with equivalent in-plane lattice constants, the calculated Young's modulus is $E = 77$ N/m, and the Poisson ratio $\nu = 0.279$ [46]. Note that this stiffness is greater than those of the elemental Si and Bi monolayers silicene and bismuthene.

10.2.5.4 Physics and Applications

First-principles calculations, which included relativistic effects, clarified the influence of an electric field, layer thickness, and mechanical strain on the electronic structure of α-SiBi nanosheets [47]. A strong electric field induces a semiconductor-to-metal transition in α-SiBi bilayers. A general effect of layer thickness is the reduction in the band gap with increasing number of layers. Under uniaxial and biaxial strains (in-plane and out-of-plane), the band gap decreases monotonically, and an indirect-to-direct band gap transition occurs already at a low strain of $\varepsilon = 0.03$.

The spiral Boltzmann transport theory, within the constant relaxation time approximation and the rigid band approach, as implemented in the BoltzTraP2 code, elucidated the thermoelectric properties of single-layer hexagonal α SiBi [46]. Specifically, the package revealed the electronic structure, Seebeck coefficient, electrical conductivity, and power factor. The results indicate a high potential of α-SiBi monolayers in terms of its thermoelectric coefficient, which increases at elevated temperatures.

10.3 Ge–VA Monolayers (GeN, GeP, GeAs, GeSb, and GeBi)

10.3.1 GeN Monolayer

10.3.1.1 Chemistry and 2D Bonding

The group of Ge–VA monolayers includes covalent-ionic bonding of Ge atoms with the pnictogens N, P, As, Sb, and Bi. In germanium nitrogen monolayers, the four valence electrons of Ge atoms ($4s^2p_x^2$), and the five valence electrons of N atoms ($2s^2p_x^2p_y^1$) form various related 2D structures, as already discussed for the carbon and silicon pnictogenides. The hexagonal four-layer structure of the α-GeN monolayer is one of the best studied phases of the group IVA–VA monolayers with space group *P*-6*m*2. The different electronegativities of Ge (2.01) and N (3.04) induce polar bonds with substantial ionic contributions. In the monolayer assembly with the N–Ge–Ge–N configuration, a large charge transfer of $\delta q = 2.77$ e occurs from the inner germanium atoms to the outer N atoms, inducing a significant fractional ionic character of 23.3% [1]. The cohesive energy of $E_c = -4.28$ eV/atom belongs to the strongest binding energies in this group of Ge pnictogens, together with the structurally related hexagonal β-GeN monolayer with the space group *P*-3*m*1 and a cohesive energy of $E_c = -4.44$ eV/atom, signifying a higher energetic stability [48].

To date, all synthesized GeN structures have monoclinic symmetry and belong to the space group *C*2/*m* with a large unequal unit cell with $b \approx 6a$ that contains 12 Ge and 12 N atoms [48]. This structure basically consists of a motif with hexagons located between two pentagons. Each Ge atom is bonded to one Ge atom and three N atoms, and each N atom is threefold coordinated to Ge atoms. The cohesive energy of this distorted low-symmetry structure with $E_c = -4.30$ eV/atom is comparable to that of the symmetric α phase. The study investigated the possibility of transforming the low-symmetry *C*2/*m* structure into the more symmetric α phase (*P*-6*m*2) or β phase (*P*-3*m*1), e.g., by the application of mechanical strain. For a transition into the β phase, the tensile strength and in-plane stiffness double and similar band structure profiles with narrowed band gaps appear.

In a comprehensive work, the lattice parameters and band gaps of 25 group IVA–VA monolayers are presented [6]. Note the deviating nomenclature in this study, which covers, for example, the four-layer mixed configuration Ge–N–Ge–N, which is proposed as the most energetically favorable structure in this work [6]. Note that this structure has three different Ge–N bond lengths. The strong ability of nitrogen to bind electrons is invoked to explain the derived extreme cohesive energy of $E_c = -6.23$ eV/atom.

10.3.1.2 2D Structures

The hexagonal α phase of the GeN monolayer with *P*-6*m*2 symmetry has the following lattice parameters: a = 0.310 nm, $d_{Ge–Ge}$ = 0.257 nm, $d_{Ge–N}$ = 0.191 nm, d_{NN} = 0.390 nm, θ_{GeGeN} = 110.40° and θ_{NGeN} = 108.53° [1]. Adding the van der Waals diameter of N atoms d_w = 0.320 nm and the distance between N atoms h_{NN} = 0.390, yields an estimate of the monolayer thickness of d_{w+h} = 0.710 nm. The calculated band gaps of the indirect semiconductor are E_g = 1.17 eV (PBE), 1.17 eV (PBE + SOC), and 2.25 eV (HSE06) [1].

Owing to similar bonding types and arrangements, the lattice parameters, cohesive energies, and band gap of other stable phases of GeN monolayers are comparable. The lattice parameters of the related β-GeN monolayer with *P*-3*m*1 symmetry of a = 0.310 nm, b = 1.610 nm, $d_{Ge–Ge}$ = 0.254 nm, and $d_{Ge–N}$ = 0.191 nm clearly demonstrate this similarity. The calculated electronic band gaps are 1.11 eV (PBE) and 2.19 eV (HSE) [48]. As discussed below, these band gaps with efficient absorption qualify the allotropes for visible-light photocatalysis [49, 50].

The lattice parameters of the *C*2*m* phase of the monoclinic GeN monolayer are a = 0.310 nm, b = 1.882 nm, $d_{1,Ge–Ge}$ = 0.246 nm, $d_{2,Ge–Ge}$ = 0.247 nm, and $d_{Ge–N}$ = 0.194 nm. The calculated values of the electronic band gap are 1.35 eV (PBE) and 2.36 eV (HSE) [48, 51]. Note that the polarization of the Ge–N bonds is much smaller than that of the α phase, with a charge transfer from the Ge atom to the N atom of only δq = 1.34 e in this low-symmetry structure [51].

10.3.1.3 Mechanics

The linear and nonlinear mechanical properties of the *C2/m* structure of low-symmetry GeN monolayers show a remarkable anisotropy of E_{ac} = 86.5 N/m and E_{zz} = 126.4 N/m for the elastic moduli and ν_{ac} = 0.18 and ν_{zz} = 0.26 for the Poisson ratios [51]. The nonlinear behavior exhibits an even more pronounced anisotropy of σ_{ac} = 4.66 N/m and σ_{zz} = 10.2 N/m for the ultimate tensile strengths and for the ultimate tensile strains ε_{ac} = 0.19 and ε_{zz} = 0.26 [51]. Additionally, this work presents the ultimate compressive strength, the ultimate compressive toughness, and the maximum compressive strain of the GeN monolayers.

The corresponding values for the isotropic linear mechanical properties of the β-GeN structure are E = 158.8 N/m for the stiffness and ν = 0.29 for the Poisson ratio [48]. Unexpectedly, the ultimate strength values of this phase of σ_{ac} = 15.2 N/m and σ_{zz} = 16.4 N/m are only slightly nonlinear, whereas the ultimate tensile strains at ε_{ac} = 0.193 and ε_{zz} = 0.258 agree well with the distinct anisotropy given above for the Poisson ratios [48]. These results indicate an isotropic behavior of Young's modulus with a large stiffness and critical fracture stress of the hexagonal β phase.

10.3.1.4 Physics and Applications

According to theory, the *C2/m* phase can be transformed into the hexagonal β configuration (*P*-3*m*1) with a lower binding energy, employing a large compressive strain of ~ 12% along the armchair direction [48]. Several factors are responsible for this phase transition, such as the similarity of the geometry of the two structures, which differ only in the armchair direction; the existence of weaker Ge–Ge bonds, which preferably deform during the transition process; and the small size of the nitrogen atom. In the monoclinic GeN monolayer, the stress concentrates in the weaker Ge–Ge bonds, and the more uniform and mechanically superior structure of β GeN shows a more homogeneous distribution. Note that a transition of the *C2/m* phase to the *P*-6*m*2 symmetry is unfeasible.

First-principles calculations revealed the structural, electronic, optical, and catalytic properties of GeN monolayers with C2/m symmetry [49]. The monolayer exhibits intense light absorption > 10^5 cm^{-1} in the visible and ultraviolet spectral ranges. Furthermore, thorough investigations of strain engineering and photocatalytic properties of GeN monolayers are available [49]. The band gap decreases monotonically under tensile strain, whereas the application of compressive biaxial strain increases the gap. The tunable band gap of the indirect semiconductor is 2.21 eV, and the band edges are suitable for producing hydrogen and oxygen over an extensive range of pH = 0–10 with a maximum absorbance of ~ 25%. Together with their mechanical stability and excellent light absorption, nanosheets may find applications in the areas of photodetectors, photocatalysis, energy conversion devices, FETs, flexible electronics, ion-batteries, and sensors [50].

10.3.1.5 Synthesis

It can be expected that first the monoclinic low-symmetry phase of the CN monolayers and then the other high-symmetry structures may be fabricated [48]. Since the transformation of the monoclinic phase into the energetically more favorable β phase, with its significant activation barrier of ~ 2.9 eV, is not reversible, this process could possibly allow the synthesis of the high-symmetry β phase of CN monolayers.

10.3.2 GeP Monolayer

10.3.2.1 Chemistry and 2D Bonding

In the germanium phosphorus monolayer GeP, the Ge atom has the valence electrons ($4s^2p_x{}^2$), and the P atom ($3s^2p_x{}^2p_y{}^1$). This electron configuration allows the formation of covalently bonded four-layer allotropes with a small ionic contribution to bonding due to the only marginally different electronegativities of Ge (2.01) and P (2.19). According to first-principles calculations, the hexagonal α-GeP monolayer with *P-6m2* symmetry is the most stable monolayer. In the four layer P–Ge–Ge–P configuration, a charge transfer of $\delta q = 1.52$ e occurs from the inner Ge atoms to the outer P atoms, introducing a low fractional ionic character of 0.8% [1]. The cohesive energy of the α-GeP monolayer at $E_c = -3.60$ eV/atom indicates a considerably weaker binding energy than that of α GeN, primarily due to the increased atom size of the P atoms, which reduces orbital overlap and thus the covalent bonding strength.

Calculations indicate a better stability of the *P-6m2* symmetry for SiP and GeP monolayers, whereas for the Si and Ge pnictides with As, Sb, and Bi, the lower *Cm* symmetry of the bulk structure is energetically favored [33]. It is important to mention that in several cases the difference in the formation energy is less than 10 meV/atom, pointing to the occurrence of polymorphism, and difficulties in the definitive identification of the most stable allotrope. Furthermore, in this situation serious problems are encountered to prepare the pure phase of the nanomaterial.

Additionally, the monoclinic structure of the GeP monolayer with *C2/m* symmetry forms stable monolayers, with hexagons and pentagons arranged in a large unit cell, as described for GeN [51]. Due to the different electronegativities, a charge transfer of $\delta q = 0.50$ e occurs from each Ge atom to each P atom. The cohesive energy of $E_c = -3.84$ eV/atom calculated for this low-symmetry structure indicates stronger binding than for the α-GeP monolayers with higher symmetry [51].

10.3.2.2 2D Structures

The stable α phase of GeP monolayers has the lattice parameters a = 0.366 nm, $d_{Ge\text{–}Ge} = 0.251$ nm, $d_{Ge\text{–}P} = 0.237$ nm, $h_{PP} = 0.465$ nm, $\theta_{GeGeP} = 116.91°$ and $\theta_{PGeP} = 101.12°$ [1]. Adding the van der Waals diameter of P atoms $d_w = 0.390$ nm and distance between P atoms h_{PP}, yields an estimate of the monolayer thickness of $d_{w+h} = 0.855$ nm. The calculated band gaps of the indirect semiconductor are $E_g = 1.35$ eV (PBE), 1.34 eV (PBE + SOC), and 2.05 eV (HSE06) [1].

For the lattice parameters of the monoclinic P2/m phase of GeP monolayers, first-principles calculations yielded a = 0.363 nm, b = 2.125 nm, $d_{1Ge\text{–}Ge} = 0.246$ nm, $d_{2Ge\text{–}Ge} = 0.248$ nm, $d_{Ge\text{–}P} = 0.237$ nm, and $h_{PP} = 0.498$ nm [51]. An estimate of the layer thickness of $d_{w+\Delta z} = 0.888$ nm can be obtained by adding the van der Waals diameter $d_w = 0.390$ nm of P and the P–P distance. The authors consider GeP to be a quasidirect semiconductor with an energy difference of only 60 meV between direct and indirect gaps of $E_g = 1.63$ eV (PBE, PBE-SOC) and 2.35 eV (HSE, HSE-SOC). Previous calculations of the lattice parameters of the low-symmetry hexagon-pentagon structure of GeP with a = 0.365 nm, b = 2.139 nm, $d_{1Ge\text{–}Ge} = 0.248$ nm, $d_{2Ge\text{–}Ge} = 0.236$ nm, $d_{Ge\text{–}P} = 0.236$ nm, and $h_{PP} = 0.498$ not only agree with those given for GeP above but are also nearly identical to those of the α phase [35]. The band gap values of the monoclinic GeP monolayer of 1.61 eV (GGA) and 2.31 eV (HSE) also agree well with recent calculations [35].

10.3.2.3 Mechanics

Strongly anisotropic mechanical behavior was found for the low-symmetry hexagon-pentagon phase of GeP monolayers by the DFT-GGA-PBE approach [36]. The elastic moduli of $E_{ac} = 65.6$ N/m and $E_{zz} = 82.1$ N/m indicate a strongly varying in-plane stiffness, while the Poisson ratios of $\nu_{ac} = 0.15$ and $\nu_{zz} = 0.16$ are nearly identical [36]. As expected, the nonlinearity increases for the ultimate strengths with the values $\sigma_{ac} = 6.0$ N/m and $\sigma_{zz} = 10.1$ N/m, and for the ultimate strains at $\varepsilon_{ac} = 0.15$ and $\varepsilon_{zz} = 0.31$. Other theoretical calculations predict a larger isotropic critical tensile stress of 13.2 N/m (14.9 GPa) and a large critical strain of 0.28 [52].

The stiffness values are in acceptable agreement with recent calculations of the mechanical properties of monoclinic GeP monolayers, whereas larger deviations are observed for the ultimate strength [51]. In addition to the anisotropic stiffnesses of $E_{ac} = 58.1$ N/m and $E_{zz} = 78.5$ N/m and the Poisson ratios $\nu_{ac} = 0.13$ and $\nu_{zz} = 0.18$, the authors present low ultimate tensile strengths of $\sigma_{ac} = 3.75$ N/m and $\sigma_{zz} = 6.51$ N/m and anisotropic ultimate tensile strains of $\varepsilon_{ac} = 0.19$ and $\varepsilon_{zz} = 0.30$. The also determined compressive strengths are $\sigma_{c,ac} = 5.63$ N/m and $\sigma_{c,zz} = 11.38$ N/m, and the compressive strains $\varepsilon_{c,ac} = \varepsilon_{c,zz} = -0.17$ [51].

10.3.2.4 Physics and Applications

The indirect band gap of low-symmetry GeP monolayers can be reduced from ~ 2.28 eV to 0.43 eV and it can be transformed into a direct band gap by applying axial strain [52]. The carrier mobility is strongly directional, with a high electron mobility of up to 1242.1 $cm^2\ V^{-1}\ S^{-1}$ in the armchair direction, which qualifies these 2D materials for applications in flexible FETs and optoelectronic devices. An experimental study using GeP flakes as photodetectors showed highly anisotropic behavior for the photoresponsivity [53].

First-principles calculations and semiclassical Boltzmann transport theory allowed the investigation of the thermoelectric properties of GeP monolayers [54]. Both phonon and electron transport are much smaller in the armchair direction, with a lattice thermal conductivity of 9.48 W/m K at 300 K in this direction. The almost isotropic Seebeck coefficient for monolayers of GeP is 2.5 mV/K at room temperature, and the maximum thermoelectric figure of merit with n-type doping approaches $ZT = 0.78$ at 700 K. The presented values of the thermal conductivity and the Seebeck coefficient are slightly lower than those calculated for SiP monolayers.

To study the ability of monoclinic GeP monolayers to sense small molecules (CO, CO_2, O_2, H_2O, NH_3, NO, and NO_2), the adsorption energy, adsorption distance, and charge transfer were calculated [55]. While all gas molecules act as acceptors, the highest sensitivities were found for NO and NO_2, where NO_2 had a smaller adsorption energy. The different work function changes allowed the identification of NO and NO_2. A small biaxial compressive strain enhances the sensitivity of the work function-type sensor to NO_2. Similarly, an external vertical electric field can increase the adsorption strength, improve charge transfer, and change the work function of the GeP-NO_2 system.

Hexagonal α GeP and monoclinic GeP monolayers are possible candidates for use as anode materials in lithium-ion batteries [56, 57]. Calculations of the activation barriers revealed anisotropic Li diffusion paths on GeP monolayers and in the interlayer region of bilayers, where Li atoms can diffuse ~ 50 times faster than in graphene bilayers. Homojunctions of low-symmetry GeP monolayers with structural similarities to the *C2/m* phase, suitable band alignments, and natural type-II homojunctions are appropriate conditions for sunlight-driven water splitting at pH = 0 and pH = 7 [58].

10.3.2.5 Synthesis

Several already synthesized layered bulk crystals of group IVA–VA compounds have monoclinic *C2/m* symmetry with weak interlayer interactions that, in principle, should allow the exfoliation of 2D sheets [51]. In fact, monoclinic GeP and GeAs nanosheets have been exfoliated by mechanical and liquid-phase methods and have already been tested in photodetectors [51, 53]. These photodetectors exhibited highly anisotropic behavior with an anisotropy factor of 1.52 for conductance and 1.83 for photoresponsivity.

10.3.3 GeAs Monolayer

10.3.3.1 Chemistry and 2D Bonding

In the germanium arsenic monolayer GeAs, the Ge atom has the valence electrons ($4s^2p_x^2$), and the As atom ($4s^2p_x^2p_y^1$). The covalent bonds formed by these valence electrons are slightly ionic owing to the small difference in the electronegativities of Ge (2.01) and As (2.18). One of the well-studied phases is the hexagonal structure of α-GeAs monolayers, where each Ge atom binds four atoms and each As atom forms three bonds with Ge atoms. The difference in the electronegativities causes a charge transfer of $\delta q = 1.39$ e from the inner Ge atoms to the negatively charged outer As atoms of the four-layer configuration, inducing a low fractional ionic character of 0.7% [1]. The cohesive energy of hexagonal α-GeAs monolayers of $E_c = -3.36$ eV/atom indicates a further decrease in binding energy due to the increase in atomic size from the P atom to the As atom.

In addition to the high-symmetry α-GeAs phase, the low-symmetry monoclinic structure of GeP monolayers with C2/m symmetry is predicted to form stable lattices, with hexagons and pentagons arranged in a large unit cell, as described for GeP [51]. A Bader analysis predicts a charge accumulation on the As atoms, which is partly due to a charge transfer of $\delta q = 0.30$ e from each Ge atom to each As atom. The cohesive energy calculated for the monoclinic structure is $E_c = -3.62$ eV/atom, which suggests an improved energetic stability of the low-symmetry GeAs monolayer structure as mentioned before for GeP [51].

10.3.3.2 2D Structures

The stable α phase of GeAs monolayers has the lattice parameters a = 0.382 nm, $d_{Ge–Ge}$ = 0.250 nm, $d_{Ge–As}$ = 0.249 nm, d_{As-As} = 0.480 nm, θ_{GeGeAs} = 117.54° and θ_{AsGeAs} = 100.33° [1]. Adding the van der Waals diameter of As atoms of d_w = 0.410 nm and the distance between As atoms h_{AsAs}, yields an estimate of the monolayer thickness of d_{w+h} = 0.890 nm. The calculated band gaps of the indirect semiconductor are E_g = 1.20 eV (PBE), 1.08 eV (PBE + SOC), and 1.81 eV (HSE06) [1].

The lattice parameters presented for the monoclinic phase of GeAs are a = 0.380 nm, b = 2.213 nm, $d_{Ge–Ge}$ = 0.248 nm, $d_{Ge–As}$ = 0.247–0.250 nm, and h_{AsAs} = 0.510 nm [35]. Remarkably, the lattice parameters are nearly identical to those of the hexagonal α phase. The reason for this agreement is the similar bonding situation. The exfoliation energy of SiP is –0.26 J/m^2 and that of GeAs is –0.37 J/m^2, which should allow exfoliation of these monolayers from the layered bulk material. Due to the structural and energetic similarity of the polymorphs, bulk exfoliation and synthesis via deposition on substrate surfaces may lead to mixed phases.

A recent study of the lattice parameters of the monoclinic phase of GeAs monolayers using first-principles calculations, yielded a = 0.378 nm, b = 2.196 nm, $d_{Ge1-Ge1}$ = 0.247 nm, $d_{Ge2-Ge2}$ = 0.248 nm,, $d_{Ge–As}$ = 0.248 nm, and h_{AsAs} = 0.511 nm in good agreement with previous results [51]. The authors consider GeAs as a quasi-direct semiconductor with an energy difference of ~ 1 meV between the direct and indirect band gaps of 1.45 eV (PBE) and 2.11 eV (HSE).

10.3.3.3 Mechanics

Strongly anisotropic mechanical behavior was found for the monolayer of the low-symmetry hexagon-pentagon phase of GeAs by the DFT-GGA-PBE approach [36]. The values of E_{ac} = 60 N/m and E_{zz} = 72.2 N/m indicate a strongly varying in-plane stiffness, whereas an isotropic Poisson ratio of ν = 0.16 is found. As expected, nonlinearity increases for the ultimate strengths at σ_{ac} = 5.6 N/m and σ_{zz} = 8.7 N/m and ultimate strains at ε_{ac} = 0.15 and ε_{zz} = 0.31 [36].

The stiffness values are in acceptable agreement with recent calculations, however, a larger variation is observed for the ultimate strength values of the anisotropic GeAs monolayers [51]. In addition to Young's moduli of E_{ac} = 49.6 N/m and E_{zz} = 66.9 N/m and slightly anisotropic Poisson's ratios of ν_{ac} = 0.14 and ν_{zz} = 0.18, the ultimate tensile strength values of σ_{ac} = 3.62 N/m and σ_{zz} = 5.41 N/m and critical tensile strains of ε_{ac} = 0.19 and ε_{zz} = 0.32 indicate strong anisotropy. Furthermore, the work provides the corresponding strongly anisotropic ultimate compressive strengths $\sigma_{c,ac}$ = 4.73 N/m and $\sigma_{c,zz}$ = 8.38 N/m and the critical compressive strains $\varepsilon_{c,ac}$ = – 0.17 and $\varepsilon_{c,zz}$ = – 0.16.

10.3.3.4 Physics and Applications

The anisotropy of the in-plane optical properties of monoclinic GeAs few-layer flakes was characterized by polarization-resolved absorption spectroscopy (400–2000 nm) employing polarization-sensitive photodetectors [59]. The nature of in-plane anisotropy was studied by Raman spectroscopy with cross- and parallel-configuration modes and an angular dependent mode. The flakes exhibit anisotropic absorption with linear dichroism and maximum absorption direction reversal. DFT calculations were employed to analyze the reverse phenomenon and the Schottky barrier that dominates photocurrent generation.

The experimental value of the band gap of liquid-phase exfoliated few-layer sheets of monoclinic GeAs was ~ 2.1 eV in comparison to ~1.5 eV of the bulk [60]. This value should be compared with first-principles calculations (HSE06) of mono-, bi, tri-, and tetra-layer assemblies with direct or quasidirect band gaps of 2.125 eV, 1.339 eV, 1.112 eV, and 1.017 eV, respectively. The 2D carrier transport properties and large photocurrents measured for GeAs nanosheets suggest their potential use in high-performance optoelectronic devices.

First-principles computations elucidated the structural, electronic, and optical properties of isovalent N/P-doped GeAs and $GeAs_2$ monolayers [61]. Interestingly, substitution of N/P on As sites with similar valence electron configuration produces fewer lattice distortions, stabilizes the structure, and increases the ionic character of bonding. Notably, tuning of the bandgap and the carrier effective mass of GeAs and $GeAs_2$ monolayers is possible by N/P doping. Furthermore, the doped monolayers exhibit strong visible light absorption > 10^5 cm^{-1}.

10.3.3.5 Synthesis

Exfoliated nanosheets of monoclinic GeAs, obtained by liquid-phase exfoliation methods from layered bulk materials, had a band gap of ~ 2.1 eV in good agreement with the value calculated for the monolayer [60]. These GeAs nanosheets deposited on n-type silicon nanowire arrays, exhibited a promising photoelectrochemical water splitting effect under visible light irradiation.

10.3.4 GeSb Monolayer

10.3.4.1 Chemistry and 2D Bonding

In the germanium antimony monolayer GeSb, the Ge atom has four valence electrons ($4s^2p_x{}^2$), and the Sb atom five valence electrons ($5s^2p_x{}^2p_y{}^1$). These valence electrons form several stable covalently bonded GeSb lattices. The structure of the hexagonal α-GeSb monolayer follows chemical experience concerning coordination and covalent bonding [1]. Owing to the quite similar electronegativities of Ge (2.01) and Sb (2.05), the bonding has a marginally ionic character, and the charge transferred from Ge atom to Sb atoms is small at $\delta q = 0.40$ e. The cohesive energy of the α phase of GeSb is $E_c = -3.12$ eV/atom, as derived by the DFT-GGA-PBE approach [1]. The low-symmetry monoclinic structure of GeSb also forms stable GeSb monolayers, with hexagons and pentagons arranged in the typical large unit cell [51]. The cohesive energy calculated for this structure is $E_c = -3.40$ eV/atom and thus indicates stronger bonding than in α GeSb with higher symmetry, as described above for GeAs.

The construction of a unit cell with hexagonal symmetry (2H) and space group P6/mmm resulted in a metallic g-GeSb monolayer with a graphene-like configuration [62]. The bond length of $d_{Ge-Sb} = 0.266$ nm calculated for this novel structure agrees well with those presented for the other phases of GeSb. However, the cohesive energy of $E_c = -3.91$ eV/atom indicates much stronger bonding of this monolayer than reported for the other allotropes. The bonding energy increases further by substitutional doping.

10.3.4.2 2D Structures

The stable hexagonal α phase of GeSb monolayers has the lattice parameters $a = 0.412$ nm, $d_{Ge-Ge} = 0.250$ nm, $d_{Ge-Sb} = 0.269$ nm, $h_{SbSb} = 0.501$ nm, $\theta_{GeGeSb} = 117.84°$ and $\theta_{SbGeSb} = 99.95°$ [1]. Adding the van der Waals diameter of Sb atoms of $d_w = 0.440$ nm and distance between Sb atoms h_{SbSb}, yields an estimate of the monolayer thickness of $d_{w+h} = 0.941$ nm. The calculated band gaps of the indirect semiconductor are $E_g = 0.65$ eV (PBE), 0.43 eV (PBE + SOC), and 1.15 eV (HSE06) [1].

For the lattice parameters of the monoclinic phase of GeSb monolayers, first-principles calculations yielded $a = 0.408$ nm, $b = 2.335$ nm, $d_{Ge1\text{-}Ge1} = 0.247$ nm, $d_{Ge2\text{-}Ge2} = 0.248$ nm, $d_{Ge-Sb} = 0.268$ nm, and $h_{SbSb} = 0.527$ nm [51]. An estimate of the layer thickness of $d_{w+\Delta z} = 0.967$ nm is found by adding the Sb–Sb distance of 0.527 nm to the van der Waals diameter of Sb of $d_w = 0.440$ nm. The GeSb monolayer is a direct semiconductor with calculated band gaps of 1.17 eV (PBE) and 1.72 eV (HSE) without SOC [51].

The lattice parameters of the constructed graphene-like g-GeSb monolayer are $a = 0.819$ nm, $d_{Ge-Sb} = 0.266$ nm, and $\theta_{GeSbGe} = 101°$ [62]. Calculations of the electronic band structure and density of states (DOS) reveal the metallic character of the lattice. In addition, this work presents structural data for As- and Bi-doped GeSb monolayers and the structural changes after hydrogen adsorption on the Ge-top, hollow, and Sb-top sides to understand the performance of GeSb monolayers as electrocatalysts for the hydrogen evolution reaction (HER), as discussed in more detail below.

10.3.4.3 Mechanics

The stiffness values of monoclinic GeSb monolayers, as determined by recent calculations, correlate with those of the other group members [51]. In addition to Young's moduli at $E_{ac} = 38.9$ N/m and $E_{zz} = 51.2$ N/m and Poisson's ratios at $\nu_{ac} = 0.15$ and $\nu_{zz} = 0.20$, the study presents the critical tensile strengths $\sigma_{ac} = 3.42$ N/m and $\sigma_{zz} = 4.03$ N/m and critical tensile strains $\varepsilon_{ac} = 0.19$ and $\varepsilon_{zz} = 0.29$. Moreover, the corresponding critical compressive strengths $\sigma_{c,ac} = 3.56$ N/m and $\sigma_{c,zz} = 4.44$ N/m and critical compressive strains $\varepsilon_{c,ac} = -0.14$ and $\varepsilon_{c,zz} = -0.20$ are provided in this report. The stiffness and ultimate strength values are in reasonable agreement with Griffith's rule.

10.3.4.4 Physics and Applications

The effects of the adsorption of As or Bi atoms on the catalytic performance of the Sb sites in the pristine structures of graphene-like monolayers (g-GeSb) and monolayers modified by substitutional doping were studied [62]. Generally, a moderate adsorption energy and Gibbs free energy characterize a good catalyst. The physisorption of hydrogen over the Sb-top site on As-doped GeSb has the lowest Gibbs free energy of 0.31 eV, compared with hydrogen adsorption over the Sb-top site on pristine GeSb monolayers [62]. For hydrogen the characteristic adsorption energy evaluated for the Sb-top side is 0.19 eV. According to a Löwdin charge analysis, charge transfer occurs from the monolayer to the hydrogen atom in all studied adsorption systems, and a large amount of charge accumulates on the hydrogen owing to the significant electronegative nature of the hydrogen atom. Overall, by electron density modulation of g-GeSb, a hydrogen evolution reaction (HER) activity of 72% is achieved under As-doped conditions.

10.3.5 GeBi Monolayer

10.3.5.1 Chemistry and 2D Bonding

In the germanium bismuth monolayer GeBi, the Ge atom has the valence electrons ($4s^2p_x{}^2$), and the Bi atom the valence electrons ($6s^2p_x{}^2p_y{}^1$) in the outer shell. First-principles calculations predict the formation of a stable hexagonal α-GeBi monolayer consisting of four covalently bonded sublayers with the sequence Bi–Ge–Ge–Bi. The small difference in the electronegativity between Ge (2.01) and Bi (1.90, 2.02), which has a large uncertainty, results in an insignificant ionic character of covalent bonding. Theory predicts a charge transfer of $\delta q = 0.44$ e from the Ge atoms to the negatively charged outer Bi sublayers, with a fractional ionic character of approximately zero [1]. The cohesive energy of the α-GeBi monolayer of $E_c = -2.98$ eV/atom designates the weakest bonding within the group of Ge–VA compounds. However, as seen before, other calculations predict substantially stronger bonding with a cohesive energy of – 4.32 eV/atom [46].

According to theory, the monoclinic monolayer of GeBi, with its typical large unit cell, is stable [51]. Negligible charge transfer between Ge and Bi atoms limits electron localization and polarization within the monolayer. The cohesive energy of $E_c = -3.25$ eV/atom is greater than that of the high-symmetry α-GeBi monolayer, a situation previously observed before for GeSb and GeAs [51].

10.3.5.2 2D Structures

The stable hexagonal α-GeBi monolayer has the lattice parameters $a = 0.426$ nm, $d_{Ge–Ge} = 0.249$ nm, $d_{Ge–Bi} = 0.278$ nm, $h_{BiBi} = 0.509$ nm, $\theta_{GeGeBi} = 117.85°$ and $\theta_{BiGeBi} = 99.94°$ [1]. Adding the van der Waals diameter of Bi atoms of $d_w = 0.460$ nm and the distance between Bi atoms h_{BiBi}, yields an estimate of the monolayer thickness of $d_{w+h} = 0.969$ nm. The calculated band gaps of the indirect semiconductor are 0.22 eV (PBE) and 0.67 eV (HSE06), whereas the inclusion of SOC (PBE + SOC) predicts a metallic character [1]. Note that the study with the larger binding energy predicts a smaller lattice constant and somewhat smaller bond lengths of $a = 0.415$ nm, $d_{Ge–Ge} = 0.243$ nm, and $d_{Ge–Bi} = 0.274$ nm, and as bond angles $\theta_{GeGeBi} = 118.86°$ and $\theta_{GeBiGe} = 98.65°$, consistent with the improved energetic stability [46].

The structural parameters of the monoclinic GeBi monolayer with hexagons and pentagons are $a = 0.421$ nm, $b = 2.403$ nm, $d_{Ge1\text{-}Ge1} = 0.247$ nm, $d_{Ge2\text{-}Ge2} = 0.248$ nm, $d_{Ge–Bi} = 0.276$ nm, and as the distance between the Bi atoms of the outer sublayers $h_{BiBi} = 0.533$ nm [51]. This yields a slightly larger thickness of $d_{w+\Delta z} = 0.993$ nm for this low-symmetry monolayer. The calculated direct electronic band gaps are 0.78 eV (PBE), 0.43 eV (PBE-SOC), and 1.25 eV (HSE).

10.3.5.3 Mechanics

The lower binding energy affects the rigidity and strength of GeBi monolayers, which has the lowest mechanical performance of the group. A comprehensive study of the mechanical properties of GeBi monolayers revealed in addition to having anisotropic Young's moduli of $E_{ac} = 33.2$ N/m and $E_{zz} = 43.0$ N/m and Poisson's ratios of $\nu_{ac} = 0.18$ and $\nu_{zz} = 0.23$, the tensile strengths $\sigma_{ac} = 3.21$ N/m and $\sigma_{zz} = 3.67$ N/m and tensile strains $\varepsilon_{ac} = 0.20$ and $\varepsilon_{zz} = 0.28$ in reasonable agreement with Griffith's rule [51]. In addition, the study provides the corresponding compressive strengths $\sigma_{c,ac} = 3.13$ N/m and $\sigma_{c,zz} = 3.48$ N/m, as well as the compressive strains $\varepsilon_{c,ac} = -0.12$ and $\varepsilon_{c,ac} = -0.19$.

10.3.5.4 Physics and Applications

To study the suitability of α-GeBi nanosheets for gas sensing applications, DFT and nonequilibrium Green's function calculations were performed to elucidate the adsorption behavior of the toxic and nontoxic gas molecules NO_2, SO_2, NO, O_2, CO_2, and H_2S [63]. The adsorption sites, binding energies, charge transfer processes, and current–voltage characteristics were elucidated for the adsorption of these molecules on GeBi nanosheets. Importantly, GeBi nanosheets possess superior sensing efficiency that surpasses that of many other 2D materials, such as graphene or bismuthene. While NO_2 and O_2 chemisorb on GeBi nanosheets, SO_2, NO, CO_2, and H_2S are weakly physisorbed with low sensitivity toward CO_2 and H_2S. This work demonstrated that 2D GeBi nanosheets provide a reusable gas sensor that works as a charge acceptor.

10.4 Sn–VA Monolayers (SnN, SnP, SnAs, SnSb, and SnBi)

10.4.1 SnN Monolayer

10.4.1.1 Chemistry, 2D Bonding, Structure, and Applications

In the tin nitrogen monolayer SnN, the Sn atom has the four valence electrons ($5s^2p_x{}^2$), and the N atom the five valence electrons ($1s^2 2s^2 p_x{}^2 p_y{}^1$) in the outer shell. The valence electrons form a hexagonal high-symmetry structure with four sublayers bonded by strong covalent and ionic interactions. As the (only) stable structure, the hexagonal α-SnN monolayer has been fully characterized by theory [1]. Owing to the substantially deviating electronegativities of Sn (1.96) and N (3.04), ionic forces make a major contribution to bonding of the conceivable 2D structures. The large ionicity is due to an enormous charge transfer of $\delta q = 3.52$ e from the inner Sn atoms to the negatively charged outer N sublayers, which is responsible for a fractional ionic character of 25.3% [1]. The currently available considerably deviating cohesive energies of $E_c = -3.78$ eV/atom [1] and -5.36 eV/atom [2] characterize the 2D lattice with the strongest bonding in this group of Sn–VA monolayers.

The stable high-symmetry hexagonal α-SnN monolayer has a lattice constant of $a = 0.342$ nm, a Sn–Sn bond length of $d_{Sn\text{-}Sn} = 0.297$ nm, a Sn–N bond length of $d_{Sn\text{-}N} = 0.211$ nm, a separation of N atoms of $h_{NN} = 0.444$ nm, and the bond angles $\theta_{SnSnN} = 110.30°$ and $\theta_{NSnN} = 108.63°$ [1, 2]. Adding the van der Waals diameter of N atoms $d_w = 0.320$ nm and distance between N atoms h_{NN}, yields an estimate of the monolayer thickness of $d_{w+\Delta z} = 0.764$ nm. The calculated small band gaps of the indirect semiconductor vary between $E_g = 0.12$ eV (PBE), 0.13 eV (PBE + SOC), and 0.89 eV (HSE06) [1].

The elastic constants c_{11} and c_{12} calculated by the DFT-GGA-PBE approach for the α-SnN monolayer yield a relatively small Young's modulus of $E = 67.8$ GPa (32 N/m) and Poisson's ratio of $\nu = 0.33$ [2].

Monolayered hexagonal α-SnN semiconductors exhibit an ultrahigh electron mobility of > 10^4 cm^2 V^{-1} s^{-1} and an abnormally high absorption of out-of-plane polarized light in the visible region, which is of interest in optoelectronics. Furthermore, ferromagnetism can be controlled by electronic doping under experimentally achievable conditions, which may be of interest in electronics and spintronics [64].

A medium exfoliation energy of $E_f = 0.69$ J/m^2 for SnN monolayers that is smaller than the exfoliation energies reported for other atomic-layered pnictides, implies the possibility of cleavage from the bulk phase [64].

10.4.2 SnP Monolayer

10.4.2.1 Chemistry and 2D Bonding

In the tin phosphorus monolayer SnP, the Sn atom has the valence electrons ($5s^2p_x{}^2$), and the P atom ($3s^2p_x{}^2p_y{}^1$) in the outer valence shell. These valence electrons can form various covalently bonded stable monolayers and few-layer structures. Owing to the similar electronegativities of Sn (1.96) and P (2.06), the bonds are slightly ionic. For the hexagonal α-SnP monolayer, a surprisingly large charge transfer of $\delta q = 2.57$ e from the inner Sn atoms to the negatively charged outer P sublayers has been predicted, which is responsible for a surprisingly low fractional ionic character of 1.3% [1]. Compared with α-SnN monolayers, the cohesive energies of α-SnP monolayers of $E_c = -\ 3.28$ eV/atom [1] and – 4.27 eV/atom [2] indicate a reduced bonding strength. Another recent study confirms stronger bonding for the two hexagonal allotropes α SnP and β SnP, predicting cohesive energies of $E_c = -\ 4.32$ eV/atom and – 4.30 eV, respectively [65].

10.4.2.2 2D Structures

The stable hexagonal α-SnP monolayer has the lattice parameters $a = 0.395$ nm, $d_{Sn-Sn} = 0.289$ nm, $d_{Sn-P} = 0.256$ nm, $h_{PP} = 0.522$ nm, $\theta_{SnSnP} = 117.05°$ and $\theta_{PSnP} = 100.94°$ [1, 2, 65]. Adding the mean van der Waals diameter of P atoms of $d_w = 0.390$ nm and the distance between P atoms h_{PP}, yields an estimate of the monolayer thickness of $d_{w+h} = 0.812$ nm. The calculated band gaps of the indirect semiconductor vary between $E_g = 1.29$ eV (PBE), 1.28 eV (PBE-SOC), and 1.91 eV (HSE06) [1].

Due to the similarity of their hexagonal structures, it is not surprising that the structural parameters of the hexagonal β-SnP monolayer are almost identical to those of the α phase. These lattice parameters are $a = 0.391$ nm, $d_{Sn-Sn} = 0.288$ nm, $h_{PP} = 0.524$ nm, and the already mentioned cohesive energy is $E_c = 4.30$ eV [65].

Based on first-principles calculations a low-symmetry structure with space group *C2/m* has been predicted for SnP and SnAs monolayers [66]. The semiconductors with a direct band gap are dynamically and thermally stable. They exhibit a high cohesive energy, excellent stability, good flexibility, and a high carrier mobility of 10^3–10^4 cm^2 V^{-1} s^{-1}.

10.4.2.3 Mechanics

Derivation of the linear mechanical properties of α-SnP monolayers from the published elastic constants $c_{11} = 84.85$ N/m and $c_{12} = 18.51$ N/m yields an elastic modulus of E = 80.8 N/m and a Poisson ratio of ν = 0.218 for the hexagonal α-SnP monolayer [65]. The mechanical stiffness is not in agreement with the lattice parameters E = 38.8 GPa (32 N/m) and ν = 0.23 [2]. A comparable stiffness of E = 83.3 N/m and similar Poisson ratio of ν = 0.229 were obtained from the elastic constants $c_{11} = 87.86$ N/m and $c_{12} = 20.08$ N/m for the hexagonal β phase, consistent with the nearly identical binding energies [65].

10.4.2.4 Physics and Applications

Using DFT, the electronic and spin properties of SnX (X = P, As, Sb, and Bi) monolayers were studied under biaxial strain [65]. For the hexagonal α-SnX and the β-SnX lattices, the band structures with and without SOC were compared to clarify the importance of spin–orbit interactions on the band structure. The band gap reaches a maximum at small tensile and compressive strains and decreases for large tensile and compressive strains. In the presence of strain, SOC modifies the band gap. The structures with α symmetry have mirror symmetry, and those with β symmetry have inversion symmetry. The determination of spin splitting in the conduction and valence bands in the presence of strain for the α structure with mirror symmetry helped to clarify the origin of the large spin splitting effect. Band and spin splitting increase with increasing SOC for the heavy elements.

The low-symmetry phases of SnP and SnAs with *C2/m* symmetry reveal good transparency in the infrared and visible regions, with the primary response located in the ultraviolet region. The monolayers exhibit high absorbance and the optical properties vary with the direction [66]. The suitable band gaps and high carrier mobilities qualify these monolayers for FETs, integrated circuits, blue light-emitting diodes (LEDs), and hole transport layers in solar cells.

10.4.3 SnAs Monolayer

10.4.3.1 Chemistry, 2D Bonding, Structure, Mechanics, and Applications

In the tin arsenic monolayer SnAs, the Sn atom has the four valence electrons ($5s^2p_x{}^2$), and the As atom the five valence electrons ($4s^2p_x{}^2p_y{}^1$). In principle, the valence electrons can form various monolayer and few-layer structures by covalent-ionic bonding. The best studied monolayers are the hexagonal α phase and β phase of SnAs. The small difference in the electronegativities of Sn (1.96) and As (2.18) introduces a moderate polarization of the covalent bonds. Nevertheless, a surprisingly large charge transfer of $\delta q = 2.39$ e from the inner Sn sublayers to the negatively charged outer As sublayers has been reported for the α-SnAs monolayer, inducing a small fractional ionic character of 1.2% [1]. The published cohesive energies of α-SnAs monolayers of $E_c = -3.10$ eV/atom [1] and − 4.00 eV/atom [2] indicate a slightly reduced bonding strength owing partly to the larger atom size of As in comparison to P atoms. Another study gives a cohesive energy of − 4.05 eV/atom that supports a stronger binding energy [65].

The dynamically and energetically stable α phase of SnAs has the lattice parameters a = 0.409 nm, $d_{Sn-Sn} = 0.288$ nm, $d_{Sn-As} = 0.267$ nm, $h_{AsAs} = 0.537$ nm, $\theta_{SnSnAs} = 117.74°$ and $\theta_{AsSnAs} = 100.04°$ [1, 2]. The bond lengths and distance between As atoms are in excellent agreement with recently published structural parameters of the α-SnAs monolayer [65]. Adding the van der Waals diameter of As atoms of $d_w = 0.410$ nm and the distance between As atoms h_{AsAs}, yields an estimate of the monolayer thickness of $d_{w+\Delta z} = 0.947$ nm. The calculated band gaps of the indirect semiconductor vary between $E_g = 1.14$ eV (PBE), 1.05 eV (PBE + SOC), and 1.72 eV (HSE06) [1].

For the hexagonal β phase with inversion symmetry, the lattice parameters are a = 0.399 nm, $d_{Sn-Sn} = 0.287$ nm, and $h_{AsAs} = 0.544$ nm. Adding the van der Waals diameter of As atoms of $d_w = 0.410$ nm and the distance between As atoms, yields a slightly largerestimate of the monolayer thickness of $d_{w+h} = 0.954$ nm for the β phase. The cohesive energy of $E_c = -4.02$ eV/atom presented for β-SnAs is consistent with the bonding behavior estimated by these authors for the α-SnAs monolayer [65]. As described above, SnAs also forms a stable low-symmetry structure consisting of a hexagon surrounded by two pentagons with a large unit cell [66]. The physicochemical properties of monoclinic SnAs follow widely those already discussed for monoclinic SnP monolayers.

Derivation of the linear mechanical properties of α SnAs from the calculated elastic constants $c_{11} = 72.87$ N/m and $c_{12} = 17.20$ N/m, yielded an elastic modulus of E = 68.8 N/m and a Poisson ratio of $\nu = 0.236$ [65]. However, a much smaller stiffness at E = 32.9 GPs (31.2 N/m) and a comparable Poisson ratio at $\nu = 0.25$ have been reported by others [2]. Similar to the situation for β SnP, the stiffness at E = 77.7 N/m and the Poisson ratio at $\nu = 0.244$ calculated for β-SnAs monolayers are larger than those of the α-SnAs monolayer, supporting stronger binding.

An extensive work explored the suitability of XAs (X = Si, Ge, and Sn) monolayers as photocatalysts for water splitting [67]. The exceptionally high anisotropic carrier mobilities support fast migration and separation of the photogenerated electron–hole pairs. Furthermore, heterobilayers of XAs and GaS or GaSe monolayers, possessing type-II alignment, enhance optical absorption and provide a strong driving force for the hydrogen evolution reaction (HER) by optimizing the CBM position. The study unravels the micro-mechanism of the photocatalytic hydrogen production process.

Searching novel photocatalysts for water splitting, theory uncovered penta-XY_2 (X = Si, Ge, and Sn; Y = P, As, and Sb) compounds as suitable monolayers [68]. These semiconductors have band gaps between E_g = 2.07 and 2.51 eV and the positions of the conduction and valence band edges match well with the reaction potentials of H^+/H_2 and O_2/H_2O. The band gaps and band edges of SnP_2, $SnAs_2$, and GeAs can be tuned by tensile and compressive strains to improve the photocatalytic performance.

The structural and electronic properties of SnAs monolayers chemically functionalized by chlorine, bromine, and iodine SnAsX (X = Cl, Br, and I) were investigated by the DFT approach (see Fig. 10.14) [69]. While pristine 2D SnAs is a metal, the functionalized non-centrosymmetric hexagonal SnAsX monolayers with the space group *P3m1* are dynamically and mechanically stable semiconductors with a direct band gap. Especially the SnAsI monolayer exhibits a strong influence of SOC, inducing a huge spin–orbit splitting of 0.56 eV. The application of biaxial compressive strains induces a change of the band gap and a direct-to-indirect band gap transition. The electron and hole mobilities in the x and y direction are almost isotropic with a much larger electron than hole mobility.

Chlorinated SnAs monolayers were employed to construct SnAs/SnAsCl heterostructures for controlling and tuning the electronic properties [70]. Depending on the stacking pattern of the heterostructure either an Ohmic contact or Schottky contact is found. Noteworthy is the ultra-low Schottky barrier at 0.03 eV, and the improved charge transport and device performance of the heterostructure.

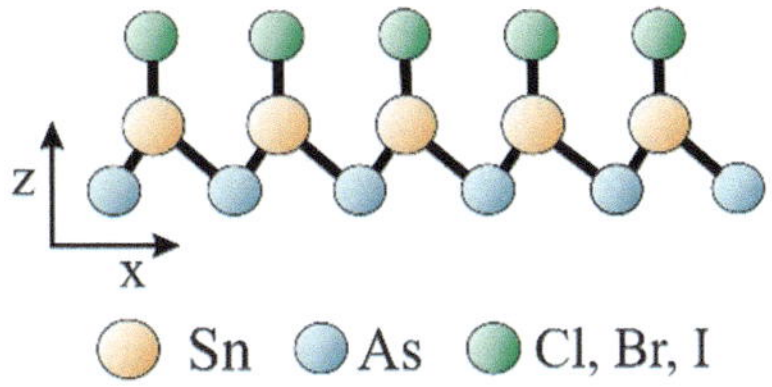

Fig. 10.14 Chemical functionalization of metallic SnAs monolayers by the elements X = Cl, Br, and I to the semiconductors SnAsX

10.4.4 SnSb Monolayer

10.4.4.1 Chemistry, 2D Bonding, Structure, Mechanics, and Applications

In the tin antimonide monolayer SnSb, the Sn atom has four valence electrons ($5s^2p_x^{\,2}$), and the Sb atom has five valence electrons ($5s^2p_x^{\,2}p_y^{\,1}$) in the outer shell. Owing to the small difference in the electronegativities of Sn (1.96) and Sb (2.05), the bonds are only slightly polar. In the stable hexagonal α-SnSb monolayer, a surprisingly large charge transfer of $\delta q = 1.75$ e occurs from the inner Sn atoms to the outer negatively charged Sb atoms, inducing a negligible fractional ionic character of only 0.2% [1]. The previously published cohesive energy of α-SnSb monolayers is $E_c = -2.89$ eV/atom [1]. The resulting binding strength is much smaller than that of the recently proposed cohesive energies of – 3.70 eV/atom [65], and – 3.65 eV/atom [2].

The stable structure of α-SnSb monolayers has the lattice parameters a = 0.439 nm, $d_{Sn-Sn} = 0.287$ nm, $d_{Sn-Sb} = 0.287$ nm, $h_{SbSb} = 0.558$ nm, $\theta_{SnSnSb} = 118.13°$ and $\theta_{SbSnSb} = 99.59°$ [1, 8]. Adding the van der Waals diameter of Sb atoms of $d_w = 0.440$ nm and distance between Sb atoms h_{SbSb}, yields an estimate of the monolayer thickness of $d_{w+h} = 0.998$ nm. The calculated band gaps of the indirect semiconductor vary between $E_g = 0.80$ eV (PBE), 0.54 eV (PBE + SOC), and 1.28 eV (HSE06) [1].

The lattice parameters available for the β phase of the hexagonal SnSb monolayer are a = 0.438 nm, $d_{Sn-Sn} = 0.286$ nm, and $h_{SbSb} = 0.556$ nm. Remarkably, these values are nearly identical to those of the α-SnSb monolayer despite the lower symmetry [65]. The cohesive energy of $E_c = -3.68$ eV/atom supports the stronger cohesion values suggested for α-SnSb monolayers [2].

Derivation of the linear mechanical properties of α-SnSb monolayers from the elastic constants $c_{11} = 57.84$ N/m and $c_{12} = 15.34$ N/m resulted in an isotropic elastic modulus of 53.8 N/m and a Poisson ratio of $\nu = 0.265$ [65]. These values deviate not much from E = 52.2 N/m and $\nu = 0.254$, derived for β-SnSb monolayers [65]. From the elastic constants $c_{11} = 57.88$ N/m, $c_{22} = 58.60$ N/m, and $c_{12} = 13.98$ N/m of α SnSb, the slightly anisotropic elastic moduli $E_{ac} = 53.9$ N/m and $E_{zz} = 54.6$ N/m and the Poisson ratios $\nu_{ac} = 0.238$ and $\nu_{zz} = 0.242$ were extracted, which are in reasonable agreement with the isotropic values given above for α SnSb [23]

Tuning of electronic properties, such as the band gap, deformation potential, and mobility in the zigzag and in the armchair directions of α-SnSb monolayers was studied by employing biaxial strain engineering [23]. For small tensile strains, the band gap reaches a maximum, while for lower compressive strains and larger tensile strains the band gap vanishes, and a semiconductor-to-metal transition is induced. Carbon-, silicon-, germanium-, and tin-antimonide monolayers exhibit this behavior. Their effective masses and the deformation potential affect the charge mobilities. Moreover, the computations provided the work function and electron affinity of the monolayers.

The epitaxial growth of atomically thin SnSb sheets with four-layer Sb–Sn–Sn–Sb configuration (H phase) was achieved on a Cu_2Sb surface alloy under ultrahigh-vacuum (UHV) conditions [71]. The structural analysis by low-temperature scanning tunneling microscopy (STM) and first-principles calculations identified the H phase, displaying a commensurate lattice growth mode on the $Cu_2Sb(111)$ surface, however, with weak interfacial interaction preserving the intrinsic properties.

10.4.5 SnBi Monolayer

10.4.5.1 Chemistry, 2D Bonding, Structure, Mechanics, and Applications

In the tin bismuth monolayer SnBi, the Sn atom has four valence electrons ($5s^2p_x^2$), and the Bi atom has five valence electrons ($6s^2p_x^2p_y^1$) in the outer shell. Owing to the small difference in the electronegativities of Sn (1.72, 1.96) and Bi (1.90), the Sn–Bi bonds are only slightly polar. In the stable hexagonal α-SnBi monolayer, the bonds are polarized by a considerable charge transfer of $\delta q = 1.48$ e from the inner Sn atoms to the negatively charged outer Bi sublayers. However, only a negligible fractional ionic character of 0.1% is induced [1]. The cohesive energy of α-SnBi monolayers of $E_c = -2.78$ eV/atom signifies the lowest binding strength obtained for Sn–VA compounds [1]. However, other studies predict much stronger binding with cohesive energies at – 3.49 eV/atom [2], – 3.54 eV/atom [65], and even – 4.06 eV/atom [46].

The hexagonal α-SnBi monolayer has the lattice parameters $a = 0.451$ nm, $d_{Sn-Sn} = 0.286$ nm, $d_{Sn-Bi} = 0.296$ nm, $h_{BiBi} = 0.566$ nm, $\theta_{Sn-Sn-Bi} = 118.26°$ and $\theta_{BiSnBi} = 99.43°$ [1, 8]. Adding the van der Waals diameter of Bi atoms of $d_w = 0.460$ nm and the distance between Bi atoms h_{BiBi}, yields an estimate of the monolayer thickness of $d_{w+\Delta z} = 1.026$ nm. The calculated band gaps of the indirect semiconductor vary between $E_g = 0.48$ eV (PBE) and 0.92 eV (HSE06), whereas the inclusion of SOC (PBE + SOC) predicts a metallic character of the monolayer [1].

The α-SnBi lattice parameters do not deviate much from those given for the hexagonal β phase of SnBi monolayers that are $a = 0.443$ nm, $d_{Sn-Sn} = 0.285$ nm, and $h_{BiBi} = 0.571$ nm [65]. This study predicts a band gap of 0.40 eV (PBE), which is consistent with the results given above and metallic behavior for PBE + SOC analysis (– 0.36 eV) [65]. For the lattice parameters of α-SnBi monolayers, this work reports nearly the same lattice parameters as those reported for the β phase.

Derivation of the linear mechanical properties of α-SnBi monolayers, using the elastic constants $c_{11} = 52.56$ N/m and $c_{12} = 14.07$ N/m, yields an elastic modulus of 48.8 N/m and a Poisson ratio of 0.268 [65]. The stiffness of β-SnBi monolayers at $E = 48.9$ N/m and Poisson ratio at $\nu = 0.277$ are only marginally larger than those of the α phase, consistent with the similarity of the structures and binding energies of the two hexagonal phases.

For the whole group of SnX (X = P, As, Sb, and Bi) monolayers, the electronic and spin properties were studied via DFT calculations by considering the mirror symmetry (α phase) and inversion symmetry (β phase) [65]. By comparing the band structures with and without spin–orbit interaction, it was demonstrated that the decrease of the band gap is more effective for heavy compounds such as SnBi. The calculations revealed the location of the valleys in the conduction and valence bands for different strains. For mirror symmetry, spin splitting in the conduction and valence bands was determined in the presence of strain.

A detailed theoretical study investigated the optical and solar energy conversion capabilities of group IVA–VA monolayers [2]. The study evaluated as solar parameters the optical band gap, short-circuit current density, open-circuit voltage, and fill factor of 17 stable IVA–VA compounds. The tabulated values revealed an outstanding solar power conversion efficiency of $\eta = 27.25\%$ for SnBi monolayers that comes from an unusually high short-circuit current density and a low open-circuit voltage. This seems to be one of the highest conversion efficiencies predicted for 2D compounds [2].

10.5 Pb–VA Monolayers (PbN, PbP, PbAs, PbSb, and PbBi)

10.5.1 PbN Monolayer

10.5.1.1 Chemistry, 2D Bonding, Structure, and Mechanics

In the plumbum nitrogen monolayer PbN, the Pb atom has four valence electrons ($6s^2p_x^2$) and the N atom five valence electrons ($1s^22s^2p_x^2p_y^1$) in the outer shell. A detailed study of 2D bonding and structure is available only for the hexagonal α phase with the four-layer configuration N–Pb–Pb–N. Owing to the strongly deviating electronegativities of Pb (1.80) and N (3.04), bonding has a significant ionic contribution with a large charge transfer of $\delta q = 2.90$ e from the inner Pb atoms to the negatively charged outer N sublayers, generating a significant fractional ionic character of 11.8% [1]. The available strongly deviating cohesive energies of $E_c = -3.04$ eV/atom [1] and -4.62 eV/atom [2] denote the lowest binding energies in this group of Pb–VA monolayers [1]. The α-PbN monolayer has the lattice parameters $a = 0.363$ nm, $d_{Pb-Pb} = 0.317$ nm, $d_{Pb-N} = 0.224$ nm $h_{NN} = 0.478$ nm, $\theta_{PbPbN} = 110.92°$ and $\theta_{NPbN} = 107.98°$ [1]. Adding the van der Waals diameter of N atoms of $d_w = 0.320$ nm and the distance between N atoms h_{NN}, yields an estimate of the monolayer thickness of $d_{w+h} = 0.798$ nm. The PbN monolayer has metallic character. The only presently available mechanical properties are the elastic modulus of 67.4 N/m and Poisson ratio of 0.15 [72].

10.5.2 PbP Monolayer

10.5.2.1 Chemistry, 2D Structure, and Applications

In the plumbum phosphorus monolayer PbP, the Pb atom has four valence electrons ($6s^2p_x^2$), and the P atom five valence electrons ($3s^2p_x^2p_y^1$) in the valence shell. The only structure studied in detail is the hexagonal α phase of the PbP monolayer with four sublayers. Owing to the relatively small difference between the electronegativities of Pb (1.80) and P (2.19) the bonds are slightly polar, with a surprisingly large charge transfer of $\delta q = 1.81$ e from the inner Pb atoms to the outer P atoms, however, a rather small fractional ionic character of 0.5% [1]. The currently available strongly deviating cohesive energies of the α-PbP monolayer are $E_c = -2.83$ eV/atom [1] and -3.82 eV/atom [2].

The energetically and dynamically stable α phase of PbP has the lattice parameters $a = 0.412$ nm, $d_{Pb–Pb} = 0.306$ nm, $d_{Pb–P} = 0.267$ nm, $h_{PP} = 0.549$ nm, $\theta_{PbPbP} = 117.08°$ and $\theta_{PPbP} = 100.91°$ [1]. Adding the van der Waals diameter of the P atoms of $d_w = 0.390$ nm and the distance between P atoms, yields an estimate of the monolayer thickness of $d_{w+h} = 0.939$ nm. The estimated band gaps of the indirect semiconductor vary between $E_g = 0.40$ eV (PBE), 0.38 eV (PBE + SOC), and 0.76 eV (HSE06).

DFT and DFPT calculations were used to study the thermoelectric performance of α-, β-, and γ-PbP (mixed four-layer structure with symmetry $P3m1$) monolayers under consideration of full mode-, energy-, and momentum-resolved electron–phonon interactions (EPI) [73]. The study revealed that intervalley scattering makes nontrivial contributions to the total carrier relaxation time and that four-phonon scattering dominates the phonon relaxation processes, because three-phonon scattering is suppressed by the large acoustic-optical phonon band gap of the three phases. The thermoelectric performance of the α-, β-, and γ-PbP monolayers has been characterized by the *ZT* values 0.90, 0.24, and 1.25, respectively.

10.5.3 PbAs Monolayer

10.5.3.1 Chemistry, 2D Bonding, and Structure

In the plumbum arsenic monolayer PbAs, the Pb atom has four valence electrons ($6s^2p_x^2$), and the As atom has five valence electrons ($4s^2p_x^2p_y^1$) in the outer shell. The hexagonal α-PbAs monolayer has slightly polar bonds due to the deviating electronegativities of Pb (1.80) and As (2.18). This difference induces a charge transfer of $\delta q = 1.63$ e that occurs from the inner Pb atoms to the outer As atoms, leading to a fractional ionic character of 0.6% [1]. An increase in the atom size from the P to the As atom increases the calculated cohesive energy of α-PbAs monolayers to $E_c = -2.72$ eV/atom [1] and -3.60 eV/atom [2]. The stable α phase of PbAs has the lattice parameters $a = 0.425$ nm, $d_{Pb–Pb} = 0.305$ nm, $d_{Pb–As} = 0.277$ nm, $h_{AsAs} =$

0.562 nm, $\theta_{PbPbAs} = 117.66°$ and $\theta_{AsPbAs} = 100.18°$ [1]. Adding the van der Waals diameter of the As atom of $d_w = 0.410$ nm and the distance between As atoms h_{AsAs}, yields an estimate of the monolayer thickness of $d_{w+h} = 0.972$ nm. The calculated band gaps of the indirect semiconductor vary between $E_g = 0.36$ eV (PBE), 0.25 eV (PBE + SOC), and 0.67 eV (HSE06) [1].

10.5.4 PbSb Monolayer

10.5.4.1 Chemistry, 2D Bonding, and Structure

In the plumbum antimony monolayer PbSb, the Pb atom has four valence electrons ($6s^2p_x{}^2$), and the Sb atom has five valence electrons ($5s^2p_x{}^2p_y{}^1$) in the outer shell. The bonds of the stable hexagonal α-PbSb monolayer are polar due to the difference in electronegativities of 1.80 (Pb) and 2.05 (Sb). This difference induces a charge transfer of $\delta q = 0.93$ e from the inner Pb atoms to the outer Sb sublayers, generating a fractional ionic character of 1.9% [1]. Compared with the binding energy of the lighter compounds in this group, the cohesive energy of the α-PbSb monolayer is further increases to $E_c = -2.58$ eV/atom [1] and − 3.33 eV/atom [2]. The lattice parameters of α-PbSb monolayers are a = 0.453 nm, $d_{Pb–Pb} = 0.303$ nm,, $d_{Pb–Sb} =$ 0.296 nm, $h_{SbSb} = 0.582$ nm, $\theta_{PbPbSb} = 118.01°$ and $\theta_{SbPbSb} = 99.75°$ [1]. Adding the van der Waals diameter of the Sb atoms of $d_w = 0.440$ nm and distance between Sb atoms h_{SbSb}, yields an estimate of the monolayer thickness of $d_{w+h} = 1.022$ nm. The calculated band gaps of the indirect semiconductor vary between $E_g = 0.28$ eV (PBE) and 0.51 eV (HSE06), whereas the inclusion of SOC (PBE + SOC) predicts a metallic α phase of PbSb monolayers.

An effective approach to achieve giant and ideal spin–orbit splitting (Rashba-type splitting occurring due to broken inversion symmetry) is to combine ordered surface alloying and interface engineering [74]. The suggested strategy, illustrated by first-principles calculations, involves a buckled hexagonal PbBi (SbBi) monolayer supported on an $Al_2O_3(0001)$ surface. The calculations predict ideal Rashba-type states in the interface structures with giant spin–orbit splitting characterized by energy offsets over 600 meV and momentum offsets over 0.3 Å^{-1}. Ideally, this is located inside the band gap of $Al_2O_3(0001)$. The authors suggest this strategy as an effective way of tuning spin–orbit splitting.

10.5.5 PbBi Monolayer

10.5.5.1 Chemistry, 2D Bonding, Structure, and Mechanics

In the plumbum bismuth monolayer PbBi, the Pb atom has four valence electrons ($6s^2p_x{}^2$), and the Bi atom has five valence electrons ($6s^2p_x{}^2p_y{}^1$) in the outer shell. The stable covalently bonded hexagonal α-PbBi monolayer has a weak ionic character owing to the minor difference in the electronegativities of Pb (1.80) and Bi (1.90). A modest charge transfer of $\delta q = 0.91$ e from the inner Pb atoms to the outer Bi sublayers induces ionicity with a fractional ionic character of 2.4% [1]. The calculated deviating cohesive energies of α-PbBi monolayers at $E_c = -2.50$ eV/atom [1] and −3.20 eV/atom [2] are the smallest binding energies obtained for this group of Pb–VA compounds.

The stable α-PbBi phase has the lattice parameters a = 0.463 nm, $d_{Pb–Pb} = 0.302$ nm, $d_{Pb–Bi} = 0.303$ nm, $h_{BiBi} = 0.588$ nm, $\theta_{PbPbBi} = 118.10°$ and $\theta_{BiPbBi} = 99{,}62°$ [1]. Adding the van der Waals diameter of the Bi atoms of $d_w = 0.460$ nm and the distance between Bi atoms h_{BiBi}, yields an estimate of the monolayer thickness of $d_{w+h} = 1.048$ nm. The small, estimated band gaps of the indirect semiconductor vary between $E_g = 0.06$ eV (PBE), 0.35 eV (HSE06), and 0.01 eV (PBE + SOC) [1].

An experimental and theoretical study revealed the atomic, electronic, and spin structures of single-layer PbBi grown on a silicon surface [95]. The adsorption of a half-monolayer of Bi and a half-monolayer of Pb onto a Si(111) surface induced the formation of a layer with rhombus-like motifs arranged in diverse periodical structures. Specifically, in the (2 ×2) PbBi phase, the motifs were oriented along a single direction, forming a one-dimensional metallic spin-polarized electron channel, which may be of interest in future spintronic-based applications.

10.6 Summary of Group VIA–VA Monolayers

10.6.1 Binding–Structure Relationships:

Figure 10.15 shows the cohesive energy versus bond length of the well-studied α phase of group IVA–VA monolayers. Note that the inner IVA–IVA bond lengths of the four-layer structures do not vary much for the same the type of pnictogen atom. The figure shows the dependence of the cohesive energies versus bond lengths for five groups obtained by combining each group-IVA element with all group-VA elements. The behavior of these groups follows the typical inverse relationship and allows the identification of strongly deviating results, such as the much lower cohesive energies mentioned above. The values between $E_c = -6.0$ eV/atom and −2.5 eV/atom are in the normal range expected for these compounds.

The plot documents the inverse behavior of the dominating covalent binding energy controlled by orbital overlap and the strongly varying ionic contribution as a function of bond length, as predicted by tight-binding theory. The ionic contributions to covalent bonding are small in most cases and decrease drastically going down the column of pnictogens in each group. Furthermore, the behavior of the diverse groups tends to coincide in a single dependence. Larger deviations from perfect inverse behavior can be seen, especially for nitrogen compounds with stronger bonding. This effect partly can be attributed to the significant ionicity of bonding in

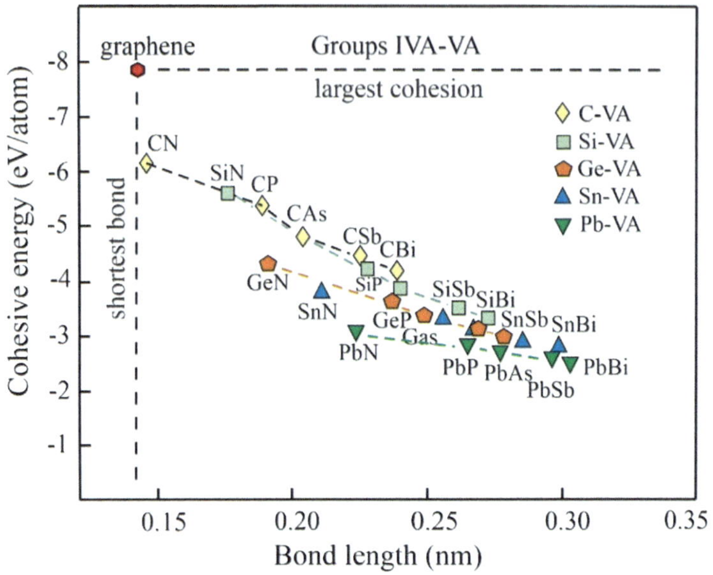

Fig. 10.15 Cohesion energies of the binary group IVA–VA monolayers as a function of bond length

these compounds, which substantially reduces the covalent bonding strength in two dimensions. Nevertheless, covalent bonding is strong due to the small size of the atoms.

10.6.2 Mechanics–Structure Relationships

Figure 10.16 shows the dependence of the calculated Young's moduli and the ultimate strengths along the stronger zigzag directions as a function of the bond lengths for various anisotropic group IVA–VA monolayers. Note that the plot presents results obtained for structures with different symmetries. The extremely sensitive dependence of the mechanical properties on the bond length and lattice direction introduces a separation of the stiffnesses of the individual groups. The monolayer stiffness changes by a factor of approximately five for a change in the bond length of less than a factor of two. The stiffness–bond length plot reveals an enormous change in the mechanical stiffness, especially for C- and Si-pnictogen monolayers.

The few calculations of ultimate strengths presently available follow only roughly Griffith's rule. The strength values are greater than one-tenth of Young's modulus for CP, SiAs, SiSb, GeN, GeP, and GeAs and lower for GeSb and GeBi. The plot demonstrates low stiffnesses and ultimate strengths, especially for monolayers with heavy elements such as Sn–VA and Pb–VA compounds. Note that a low mechanical

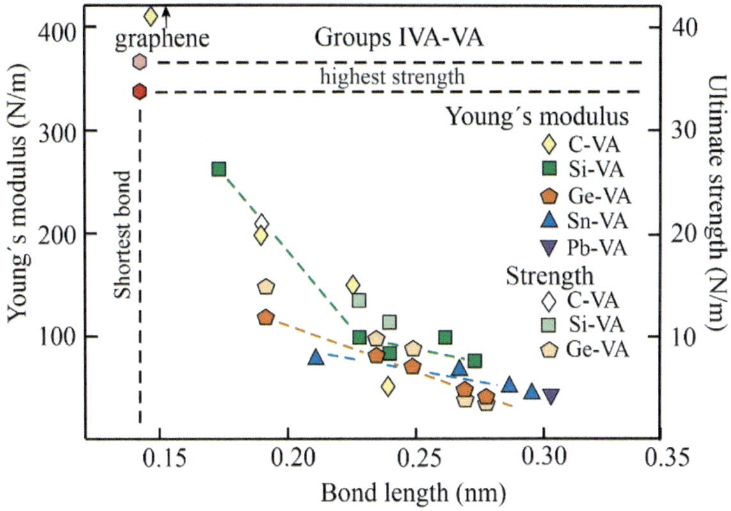

Fig. 10.16 Young's moduli and ultimate strengths of binary group IVA–VA monolayers versus bond lengths

performance resulting from weak bonding favors structural defect formation and deterioration by impurities owing to an increase in chemical reactivity.

References

1. B. Özdamar, G. Özbal, M.N. Çinar, K. Sevim, G. Kurt et al., Phys. Rev. B **98**, 045431 (2018)
2. A.K. Bhojani, H.L. Kakdada, D.K. Singh, J. Appl. Phys. **135**, 095106 (2024)
3. S. Zhang, J. Zhou, Q. Wang, P. Jena, J. Phys. Chem. C **120**, 3993 (2016)
4. H. Piao, G. Choi, X. Jin, S.-J. Hwang, Y.J. Song et al., Nano-Micro Lett. **14**, 55 (2022)
5. L. Tan, C. Nie, Z. Ao, H. Sun, T. An et al., J. Mater. Chem. A **9**, 17 (2021)
6. L. Liu, C. Yan, L. Gao, W. Shangguan, J. Dai et al., Mater. Res. Express **8**, 125010 (2021)
7. G. Wang, R. Pandey, S.P. Karna, Nanoscale **8**, 8819 (2016)
8. H. Cheng, J.-C. Zheng, Front. Phys. **16**, 43505 (2021)
9. H. Zhao, H. Yu, X. Quan, S. Chen, Y. Zhang et al., Appl. Catal. B Environ. **152–153**, 46 (2014)
10. J. Guan, D. Liu, Z. Zhu, D. Tománek, Nano Lett. **16**, 3247 (2016)
11. A.A. Kistanov, V.R. Nikitenko, O.V. Prezhdo, J. Phys. Chem. Lett. **12**, 620 (2021)
12. V. Sorkin, Y.W. Zhang, Nanotechnology **29**, 435707 (2018)
13. D. Singh, V. Shukla, P.K. Panda, Y.K. Mishra, H.-G. Rubahn et al., New J. Chem. **44**, 3777 (2020)
14. A.A. Kistanov, E.A. Korznikova, M. Huttula, W. Cao, Phys. Chem. Chem. Phys. **22**, 11307 (2020)
15. W.C. Tan, Y. Cai, R.J. Ng, L. Huang, X. Feng et al., Adv. Mater. **29**, 1700503 (2017)
16. L. Cao, Y.S. Ang, Q. Wu, L.K. Ang, Phys. Rev. B **101**, 035422 (2020)
17. W. Zhang, J. Yin, P. Zhang, X. Tang, Y. Ding, J. Mater. Chem. A **6**, 12029 (2018)
18. D. Chen, Z. Chen, X. Zhang, Z. Lu, S. Xiao et al., J. Energy Chem. **52**, 155 (2021)
19. M. Naseri, Appl. Surf. Sci. **423**, 566 (2017)
20. C. Kamal, Phys. Rev. B **100**, 205404 (2019)

21. P.-F. Liu, T. Bo, Z. Liu, O. Eriksson, F. Wang et al., J. Mater. Chem. C **6**, 12689 (2018)
22. H.G. Abbas, M.U. Rehman, Appl. Surf. Sci. **562**, 150109 (2021)
23. A. Rajabi-Maram, N. Hasani, S.B. Touski, Physica E **138**, 115065 (2022)
24. M.M. Abutalib, Chem. Phys. Lett. **708**, 188 (2018)
25. J.W. Jan, S.Y. Chen, Mater. Today Commun. **22**, 100817 (2020)
26. S.-D. Guo, Y.-T. Zhu, J. Phys. Condens. Matter **34**, 235501 (2022)
27. M. Xi, C. He, H. Yang, X. Fu, L. Fu et al., Chin. Chem. Lett. **33**, 2595 (2022)
28. H. Liu, G. Qin, Y. Lin, M. Hu, Nano Lett. **16**, 3831 (2016)
29. S. Zhang, J. Zhou, Q. Wang, X. Chen, Y. Kawazone, et al., Proc. Natl. Acad. Sci. USA **112**, 2372 (2015)
30. N. Ghobadi, S.B. Touski, J. Phys. Condens. Matter **33**, 285502 (2021)
31. R.N. Somaiya, Y.A. Sonvane, S.K. Gupta, Phys. Chem. Chem. Phys. **22**, 3990 (2020)
32. J.-H. Lin, H. Zhang, X.-L. Cheng, Y. Miyamoto, Phys. Rev. B **96**, 035438 (2017)
33. M. Ashton, S.B. Sinnott, R.G. Hennig, Appl. Phys. Lett. **109**, 192103 (2016)
34. S. Zhang, S. Guo, Y. Huang, Z. Zhu, B. Cai, et al., 2D Mater. **4**, 015030 (2017)
35. A.-Q. Cheng, Z. He, J. Zhao, H. Zeng, R.-S. Chen, A.C.S. Appl, Mater. Interfaces **10**, 5133 (2018)
36. B. Mortazavi, T. Rabczuk, Physica E **103**, 273 (2018)
37. S. Kansara, P.D. Bhuyan, Y. Sonvane, S.K. Gupta, J. Mater. Sci. **54**, 11878 (2019)
38. B. Mortazavi, M. Shahrokhi, G. Cuniberti, X. Zhuang, Coatings **9**, 522 (2019)
39. E. Jiang, X. Zhu, T. Quyang, C. Tang, J. Li et al., J. Appl. Phys. **126**, 185106 (2019)
40. T. Yu, H. Qiu, H. Liu, Z. Hu, Y. Wu, Inorg. Chem. Front. **11**, 2136 (2024)
41. L. Zhou, Y. Guo, J. Zhao, Physica E **95**, 149 (2018)
42. S. Bai, C.-Y. Niu, W. Yu, Z. Zhu, X. Cai et al., Nanoscale Res. Lett. **13**, 404 (2018)
43. X. Wang, G. Deng, W. Yu, B. Wang, X. Li et al., Europ. Phys. J. Plus **134**, 287 (2019)
44. H. Morshedi, M. Naseri, M.R. Hantehzadeh, S.M. Elahi, J. Electron. Mater. **47**, 2290 (2018)
45. H.H. Huang, X. Fan, D.J. Singh, W.T. Zheng, J. Mater. Chem. C **7**, 10652 (2019)
46. A. Bafekry, M. Yagmurcukardes, B. Akgenc, M. Ghergherehchi, Phys. Chem. Chem. Phys. **23**, 12471 (2021)
47. A. Bafekry, F. Shojaei, M.M. Obeid, M. Ghergherehchi, C. Nguyen et al., RSC Adv. **10**, 31894 (2020)
48. M. Abboud, D.H. Ozbey, E. Durgun, Appl. Surf. Sci. **567**, 150793 (2021)
49. K. Zhang, N. Li, RSC Adv. **10**, 14225 (2020)
50. X. Gao, Y. Shen, Y. Ma, S. Wu, Z. Zhou, Inorg. Chem. **58**, 12053 (2019)
51. M. Aboud, D.H. Ozbey, M.E. Kilic, E. Durgun, J. Phys. D Appl. Phys. **55**, 185302 (2022)
52. J. Sun, J. Leng, Phys. Lett. A **383**, 125856 (2019)
53. L. Li, W. Wang, P. Gong, X. Zhu, B. Deng et al., Adv. Mater. **30**, 1706771 (2018)
54. E. Jiang, X. Zhu, T. Ouyang, C. Tang, J. Li et al., J. Appl. Phys. **126**, 185106 (2019)
55. C. Niu, T. Lan, D. Wang, J. Pan, J. Chu et al., Appl. Surf. Sci. **520**, 146257 (2020)
56. F. Shojaei, H.S. Kang, Phys. Chem. Chem. Phys. **18**, 32458 (2016)
57. F. Shojaei, H.S. Kang, J. Phys. Chem. C **120**, 23842 (2016)
58. N. Jiao, P. Zhou, C. He, J. He, X. Liu et al., Phys. Stat. Solid. **13**, 1900470 (2019)
59. Z. Zhou, M. Long, L. Pan, X. Wang, M. Zhong et al., ACS Nano **12**, 12416 (2018)
60. C.S. Jung, D. Kim, S. Cha, Y. Myung, F. Shojaei et al., J. Mater. Chem. A **6**, 9089 (2018)
61. L. Feng, J. Lu, X. Zhang, G. Xiang, Physica B **614**, 413033 (2021)
62. M.H. Dalsaniya, T.K. Gajaria, N.N. Som, P.K. Jha, Phys. Chem. Chem. Phys. **22**, 19823 (2020)
63. V. Kumar, K. Rajput, D.R. Roy, Appl. Surf. Sci. **606**, 154741 (2022)
64. S. Zhang, X. Zou, J. Mater. Chem. C **8**, 6396 (2020)
65. N. Hasani, A. Rajabi-Maram, S.B. Touski, J. Phys. Chem. Solids **174**, 111131 (2022)
66. Q.-Y. Chen, F.-J. Huang, J.-Q. Ruan, T. Ma, K. Xiond et al., Phys. Rev. Mater. **6**, 104005 (2022)
67. R. Meng, X. Sun, D. Yang, J. Bao, X. Chen, Appl. Mater. Today **13**, 276 (2018)
68. L. Liu, X. Kang, S. Gao, X. Duan, Nanoscale **15**, 9123 (2023)
69. V.T.T. Vi, C.Q. Nguyen, B.D. Hoi, H.V. Phuc, C.V. Nguyen et al., J. Phys. D Appl. Phys. **55**, 505302 (2022)

70. P.H. Nha, C.Q. Nguyen, C.V. Nguyen, Physica B: Condens. Matter **671**, 415392 (2023)
71. H. Li, D. Zhou, Q. He, N. Si, B. Xin et al., ACS Nano **15**, 16335 (2021)
72. Z. Cheng, Y. Wang, R. Zheng, W. Mu, Photcatal. Photochem. **12** (2024). https://doi.org/10.3389/fchem.2024.1382850
73. A. Wu, Y. Zhang, Y. Xia, L. Peng, H. Zhu, arX: 2206.02188v1 (2022)
74. M. Chen, F. Liu, Nat. Sci. Rev. **8**, nwaa341 (2021)
75. A.N. Mihalyuk, L.V. Bondarenko, A.Y. Tupchaya, T.V. Utas, Y.E. Vekovshinin et al., Phys. Rev. B **104**, 125413 (2021)

Chapter 11
Elemental Group–VIA Monolayers

Abstract This chapter covers the elemental monolayers of Se, Te, and Po, which belong to the main group-VIA elements and evolve from the insulators O and S to the metal Po. The most stable bulk structures of the chalcogens selenium and tellurium consist of quasi 1D helical chains with only two chemical bonds per atom. Consistent with the increasing metallic character in this group, bulk polonium shows enhanced metallic behavior with a simple cubic structure. The inherent multivalency of these elements with variable oxidation states (+ 6 and − 2) extends the covalent bonding types of Se and Te from low-dimensional chains to the formation of various metastable tetragonal, rhombic, and monoclinic 2D structures, usually derived by global structural searching combined with first-principles calculations. Note that tellurene is a 2D material first proposed by computational simulations. The most stable structure of poloniumene seems to be a trigonal three-layer assembly. In general, 2D bonding is weak in this group, and conflicting discussions concerning the existence and the thermodynamic, dynamic, and thermal stability of some postulated phases are going on. Liquid-phase exfoliation and physical and chemical vapor deposition provide few-layer sheets of selenene and tellurene with the trigonal bulk structure, whereas molecular beam epitaxy delivers monolayers of helical chain α tellurene and tetragonal β tellurene. The relevant features of multifunctional chalcogen monolayers and few-layer assemblies are tunable band gaps, high carrier mobilities, low thermal conductivity, strong angle-dependent response, and ambient stability. Potential applications of 2D chalcogens cover a wide range of possibilities, including batteries, photodetectors, modulators, as well as photothermal and medical applications. Both selenene and tellurene are suitable materials for application in field-effect transistors and in ultrascaled complimentary metal–oxide–semiconductor circuits. Furthermore, some monolayers are 2D topological insulators with gapped interior states and gapless metallic edge states, which are of interest in spintronics, optoelectronics, and thermoelectrics.

P. Hess, *Bonding, Structure, and Performance of Two-Dimensional Materials*, NanoScience and Technology, https://doi.org/10.1007/978-3-031-74161-6_11

11.1 Selenene Monolayers

11.1.1 Chemistry and 2D Bonding

The elements Se, Te, and Po of the main group–VIA of the periodic table, which form interesting elemental monolayers are displayed in Fig. 11.1.

The most stable bulk structure of the chalcogens selenium and tellurium consist of quasi 1D helical chains of covalently bonded atoms with strong directional bonding held together by weaker van der Waals forces as well as stronger covalent-like forces in an ordered hexagonal array of these chains. With only two covalent bonds, these atoms could not play a vital role in 2D materials science because they would prefer the 1D morphologies, e.g., of nanowires. However, owing to the variable oxidation states of + 6 and – 2, the bonding capabilities of these elements extends far beyond low-dimensional chains. The different oxidation states are the key factor for understanding the stability of their 2D allotropes. Note that this inherent multivalency has the largest difference in oxidation states considered thus far in 2D materials. Owing to the multivalency of group-VIA elements with metallic and nonmetallic characteristics, several stable and metastable covalent 2D phases exist, different from the usual honeycomb or buckled honeycomb structures [1]. However, there are still conflicting discussions regarding the existence, as well as the dynamic and thermal stability of various phases derived by global structural searches combined with first-principles approaches.

Unfortunately, the individual nomenclature with creek letters currently used by several authors is not consistent and introduces confusion. Therefore, the unequivocal symmetry of the monolayers or their space groups are given here to unambiguously classify the individual phases. Eventually, in parentheses, the mostly used nomenclature, e.g., based on formation energies is presented [1, 2]. Note that normally the thickness direction is defined as the z direction, and the chain direction is defined as the y direction for the chain-like and tetragonal phases, whereas for the rhombic phase, the y-direction extends along the armchair direction due to the loss of the original chain structure by reconstruction [2].

The group-VIA atoms have an electron configuration with six valence electrons in their outer shell, i.e., ($4s^2p_x{}^2p_y{}^2$) for Se, and therefore two extra electrons are needed or six electrons are provided to reach octet stability (see Fig. 11.2). As mentioned

Fig. 11.1 Periodic table of the main group elements highlighting the group–VIA elements Se, Te, and Po involved in the formation of 2D materials

H							He
Li	Be	B	C	N	O	F	Ne
Na	Mg	Al	Si	P	S	Cl	Ar
K	Ca	Ga	Ge	As	Se	Br	Kr
Rb	Sr	In	Sn	Sb	Te	I	Xe
Cs	Ba	Tl	Pb	Bi	Po	At	Rn
Fr	Ra						

above, these elements can form stable 1D-like chain structures with two covalent bonds. Owing to the large separation of the 4*s* and 4*p* states, the two 4*s* electrons behave as core states and are not involved in covalent bonding. Consequently, the four $4p^4$ valence electrons form two covalent bonds with their direct neighbors arranged in a chain and in addition a lone electron pair is generated. Note that this configuration of valence electrons is threefold degenerate ((p_x^2, p_y, p_z), (p_x, p_y^2, p_z), and (p_x, p_y, p_z^2)) and forms helical chains with a helical turn of 120° (see Fig. 11.2).

Figure 11.3 displays the characteristic bonding of these group-VIA elements in the bulk phase. The helical chains, extending along the c axis in a hexagonal pattern, exhibit equilateral triangles in the plane perpendicular to the axis (see Fig. 11.4). Both the interaction of the lone electron pairs and interchain van der Waals interactions control the symmetric arrangement of the chains. Some researchers call the strong interchain interactions in tellurene 'covalent-like' interactions because they are much stronger than pure van der Waals forces. Cleaving the bulk structure either perpendicular or parallel to the chain direction by cutting through the low-index planes, e.g., (0001) or (10–10), yields two stable monolayers and the corresponding few-layer systems (Fig. 11.3).

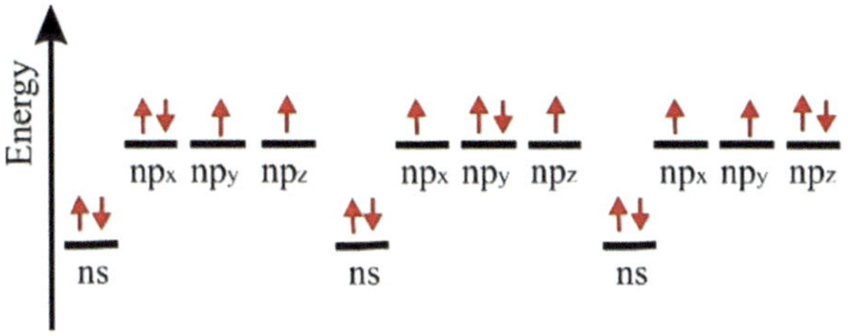

Fig. 11.2 Scheme of the degenerate electron configurations of group-VIA elements with six valence electrons in the outer valence shell

Fig. 11.3 Bulk structure of Se and Te with helical chains oriented in the c direction assembled in a hexagonal arrangement. Two low-index planes illustrate possible cuts for separating monolayers. Flat-lying parallel helical chains of the (10–10) surface expose the (1 ×1) facets of the bulk crystal [3]. Reproduced with permission from Ref., Copyright 2018, Royal Society of Chemistry

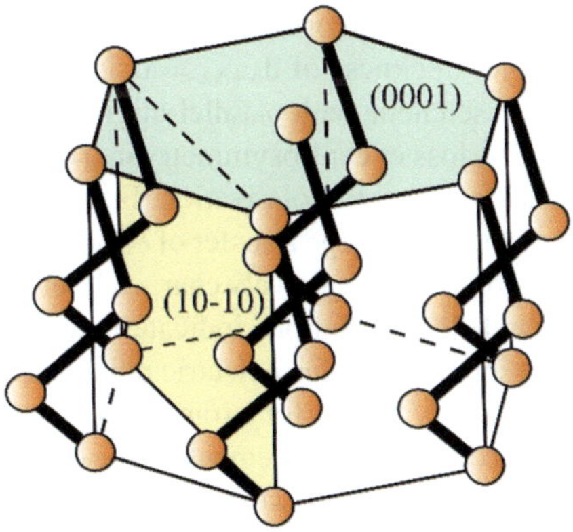

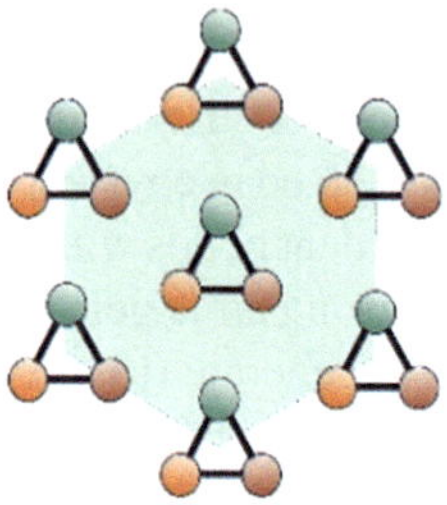

Fig. 11.4 Top view of bulk Se (α phase) and Te (α phase) with helical stacking into a hexagonal structure. The interaction of the lone-pair electrons and interchain van der Waals interactions (weak 'covalent-type' interaction) controlling the symmetric arrangement of the chains are indicated schematically in green

11.1.2 2D Structures

Helical chain selenene (α selenene): This phase consists of parallel-aligned tiled 1D helical chains arranged in a plane and held together by van der Waals forces, as displayed for a bilayer in Fig. 11.5. One stable monolayer structure derives from the trigonal bulk α Se structure with threefold screw symmetry that breaks the spatial inversion symmetry. The lattice constants are a = 0.499 nm and b = 0.413 nm, and the bond lengths within the chain are d_1 = 0.240 nm and d_2 = 0.237 nm [4]. The Se–Se distance is > 0.31 nm for atoms of adjacent chains or rings [5]. Independent calculations revealed similar lattice constants a = 0.498 nm and b = 0.401 nm, a distance between the upper and lower layers of Δz = 0.176 nm, and a cohesive energy of E_c = – 2.78 eV/atom [5]. The cohesive energies and lattice parameters of Se, Te, and Po are collected in Table 11.1. The formation energy of α selenene, which is defined as the difference between the energy per atom of α selenene and the energy per atom of the bulk material, is E_f = 0.122 eV/atom [4]. The 1D helical chain structure is an indirect semiconductor with a band gap of E_g = 0.71 eV (HSE06) [4]. Adding the corrugation and van der Waals diameter of d_w = 0.380 nm [6] yields a layer thickness of $d_{w+\Delta z}$ = 0.556 nm. Figure 11.5 displays the bilayer of helical chain selenene with parallel-aligned 1D chains.

The loss of centrosymmetry in the helical chain structure is the reason for the existence of nonzero polarization. For this noncentrosymmetric structure, a Bader analysis yields a charge transfer of δq = 0.22 e from the positively charged pseudo-central Se atom to the negatively charged boundary atoms of Se, visible in the cross-sectional view [4]. Calculations indicate that this chain structure with twofold coordination and spontaneous ferroelectric polarization is the 2D structure with the lowest energy. The energy barrier for a transition to the nonpolar state is 271 meV per unit cell, and the necessary displacement of the central atom is 0.059 nm (see Fig. 11.6) [4]. The figure displays the two possible polarizations and the nonpolar transition state. The magnitude of polarization in the y direction is 2.68×10^{-10} C/m per layer [4]. The theoretically predicted dynamic stability of the monolayer and few-layer sheets was confirmed by the recently achieved synthesis of this phase.

Table 11.1 Symmetry, space group, cohesive energy, lattice constants, bond lengths, corrugation, van der Waals diameter, and corrugation corrected van der Waals diameter of group–VIA monolayers

Monolayer, structure	Space group	Cohesive energy, E_c (eV/atom)	Lattice constants a, b (nm)	Bond lengths d_1, d_2, d_{int} (nm)	Corrugation, buckling Δz (nm)	Van der Waals diameter d_w [6], $d_{w+\Delta z}$ (nm)
Selenene						
Trigonal, chain	α selenene	− 2.78 [5]	0.499. 0.413 [4]	0.240 [1, 4], > 0.31 [5]	0.176 [5]	0.380 [6], 0.556
Tetragonal	β selenene	− 2.77 [7]	0.372, 0.492 [7]	0.265–0.388 [7]	0.177 [7]	0.380 [6], 0.557
Rhombic	*P*-3*m*1 [10]	− 2.72 [7]	0.374 [4], 0.365 [10]	0.267 [4], 0.262 [10]	0.311 [10]	0.380 [6], 0.691
Monoclinic	*Cm* [5]	− 2.65 [5]	0.365 [4, 5]	0.240, 0.271 [4]	0.077 [5]	0.380 [6], 0.457
Tellurene						
trigonal, chain bulk	$P3_121$, $P3_221$ $P2$, $P2_1$ [2, 25]	− 2.73 [8] − 2.79 [9]	~ 0.433, ~ 0.578 [1] 0.433 [9]	0.282, 0.291 [2] 0.290 [9]		0.420 [6]
Tetragonal	*P*2/*m*, *P*21/*m* [2]	− 2.56 [9]	0.417, 0.549 [9]	0.275, 0.302 [9]	0.216 [9]	0.420 [6], 0.636
Rhombic	*P*-3*m*1 [2, 10]	− 2.62 [9]	0.415 [9]	0.302 [9]	0.367 [9]	0.420 [6], 0.787
Monoclinic	*Cm* [5]	− 2.51 [5]	0.418 [12]	0.285, 0.309 [12]	0.088 [12]	0.420 [6], 0.508
Poloniumene						
	square	− 2.58 [44]	0.326 [44]			
	P-3*m*1 [45], trig	− 3.25 [45]	0.443 [45]	0.320 [45]		

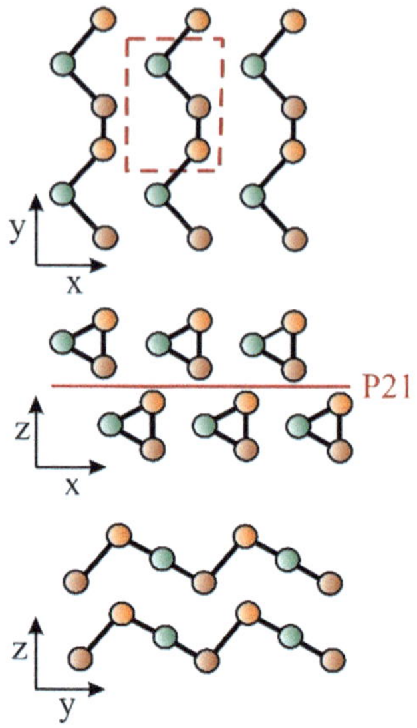

Fig. 11.5 Top and side views of the helical chains derived from the trigonal bulk structure by cutting a sheet parallel to the (10–10) plane [2]. Reproduced with permission from Ref., Copyright 2019, Elsevier

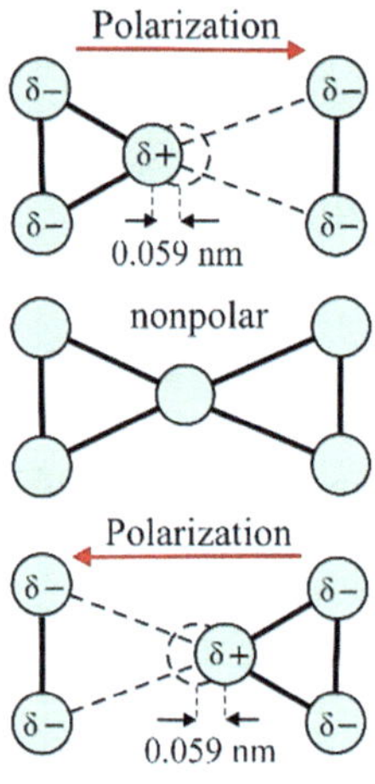

Fig. 11.6 Scheme of spontaneous polarization according to Bader charge transfer. Movement of the central atom of the tetragonal nonpolar structure in the middle to the two polarization states on the left and right positions by spontaneous polarization

Tetragonal mirror-plane selenene (β selenene): It is important to note that the incomplete set of structural data, given by two studies postulating the existence of a tetragonal phase for selenene, resembles the data presented above for the helical chain phase [7, 8]. This structure, originating from a helix-to-layer transformation process is composed of deformed four-membered and deformed six-membered rings with coordination numbers four for the central and three for the outer atoms, as shown for the bilayer displayed in Fig. 11.7. The structural properties given for the tetragonal phase are a = 0.372 nm and b = 0.492 nm, Δz = 0.177 nm, d = 0.265–0.338 nm, and E_c = –2.77 eV/atom [7]. The bond lengths presented in this study would fit better to a structure with separated helical chains. These values should be compared

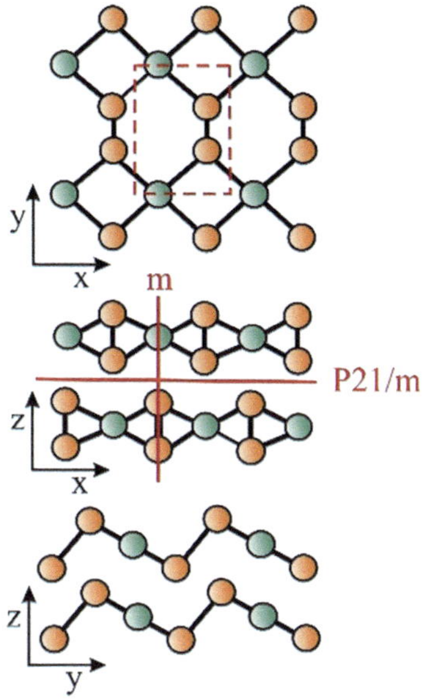

Fig. 11.7 Top and side views of the tetragonal mirror-plane phase after structural relaxation to the nonpolar symmetric position. The rectangular primitive cell contains three atoms [2]. Reproduced with permission from Ref., Copyright 2019, Elsevier

with a = 0.420 nm and b = 0.499 nm, Δz = 0.175 nm, and E_c = – 2.71 eV/atom [8]. Only a very small Bader charge transfer of δq = 0.04 e was postulated for this structure [8]. Unfortunately, in this latter report, the crucial bond lengths are not given [8]. Therefore, it is an open question whether the postulated tetragonal phase truly exists for selenene or whether the delocalization of electrons between the individual chains is not sufficient to induce the described reconstruction to a tetragonal phase with a mirror plane [5]. The calculated stiffness of the β-Se monolayer is extremely anisotropic at E_{ac} = 31 N/m and E_{zz} = 5.4 N/m [8].

Rhombic selenene (γ selenene): In this well-characterized three-layer structure, the coordination number of the central atom is 6, whereas the coordination number of atoms in the lower and upper sublayers is only 3, as elucidated in Fig. 11.8 [4, 9, 10]. General agreement has been achieved on the dynamic stability of the monolayer with centrosymmetric rhombic structure (1T-MoS_2-like structure), a lattice constant of a = 0.374 nm, and a bond length of d = 0.267 nm [4]. Remarkably, in the 1T-MoS_2-like structure the Se atoms replace both the nonmetallic S and the metallic Mo atoms. Rhombic selenene is an indirect semiconductor with a band gap of E_g = 0.71 eV (HSE06) [4].

Due to its multivalent nature, the trilayer network with the space group *P*-3*m*1 forms two distinct bonding types [10]. The presented structural data agree well with a = 0.365 nm, d = 0.262 nm, and a distance of the outer layer atoms at h = 0.311 nm [10]. The published cohesive energies are E_c = – 2.72 eV/atom [7], – 2.62 eV/atom [9], and a much larger value of – 3.35 eV/atom [11]. The formation energy of rhombic selenene is E_f = 0.236 eV/atom [4]. The van der Waals diameter of d_w = 0.380 nm [6] leads to a thickness of t = 0.691 nm, when the layer distance is taken into account. There is no agreement on the charge transferred from the central to the

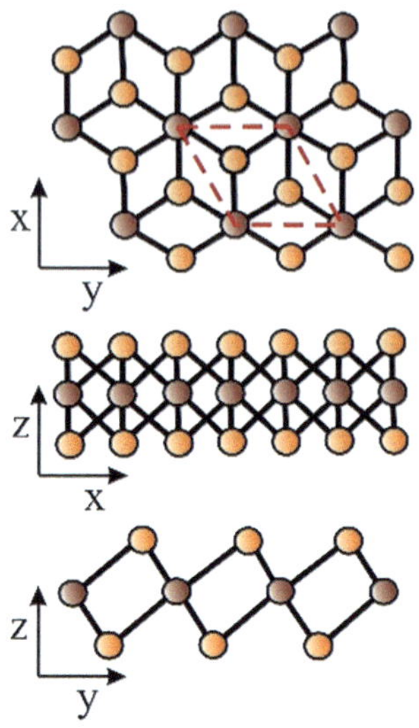

Fig. 11.8 Top and side views of the trilayer γ phase with six bonds of the central atom and three bonds of the atoms in the outer layers. The rhombic unit cell consists of three atoms [2]. Reproduced with permission from Ref., Copyright 2019, Elsevier

outer atoms, as visible in the strongly deviating values of $\delta q = 0.18$ e [8] and 0.41 e [9]. According to theory, rhombic selenene is dynamically stable in its free-standing form [4, 8].

Monoclinic 'square selenene' (Cm [5]*)*: This phase results in the formation of a chair-like buckled structure similar to silicene, however, it involves a square unit cell instead of a hexagonal one (see Fig. 10.9) [5]. The lattice constant is a = 0.365 nm, the corrugation distance between sublayers is $\Delta z = 0.077$ nm, and the cohesive energy is $E_c = -2.65$ eV/atom [5]. A comparable formation energy of square selenene of $E_f = 0.264$ eV/atom has been suggested [4]. Independent calculations of the lattice parameters yielded a = 0.365 nm, $d_1 = 0.240$ nm and $d_2 = 0.271$ nm [4]. Including the buckling effect, the corresponding van der Waals-diameter-based thickness is $d_{w+\Delta z} = 0.457$ nm, which is in very good agreement with t = 0.457–0.469 nm estimated by different approaches [12]. The dynamic and thermal stability of the free-standing monolayer of this phase has been confirmed at 300 K by molecular dynamics (MD) simulations [5]. The band gap opened by spin–orbit coupling (SOC) of ~ 0.1 eV endows square selenene and tellurene moieties with topological insulator (TI) properties, which host nontrivial edge states. The authors consider the low-symmetry square structure as a new template for studying highly anisotropic band dispersions. While previous models of TIs are limited to *s*-state electrons the low-symmetry square lattice involves *p*-state hybridizations around the Fermi level. This gives access to novel Dirac and semi-Dirac semimetals and TIs. Thus, the square phase provides an interesting platform for exploring novel electronic and spintronic applications (Fig. 11.9) [5].

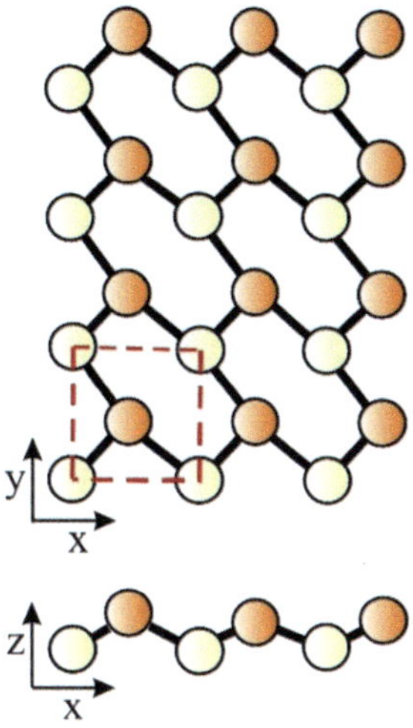

Fig. 11.9 Top and side views of the buckled monolayer of the monoclinic (square) phase. The square unit cell of this low-symmetry structure contains only two atoms [5]. Reproduced from Ref., Copyright 2017, IOP Publishing Ltd

11.1.3 Mechanics

In the buckled tetragonal structure, strain in the armchair direction initially allows stretching without extending covalent bonds because the structure flattens under low-strain conditions. The chemical connection of adjacent chains to the mid atoms leads to the superior mechanical performance of tetragonal selenene compared with that of the helical-chain phase. In contrast, any in-plane strain causes an extension of chemical bonds in the rhombic phase, which generates a much greater stiffness and strength.

The linear mechanical response of rhombic selenene was determined by using the stress–strain relationship. The various reported Young´s moduli of the rhombic phase are E = 59 N/m [7], E = 51 N/m [8], and E = 55.1 N/m [10] neglecting weak anisotropy. The same is assumed for the Poisson ratio of $\nu = 0.27$ [7]. Elastic moduli with a negligible anisotropy were also reported for the rhombic phase at $E_{ac} = 31.6$ N/m and $E_{zz} = 31{,}4$ N/m [11]. An extensive study considered the anisotropy of the elastic modulus, ultimate strength, and Poisson ratio for rhombic selenene with Young´s moduli at $E_{ac} = 123.7$ GPa (86 N/m) and $E_{zz} = 135.1$ GPa (93 N/m) that are much larger than the former isotropic values [11]. The transformation of 3D units to 2D units given in brackets is based on a layer thickness of $d_{w+h} = 0.691$ nm. A Poisson ratio of $\nu \approx 0.25$ was calculated for both directions, in good agreement with the value given above [11]. The ultimate strengths of $\sigma_{ac} = 6.61$ GPa (4.6 N/m) and $\sigma_{zz} = 9.51$ GPa (6.6 N/m) are much lower than expected from Griffith´s rule and would in fact fit better to the lower stiffness values around 55 N/m. The linear and nonlinear mechanical properties of Se and Te monolayers are presented in Table 11.2.

Table 11.2 Structure, Young's modulus, Poisson's ratio, ultimate strength, and ultimate strain of group–VIA monolayers

Monolayer, structure	Young´s modulus E_{ac}, E_{zz} (N/m)	Poisson´s ratio ν_{ac}, ν_{zz}	Intrinsic strength σ_{ac}, σ_{zz} (N/m)	Ultimate strain ε_{ac}, ε_{zz}
Selenene				
Tetragonal	31, 5.4 [8]			
Rhombic	55.1 [10], 59 [7], 31.5 [11]	0.27 [7], ~ 0.25 [11]	4.6, 6.6 [11]	
Tellurene				
Tetragonal	11, 24.8 [27], 11.6, 20.4 [31]	0.26, 0.59 [27]	4.5, 7.2 [30]	0.36, 0.35 [30]
	17, 35 [25]	0.29, 0.49 [31]		
Rhombic	44.5 [7], 45.1 [30]	0.29 [7], 0.28 [30]	5.3, 5.6 [30]	0.26, 0.30 [30]
	32.8, 31.5 [25]	0.41, 0.39 [25]	2.6, 3.5 [25]	0.12, 0.24 [25]

11.1.4 Physics and Applications

The thermoelectric properties of 2D Se and Te have become attractive topics of research. Both monolayers exhibit three-phonon-limited thermal conductivity. According to density functional theory (DFT) calculations, anharmonic phonon scattering is responsible for the particularly low lattice thermal conductivity calculated for tellurene [12]. The square unit cell results in a highly anisotropic electronic structure, and the larger effective mass and Seebeck coefficient in the armchair direction provide better thermal performance in this direction. The conclusion that tellurene has the lowest lattice thermal conductivity of all the elemental 2D monolayers is based on the higher 'umklapp' scattering of tellurene [12]. Interestingly, for rhombic selenene, a much smaller isotropic in-plane thermal conductivity was derived from first-principles calculations than for rhombic tellurene at room temperature, originating from the greater anharmonicity of the acoustic phonon branch [10]. The authors studied the size and boundary effects, the total phase space for the three-phonon process, the phonon group velocity, and the relaxation time and concluded that the latter is the key ingredient of thermal transport in selenene.

Up-to-date comprehensive treatments for potential device applications of chalcogens have been published [1, 13]. Potential applications of selenene (tellurene) can be envisaged in the fields of optoelectronics, thermoelectric materials, devices employing ferroelectric materials, and sensors based on tunable band gaps, high carrier mobilities, good thermoelectric properties, ambient chemical stability, and strong angle-dependent photoresponses [14]. The optical gaps of monolayer selenene and $TeSe_2$ lie in the ideal window of 1.0–1.5 eV with the maximum photoconversion efficiency. The rhombic structure has strong optical absorbance in the whole visible region, and changing the number of layers in a stack modifies the absorption edge over a wide range [15].

Following the fabrication of air-stable selenium nanosheet field-effect transistors (FETs) with a gate length of 5 nm, an ab initio quantum transport approach was employed to simulate a sub-5 nm gate-length double-gate monolayer selenene FET [16]. This study showed that selenene has the potential to deliver a channel material that can scale Moore´s law down to a gate length of only 3 nm. These devices can meet the 2013 ITRS standards for high-performance applications in the armchair and zigzag directions in the 2028 horizon.

Functionalization gives rise to topologically nontrivial dynamically stable honeycomb lattice structures of Se and Te with large gaps [17]. A theoretical study investigated oxygen-functionalized Se (selenene oxide SeO) and sulfur-functionalized Se (selenene sulfide SeS) and oxygen-functionalized Te monolayers (tellurene oxide TeO), which break the spatial inversion symmetry. The authors concluded that functionalization allows the realization of exotic quantum properties that are robust against degradation, such as SOC-induced large bandgaps with isolated Dirac cones along the edges, suggesting applications in room temperature spin-transport devices.

A concise review addresses the biomedical applications of Se and Te nanomaterials [18]. These 2D materials have several advantages over conventional nanoparticle-based systems, namely, better photothermal conversion, excellent electrical conductivity, and larger surface area. The latter endows them with a higher loading capacity for therapeutic applications and fluorescent molecules in drug delivery platforms. Furthermore, they can be employed as diagnostic agents in computer tomography (CT), photoacoustic imaging, fluorescence imaging, and Xene-based theragnostic nanoplatforms [18]. The future application of 2D Xenes in the biomedical field faces unprecedented opportunities, however, also unprecedented challenges.

A well referenced review of 2D selenium and its composites details the various methods of controllable, low-cost, and eco-friendly synthesis, and advanced 2D properties including size-dependent band gaps, environmental robustness, strong photoluminescence, anisotropic thermal conductivity, and high photoconductivity. Furthermore, potential device applications in batteries, photodetectors, modulators, photothermal and medical applications are discussed in detail [19].

11.1.5 Synthesis

To date, only few-layer assemblies of selenene could be synthesized, avoiding the strong tendency of selenene to grow in 1D nanowire forms [20, 21]. By liquid-phase exfoliation, 2D Se sheets of the trigonal phase with 40–120 nm lateral dimensions and 3–6 nm thicknesses could be exfoliated by sonication, revealing the stability of the bulk phase despite dimensional reduction [20]. Trigonal crystalline nanosheets exhibit strong photoluminescence and robust chemical stability under ambient conditions. The physical vapor deposition (PVD) method allows the deposition of large-size high-quality Se nanosheets with a minimum thickness of 5 nm [21]. These selenene sheets exhibit strong in-plane anisotropy, as determined by angle-resolved Raman spectroscopy [21]. Fabricated back-gated FETs exhibit p-type transport behavior. Seed-assisted chemical vapor transport (CVT) growth of selenene, oriented in the (0001) direction, induced the growth of bilayers of ultrathin triangular flakes of highly crystalline trigonal selenene with a lateral size > 30 μm [22]. Polarization angle-resolved Raman spectra showed in-plane isotropic properties due to the high symmetry originating from the unique growth orientation. DFT calculations confirmed the structure and stability of the as-grown selenene sample. This growth strategy can be applied to other elemental 2D materials. A review provides a detailed discussion of the methods used for the synthesis of group VIA elemental 1D and 2D structures, their basic properties, and diverse device applications [23].

11.2 Tellurene Monolayers

11.2.1 Chemistry and 2D Bonding

Based on the valence electrons of Te ($5s^2p_x{}^2p_y{}^2$), tellurium forms atomic chains and 2D layered structures as already described for the chemically related selenium. A significant difference between the two metalloid elements is the more metallic nature of Te according to the transition from the nonmetal S to the metal Po in the VIA group. This transition is responsible for the stronger delocalization of lone electron pairs in the interchain region of tellurene, which increases the metallic bonding character. An in-depth study of the interchain interactions in the bulk, few-layer, and monolayer tellurene structures revealed the specific nature of bonding between chains [24]. In the bulk, each helical chain has six neighbors, and the intra- and interchain distances are $d_l = 0.290$ nm and $d_{int} = 0.351$ nm, respectively. In the bilayer, these distances decrease to $d_l = 0.287$ nm and $d_{int} = 0.337$ nm. Note that in the monolayer with two neighboring chains, the bond length and interaction distance decrease substantially to $d_l = 0.278$ nm and $d_{int} = 0.304$ nm [24]. Thus, the delocalization of lone electron pairs in the interchain region adds metallic-type bonding to typical van der Waals interaction forces, though, this enhanced interaction is still not as strong as covalent intrachain bonding. Going down the column in the periodic table a further decrease in covalent bonding strength and complete electron delocalization generate the Po metal.

11.2.2 2D Structures

Helical chain tellurene (α-tellurene [1]*)*: According to theory, few-layer systems of α Te should exfoliate easily from layered bulk samples owing to the small interlayer coupling energy of only 26 meV/Å^2, which is in the same range as that of MoS_2 (21 meV/Å^2) [9]. Cutting the bulk structure parallel to the (10–10) plane leads to a stable 2D Te slab. The most stable tellurene phase at equilibrium is a few-layer assembly of helical chains identical to a dimensionally reduced bulk structure. In fact, experimentally realized few-layer tellurene sheets revealed such a bulk-like structure. The helical bulk structure is a chiral-type crystal owing to the lack of mirror symmetry and inversion center. This bulk structure has the space groups $P3_121$ and $P3_221$ for the inequivalent right- and left-handed structure, respectively, whereas the space group is $P2_1$ for the few-layer system with an even number of layers and $P2$ for those with an odd number of layers (see Fig. 11.5) [2, 25]. Structural relaxation is expected for thicker slabs in the surface region, resulting in quasi-bulk properties of thicker slabs. These slabs approach the bulk lattice constant $a = b = 0.433$ nm, the bond length $d = 0.290$ nm and the cohesive energy $E_c = -2.79$ eV/atom [9].

More accurate values are a = 0.433 – 0.436 nm and b = 0.578 –0.581 nm [1] and d_1 = 0.282 nm and d_2 = 0.289–0.291 nm [2].

Notably, a monolayer with a chain-like structure seems not to be stable and transforms without an activation barrier to a phase with an additional mirror plane due to stronger interchain interactions in tellurene, compared to selenene that has a less metallic character (see Fig. 11.7) [26]. This stronger interchain interaction originates from the mid-Te atom that moves to the symmetric position to generate the mirror plane, a transformation that is absent in selenene (see Fig. 11.6) [27]. A detailed study investigating the microscopic mechanism of helix-to-layer transformation in elemental group-VIA solids, revealed as explanation the release of the activation barrier [28]. The transition from the helical chain phase to the tetragonal phase discussed in more detail below is also promoted by a compression of this phase in any in-plane direction [2]. Note that a change in the chiral structure (α phase), e.g., from the left-handed to the right-handed structure, can occur through the intermediate tetragonal structure (the β phase) because the energy difference between the two phases is comparable to the thermal energy. Due to the chirality of the helical chain structure with two polarization directions, few-layer systems exhibit a piezoelectric effect owing to strain-induced α–β transitions, a ferroelectric effect based on transitions between the two chiral directions, and electrostrictive phenomena due to charge-induced α–β transitions [26].

Tetragonal mirror-plane tellurene (β tellurene [1]*)*: This semiconducting three-layer structure consists of planar four-membered and chair-like six-membered rings, as displayed in Fig. 11.7. The central atoms have a coordination number of four, and for the atoms of the upper and lower sublayers, it is three. The outer atoms are covalently bonded with a shorter bond length of d_1 = 0.275 nm, while the metal–ligand-like bonding of the center atoms with the outer atoms has a longer bond length of d_2 = 0.302 nm [9]. The lattice constants are a = 0.417 nm and b = 0.549 nm, the cohesive energy is E_c = – 2.56 eV/atom, and the charge transfer from the center to the outer atoms is δq = 0.11 e [9]. These properties have been confirmed by a = 0.413 nm and b = 0.543 nm, the bond lengths d_1 = 0.270 nm and d_2 = 0.298 m, and the cohesive energy E_c = – 2.57 eV/atom for single-layers of β Te [29]. The formation energy of β tellurene is E_f = 0.228 eV/atom [27]. Note that data obtained for a structure called 'rectangular tellurene' obviously describe this structure [5]. As the distance between the upper and lower sublayers increases to Δz = 0.216 nm [9] and the van der Waals diameter increases to d_w = 0.420 nm, the thickness reaches $d_{w+\Delta z}$ = 0.636 nm. The step height measured by scanning tunneling microscopy (STM) between two consecutive layers was d_s = ~ 0.40 nm [9], which agrees with the calculated interlayer spacing of d_s = 0.40 nm [30]. The step height is expected to be smaller than the monolayer thickness due to the missing interlayer interactions.

As described above, a precursor of the β phase appears by thinning of bulk chain-like tellurene to a single layer in the first step. Then, the β phase is generated from the helical-chain phase in an 'gedanken' experiment by shifting the mid-layer Te atoms toward their closest neighboring helical chains to generate periodic mirror planes (m) perpendicular to the x direction (helix-to-layer reconstruction) [2]. The space group of the derived few-layer systems is *P*21/*m* for even layers and *P*2/*m* for odd layers. The stable monolayer is centrosymmetric, whereas bilayers of the tetragonal phase are unstable against a transition to the bilayer helical phase with broken centrosymmetry. The strong repulsion of the lone electron pairs in bilayers may induce this reconstruction, to reduce interlayer lone-pair interactions. Theory supports the existence of both a few-layer helical chain and a tetragonal phase, with the difference that β Te can be more easily converted to a metal due to the small direct band gap [27]. Strain-induced band gap engineering in few-layer tellurene is an important tool for fabricating FETs, photodetectors, sensors, and memristors.

The rearrangement of atoms within the chain leads to the noncentrosymmetric α phase structure, and electronically induced distortions in the charge density generate an in-plane ferroelectric polarization that is mostly based on a contribution originating from the displacement of ions [3]. Such spontaneous polarizations in an elemental material is astonishing because structures usually assume high symmetry if no difference in electronegativity exists. Such a spontaneous nonzero electrical polarization also occurs in multilayers and can be stable at elevated temperatures up to 600 K. Interestingly, in binary SnTe, which is a group IVA–VIA compound, for example, only odd layers show polarization due to opposite polarizations of neighboring layers.

Rhombic tellurene: (γ tellurene [1]): Fig. 11.8 shows the dynamically stable semiconducting 1T-MoS_2-like trilayer structure with the space group *P*-3*m*1 [2]. The unit cell contains three atoms, and the central atoms have a coordination number of 6, whereas the outer-layer atoms have only three. The lattice constant is a = 0.415 nm, the bond length is d = 0.302 nm, the distance between upper and lower layer Δz = 0.367 nm, the cohesive energy E_c = – 2.62 eV/atom, and the Bader charge transfer from the center to the outer atoms δq = 0.41 [9]. These values partly differ from other calculations at a = 0.423 nm, d = 0.364 nm, and E_c = – 3.05 eV/atom [11]. The structure can be determined by the following 'gedanken' experiment. Cleavage of the bulk structure along the (0001) plane breaks intrachain covalent bonds, introducing dangling bonds in the 2D structure, which induces relaxation that destroys the original helical chain structure. The relaxed structure of a few-layers thick slab removed from the bulk structure depends on the number of layers. Interestingly, sheets with 3, 6, 9, 12, and 15 layers transform to the respective layered structures of rhombic tellurene, e.g., one trilayer structure for three chain layers, whereas a slab with 4 layers, for example, will maintain its chain-like structure [9]. Note that monolayer rhombic tellurene is energetically the most stable phase [9]. With the sum of the distance between upper and lower layer of Δz = 0.367 nm and the van der Waals diameter of d_w = 0.420 nm we find a thickness of $d_{w+\Delta z}$ = 0.787 nm [9].

Monoclinic square tellurene (Cm [5]): Important structural data of square tellurene are a = 0.408 nm, Δz = 0.092 nm, and E_c = – 2.51 eV/atom (see Fig. 11.9) [5]. These values are in good agreement with a = 0.418 nm, d_1 = 0.285 nm, d_2 = 0.309 nm, and Δz = 0.088 nm [12]. Based on the presence of only two atoms in the unit cell and the low-symmetry with space group *Cm* for the square lattice of tellurene the authors predict the properties of TIs, hosting nontrivial edge states [5]. Using the van der Waals diameter of d_w = 0.420 nm, the estimated thickness is $d_{w+\Delta z}$ = 0.508 nm.

δ phase and ε phase: The δ phase consists of helical chains; however, they are not as tightly packed on the surface as the α phase [2]. This leads to a high surface energy and the highest formation energy of all phases for a thickness of less than five layers. With a nanosheet of approximately 7 layers, the formation energy becomes comparable to that of the α and β phases. The space group of the δ structure is *P*2 for an odd and even number of layers.

State-of-the-art global structural searches combined with first-principles calculations played a significant role in the discovery of group–VIA allotropes, including new phases of tellurene, such as the metastable ε phase with a 2H-MoS_2-like structure and a comparable cohesive energy of $E_c = -2.46$ eV/atom. This phase of the metalloid Te has metallic characteristics, and ab initio MD simulations predict kinetic stability up to 200 K [9].

11.2.3 Mechanics

With the monolayer thickness of the β phase of tetragonal tellurene $d_{w+\Delta z} = 0.636$ nm, the published elastic constants deliver for Young´s modulus the values $E_{ac} =$ 17.5 N/m and $E_{zz} = 38.8$ N/m, and for and Poisson´s ratios $\nu_{ac} = 0.263$ and $\nu_{zz} =$ 0.585 [25]. For the ultimate strengths this study predicts $\sigma_{ac} = 7.3$ GPa (4.6 N/m), $\sigma_{zz} = 11.4$ GPa (7.3 N/m), and for the ultimate strains $\varepsilon_{ac} = 0.36$ and $\varepsilon_{zz} = 0.35$. The large anisotropy of the mechanical properties of tetragonal tellurene was confirmed by the elastic moduli of $E_{ac} = 11.1$ N/m and $E_{zz} = 24.8$ N/m and the Poisson ratios $\nu_{ac} = 0.26$ and $\nu_{zz} = 0.59$, however, with much lower stiffnesses [27]. The tetragonal monolayer and few-layer structures deform more easily in the armchair direction. Compared to the helical-chain α structure, the β structure has a larger stiffness and Poisson ratio along the zigzag direction owing to the strong interchain bonding of the mid-Te atoms. Another study supports the low stiffness values $E_{ac} = 11.6$ N/m and $E_{zz} = 20.4$ N/m, and $\nu_{ac} = 0.29$ and $\nu_{zz} = 0.49$ for the Poisson ratios, based on uniaxial loading of the tetragonal structure [31].

An extensive study covers the elastic constants, Young´s moduli, ultimate strengths, and critical strains of several phases. For the tetragonal phase, the strongly anisotropic stiffnesses under uniaxial tensile strain are $E_{ac} = 27$ GPa (17 N/m) and $E_{zz} = 55$ GPa (35 N/m), and the ultimate strengths $\sigma_{ac} = 7.23$ GPa (4.5 N/m) and $\sigma_{zz} = 11.4$ GPa (7.2 N/m), employing a thickness of $d_{w+\Delta z} = 0.636$ nm [30]. Note that these stiffness values support the large elastic moduli given above. The critical strains obtained in this study are $\varepsilon_{ac} = 0.36$ and $\varepsilon_{zz} = 0.35$. Tetragonal tellurene exhibits a negative Poisson ratio. When stretched in the zigzag direction, this phase shows an auxetic effect in the out-of-plane direction. The elastic constants fulfill the Born-Huang conditions necessary for the mechanical stability of the monolayer.

For the rhombic Te monolayers, the nearly isotropic Young´s moduli are $E_{ac} =$ 41.6 GPa (32.8 N/m) and $E_{zz} = 39.9$ GPa (31.5). The stiffness values in 2D unit were calculated using a layer thickness of t = 0.789 nm [25]. The Poisson ratios of ν_{ac} = 0.41 and $\nu_{ac} = 0.39$ were estimated from the elastic constants. For the ultimate strengths, these authors obtained $\sigma_{ac} = 3.25$ GPa (2.6 N/m) and $\sigma_{zz} = 4.49$ GPa (3.5 N/m), and for and ultimate strains $\varepsilon_{ac} = 0.12$ and $\varepsilon_{zz} = 0.24$ [25]. Other calculations revealed a greater isotropic stiffness for the rhombic phase at E = 57.1 GPa (45.1 N/m), a Poisson ratio at $\nu = 0.28$, ultimate strengths at $\sigma_{ac} = 6.7$ GPa (5.3 N/m) and $\sigma_{zz} = 7.12$ GPa (5.6 N/m), and critical strains at $\varepsilon_{ac} = 0.26$ and $\varepsilon_{zz} = 0.30$, using t = 0.789 nm [30]. Additional mechanical properties reported for rhombic tellurene include E = 44.5 N/m [7], E = 35.2 N/m [10], and $\nu = 0.29$ [7].

Several tellurene phases exhibit a linear increase in the stiffness of the first four stacked monolayers when biaxial strain is applied. The rhombic and tetragonal phases have the largest elastic moduli owing to covalent bonding in the 2D plane. For the rhombic phase, the elastic modulus increases from approximately 50 N/m to approximately 100 N/m from the monolayer to the bilayer. The corresponding moduli of the tetragonal phase possess roughly half these stiffnesses [2]. The reason for the high modulus of the rhombic phase is that in this 2D network, any in-plane strain causes a direct extension of chemical bonds, while in the tetragonal phase, the buckled structure allows an initial stretching in the armchair direction without extending covalent bonds. Flattening of the structure is possible under low-strain conditions. The stiffness decreases further for chain-like phases with weak van der Waals interchain forces.

11.2.4 Physics and Applications

Current interest in the group–VIA chalcogens selenium and tellurium is based on their favorable properties, such as strong optical absorptions, high carrier mobilities, excellent thermoelectric performance, piezoelectricity, high stretchability, good environmental and mechanical stabilities, and suitability for spintronics, and quantum devices [31–33]. Strain engineering of monolayers and few-layer systems is an important technique to tailor the electrical and optical properties of 2D materials. An in-depth study of the effects of in-plane and out-of-plane strains on helical (α) and tetragonal (β) few-layer assemblies, revealed variations in the band structures and semiconductor-to-metal transitions [27]. The mechanism of helix-to-layer transformation and the temperature-, strain-, and charge-induced dynamic structural transformations of the selenene and tellurene phases were studied theoretically [28].

The noncentrosymmetric phases of tellurene show outstanding piezoelectricity. In a theoretical study the mechanical and piezoelectric properties of a puckered structure with alternate planar four-membered and chair-like six-membered rings with the space group *P2*/*m* were compared with those of an optimized noncentrosymmetric β' structure with the monoclinic space group *P*2(3) that may be realized by strain engineering [31]. The monoclinic phase of tellurene exhibits a large thermal expansion, a negative Poisson ratio of $\nu = -0.024$ along the zigzag direction, and large in-plane piezoelectric coefficients that are a factor of 23 and 12 greater than those of the widely used MoS_2 monolayer under biaxial and uniaxial strain in the zigzag direction, respectively. The observation of giant in-plane piezoelectric coefficients of -83.9×10^{-10} C/m and -42.6×10^{-10} C/m was possible under biaxial and uniaxial strains, respectively [31].

Despite several structural and electrical advantages of tellurene for advanced complementary metal-oxide-semiconductor (CMOS) technology, the low bulk band gap poses problems, which can be solved, however, in few-layer sheets with a suitable gap of > 1 eV [33]. In view of the progress made in the synthesis of 2D tellurene sheets, their application as FET and potential use in ultrascaled CMOS circuits are discussed. Exact ab initio quantum transport simulations provided the performance limits of sub-5 nm monolayer tellurene metal-oxide-semiconductor field-effect transistors (MOSFETs) [16]. The authors concluded that monolayer tellurene may offer the possibility of extending Moore´s law to a limit of 3 nm.

The investigation of binary 2D compounds of group–VIA elements is still in its infancy. Due to the similar chemical nature of Se and Te, they easily form solutions. Among the large number of ordered polymorphic crystals and disordered structures, the three mixed-crystal structures of TeSe, $TeSe_2$, and Te_2Se were characterized through a global structure search based on swarm intelligence [34]. This work evaluated the structural parameters, the formation energy, the Bader charge analysis, and the band gaps of the stable semiconductors with gaps between $E_g = 0.12$ eV and 1.55 eV. A change in stoichiometry and polymorphism add additional degrees of freedom and make the semiconducting properties highly tunable.

The liquid-phase exfoliation method employing anisotropic ultrasonication yields nonlayered helical-chain (α-phase) tellurene monolayers with thicknesses less than 0.6 nm [35]. This phase is of particular interest because theory predicts a topological interior insulating state and metallic edge states for this structure (see Fig. 11.10). The built-in strain upon exfoliation of α tellurene to nanoflakes, revealed through a shift in the Raman modes, explains the occurrence of the nontrivial band structure. In the inner part of the monolayer, STS measurements revealed the valence band (VB) and conduction band (CB) energies that along the periphery of the deposited nanoflakes exhibit gapless metallic states with a Dirac point. The topologically protected edge extended approximately 3 nm into the interior of the nanoflake. Thus, the nontrivial TI phase of tellurene provides a model candidate for quantum computation, spintronics, and nonlinear photonic devices.

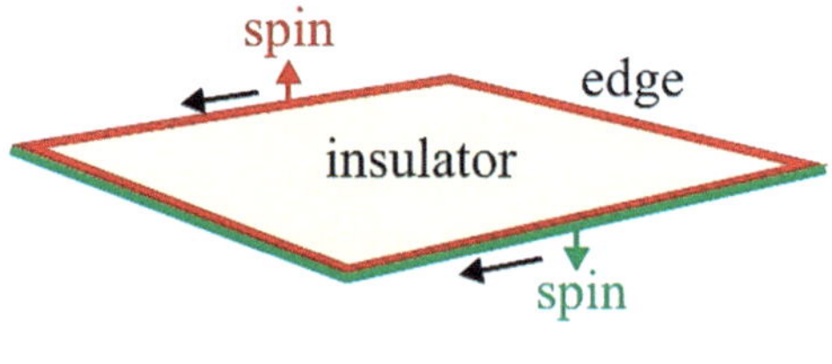

Fig. 11.10 Scheme of a 2D topological insulator presented schematically by quantum-spin Hall layers with opposite chirality presenting the interior semiconducting part and the metallic edge states with counterpropagating opposite spins

11.2.5 Synthesis

Like for selenene, diverse methods, such as liquid-phase exfoliation (LPE) and physical and chemical vapor deposition (PVD, CVD), have been applied to grow few-layer sheets of tellurene. As an exception, only molecular beam epitaxy (MBE) on graphene/6H-SiC(0001) allowed the growth of monolayers and few-layer tellurenes. In situ STM and STS analysis revealed flat-lying and parallel arranged helical chains on the graphene surface, exposing the (1 $\times$ 1) facets of the (10–10) surface of the bulk crystal [36]. Another experimental study employed MBE for deposition on highly oriented pyrolytic graphite (HOPG), using STM and STS for analysis [37]. For ultrathin layers, the cell size was consistent with the predicted rectangular surface cells of the tetragonal phase, whereas for thicker films, the detected structure agreed better with the (10–10) surface of the trigonal bulk crystal (α-Se) [37].

A free-standing, high-quality large-area chiral-chain van der Waals structure of tellurene was fabricated in a substrate-free hydrothermal solution process with a tunable thickness ranging from monolayers to tens of nanometers and lateral sizes reaching 100 μm [38]. The structure, composition, and quality of these tellurene sheets were thoroughly analyzed via high-angle annular dark-field scanning transmission microscopy (HAADF-STM), high-resolution transmission electron microscopy (HRTEM), energy dispersive X-ray spectroscopy (EDS), and X-ray diffraction (XRD). A tellurene field-effect transistor with an on/off ratio of approximately 10^6 and a field-effect mobility of approximately 700 $cm^2\ V^{-1}\ s^{-1}$ showed air-stable performance for more than two months.

The PVD method made the synthesis of subnanometer trilayer rhombic tellurene possible, and pulsed laser deposition (PLD) allowed the growth of uniform and contiguous sub-7 nm films with a centimeter square size [39]. DFT calculations predict the coexistence of three polytypes, including the observed rhombic structure, for subnanometer flakes. In addition to detecting flat regions, high-resolution scanning tunneling electron microscopy (STEM) revealed the existence of rippled regions. Their occurrence can be explained by the low stiffness of $E = 44.5$ N/m for rhombic tellurene.

In addition to the usual problem of controlling the thickness of the growing deposit, the inherent multivalency feature makes it difficult to avoid the growth of several crystal faces and morphologies of selenene and tellurene. A recent review covers in detail the present stage of tellurene preparation techniques, the role of tellurium in 2D electronic and optoelectronic devices, and the fabrication of 1D to 3D Te photodetectors consisting of van der Waals heterostructures [40].

A rectangular phase of tellurene monolayers was predicted by first-principles calculations and realized by MBE on a Ni(111) substrate, using STM to analyze the deposited structure [41]. Strong coupling of the deposit with the surface and good lattice registry worked as driving forces for tellurene formation. With continued growth, the Te/Ni(111) system experienced a unique morphological transition from the rectangular tellurene monolayer to periodic striped patterns, and then to thick striped patterns.

First experimental realization of the structurally complex α-tellurene monolayer was achieved by a novel self-assemble approach [42]. The epitaxial growth of the monolayers on a Sb_2Te_3 substrate was achieved by using Te chains consisting of three atoms as building blocks, derived from the dynamic nonequilibrium epitaxial growth of an a-Si:Te alloy, consisting of Si atoms linked with two-fold coordinated Te chains. A characteristic (1 ×1) structure of the hexagonal lattice of α-tellurene monolayers was identified by STM measurements. The authors recommend the self-assembly-based synthesis strategy for the growth of other elemental artificial 2D materials. Furthermore, STS measurements revealed a band gap of $E_g = 0.36$ eV of the α-tellurene monolayer grown on Sb_2Te_3.

A comprehensive review covers the diverse fabrication methods used for growing few-layer sheets and various possibilities of potential applications. Based on the electrical, optical, and thermoelectric properties the work covers practical use of tellurene in photodetectors, FETs, piezoelectric devices, modulators, energy harvesting devices, logic gates and circuits, and stability issues, together with a comparison of performance with that of competing materials [43].

11.3 Poloniumene Monolayers

11.3.1 Chemistry, 2D Bonding, and Structure

In elemental metals face-centered cubic, body-centered cubic, and hexagonal close-packed lattices have the highest energetic stability. However, in 2D polonium (poloniumene), the square lattice structure forms the ground state, which is the 2D analog of the simple cubic structure of bulk polonium. First DFT calculations treating the effects of exchange and correlation by the local density approximation (LDA) and the generalized gradient approximation (GGA), identified the square structure as the ground state of monolayer Po [44]. The calculations included the *d* electrons in the electronic configuration of Po (Xe|$5d^{10}6s^2p_x{}^2p_y{}^2$). The cohesive energy of the simple cubic lattice of bulk polonium is $E_c = -2.73$ eV/atom. Without SOC, the cohesive energy of poloniumene changes to $E_c = -2.58$ eV/atom. SOC increases the electron density at the nuclei by 0.033 a.u. and changes electron density at the middle of the bonds by 0.005 a.u.. This effect weakens directional covalent bonding and elongates the distance between the Po atoms. Due to the reduction in the coordination number in comparison with that of the hexagonal structure, the lattice constant decreases to a = 0.326 nm in the square structure in comparison to that in the cubic structure at a = 0.336 nm. Moreover, SOC plays a crucial role in suppressing the Peierls instability and generates dynamical stability by preventing the occurrence of imaginary phonon frequencies over the Brillouin zone.

Contrary to these results, first-principles calculations combined with the tight-binding method indicated that a trigonal 1T-MoS_2-like structure with three sublayers is more stable than the square phase [45]. The space group of this structure is *P*-3*m*1.

A cohesive energy of $E_c = -3.25$ eV/atom indicates stronger covalent bonding of trigonal poloniumene. The calculated bond length is d = 0.320 nm, and the optimized lattice constant is a = 0.443 nm. A Bader charge transfer of $\delta q = 0.21$ e per atom occurs to the upper (lower) layer. The central layer of Po atoms tends to be metal like, whereas the outer layer atoms prefer a semiconductor-like behavior. The phonon spectra confirm dynamic stability, and inherent multivalency enhances structural stability. Moreover, ab initio MD simulations predict thermal stability at room temperature. Importantly, trigonal multivalent poloniumene exhibits an intrinsic QSH effect with the advantages of improved experimental feasibility and the absence of scattering from extra atoms [45]. As rationalized by the tight-binding model, the $p_{x,y}$-p_z band inversion mechanism explains the transition from the normal insulator to the QSH insulator.

Additionally, first-principle DFT calculations and ab initio MD simulations confirmed the existence of the stable trigonal structure of monolayer Po [46]. In addition to the metallic cubic phase, the trigonal phase of bulk metallic Po becomes stable at higher temperatures. The authors showed that magic thicknesses of N = 3, 6, 9, up to 24 atomic layers are energetically preferred. These assemblies spontaneously transform to a monolayer or a multilayer with a 1T-MoS_2 structure with weaker interlayer interactions between its three covalently bonded sublayers. The nanosheets with magic thicknesses are semiconducting, while the other sheets remain metallic. The band gap of the semiconducting monolayer of poloniumene is $E_g \sim 0.9$ eV (HSE06) and decreases to ~ 0.1 eV (HSE06) at a thickness of four monolayers. The spontaneous phase transition of nanosheets with magic thicknesses offers a new preparation method, resulting in van der Waals bonded few-layer assemblies with an interlayer interaction of ~ 32 meV/$Å^2$. Driven by strong quantum size effects, Peierls-like distortion induces such a phase transition.

11.4 Summary of Elemental Group–VIA Monolayers

11.4.1 Binding–Structure Relationships

Figure 11.11 displays the cohesive energies of four allotropes of selenene and tellurene and two of poloniumene versus their bond lengths [45]. Surprisingly, the trigonal, tetragonal, rhombic, and monoclinic structures of selenene, tellurene, and poloniumene with different bonding types have comparable cohesive energies of approximately $E_c = -2.7 \pm 0.2$ eV/atom, except for the trigonal phase of poloniumene, which has a surprisingly large binding energy of – 3.25 eV/atom (see Table 11.1). The longer covalent bonds of the tellurenes are due to the larger size of the Te atoms. The anisotropic structures of bulk and few-layer assemblies of helical-chain selenene and tellurene contain two strongly deviating spacings between atoms. For the trigonal Se bulk structure, the cohesive energy is $E_c = -2.88$ eV/atom, the length of covalent intrachain bonds is $d_1 = 0.243$ nm, and for interchain interactions

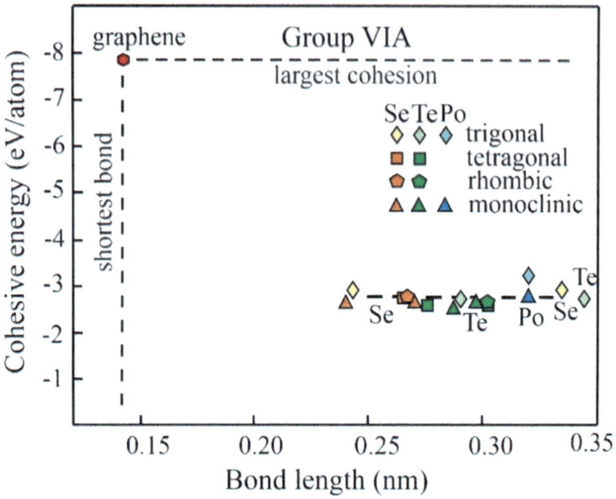

Fig. 11.11 Plots of the cohesive energies versus bond lengths of the trigonal, tetragonal, rhombic, and monoclinic phases of selenene and tellurene and of the trigonal bulk and monoclinic phases of poloniumene [45]. Adapted from Ref., Copyright 2021, Royal Society of Chemistry

between neighboring chains, it is $d_{int} = 0.334$ nm [46]. In the case of the trigonal Te bulk structure, the bond length is $d_l = 0.290$ nm, the atom distance is $d_{int} = 0.342$ nm, and the cohesive energy is – 2.71 eV/atom [46]. As mentioned above, a single layer of the trigonal Te phase seems not to be stable, and the bonding shrinks to that of the tetragonal monolayer [1]. In the layered tetragonal and square phases, these large differences in the spacing of atoms are reduced to two comparable bond lengths, and in the rhombic phase, only a single bond length exists.

The observation that monolayer and bulk phases, with strongly varying bonding types, exhibit a similar cohesive energy within the errors involved in such calculations domonstrates a unique behavior of the chalcogenides Se, Te, and Po. The increasing metallic character induced by the interchain interaction of delocalized lone electron pairs may be the reason for the comparable cohesion of the Se and Te phases with 1 D-like helical chains. In the covalently bonded networks of tetragonal, rhombic, and monoclinic monolayers, the multivalency of the two elements obviously allows the realization of distinct structures with similar cohesive energies ('multiple structures with similar cohesion'). This fascinating effect eliminates the strong dependence of bond energy on bond length over a range of bond extensions. Furthermore, it seems that the activation barrier for the transition from the helix to layer phases creates no problems [28]. Comparable cohesive energies make transformation between phases possible even when using thermal energy and the adsorbed state. Thus, stable and metastable allotropes may coexist at higher temperatures. Moreover, the application of mechanical strain promotes the helix-to-layer transformation process [2].

11.4.2 Mechanics–Structure Relationship

Figure 11.12 shows the dependence of Young´s modulus and ultimate strength on the bond length of the tetragonal and rhombic phases of selenene and tellurene [45]. While the cohesive energies of the corresponding structures decrease only slightly from selenene to tellurene, the stiffnesses of the tellurene allotropes are substantially smaller than those of selenene. The figure demonstrates the extremely reduced mechanical performance of the strongest rhombic phases with covalent bonds in the in-plane directions in comparison to that of graphene.

Unfortunately, only limited information is available on the mechanical properties of group-VIA monolayers. The elastic moduli of anisotropic tetragonal selenene with $E_{ac} = 5.4$ N/m and $E_{zz} = 31$ N/m are smaller than that of rhombic selenene (55.1 N/m) [8, 10]. The intrinsic strengths of the rhombic phase mostly follow Griffith´s rule. Similarly, the stiffnesses of the anisotropic tetragonal tellurene phases of $E_{ac} = 17$ N/m and $E_{zz} = 35$ N/m and that of the rhombic phase at $E = 44.5$ N/m follow this behavior [30]. For the rhombic phase the ultimate strengths agree with the E/10 rule, which is not the case for tetragonal tellurene. Due to its lower stiffness and larger Poisson ratio, rhombic tellurene is softer and more flexible than rhombic selenene (see Table 11.2). Notably, rhombic selenene, with its 1T-MoS_2-like structure, is approximately a factor of three softer than the well-studied 1T-MoS_2 monolayer. Importantly, the elastic moduli of the two rhombic phases reach the mechanical performance of their group-VA neighbors in the periodic table with graphene-like networks.

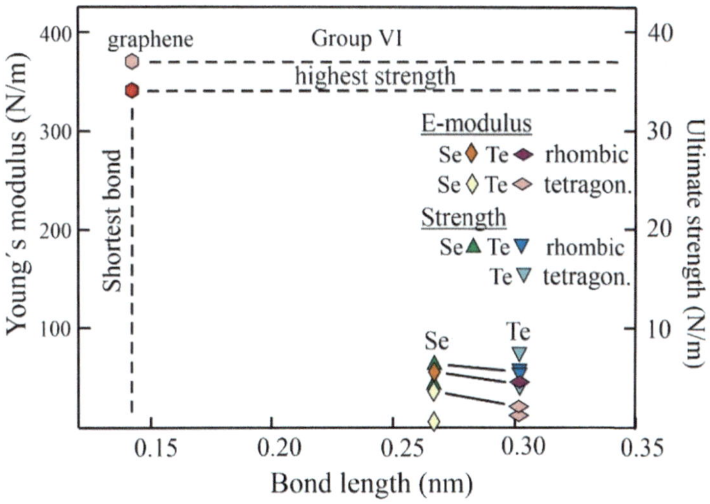

Fig. 11.12 Young´s moduli and ultimate strengths of the tetragonal and rhombic phases of selenene and tellurene as a function of bond length [45]. Adapted from ref., Copyright 2021, Royal Society of Chemistry

References

1. L. Su, X. Fan, C. Wang, Q. Wu, Y. Li et al., Nanophotonics **9**, 1621 (2020)
2. Y. Xiang, S. Gao, R.-G. Xu, W. Wu, Y. Leng, Nano Energy **58**, 202 (2019)
3. P.K. Sachdeva, S. Gupta, C. Bera, Nanoscale Adv. **3**, 3279 (2021)
4. C. Liu, T. Hu, Y. Wu, H. Gao, Y. Yang. et al., J. Phys. Condens. Matter **31**, 235702 (2019)
5. L. Xian, A.P. Paz, E. Bianco, P.M. Ajayan, A. Rubio, 2D Materials **4**, 041003 (2017)
6. S.S. Batsanov, Inorg. Mater. **37**, 871 (2001)
7. L. Ramirez-Montes, W. López-Pérez, R. González-Hernández, C. Pinilla, Int. J. Quantum Chem. e26267 (2020)
8. D. Wang, L.-M. Tang, X.-X. Jiang, J.-Y. Tan, M.-D. He et al., Adv. Electron. Mater. **5**, 1800475 (2019)
9. Z. Zhu, X. Cai, S. Yi, J. Chen, Y. Dai et al., Phys. Rev. Lett. **119**, 106101 (2017)
10. G. Liu, Z. Gao, G.-L. Li, H. Wang, J. Appl. Phys. **127**, 065103 (2020)
11. C. Lin, W. Cheng, G. Chai, H. Zhang, Phys. Chem. Chem. Phys. **20**, 24250 (2018)
12. Z. Lin, C. Wang, Y. Cai, Small **16**, 2003319 (2020)
13. P.-Y. Liao, J.-K. Qin, G. Qiu, Y. Wang, W. Wu, et al., in Woodhead Publishing Series in Electronic and Optical Materials, Xenes, 2D Synthetic Materials Beyond Graphene, Elsevier, pp. 197–224 (2022)
14. J. Min, C. Zhao, Z. Zeng, Y. Jia, Z. Du, Phys. Rev. B **100**, 085402 (2019)
15. Q. Li, X. Tan, Y. Yang, X. Xiong, T. Zhang et al., Molecules **28**, 5390 (2023)
16. R.M. Sattigeri, P.K. Jha, J. Phys. Condens. Matter **34**, 08LT01 (2021)
17. P. Gao, Y. Xiao, Y. Wang, L. Li, W. Li et al., J. Nanobiotechnol. **19**, 96 (2021)
18. Z. Shi, H. Zhang, K. Khan, R. Cao, K. Xu et al., Nano Res. (2021). https://doi.org/10.1007/s12274-021-3493-x
19. C. Xing, Z. Xie, Z. Liang, W. Liang, T. Fan, et al., Adv. Opt. Mater. **5**, 1700884 (2017)
20. P.V. Sarma, R. Nadarajan, R. Kumar, R.M. Patinharayil, N. Biju, et al., 2D Mater. **9**, 045004 (2022)
21. A.K. Tareen, K. Khan, M. Aslam, H. Zhang, X. Liu, Nanoscale **13**, 510 (2021)
22. Y. Liu, W. Wu, W.A. Goddard, J. Am. Chem. Soc. **140**, 550 (2018)
23.] X. Cai, X. Han, C. Zhao, C. Niu, Y. Jia, J. Semicond. **41**, 081002 (2020)
24. J. Qiao, Y. Pan, F. Yang, C. Wang, Y. Chai et al., Science Bulletin **63**, 159 (2018)
25. G. Wang, Y. Ding, Y. Guan, Y. Wang, L. Yang, Nanomaterials **12**, 875 (2022)
26. D. Liu, X. Lin, D. Tománek, Nano Lett. **18**, 4908 (2018)
27. B. Wu, X. Liu, J. Yin, H. Lee, Mater. Res. Express **4**, 095902 (2017)
28. Y. Dong, B. Zeng, X. Zhang, D. Li, J. He et al., J. Appl. Phys. **125**, 064304 (2019)
29. Y. Wang, C. Xiao, M. Chen. C. Hua, J. Zou, et al., Mater. Horiz. **5**, 521 (2018)
30. A. Apte, E. Bianco, A. Krishnamoorthy, S. Yazdi, R. Rao, et al., 2D Mater. **6**, 015013 (2019)
31. G. Qiu, A. Charnas, C. Niu, Y. Wang, W. Wu, et al., NPJ 2D Mater. Appl. **6**, 17 (2022)
32. Q. Zheng, C. Liu, H. Guo, F. Jia, J. Hong et al., Mater. Today Commun. **31**, 103409 (2022)
33. S. Khatun, A. Banerjee, A.J. Pal, Nanoscale **11**, 3591 (2019)
34. X. Huang, J. Guan, B. Liu, S. Xing, W. Wang et al., J. Nano Lett. **17**, 4619 (2017)
35. J. Chen, Y. Dai, Y. Ma, X. Dai, W. Ho et al., Nanoscale **9**, 15945 (2017)
36. Y. Wang, G. Qiu, R. Wang, S. Huang, Q. Wang et al., Nat. Electron. **1**, 228 (2018)
37. T. Zhu, Y. Zhang, X. Wei, M. Jiang, H. Xu, Front. Phys. **18**, 33601 (2023)
38. W. Quan, C. Hong, S. Pan, J. Hu, Q. Wu et al., ACS Appl. Mater. Interfaces **15**, 16144 (2023)
39. W. Wu, G. Qiu, Y. Wang, R. Wang, P. Ye, Chem. Soc. Rev. **47**, 7203 (2018)
40. X. Huang, R. Xiong, C. Hao, W. Li, B. Sa, Adv. Mater. **36**, 2309023 (2024)
41. Z. Shi, R. Cao, K. Khan, A.K. Tareen, X. Liu et al., Nano-Micro Lett. **12**, 99 (2020)
42. S. Ono, Sci. Rep. **10**, 11810 (2020)
43. H. Bao, B. Zhao, J. Zhang, Y. Xue, H. Huan et al., Sci. Rep. **12**, 2129 (2022)

44. C. Zhao, X. Cai, X. Liu, J. Wang, W. Chen et al., Phys. Chem. Chem. Phys. **24**, 7512 (2022)
45. P. Hess, Nanoscale Horiz. **6**, 856 (2021)
46. S. Demirci, H.H. Gürel, S. Jahangirov, S. Ciraci, Nanoscale **12**, 3249 (2020)
47. J. Singh, P. Jamdagni, M. Jakhar, A. Kumar, Phys. Chem. Chem. Phys. **22**, 5749 (2020)
48. J. Qin, G. Qiu, J. Jian, H. Zhou, L. Yang, et al., ACS Nano **11**, 10222 (2017)

Chapter 12
Group IIIA–VIA Monolayers

Abstract The group of IIIA–VIA monolayers with the metals M = B, Al, Ga, In, and Tl and the chalcogens X = O, S, Se, Te, and Po contains 25 binary combinations; however, these compounds have not all received the same attention. The best studied hexagonal 2H and 1T structures consist of four covalently bonded sublayers arranged in the form of X–M–M–X with polarized bonds owing to the different electronegativities of the constituents. Many of these compounds are energetically, dynamically, thermally, and mechanically stable. Most materials are indirect semiconductors with wide band gaps in the VIS-UV range offering the possibility of band gap tuning by strain, an external electric field, or by changing the number of layers in a few-layer sheet. The remarkable properties of these 2D materials include superior optical performance, promising photoelectrochemical performance, and topological and quantum features. Some of the compounds have attracted particular attention as potential thermoelectric materials owing to their moderate band gap of ~ 2 eV and large electrical conductivity, which are based on their unique density of states (DOS). Estimates of the power factor and figure of merit employing Boltzmann transport theory and first-principles calculations suggest GaS, GaSe, and InSe monolayers as promising good candidates for thermoelectric applications. Photocatalysis is another important field of potential applications. One reason for their use in water splitting is the optimal band gap required for efficient absorption in the visible spectrum. Moreover, band gap engineering allows the realization of the band edge positions necessary to straddle the oxidation potential of H_2O and the reduction potential of H_2. Furthermore, the carrier mobilities are high, and the photogenerated electrons and holes tend to separate spatially in heterostructures with type-II band alignment, where the valence and conduction bands belong to different monolayers.

P. Hess, *Bonding, Structure, and Performance of Two-Dimensional Materials*, NanoScience and Technology, https://doi.org/10.1007/978-3-031-74161-6_12

12.1 Boron Monochalcogenide Monolayers (BO, BS, BSe, BTe, and BPo)

12.1.1 BO Monolayer

12.1.1.1 Chemistry and 2D Bonding

The elements forming the binary monolayers of group of IIIA–VIA compounds are displayed in Fig. 12.1.

The two second row elements, B and O, with three valence electrons of B atoms ($2s^2p_x{}^1$) and six valence electrons of O atoms ($2s^2p_x{}^2p_y{}^2$), have the smallest atom sizes and thus the strongest and shortest covalent bonds in this group, due to efficient orbital overlap. Covalent bonding in the stable hexagonal 2H structure consists of four sublayers arranged as O–B–B–O with sp^3 hybridized chalcogen atoms, where three of these orbitals form B–O bonds and the lone pair of electrons occupies the remaining p_z orbital (Fig. 12.2) [1]. As shown in this figure, the monolayer consists, in principle, of two buckled triangular sublayers bonded together by covalent B–B bonds (see IVA-VA). The short bond length between B atoms indicates strong covalent bonding that is basically independent of the type of chalcogen atom.

This experimentally verified chemically-bonded 'double' structure has a B_2O_2 stoichiometry and a unit cell with four atoms, two B atoms and two O atoms. Due to the enormous difference in the electronegativities of B (2.04) and O (3.44), bonding has a strong ionic contribution with a charge transfer of $\delta q = 1.52$ e from the B atom to the O atom. Since the electronegativity decreases from oxygen to tellurium, the charge transfer from the group–IIIA atom to the chalcogen atom decreases going down the column of chalcogen constituents. A perspective view of the hexagonal 2H structure illustrates the geometrical arrangement of bonds between inner metal atoms and outer chalcogen atoms (see Fig. 12.3). Since the two outer and two inner sublayers have the same electronegativity, the charge distribution is uniform in the four-sublayer system. The cohesive energy of the 2H phase of $E_c = -6.65$ eV/atom is the largest binding energy in the group of B–VIA monolayers and indicates considerably stronger binding than in TlTe monolayers, for example, with one of the highest cohesive energies of $E_c = -2.54$ eV/atom [2]. The small atom size and large electronegativity of oxygen allow strong covalent-ionic bonding via efficient

H							He
Li	Be	B	C	N	O	F	Ne
Na	Mg	Al	Si	P	S	Cl	Ar
K	Ca	Ga	Ge	As	Se	Br	Kr
Rb	Sr	In	Sn	Sb	Te	I	Xe
Cs	Ba	Tl	Pb	Bi	Po	At	Rn
Fr	Ra						

Fig. 12.1 Periodic table of the main group elements highlighting the elements of group IIIA and group VIA involved in the formation of binary group IIIA–VIA monolayers

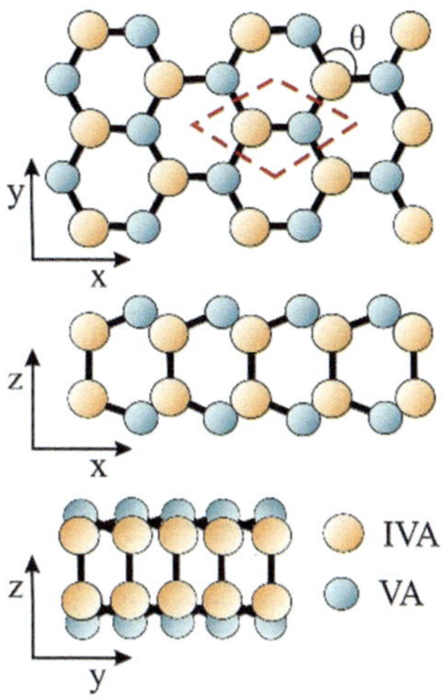

Fig. 12.2 Top view, front view, and side view of the 1H structure of group IIIA–VIA monolayers. The dashed box indicates the unit cell with four atoms

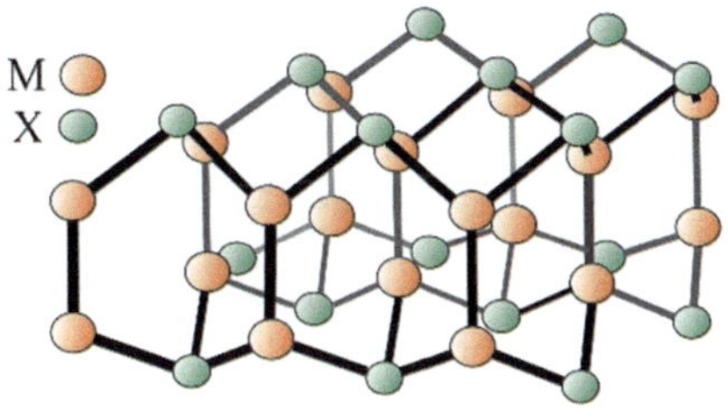

Fig. 12.3 Perspective view of the hexagonal 2H structure of group IIIA–VIA monolayers. Adapted with permission from Ref. [3], Copyright 2014, American Physical Society

orbital overlap and significant ionic interactions leading to strong polarization of the heteroatomic bonds.

12.1.1.2 2D Structures and Mechanics

<u>B_2O_2</u>: The lattice constant of 2H–BO monolayers of a = 0.245 nm, bond lengths of $d_{B\text{–}O} = 0.152$ nm and $d_{B\text{–}B} = 0.177$ nm, and O–O distance $h_{OO} = 0.292$ nm are the smallest values in the whole group of 25 IIIA–VIA compounds. The bond angles $\theta_{BBO} = 112.11°$ and $\theta_{OBO} = 106.71°$ do not vary much in the other groups of IIIA–VIA monolayers considered in this chapter [1, 2]. The mean van der Waals diameter of BO is $d_w = 0.335$ nm [4]. Adding the van der Waals diameter of O $d_w = 0.310$ nm [4] and the distance between O atoms yields a monolayer thickness of $d_{w+h} = 0.602$ nm. The space group of the 2H phase is *P*-6*m*2. The lattice parameters based on density functional theory (DFT) calculations using the generalized gradient approximation (GGA) and the Perdew, Burke, and Ernzerhof (PBE) functionals agree well with those obtained by molecular dynamics (MD) simulations employing a Stillinger–Weber potential [5]. The cohesive energies and lattice parameters of the IIIA–VIA monolayers are presented in Table 12.1.

Table 12.1 Symmetry, cohesive energy, lattice constant, distance between X atoms, bond lengths, bond angle, and van der Waals diameter of group IIIA–VIA monolayers

Monolayer	Symmetry	Cohesive energy, E_c (eV/atom)	Lattice constant a (nm)	Distance h_{XX} (nm)	Bond length d_{MX} (nm)	Bond length d_{MM} (nm)	Bond angle θ_{XMX} (°)	Van der Waals diameter d_w (nm)
BO B_2O	2H *Cmmm*	– 6.65 [2] – 6.80 [6]	0.245 [1] 0.393 [6]	0.292 [1]	0.153 [1] 0.134 [6]	0.177 [1] 0.171 [6]	106.71 [1] 107.8 [6]	0.335 [4]
BS	2H 1T	– 5.22 [2], – 5.81 [11] – 5.79 [10], – 5.82 [11]	0.304 [1] 0.306 [10]	0.342 [1] 0.381 [10]	0.195 [1] 0.195 [10]	0.172 [1] 0.170 [10]	102.47 [1]	0.360 [4]
BSe	2H 1T	– 4.71 [2], – 5.23 [10] – 5.24 [10], – 5.26 [11]	0.325 [1] 0.327 [10]	0.361 [1] 0.402 [10]	0.210 [1] 0.211 [10]	0.171 [1] 0.169 [10]	101.33 [1]	0.370 [4]
BTe	2H 1T	– 4.24 [2], – 4.69 [10] – 4.72 [10], – 4.75 [11]	0.356 [1] 0.359 [10]	0.383 [1] 0.428 [10]	0.231 [1] 0.232 [10]	0.171 [1] 0.168 [10]	100.74 [1]	0.390 [4]
BPo	2H		0.372 [1]	0.396 [1]	0.243 [1]	0.170 [1]	100.13 [1]	0.377 [4]
AlO	2H	– 5.73 [2], – 5.67 [18]	0.297 [1]	0.394 [1]	0.184 [1]	0.262 [1]	107.81 [1]	0.365 [4]
AlS	2H	– 4.26 [2], – 4.22 [18]	0.358 [1]	0.473 [1]	0.233 [1]	0.260 [1]	100.65 [1]	0.390 [4]
AlSe	2H	– 3.84 [2], – 3.87 [18]	0.378 [1]	0.492 [1]	0.247 [1]	0.259 [1]	99.60 [1]	0.400 [4]
AlTe	2H	– 3.38 [2], – 3.28 [18]	0.412 [1]	0.513 [1]	0.270 [1]	0.257 [1]	99.32 [1]	0.420 [4]
AlPo	2H		0.425 [1]	0.528 [1]	0.280 [1]	0.257 [1]	98.67 [1]	0.407 [4]
GaO	2H	– 4.52 [2]	0.312 [1]	0.400 [1]	0.194 [1]	0.252 [1]	106.32 [1]	0.365 [4]

(continued)

Table 12.1 (continued)

Monolayer	Symmetry	Cohesive energy, E_c (eV/atom)	Lattice constant a (nm)	Distance h_{XX} (nm)	Bond length d_{MX} (nm)	Bond length d_{MM} (nm)	Bond angle θ_{XMX} (°)	Van der Waals diameter d_w (nm)
GaS	2H	– 3.62 [2], – 4.16 [27]	0.364 [1]	0.466 [1]	0.237 [1]	0.248 [1]	100.45 [1]	0.390 [4]
GaSe	2H	– 2.81 [2], – 3.80 [27]	0.382 [1]	0.483 [1]	0.250 [1]	0.247 [1]	99.68 [1]	0.400 [4]
GaTe	2H	– 2.96 [2], – 3.47 [29]	0.413 [1]	0.503 [1]	0.271 [1]	0.247 [1]	99.56 [1]	0.420 [4]
GaPo	2H		0.426 [1]	0.514 [1]	0.280 [1]	0.247 [1]	99.14 [1]	0.407 [4]
InO	2H	– 3.98 [2], – 5.29 [56]	0.347 [1]	0.447 [1]	0.216 [1]	0.286 [1]	106.89 [1]	0.375 [4]
InS	2H	– 3.32 [2], – 3.85 [27]	0.394 [1]	0.520 [1]	0.257 [1]	0.282 [1]	100.24 [1]	0.400 [4]
InSe	2H	– 2.57 [2], – 3.54 [27]	0.409 [1]	0.539 [1]	0.269 [1]	0.282 [1]	99.10 [1]	0.410 [4]
InTe	2H	– 2.77 [2], – 2,97 [3]	0.439 [1]	0.560 [1]	0.289 [1]	0.282 [1]	98.81 [1]	0.430 [4]
InPo	2H		0.450 [1]	0.573 [1]	0.298 [1]	0.282 [1]	98.23 [1]	0.417 [4]
TlO	2H		0.369 [1]	0.463 [1]	0.229 [1]	0.292 [1]	107.04 [1]	0.375 [4]
TlS	2H	– 3.04 [29]	0.412 [1]	0.531 [1]	0.267 [1]	0.287 [1]	100.80 [1]	0.400 [4]
TlSe	2H	– 2.89 [29]	0.427 [1]	0.548 [1]	0.279 [1]	0.287 [1]	99.88 [1]	0.410 [4]
TlTe	2H	– 2.54 [29]	0.456 [1]	0.568 [1]	0.298 [1]	0.288 [1]	99.69 [1]	0.430 [4]
TlPo	2H		0.465 [1]	0.579 [1]	0.305 [1]	0.288 [1]	99.21 [1]	0.417 [4]

DFT-GGA-PBE calculations deliver a large isotropic Young's modulus of E = 350 N/m, comparable to that of graphene, and a Poisson ratio of $\nu = 0.20$ [2]. In comparison to these DFT calculations, MD simulations allow the treatment of larger system sizes, inclusion of the effect of temperature, and estimation of the influence of defects on mechanical performance [5]. The slightly anisotropic Young's moduli of BO derived by MD simulations are $E_{ac} = 305.5$ N/m and $E_{zz} = 302.8$ N/m at 1 K and $E_{ac} = 301.2$ N/m and $E_{zz} = 300.8$ N/m at 300 K. The results obtained for the ultimate fracture strengths are $\sigma_{ac} = 58$ N/m and $\sigma_{zz} = 56$ N/m at 1 K and $\sigma \approx 44$ N/m at 300 K. The corresponding ultimate strains of $\varepsilon_{ac} = 0.194$ and $\varepsilon_{zz} = 0.247$ at 1 K decrease substantially to $\varepsilon_{ac} = 0.117$ and $\varepsilon_{zz} = 0.134$ at 300 K. The temperature effect on the stiffness is relatively small with a decrease of 4–5% in the covered region from 1 to 900 K. Regarding the influence of defects, the crack length has a considerable impact on the fracture strength, whereas its effect on the elastic modulus is negligible. Notably, the authors discuss the relevance and limitation of Griffith's criterion for identifying nanocracks. Mono- and bi-vacancies reduce the fracture strength significantly if patterned perpendicular to the loading direction. The mechanical properties of the IIIA–VIA monolayers are listed in Table 12.2.

In addition to the extreme in-plane stiffness of BO monolayers, the bending rigidity of 30 eV is roughly an order of magnitude greater than that of graphene [2]. This enormous resistance against out-of-plane distortions is due to the compact four-layer configuration with a strong B–B bond perpendicular to the plane. Owing to their structural similarity to BO, the group IIIA–VIA compounds have substantial bending rigidities.

B_2O: A global minimum search adopting particle swarm optimization (PSO) revealed a B_2O structure with a global minimum binding energy of $E_c = -6.80$ eV/atom [6]. Figure 12.4 shows the orthorhombic structure of B_2O with the space group *Cmmm*. Note that in this structure, both B and O atoms possess their natural coordination of 3 and 2, respectively. The lattice constant is a = 0.393 nm, the B–B bond length is $d_{B-B} = 0.171$ nm, and the B–O bond length is $d_{B-O} = 0.134$ nm, indicating strong covalent bonding with the two bond angles 107.8° and 126.1° adopted by the slightly distorted trigonal coordinated B atoms (see Fig. 12.4). Furthermore, first-principles calculations provided the electronic, phonon, and mechanical properties of the metallic B_2O monolayer. Note that B_2O is not only a metallic monolayer but also an intrinsic superconductor described in more detail below [6]. Young's modulus is extremely anisotropic, with extremely deviating values of $E_{ac} = 23.4$ N/m and $E_{zz} = 130.4$ N/m, and the same is true for Poisson's ratios, which are $\nu_{ac} = 0.29$ and $\nu_{zz} = 1.59$ [6].

B_xO: The incorporation of oxygen into the basal plane of electron-deficient borophene follows chemical intuition and leads to various stable stoichiometric 2D boron oxide structures. First-principles global optimization studies confirmed this expectation and proposed B_xO_y systems such as B_4O, B_5O, B_6O, B_7O, and B_8O with well-defined degrees of oxidation, which contain motifs such as B–O–B and the OB_3 tetrahedron [6, 7]. The predicted properties include strong linear dichroism, Dirac node-line semimetallicity, and negative differential resistance. Electronic structure

Table 12.2 Structure, Young's modulus, Poisson's ratio, ultimate strength, and ultimate strain of group IIIA–VIA monolayers

Monolayer	Structure	Young's modulus E_{ac}, E_{zz} (N/m)	Poisson's ratio ν_{ac}, ν_{zz}	Intrinsic strength σ_{ac}, σ_{zz} (N/m)	Ultimate strain ε_{ac}, ε_{zz}
BO	2H	350 [2], 306, 303 [5]	0.20 [2]	58, 56 [5]	0.194, 0.247 [5]
BS	2H	212 [2], 203, 201 [11]	0.12 [2], 0.13, 0.12 [11]	15.2, 20.3 [11]	0.12, 0.26 [11]
BSe	2H	172 [2], 162, 159 [11]	0.15 [2], 0.17, 014 [11]	12.7, 14.9 [11]	0.13, 0.18 [11]
BTe	2H	136 [2], 131, 130 [11]	0.15 [2], 0.17, 0.14 [11]	11.4, 14.0 [11]	0.15, 0.30 [11]
AlO	2H	149 [2], 155 [18]	0.37 [2], 0.36 [18]		
AlS	2H	80 [2], 79.9 [18]	0.25 [2], 0.25 [18]		
AlSe	2H	66 [2], 68.3 [18]	0.24 [2], 0.24 [18]		
AlTe	2H	54 [2], 55.0 [18]	0.23 [2], 0.24 [18]		
GaO	2H	130 [2]	0.40 [2]		
GaS	2H	73 [2], 91 [28]	0.22 [2], 0.26 [28]	5.6 [28]	0.24 [28]
GaSe	2H	67 [2], 77 [28]	0.24 [2], 0.25 [28]	4.8 [28], 8.6, 10.5 [39]	0.23 [28], 0.24 [39]
GaTe	2H	55 [2], 66 [31]	0.20 [2]		
InO	2H	75 [2], ~ 100, ~ 104 [57]	0.44 [2]		
InS	2H	50 [2], 55.5 [29]	0.31 [2], 0.26 [29]		
InSe	2H	42 [2], 44.1, 41.3 [64]	0.29 [2], 0.27 [29]	9.45, 12.4 [64]	0.362, 0.433 [64]
InTe	2H	39 [2], 37 [29]	0.23 [2], 0.26 [29]		
TlS	2H	40.2 [29]	0.29 [29]		
TlSe	2H	35.2 [29],	0.28 [29]		
TlTe	2H	28.2 [9]	0.25 [29]		

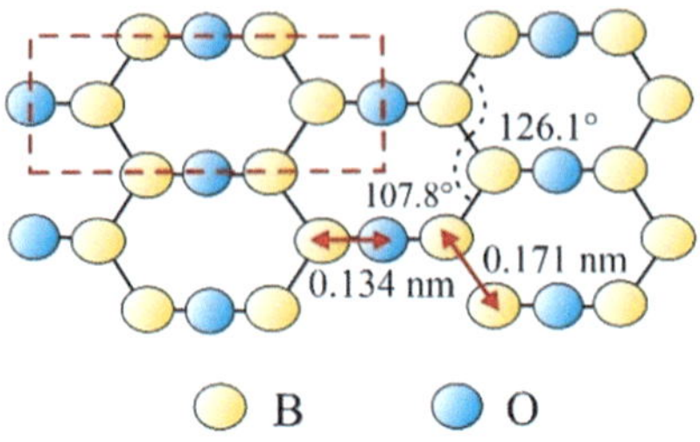

Fig. 12.4 Structure of the B_2O monolayer. The dashed red lines indicate the rhombic primitive cell. Note the short bonds, especially the B–O bond length, and the distorted trigonal coordination. Reproduced from Ref. [6], Copyright 2020, Springer Nature

engineering is possible by tuning the stoichiometry, where the O content provides an important chemical degree of freedom.

ph-BO: The planar porous hexagonal boron oxide structure (ph-BO), consisting of six hexagons with B–O bonds enclosing a hole with a diameter of 0.627 nm, demonstrates the variability of B–O bonding (see Fig. 12.5) [8]. The structure presents the global minimum among the boron monoxide structures with a cohesive energy of $E_c = -6.68$ eV/atom. This phase, named ph-BO, has a large lattice constant of $a = 0.782$ nm and short bond lengths of $d_{B\text{–}O} = 0.139$ nm and $d_{B\text{–}B} = 0.172$ nm, which are shorter than those of the hexagonal 2H structure. The electron localization function (ELF) analysis confirmed the presence of electrons in the middle of the B–B bonds and the B–O bonds, and, of course, around O atoms. Interestingly, ph-BO is a semiconductor with wide band gaps of $E_g = 3.59$ eV (PBE) and 5.23 eV (Heyd-Scuseria-Ernzerhof (HSE06)) and thermodynamic, dynamic, and thermal stability up to 1500 K. Young's modulus of this porous lattice is only $E = 71.8$ N/m. These properties endow the structure with great promise for use in high-power electronics, deep-UV emitters, and blue-violet laser diode applications.

Fig. 12.5 Top view of the planar hexagonal ph-BO monolayer with porous geometry. Reproduced with permission from Ref. [8], Copyright 2018, Elsevier

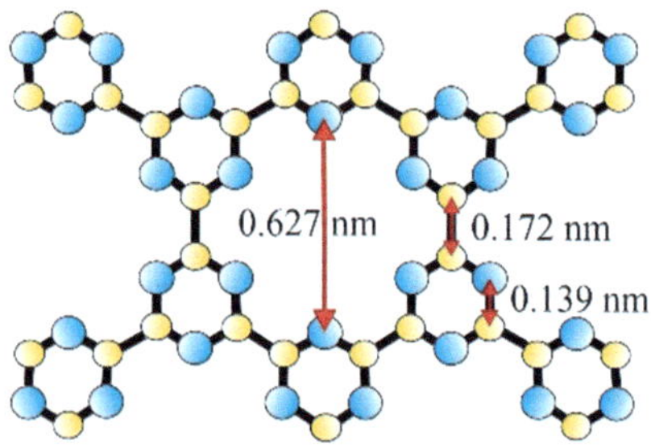

12.1.1.3 Physics and Applications

While the synthesis of BO monolayers and few-layer sheets is still lacking, these 2D materials offer a versatile platform for engineering electronic structures by oxidation. Unfortunately, it is difficult to accurately control the exact degree of oxidation. A well-defined previously studied stoichiometric material is the honeycomb borophene oxide B_2O, which exhibits intrinsic superconducting features that are tunable by strain [6]. Within the framework of the Bardeen–Cooper–Schrieffer (BCS) theory, B_2O monolayers exhibit intrinsic superconducting behavior with a transition temperature of $T_c \approx 10.3$ K. The superconductivity is tunable by strain to an optimal temperature of $T_c \approx 14.7$ K under a tensile strain of 1%. The specific features originate from the out-of-plane soft mode vibrations of the monolayer, which are enhanced by the light O atoms.

A systematic global search was performed to identify the lowest-energy structures of 2D boron oxides by using the PSO algorithm combined with first-principles methods. Several interesting structures were identified [9]. For example, B_6O shows a Dirac loop near the Fermi level and a maximum Fermi velocity comparable to that of graphene. The stable B_4O monolayer is a direct semiconductor with an extreme anisotropy in the carrier mobility that is approximately five orders of magnitude greater in the armchair direction than in the zigzag direction; such an anisotropy was not previously observed in 2D materials.

12.1.2 BS Monolayer

12.1.2.1 Chemistry and 2D Bonding

In addition to the most extensive data collection on group IIIA–VIA monolayers providing 2D bonding, structural, and electronic properties of 25 compounds [1], two subsequent studies cover both the 2H and 1T structures of the boron monochalcogenides BS, BSe, and BTe, one concentrating on their thermoelectric properties [10] and one on their mechanical properties [11]. Usually the structural parameters of these calculations agree quite well, whereas the published cohesive energies, reported by different groups, sometimes deviate substantially (see Table 12.1).

In the BS monolayer, boron with the valence electrons ($2s^2p_x^{\ 1}$) contributes three electrons, and S with the valence electrons ($3s^2p_x^{\ 2}p_y^{\ 2}$) contributes six valence electrons to the outer shell. In the stable 2H structure, the B atom forms three covalent B–S bonds, each S atom is also involved in three covalent B–S bonds, and in addition the S atom has a lone pair of electrons in the z direction. According to a Bader analysis, the different electronegativities of B (2.04) and S (2.58) induce a charge transfer of $\delta q = 0.97$ e from the B atom to the S atom in the 2H structure [2]. Note that a much larger Bader charge transfer of $\delta q = 1.88$ e is postulated elsewhere from a single B atom to a chalcogen atom [11]. An analysis of the type of bonding by the electron localization function (ELF) resulted in ELF values > 0.75 around the centers of all the bonds, confirming the dominant covalent bonding in this structure [11]. The first published value of the binding energy of 2H–BS monolayers of $E_c = -5.22$ eV/atom (PBE) is lower than the binding energies reported afterwards by other groups [2]. Two recent publications give for the 2H– and 1T–BS monolayers larger, nearly identical cohesive energies of $E_c = -5.81$ eV/atom and $E_c = -5.82$ eV/atom [11] as well as $E_c = -5.78$ eV/atom and $E_c = -5.79$ eV/atom [10], respectively. The nearly identical values for the two phases are due to the similarity of chemical bonding in these two phases [10, 11]. Note that the unit cell of the 2H and 1T structure contains four atoms.

12.1.2.2 2D Structures

The two phases are not only energetically but also dynamically, and thermally stable structures of BS monolayers. Theory suggests that in both hexagonal structures, two sublayers of boron are sandwiched between two chalcogen sublayers. Thus, the 2H phase of the honeycomb monolayer is arranged in the stacking order S–B–B–S. The 2H structure consists of an assembly of two buckled hexagonal layers (B_2S_2) combined such that boron atoms are located on boron atoms and chalcogen atoms on chalcogen atoms (see top view in Fig. 12.2). The same order describes the second 1T structure with a slightly lower binding energy. In the 1T phase the chalcogen atoms of one sublayer are moved to the hollow center of the hexagonal lattice, either on the bottom or top side (see Fig. 12.6) [10, 11]. Compare with IVA-VA monolayers. In the 1T structure, boron atoms are situated on boron atoms; however, in this configuration the chalcogen atoms are located at the hexagon centers (see Fig. 12.6).

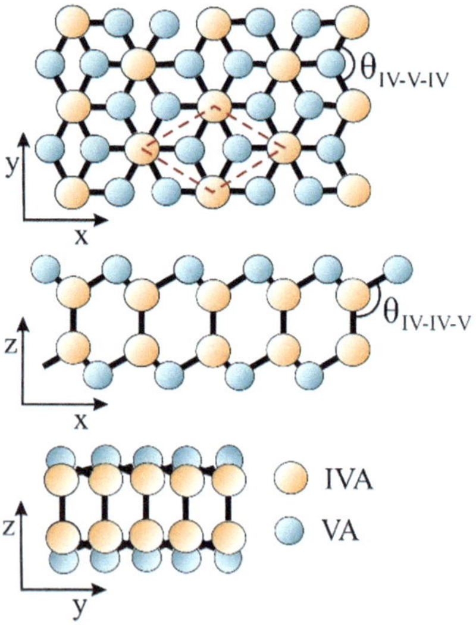

Fig. 12.6 Top, front, and side view of the 1T structure of group IIIA–VIA monolayers. The dashed box indicates the unit cell with four atoms

First-principles DFT calculations approximating the exchange correlation potential by GGA with the PBE functional showed that the hexagonal four-layer 2H-structure (α-phase) of BS monolayers is strongly bonded with the structural parameters a = 0.304 nm, $d_{B\text{–}S}$ = 0.195 nm, $d_{B\text{–}B}$ = 0.172 nm, h_{SS} = 0.342 nm, θ_{BBS} = 115.80°, and θ_{SBS} = 102.47° [1]. Adding the van der Waals diameter of S atoms d_w = 0.36 nm and the distance between S atoms, yields an estimate of the monolayer thickness of d_{w+h} = 0.702 nm. The dynamically and thermally stable 2H-structure has a *P*-6*m*2 symmetry. For the indirect semiconductor theory predicts a large band gap of E_g = 2.80 eV, based on the PBE approximation, and as expected, for the 2H phase, a much larger gap of 3.85 eV by the HSE06 approximation [11].

The hexagonal four-layer 1T structure (β-phase) has as space group *P*-3*m*2. The published lattice parameters of the 1T structure of BS monolayers deviate only slightly from those of the corresponding 2H structure due to the similarity in the type of bonding (see Table 12.1) [10, 11]. Owing to the same type of bonds in the 2H and 1T structures, the cohesive energy of the latter phase of E_c = – 5.79 eV/atom [10] (– 5.82 eV/atom [11]) agrees with that of the 2H phase within the errors involved in such calculations. Moreover, the band gaps estimated by different approaches of E_g = 2.87 eV (PBE) and 4.0 eV (HSE06) are only slightly larger than those of the 2H structure [11].

A theoretical study combining ab initio calculations and a structure search method identified two metastable orthorhombic monolayers, γ BS with a cohesive energy of E_c = – 5.11 eV and space group *C*2/*m* and δ BS with the lowest binding energy of – 4.92 eV and the space group *Pmma* [12]. The γ-BS monolayer with the lattice constants a = 0.738 nm and b = 0.306 nm is a semiconductor with a direct band gap at E_g = 2.94 eV (HSE06) and strongly anisotropic stiffness values at E_{ac} = 81 N/m and E_{zz} = 140 N/m, and Poisson ratios at ν_{ac} = 0.02 and ν_{zz} = 0.03 (see Fig. 12.7). The δ-BS monolayer is a superconducting metallic polymorph with large electron–phonon coupling and one of the highest critical superconducting temperatures of T_c ≈ 21.56 K [12]. The lattice parameters of this monolayer are a = 0.307 nm and b = 0.259 nm, and the anisotropic Young's moduli E_{ac} = 198 N/m and E_{zz} = 166 N/m and Poisson's ratios ν_{ac} = 0.14 and ν_{zz} = 0.11.

12.1.2.3 Mechanics

The large binding energies of the 2H- and 1T-BS networks imply a large stiffness and high fracture strength for both structures. The early published elastic modulus of the 2H phase at E = 212 N/m and Poisson ratio at ν = 0.12 [1] are in good agreement with the slightly anisotropic stiffness values of E_{ac} = 203 N/m and E_{zz} = 201 N/m and Poisson ratios of ν_{ac} = 0.13 and ν_{zz} = 0.12 (see Table 12.2) [11]. The nearly isotropic elastic response in the linear regime of the stress–strain curve is a consequence of the trigonal symmetry. Of interest is the large bending rigidity of D = 22 eV, which is a consequence of the strong B–B bonds perpendicular to the plane formed by the two buckled sublayer assemblies [2]. The anisotropy increases for the ultimate strengths of σ_{ac} = 15.2 N/m and σ_{zz} = 20.3 N/m, where the larger strength

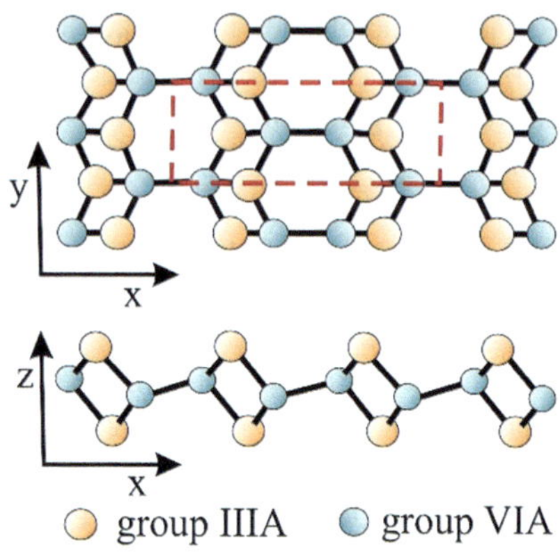

Fig. 12.7 Top and side view of γ-BS monolayers. The red dashed line indicates the unit cell. Reproduced with permission from Ref. [12], Copyright 2020, AIP Publishing

in the zigzag direction fulfills Griffith's rule quite well. The ultimate strains of $\varepsilon_{ac} = 0.12$ and $\varepsilon_{zz} = 0.26$ indicate much greater flexibility in the zigzag direction [11]. In addition, the latter study investigated the linear and nonlinear mechanical properties of the 1T phases of the BS, BSe, and BTe monolayers [11]. The 1T structures of these compounds exhibit a marginally smaller stiffness, similar Poisson ratios, and slightly smaller ultimate strengths and strains due to the somewhat lower symmetry of the 1T structure [11].

12.1.2.4 Physics and Applications

Boron monochalcogenides possess interesting physicochemical properties, especially for potential thermoelectric applications. The intriguing electronic band structure and transport properties of these compounds led to a systematic study of the ballistic transport and thermoelectric properties of the 25 group IIIA–VIA MX compounds (M = B, Al, Ga, In, and Tl; X = O, S, Se, Te, and Po) [1]. In particular, for the boron chalcogenides BS, BSe, and BTe, superior carrier mobilities were predicted due to the small effective mass of electrons [10]. In the temperature range of 200–450 K, an electronic figure of merit of $ZT \approx 1$ was found. High-performance 2D thermoelectric materials have attracted significant interest for transforming automotive radiator heat to electricity, in thermally powered night vision systems, in electric refrigeration systems, as coolers for diode lasers, and as low-voltage power generators.

Based on state-of-the-art first-principles calculations, the possible experimental realization of the four 2D BS sheets named α-BS, β-BS, γ-BS, and δ-BS was proposed, all of which have thermal and dynamical stability [12]. According to theory, δ-BS with an orthorhombic lattice belonging to the space group *Pmma* is a superconductor with a large electron–phonon coupling and a high critical temperature. For the 2H structure (α-phase) and the 1T structure (β-phase) these authors give somewhat lower binding energies at $E_c = -5.23$ eV/atom and -5.22 eV/atom [12], respectively, in excellent agreement with $E_c = -5.22$ eV/atom [2] for the 2H

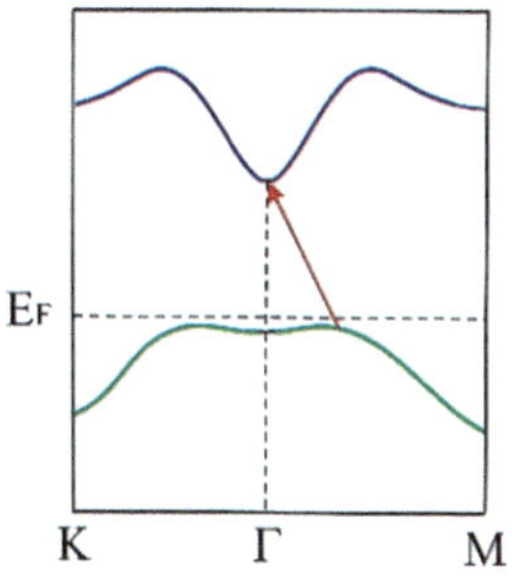

Fig. 12.8 Scheme of nearly flat VBM with Mexican-hat-shaped band edge and parabolic CBM. This causes a high DOS at the top of the valence band and an almost constant DOS at the bottom of the conduction band

phase. The top valence bands of the 2H and the 1T BS sheets are nearly flat around the Γ point with Mexican-hat-shaped valence band edges, leading to sharp peaks in the DOS and strong van Hove singularities near the Fermi level (see Fig. 12.8) [12]. These shapes of the valence bands are attributed to the strongly overlapping 2*p* orbitals of the B and the S atoms, which is the source of efficient covalent bonding.

12.1.2.5 Synthesis

Notably, BS monolayers belong to the few synthesized group IIIA–VIA monolayers. Inspired by the medium exfoliation energy of $E_f = 0.96$ N/m proposed for the 2H phase [12], it has been suggested that the most stable mono- and few-layer 1T structures may be obtained from their bulk counterparts [11]. In fact, physical exfoliation of BS nanosheets can be achieved from bulk rhombohedral boron monosulfide (r-BS) [13]. The synthesis of single-phase r-BS with the space group *R*-3*m* was revealed by X-ray diffraction (XRD), X-ray photoelectron spectroscopy (XPS), scanning electron microscopy (SEM), electron probe microanalysis (EPMA), and Raman spectroscopy. Acetonitrile allowed the separation of a mixture of BS nanosheets from r-BS powder. The authors assumed that the nanosheets preserved the trigonal structure of bulk r-BS with the space group *R*-3*m*.

12.1.3 BSe Monolayer

12.1.3.1 Chemistry and 2D Bonding

The boron atom has three valence electrons ($2s^2p_x^1$) and the Se atom has six valence electrons ($4s^2p_x^2p_y^2$) in the outer shell. These nine bonding electrons form in the four-layer 2H and 1T structures one B–B, three B–S, and one lone electron pair. Due to the difference in the electronegativities of B (2.04) and Se (2.55), a Bader charge transfer of $\delta q = 0.47$ e [2] (0.68 e [11]) occurs from the B atom to the Se atom, inducing ionic interactions in the heteroatomic B–Se bonds of the 2H structure. Like for BS monolayers, an ELF value > 0.75 was observed for B–Se bonds, confirming strong covalent bonding [11]. The first published cohesive energy of the 2H phase of $E_c = -4.71$ eV/atom (PBE) is in reasonable agreement with the binding energy of related compounds [2]. Two recent publications give for both the 2H- and 1T-BSe monolayer smaller, but nearly identical cohesive energies of $E_c = -5.24$ eV/atom and -5.26 eV/atom [11], as well as $E_c = -5.23$ eV/atom and -5.24 eV/atom [10], respectively, indicating stronger bonding. Note that the unit cells of the 2H- and 1T-BSe structures contain four atoms despite different symmetries.

12.1.3.2 2D Structures

The 2H and 1T structures of BSe monolayers are energetically, dynamically, and thermally stable. The lattice parameters of the 2H phase of BSe monolayers are a $= 0.325$ nm, $d_{B–Se} = 0.210$ nm, $d_{B–B} = 0.171$ nm, the distance of the Se atoms is $h_{SeSe} = 0.361$ nm, and the bond angles are $\theta_{BBSe} = 116.73°$ and $\theta_{SeBSe} = 101.33°$ [1]. Adding the van der Waals diameter of Se atoms $d_w = 0.38$ nm and the distance between S atoms, yields an estimate of the monolayer thickness of $d_{w+h} = 0.741$ nm. The band gaps of the 2H structure calculated by different approaches are $E_g = 2.07$ eV (PBE) and 2.96 eV (HSE06) [11]. The 2H structure belongs to the space group *P*-6*m*2.

As shown in Table 12.1, the results for the 1T structure of BSe monolayers do not deviate much from those for the structurally similar 2H phase [10, 11]. The band gaps derived by different approaches at $E_g = 2.50$ eV (PBE) and 3.43 eV (HSE06) are slightly larger than those of the 2H phase [11]. The symmetry of the 1T structure is *P*-3*m*2.

12.1.3.3 Mechanics

The combination of boron with the larger Se atom, compared to the S atom, reduces orbital overlap; thus, the binding energy and stiffness of the 2H- and 1T-BSe monolayers decrease. The first reported value of the isotropic Young's modulus is $E = 172$ N/m and of the Poisson's ratio it is $\nu = 0.15$ for the 2H structure [2]. These values are in reasonable agreement with the slightly lower anisotropic elastic moduli of $E_{ac} = 162$ N/m and $E_{zz} = 159$ N/m and Poisson ratios of $\nu_{ac} = 0.17$ and $\nu_{zz} = 0.14$ [11]. As expected, the anisotropy increases for the nonlinear mechanical properties, e.g., for the ultimate strengths at $\sigma_{ac} = 12.7$ N/m and $\sigma_{zz} = 14.9$ N/m and the ultimate strains at $\varepsilon_{ac} = 0.13$ and $\varepsilon_{zz} = 0.18$, where the larger strength in the zigzag direction fulfills Griffith's rule reasonably well [11]. The bending rigidity decreases from 22 to 16 eV from the 2H-BS to the 2H-BSe monolayers [2].

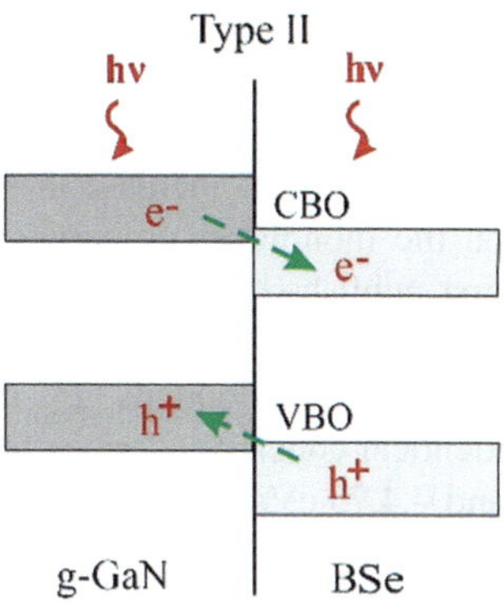

Fig. 12.9 Comparison of the VBM and CBM of the heterostructure g-GN/BSe. In type-II band alignment, electron transfer occurs between the CBs from g-GaN to BSe, and hole transfer occurs between the VBs from BSe and g-GN controlled by the conduction band offset (CBO) and the valence band offset (VBO)

12.1.3.4 Physics and Applications

A detailed investigation of the optical properties of 2H-BSe monolayers based on the DFT framework included optical parameters such as the imaginary and real components of the dielectric function, the refractive index, and the absorption coefficients using the random phase approximation method (RPA) [14]. Owing to the absorption peaks at 8.72 eV and 19.72 eV, BSe is a strong absorber of UV radiation.

The suitability of 2D heterostructures such as g-GaN/BSe for high-efficiency photocatalytic water splitting was investigated theoretically by characterizing the structural, electronic, optical, and interfacial properties of these van der Waals heterostructures [15]. The calculations showed that the indirect semiconductor with a band gap of $E_g = 2.27$ eV has suitable band edges for the redox potential of water splitting at pH = 0 and 7, and type-II band alignment that helps to prevent the efficient recombination of photogenerated electron–hole pairs. As shown in Fig. 12.9, in the type-II band alignment, the valence band and the conduction band belong to different compounds and thus promote the separation of photogenerated electrons and holes. In type-II band alignment, for example, electron transfer occurs from the conduction band of g-GaN to BSe, and hole transfer occurs from the valence band of BSe to g-GaN controlled by the conduction band offset (CBO) and the valence band offset (VBO). In addition, the considered heterostructure has high carrier mobilities and excellent absorption abilities under sunlight, based on an appropriate band gap.

12.1.4 BTe Monolayer

12.1.4.1 Chemistry and 2D Bonding

The valence electrons of B atoms ($2s^2p_x{}^1$) and those of Te atoms ($5s^2p_x{}^2p_y{}^2$) deliver the bonding electrons needed for the formation of stable 2D lattices such as the stable 2H and 1T structures by covalent bonding. The electronegativities of B (2.04) and Te (2.10) are almost identical and therefore the ionic contribution to bonding is marginal. Note that a Bader analysis formally suggests a negligible reverse charge transfer of $\delta q = 0.03$ e [2] (0.1 e [11]) from the Te atom to the B atom in the 2H-BTe monolayer. The increase in the atom size from Se to Te has a dominant influence on the monolayer properties such as bonding and mechanical performance. The first published value of the cohesive energy of the 2H–BTe monolayer of $E_c = -4.24$ eV/atom (PBE) is in agreement with the behavior of the binding energies of the related compounds [2]. Two following publications present smaller, but nearly identical cohesive energies for the 2H- and 1T-BTe phase of $E_c = -4.72$ eV/atom and – 4.75 eV/atom [11], as well as $E_c = -4.69$ eV/atom and – 4.72 eV/atom [10], respectively, suggesting stronger bonding. In any case, these values denote a decrease in the binding strength from BSe to BTe monolayers.

12.1.4.2 2D Structures

The lattice parameters of the 2H–BTe monolayer are a = 0.356 nm, $d_{B-Te} = 0.231$ nm, $d_{B-B} = 0.171$ nm, $h_{TeTe} = 0.383$ nm, $\theta_{BBTe} = 117.22°$, and $\theta_{TeBTe} = 100.74°$ [1, 2]. As discussed before, there is good agreement among the published data on the geometrical structure. Adding the van der Waals diameter of Te atoms $d_w = 0.42$ nm and the distance between Te atoms, yields an estimate of the monolayer thickness of $d_{w+h} = 0.803$ nm. The reported values of the reduced band gaps of the indirect semiconductor vary between $E_g = 1.52$ eV (HSE06)[1], 1.49 eV (PBE), and 2.07 eV (HSE06) [11]. Table 12.1 presents the lattice parameters and cohesive energies of both the 2H and 1T structure.

12.1.4.3 Mechanics

The isotropic Young's modulus and Poisson's ratio of E = 136 N/m and $\nu = 0.15$, respectively, for the 2H structure are comparable with the slightly anisotropic elastic moduli at $E_{ac} = 131$ N/m and $E_{zz} = 130$ N/m and the Poisson ratios at $\nu_{ac} = 0.17$ and $\nu_{zz} = 0.14$ [2, 11]. The bending rigidity at D = 15 eV remains nearly constant from the 2H-BS to the 2H-BTe monolayer since the crucial B–B bonds have essentially identical binding strengths and bond lengths of $d_{B-B} \approx 0.171$ nm [2]. As usual, the nonlinear mechanical properties exhibit greater anisotropy for the ultimate strengths at $\sigma_{ac} = 11.4$ N/m and $\sigma_{zz} = 14.0$ N/m, in reasonable agreement with Griffith's rule, and the ultimate strains at $\varepsilon_{ac} = 0.15$ and $\varepsilon_{zz} = 0.30$ [11], indicating strongly anisotropic behavior.

12.1.4.4 Physics and Applications

Based on first-principles calculations, the already described γ-BS structure of the direct band gap semiconductor has been extended to several B-VIA compounds, namely, γ-BSe and γ-BTe monolayers [16]. Owing to the moderate direct band gap, excellent transport properties, dramatically high absorption from the visible to the ultraviolet spectral region, and extraordinary excitonic behavior of γ BTe, this structure is considered to provide a promising 2D material for optoelectronic and photovoltaic devices. For a combination of γ-BTe monolayers with a monolayer of GeP_2 as an acceptor material, first-principles calculations predict for the heterostructure γ-BTe/GeP_2 a power conversion efficiency of up to ~ 21.8 % and ~ 23 % under strain, which is comparable to that of typical heterolayer solar cells.

As discussed for the g-GaN/BSe heterostructure in more detail above [15], first-principles calculations elucidated the structural, electronic, optical, and charge transfer properties of the van der Waals heterostructure g-GaS/BTe under an applied electric field [17]. A positive external transverse electric field modulates the electronic band gaps by transforming the type-II band alignment to type-I or type-III band alignment. At an external electric field between + 0.6 and + 0.8 V/Å, the type-II band alignment transforms to type-I band alignment and a transition of the indirect-to-direct band gap occurs (see Fig. 12.10a). A larger external electric field of + 0.9 V/Å induces the typical broken-gap type-III band alignment with metallic character (see Fig. 12.10b). Importantly, an external electric field enhances the efficiency of the hydrogen evolution reaction (HER) and improves the optical performance in the visible region.

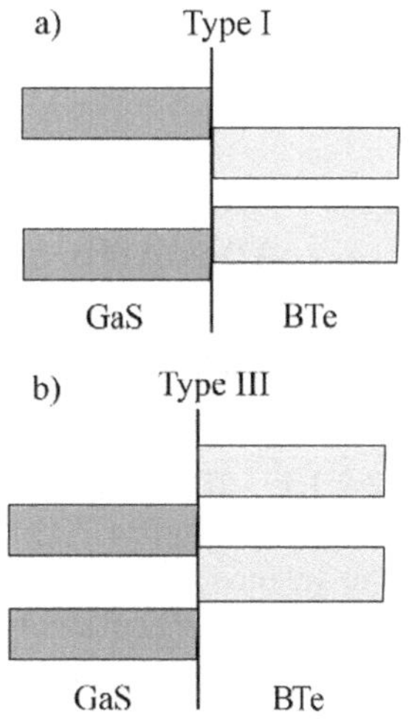

Fig. 12.10 **a** Type-I band alignment of the van der Waals heterostructure g-GaN/BTe induced by a low external transverse electric field. **b** Type-III band alignment of the heterostructure g-GaN/BSe induced by a medium external transverse electric field

12.1.5 BPo Monolayer

12.1.5.1 2D Bonding and Structure

Considering the valence electrons of B atoms ($2s^2p_x^{\,1}$) and Po atoms ($6s^2p_x^{\,2}p_y^{\,2}$), the available nine bonding electrons are expected to form stable hexagonal covalently bonded 2H and 1T-BPo structures, following the previously described chemical behavior of the homologues. The currently known electronegativities of 2.04 (B) and ~ 2.0 (Po) indicate a negligible polarization of the covalent B–Po bonds. Owing to the uncertainty in the electronegativity of Po, a prediction of the direction of charge transfer presently is not possible.

The lattice parameters of the 2H phase of BPo monolayers are a = 0.372 nm, $d_{B\text{–}Po}$ = 0.243 nm, $d_{B\text{–}B}$ = 0.170 nm, h_{PoPo} = 0.396 nm, θ_{BBPo} = 117.70°, and θ_{PoBPo} = 100.13°, and the symmetry of the space group is *P*-6*m*2 (see Table 12.1) [1]. Adding the van der Waals diameter of Po atoms d_w = 0.394 nm and the distance between S atoms, yields an estimate of the monolayer thickness of d_{w+h} = 0.790 nm. A comparison of the lattice parameters of 2H BPo with those of the 2H BTe monolayers reveals an increase in the lattice constant, the B–Po bond length, and the distance between the Po atoms, indicating the expected increase in the thickness of the BPo monolayer. The 2H–BPo monolayer is a semiconductor with a band gap of E_g = 0.90 eV (PBE) [1]. To date, this study seems to be the only one that provides fundamental information on the intrinsic properties of BPo monolayers.

12.2 Aluminum Monochalcogenide Monolayers (AlO, AlS, AlSe, AlTe, and AlPo)

12.2.1 AlO Monolayer

12.2.1.1 2D Bonding, Structure, and Applications

The valence electrons of Al atoms ($3s^2p_x^{\,1}$) and O atoms ($2s^2p_x^{\,2}p_y^{\,2}$) initiate the formation of 2D covalent networks, consisting of stable 2H and 1T four-layer structures with Al–Al bonds, Al–O bonds, and lone electron pairs. Owing to the substantial difference in the electronegativities between O (3.44) and Al (1.61), a large charge transfer of δq = 1.69 e occurs from each Al to each O atom [18]. Thus, in addition to covalent bonding based on sp^2 hybrids of Al atoms and sp^3 hybridization of O atoms, ionic interactions play an essential role in the strongly polar Al–O bonds. In comparison to the B–B bonds in the boron compounds the Al–Al bond is much weaker, however, still independent of the type of chalcogen atom. In comparison with bonding in BO monolayers, the reported cohesive energies of the 2H phase of AlO monolayers are only E_c = – 5.73 eV/atom [2] (PBE) and – 5.67 eV/atom [18] (PBE).

The estimated lattice constant of 2H AlO is a = 0.287 nm, the bond lengths are $d_{Al\text{–}O} = 0.180$ nm, $d_{Al\text{–}Al} = 0.261$ nm, the distance between O atoms is $h_{OO} = 0.401$ nm, and the bond angle is $\theta_{AlAlO} = 105.9$ [18]. The lattice parameters agree well with those of other DFT-GGA-PBE calculations, presented in Table 12.1 [1]. Adding the van der Waals diameter of O atoms $d_w = 0.31$ nm and the distance between O atoms yields an estimate of the monolayer thickness of $d_{w+h} = 0.711$ nm. The energetic and dynamic stabilities of the 2D material were confirmed by DFT calculations and ab initio molecular dynamics (AIMD) simulations [18, 19]. As linear elastic properties, the first published Young's modulus of E = 149 N/m and Poisson's ratio of $\nu = 0.37$ are in good agreement with those of succeeding studies giving E = 155 N/m and $\nu = 0.36$ [2, 18].

The calculated band gap of the indirect semiconductor depends on the approach employed and may vary substantially from publication to publication, as demonstrated by $E_g = 1.06$ eV (PBE) and 1.61 eV (HSE06) [18] in comparison with $E_g = 1.34$ eV (PBE) and 1.85 eV (HSE06) for the 2H phase [19]. In addition to AlO, these studies cover the AlS, AlSe, and AlTe monolayers and characterize their electronic band gaps and optical performance, including the complex refractive index, absorption coefficient, and energy loss function by means of the complex dielectric function [19]. All the compounds possess high absorption coefficients in the visible and UV regions.

The band edge positions straddle the reduction potential of H_2 and oxidation potential of H_2O. Furthermore, the presence of an electric field in AlO monolayers significantly reduces the Gibbs free energy for the hydrogen evolution reaction (HER), approaching an extremely low value of $\Delta G_{H*} = -0.02$ eV. This suggests that AlO may be a promising catalyst for hydrogen production.

In the group of IIIA–VIA compounds, the derivation of ternary monolayers has attracted increasing amounts of attention; for example, by substituting half of the Al atoms in Al_2O_2 with B atoms, Janus monolayers with the stoichiometry $BAlO_2$ are formed. A series of such Janus structures, $BAlX_2$ (X = O, S, Se, Te), function as model systems for studying the piezoelectric behavior and structural, electromechanical, and phonon dispersion properties of these 2D materials [20]. DFT and density functional perturbation theory (DFPT) calculations showed that all the monolayers are dynamically stable semiconductors with outstanding piezoelectric properties. The $BAlO_2$ monolayer exhibits the strongest piezoelectric performance with a relaxed-ion piezoelectric coefficient of $d_{11} = 15.89$ pm V^{-1}. The large in-plane and out-of-plane piezoelectric coefficients are comparable to or larger than those of previously described non-Janus and Janus monolayer structures.

12.2.2 AlS Monolayer

12.2.2.1 2D Bonding, Structure, and Applications

In the column of VIA elements, the van der Waals diameter of chalcogen atoms increases from O to Te [4]. The resulting larger size of AlS strongly influences the cohesive energy, structural parameters and especially the stiffness, which decreases drastically in comparison to AlO monolayers. The valence electrons of Al atoms ($3s^2p_x^1$) and S atoms ($3s^2p_x^2p_y^2$) form the already characterized 2H-structure of AlS. Owing to the medium difference in the electronegativities of Al (1.61) and S (2.58), the charge transfer from Al to S atoms is $\delta q = 1.49$ e, with a slightly diminished ionic contribution to bonding compared with that of AlO [17]. The currently available binding energies of the 2H structure are $E_c = -4.26$ eV/atom [2] (PBE) and -4.22 eV/atom [18] (PBE).

The lattice constant of the 2H–AlS monolayer increases to a = 0.358 nm, and the same is true for the bond lengths $d_{Al-S} = 0.232$ nm and $d_{Al-Al} = 0.260$ nm, and the distance between S atoms $h_{SS} = 0.473$ nm compared with the AlO monolayer, whereas the bond angle decreases to $\theta_{AlAlS} = 100.6°$ [1, 18]. Adding the van der Waals diameter of S atoms $d_w = 0.36$ nm and the distance between S atoms, yields an estimate of the monolayer thickness of $d_{w+h} = 0.833$ nm. Subsequent calculations confirmed the lattice parameters and yielded band gaps at $E_g = 2.09$ eV (PBE) and 2.88 eV (HSE06) [18]. The elastic modulus of AlS monolayers was estimated to be E = 80 N/m, indicating a strongly reduced stiffness, and the Poisson ratio is diminished to $\nu = 0.25$ [2], in good agreement with E = 79.9 N/m and $\nu = 0.25$ [18]. The bending rigidity of D = 16 eV is large for the 2H structure [2].

Substitution of one S sublayer by an O sublayer leads to the formation of a Janus aluminum oxysulfide (Al_2OS) monolayer [21]. In this structure, the lattice constant increases to a = 0.378 nm and the Al–Al and Al–S bonds agree with those of the AlS monolayer, whereas the thickness decreases to t = 0.438 nm, with an Al–O bond length at 0.195 nm. Remarkably, the reported direct band gaps decrease to $E_g = 1.58$ eV (PBE) and 2.11 eV (HSE06) [21]. Compared with the parental Al_2S_2 monolayer, the Janus compound exhibits suitable band edges for the HER process and enhanced absorption of visible light based on a direct band gap. Figure 12.11 shows a schematic representation of such a Janus monolayer with a suitable VBM, CBM, and redox potential for the water splitting reaction. The scheme shows the two electrochemical half-reactions consisting of the oxidation of H_2O by photogenerated holes to O_2 and the reduction of protons by photogenerated electrons to H_2. Importantly, the dipole moment generated by the intrinsic electric field along the z direction further helps to separate the photoinduced charges into different surfaces [21]. Moreover, the Gibbs free energy decreases from $\Delta G_{H*} = 0.97$ eV to 0.69 eV under an external electric field, which enhances the hydrogen adsorption energy.

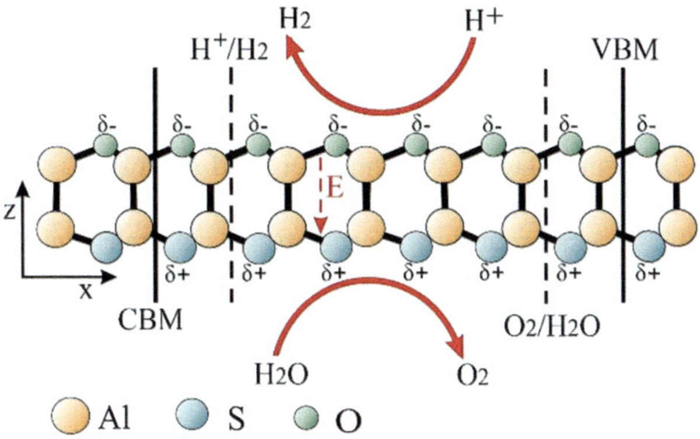

Fig. 12.11 Scheme of the photocatalytic activity of Janus monolayer OAl_2S indicating the VBM, CBM, and redox potential. Due to the intrinsic electric field, photogenerated electrons transfer to the top surface and contribute to the HER. Adapted from Ref. [21], Copyright 2022, Elsevier

12.2.3 AlSe Monolayer

12.2.3.1 2D Bonding, Structure, and Applications

As described above, the nine valence electrons of Al atoms ($3s^2p_x^1$) and Se atoms ($4s^2p_x^2p_y^2$) can form two types of covalent bonds, an Al–Al and an Al–Se bond with a remaining lone electron pair, which can arrange to various stable 2D structures. With further increase in the size of the Se atom by moving down in the column of chalcogen elements, the mean van der Waals diameter of AlSe increases to $d_w = 0.40$ nm [4], and the binding energy decreases accordingly. For the comparable electronegativities of Se (1.39) and S (1.49) a surprisingly large charge transfer of $\delta q = 1.36$ e is proposed from Al to Se atoms, inducing ionicity in the Al–S bonds [17]. Nearly identical cohesive energies of $E_c = -3.84$ eV/atom [2] (PBE) and -3.87 eV/atom [18] (PBE) have been reported for the 2H structure, based on the same approach.

The lattice parameters of the 2H-AlSe monolayer expand to $a = 0.378$ nm, $d_{Al\text{-}Se} = 0.247$ nm, $d_{Al\text{-}Al} = 0.258$ nm, $h_{SeSe} = 0.491$ nm, and $\theta_{AlAlSe} = 99.6°$ [18]. These lattice parameters are in excellent agreement with previous calculations (see Table 12.1) [1]. Adding the van der Waals diameter of Se atoms $d_w = 0.38$ nm and the distance between Se atoms, yields an estimate of the monolayer thickness of $d_{w+h} = 0.871$ nm. Compared with the stiffness of AlS, the elastic modulus decreases for 2H-AlSe monolayers to 66 N/m and the Poisson ratio is $\nu = 0.24$ [2], in good agreement with $E = 68.3$ N/m and $\nu = 0.24$ [18]. Surprisingly, the bending rigidity increases to $D = 20$ eV [2]. The values reported for the band gap of the indirect semiconductor are $E_g = 2.00$ eV (PBE) and 2.73 eV (HSE06) [18]. The 2H phase is energetically and dynamically stable, however, could not be synthesized so far.

DFT and AIMD calculations confirmed the dynamic and thermal stability of the Janus structure Al_2SSe, which exhibits lattice parameters and optical properties between those of the two parental monolayers of AlS and AlSe [22]. This is not surprising because these two monolayers constitute the four-layer structure of Al_2SSe. The Janus monolayer has an indirect band gap of $E_g = 2.08$ eV, to be compared with 2.13 eV (AlS) and 2.07 eV (AlSe), calculated with the HSE06 hybrid functional [22]. As demonstrated by the authors, biaxial strain engineering notably alters the band gap, while an electric field has only a negligible influence on the electronic and optical properties. The Al_2SSe monolayer absorbs strongly in the near-ultraviolet region, with a first absorption peak at 5.27 eV.

12.2.4 AlTe Monolayer

12.2.4.1 2D Bonding, Structure, and Applications

The valence electrons of the Al atoms ($3s^2p_x{}^1$) and Te atoms ($4s^2p_x{}^2p_y{}^2$) provide the nine bonding electrons needed to form stable 2D covalent structures, as discussed above for the AlSe monolayer. This includes the two hexagonal 2H and 1T phases. Due to the difference in the electronegativities of 1.61 (Al) and 2.10 (Te), a charge transfer of $\delta q = 1.19$ e (1.22 e) [18]) occurs from each Al atom to each Te atom, introducing ionic character to bonding [2]. The comparable cohesive energies reported for the 2H-AlTe monolayer are $E_c = -3.38$ eV/atom [2] (PBE) and -3.28 eV/atom [18] (PBE).

The lattice parameters of 2H-AlTe monolayers are $a = 0.411$ nm, $d_{Al\text{–}Te} = 0.270$ nm, $d_{Al\text{–}Al} = 0.257$ nm, $h_{TeTe} = 0.513$ nm, and $\theta_{AlAlTe} = 99.4°$ [1, 18]. Adding the van der Waals diameter of Te atoms $d_w = 0.42$ nm and the distance between Te atoms, yields an estimate of the monolayer thickness of $d_{w+h} = 0.933$ nm. Independent calculations presented for the indirect band gap of the stable semiconductor the values $E_g = 1.84$ eV (PBE) and 2.48 eV (HSE06) [18, 19]. The small binding energy reduces the stiffness to $E = 54$ N/m, whereas the Poisson ratio of $\nu = 0.23$ is nearly identical to that of AlSe [2], in very good agreement with $E = 55$ N/m and $\nu = 0.24$ [18]. The AlTe monolayer has the smallest bending rigidity of 14 eV in this group of compounds, due to the weak Al–Al bond in the center [2].

The thermoelectric performance of AlX (X = S, Se, and Te) monolayers was systematically studied by first-principles calculations combined with Boltzmann transport theory [23]. The sharp peak at the VBM and almost constant CBM in the DOS qualify these 2D materials for thermoelectric applications. The figures of merit *ZT* of the three compounds increases with temperature, and the optimized values are 0.52, 0.59, and 0.26 at room temperature and 0.67, 0.74, and 0.73 at 700 K, determined under moderate electronic concentrations for the AlS, AlSe, and AlTe monolayers, respectively. The corresponding optimized power factors are 23.4, 54.4, and 10.4 mW m^{-1} K^{-2} at 700 K, respectively.

For the two Janus monolayers Al (Ga, In) STe, computations performed within the framework of the DFT and the AIMD approach, predicted the structural, optoelectronic, photocatalytic, and stability properties [24]. The two monolayers are thermodynamically stable and exhibit semiconducting characteristics with direct and indirect gaps of $E_g = \sim 2.23$ eV and ~ 2.69 eV (HSE06) for AlGaSTe and AlInSTe, respectively. Strong optical absorption occurs in the spectral range between 3 eV and 18 eV with an absorption coefficient > 10^5 cm^{-1}. In addition, the adequate band edge positions together with the high optical absorption make the two Janus monolayers suitable for photocatalytic water splitting.

12.2.5 AlPo Monolayer

12.2.5.1 2D Bonding and Structure

Based on the valence shell of Al atoms ($3s^2p_x^1$) and Po atoms ($6s^2p_x^2p_y^2$) with nine bonding electrons, covalent bonding leads to a stable hexagonal 2H–AlPo structure. Owing to the different electronegativities of 1.61 (Al) and ~ 2.0 (Po), the Al–Po bond is polarized with expected charge transfer from the Al atom to the Po atom. The lattice parameters of the 2H-phase of AlPo monolayers are a = 0.425 nm, $d_{Al\text{–}Po}$ = 0.280 nm, $d_{Al\text{–}Al}$ = 0.257 nm, h_{PoPo} = 0.528 nm, θ_{AlAlPo} = 118.85°, and θ_{PoAlPo} = 98.67, and the space group is *P*-6*m*2 (see Table 12.1) [1]. Adding the van der Waals diameter of Po atoms d_w = 0.394 nm and the distance between Po atoms, yields an estimate of the monolayer thickness of d_{w+h} = 0.922 nm.

A comparison of the AlPo lattice parameters with those of the AlTe monolayer reveals an increase in the lattice constant, the Al–Po bond length, and the distance between the Po atoms, indicating an extended lattice of the 2H–AlPo monolayer. Note that the currently proposed van der Waals radius of Po d_w = 0.197 nm is smaller than the radius of Te (0. 210 nm), resulting in a decrease in the mean van der Waals diameter from AlTe to AlPo. The 2H–AlPo monolayer is a semiconductor with a band gap of E_g = 1.52 eV (HSE06) [1]. To date, this work seems to be the only detailed study on the structural and electronic properties of AlPo monolayers.

12.3 Gallium Monochalcogenide Monolayers (GaO, GaS, GaSe, GaTe, and GaPo)

12.3.1 GaO Monolayer

12.3.1.1 2D Bonding, Structure, and Mechanics

The three valence electrons of Ga atoms ($4s^2p_x^1$) and the six valence electrons of O atoms ($2s^2p_x^22p_y^2$) form stable covalently bonded four-layer 2H–GaO monolayers. The large difference between the electronegativities of Ga (1.81) and O (3.44) induces strong covalent-ionic bonding with a surprisingly small Bader charge transfer of $\delta q = 1.08$ e from the Ga atom to the O atom with its much greater electronegativity [2]. The cohesive energy of the stable hexagonal 2H–GaO monolayer is $E_c = -4.52$ eV/atom (PBE) [2].

The lattice parameters of the 2H–GaO phase are a = 0.312 nm, $d_{Ga–O} = 0.194$ nm, $d_{Ga–Ga} = 0.252$ nm, $h_{OO} = 0.399$ nm, $\theta_{GGaO} = 112.46°$, and $\theta_{OGaO} = 106.32°$ [1, 2]. Adding the van der Waals diameter of O atoms $d_w = 0.310$ nm and the distance between O atoms, yields an estimate of the monolayer thickness of $d_{w+h} = 0.709$ nm. The 2H–GaO monolayers are energetically and dynamically stable. For the semiconducting monolayer with an indirect band gap nearly identical gaps of $E_g = 1.56$ eV [1] (PBE) and 1.54 eV [2] (PBE) have been presented, based on the same theoretical approach. As linear mechanical properties, theory predicts an elastic modulus of E = 130 N/m, a Poisson ratio of $\nu = 0.40$, and a bending rigidity of 26 eV [2].

First-principles calculations elucidated the energetic, dynamic, and thermal stability, electronic properties, and hole-doping-induced magnetic properties of GaO (InO) monolayers [25]. Hole doping can switch the intrinsically non-magnetic stable semiconductor with Mexican-hat-like band edges in the valence band, to a spontaneous ferromagnetic half-metal phase over a wide range of hole densities. The Curie temperature, estimated by Monte Carlo simulations, can reach ~ 125 K for 2H-GaO monolayers.

12.3.2 GaS Monolayer

12.3.2.1 Chemistry and 2D Bonding

The three valence electrons of Ga atoms ($4s^2p_x^1$) and the six valence electrons of S atoms ($3s^2p_x^2p_y^2$) form the two stable hexagonal 2H and 1T structures. Both phases consist of Ga–Ga bonds, Ga–S bonds, and lone electron pairs, however, with the already discussed symmetries. Therefore, the two 2D structures have similar cohesive energies and physicochemical properties. Ionic interactions play a vital role owing to the difference in the electronegativities of Ga (1.81) and S (2.58). In the four-layer

2H phase of GaS monolayers, each Ga atom donates a charge of $\delta q = 0.75$ e to each S atom. Therefore, the binding electrons of the polar bonds concentrate at the S atoms [2]. The cohesive energy published for the 2H-GaS monolayer is $E_c = -3.62$ eV/atom [2] (PBE) (PBE + D2). The increase in the mean van der Waals diameter from $d_w = 0.365$ nm for GaO to $d_w = 0.390$ nm for GaS is mainly responsible for the lower binding strength of the polar Ga-S bonds [4].

12.3.2.2 2D Structures

For the structural parameters, the most complete study gives a = 0.364 nm, $d_{Ga\text{–}S} = 0.237$ nm, $d_{Ga\text{–}Ga} = 0.248$ nm, $h_{SS} = 0.466$ nm, $\theta_{GaGaS} = 117.44°$, and $\theta_{SGaS} = 100.45°$ [1]. These properties agree well with those of other sources [2]. Adding the van der Waals diameter of S atoms $d_w = 0.36$ nm and the distance between S atoms, yields an estimate of the monolayer thickness of $d_{w+h} = 0.826$ nm. The space group of the best studied 2H structure is *P*-6*m*2. The structure is energetically, dynamically, and thermally stable and has reported indirect band gaps of $E_g = 2.568$ eV (PBE + D2) [26] and 2.35 eV (PBE) [1, 2].

Engineering of the electronic properties of the 2H phase of GaS monolayers under biaxial strain and an electric field was systematically studied via DFT calculations [26]. The energy gap and especially the conduction band depend strongly on biaxial tensile strain, whereas the effect of compressive strain is quite weak. Importantly, a negative external electric field initiates a semiconductor-to-metal transition, while a positive electric field has only a weak effect on the band structure.

Global structural searches based on the PSO algorithm identified a polymorph of the most stable 2H phase [27]. This polymorph consists of a monoclinic lamellae phase with a large unit cell of eight atoms. According to a phonon spectrum analysis and ab initio MD simulations, the polymorph is dynamically and thermally stable. Considering the C_{2h} symmetry, the predicted large binding energy of $E_c = -4.11$ eV/atom for these GaS monolayers is surprising. Moreover, this polymorph has lattice constants of a = 0.981 nm and b = 0.365 nm, and a direct band gap of $E_g = 2.84$ eV with a strong optical absorption at ~ 10^5 cm^{-1}. Interestingly, the polymorphic monoclinic lamella phase of several Ga and In monochalcogenides with C_{2h} symmetry have nearly the same binding energy as the 2H honeycomb lattice, as predicted through an artificial swarm intelligence computational structure search using crystal structure analysis by particle swarm optimization (CALYPSO) [26]. For the direct semiconductor, higher carrier mobilities were found than for the 2H phase. This structure is a potential candidate for high-efficiency solar cells and optoelectronics owing to the strong anisotropic optical absorption in the VIS and UV regions.

12.3.2.3 Mechanics

As linear elastic properties of the 2H-GaS monolayer, the DFT-GGA-PBE approach delivered an in-plane stiffness of E = 91 N/m, and the Poisson ratio $\nu = 0.26$ [28], which are in reasonable agreement with E = 79.3 N/m and $\nu = 0.24$ [29]. A subsequent publication reported lower values of E = 73 N/m and $\nu = 0.22$ [2]. These quantities characterize the in-plane rigidity of the network and its sensitivity to an applied uniaxial strain. The ultimate strength extracted from the fracture point is $\sigma = 12$ GPa or 5.6 N/m by using a thickness of t = 0.466 nm [28]. An ultimate strain of $\varepsilon = 0.26$ indicates a higher ductility, as for example for graphene with an ultimate strain of 0.20 [28]. The maximum stress point of the calculated stress–strain curve determines the fracture strain, where the structure becomes elastically unstable by failure. There is no guarantee, however, that the material remains dynamically stable up to this point of elastic instability. The calculated phonon frequencies of GaS indicate that the GaS crystal is no longer stable at a biaxial strain of $\varepsilon = 0.17$ due to out-of-plane soft modes inducing instability before elastic failure of the lattice.

12.3.2.4 Physics and Applications

Switching the atomic positions of the group –IIIA and group–VIA atoms in the 2H structure to M-X–X–M was proposed as a possible strategy to achieve a drastic reduction in the thermal conductivity of hexagonal MX (M = Ga, In, and Tl; X = S, Se, and Te) monolayers [29]. The authors explain this surprising effect by showing that increased phonon anticrossing leads to enlarged anharmonic scattering and repulsion of the lone-pair electrons between the two chalcogen atoms representing the two middle sublayers of the four-layer arrangement.

As revealed by first-principles calculations, the noncentrosymmetric 2H phases of GaS, GaSe, and InSe exhibit piezoelectric coefficients on the same order of magnitude as those of other 2D materials, such as MoS_2, with the same space group of *P*-6*m*2 [30]. These 2D semiconductors may find use in applications, where excellent photosensitivity and electromechanical coupling are an issue, e.g., in integrating stress sensors, photodetectors, and nanoscale transducers. The 2D materials may open the door to flexible photodetectors in which electrical sensing of both mechanical and optical signals becomes possible using the same active material.

To investigate the suitability of single-layer group–IIIA monochalcogenides MX (M = Ga and In; X = S, Se, and Te) as photocatalysts for hydrogen generation by photocatalytic water splitting, the valence and conduction band edge positions were calculated by using the density-functional calculations and the quasiparticle many-body G_0W_0 approximation [31]. The band gaps of 2D GaS and GaSe are not optimal to absorb a significant amount of solar energy. However, tuning the band gaps, band edge positions, and optical absorption by mechanical strains improves the matching between the solar spectrum and the redox potentials of water. Calculations of the enthalpies of solvation indicate the stability of single-layer monochalcogenides in aqueous solution.

12.3.2.5 Synthesis

Group IIIA–VIA monolayers have gained enormous attention as alternative 2D semiconductors due to their superior optical and electronic properties. A comprehensive review describes the strategies employed for the synthesis of monolayers and few-layer assemblies of these compounds, such as top-down, bottom-up, and template-based chemical conversion methods [32]. Most of the experimental work has been performed on Ga-monochalcogenides, which has resulted in the successful growth of few-layer sheets [33]. Optical microscopy, AFM, and Raman spectroscopy combined with Raman imaging allowed the determination of the thickness of the ultrathin GaS and GaSe flakes. Efficient ambient pressure chemical vapor deposition (CVD) allowed the synthesis of high-quality GaS nanosheets employing hydrogen carrier gas for the reduction of the single-source precursor Ga_2S_3 at atmospheric pressure [34]. The films grown on a SiO_2/Si substrate contained monolayer domains and multilayer flakes with thicknesses varying between 1 and 15 nm and a uniform coverage of 0.7 cm^2. The authors used energy-dispersive X-ray spectroscopy (EDX), atomic force microscopy (AFM), Raman spectroscopy, photoluminescence (PL) spectroscopy, and transmission electron microscopy (TEM) to analyze the chemical composition, thickness, film quality, and crystallinity.

Single-layer and few-layer GaS nanosheets were obtained by mechanical cleavage and subsequent transfer to a SiO_2/Si wafer or a flexible polyethylene terephthalate (PET) substrate [35]. VIS-UV photodetectors produced from ultrathin GaS sheets, exfoliated from bulk crystals, showed an enhanced photoresponse, exceeding that of other 2D materials, e.g., that of the currently exploited InGaAs photodetector. Similarly, a micromechanical cleavage technique was employed to fabricate photodetectors containing GaS sheets by repeatedly peeling the bulk material and transferring the sheets to a SiO_2/Si substrate using adhesive tape [36]. In this study, the authors measured the photoresponse in different gas environments, such as NH_3, O_2, and air, to determine the adsorption behavior.

GaS nanosheets were also prepared by liquid exfoliation of layered GaS powder via sonification of the powder in a variety of suitable solvents [37]. Centrifugation allowed the removal of nonexfoliated material and size selection of the dispersions. The sheets had lateral sizes of 50–1000 nm and thicknesses of 3–80 layers. The nanosheet dispersions with known concentrations and lateral dimensions served as electrodes for hydrogen evolution catalysis. The observed size effect of better catalytic performance by the smaller sheets suggested that the catalytic sites were preferentially located at the sample edges.

12.3.3 GaSe Monolayer

12.3.3.1 Chemistry and 2D Bonding

The larger size of Se atoms reduces orbital overlap in GaSe monolayers compared to GaS; thus, the binding energy decreases substantially from $E_c = -3.62$ eV/atom (GaS) to -2.81 eV/atom (GaSe) [2]. Moreover, the contribution of ionic interactions to covalent bonding is small due to the modest difference in the electronegativities of Ga (1.81) and Se (2.55). The Bader analysis predicts a charge transfer of $\delta q = 0.61$ e from each Ga atom to each Se atom in the four-layer 2H–GaSe structure [2]. Notably, based on the DFT-GGA-PBE approach, also a binding energy of $E_c = -3.80$ eV/atom was derived for GaSe that is greater than that given above for GaS [27].

Since the valence electrons of Ga atoms ($4s^2p_x^{\ 1}$) and those of S, Se, and Te atoms ($ns^2p_x^{\ 2}p_y^{\ 2}$, n = 3–5) are compatible, and the increase of atom size is relatively small, chalcogen atoms can be mixed in hexagonal 2H-structures to form, for example, Ga_2SSe with the S–Ga–Ga–Se configuration. The valence electrons provide one Ga–Ga bond, three Ga–S bonds with the upper layer, three Ga–Se bonds with the lower layer, and two lone electron pairs. Owing to the slightly varied sizes of S atoms (3.6 nm) and Se atoms (3.8 nm) [4], bonding in the upper and lower two sublayers will be different, however, due to the nearly identical electronegativities of S (2.58) and Se (2.55) the difference in polarization will be small.

12.3.3.2 2D Structures

The most complete study presents as structural parameters of 2H-GaSe monolayers $a = 0.382$ nm, $d_{Ga\text{–}Se} = 0.250$ nm, $d_{Ga\text{–}Ga} = 0.247$ nm, $h_{SeSe} = 0.483$ nm, $\theta_{GaGaSe} = 118.12°$, and $\theta_{SeGaSe} = 99.68°$ [1]. These properties agree well with those of another source [2]. Adding the van der Waals diameter of Se atoms $d_w = 0.38$ nm and the distance between Se atoms, yields an estimate of the monolayer thickness of $d_{w+h} = 0.863$ nm. The space group of the best studied energetically, dynamically, and thermally stable 2H structure is *P*-6*m*2. The monolayer of GaSe is an indirect semiconductor with calculated band gaps of $E_g = 1.88$ eV (PBE) (1.80 eV [2] PBE)) and 2.73 eV (HSE06) [38].

For the monoclinic GaSe monolayer with the point group C_{2h}, the CALYPSO approach predicts a cohesive energy of $E_c = -3.75$ eV/atom, lattice constants of $a = 0.997$ nm and $b = 0.382$ nm, and a direct band gap of $E_g = 2.39$ eV with a strong optical absorption at ~ 10^5 cm^{-1} [27]. Like GaS, the GaSe monolayer exhibits an anisotropic carrier mobility that is greater than that of its counterpart with the hexagonal 2H structure.

12.3.3.3 Mechanics

The mechanical properties, such as the mechanical stiffness, stability, and nonlinear elastic properties of monolayer gallium selenide under tension were determined by first-principles DFT calculations. The enormous elastic modulus of E = 335 N/m and Poisson ratio of $\nu = 0.18$ obtained under small tension are comparable with the stiffness and Poisson ratio of graphene, however, seem not to be consistent with other results reported for the GaSe monolayer [39]. On the other hand, the ultimate strengths of $\sigma_{ac} = 8.6$ N/m and $\sigma_{zz} = 10.5$ N/m and the ultimate strains $\varepsilon_{ac} = 0.244$ and $\varepsilon_{zz} = 0.473$, presented by these authors, seem to be reasonable for GaSe monolayers [39]. Importantly, a much lower stiffness of E = 67 N/m and larger Poisson ratio of $\nu = 0.24$ confirm the results reported by other groups and are in much better agreement with Griffith's rule [2]. The much lower Young's modulus and larger Poisson's ratio are supported by E = 77 N/m and $\nu = 0.25$ [28], and by E = 67.6 N/m and $\nu = 0.23$ [29]. The ultimate strength of $\sigma = 4.8$ N/m (10 GPa), ultimate strain of $\varepsilon = 0.23$, and fracture strain of $\varepsilon_f = 0.30$ seem to provide a more realistic description of the mechanical behavior of 2H–GaSe monolayers [28].

12.3.3.4 Physics and Applications

A review covers the specific physical properties of 2D GaSe and InSe and their potential applications based on the direct band gaps, high intrinsic mobilities, and high optical absorbances, which enable the construction of fast response photodetectors on rigid and flexible substrates for a wide spectral range from near infrared (NIR) to UV light [40]. Furthermore, due to the lack of inversion symmetry, 2D GaSe allows the study of second-order nonlinearity in single-photon light emitters, second harmonic generation, and terahertz sources [40]. A property of special interest is the high intrinsic mobility because understanding 2D charge transport is a prerequisite for integration into technological devices. Moreover, the review covers encapsulation techniques to prevent degradation of GaSe (InSe) layers under ambient conditions.

As revealed by Raman spectroscopy, oxidation is a severe problem that needs to be solved for technological applications of GaSe sheets. Exposure to air leads to almost immediate oxidation, which markedly changes the electrical and optical properties through morphological changes [41]. Exposure to the environment and moisture leads to an oxide layer by sorption of O_2 and H_2O on the Se vacancies and generates the reaction products Ga_2O_3, Ga_2Se_3, and amorphous Se. Various protection and encapsulation procedures, such as inert top/bottom layers of h-BN, atomic layer deposition (ALD) of Al_2O_3, or an argon atmosphere were tested to minimize contamination effects [40].

A systematic investigation of the water splitting reaction using GaSe, GaTe, InSe, and InTe as photocatalysts revealed that few-layer GaSe is the most appropriate catalyst for several reasons [42]. As the number of layers in a few-layer assembly increases, GaSe sheets tend to transform into a direct band gap semiconductor. Few-layer GaSe sheets have the optimal band gap of ~ 2 eV required for efficient absorption in the visible spectrum. In GaSe sheets, it is possible to tune the band edge positions necessary to straddle the oxidation potential of H_2O and the reduction potential of H_2. In addition, the carrier mobilities of GaSe sheets are high, and the photogenerated electrons and holes tend to separate spatially.

Another systematic first-principles study elucidated the HER activity of GaX (X = O, S, Se, and Te) and that of the Janus monolayers Ga_2XY (X $\neq$ Y, X/Y = O, S, Se, and Te) [43]. The lowest Gibbs free energy of $\Delta G_{H^*} = 0.19$ eV was found for Ga_2OSe by applying minimal strain of $\varepsilon = 0.03$. The reasons for the enhanced activity of Ga_2OSe include the presence of the most electronegative O atom, the larger lattice parameter of Ga_2OSe with respect to GaO, and the enhancement of the built-in electric field after H_2 adsorption. The enhanced chemisorption of H with lattice stretching is due to the weak Ga–O bonds and reduced electron filling of antibonding molecular orbitals formed by H 1s and O $2p_z$ orbitals after H adsorption. With the appropriate band alignment, an optical absorption coefficient of ~ 10^5 cm^{-1}, and a low exciton binding energy of 0.71 eV, the Ga_2OSe monolayer is a suitable photocatalyst for hydrogen production in the near-infrared region.

12.3.3.5 Synthesis

The monolayer of gallium selenide is a prominent IIIA–VIA monochalcogenide. Therefore, the synthesis of high-quality mono- and few-layer systems has attracted enormous interest. Mechanical exfoliation and micromechanical cleavage are low-cost methods that cause lesser destruction than other methods and are therefore the preferably used techniques for preparing high-quality GaSe layers from van der Waals-bonded bulk materials. For example, high-performance photodetectors were derived from layered bulk crystals using ultrathin GaSe sheets synthesized by mechanical cleavage and solvent exfoliation [44]. Large atomically thin single-crystal GaSe sheets were grown directly on insulating substrates by the vapor phase mass transport method to measure their photoresponse [45, 46]. Note that these methods primarily developed for research purposes are difficult to scale up for industrial applications.

Promising techniques for the scalable and quality-controllable production of GaSe monolayers include physical vapor deposition (PVD), pulsed laser deposition (PLD), and CVD. A major step forward in the realization of large-scale methods for the fabrication of single crystals is the proposed multistep growth of 2D GaSe on 200 mm Si(111) wafers in a complementary metal-oxide semiconductor (CMOS) compatible platform [47]. Vicinal Si(111) wafers, slightly misoriented toward (−1–12), allowed step-edge guided growth of preferentially single-oriented 2D-GaSe crystals. By the application of an optimized growth-sublimation-regrowth process, it was possible to remove most of the screw dislocations that formed at the step edges.

12.3.4 GaTe Monolayer

12.3.4.1 Chemistry and 2D Bonding

The valence electrons of Ga atoms ($4s^2p_x^1$) and Te atoms ($5s^2p_x^2p_y^2$) form not only high-symmetry hexagonal structures but also low-symmetry monoclinic monolayers. In these stable covalently bonded GaTe monolayers, the same type of chemical bonds is geometrically arranged in a different way. For this reason, the cohesive energies of the hexagonal and monoclinic phases do not substantially differ. The 2H phase of the GaTe monolayer has the smallest binding energy of $E_c = -2.96$ eV/atom (PBE) in this group, excluding the less studied GaPo monolayers [2]. The reduction of the cohesive energy is due to decreasing orbital overlap caused by the large van der Waals diameter of Te atoms of $d_w = 4.20$ nm [4] and the small ionic contribution to bonding caused by the small difference in the electronegativities of Ga (1.81) and Te (2.10). Accordingly, the charge transfer from the Ga atom to the Te atom is reduced to $\delta q = 0.41$ e [2].

12.3.4.2 2D Structures

The lattice parameters of the 2H phase of GaTe monolayers are $a = 0.413$ nm, $d_{Ga-Te} = 0.271$ nm, $d_{Ga-Ga} = 0.247$ nm, $h_{TeTe} = 0.503$ nm, $\theta_{GaGaTe} = 118.16°$, and $\theta_{TeGaTe} = 99.56°$ [1]. These values are in good agreement with those of another publication [2]. Adding the van der Waals diameter of Te atoms $d_w = 0.42$ nm and the distance between Te atoms, yields an estimate of the monolayer thickness of $d_{w+h} = 0.923$ nm. The absence of imaginary frequencies in the phonon dispersion curve indicates the dynamic stability of GaTe monolayers [1, 2]. The calculated band gap of the indirect semiconductor is $E_g = 1.43$ eV (PBE) [1, 2]. The observed decrease in the band gap from GaS to GaTe is due to the decreasing ionicity of the homologues going down the column of chalcogenides in the periodic table.

As mentioned already, the hexagonal 2H phase of GaTe (InTe) monolayers form a monoclinic structure that deviates completely from the hexagonal structure of GaTe (see Fig. 12.12). This structure derives from the most stable hexagonal phase by twisting every third M–M bond in the in-plane direction and by changing the in-plane bond length slightly compared to the out-of-plane bonds [48]. The lattice parameters of this low-symmetry structure are $a = 0.414$ nm and $b = 2.397$ nm. The resulting large unit cell contains 12 Ga and 12 Te atoms. The structure is thermodynamically stable with a formation energy of $E_f = -0.371$ eV [48]. The low-symmetry structure is a semiconductor with estimated direct band gaps at $E_g = 1.4$ eV (PBE) and 2.16 eV (HSE06).

By using substrate-induced strain and precisely controlling the growth temperature, a phase-controlled synthesis of monoclinic and hexagonal GaTe flakes could be achieved via PVD [49]. Stabilization by tensile strain on a mica substrate allowed the growth of hexagonal 2D GaTe. Interestingly, with increasing temperature, the

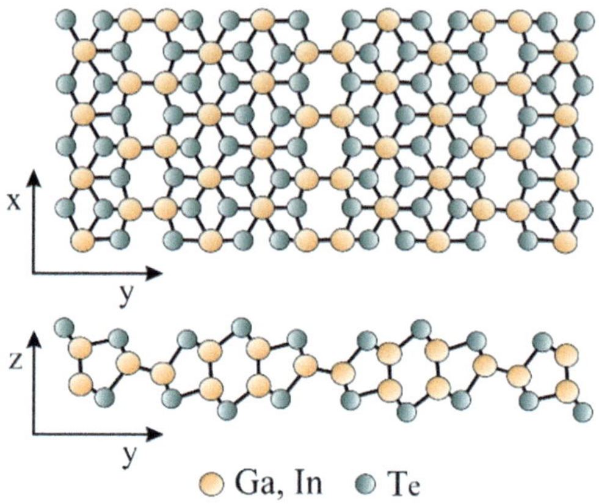

Fig. 12.12 Top and front views of the low-symmetry monoclinic structure with a large unit cell, e.g., of group IIIA telluride monolayers. Reproduced from Ref. [48], Copyright 2017, Royal Society of Chemistry

structure of GaTe tends to transform to the monoclinic phase owing to a decrease in substrate interactions. Moreover, the hexagonal phase is transformed to a stable monoclinic phase by femtosecond laser irradiation via a novel selected-area phase transition process [49].

12.3.4.3 Mechanics

As expected for the low binding energy of the 2H phase, the stiffness of GaTe monolayers is only 55 N/m, and the Poisson ratio decreases to $\nu = 0.20$ [2], in good agreement with E = 51.5 N/m and $\nu = 0.22$ [29]. These elastic moduli of GaTe are consistent with the behavior of its stronger homologues. The PBE approach delivered a stiffness of 66 N/m, whereas the HSE06 functionals yielded a much greater stiffness of 84 N/m [31]. This larger elastic modulus is likely due to the smaller lattice parameters predicted by the HSE06 approach. Weaker bonding not only reduces the overall mechanical stability but also leads to easier defect formation and the incorporation of impurities in the lattice. Nanoindentation with a Berkovich tip served as a tool for studying the mechanical behavior of substrate-supported and suspended GaTe multilayers [50]. Surprisingly, the loading curve showed unusual pop-in and load-drop events. Theory identified an interlayer-sliding-mediated layer-by-layer fracture mechanism in GaTe multilayers as the source of these unusual processes.

12.3.4.4 Physics and Applications

A comparative study of the homologues GaS, GaSe, and GaTe by DFT and the semiclassical Boltzmann transport equation suggested future applications of these monolayers as low-temperature thermoelectric materials [29]. A comparison of the band edge positions with the redox potentials of water indicated the suitability of the three gallium monochalcogenides as photocatalysts for water splitting [31]. Furthermore, the band gaps, band edge positions, and optical absorption can be tuned by biaxial strain to improve matching with the solar spectrum and redox potentials. The low mobility of charge carriers and short lifetime of photo excitations currently hamper the application of these materials. Van der Waals heterostructures of GaTe monolayers and graphene sheets may provide a solution to this problem and simultaneously combine high photoabsorption of the former and ballistic transport by the latter [48].

DFT calculations provided information on the modulation of the electronic and optical properties under biaxial strain and an external electric field [51]. Compressive strain significantly increases the absorption in the ultraviolet region, while compressive strain increases the energy gap only slightly. The gap in monolayer GaTe decreases under tensile strain, and an external electric field induces a semiconductor-to-metal transition.

Importantly, a comprehensive study of the photoconductivity and photodetection of GaTe nanowires and nanosheets, grown by the physical vapor transport (PVT) method, affirms high carrier mobility and long carrier lifetime, contrary to a former theoretical result [52]. According to first-principles calculations, GaTe monolayers are potential gas sensors for NO_2 molecules with moderate adsorption energy and for SO_2 with strong bonding to the GaTe monolayer [53]. The adsorption of these molecules changes the band gap energy, whereas NO and H_2S are weakly physisorbed.

12.3.4.5 Synthesis

According to the formation energies of single-layer Ga monochalcogenides, the preparation of monoclinic GaTe should be easily possible by cleavage because it has the lowest formation energy. Layered GaTe sheets were first obtained by mechanical exfoliation and used to study their performance in optoelectronic devices and field-effect transistors (FETs) [32]. On a mica substrate stabilized hexagonal GaTe grows under additional stabilization by tensile strain [32]. Both monoclinic and hexagonal GaTe can be selectively realized by phase-controlled synthesis. This method uses the vapor phase deposition method under precisely controlled growth temperature [49]. A novel method called selected-area phase transition induces the transition from the hexagonal phase to monoclinic GaTe via fs-laser irradiation.

High-quality crystalline nanosheets of GaTe with up to 100 μm in size were synthesized by the CVD method and van der Waals epitaxial growth on a mica substrate [54]. Under ambient conditions, adsorbates strongly deteriorated the performance of a fabricated GaTe photodetector. Removal of the adsorbates under vacuum ($\sim 7 \times 10^{-5}$ torr) yields a flexible photodetector with a fast response and an excellent linear input–output relationship.

Interestingly, the growth of high-quality hexagonal h-GaTe nanosheets on the (001) surface of hexagonal ZnO has been achieved by CVD under ambient pressure [55]. SEM, XPS, XRD, and high-resolution transmission electron microscopy

(HRTEM) were used to characterize the flower-like morphology and the hexagonal structure. The lattice constants were a = 0.406 nm and c = 1.696 nm, and the monolayer thickness ~ 0.80 nm. The space group of this structure is *P63/mmc*. The h-GaTe nanosheet is a p-type semiconductor with an indirect band gap. These p-type GaTe sheets can form heterostructure p-n junctions by combination with n-type layers (e.g., n-type ZnO), which are applicable as photodiode or catalyst for heterojunction-based water splitting.

12.3.5 GaPo Monolayer

12.3.5.1 2D Bonding and Structure

The valence electrons of Ga atoms ($4s^2p_x^{\ 1}$) and Po atoms ($6s^2p_x^{\ 2}p_y^{\ 2}$) provide the nine electrons needed to form 2D structures by covalent Ga–Ga and Ga–Po bonds and lone electron pairs, as proposed theoretically for the hexagonal GaPo structures. Assuming electronegativities of 1.81 (Ga) and ~ 2.0 (Po), the Ga–Po bond is slightly polar with minor charge transfer from the Ga atoms to the Po atoms. The calculated lattice parameters of the stable 2H phase of GaPo monolayers with space group *P*-6*m*2 are included in Table 12.1 [1]. A comparison with the lattice parameters of GaTe monolayers shows an increase in the lattice constant, an extended Ga–Po bond length, and a greater distance between the Po atoms, indicating an extended thickness of the GaPo monolayer despite a decrease in the van der Waals radius from Te (0.210 nm) to Po (~0.20 nm). The 2H phase of the GaPo monolayer is an indirect semiconductor with a band gap of E_g = 0.90 eV (PBE). To date, the cited theoretical study seems to be the only extensive work predicting properties of the GaPo monolayer [1].

12.4 Indium Monochalcogenide Monolayers (InO, InS, InSe, InTe, and InPo)

12.4.1 InO Monolayer

12.4.1.1 2D Bonding, Structure, Mechanics, Applications, and Synthesis

The valence electrons of In atoms ($5s^2p_x^{\ 1}$) and O atoms ($2s^2p_x^{\ 2}p_y^{\ 2}$) provide the nine electrons necessary to form hexagonal covalently bonded four-layer phases of InO monolayers. DFT calculations indicate the stability of the 2H structure of InO monolayers with the lattice parameters a = 0.347 nm, $d_{In\text{–}O}$ = 0.216 nm, $d_{In\text{–}In}$ = 0.286 nm, h_{OO} = 0.447 nm, θ_{InInO} = 111.94° and θ_{OInO} = 106.89° [1, 2]. Adding the van der Waals diameter of O atoms d_w = 0.31 nm and the distance between O atoms, yields an estimate of the monolayer thickness of $d_{w+h} = 0.757$ nm. Owing to its larger size, the binding energy of the 2H-InO monolayer decreases to $E_c = -3.98$ eV/atom in comparison to that of GaO [2]. Note that also a cohesive energy of -5.29 eV/atom has been derived by the GGA/PBE approach [56]. The negative formation energy $E_f = -1.10$ eV/atom supports the thermodynamic stability of the hexagonal phase, and implies a good chance for synthesis [57]. Despite a large difference in the electronegativities of In (1.78) and O (3.44), only a moderate Bader charge transfer of $\delta q = 1.01$ e occurs from each inner In atom to an outer O atom of the quadruple layer, which is comparable with the charge transfer estimated for GaO. The energetically and dynamically stable InO monolayer is an indirect semiconductor with a band gap of $E_g = 0.45$ eV (PBE) [2].

The DFT-PBE approach gives as Young's modulus E = 75 N/m and as Poisson's ratio $\nu = 0.44$ for the 2H phase of InO monolayers [2]. Another study obtained slightly anisotropic and substantially larger elastic moduli at $E_{ac} = 100.4$ N/m and $E_{zz} = 103.5$ N/m [57].

In a group of 18 metal-oxide-based monolayers with excellent oxidation resistance and notable absorption in the UV region, the InO monolayer showed the largest carrier mobility of 8540 $cm^2\ V^{-1}\ s^{-1}$ [57]. Such a high carrier mobility can be attributed to a small effective mass, large stiffness, and small deformation potential constant. Therefore, 2D InO offers a wide range of possibilities for advanced applications in high-speed electronics and optoelectronics.

Targeted intercalation experiments provided experimental evidence for the formation of 2D InO [58]. To initiate the formation of InO nanosheets, it is necessary to employ a robust metal organic chemical vapor deposition (MOCVD) process using $(CH_3)_3In$ as a precursor to intercalate In atoms at a graphene/SiC interface. The two In sublayers, in 'atop' position in the quadruple layer in confinement between graphene and SiC, were imaged by scanning transmission electron microscopy (STEM). Electron energy loss spectroscopy (EELS) served as a tool for detecting the bonding of oxygen to indium atoms. Conductive AFM measurements revealed a wide band gap of $E_g = 4.1$ eV.

Optimization of the hexagonal structure of InO yields another minimum on the potential energy hypersurface, namely, a mixed O–In–O–In four-layer configuration [56]. The calculations delivered a surprisingly low cohesive energy of $E_c = -5.21$ eV/atom for this metastable structure, comparable with the binding energy of $E_c = -5.29$ eV/atom calculated by these authors for the 2H phase. Interestingly, the metastable mixed-layer structure appears gapless and thus is a metal.

First-principles calculations elucidated the energetic, dynamic, and thermal stability, the electronic properties, and hole-doping-induced magnetic properties of H-InO monolayers [25]. As discussed for GaO, hole doping can convert the intrinsically non-magnetic stable semiconductor to a spontaneous ferromagnetic half-metal phase over a wide range of hole densities. The Curie temperature estimated by Monte Carlo simulations can reach ~ 107 K for the H-InO monolayer.

12.4.2 *InS Monolayer*

12.4.2.1 Chemistry and 2D Bonding

The nine valence electrons of the In ($5s^2p_x{}^1$) and S atoms ($3s^2p_x{}^2p_y{}^2$) form stable covalently bonded 2H and 1T quadruple structures. Each In atom of the central In–In dimer connects to three S atoms, and vice versa. The different electronegativities of In (1.78) and S (2.58) lead to a polarization of the In–S bonds with a charge transfer of $\delta q = 0.73$ e from the inner In atoms to the outer S atoms [2]. Owing to the increasing atom size from Ga (0.420 nm) to In (0.440 nm), bonding in In-monochalcogenide monolayers is generally weaker than that in the Ga group. The published cohesive energies of the 2H phase of InS vary between $E_c = -3.32$ eV/atom [2] (PBE) and – 3.85 eV/atom [26] (HSE06). Calculations using different approaches to study the hexagonal polytypes obtained $E_c = -3.46$ eV/atom (PBE) for the 2H and 1T phase and $E_c = -4.04$ eV/atom (LDA) for these two hexagonal phases [3].

12.4.2.2 2D Structures

The lattice parameters of the 2H phase of InS monolayers are a = 0.394 nm, $d_{In–S}$ = 0.257 nm, $d_{In–In}$ = 0.282 nm, h_{SS} = 0.520 nm, θ_{InInS} = 117.61° and θ_{SInS} = 100.24° [1, 2], in excellent agreement with subsequently published results [48]. Adding the van der Waals diameter of S atoms d_w = 0.36 nm and the distance between S atoms, yields an estimate of the monolayer thickness of d_{w+h} = 0.880 nm. The energetically and dynamically stable hexagonal lattice belongs to the space group *P*-6*m*2 [59]. The band gap of the hexagonal semiconducting monolayer is E_g = 1.74 eV (PBE), and the systematically larger gap of the HSE06 approach is 2.71 eV [31], in good agreement with other results with values at 1.77 (PBE) and 2.62 (HSE06) [38].

Note that for InS, InSe, and InTe monolayers, the 1T structure (β-phase) also exists with comparable binding energies. Importantly, a simple shift of one S (Se, Te) layer in the 2H structure with respect to the other S (Se, Te) layer transforms the 2H into the 1T structure. This minor change in the arrangement of the four-layer system breaks the mirror symmetry and establishes inversion symmetry. Note that the 1T phase of the InS (InSe, InTe) monolayer has essentially the same lattice parameters and cohesive energy as the 2H phase, however, different sets of Raman-active phonon modes, which allow identification of the two phases [3]. This study presents first-principles calculations of the electronic band structure, phonon dispersions, and optical absorption spectra of the three monolayers InS, InSe, and InTe.

12.4.2.3 Mechanics

The weaker binding energies reduce the stiffness and ultimate strength of the In-VIA monolayers in comparison with those of the corresponding Ga-VIA monolayers.

Furthermore, the stiffness decreases drastically with increasing atom size of the chalcogenide. For the hexagonal 2H-InS monolayer, the calculated Young's modulus is $E = 50$ N/m and Poisson's ratio is $\nu = 0.31$ [2], in good agreement with $E = 55.5$ N/m and $\nu = 0.26$ [10]. No nonlinear mechanical properties could be found in the literature for InS monolayers.

12.4.2.4 Physics and Applications

The first-principles solution of the phonon Boltzmann transport equation revealed the effects of nonanalytical corrections of the phonon transport properties of InS, InSe, and InTe [59]. Note that at room temperature, the lattice thermal conductivities are 57.1 W/mK, 44.4 W/mK, and 33.1 W/mK, respectively, where the value for InS monolayers, for example, is 30.2% greater under nonanalytical corrections. The longitudinal acoustic branch is the main contributor to the behavior of the lattice thermal conductivity.

Concerning the efficiency of solar energy conversion, the CBM of InS monolayers is too close to the redox potential of water. Band gap engineering by applying strains solves this problem. While the synthesis of 2D InSe has found enormous interest due to its band gap matching of the visible region, no such effort is visible for mono- and few-layer InS sheet fabrication.

A systematic study investigated the suitability of van der Waals heterostructures of six group–IIIA monochalcogenides (MX: M = Ga and In; X = S, Se, and Te) for use in photocatalysis and 2D excitonic solar cells by addressing the important aspects of stability, energetics, lattice dynamics, and mechanics of heterobilayers [60]. The heterobilayers InS/InSe and GaTe/InS are direct semiconductors with band gaps of $E_g = 3.31$ eV and 1.28 eV, respectively, although the individual monolayers have indirect gaps. The InS/InSe and GaTe/InTe heterostructures reach power conversion efficiencies of 12.7% and 13.2%, respectively, which are higher than those of organic solar cells. With a solar energy absorbance of ~ 10^4 cm^{-1}, GaTe/InTe and GaSe/InS exhibit the ability to perform photocatalytic water splitting over a large pH range.

Oxygen functionalization is an efficient tool for dramatically changing the electronic band structure [38]. The chemisorption of oxygen on the 2H structure is possible on one side (e.g., InSO) or both sides (e.g., OInSO). This process reduces or even closes the band gap, as studied for the InS, InSe, and InTe monolayers. The typical band structure of group–IIIA monochalcogenides consists of a flat band (Mexican hat-shaped or sombrero-like) at the top of the valence band and a parabolic band at the bottom of the conduction band (see Fig. 12.8). This causes a high DOS at the top of the valence band and an almost constant DOS at the bottom of the conduction band. Single-sided oxidation transforms the indirect semiconductor into a direct semiconductor with a reduced gap (see Fig. 12.13). Moreover, the polarity of the surface increases, and the carrier mobility is enhanced. Interestingly, oxygen functionalization of InS, InSe, and InTe on both sides creates a 2D topological insulator (TI). These fully functionalized nanosheets are protected against further oxidation and provide a new platform for realizing the quantum-spin Hall (QSH) effect at room

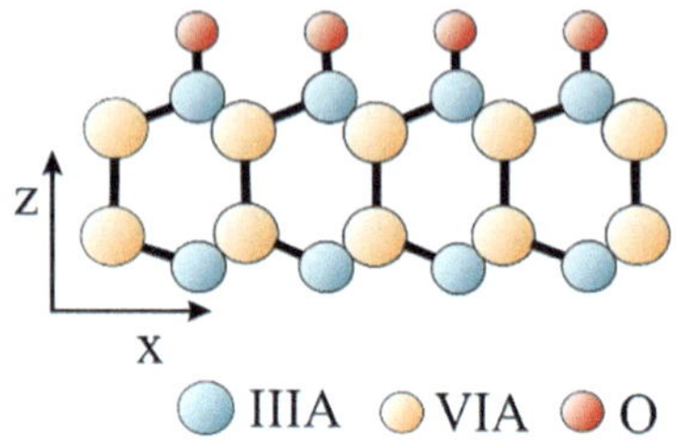

Fig. 12.13 Side views of oxygen functionalized IIIA–VIA monolayer (e.g., InSeO)

temperature. The $InXO_2$ (X = S, Se, and Te) compounds are TIs with a bulk band gap of up to 0.21 eV.

12.4.3 InSe Monolayer

12.4.3.1 Chemistry and 2D Bonding

Like in the other group IIIA–VIA monolayers, the valence electrons of In atoms ($5s^2p_x{}^1$) and Se atoms ($4s^2p_x{}^2p_y{}^2$) provide the electrons needed to form the energetically preferred dynamically stable hexagonal 2H- and 1T-InSe monolayers. Despite a negligible decrease in the electronegativity of Se (2.55), compared with that of S (2.58), the Bader charge transferred from the In atom to the Se atoms is $\delta q = 0.62$ e, inducing the ionic character of bonding in InSe monolayers. Furthermore, due to the larger size of the Se atom ($d_w = 0.38$ nm) compared with the S atom ($d_w = 0.36$ nm), the orbital overlap decreases, and the cohesive energy of the 2H phase of the InSe monolayer is only $E_c = -2.57$ eV/atom (PBE) [2]. As mentioned before, there is no general agreement on the binding energy, and much stronger bonding of $E_c = -3.22$ eV/atom [3] (PBE) and – 3.54 eV/atom [27] has been reported. For the 1T phase of the InSe monolayer, the authors propose the same value of $E_c = -3.22$ eV/atom as for the 2H phase [3].

12.4.3.2 2D Structures

The lattice parameters of the 2H structure of InSe monolayers are a = 0.409 nm, $d_{In\text{-}Se} = 0.269$ nm, $d_{In\text{–}In} = 0.282$ nm, $h_{SeSe} = 0.539$ nm, $\theta_{InInSe} = 118.52°$ and $\theta_{SeInSe} = 99.10°$ [1]. The lattice constant and bond lengths agree with those of subsequent publications [2, 26, 59]. Adding the van der Waals diameter of Se atoms $d_w = 0.38$ nm and the distance between Se atoms, yields an estimate of the monolayer thickness of $d_{w+h} = 0.919$ nm. The energetically and dynamically stable 2H-structure with four atoms in the unit cell belongs to the space group *P*-6*m*2. The band gap of the hexagonal semiconducting InSe monolayer is $E_g = 1.49$ eV (PBE), and for the systematically larger gap of the HSE06 approach it is 2.37 eV [31], in good agreement with other results at 1.52 eV (PBE) and 2.24 eV (HSE06) [38]. By analyzing the orbital contributions to the energy band and the evolution of the band edges, indirect-direct–indirect band gap transitions were revealed in monolayer InSe as the strength of a perpendicular electric field is increased continuously [61]. Simultaneously, a semiconductor-to-metal transformation occurs because the global band gap is gradually suppressed to zero. Alternatively, tuning of the band gap is possible by exerting stress or strain, where compressive stress causes a blueshift of the optical spectrum and the application of strain a redshift of the optical spectrum [62].

12.4.3.3 Mechanics

Young's modulus calculated for the 2H phase of InSe monolayers is $E = 42$ N/m, and Poisson's ratio $\nu = 0.29$ [2], in good agreement with $E = 45.7$ N/m and $\nu = 0.27$ [29]. These DFT-based calculations are in agreement with classical MD simulations, yielding the slightly anisotropic Young's moduli $E_{ac} = 45.7$ N/m and $E_{zz} = 44.8$ N/m , the ultimate tensile strengths $\sigma_{ac} \approx 9.8$ N/m (1K) and $\sigma_{zz} \approx 12$ N/m (1 K), and the ultimate strains $\varepsilon_{ac} = 0.36$ and $\varepsilon_{zz} = 0.44$ [63]. The results are confirmed by MD simulations using the large-scale atomic/molecular massively parallel simulator (LAMMPS) software package [64]. Other published nearly identical results are $E_{ac} = 44.1$ N/m, $E_{zz} = 41.3$ N/m, $\sigma_{ac} \approx 9.45$ N/m (1 K), $\sigma_{zz} \approx 12.4$ N/m (1 K), $\varepsilon_{ac} = 0.362$, and $\varepsilon_{zz} = 0.433$. In addition to the properties at 1 K, both MD studies present the properties at 300 K with reduced stiffnesses and strengths. Moreover, the latter work investigates the fracture mechanism of precracked and defected InSe monolayers [64]. Significant differences appear between the MD-predicted fracture strength and the brittle failure mechanism of perfect structures described by Griffith's rule.

To date, most mechanical properties of free-standing few-layer sheets were measured via indentation using an AFM probe [65]. According to the empirical evaluation of indentation measurements of InSe flakes (>6 ML), Young's modulus is $E \sim 100$ Ga (~79 N/m), Poisson's ratio is $\nu = 0.40$, the breaking strength is $\sigma = 8.7$ GPa (6.9 N/m), and the maximum strain is $\varepsilon = 0.086$, based on a thickness $t = 0.80$ nm [65]. The authors explain the insensitivity of the intrinsic elastic properties of perfect layered InSe sheets to thickness changes by the very small difference in the charge density distribution between monolayer and bulk InSe. However, this explanation neglects the effect of interlayer sliding. In contrast to these results, measurements using a buckling-based method produced a surprisingly low Young's modulus of 23.1 GPa (18.5 N/m) for thin InSe flakes [66]. This method measures the buckling of flakes deposited on a very compliant substrate in competition with adhesion, when subjected to uniaxial compression.

12.4.3.4 Physical Properties

2D InSe has attracted particular attention as a potential thermoelectric material owing to its moderate band gap of approximately 2 eV and its large electrical conductivity, which is based on its unique DOS [67]. Estimates of the power factor and the figure of merit of monolayer InSe employing the Boltzmann transport theory and first-principles calculations, support the supposition that in addition to GaS and GaSe, InSe is a suitable candidate for thermoelectric applications [67]. In the next step, the influence of uniaxial strain on the figure of merit was studied [68]. As revealed by these calculations, convergence of the conduction band occurs at a compressive strain of – 6% along the zigzag direction, inducing an increase in *ZT* for p-type InSe at room temperature. A compressive strain of – 7% yields the maximum value of *ZT* = 0.46, with the convergence of the valence band.

DFPT revealed the unusual phonon properties and low thermal conductance of InSe monolayers [69]. The authors noted the occurrence of strong long-range dipole–dipole interactions in polar InSe, which stabilize the structure. These long-range interactions induce a significant frequency shift of the out-of-phase vibrations with two isolated frequency regimes, each with a single phonon mode. Without these interactions, the acoustic vibrations would be divergent and destroy the 2D structure. The extremely low thermal conductance originates from the low cutoff frequency, the low phonon group velocities, and the presence of a large frequency gap.

The potential of van der Waals heterostructures consisting of a combination of group IIIA–VIA MX (M = Al, Ga, and In; X = S, Se, and Te) monolayers and of elemental group–VIA Se or Te monolayers for solar cell applications was screened by first-principles investigations [70]. Promising results were obtained, for example, for the InSe/Te van der Waals heterostructure consisting of an InSe monolayer as a window layer with a large band gap and high transparency and a tellurene layer as a strong absorber with a direct band gap of $E_g = 1.14$ eV. The heterostructure has a type-II band structure with a direct band gap of $E_g = 0.88$ eV. Based on a remarkable light absorption, the maximum power conversion efficiency (PCE) of the heterostructure reaches 13.4%.

First-principles calculations suggest Janus indium-mixed-chalcogenide monolayer In_2XY (X, Y = S, Se, and Te, X $\neq$ Y) monolayers as potential photocatalysts for water splitting [71]. The band gaps are in the visible regime, and the band edges straddle the H_2O redox potential, except for In_2STe. The valence and conduction band edges shift to higher energies under compressive strain, the latter band moves above the hydrogen reduction potential, with the band gap still in the visible range. Simultaneously, the carrier mobility increases, and the exciton binding energy decreases. The internal electric field generated by the different electronegativities of the In and Te atoms breaks up the exciton pair, creating free electrons and holes that diffuse in opposite directions.

A first-principles study investigated several structural parameters, electronic properties, band alignments of GaS, GaSe, GaTe, InS, InSe, and InTe monolayers, and the possibility of forming type-I and type-II heterostructures by combining these

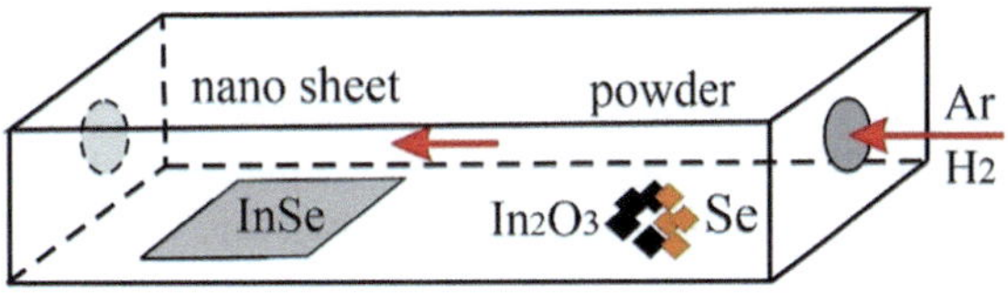

Fig. 12.14 Synthesis of InSe monolayers by CVD, using In_2O_3 and Se powder as precursors and argon and hydrogen as carrier gas and reducing agent, respectively

compounds with 2D semiconductors [72]. All the compounds are potential photocatalysts for water splitting and can form heterostructures with type-II alignment in combination with various other 2D materials. Furthermore, the band convergence indicates the potential of these materials for use as a 2D thermoelectric material. This work supports the application of group–IIIA monochalcogenides in thermoelectric devices, photocatalysis, photovoltaics, and nanoelectronics.

12.4.3.5 Synthesis

The remarkable electronic properties of 2D InSe triggered experimental work on the synthesis of monolayers and few-layer sheets. Vapor-phase selenization of In_2O_3 in a CVD system resulted in large-area monolayer flakes or thin continuous and uniform films with an area in the range of ~ cm^2 (see Fig. 12.14) [73]. The two precursors In_2O_3 and Se powder are deposited in the CVD chamber and a mixture of argon and hydrogen serves as carrier gas and reducing agent, respectively, to initiate the vapor-phase reaction.

Furthermore, the synthesis of InSe monolayers was achieved by physical vapor deposition (PVD) [74]. The quality of the samples was analyzed by Raman spectroscopy, AFM, and high-resolution annular dark field scanning transmission electron microscopy (ADF-STEM). A review article covers the synthesis, properties, and applications of metallic group–IIIA chalcogenides (metals: In and Ga; chalcogens: S, Se, and Te) in electronics and optoelectronics [75]. Moreover, the comprehensive work extensively discusses the construction of hybrid heterostructures of these compounds and their potential applications.

12.4.4 InTe Monolayer

12.4.4.1 Chemistry and 2D Bonding

The valence electrons of In atoms ($5s^2p_x^1$) and Te atoms ($5s^2p_x^2p_y^2$) imply similar bonding to stable covalent 2H and 1T quadruple structures, as described for the related compounds. The ionic character of the bonds is small due to the minor difference in the electronegativities between Te (2.1) and In (1.78). This difference yields a Bader charge transfer from an In atom to a Te atom of $\delta q = 0.45$ e [2]. As the van der Waals diameter increases further from Se to Te, the cohesive energies at $E_c = -2.77$ eV/

atom [2] and – 2.97 eV/atom [3] for the 2H and 1T phases, respectively, designate the weakest bonding in this group. Note that except for the nonmetallic boron, all group–IIIA elements have metallic character; similarly, the behavior of chalcogen atoms shows a downward trend to metallic tellurium.

12.4.4.2 2D Structures

The 2H-InTe monolayer has the largest lattice parameters of a = 0.439 nm, d_{In-Te} = 0.289 nm, d_{In-In} = 0.282 nm, h_{TeTe} = 0.560 nm, θ_{InInTe} = 118.82° and θ_{TeInTe} = 98.81° [1, 2], in reasonable agreement with the results of other studies [59]. Adding the van der Waals diameter of Te atoms d_w = 0.42 nm and the distance between Te atoms, provides an estimate of the monolayer thickness of d_{w+h} = 0.702 nm. The structure of the 2H-InTe monolayer is dynamically stable, as shown by the analysis of the phonon spectrum [60].

According to its low binding energy, DFT calculations give a small stiffness at E = 39 N/m and a Poisson ratio at ν = 0.23 for InTe monolayers [2]. These linear elastic properties have been confirmed by E = 37 N/m and ν = 0.26 [29], and E = 39 N/m (PBE), while the HSE06 approach yields 63 N/m [31].

The hexagonal 2H phase of the InTe monolayer is an indirect semiconductor with calculated band gaps of E_g = 1.40 eV (PBE) and 2.20 eV (HSE06) [31], in good agreement with 1.46 eV (PBE) and 2.04 eV (HSE06) [38]. The tunability of the electronic and optical properties of InTe monolayers by biaxial strain and an external electric field was studied via DFT calculations [76]. The electronic properties of InTe vary strongly in response to strain engineering and an applied electric field, resulting in significant changes in energy, the dispersion, and band gap. A large electric field induces a semiconductor-to-metal transformation. Furthermore, the optical absorption of a novel graphene-like InTe monolayer extends from the ultraviolet to the visible region, reaching an absorption coefficient of up to 10^{-5} cm^{-1} in the ultraviolet region. This may be of interest for various applications in optoelectronics, such as ultraviolet detectors and photovoltaic devices [77].

Hybrid-DFT calculations provided information on the lattice dynamics and thermodynamic properties of the bulk and monolayer phases of InTe and GaTe [78]. Different monolayer structures are distinguished by criteria based on the vibrational frequencies. The sum of the vibrational contributions over the corresponding Brillouin zone gives the Helmholtz free energy. The calculated Helmholtz free energy of monolayer formation confirmed the greater stability of the hexagonal phase. The calculated relative stabilities of the polymorphs of GaTe and InTe at different temperatures indicate that an increasing temperature favors monolayer formation of GaTe but prevents monolayer formation of InTe.

12.4.4.3 Physics and Applications

The investigation of the ballistic thermoelectric properties by DFT calculations and the nonequilibrium Green's function method elucidated the thermoelectric performance of the Ga and In monochalcogenides [79]. InTe monolayers show the highest figure of merit of $ZT = 2.03$ at 300 K owing to the low thermal conductance. The application of biaxial strain between – 10% (compressive) and 10% (tensile) yields a maximal ZT value of 2.7 under a tensile strain of 8%.

Using self-consistent phonon (SCP) theory to determine thermodynamic parameters, reveals the considerable influence of lattice anharmonicity on both the lattice dynamics and thermal conductivity [80]. Determination of the lattice thermal conductivity by the standard Boltzmann transport equation (BTE) yields 0.30 W m^{-1} K^{-1}. A value of 3.58 W m^{-1} K^{-1} results from a combined approach based on SCP and BTE theory. These results demonstrate the enormous effect of anharmonicity on the thermal conductivity.

In monolayers with broken inversion symmetry, spin–orbit coupling (SOC) lifts the spin degeneracy, and spin splitting in the momentum direction is termed Rashba spin splitting. Monolayers with large Rashba effects are key materials for applications in spintronics and optoelectronics, including spin field-effect transistors. A first-principles study revealed large Rashba spin splitting in hexagonal fluorinated-InTe monolayers, which has a magnitude of 1.08 eV Å [81]. InTeF forms a buckled hexagonal structure with space group $P3m1$, where the F atoms bond on top of the In atoms (see Fig. 12.15). Each In atom connects to three Te atoms and one F atom, forming a tetrahedron. Remarkably, the Bader charges at In (0.038 e), Te (8.07 e), and F (7.89 e) do not change noticeably under biaxial strains. This is due to strong ionic bonding normal to the plane and explains the small tunability of the Rashba effect by strain. The cohesive energy of $E_c = -5.13$ eV/atom indicates strong bonding. The fluorinated monolayer has a direct band gap of $E_g = 2.48$ eV (HSE06) and is energetically and dynamically stable, as derived from the calculated cohesive energy and phonon dispersion. Concerning the effect of substrates, the study indicates that neither the Rashba effect nor the direct band gap change upon adsorption on a surface.

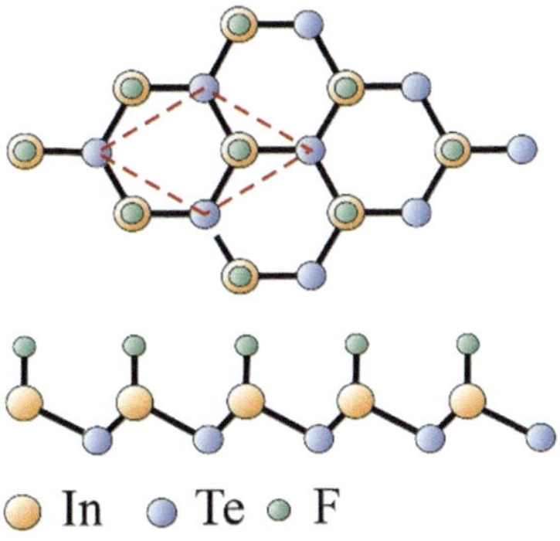

Fig. 12.15 Top and side views of the InTeF monolayer. The dashed line indicates the unit cell with three atoms

First-principles calculations supplied several structural parameters, electronic properties, and band alignments of GaS, GaSe, GaTe, InS, InSe, and InTe [65]. The results obtained for the band structure and band edges indicate that all these compounds are potential photocatalysts for water splitting and can form heterostructures realizing type II alignment with various other 2D materials. Furthermore, band convergence indicates the potential of the material for use as a 2D thermoelectric material. This extensive work recommends the application of group–IIIA monochalcogenides in thermoelectrics, photocatalysis, photovoltaics, and nanoelectronics.

12.4.5 InPo Monolayer

12.4.5.1 2D Bonding and Structure

Based on the valence electrons of In atoms ($5s^2p_x^{\ 1}$) and Po atoms ($6s^2p_x^{\ 2}p_y^{\ 2}$), the available nine electrons produce a stable quadruple 2H-InPo structure by covalent bonding. Assuming electronegativities of 1.78 (In) and ~ 2.0 (Po), the In–Po bond is only slightly polar with minor charge transfer from the In atom to the Po atom. The calculated lattice parameters of the 2H phase of InPo monolayers with space group *P*-6*m*2 are included in Table 12.1 [1]. A comparison with the lattice parameters of InTe monolayers shows an increase in the lattice constant, an extension of the In–Po bond length, and a larger distance between the Po atoms, indicating a greater thickness of the InPo monolayer despite a decrease in the van der Waals radius from Te (0.210 nm) to Po (0.197 nm). The 2H phase of the InPo monolayer is a semiconductor with a band gap of $E_g = 0.87$ eV (PBE) [1]. To date, this study seems to be the only work providing detailed properties for GaPo monolayers.

12.5 Thallium Monochalcogenide Monolayers (TlO, TlS, TlSe, TlTe, and TlPo)

12.5.1 TlO Monolayer

12.5.1.1 Chemistry and 2D Bonding

The three valence electrons of Tl atoms ($6s^2p_x^{\ 1}$) together with the six valence electrons of O atoms ($2s^2p_x^{\ 2}2_y^{\ 2}$) imply covalent 2D bonding, as discussed for the other group IIIA–VIA monolayers. In addition to the stable hexagonal four-layer 2H-TlO structure with stoichiometry Tl_2O_2, a Tl_2O structure has recently received increasing attention [82]. The significant difference in the electronegativities of Tl (1.8) and O

(3.44) polarizes the Tl–O bond by substantial charge transfer from the Tl atom to the O atom, producing a strong ionic character of bonding.

In the hexagonal three-layer T and H structures with Tl_2O stoichiometry, the O sublayer is sandwiched between two sublayers of Tl atoms (see Fig. 12.16) [82]. In the three-layered 1T structure, one Tl sublayer is shifted, reducing the symmetry of the structure. The unit cell contains one O atom and two Tl atoms. The T and H phases with Tl_2O stoichiometry are energetically and dynamically stable [82]. Determination of the charge distribution by ELF analysis indicates for the T phase almost no electron localization between Tl and O atoms and between Tl atoms, reflecting strong ionic bonding [83]. The ionicity is based on the donation of Tl 6*p* electrons to the oxygen atoms. Electronic orbital analysis indicates that the VBM is formed by O 2*p* and Tl 6*s* orbitals, whereas the CBM is dominated by Tl 6*p* orbitals.

The investigation of the Tl_2O stoichiometry has been extended to more compositions M_2X (M = Tl and In; X = O, S, and Se) monolayers. These metal-shrouded 2D

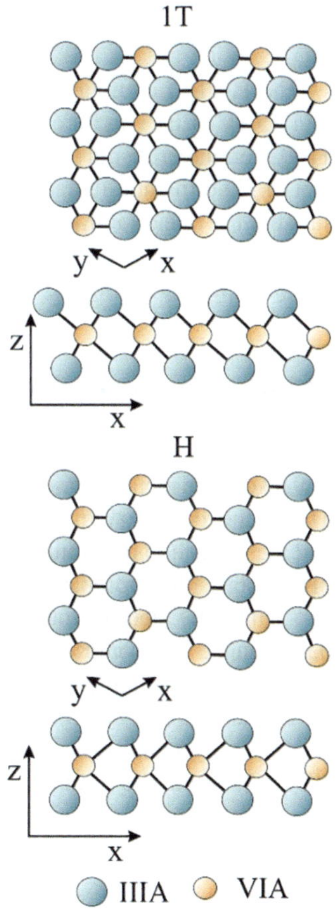

Fig. 12.16 Comparison of the hexagonal three-layer M_2X structure with 1T and H symmetry

semiconductors can form low-resistance contacts with electrode materials, except for 1T Tl_2O and H Tl_2O [82].

12.5.1.2 2D Structures

Table 12.1 displays the lattice parameters of the four-layer hexagonal 2H phase of TlO. The lattice parameters are a = 0.369 nm, $d_{Tl–O}$ = 0.229 nm, $d_{Tl–Tl}$ = 0.292 nm, h_{OO} = 0.463 nm, θ_{TlTlO} = 111.80°, and θ_{OTlO} = 107.04° [1]. This structure is a semimetal that up to now has not received much attention [2]. Adding the van der Waals diameter of O atoms d_w = 0.31 nm and the distance between O atoms, yields an estimate of the monolayer thickness of d_{w+h} = 0.773 nm.

The lattice constant of the 1T-Tl_2O monolayer is a = 0.359 nm and the Tl–O bond lengths in the Tl_6O octahedron are $d_{Tl–O}$ = 0.257 nm [83]. The 1T-Tl_2O phase is a direct semiconductor with a high carrier mobility of 4.3×10^3 cm^2 V^{-1} s^{-1} and a band gap of 1.56 eV that finally decreases to ~ 1.0 eV with increasing layer thickness at the bulk state of Tl_2O [83]. The band gap is similar in size to that of silicon (~1.4 eV).

The H-Tl_2O monolayer is an indirect semiconductor with a smaller band gap of 1.15 eV (PBE) and a carrier mobility of 9.8×10^3 cm^2 V^{-1} s^{-1} [84]. The H–Tl_2O monolayer displays large valley spin splitting (VSS), significantly larger than that of 2D transition-metal dichalcogenides (TMDCs), and a finite band gap that is suitable for integration of valleytronics and spintronics. The VSS is pronounced in the conduction band (0.61 eV) and negligible in the valence band (21 meV), opening a way for manipulating the coupled spin and valley physics.

12.5.1.3 Physics and Applications

The 1T-Tl_2O monolayer is a natural semiconductor with the special potential to form low resistance metal–semiconductor contacts with selected electrode materials [83]. This is due to the small lattice mismatch between the (001) surface of metallic hcp Tl and the surface of Tl_2O monolayers. Note that different from other metal-shrouded 2D crystals, the monolayer of Tl_2O is a natural semiconductor with direct band gap. Thus, the metal-shrouded Tl_2O monolayer is an excellent candidate for high-performance electronics and owing to strong optical absorption of ~ 10^5 cm^{-1} and high mobility of photogenerated electrons it qualifies for optoelectronic applications.

The lattice thermal conductivity and the electronic transport coefficients were calculated by semiclassical Boltzmann transport theory. The study revealed that the 1T-Tl_2O monolayer exhibits an ultralow thermal conductivity of 0.17 W/mK at 300 K and therefore may be a suitable candidate for thermoelectric applications [85].

12.5.1.4 Synthesis

Based on the low cleavage energy of experimentally available layered Tl_2O bulk crystals, which is comparable to that of graphite and MoS_2, the mechanical exfoliation of monolayers and few-layer sheets of Tl_2O should be possible from available bulk samples and may be achieved soon [83].

12.5.2 TlS Monolayer

12.5.2.1 Chemistry, 2D Bonding, Structure, Mechanics, Synthesis, and Applications

With similar valence electrons for S atoms ($3s^2p_x{}^2p_y{}^2$) as for O atoms and those of Tl atoms, the chemistry of 2D-TlS compounds has many similarities with that of 2D-TlO monolayers. A smaller electronegativity of the S atom (2.58) leads to a substantially lower ionicity of the Tl–S bond than of the Tl–O bond. The Tl_2S monolayer with its peculiar anti-MoS_2 structure, where the chalcogen is sandwiched between metal layers exhibits several unusual features (see Fig. 12.17).

Table 12.1 shows the confirmed lattice parameters of the stable four-sublayer 2H-TlS monolayer, which are a = 0.412 nm, $d_{Tl–S}$ = 0.267 nm, $d_{Tl–Tl}$ = 0.287 nm, h_{SS} = 0.531 nm, θ_{TlTlS} = 117.16°, and θ_{STlS} = 100.80° [1]. Adding the van der Waals diameter of S atoms d_w = 0.36 nm and the distance between S atoms, yields an estimate of the monolayer thickness of d_{w+h} = 0.891 nm. The band gaps of the indirect semiconductor calculated by two different approaches are E_g = 0.91 eV (PBE) and 1.50 eV (HSE06) [29]. The elastic modulus is only E = 40.2 N/m, and the Poisson ratio is $\nu = 0.29$ [29].

More importantly, according to first-principles calculations, the H phase of the Tl_2S monolayer is thermally and dynamically stable, with a modulable band gap of 1.89–2.31 eV, intense light absorption, and efficient conversion of photon energy into electricity [86]. The 2D metal-covered semiconductor is thermally and dynamically stable. The implications of the anti-structure for sensing, energy generation, and specifically the electrocatalytic performance toward HER were studied in detail [87].

The unique metal-shrouded 2D structure endows the semiconducting Tl_2S monolayer with the potential to form low-resistance contacts with electrode materials. A strong SOC and the loss of inversion symmetry produce a large conduction band

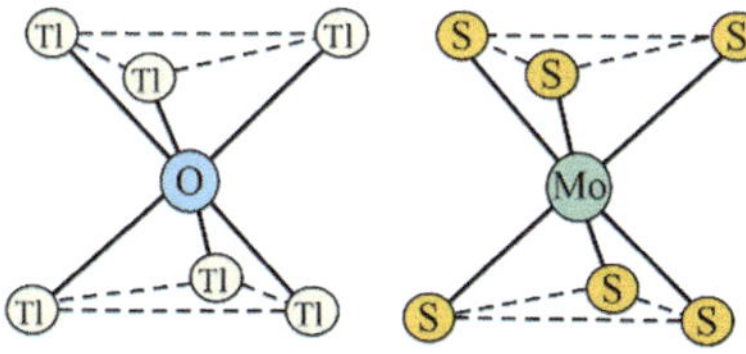

Fig. 12.17 Scheme of anti-Tl_2O structure in comparison with normal MoS_2 structure

spin-valley coupling in the H-Tl_2S monolayer. Like in Tl_2O, exfoliation of Tl_2S monolayers and 2D sheets from layered bulk Tl_2S crystals should be possible and is expected to be realized soon.

12.5.3 *TlSe Monolayer*

12.5.3.1 Chemistry, 2D Bonding, Structure, Mechanics, and Applications

With the valence electrons of Tl atoms ($6s^2p_x{}^1$) and Se atoms ($4s^2p_x{}^2p_y{}^2$), the TlSe monolayers form the same covalently bonded stable quadruple 2H structure as the TlO and TlS phases. Moreover, the Tl–Se bond has a medium ionicity owing to a similar difference in the electronegativities of Tl (1.8) and Se (2.55) as for Tl and S atoms.

Table 12.1 presents the lattice parameters of the hexagonal 2H-phase, which are $a = 0.427$ nm, $d_{Tl-Se} = 0.279$ nm, $d_{Tl-Tl} = 0.287$ nm, $h_{SeSe} = 0.548$ nm, $\theta_{TlTlSe} = 117.90°$, and $\theta_{SeTlSe} = 99.88°$ [1]. Adding the van der Waals diameter of Se atoms $d_w = 0.38$ nm and the distance between Se atoms, yields an estimate of the monolayer thickness of $d_{w+h} = 0.928$ nm. As band gap, different theoretical approaches yield values between $Eg = 0.69$ eV (PBE) and 1.21 eV (HSE06) [28]. The elastic modulus of the H-TlSe monolayer is further reduced to $E = 35.2$ N/m, and the Poisson ratio is $\nu = 0.28$ [29].

Three hexagonal 2D phases of TlSe studied by first-principles calculations exhibited low lattice thermal conductivities between 0.088 and 0.429 W m^{-1} K^{-1} at 900 K [88]. The resulting figure of merit varied between $ZT \approx 0.5$ and 1.94 in the temperature range of 300 K to 900 K. Moreover, this work investigated the way in which states of a topological insulator (TI) transform from the monolayer to the bulk in view of potential applications in spintronics.

12.5.4 *TlTe Monolayer*

12.5.4.1 Chemistry, 2D Bonding, Structure, Mechanics, and Applications

The similarity of the valence electrons of Tl atoms ($6s^2p_x{}^1$) and Te atoms ($5s^2p_x{}^2p_y{}^2$) with those of the other Tl-chalcogenide monolayers points to the related stable covalent 2D structures of TlTe monolayers. The decrease in electronegativity within the group of chalcogenides to 2.1 for the Te atom is responsible for a smaller ionic contribution to bonding in comparison with those of the lighter compounds.

The lattice parameters of 2H–TlTe monolayers are $a = 0.456$ nm, $d_{Tl-Te} = 0.298$ nm, $d_{Tl-Tl} = 0.288$ nm, $h_{TeTe} = 0.568$ nm, $\theta_{TlTlTe} = 118.05°$, and $\theta_{TeTlTe} = 99.69°$

[1]. Adding the van der Waals diameter of Te atoms $d_w = 0.42$ nm and the distance between Te atoms, yields an estimate of the monolayer thickness of $d_{w+h} = 0.988$ nm. The resulting reduction in the indirect band gap of the 2H-phase of TlTe monolayers as confirmed by different theoretical approaches are $E_g = 0.49$ eV (PBE) and 0.88 eV (HSE06) [28]. The elastic modulus of 2H-TlTe monolayers is further reduced to E = 28.2 N/m, and the Poisson ratio is $\nu = 0.25$ [29].

In a first-principles study on layer-separable and gap-tunable TIs, the authors predict the realization of layered TlTe crystals with weak interlayer coupling, which can be easily exfoliated for applications in spintronics [89]. Topological effects can be induced by doping with In atoms and the application of lateral strains.

12.5.5 TlPo Monolayer

12.5.5.1 Chemistry, 2D Bonding, and Structure

To date, the available information on TlPo monolayers is limited. Based on the valence electrons of Tl atoms ($6s^2p_x{}^1$) and Po atoms ($6s^2p_x{}^2p_y{}^2$), with the same set of bonding electrons as the already discussed compounds, the hexagonal 2H structure should be stable. This was confirmed by first-principles calculations, which predict for the lattice parameters of the 2H-phase of TlPo monolayers a = 0.465 nm, $d_{Tl–Po} = 0.305$ nm, $d_{Po–Po} = 0.288$ nm, $h_{PoPo} = 0.579$ nm, $\theta_{TlTlPo} = 118.43°$, and $\theta_{PoTlPo} = 99.21°$ [1]. Adding the van der Waals diameter of Po atoms $d_w = 0.394$ nm and the distance between Po atoms, yields an estimate of the monolayer thickness of $d_{w+h} = 0.973$ nm. In comparison with the geometrical data of TlTe, the monolayer of TlPo has a larger lattice constant, an extended distance between the Po atoms, and longer Tl–Po bonds, indicating an increase in the thickness of the monolayer. Notably, the mean van der Waals diameter decreases slightly from 0.43 nm (TlTe) to 0.42 nm (TlPo). The Tl–Po bonds are marginally polarized due to the small difference in the electronegativities of Tl (1.8) and Po (~2.0). Regarding the cohesive energy, the weakest bonding in this group is expected due to the size of Po atoms and the lattice extension. According to band structure calculations, the semiconducting TlPo monolayers have a narrow band gap of $E_g = 0.24$ eV (HSE06) [1].

12.6 Summary of Group IIIA–VIA Monolayers

12.6.1 Binding–Structure Relationships

Figure 12.18 shows the cohesive energy versus bond length for the M–X bonds of IIIA–VIA monolayers. Note that in the hexagonal four-layer structure the M–M bond lengths do not vary much when the chalcogen atom is changed. Surprisingly, the use of the DFT-GGA-PBE approach leads to significant differences in the estimated cohesive energies. The reasons for these deviations are not thoroughly discussed in the literature. On the contrary, the lattice parameters of diverse sources are in most cases in good agreement. The data presented for the group of B-VIA monolayers originate from two extensive collections of cohesive energies and structural data [1, 2]. Table 12.1 indicates the strongest bonding for BO, BS, BSe, and BTe. The literature values of cohesive energies are in satisfactory agreement for the group of Al-VIA monolayers. For the Ga-chalcogenides, the most complete data collection gives small bonding energies [2]. Therefore, Fig. 12.18 displays the larger binding energies reported for GaS, GaSe, and GaTe [27, 29]. Too large deviations of the binding energies are also found for the group In-VIA monolayers. Figure 12.18 displays the lower binding energies of the In-VIA monolayers taken from the complete data set [2], despite the much larger binding energies estimated for InS, InSe, and InTe monolayers [27].

The plot documents the general inverse behavior of the selected cohesive energies versus the bond lengths of the covalent-ionic M–X bonds, as predicted by tight-binding theory. Most results presently available for the five considered groups follow surprisingly well the same inverse dependence and may allow the identification of strongly deviating results. One reason for the virtual coincidence of the cohesive energies of the five groups in an approximately uniform dependence is the limited variation of the cohesive energies and the length of the chemical bonds. Note that this

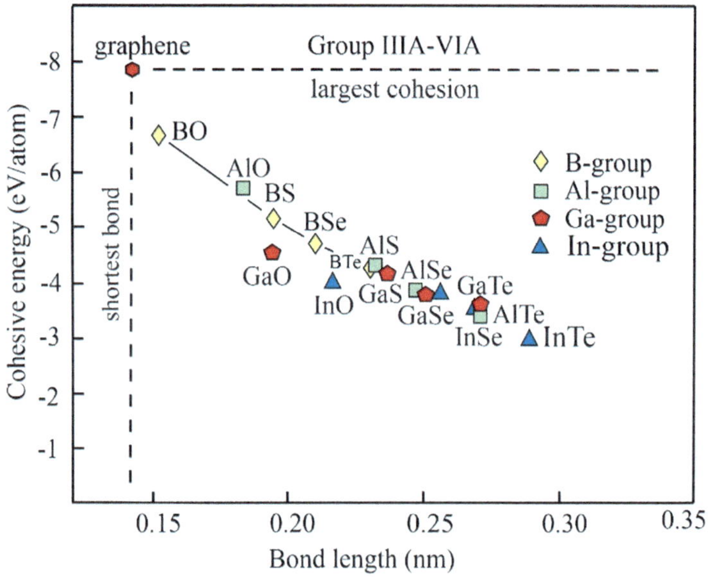

Fig. 12.18 Plot of the cohesive energies versus bond lengths of the hexagonal 2H structure of group IIIA–VIA monolayers

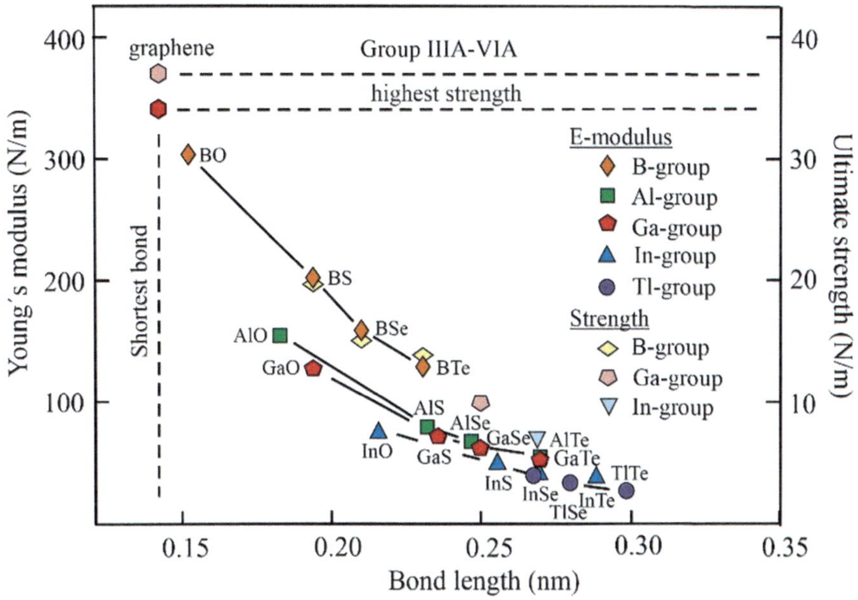

Fig. 12.19 Young's moduli and ultimate strengths versus bond lengths of the hexagonal 2H-structures of group IIIA–VIA monolayers

observation is no longer true for linear and nonlinear mechanical properties because of their sensitive dependence on the change of bond length.

12.6.2 *Mechanics–Structure Relationships*

Figure 12.19 shows the calculated Young's moduli and for few monolayers the ultimate strengths as a function of the M–X bond lengths. The extremely sensitive dependence of the mechanical properties on bond lengths introduces a separation of the stiffnesses of the individual groups. In fact, the stiffness changes by a factor of approximately eight for a change in bond length of less than a factor of two. There is an enormous variation in the mechanical stiffness of the B-chalcogenide monolayers, whereas the IIIA-chalcogenide groups of the higher homologues move together forming a broad inverse dependence in the stiffness–bond length plot.

The few calculations of ultimate strengths presently available follow Griffith's rule only for the B-chalcogenide group, whereas the strengths are much greater than one-tenth of Young's modulus in the other cases. The plot demonstrates the low stiffness and critical strength of selenides and tellurides, which are of special interest for potential future applications. Note that low mechanical performance favors structural defect formation and deterioration by impurities owing to an increasing chemical reactivity of the lattice atoms.

References

1. M.N. Çinar, G.Ö. Sargin, K. Sevim, B. Özdamar, G. Kurt et al., Phys. Rev. B **103**, 165422 (2021)
2. S. Demirci, N. Avazli, E. Durgun, S. Cahangirov, Phys. Rev. B **95**, 115409 (2017)
3. V. Zólyomi, N.D. Drummond, V.I. Fal'ko, Phys. Rev. B **89**, 205416 (2014)
4. S.S. Batsanov, Inorg. Mater. **37**, 871 (2001)
5. M. Saaoud, K. Sadki, L.B. Drissi, M. Abdel-Hafiez, Mater. Sci. Engineer. B **283**, 115758 (2022)
6. L. Yan, P.-F. Liu, H. Li, Y. Tang, J. He, et al., NPJ Comput. Mater. **6**, 94 (2020)
7. R. Zhang, Z. Li, J. Yang, J. Phys. Chem. Lett. **8**, 4347 (2017)
8. S. Lin, J. Gu, H. Zhang, Y. Wang, Z. Chen, Flat Chem **9**, 27 (2018)
9. R. Zhang, J. Yang, J. Phys. Chem. Lett. **8**, 4347 (2017)
10. P. Mishra, D. Singh, Y. Sonvane, R. Abuja, Sust. Energy Fuels **4**, 2363 (2020)
11. B. Mortazavi, T. Rabczuk, Energies **11**, 1573 (2018)
12. D. Fan, S. Lu, C. Chen, M. Jiang, X. Li, Appl. Phys. Lett. **117**, 013103 (2020)
13. H. Kusaka, R. Ishibiki, M. Toyoda, T. Fujita, T. Tokunaga et al., J. Mater. Chem. A **9**, 24631 (2021)
14. P. Mishra, Y. Sonvane, Mater. Res. Proceed. **22**, 65 (2022)
15. K. Ren, Y. Luo, S. Wang, J.-P. Chou, J. Yu et al., ACS Omega **4**, 21689 (2019)
16. Y. Xu, Y. Liu, Y. Chen, Y. Zhang, C. Ma et al., ACS Appl. Mater. Interfaces **12**, 58349 (2020)
17. D. Singh, N. Khossossi, A. Ainane, R. Ahuja, Catal. Today **370**, 14 (2021)
18. M. Demirtas, M.J. Varjovi, M.M. Cicek, E. Durgun, Phys. Rev. Mater. **4**, 114003 (2020)
19. Z. Haman, N. Khossossi, M. Kibbou, I. Bouziani, D. Singh et al., Appl. Surf. Sci. **556**, 149561 (2021)
20. S. Choopani, M.M. Alyörük, J. Phys. D Appl. Phys. **55**, 155301 (2022)
21. Z. Haman, N. Khossossi, M. Kibbou, I. Bouziani, D. Singh et al., Appl. Surf. Sci. **589**, 152997 (2022)
22. L.C. Nhan, C.Q. Nguyen, N.V. Hieu, H.V. Phuc, C.V. Nguyen et al., Optik **238**, 166761 (2021)
23. X. Chen, Y. Huang, J. Liu, H. Juan, H. Chen, ACS Omega **4**, 17773 (2019)
24. S. Bahti, M. Kibbou, N. Khossossi, I. Essaoudi, A. Ainane et al., Physica E **142**, 115229 (2022)
25. R. Meng, M. Houssa, K. Iordanidou, G. Pourtois, V. Afanasiev et al., Phys. Rev. Materials **4**, 074001 (2020)
26. K.D. Pham, V.T.T. Vi, D.V. Thuan, N.V. Hieu, C.V. Nguyen et al., Chem. Phys. **524**, 101 (2019)
27. T. Hu, C. Xu, A. Zhang, P. Yu, Mater. Adv. **3**, 2213 (2022)
28. M. Yagmurcukardes, R.T. Senger, F.M. Peeters, H. Sahin, Phys. Rev. B **94**, 245407 (2016)
29. A. Majumdar, S. Chowdhury, R. Ahuja, Nano Energy **88**, 106248 (2021)
30. W. Li, J. Li, Nano Res. **8**, 3796 (2015)
31. H.L. Zhuang, R.G. Hennig, Chem. Mater. **25**, 3232 (2013)
32. H. Cai, Y. Gu, Y.-C. Lin, Y. Yu, D.B. Geohegan et al., Appl. Phys. Rev. **6**, 041312 (2019)
33. D.J. Late, B. Liu, H.S.S.R. Matte, C.N.R. Rao, V.P. Dravid, Adv. Funct. Mater. **22**, 1894 (2012)
34. X. Wang, Y. Sheng, R.-J. Chang, J.K. Lee, Y. Zhou et al., ACS Omaga **3**, 7897 (2018)
35. P. Hu, L. Wang, M. Yoon, J. Zhang, W. Feng et al., Nano Lett. **13**, 1649 (2013)
36. S. Yang, Y. Li, X. Wang, N. Huo, J.-B. Xia et al., Nanoscale **6**, 2582 (2014)
37. A. Harvey, C. Backes, Z. Gholamvand, D. Hanlon, D. McAteer et al., Chem. Mater. **27**, 3483 (2015)
38. S. Zhou, C.-C. Liu, J. Zhao, Y. Yao, NPJ Quant. Mater. **3**, 16 (2018)
39. G. Liu, S. Xia, B. Hou, T. Gao, R. Zhang, Modern Phys. Lett. B **29**, 1550049 (2015)
40. H. Arora, A. Erbe, InfoMat **3**, 662 (2021)
41. M. Rahaman, R.D. Rodriguez, M. Monecke, S.A. Lopez-Rivera, D.R.T. Zahn, Semicond. Sci. Technol. **32**, 105004 (2017)
42. Y. Cui, L. Peng, Q. Qian, Y. Huang, J. Mater. Chem. A **6**, 22768 (2018)
43. Y. Cui, M. Li, X. Zhang, S. Wang, Y. Huang, Green Energy Environ. **7**, 1045 (2022)
44. P. Hu, Z. Wen, L. Wang, P. Tan, K. Xiao, ACS Nano **6**, 5988 (2012)

45. S. Lei, L. Ge, Z. Liu, S. Najmaei, G. Shi et al., Nano Lett. **13**, 2777 (2013)
46. X. Li, M.-W. Lin. A.A. Puretzky, J.C. Idrobo, C. Ma, et al., Sci. Rep. **4**, 5497 (2014)
47. M. Martin, P. Pochet, H. Okuno, C. Alvarez, E. Bellet-Amalric et al., APL Mater. **10**, 051106 (2022)
48. J.A. Olmos-Asar, C.R. Leão, A. Fazzio, RSC Adv. **7**, 32383 (2017)
49. Y. Yu, M. Ran, S. Zhou, R. Wang, F. Zhou et al., Adv. Funct. Mater. **29**, 1901012 (2019)
50. Y. Zhou, S. Zhou, P. Ying, Q. Zhao, Y. Xie et al., J. Phys. Chem. Lett. **13**, 3831 (2022)
51. V.T.T. Vi, N.N. Hieu, B.D. Hoi, N.T.T. Binh, T.V. Vu, Superlatt. Microstruct. **140**, 106435 (2020)
52. L.-C. Tien, Y.-C. Shih, C.-Y. Chen, Y.-T. Huang, R.-S. Chen, J. Alloys Comp. **876**, 160195 (2021)
53. S.S.A. Al-Abbas, M.K. Muhsin, H.R. Jappor, Superlatt. Microstruct. **135**, 106245 (2019)
54. Z. Wang, M. Safdar, M. Mirza, K. Xu, Q. Wang et al., Nanoscale **7**, 7252 (2015)
55. Q. Chen, Y. Chen, J. Wang, M. Liu, Z. Chen, Crystals **12**, 627 (2022)
56. R. Batista dos Santos, R. Rivelino, G.K. Gueorguiev, A. Kakanakova-Georgieva, Cryst. Eng. Commun. **23**, 6661 (2021)
57. Y. Guo, L. Ma, K. Mao, M. Ju, Y. Bai et al., Nanoscale Horiz. **4**, 592 (2019)
58. A. Kakanakova-Georgieva, F. Giannazzo, G. Nicotra, I. Cora, G.K. Gueorguiev et al., Appl. Surf. Sci. **548**, 149275 (2021)
59. A. Shafique, Y.-H. Shin, Sci. Rep. **10**, 1093 (2020)
60. A. Rawat, R. Ahammed, Dimple, N. Jena, M.K. Mohanta, et al., J. Phys. Chem. C **123**, 12666 (2019)
61. X.-B. Xiao, Q. Ye, Z.-F. Liu, Q.-P. Wu, Y. Li et al., Nanoscale Res. Lett. **14**, 322 (2019)
62. J. Jalilian, M. Safari, Phys. Lett. A **381**, 1313 (2017)
63. X. Chang, H. Li, G. Tang, Comp. Mater. Sci. **158**, 340 (2019)
64. F. Jamil, S.A.M. Jony, T.S. Akash, R.A.S.I. Subad, M. Islam, Comp. Condens. Matter **30**, e00637 (2022)
65. Y. Li, C. Yu, Y. Gan, Y. Kong, P. Jiang et al., Nanotechnology **30**, 335703 (2019)
66. Q. Zhao, R. Frisenda, T. Wang, A. Castellanos-Gomez, Nanoscale **11**, 9845 (2019)
67. N.T. Hung, A.R.T. Nugraha, R. Saito, Appl. Phys. Lett. **111**, 092107 (2017)
68. L.W. Sprague Jr., C. Huang, J.-P. Song, B.M. Rubinstein, J. Phys. Chem. C **123**, 25437 (2019)
69. H. Zhou, Y. Cai, G. Zhang, Y.-W. Zhang, Nanoscale **10**, 480 (2018)
70. Z. Ma, R. Li, R. Xiong, Y. Zhang, C. Xu et al., Materials **14**, 3768 (2021)
71. Z. Wang, G. Zhou, J. Phys. Chem. C **124**, 167 (2020)
72. C. Ren, S. Wang, H. Tian, Y. Luo, J. Yu et al., Sci. Rep. **9**, 13289 (2019)
73. H.-C. Chang, C.-L. Tu, K.-I. Lin, J. Pu, T. Takenobu et al., Small **14**, 1802351 (2018)
74. J. Zhou, J. Shi, Q. Zeng, Y. Chen, L. Niu, et al., 2D Mater. **5**, 025019 (2018)
75. W. Huang, L. Gan, H. Li, Y. Ma, T. Zhai, Cryst. Eng. Commun. **16**, 3968 (2016)
76. T.-N. Do, V.T.T. Vi, N.T.T. Binh, N.N. Hieu, N.V. Hieu, Superlatt. Microstruct. **151**, 106816 (2021)
77. Y. Liao, H. Liu, G. Yuan, Z. Huang, X. Qi, Crystal Res. Technol. **55**, 2000102 (2020)
78. A.V. Bandura, A.V. Kovalenko, D.D. Kuruch, R.A. Evarestov, Eur. J. Inorg. Chem. **2021**, 126 (2020)
79. M.-S. Li, K.-X. Chen, D.-C. Mo, S.-S. Lyu, Phys. Chem. Chem. Phys. **21**, 24695 (2019)
80. H. Alqurashi, A. Pandit, B. Hamad, Mater. Sci. Engin. B **283**, 115823 (2022)
81. K. Li, X. Xian, J. Wang, N. Yu, Appl. Surf. Sci. **471**, 18 (2019)
82. H. Li, S. Hu, S. Zhao, C. Lan, Chem. Phys. Lett. **749**, 137404 (2020)
83. Y. Ma, A. Kuc, T. Heine, J. Am. Chem. Soc. **139**, 11694 (2017)
84. Y. Ma, L. Kou, A. Du, B. Huang, Y. Dai et al., Phys. Rev. B **97**, 035444 (2018)
85. M. Sajjad, N. Singh, S. Sattar, S. De Wolf, U. Schwingenschlögl, A.C.S. Appl, Energy Mater. **2**, 3004 (2019)
86. S. Shen, Y. Liang, Y. Ma, B. Huang, W. Wei et al., Phys. Chem. Chem. Phys. **20**, 14778 (2018)
87. X. Chia, A. Ambrosi, Z. Sofer, J. Luxa, D. Sedmidubsky et al., ACS Nano **10**, 112 (2016)
88. A. Majumdar, S. Chowdhury, R. Ahuja, A.C.S. Appl, Energy Mater. **3**, 9315 (2020)
89. K.-W. Chang, W. Ji, C.C. Kaun, Phys. Chem. Chem. Phys. **19**, 3932 (2017)

Chapter 13
Group IVA–VIA Monolayers

Abstract The elements of group IVA–VIA compounds or group–IVA monochalcogenides with four valence electrons (n$s^2p_x^2$) for the IVA elements and six valence electrons (n$s^2p_x^2p_y^2$) for the VIA elements are direct neighbors of group–VA elements. Therefore, these binary compounds are their isoelectronic counterparts. They form the monolayers MX (M = C, Si, Ge, Sn, and Pb; X = O, S, Se, Te, and Po). The most stable structures of these monolayers are the puckered and buckled configurations, where the group–IVA atoms form three bonds with group–VIA atoms and vice versa, and both the group–IVA atoms and the chalcogen atoms form sp^3-like hybrids and the chalcogen atoms possess a lone electron pair. Owing to the different electronegativities and atom sizes of the constituents, the binary monolayer structures lack centrosymmetry. The stereo-chemically active lone electron pairs are the source of structural distortions leading to the anisotropy of thermal expansion and the distinctive elastic, piezoelectric, optical, and optoelectronic properties of this group of compounds. Of enormous practical interest are the distorted orthorhombic crystals consisting of an asymmetric washboard-like (aw) structure with the space group $Pmn2_1$. Up to now, the large exfoliation energy has hindered the realization of single layers by versatile mechanical exfoliation methods. Only bottom-up techniques, such as molecular beam epitaxy (MBE) and chemical vapor deposition (CVD), allowed the successful synthesis of selected monolayers such as SnSe, SnTe, and PbTe. The aw phases of the four compounds GeS, GeSe, SnS, and SnSe have attracted enormous attention owing to their high mechanical flexibility, high carrier mobility, low lattice thermal conductivity, and strong absorbance of visible light. These properties qualify the 2D materials for potential applications in optoelectronics, thermoelectrics, piezoelectric devices, photodetectors, and gas sensors. Remarkable phenomena related to in-plane strain include giant piezoelectricity and multiferroicity, combining ferroelastic and ferroelectric behavior.

P. Hess, *Bonding, Structure, and Performance of Two-Dimensional Materials*,
NanoScience and Technology, https://doi.org/10.1007/978-3-031-74161-6_13

13.1 Carbon Monochalcogenide Monolayers (CO, CS, CSe, and CTe)

13.1.1 CO Monolayer

13.1.1.1 Chemistry, 2D Bonding, and Structure

The elements, which form the monolayers MX (M = C, Si, Ge, Sn, and Pb; X = O, S, Se, Te, and Po) are highlighted in Fig. 13.1.

Based on the experience with monolayers containing an element from the second period of the periodic table, one might expect that compounds such as CO, CS, CSe, and CTe, which are composed of an element from the second period, could form stable planar 2D structures. The structural data derived for these monolayers from density functional theory (DFT) calculations employing the generalized gradient approximation (GGA) given by the Perdew–Burke–Ernzerhof (PBE) functionals indicate that the puckered (asymmetric washboard (aw)) and hexagonal buckled (hb) phases form the most stable structures [1]. In a dynamically stable puckered four-layer structure with two different atoms, centrosymmetry is lost (see Fig. 13.2). While the only slightly decreasing negative cohesive energies indicate the energetic stability of all four carbon monolayers, to date, only aw CSe (α CSe) has attracted significant attention owing to the direct band gap of the puckered configuration, whereas the remaining compounds are indirect semiconductors. The two-layer hexagonal buckled hb phase (ß phase) of CO monolayers has a larger lattice constant and C–O bond length and is energetically less stable than the puckered phase. The buckled two-layer structure of hb-CO monolayers is displayed in Fig. 13.3.

With the valence electrons of C atoms ($2s^2p_x^2$) and O atoms ($2s^2p_x^2p_y^2$), CO is isoelectronic to nitrogen and group–VA elements with five valence electrons. Therefore, many similarities exist between the physicochemical and electronic properties of elemental group VA and binary group IVA–VIA monolayers. The ten valence electrons of CO form chemically inactive but sterically active lone electron pairs, which influence the three bonds of the group–IVA atoms with the chalcogen atoms and therefore the crystal structure. The chalcogen atoms form sp^3-like hybrids with three bonds to group–IVA atoms and a lone electron pair. This lone electron pair in each chalcogen atom pushes the three covalently bonded atoms toward tetrahedral

H							He
Li	Be	B	C	N	O	F	Ne
Na	Mg	Al	Si	P	S	Cl	Ar
K	Ca	Ga	Ge	As	Se	Br	Kr
Rb	Sr	In	Sn	Sb	Te	I	Xe
Cs	Ba	Tl	Pb	Bi	Po	At	Rn
Fr	Ra						

Fig. 13.1 Periodic table of the main group elements, highlighting groups IVA and VIA involved in group–IVA monochalcogenide monolayers

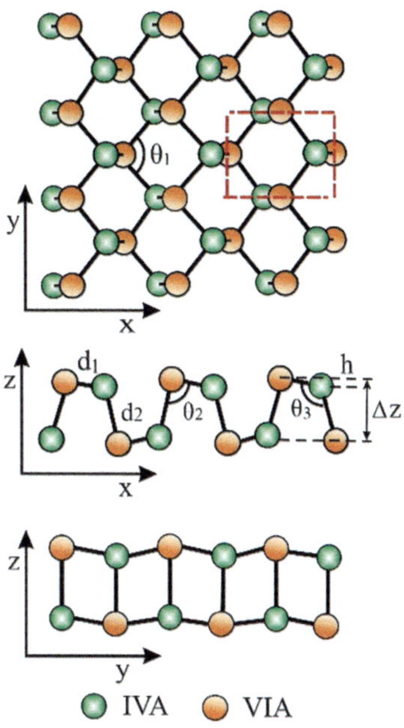

Fig. 13.2 Top view and side views of the asymmetric washboard (aw) or puckered structure with four sublayers of group IVA–VIA monolayers. The unit cell is indicated by dashed lines

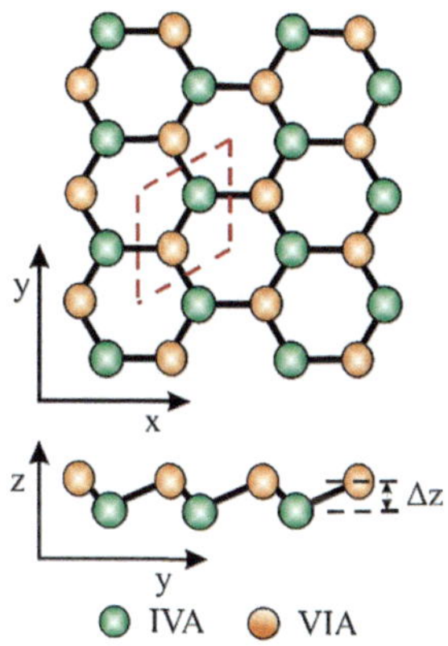

Fig. 13.3 Top and side views of the hexagonal buckled (hb) structure of group IVA–VIA monolayers. The unit cell is indicated by dashed lines

coordination. Note that covalent bonding is more complicated in this case than classical hybridization schemes would suggest. The energetically more stable aw-CO monolayer has a considerably smaller cohesive energy at $E_g = -5.07$ eV/atom with stronger bonding than the hb phase with -4.34 eV/atom [1].

The lattice constants of the aw-CO monolayer are a = 0.218 nm, b = 0.401 nm, the short bond length is d = 0.138 nm, and the favorable cohesive energy indicates strong covalent-ionic bonding of this monolayer (see Table 13.1) [1]. Owing to the difference in the electronegativities of C (2.55) and O (3.44), the C–O bond is polarized with a large charge transfer from the C atom to the more electronegative

Table 13.1 Space group, cohesive energy, lattice constants, bond lengths, corrugation effect, and van der Waals diameter of group IVA–VIA monolayers

Mono-layer	Space group	Cohesive energy, E_c (eV/atom)	Lattice constants a, b (nm)	Bond lengths d_1, d_2 (nm)	Pucker, buckle Δz (nm)	Bond angles $\theta_1, \theta_2, \theta_3$ (°)	Van der Waals diameter d_w [2], $d_{w+\Delta z}$ (nm)
CO	$Pmn2_1$ [1] $P3m1$ [1]	− 5.07 [1] − 4.34 [1]	0.218, 0.401 [1] 0.245 [1]	0.138 0.164 [1]			0.325 [2]
CS	$Pmn2_1$ [1] $P3m1$ [1]	− 4.37 [1] − 4.08 [1]	0.280, 0.432 [1] 0.284 [1]	0.176, 0.185 [1] 0.188 [1]			0.350 [2]
CSe	$Pmn2_1$ [1] $P3m1$ [1]	− 3.84 [1], − 3.91 [3] − 3.64 [1], − 3.79 [4]	0.303, 0.430 [1] 0.306 [1]	0.196, 0.201 [1] 0.206 [1]	0.184 [5] 0.104 [4]	110.9 (θ_2), 98.2 (θ_3) [5] 96.45 [4]	0.360 [2], 0.564 0.360 [2], 0.464
CTe	$Pmn2_1$ [1] $P3m1$ [1]	− 3.68 [1] − 3.43 [1]	0.436, 0.425 [5] 0.335 [1]	0.216, 0.218 [5] 0.223 [1]	0.208 [5]	109.3 (θ_2), 90.1 (θ_3) [5]	0.380 [2], 0.628
SiO	$Pmn2_1$ [1] $P3m_1$ [1]	− 5.27 [1] − 5.02 [1]	0.274, 0.470 [1] 0.282 [1]	0.184, 0.186 [1] 0.188 [1]	0.131 [5]	94.2 (θ_2), 132.5 (θ_3) [5]	0.365 [2], 0.441
SiS	$Pmn2_1$ [1] $P3m_1$ [1]	− 3.91 [1] − 3.92 [1], − 3.95 [9]	0.335, 0.477 [1] 0.330 [1, 9]	0.230, 0.234 [1] 0.232 [1, 9]	0.217 [5] 0.133 [9]	96.3 (θ_2), 111.1 (θ_3) [5]	0.390 [2], 0.577 0.390 [2], 0.523
SiSe	$Pmn2_1$ [1] $P3m1$ [1]	− 3.53 [1] − 3.55 [1], − 3.60 [9]	0.374, 0.440 [1] 0.352 [1, 9]	0.252, 0.245 [1] 0.248 [1, 9]	0.238 [5] 0.142 [9]	97.0 (θ_2), 101.5 (θ_3) [5] 76.62 [6]	0.400 [2], 0.618 0.400 [2], 0.542
SiTe	$Pmn2_1$ [1] $P3m1$ [1]	− 3.23 [1], − 3.30 [17] − 3.20 [1], − 3.27 [17]	0.411, 0.429 [1] 0.384 [1]	0.264, 0.278 [1] 0.269 [1]	0.272 [5] 0.153 [9]	95.6, 100.3, 87.8 [5, 17] 90.9 [17]	0.420 [2], 0.692 0.420 [2], 0.573
GeO	$Pmn2_1$ [1] $P3m1$ [1]	− 4.67 [1] − 4.37 [1]	0.305, 0.480 [1] 0.312 [1], 0.305 [6]	0.196, 0.199 [1] 0.203 [1], 0.266 [6]	0.200 [5]	69.88 [6]	0.365 [2]
GeS	$Pmn2_1$ [1] $P3m1$ [1]	− 3.66 [1], − 3.76 [20] − 3.64 [1], − 3.74 [20]	0.364, 0.449 [1] 0.349 [1], 0.351 [6]	0.246, 0.242 [1] 0.243 [1], 0.285 [6]	0.261 [20] 0.136 [20]	95.17, 95,24, 104.1 [21] 76.84 [6]	0.390 [2], 0.595 0.390 [2], 0.526

(continued)

Table 13.1 (continued)

Mono-layer	Space group	Cohesive energy, E_c (eV/atom)	Lattice constants a, b (nm)	Bond lengths d_1, d_2 (nm)	Pucker, buckle Δz (nm)	Bond angles θ_1, θ_2, θ_3 (°)	Van der Waals diameter d_w [2], $d_{w+\Delta z}$ (nm)
GeSe	*Pmn*2$_1$ [1] *P*3*m*1 [1]	− 3.37 [1], − 3.88 [20] − 3.36 [1], − 3.85 [20]	0.397, 0.430 [1] 0.368 [1]	0.266, 0.254 [1] 0.257 [1]	0.249 [20] 0.141 [20]	96.26, 93.15, 97.86 [21] 78.26° [6]	0.400 [2], 0.633 0.400 [2], 0.541
GeTe	*P*3*m*1 [1] *P*3*m*1 [1]	− 3.09 [1] − 3.06 [1]	0.424, 0.438 [1] 0.394 [1]	0.274, 0.288 [1] 0.277 [1]	0.305 [50] 0.156 [50]	99.9 (θ_2), 87.3 (θ_3) [5] 91.5 [53]	0.420 [2], 0.725 0.420 [2], 0.576
SnO	*Pmn*2$_1$ [1] *P*3*m*1 [1]	− 4.45 [1] − 4.17 [1]	0.340, 0.476 [1] 0.344 [1]	0.213, 0.216 [1] 0.220 [1]			0.380 [2]
SnS	*Pmn*2$_1$ [1] *P*3*m*1 [1]	− 3.55 [1], − 3.67 [20] − 3.47 [1], − 3.58 [20]	0.405, 0.435 [1] 0.376 [1]	0.273, 0.260 [1] 0.262 [1]	0.261 [20] 0.146 [20]	95.53,89.42, 100.0 [21] 79.23° [6]	0.405 [2], 0.666 0.405 [2], 0.546
SnSe	*Pmn*2$_1$ [1] *P*3*m*1 [1]	− 3.28 [1], − 4.57 [20] − 3.22 [1], − 4.56 [20]	0.426, 0.445 [1] 0.392 [1]	0.289, 0.273 [1] 0.275 [1]	0.259 [20] 0.153 [20]	94.53, 93.56, 92.50 [21] 80.55 [6]	0.415 [2], 0.639 0.415 [2], 0.568
SnTe	*Pmn*2$_1$ [1] *P*3*m*1 [1]	− 2.99 [1] − 2.93 [1]	0.454, 0.458 [1] 0.415 [1]	0.293, 0.316 [1] 0.295 [1]		92.1 (θ_1), 95.5 (θ_2) [29]	0.435 [2]
PbS	sb− phase *P*3*m*1		0.374 [80] 0.394 [10]	0.284 [80]	0.197 [80] 0.147 [10]	21.6 [80]	0.410 [2], 0.607 0.410 [2], 0.557
PbSe	sb-phase *P*3*m*1		0.382 [80] 0.406 [10]	0.296 [80]	0.231 [80] 0.159 [10]	24.4 [80]	0.420 [2], 0.651 0.420 [2], 0.579
PbTe	sb-phase *P*3*m*1		0.401 [80] 0.432 [10]	0.316 [80]	0.265 [80] 0.172 [10]	26.3 [80]	0.440 [2], 0.705 0.440 [2], 0.612

O atom of δq = 1.84 e [1]. The ionic interaction results in an indirect band gap of E_g = 1.37 eV [1]. First-principles calculations of the corresponding properties of the buckled and planar CO monolayers indicate longer bonds and weaker bonding for these monolayers, especially for the planar structure [1]. Unfortunately, no values for the pucker effect and the buckling effect are available, which allow an estimate of the layer thickness by adding the effective van der Waals diameter [2] to the corrugation. The space group of the puckered aw structure is $Pmn2_1$, and that of the buckled hb phase $P3m1$. The space group, cohesive energy, lattice constant, bond length, corrugation, bond angles, van der Waals diameter, and corrugation corrected van der Waals diameter of group IVA–VIA monolayers are presented in Table 13.1.

It is important to note that the band structure of CO monolayers, with their incomplete 2D-like geometric structure, differs completely from that of the other oxides of group–IVA elements [1]. CO does not form a real 2D structure but rather linear chains, while the optimized geometric structures of the oxides SiO, GeO, and SnO show puckered configurations, which are slightly distorted [1]. In fact, the oxides behave differently from the behavior of the sulfides, selenides and tellurides.

13.1.2 CS Monolayer

13.1.2.1 Chemistry, 2D Bonding, and Structure

With the valence electrons of C atoms ($2s^2p_x^2$) and S and ($3s^2p_x^2p_y^2$), CS is isoelectronic to group-VA elements. As already discussed for CO, the aw phase of the CS monolayer has a greater stability than the hb phase. The aw-CS monolayers entail a four-layer structure O–C–C–O. In general, the group–IVA atoms form the two inner sublayers and the group–VIA atoms the two outer sublayers. In this case, the lone electron pairs extend perpendicular to the monolayer surface in sp^3-like hybrids. In fact, the cohesive energy of E_c = − 4.37 eV/atom signifies stronger bonding for aw CS than for hb CS with a cohesive energy at − 4.08 eV/atom or the planar phase at − 3.58 eV/atom [1]. Note that despite the somewhat larger electronegativity of S atoms (2.58) compared with that of C atoms (2.55), as given by the Pauling scale, a Bader net charge transfer of δq = 0.5 e (0.454 e [1]) occurs from the S atom to the C atom. As indirect band gaps PBE and HSE calculations delivered E_g = 1.13 eV and 2.02 eV, respectively.

The lattice constants of the aw-CS monolayer are a = 0.280 nm and b = 0.432 nm, and the bond lengths are d_1 = 0.176 nm and d_2 = 0.185 nm [1]. The space group of the puckered aw structure is $Pmn2_1$.

The buckled hb-CS monolayer has a slightly larger lattice constant of a = 0.284 nm and bond length of d = 0.188 nm, consistent with the weaker bonding of the buckled CS monolayer [1]. The buckled phase of CS monolayers belongs to the space group $P3m1$.

13.1.3 CSe Monolayer

13.1.3.1 Chemistry and 2D Bonding

With the same number and type of valence electrons resulting from the electronic configurations of C atoms ($1s^2 2s^2 p^2$) and Se atoms ($Ar|3d^{10}4s^2p^4$), the stable monolayer phases of CSe follow the bonding types and structural arrangements of the CS monolayers. However, an increase in the atom size from S to Se reduces the orbital overlap and thus leads to weaker covalent bonding in CSe monolayers. Despite the quasi identical electronegativities of C (2.57) and Se (2.55) atoms, a Bader charge transfer of $\delta q = 0.216$ e is postulated from the Se atom to the C atom [3]. DFT calculations confirmed the energetic stability of the monolayer, revealing a slightly greater energetic stability of the puckered phase than that of the buckled phase, with cohesive energies of $E_c = -3.84$ eV/atom [1] and − 3.91 eV/atom [3] for aw CSe and $E_c = -3.64$ eV/atom [1] and − 3.79 [4] for hb-CSe monolayers.

13.1.3.2 2D Structures

The published lattice constants of the aw phase of CSe monolayers of a = 0.303 nm and b = 0.430 nm and bond lengths of $d_1 = 0.196$ nm and $d_2 = 0.201$ nm, obtained from various studies, agree quite well [1, 3], and the bond angles are $\theta_2 = 110.9$ and $\theta_3 = 98.2$ (see Fig. 13.2) [5]. Compared with the well-known symmetric washboard (sw) structure of phosphorus with two sublayers the aw phase consists of four sublayers with a tiny distance effect h between the two outer sublayers, as indicated in Fig. 13.2. Accordingly, a recent study of the structure of CSe reported a total pucker effect of $\Delta z = 0.184$ nm with a small buckling distance of the outer two sublayers of h = 0.023 nm [5]. In addition to the energetic stability, first-principles DFT-PBE calculations also confirmed the dynamic stability of aw CSe (α CSe) [3]. This phase of the CSe monolayer is a semiconductor with a direct band gap of $E_g = 0.905$ eV based on the PBE and 1.59 eV based on the Heyd–Scuseria–Ernzerhof (HSE) approach [3]. With the pucker distance and the van der Waals diameter of Se $d_w = 0.380$ nm, an estimate of the corrugated monolayer thickness of $d_{w+\Delta z} = 0.564$ nm is found [5].

For the buckled hb-CSe (ß-CSe) monolayer, the calculated structural parameters are a = 0.307 nm (0.306 nm [1]) and b = 0.522 nm as the lattice constants; d = 0.206 nm (0.206 nm [1]) as the bond length; $\Delta z = 0.104$ nm as the buckling height; and $\theta = 96.45°$ as the bond angle [4]. The cohesive energy of $E_c = -3.64$ eV/atom is less favorable for hb CSe, which is consistent with the extended bond length. The monolayer is not only energetically but also dynamically stable. The approach-dependent indirect bandgaps at $E_g = 1.54$ eV (PBE) and 2.37 eV (HSE) are sensitive to in-plane strain [4]. With a mean van der Waals diameter of CSe of $d_w = 0.360$ nm [2], the corrugated layer thickness is $d_{w+\Delta z} = 0.464$ nm.

13.1.3.3 Mechanics

Reported Young's moduli of the anisotropic hb structure of CSe monolayers are $E_{ac} = 86.1$ N/m and $E_{zz} = 83.5$ N/m, and the corresponding Poisson ratios are $\nu_{ac} = 0.14$ and $\nu_{zz} = 0.16$ [4]. The slightly anisotropic ultimate strengths at $\sigma_{ac} = 5.6$ N/m and $\sigma_{zz} = 5.9$ N/m are surprisingly small with respect to Griffith's rule. The anisotropic ultimate strains are equal to $\varepsilon_{ac} = 0.10$ and $\varepsilon_{zz} = 0.14$ [4]. The structure, Young's modulus, Poisson's ratio, ultimate strength, and ultimate strain of group IVA–VIA monolayers are presented in Table 13.2.

13.1.3.4 Physics and Applications

The aw-CSe monolayer is one of the very few 2D compounds with a similar band structure and a direct band gap as determined for phosphorene. A gap of $E_g = 0.905$ eV was estimated by the standard DFT-PBE functional, and a larger gap of 1.59 eV by the HSE06 approach [3]. This similarity is because the electronegativities of C and Se are almost identical. With these electronic properties, aw CSe may be an alternative semiconductor to phosphorene with the space group $Pmn2_1$ [3]. The CSe monolayer is a good absorber in the UV region and therefore qualifies as a window layer in solar cells and a top junction in tandem solar cells.

Tailoring the electronic and optical properties of aw-CSe monolayers is possible by doping with group–VIA elements [3]. The effects depend on the electronegativity and size of the doping atom and increases with doping concentration. For example, substitution of Se atoms by O atoms induces a transition from a direct to an indirect band gap of CSe. Furthermore, the red-shift phenomenon increases with the radius and concentration of doping atoms.

In aw-CSe/hb-CSe heterojunctions, charge transfer from the aw phase to the hb phase causes spatial separation of photogenerated carriers. Therefore, such a junction may serve as a metal-free photocatalyst for the water splitting reaction [4]. Similarly, electron charge is transferred in a variety of heterostructures consisting of group IVA–VIA monolayers, e.g., from GeO to the CSe layer at the interface of the van der Waals heterostructure GeO/CSe [6]. This semiconducting bilayer structure exhibits narrow indirect type-II band alignment. In addition, this work provides the transport coefficients for these heterobilayers, including Seebeck coefficients, electrical conductivity, and power factors versus chemical potential using the Boltzmann transport theory that is implemented in the BoltzTrap code [6].

Table 13.2 Structure, Young's modulus, Poisson's ratio, ultimate strength, and ultimate strain of group IVA–VIA monolayers

Monolayer	Structure	Young's modulus E_{ac}, E_{zz} (N/m)	Poisson's ratio ν_{yx}, ν_{xy}; ν_{zx}, ν_{zy}	Ultimate strength σ_{ac}, σ_{zz} (N/m)	Ultimate strain ε_{ac}, ε_{zz}
CSe	hb	86.1, 83.5 [4]	0.14, 0.16 [4]	5.6, 5.9 [4]	0.10, 0.14 [4]
SiS	hb	51.3 [9], 46 [11]	0.13 [9]	5.1, 4.9 [11]	0.21, 0.24 [11]
SiSe	hb	41.5 [9]	0.17 [9]		
SiTe	aw hb	20.8, ~ 62.9 [17] 34.4 [9], 34 [11]	-0.20, ~ 0.60 [17] 0.18 [9], 0.18 [17]		
GeS	aw	4.2, 15.9 [23] 15.5, 33.3 [29]	0.42, 1.40; – 0.21, 0.41 [30]	1.4, 5.1 [25]	0.45, 0.28 [25]
GeSe	aw hb	12.3, 30.4 [23] 12.5, 30.6 [29] 26.0 [12]	0.39, 1.04; 0.583, – 0.433 [30]	1.4, 5.0 [25] 3.3, 3.1 [12]	0.55, 0.23 [25] 0.14, 0.15 [12]
GeTe	aw	12.0, 21.1 [29]			
SnS	aw	14.2, 26.0 [29] 14.5, 25.2 [23]	0.422, 0.961; – 0.004, 0.404 [30]	1.1, 4.72 [25]	0.61, 0.23 [25]
SnSe	aw	11.0, 24.9 [29] 14.0, 26.6 [23]	0.423, 0.851; – 0.210, 0.352 [30]	1.1, 4.53 [25]	0.63, 0.22 [25]
SnTe	aw	5.57, 12.9 [29]			
PbS	hb	19.6 [10]	0.284 [10]		
PbSe	hb	15.4 [10]	0.310 [10]		
PbTe	hb	16.5 [10]	0.268 [10]		

13.1.4 CTe Monolayer

13.1.4.1 Chemistry, 2D Bonding, and Structure

A further increase in the atom size from Se to Te further reduces orbital overlap, and with the resulting weaker covalent bonding in CTe, the cohesive energy amounts to $E_c = -3.68$ eV/atom for aw CTe and -3.43 eV/atom for hb CTe [1]. Compared with the ionicity of aw-CSe monolayers, the polarity induced by the different electronegativities of C (2.55) and Te (2.10) increases slightly with the greater difference, inducing a Bader charge transfer of $\delta q = 1.0$ e from the Te atom to the C atom [7].

The information presently available for aw-CTe monolayers is limited to the lattice constants a = 0.436 nm and b = 0.425 nm, the bond lengths $d_1 = 0.216$ nm and $d_2 = 0.218$ nm, the pucker distance $\Delta z = 0.208$ nm with a small buckling distance of h = 0.037 nm, and the two bond angles $\theta_2 = 109.3°$ and $\theta_3 = 90.1°$ [5]. With the van der Waals diameter of Te of $d_w = 0.420$ nm [2], the estimated monolayer thickness of the aw phase increases to $d_{w+\Delta z} = 0.628$ nm if corrugation is added. The indirect band gaps calculated by the PBE and HSE approach are $E_g = 0.56$ eV and 1.07 eV, respectively [7]. The space group for the aw structure is $Pmn2_1$.

The structural parameters of hb-CTe monolayers are a = 0.335 nm and d = 0.223 nm, supporting weaker bonding due to longer C–Te bonds [1]. The buckled phase belongs to the space group *P*3*m*1. While energetic stability is confirmed by the negative cohesive energy, thermodynamic, thermal, and mechanical stability still must be approved.

13.2 Silicon Monochalcogenide Monolayers (SiO, SiS, SiSe, and SiTe)

13.2.1 SiO Monolayer

13.2.1.1 Chemistry, 2D Bonding, and Structure

The electronic configuration of SiO, with the valence electrons of the Si atom of $(3s^2p_x^{\,2})$ and the O atom of $(2s^2p_x^{\,2}p_y^{\,2})$, is isoelectronic to group–VA elements with five valence electrons. Therefore, the Si monochalcogenide compounds have electronic properties like those of the carbon monochalcogenides. Therefore, the 2D chemical bonding types and structure formation show many similarities. The energetically most stable aw phase of SiO exhibits the strongest bonding in this group, with a cohesive energy of $E_c = -5.27$ eV/atom. A comparable cohesive energy of -5.02 eV/atom has been reported for the hb phase [1]. The S–O bonds are strongly polarized due to the large difference in the electronegativities of Si (1.90) and O (3.44), which induces a charge transfer of $\delta q = 2.36$ e from the Si atom to the O atom [1].

The structural parameters of the aw phase of SiO monolayers are a = 0.274 nm, b = 0.470 nm, d_1 = 0.184 nm, and d_2 = 0.186 nm [1]. Similar values, together with the pucker effect of Δz = 0.131 nm, buckling amplitude of h = 0.077 nm, and the two bond angles θ_2 = 94.2° and θ_3 = 132.5°, have been reported elsewhere [5]. The van der Waals diameter of O atoms of d_w = 0.31 nm [2] and the pucker effect yield a corrugated layer thickness of $d_{w+\Delta z}$ = 0.441 nm. The aw-SiO monolayer has been described as a semiconductor with a direct band gap of E_g = 1.45 eV (PBE) [1]. Note that other authors designated SiO as an indirect semiconductor with calculated band gaps of 1.37 eV (PBE) and 2.06 eV (HSE) [7].

The hb phase of SiO monolayers has not only a slightly smaller cohesive energy of E_c = − 5.02 eV/atom but consistent with this observation a marginally longer Si–O bond of d = 0.188 nm [1]. The lattice constant of the semiconductor with an indirect band gap of E_g = 0.706 eV is a = 0.282 nm [1]. The space group of the hb-SiO monolayer is *P*3*m*1.

13.2.2 SiS Monolayer

13.2.2.1 Chemistry and 2D Bonding

With the valence electrons of Si atoms (Ne|$3s^2p^2$) and S atoms (Ne|$3s^2p^4$), bonding in the SiS monolayer follows the general 2D bonding scheme described above. The structural distortions caused by the additional electrons in the lone electron pairs are responsible, for example, for the generation of the characteristic buckled or puckered structure with strong interlayer interactions and increased chemical reactivity. The lone electron pairs push the three covalently bonded atoms toward a tetrahedral coordination. Due to the difference in the electronegativities of Si (1.8) and S (2.5), a significant Bader charge transfer of δq = 2.48 e occurs from each Si atom to each S atom, inducing a sizable ionic contribution to covalent bonding [1]. The cohesive energy of the aw-SiS monolayer at E_c = − 3.91 eV/atom [1] (PBE) essentially agrees within the error involved in these calculations, with the nearly identical binding energies reported for hb-SiS monolayers at E_c = − 3.92 eV/atom [1] (PBE) and − 3.91 eV/atom [3] (PBE).

13.2.2.2 2D Structures

The existence of monolayers of SiS was predicted by theory as the first stable compound of the 2D group IVA–VIA family [8]. This report proposed the stability of aw- and hb-SiS monolayers based on their similarity with the isoelectric counterparts of black and blue phosphorene, respectively. The derived lattice constants of puckered aw SiS of a = 0.340 nm and b = 0.476 nm [8] are in good agreement with a = 0.335 nm and b = 0.477 nm [1] (see Table 13.1). The bond lengths are d_1 = 0.230 nm and d_2 = 0.234 nm [1]. The pucker effect is Δz = 0.217 nm and the two available bond angles are $\theta_2 = 96.3°$ and $\theta_3 = 111.1°$ [5]. When the mean van der Waals diameter of S of $d_w = 0.360$ nm [2] is added the pucker effect of $\Delta z = 0.217$ nm, a monolayer thickness of $d_{w+\Delta z} = 0.577$ nm is obtained. The space group of aw-SiS monolayers is $Pmn2_1$ [1].

The short lattice constant of a = 0.335 nm for the buckled hb-SiS lattice supports the assumption of a slightly more energetically stable buckled hb phase [8]. Previous calculations reported for the lattice parameters a = 0.330 nm, d = 0.232 nm [1, 9], and $\Delta z = 0.133$ nm for the buckling height [6]. For hb-SiS monolayers the calculated band gaps at 2.20 eV (PBE) and 3.00 eV (HSE06) exceed those of the aw phase [9]. Adding the buckling effect to the mean van der Waals diameter of SiS $d_w = 0.390$ nm [2] yields a corrugated layer thickness of $d_{w+\Delta z} = 0.523$ nm. The different electronegativities of Si and S atoms and the nonplanar configuration cause an in-plane polarization of aw-SiS and an out-of-plane polarization of hb-SiS monolayers [8].

The space group $P3m1$ of buckled honeycomb group–IVA monochalcogenides has a threefold rotational symmetry and a mirror symmetry, whereas an out-of-plane inversion symmetry is lacking [10]. Therefore, these materials possess an intrinsic out-of-plane electric polarization and an out-of-plane piezoelectric coefficient, which may be relevant for future applications. Note that the out-of-plane polarization changes with an applied in-plane strain [10].

13.2.2.3 Mechanics

Importantly, the buckled monolayers hb SiS, hb SiSe, and hb SiTe are dynamically, thermally, and mechanically stable at room temperature. The isotropic Young's modulus of hb SiS is E = 51.3 N/m, and the Poisson ratio is $\nu = 0.13$ [9]. This stiffness is in reasonable agreement with E = 46 N/m obtained by molecular dynamics (MD) simulations at 1 K, using a Stillinger–Weber potential [11]. These simulations yielded anisotropic ultimate strengths of $\sigma_{ac} = 5.1$ N/m and $\sigma_{zz} = 4.9$ N/m and as ultimate strains $\varepsilon_{ac} = 0.21$ and $\varepsilon_{zz} = 0.24$.

Simulations employing a Tersoff potential were performed for the mechanical properties of hb-SiS monolayers at 300 K, yielding for the elastic modulus E = 38.3 N/m, for the anisotropic critical stresses $\sigma_{ac} = 3.46$ N/m and $\sigma_{zz} = 3.54$ N/m, and for the ultimate strains at maximal stresses $\varepsilon_{ac} = 0.102$ and $\varepsilon_{zz} = 0.117$ [12]. In this work, uniaxial tensile stress–strain curves were analyzed not only for pristine monolayers but also for precracked samples, which exhibited almost identical brittle fracture behaviors. The mode-I fracture toughness values or critical stress intensity factors of hb-SiS monolayers, considering an ideal initial crack, are K_{Ic} t, $_{ac} = 1.17 \times 10^{-4}$ N/m$^{1/2}$ and K_{Ic} t, $_{zz} = 1.15 \times 10^{-4}$ N/m$^{1/2}$ [12].

A recent extensive study on negative Poisson's ratios of the orthorhombic structure of aw-SiS monolayers evaluated $\nu_{zx} = -0.073$ for extension in the z direction by tension in the armchair direction, $\nu_{zy} = 0.058$ for tension in the zigzag direction, and for the anisotropic in-plane values $\nu_{xy} = 1.025$ and $\nu_{yx} = 0.219$ [13]. Mixed Poisson ratios describe deformations in two given directions; for example, ν_{xy} represents the deformation in the x direction induced by tension applied in the y direction.

13.2.2.4 Physics and Applications

Several studies by the DFT-PBE approach provide insight into the electronic and optical properties needed, for example, to find efficient photocatalysts that work under visible-light irradiation. The requirements necessary for this application are a suitable band gap, a favorable band alignment, efficient light absorption, and high carrier mobility. The puckered SiS and SiSe monolayers have suitable electronic properties for photocatalytic water splitting and hydrogen evolution [14]. With their large band gaps, they fulfill the minimum energy requirement for the water splitting reaction of 1.23 eV, which is the free energy of water splitting. These compounds efficiently harvest solar radiation in the spectral window from 1.3 to 3.0 eV and exhibit a robust stability under ambient electrochemical conditions. By employing biaxial strain or by changing the number of layers or stacking order of few-layer assemblies, it is possible to tune the band gap and edge positions. The chemical composition, e.g., in Janus compounds, is another degree of freedom in group–IVA monochalcogenides for engineering the electronic band gap.

The selection of high-performance electrode materials is a critical step in the development of Li-ion batteries. Both SiS and SiSe preserve their structural integrity upon lithiation, where a semiconductor–metal transition occurs. Therefore, these 2D materials can be used as host materials for lithium [15]. As potential electrode materials for Li-ion batteries, they exhibit a low diffusion barrier, moderate to high specific capacity, low open-circuit voltages, small volume changes, and good electrical conductivity.

To date the synthesis of high-quality puckered SiS and most group IVA–VIA monolayers and few-layer assemblies is lacking, despite moderate interlayer interaction energies, which in principle should allow the mechanical, sonication-based, or liquid-phase exfoliation of these 2D materials [12].

13.2.3 SiSe Monolayer

13.2.3.1 Chemistry, 2D Bonding, and Structure

With a similar electronic configuration and number of valence electrons, Si and Se atoms form three covalent bonds with their nearest neighbors, and in addition, the Se atom has a lone electron pair orbital responsible for tetrahedral sp^3-like hybridization of the Si and Se atoms. The bonding configurations of buckled and puckered SiSe monolayers are like those discussed for SiS, except that the ionic contribution to bonding and band gap are smaller. The difference in the electronegativities of Si (1.8) and Se (2.4) induces a Bader charge transfer of $\delta q = 1.39$ e from the Si atom to the more electronegative Se atom [1]. The cohesive energy of aw SiSe of $E_c = -$ 3.53 eV/atom (PBE) is within calculation errors identical to the cohesion energies published for the energetically slightly more stable hb SiSe of $E_c = -$ 3.55 eV/atom [1] (PBE) and − 3.60 eV/atom [9] (PBE). The larger atom size of Se leads to longer bonds and smaller binding energies than those found for the SiS structures.

The results of DFT-PBE calculations for the structural properties of aw SiSe are a = 0.374 nm, b = 0.440 nm, d_1 = 0.252 nm and d_2 = 245 nm [1]. These values are nearly identical to those of other reports, which provided in addition the pucker effect of Δz = 0.238 nm [5] and the two bond angles θ_2 = 97.0° and θ_3 = 101.5° [5]. With the pucker distance and van der Waals radius of Se at d_w = 0.38 nm [2], a monolayer thickness of $d_{w+\Delta z}$ = 0.618 nm is obtained for the corrugated monolayer. The calculated band gaps of the puckered structure are 1.19 eV (PBE) and 1.81 eV (HSE) [7]. The space group of the puckered SiSe phase is $Pmn2_1$ [1].

Analogous to the situation in hb SiS, the lattice constant a = 0.352 nm and bond length d = 0.248 nm support the proposal that hb SiSe represents the minimum energy configuration, as suggested by the slightly lower cohesive energies presented above [1, 9]. A set of structural parameters with a much larger bond length exists with the values a = 0.355 nm, d = 0.286 nm, and θ = 76.62° [6]. Similar to most group IVA–VIA compounds, hb SiSe is an indirect semiconductor with estimated band gaps of E_g = 2.12 eV (PBE) and 2.9 eV (HSE06) [9]. The space group of the buckled configuration is $P3m1$ [1]. By taking the buckling effect of Δz = 0.142 nm [9] into account, we obtain an estimate of the thickness of the corrugated monolayer of $d_{w+\Delta z}$ = 0.542 nm from the mean van der Waals diameter of SiSe d_w = 0.40 nm. Compared to the stiffness of hb SiS, the elastic modulus of hb SiSe decreases to 41.5 N/m and the Poisson ratio is 0.17 [9].

Biaxial strain engineering served as a tool for studying the effect of strain on the absorption behavior and electronic properties of aw- and hb-SiSe monolayers [16]. The electronic band gap of the aw phase increases with tensile strain and decreases with compressive strain, which induces a semiconductor-to-metal transition at − 4%. Otherwise, under biaxial strain between − 6% and 6% the hb-SiSe monolayer remains a semiconductor.

13.2.4 SiTe Monolayer

13.2.4.1 Chemistry, 2D Bonding, and Structure

Bonding in aw-SiTe monolayers follows the principles described for the SiS and SiSe monolayers. Due to the small difference in the electronegativities of Si (1.8) and Te (2.1), a Bader charge transfer of only δq = 0.362 e occurs from the Si atom to the Te atom [1]. This reduces the ionic contribution to covalent bonding. Due to the larger size of Te atoms orbital overlap is reduced, which further weakens covalent bonding in comparison to SiSe monolayers. Thus, the cohesive energies of $E_c = -$ 3.23 eV/atom [1] (PBE) and − 3.30 eV/atom [17] (PBE) obtained for aw SiTe indicate the weakest bonding in this group. Note that the puckered structure has a comparable bonding strength as the buckled phase at $E_c = -3.20$ eV/atom [1] (PBE) and -3.27 eV/atom [17] (PBE).

The published structural parameters for aw-SiTe monolayers of a = 0.411 nm and b = 0.429 nm, $d_1 = 0.264$ nm and $d_2 = 0.278$ nm, $\theta_1 = 95.6°$, $\theta_2 = 100.3°$ and $\theta_3 = 87.8°$, and $\Delta z = 0.272$ nm are in good agreement [1, 5]. Besides its dynamic and thermal stability, the aw phase is energetically more stable than the hb phase. The van der Waals diameter of Te $d_w = 0.420$ nm [2] and the pucker effect of $\Delta z = 0.272$ nm provide an estimate of the thickness of the monolayer of $d_{w+\Delta z} = 0.692$ nm. The space group of puckered SiTe is $Pmn2_1$ [1]. Flexible band gap engineering allows the indirect band gap of 0.40 eV (PBE) or 0.57 eV (HSE) to be transformed into a direct band gap, and biaxial strain transforms the semiconductor into a metal [17]. Since at moderate tensile strains, the indirect band gap becomes a direct gap, such a transition may be induced even by epitaxial mismatch at the substrate surface.

Furthermore, for the puckered structure, the published Young's moduli of $E_{ac} = 20.8$ N/m, $E_{zz} \sim 42$ N/m and $E_{max} = 62.9$ N/m in the strongest direction are strongly anisotropic. Poisson's ratios show the expected anisotropic auxetic behavior and vary between the negative value of $\nu = -0.20$ and the positive value of $\nu = 0.60$ [17]. While for a perfect isotropic incompressible material, only a maximal Poisson ratio of $\nu = 0.5$ is possible, it may clearly exceed this value in certain directions for an anisotropic orthorhombic 2D structure.

The lattice constants of a = 0.384 nm [1] (0.383 nm [17]) and the bond lengths of d = 0.272 nm [1] (0.269 nm [17]) reported for hb-SiTe monolayers show good agreement. The layer thickness determined by adding the mean van der Waals diameter of SiTe dw = 0.42 nm and the buckling effect of $\Delta z = 0.153$ nm [17] is $d_{w+\Delta z} = 0.573$ nm. The space group of the buckled phase is $P3m1$. Contrary to the small band gap of aw-SiTe monolayers, calculations predict larger gaps of $E_g = 1.83$ eV (PBE) and 2.36 eV (HSE) for hb-SiTe monolayers [17]. Calculations of the linear elastic constants resulted in an isotropic Young's modulus of 34.4 N/m [9] and a Poisson ratio of 0.18 [9] in excellent agreement with E = 34.1 N/m and $\nu = 0.18$ [17].

Ab initio total energy calculations revealed for a novel structure, the lattice dynamic properties, 2D bonding behavior, and electronic structure. The hexagonal structure with the stoichiometry $(AX)_2$ consists of a monolayer with four-sublayers X–A–A–X and the elements A = Si, Ge, Sn, and Pb and X = Se and Te [18]. In this lattice, the covalent A–X bonds are much stronger than the A–A bonds with van der Waals-like bonding features. The space group of the hexagonal structure is P-3m1. While most compounds are narrow-gaped semiconductors, the dynamically stable $(SiTe)_2$ exhibits strong interactions between electrons and phonons, not observed for the related compounds. This explains the topological insulator (TI) properties of the possibly zero-gap semimetal $(SiTe)_2$ if spin–orbit coupling (SOC) is included. In addition, $(AX)_2$ monolayers with their chalcogenide elements on both sides are building blocks for 2D van der Waals heterostructures and 3D superlattices.

13.3 Germanium Monochalcogenide Monolayers (GeO, GeS, GeSe, and GeTe)

13.3.1 GeO Monolayer

13.3.1.1 Chemistry, 2D Bonding, and Structure

The electronic configurations of Ge atoms (Ar|$3d^{10}4s^2p^2$) and O atoms ($1s^22s^2p^4$) determine 2D bonding in GeO monolayers. The nature of bonding is covalent-ionic owing to a significant difference in the electronegativities of Ge (2.01) and O (3.44). This difference induces a medium Bader charge transfer of $\delta q = 1.24$ e from each Ge atom to each O atom. The negative cohesive energy of $E_c = -4.67$ eV/atom reveals the energetic stability of aw-GeO monolayers [1]. The hb-GeO monolayer has a somewhat larger cohesive and thus lower binding energy of $E_g = -4.37$ eV/atom [1].

The lattice parameters available for the aw-GeO phase are a = 0.306 nm, b = 0.480 nm, $d_1 = 0.196$ nm and $d_2 = 0.199$ nm [1], a pucker distance of $\Delta z = 0.200$ nm, and the bonding angles $\theta_2 = 91.3°$ and $\theta_3 = 126.5°$ [5]. Table 13.1 presents the lattice constants, lattice parameters, and layer thicknesses of the two stable phases. Adding the van der Waals diameter of O of $d_w = 0.31$ nm and the buckling distance Δz, yields an estimate of the layer thickness of $d_{w+\Delta z} = 0.510$ nm. The band gap of the indirect semiconductor is $E_g = 2.76$ eV (aw) [1].

As structural information of the hb-GeO structure a larger lattice constant and bond length at a = 0.312 nm and d = 0.203 nm were obtained [1]. A partly deviating set of lattice parameters suggests a = 0.305 nm, d = 0.266 nm, $\theta = 69.88°$ and $E_g = 2.17$ eV (PBE) [6]. The latter band gap of the indirect semiconductor agrees well with $E_g = 2.23$ eV and confirms the lower gap of the hb phase [1].

DFT calculations elucidated the structural, electronic, and optical properties of Janus GeSO monolayers with a 1 T structure and space group *P*3*m*1 [19]. While the additionally characterized GeO_2 and GeS_2 monolayers are indirect semiconductors, GeSO has a direct band gap and due to the lack of mirror symmetry the symmetry is lower than that of GeO_2 and GeS_2. The calculated band gaps are $E_g = 0.63$ eV and 1.55 eV at the PBE and HSE06 levels, respectively [19]. With its low stiffness at E = 59.9 N/m and large Poisson ratio at $\nu = 0.57$, the GeSO Janus monolayer can withstand large-amplitude strain and thus is mechanically more flexible than most 2D materials. Moreover, it can be easily rolled to nanotubes. Note that the corresponding stiffness of GeO_2 is E = 148.4 N/m and the Poisson ratio is $\nu = 0.29$. The GeSO monolayer exhibits high optical absorption in the visible region and especially in the near UV at approximately 9 eV.

13.3.2 GeS Monolayer

13.3.2.1 Chemistry and 2D Bonding

The chemistry of GeS monolayers derives from the five valence electrons, as described for his homologues. The lone electron pairs present in the group IVA–VIA monolayers influence the corrugation of the structure of monolayers and few-layer sheets. The most stable phase usually is a derivative of the orthorhombic structure with four atoms in the unit cell, e.g., consisting of four sublayers in the stable aw GeS (α GeS) lattice. Each covalently bonded atom has three next neighbors, forming zigzag rows of alternating elements. The lone electron pair pushes the three bonds toward a tetrahedral-type configuration. In this arrangement, each monolayer consists of six-membered rings in the so-called chair conformation. According to the Bader charge analysis, a charge transfer of $\delta q = 0.815$ e occurs from Ge (1.8) to S (2.5) owing to the difference in their electronegativities [1]. The cohesive energies of the aw phase of $E_c = -3.66$ eV/atom [1] (PBE) and -3.76 eV/atom [20] (PBE) indicate weaker bonding than in SiS monolayers due to the larger size of the Ge atoms. In addition to the aw phase, the not as well studied hb-GeS (ß-GeS) monolayer has similar cohesive energies at $E_c = -3.64$ eV/atom [1] (PBE) and -3.74 eV/atom [20] (PBE).

13.3.2.2 2D Structures

The lattice constants of aw-GeS monolayers are a = 0.364 nm and b = 0.449 nm, and the bond lengths $d_1 = 0.246$ nm and $d_2 = 0.242$ nm [21]. These values agree well with those of another study, providing the three bonding angles $\theta_1 = 95.17°$, $\theta_2 = 95.24°$, and $\theta_3 = 104.12°$ [21]. The hinge-like structure exhibits a large pucker effect of $\Delta z = 0.235$ nm [5] ($\Delta z = 0.256$ nm [22]). From the tabulated van der Waals radii, we find a diameter of $d_w = 0.360$ nm [2] for S atoms and by adding the van der Waals diameter to the pucker effect, we obtain an estimate of the layer thickness of $d_{w+\Delta z} = 0.595$ nm. This thickness of the free-standing layer is larger than the theoretical interlayer spacing of $d_s = 0.541$ nm (0.536 nm [23]), chosen as half the length of the lattice constant in the z direction of the bulk material, as expected [24]. The calculated band gaps are $E_g = 1.70$ eV (PBE) and 2.45 eV (HSE) [20]. The space group of the 2D asymmetric puckered crystal is $Pmn2_1$ [20, 25].

For few-layer systems the interlayer bonding and stacking order are important as a tool to tune electronic properties. Both interlayer distances and interlayer binding energies are available for the bilayer stacking orders AA, AB, AC, and AD (see Fig. 13.4) [26]. The most stable AB stacking order of aw-GeS bilayers shows slight changes in the lattice constants a = 0.367 nm and b = 0.436 nm, and the shortest distance between adjacent S layers of only $d_{int} = 0.279$ nm clearly points to covalent-like interactions of lone electron pairs beyond van der Waals interlayer interactions. The interlayer separation of $d_s = 0.536$ nm [23] is in good agreement with the bulk

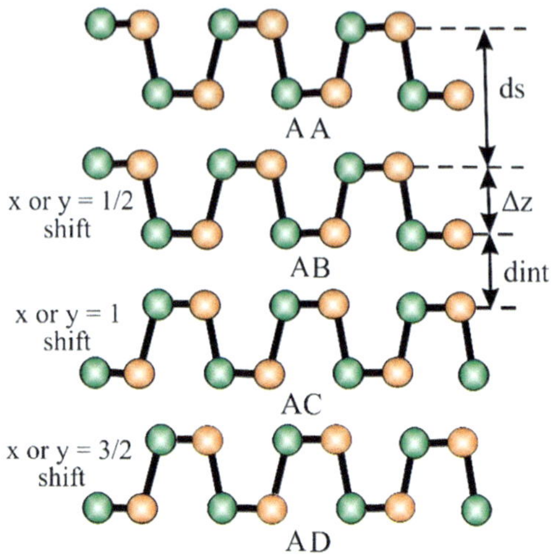

Fig. 13.4 The AA, AB, AC, and AD bilayer stacking orders and the different distances used to characterize multilayer assemblies. Adapted with permission from Ref. [26], Copyright 2018, Royal Society of Chemistry

interlayer spacing of 0.524 nm determined via X-ray diffraction (XRD) measurements [27]. Using the larger distance between adjacent layers of d_{int} = 0.346 nm found for AD stacking results in a larger interlayer spacing of d_s = 0.736 nm [26]. It is reasonable to assume that the interlayer spacing extracted from the most stable bulk structure is smaller than the thickness of the free-standing monolayer, which is approached by few-layer assemblies with decreasing layer number and interlayer interaction. This conclusion is consistent with the stronger interlayer interaction energy of E_{int} = – 29.39 meV/A^2 for AB stacking in a bilayer [26]. The existence of the strong interlayer interactions explain the difficulties faced in isolating single monolayers by mechanical exfoliation.

The calculated lattice constant of hb-GeS monolayers a = 0.349 nm is smaller than that of the aw phase, whereas the bond length d = 0.243 nm is comparable [1]. Of course, the buckling height of Δz = 0.136 nm is much smaller [20]. Another set of structural parameters with an extended bond length is a = 0.351 nm, d = 0.285 nm, and θ = 76.84° [6, 7]. The lower buckling height yields a corrugated thickness of $d_{w+\Delta z}$ = 0.526 nm. The calculated band gaps of the hb phase are E_g = 2.47 eV (PBE) and 3.01 eV (HSE), and the space group of buckled GeS monolayer is $P3m1$ [20].

13.3.2.3 Mechanics

The mechanical properties of puckered aw GeS are strongly anisotropic. To determine the values E_{ac} = 4.2 N/m (7.80 GPa) and E_{zz} = 15.9 N/m (29.7 GPa) [23], the published 3D units in brackets were transformed to 2D units using a thickness of t = 0.536 [17]. Notably, these stiffness values are only slightly lower than those at E_{ac} = 5.4 N/m and E_{zz} = 17.9 N/m [28]. However, substantially larger stiffness values of E_{ac} = 15.5 N/m and E_{zz} = 33.3 N/m were reported elsewhere [29]. The values reported for the in-plane Poisson ratios are ν_{yx} = 0.420 and ν_{xy} = 1.40 [30]. The out-of-plane Poisson ratio is negative for extension in the armchair direction (ν_{zx} = – 0.208) but positive for extension in the zigzag direction (ν_{zy} = 0.411) [18, 30]. The ideal strength of σ_{ac} = 1.4 N/m (2.59 GPa) in the accordion-like armchair direction is extremely small compared with that of σ_{zz} = 5.1 N/m (9.56 GPa) in the zigzag direction [25]. The corresponding critical strain values reflect this strong anisotropy of failure mechanics at ε_{ac} = 0.45 and ε_{zz} = 0.28 [25]. The high failure strain in the armchair direction indicates ultrahigh flexibility.

13.3.2.4 Physics and Applications

The aw-GeS monolayer is a semiconductor with an indirect and only marginally larger direct band gap and a high carrier mobility [31]. Since the diameter of the constituents determines the packing density, GeS has the most compact structure of the thoroughly studied Ge and Sn monochalcogenides. The electronic properties of aw GeS permit strong tunability by external strain and electric fields [31, 32]. A uniform electric field reduces the band gap, and ultimately a semiconductor-to-metal transition occurs. Moreover, it is possible to induce a phase transition from $Pmn2_1$ to the space group *Pmmn* of phosphorene with greater symmetry by the application of strain [25].

Multiple valence and conduction band valleys characterize the band structures of the group–IVA monochalcogenides (GeS, GeSe, SnS, and SnSe), which make these monolayers suitable for a variety of applications, such as piezotronics and ferroelectricity [33–35]. The polar character is introduced by the formation of monolayers because in the bulk, the layers have alternative electrical dipoles. According to first-principles calculations, the puckered GeS, GeSe, SnS, and SnSe monolayers possess piezoelectric coefficients approximately one to two orders of magnitude greater than those of other 2D materials [33]. The large piezoelectric effect is the result of the unique puckered symmetry and weaker chemical bonds, which are more sensitive to applied stress. These materials will allow the fabrication of ultrasensitive mechanical detectors and efficient energy conversion devices.

The four well-studied group–IVA monochalcogenides (GeS, GeSe, SnS, and SnSe) have in-plane ferroelectricity [34]. Figure 13.5 shows the top and side view of the monolayer. Remarkably, these monolayers exhibit strain-tunable in-plane ferroelectricity. The GeS phase has the best ferroelectric parameters, with a spontaneous polarization of $P_s = 4.88 \times 10^{-10}$ C/m, on par with the polarizations reported for SbN and BiP [34]. Figure 13.6 shows the side view of the two degenerate non-centrosymmetric structures (ferroelectric phase) and the centrosymmetric structure (paraelectric phase). Applying uniaxial tensile strain further improves polarization.

A structural evolution diagram established for GeS monolayers leads to a noncollinear ferroelectric δ phase with an energetically stable puckered phase [35]. The induction of ferroelectricity by phonon frustration, i.e., the competition between ferroelectric and antiferroelectric modes allows for the possibility of tuning polarity. A morphotropic phase boundary between the aw and δ phases is possible and opens the door to giant piezoelectricity for lead-free applications.

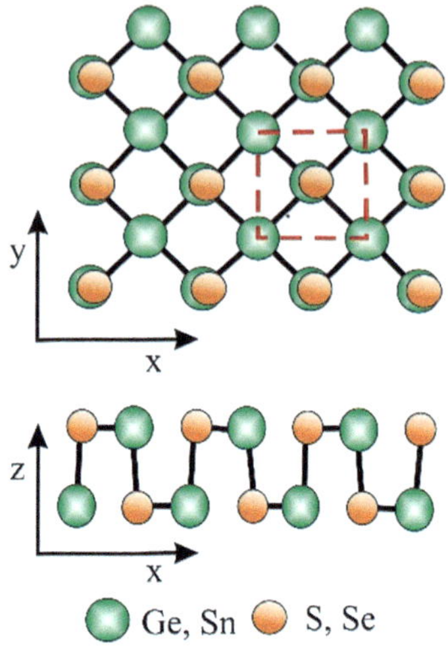

Fig. 13.5 Top and side view of MX (M = Ge, Sn; X = S, Se) monolayers. The red dashed line marks the unit cell. Adapted with permission from Ref. [34], Copyright 2022, AIP Publishing

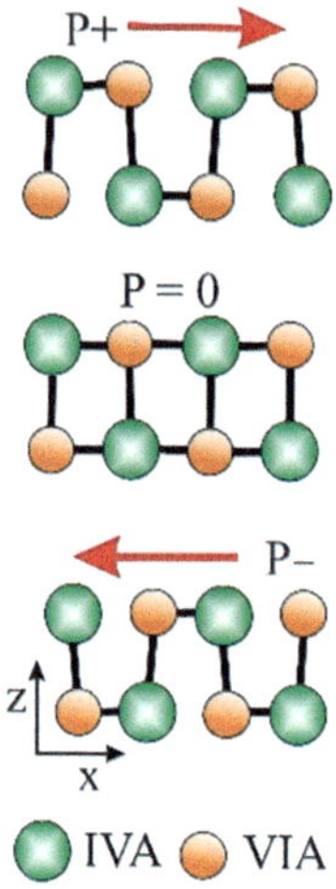

Fig. 13.6 Side view of the two degenerate non-centrosymmetric structures (ferroelectric phase P + and P−) and the centrosymmetric structure with P = 0 (paraelectric). Adapted with permission from Ref. [34], Copyright 2022, AIP Publishing

Thermal transport is a fundamental property of significance in emerging technologies. While thermal transport shows strong anisotropy along the zigzag and armchair directions of aw GeS, phonon transport is almost isotropic in the monolayers of SnS and SnSe despite similar hinge-like structures [23]. The larger the band gap is, the stronger the anisotropy. The anisotropic characteristic of bonding visualized by the electron localization function (ELF) is the chemical origin of the anisotropic behavior.

The hb-GeS phase is a suitable material for high-performance anodes in alkali metal ion batteries, as revealed by DFT calculations [36]. After the adsorption of alkali atoms, the electrical conductivity of hb GeS is enhanced owing to a semiconductor-to-metal transition. The low diffusion barrier of alkali atoms allows a fast charge/discharge rate, a low average open-circuit voltage, and a high theoretical capacity, which implies that hb GeS is a promising anode material for Na-ion batteries. A feature article treats the incorporation of nanocrystals of GeS, GeSe, SnS, and SnSe in the photoabsorbant material of photovoltaic devices [37]. The article describes the efforts made to control the size and morphology of the nanocrystals of these IVA–VIA compounds during synthesis. They are less toxic than the currently studied lead chalcogenides.

In comparison to phosphorene, GeS and the other group–IVA monochalcogenides exhibit better resistance to oxidation according to first-principles calculations. The superior oxidation resistance and better environmental stability of group IVA–VIA monolayers follow from the activation energy needed for chemisorption of O_2, which is $E_a = 0.70$ eV for phosphorene, $E_a = 1.26$ eV for GeS, and $E_a = 1.6$ eV for SnS [38]. No severe deterioration of the band structure seems to occur by chemisorption of a small amount of oxygen atoms.

A high carrier mobility is a crucial property for many microelectronic applications. Hybrid DFT computations indicate a remarkable electron mobility of the aw phase of GeS monolayers of 3680 $cm^2\ V^{-1}\ s^{-1}$, which is much greater than, for example, the electron mobility of MoS_2 monolayers [39].

13.3.2.5 Synthesis

The present status of preparation methods available for group–IVA monochalcogenides, such as mechanical exfoliation, solution-based methods, chemical vapor deposition (CVD), molecular beam epitaxy (MBE), atomic layer deposition (ALD), and two-step etching methods, has been reviewed extensively [40]. Even in the case of the best developed mechanical exfoliation techniques, it is difficult to control the yield, layer number, and size of the uniform area of few-layer nanosheets and nanoflakes. While solution-based methods allow the synthesis of micrometer-sized GeS nanosheets [41], most reported studies have used CVD techniques, e.g., for the synthesis of single-crystalline nanoribbons [9]. Further separation of already grown few-layer assemblies is difficult owing to the large exfoliation energy, especially for GeS [40].

13.3.3 GeSe Monolayer

13.3.3.1 Chemistry and 2D Bonding

The nature of bonding in GeSe monolayers deviates from that of GeS primarily because of the larger size of the Se atoms. As discussed for aw GeS, the Ge and Se atoms form three covalent bonds with their nearest neighbors and together with the lone electron pair orbital sp^3-like hybridization is realized. Owing to the difference in the electronegativities of 1.81 (Ge) and 2.55 (Se), a Bader charge transfer of $\delta q =$

0.649 e occurs from the Ge atom to the Se atom [1]. As described before, the most stable aw phase with an orthorhombic structure has only a marginally lower cohesive energy of $E_c = -3.37$ eV/atom in comparison with -3.36 eV/atom for the hb-GeSe monolayer [1]. These results should be compared with the cohesive energies of $E_c = -3.88$ eV/atom obtained for aw GeSe and -3.85 eV/atom found for the hb-GeSe monolayers [20]. First-principles calculations produced an even lower cohesive and thus larger binding energy of $E_c = -4.20$ eV/atom for the phosphorene-analog symmetric washboard (sw) structure with only two atomic sublayers of the GeSe monolayer, belonging to the space group $Pmn2_1$ [42].

13.3.3.2 2D Structures

The reported lattice constants of aw GeSe are a = 0.397 nm and b = 0.430 nm, and the bond lengths $d_1 = 0.266$ and $d_2 = 0.254$ nm [1]. The published similar pucker effects are $\Delta z = 0.249$ nm [20] and $\Delta z = 0.253$ nm [5], and the three bond angles of the puckered phase are $\theta_1 = 96.26°$, $\theta_2 = 93.15°$, and $\theta_3 = 97.86°$ [21]. The space group is $Pmn2_1$ [1]. Note that there is a change in the bonding angle associated with a change in the type of atom located in the outer sublayer. Notably, for aw-GeS, aw-SnS, and aw-SnSe monolayers, the group–IVA atoms are located at the outermost positions of the four-layer system, whereas for aw-GeSe monolayers the chalcogen atoms occupy the outer positions [30]. Thus, contrary to the normal situation encountered in group IVA–VIA monolayers, in GeS, SnS, and SnSe the Ge and Sn atoms form a tetrahedral structure with sp^3-like orbitals including a lone electron pair orbital perpendicular to the surface with the configuration M–X–X–M. For aw GeSe the Se–Ge–Ge–Se arrangement is the most stable structure of the known polymorphs, and is the only monolayer with a direct band gap in this group [21]. For the thickness of the corrugated monolayer, we find $d_{w+\Delta z} = 0.633$ nm with the van der Waals diameter of Se $d_w = 0.38$ nm [2]. The interlayer spacing estimated from the bulk-crystal data is $d_s = 0.556$ nm [23]. The latter value is in good agreement with early XRD measurements for an interlayer spacing of bulk GeSe of $d_s = 0.541$ nm [37]. For aw GeSe, direct band gaps of $E_g = 1.28$ eV (PBE) and 1.68 eV (HSE06) were postulated, which are suitable to efficiently harvest visible light from solar radiation, and in addition, this phase shows improved photocatalytic activity [14]. Note that other publications assign an indirect band gap to aw-GeSe monolayers with a similar gap size of 1.81 eV (HSE06) [43].

Inspection of the results obtained by DFT calculations based on the PBE parametrization of the GGA approach indicates that the bonding, structural, and optical properties are sensitive to the choice of approximation for the exchange–correlation functional. For this reason, the authors conducted quantum Monte Carlo (QMC) calculations with the diffusion Monte Carlo (DMC) method for aw GeSe [44]. The structural parameters obtained were a = 0.440 nm, b = 0.389 nm, $d_1 = 0.263$ nm, $d_2 = 0.249$ nm, and $\Delta z = 0.245$ nm, with the bonding angles $\theta_1 = 95.4°$, $\theta_2 = 100.2°$, and $\theta_3 = 93.4°$. A detailed comparison of band gaps calculated by different approaches revealed large direct band gaps in the range of 3.2 eV to 3.6 eV for monolayer GeSe, which are much larger than those of bulk GeSe at 1.6 eV to 2 eV.

To date, the buckled monolayer of hb GeSe has not received much attention. The structural data of hb GeSe of $a = 0.368$ nm and $d = 0.257$ nm [1] are in acceptable agreement with those of $a = 0.363$ nm and $\Delta z = 0.141$ nm [20]. A much larger bond length is supported by calculations giving $a = 0.367$ nm, $d = 0.292$ nm, and $\theta = 78.26°$ [7]. The calculated band gaps at 2.32 eV (PBE) and 3.01 eV (HSE) [20] are in good agreement with 2.27 eV (PBE) and 3.08 eV [10].

A detailed study of the structure and stability of aw-GeSe bilayers revealed similar stabilities for AA, AB, and AC stacking [26]. The AC order has not the shortest distance between adjacent layers ($d_{int} = 0.309$ nm), however, the strongest interlayer binding, $E_{int} = -28.47$ meV/A^2. A slightly shorter distance of $d_{int} = 0.300$ nm is found for AA stacks with a slightly weaker interaction $E_{int} = -27.53$ meV/A^2 [26]. The corresponding distances between the nearest adjacent layers are $d_{int} = 0.303$ nm (AA), 0.312 nm (AB), 0.340 nm (AC), and 0.330 nm (AD). Accordingly, AA stacking has the highest stability with the shortest sublayer distance. The large interlayer interaction energies clearly indicate the contribution of covalent interactions. The band gap of bilayers is indirect for AA and AC stacking and direct for AB and AD stacking [45]. Note that in-plane strain tunes the bilayer gap, and the transition between direct and indirect band gaps is also achievable. For example, it is possible to transform the indirect band gap of AA-stacked bilayers to a direct band gap by a small biaxial strain or stretching along the z direction. Furthermore, a Ge–Se–Ge bond angle of $\theta = 96.6°$ and a Se–Ge–Se bond angle of $\theta = 97.4°$ were determined for AA stacking in aw-GeSe bilayers [42].

13.3.3.3 Mechanics

The elastic moduli of the strongly anisotropic aw-GeSe monolayer are $E_{ac} = 12.3$ N/m (22.1 GPa) and $E_{zz} = 30.4$ N/m (54.6 GPa) [23] for a layer thickness of $t = 0.556$ [23]. These values agree with $E_{ac} = 12.5$ N/m and $E_{zz} = 30.6$ N/m [29]. The in-plane Poisson ratios are $\nu_{xy} = 1.039$ and $\nu_{yx} = 0.391$ for aw GeSe [30]. Different from that of GeS, the auxetic effect of $\nu_{zy} = -0.433$ ($\nu_{zx} = 0.583$) in the z direction occurs for extension in the zigzag (armchair) direction because, in the four-layer structure of aw GeSe, the Se atoms and not the Ge atoms are located at the two outermost atomic layers [30]. This monolayer exhibited failure behavior comparable to that of GeS monolayers, with ultimate strengths of $\sigma_{ac} = 1.4$ N/m (2.48 GPa) and $\sigma_{zz} = 5.0$ N/m (9.01 GPa) [25]. These values increase only slightly for multilayers. The corresponding critical strains are $\varepsilon_{ac} = 0.55$ and $\varepsilon_{zz} = 0.23$ [25].

The mechanical properties of the hb-GeSe monolayer derived by a MD study with a Stillinger–Weber potential at 300 K (1 K) were 26.0 N/m (31.5 GPa) for Young's modulus, $\sigma_{ac} = 3.27$ N/m (4.4 N/m) and $\sigma_{zz} = 3.13$ N/m (4.2 N/m) for the maximal stresses [12]. The corresponding maximal strain values are $\varepsilon_{ac} = 0.139$ (0.24) and $\varepsilon_{zz} = 0.146$ (0.28) [12]. Moreover, the authors determined the mode-I critical stress intensity factor of $K_{Ic}\,t = 1.058 \times 10^{-4}$ N/m$^{1/2}$ (ac) and $K_{Ic}\,t = 1.095 \times 10^{-4}$ N/

$m^{1/2}$ (zz) at 300 K, where zz denotes the stress in the zigzag direction and the initial crack in the armchair direction and vice versa.

13.3.3.4 Physics and Applications

The aw-GeSe monolayer with the suggested large direct band gap not only efficiently harvests visible light from solar radiation but also shows improved photocatalytic activity [14]. However, the calculated high overpotential of GeSe for the oxygen evolution reaction (OER) limits its application as a photocatalyst. Nevertheless, GeSe is suitable for the half-cell hydrogen evolution reaction (HER). First-principles calculations confirmed the suitability for oxygen water splitting through strain engineering [46]. Accordingly, biaxial strain can increase the absorption coefficient to approximately 6×10^5 cm^{-1} and the photocatalytic activity is enhanced by the application of strain. Figure 13.7 elucidates the formation of electrons and holes by the absorption of photons greater than the band gap, their spatial separation, the induction of the oxygen evolution reaction by holes, and the initiation of the hydrogen evolution reaction by electrons.

Based on the interest introduced by the environmental and air stability of GeSe multilayers, DFT calculations and the nonequilibrium Green's function formalism were combined to elucidate the potential use of these materials as gas sensing materials [47]. The calculations revealed a marked change in the carrier density due to charge transfer processes between the gas molecule and the GeSe monolayer. The zigzag direction showed a higher sensitivity for the gas molecules SO_2, NH_3, and NO_2. The decomposition components SO_2, SOF_2, SO_2F_2, H_2S, and HF from SF_6 in gas-insulated switches were detected and used as samples to evaluate the performance of GeSe monolayers as gas-sensitive detectors [48]. According to the theoretical analysis, selective detection of SO_2F_2 is possible through the changes induced in the energy band structure, density of state distribution, and gas desorption time.

Several publications stress the long-term stability of aw GeSe with strong resistance against environmental degradation in air, but surface science investigations and DFT calculations do not confirm this claim [49]. Atomically thin flakes of GeSe form

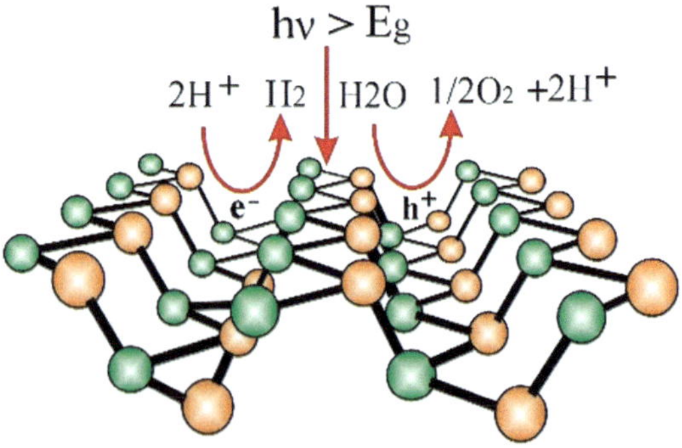

Fig. 13.7 Perspective view of the aw structure of a group IVA–VIA monolayer involved in the hydrogen evolution reaction (HER)

a self-assembled germanium oxide skin with subnanometric thickness, which leads to a decrease in the band gap. A self-assembled heterostructure consisting of a germanium oxide/germanium selenide assembly with different band gaps enables the activation of photocatalytic processes by the absorption of light at different wavelengths owing to different band gaps.

13.3.3.5 Synthesis

By sonication-assisted liquid-phase exfoliation of bulk GeSe powders, aw GeSe nanosheets with thicknesses of 4–10 nm could be prepared in different solvents and characterized by selected area electron diffraction (SAED), high-resolution transmission electron microscopy (HRTEM), and XPS [43]. An estimate of the number of layers of the synthesized few-layer GeSe sheets was based on a monolayer thickness of t = 0.54 nm. The authors used these GeSe nanosheets to examine their suitability as UV–VIS photodetectors in terms of response time, responsivity, and external quantum efficiency in optoelectronics and photovoltaics.

13.3.4 GeTe Monolayer

13.3.4.1 Chemistry and 2D Bonding

Based on the electronic configuration of Ge atoms (Ar|$3d^{10}4s^2p^2$) and Te atoms (Kr|$4d^{10}5s^2p^4$), GeTe forms a puckered and buckled monolayer, where each Ge atom binds to three Te atoms and vice versa. Owing to the increasing atom size from Se (3.8 nm) to Te (4.2 nm), the bond length increases. The Bader charge transfer from Ge atoms to Te atoms is only $\delta q = 0.372$ e because the electronegativities of Ge (2.01) and Te (2.10) are nearly identical. This denotes a small ionic contribution to covalent bonding and a minor polarity of the Ge–Te bonds. The published cohesive energies are $E_c = -3.09$ eV/atom for aw GeTe and -3.06 eV/atom for hb GeTe [1]. Note that the nomenclature follows the practice previously used in this book, different from various designations found in the literature.

13.3.4.2 2D Structures

As structural data of aw-GeTe monolayers we find a = 0.424 nm, b = 0.438 nm, $d_1 = 0.274$ nm, $d_2 = 0.288$ nm [1], and $\Delta z = 0.305$ nm [50, 51]. With the van der Waals diameter of Te $d_w = 0.420$ nm [2], the corrugated monolayer has a thickness of $d_{w+\Delta z} = 0.725$ nm. The puckered structure is dynamically, thermally, and mechanically stable. It is an indirect or quasi-direct semiconductor with calculated band gaps of $E_g = 0.90$ eV (optimized norm-conserving Vanderbilt pseudopotential, (ONCV-GGA)) and 1.30 eV (HSE) [51]. The authors argue that in the case of GeTe, the HSE method overestimates the band gap, and a lower value is more realistic. Other calculations yielded different values, such as $E_g = 1.16$ eV (HSE) and 1.08 eV (HSE + SOC), by including spin–orbital coupling (SOC) [50]. The space group of the puckered structure is $Pmn2_1$ [52]. The elastic moduli of the anisotropic aw phase are $E_{ac} =$ 12.0 N/m and $E_{zz} = 21.1$ N/m [29].

For hb-GeTe monolayers, a lattice constant of a = 0.395 nm, a bond length of d = 0.278 nm, a buckling effect of $\Delta z = 0.156$ nm (which is smaller than that for the puckered lattice), and a bonding angle of $\theta = 91.5°$ were calculated [51]. The mean van der Waals diameter of GeTe (0.42 nm) yields a layer thickness of $d_{w+\Delta z}$ = 0.576 nm. For the buckled phase, various theoretical approaches produced large indirect band gaps of $E_g = 1.81$ eV (ONCV-GGA), 1.91 eV (PBE), and 2.60 eV (HSE) [51]. The space group of the buckled structure is $P3m1$.

First-principles calculations revealed the stability, electronic structure, optical absorption, change in electronic properties as a function of the interlayer distance, and the effects of an external electric field applied to hb-GeTe bilayers [53]. The hexagonal AA stacking order is more stable than the Bernal AB stacking with an equilibrium distance of $d_s = 0.292$ nm. The van der Waals bilayer is dynamically stable and has an indirect band gap of $E_g = 0.61$ eV (HSE06). Furthermore, the band gap can be tuned by applying vertical strain and external electric fields. The bilayer exhibits distinct light absorption in the visible and infrared regions.

13.3.4.3 Physics and Applications

The monolayers of group–IVA monotellurides possess intrinsic giant piezoelectric properties. The calculated lattice structures, band structures, and elastic properties of the SiTe, GeTe, and SnTe monolayers are based on the PBE functional. Density functional perturbation theory (DFPT) delivered the piezoelectric coefficients of these compounds. The superior piezoelectricity of these materials qualifies them for flexible electronic and energy transfer devices [54].

The investigation of 2D materials suitable for photocatalytic water splitting to convert solar energy into chemical energy is an important field of current research. GeTe monolayers, with their large band gap, high electron and hole mobilities, and pronounced absorption of visible light, render GeTe monolayers promising photocatalysts [55, 56]. Four-monolayers-thick GeTe sheets served as photocatalysts in a hydrogen production experiment. The GeTe sheets were prepared by sonication-assisted liquid-phase exfoliation [56]. The layered GeTe samples behaved as indirect-gap semiconductors with an increased gap after oxidation. Figure 13.8 compares the minimum energy needed for inducing the reaction with the band gaps of two GeTe sheets, one prepared with oxygen present during preparation (O-GeTe) and one obtained by excluding oxygen (Ar-GeTe). DFT calculations indicate that oxygen atoms can easily bind to Ge atoms, forming a more stable structure. The highest hydrogen production rate of 1.13 mmol g^{-1} h^{-1}, and best cycling stability was reached by using 10% methanol as the agent and Ar-GeTe, whereas the hydrogen

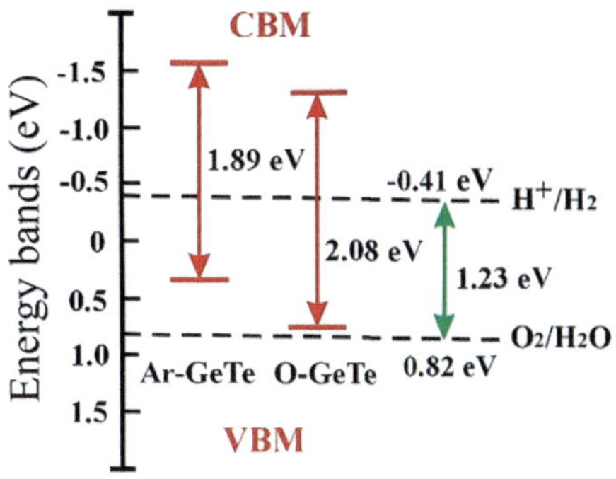

Fig. 13.8 Band structure of Ar-GeTe and O-GeTe, showing the VBM and CBM and the energies needed for the hydrogen and oxygen evolution reactions. Adapted with permission from Ref. [56], Copyright 2020, American Chemical Society

production rate decreased to 0.54 mmol g^{-1} h^{-1} for O-GeTe, which possessed a wider band gap.

As another potential application of this family, the thermoelectric properties of GeTe and Ge_2SeTe were studied using the Boltzmann semiclassical theory. The electronic band structure and DOS have been calculated for the indirect band gap semiconductor GeTe and Ge_2SeTe with a direct band gap. Moreover, the work provides the Seebeck coefficient, electrical conductivity, power factor, and electronic thermal conductivity for different temperatures and doping levels of monolayers at 300 K, 500 K, and 700 K [57].

Compared with their bulk counterparts, group–IVA monochalcogenide monolayers lose symmetry owing to a reduction in dimensionality, which favors ferroelectricity [58]. First-principles calculations revealed an in-plane spontaneous polarization of 2D GeTe with a hinge-like structure. In an orthorhombic structure the Ge atoms are displaced in the x direction with respect to Te atoms breaking the centrosymmetry. Reversal polarization occurs through a phase transition between two symmetry-equivalent ferroelectric states with opposite polarizations. For a monolayer thickness of 1 nm, the intrinsic polarization is $P_s = 3.28 \times 10^{-10}$ C/m. The stability of ferroelectricity is determined by the Curie temperature, which is 570 K in the case of 2D GeTe. The in-plane coercive field for reversing polarization is tunable by a vertical field. Importantly, for the bilayer, the ferroelectric state is the ground state.

A novel concept for energy harvesting by tribological nanogenerators was studied by the van der Waals force-induced interlayer ferroelectric-to-antiferroelectric transition via interlayer sliding in bilayer group–IVA monochalcogenides [59]. In these bilayers, AA and AC stacking prefer a ferroelectric state with parallel polarization, while AB and AD stacking favor an antiferroelectric state with an antiparallel arrangement of polarization. Tribological interlayer sliding permits reversible and hysteretic transitions between the two states. A change in the stacking order allows a change in the electric polarization with high magnitude and rate.

The topological phase of superconductors has attracted increasing attention owing to various exotic phenomena. While there is insight into the theoretical aspects of these methods, experimental realization is challenging because it is necessary to adjust several microscopic parameters simultaneously.

13.3.4.4 Synthesis

Sonication-assisted liquid-phase exfoliation of bulk GeTe powder yielded 2–4 layers and even monolayer thick GeTe sheets [60]. HRTEM and SAED analyses confirmed the crystallinity, and XPS verified the purity and elemental distribution. The large monolayer thickness of 1.6 nm, extrapolated from atomic force microscopy (AFM) measurements of few-layer assemblies, seems to be due to deterioration of the surface of the solution-processed samples. As mentioned above, GeTe nanosheets ~10 μm in size and 5–10 nm in thickness were prepared in a pure Ar atmosphere and air by sonication-assisted liquid-phase exfoliation [56]. For preparation in a pure Ar atmosphere, a better crystal structure and sample morphology could be detected by scanning electron microscopy (SEM), energy-dispersive X-ray spectroscopy (EDS), transmission electron microscopy (TEM), and AFM.

13.4 Tin Monochalcogenides (SnO, SnS, SnSe, and SnTe)

13.4.1 SnO Monolayer

13.4.1.1 Chemistry, 2D Bonding, and Structure

The electronic configurations of Sn atoms (Kr|$4d^{10}5s^2p^2$) and O atoms ($1s^22s^2p^4$) with the valence electrons of Sn ($5s^2p_x{}^2$) and O ($2s^2p_x{}^2p_y{}^2$) determine 2D bonding and formation of stable SnO monolayers. Detailed DFT-PBE level calculations are available for the puckered, buckled, and planar structures of SnO monolayers. The different electronegativities of Sn (1.96) and O (3.44) induce a Bader charge transfer of $\delta q = 1.29$ e from the inner Ge atoms to the outer O atoms. The energetic stability of aw-SnO monolayers is demonstrated by the negative cohesive energy of $E_c = -$ 4.45 eV/atom, which is the most stable phase, while the hb-SnO monolayer has a lower binding energy of − 4.17 eV/atom (PBE) [1]. The lattice parameters of the stable aw-SnO phase are a = 0.340 nm and b = 0.476 nm and $d_1 = 0.213$ nm and $d_2 = 0.216$ nm. The band gap of the indirect semiconductor decreases from 2.79 eV (aw SnO) to 1.64 eV (hb SnO). Consistent with weaker bonding the lattice constant and bond length of hb-SnO monolayers increase to a = 0.344 nm and d = 0.220 nm [1].

A recent study investigated the down-scaling of bulk structural and electronic properties of SnO to monolayers [61]. In layered bulk SnO crystals van der Waals forces play a key role, controlling structural and electronic properties. To accurately include the important contribution of lone electron pairs to interlayer interactions, in addition to employing the DFT-PBE approach dealing with the electronic exchange–correlation interaction, the DFT + D3 (Grimme) dispersion correction was included. The lattice parameters presented for SnO monolayers are a = 0.386 nm, d = 0.227 nm, and E_g = 2.85 eV (close to the experimental value of 2.7 eV) that reduces to 0.6 eV for bulk SnO. Since the growth of monolayers is energetically favorable, future applications in optoelectronics and thin film transistor devices can be expected owing to its optical transparency and ambipolar conductivity.

13.4.2 SnS Monolayer

13.4.2.1 Chemistry and 2D Bonding

The valence electrons of Sn atoms ($5s^2p^2$) and S atoms ($3s^2p_x{}^2p_y{}^2$) resulting from their electron configurations determine covalent bonding in stable SnS monolayers. As mentioned before, the lone electron pairs of the S atoms and sp^3-like hybridization are responsible for specific structural arrangements, such as corrugation with a lack of centrosymmetry, strong anisotropy of various properties, and greater chemical reactivity, e.g., stronger interlayer coupling between adjacent layers. The difference in the electronegativities of 1.96 (Sn) and 2.58 (S) is responsible for a Bader charge transfer of δq = 0.98 e from the Sn to the S atoms [1]. The cohesive energies of E_c = − 3.55 eV/atom [1] (PBE) and − 3.67 eV/atom [20] (PBE) for aw SnS and − 3.47 eV/atom [1] (PBE) and − 3.58 eV/atom [20] (PBE) for the hb-SnS monolayers indicate similar binding energies for the two stable allotropes.

13.4.2.2 2D Structures

For the aw-SnS monolayers, early calculations yielded the lattice constants a = 0.405 nm and b = 435 nm and the bond lengths d_1 = 0.273 nm and d_2 = 0.260 nm [1]. These structural parameters were confirmed and extended by a = 0.403 nm, b = 0.426 nm, d_1 = 0.272 nm, d_2 = 0.258 nm, and the three bond angles θ_1 = 95.53°, θ_2 = 89.42°, and θ_3 = 100.0° [21]. With a mean van der Waals diameter of S atoms of d_w = 0.360 nm [2] and a pucker height of 0.261 nm [20], a monolayer thickness of $d_{w+\Delta z}$ = 0.621 nm has been estimated. For this compound, cross-sectional TEM images supplied a layer spacing of d_s = 0.70 nm, which is consistent with the AFM measurements of few-layer samples yielding 0.9 nm [62, 63]. These values are larger than the theoretical and experimental values of d_s = 0.571 nm and d_s = 0.560 nm, respectively [23, 27]. The calculated band gaps are 1.45 eV (PBE) and 2.05 eV (HSE) [20]. The space group of multilayers with an odd number of layers is $Pmn2_1$, and that with an even number of layers is identical with that of the bulk structure *Pnma* [62, 63]. Note that only odd-numbered layers break the symmetry and exhibit ferroelectricity.

In aw-SnS bilayers, the AB stacking sequence has the shortest distance between the inner adjacent layers (d_{int} = 0.291 nm) and the largest interlayer interaction energy, E_{int} = − 39.5 meV/Å^2 [26]. This behavior is analogous to that discussed for 2D GeS. The bilayer lattice constant of a = 0.403 nm is slightly smaller in the armchair direction than in the zigzag direction at b = 0.413 nm.

The limited structural parameters for hb-SnS monolayers are a = 0.376 nm and d = 0.262 nm [1]. The surprisingly small lattice constant was partly confirmed by a recent study that reported a = 0.650 nm and b = 0.375 nm, and a buckling distance of Δz = 0.146 nm [20]. Another set of suggested structural parameters is a = 0.376 nm, d = 0.295 nm, and θ = 79.23° [6]. Obviously, there is agreement on the lattice constant but not on the bond length. The calculated band gaps are E_g = 2.31 eV (PBE) and 2.82 eV (HSE) [9, 20]. The space group of the buckled structure is *P*3*m*1.

13.4.2.3 Mechanics

The elastic moduli of E_{ac} = 14.5 N/m (25.3 GPa) and E_{zz} = 25.2 N/m (44,1 GPa), based on a thickness of t = 0.571 nm, demonstrate the large anisotropy of the stiffness of aw-SnS monolayers [23]. These values agree quite well with E_{ac} = 14.2 N/m and E_{zz} = 26.0 N/m [29]. The in-plane Poisson ratios are ν_{yx} = 0.422 and ν_{xy} = 0.961, and a negative Poisson ratio of ν_{zx} = − 0.004 and a positive value of ν_{zy} = 0.404 were found for stretching in the armchair direction and the zigzag direction, respectively, owing to the location of Sn atoms at the outermost two layers of the puckered four-layer structure [30]. As expected, the strongly anisotropic intrinsic strength of SnS monolayers is further weakened to σ_{ac} = 1.09 N/m (1.91 GPa) and σ_{zz} = 4.72 N/m (8.27 GPa), with similar values for the multilayers. The ultimate strengths in 2D units were derived using an effective thickness of t = 0.571 nm [25]. The critical strains of monolayers (multilayers) are ε_{ac} = 0.61 (0.40) and ε_{zz} = 0.23 (0.18) [25].

13.4.2.4 Physics and Application

The aw-SnS monolayer is a semiconductor with an indirect band gap [64]. By applying a moderate uniaxial or biaxial strain of approximately 2%, aw SnS experiences a transition to a direct semiconductor with a moderate gap < 2.7 eV [64]. In a study on band gap engineering, first-principles calculations predicted the effects of ununiformed biaxial strain of 19.1 N/m on low-dimensional phases, revealing strain-induced changes such as semiconductor-to-metal transitions, indirect-to-direct transitions, or negative Poisson's ratios [65].

Specifically, puckered SnS monolayers exhibit an in-plane ferroelectric effect, and switching occurs below the critical thickness of the SnS sheet of < 15 layers with an odd and even number of layers by changing the stacking sequence from AA to AB, even at room temperature [66]. In the monolayer, the dipole moment in the armchair direction induces spontaneous polarization. The aw-SnS sheet is an indirect semiconductor with multiferroicity, revealing piezoelectricity and pyroelectricity.

Several publications have used DFT calculations and Boltzmann transport theory to study the thermoelectric properties of group-IVA monochalcogenides [67, 68]. In addition to their energetic and dynamic stability, ab initio calculations provided insight into the thermoelectric properties of buckled SnS monolayers [67]. In sharp contrast to the bulk material, SnS monolayers exhibited enhanced performance with high values of $ZT \sim 1.36$ and $ZT \sim 5$ at 300 K and 600 K, respectively. Therefore, these monolayers are good candidates for use in room-temperature thermoelectric devices.

A critical issue in photocatalysis is fast carrier recombination. Studies on the suitability of hb-SnS/GaS heterostructures as photocatalysts for water splitting by DFT and nonadiabatic MD simulations showed that the interlayer interaction in the heterostructure generates a built-in electric field and that nonadiabatic coupling accelerates the separation of photogenerated carriers within subpicoseconds [69]. Simultaneously, the recombination of photogenerated carriers is delayed, which allows separated electrons and holes with strong redox capacities to participate in water oxidation and reduction processes on the GaS and hb-SnS monolayers, respectively.

Monolayers of aw SnS with micrometer size were grown by physical vapor deposition (PVD), by precisely controlling the deposition pressure and temperature [66]. It has been generally assumed that SnS shows an odd–even effect, restricting ferroelectricity to sheets with an odd number of layers, while sheets with an even number of layers are centrosymmetric. Importantly, room temperature ferroelectricity could be observed in SnS sheets with a critical thickness of 15 layers. The authors explain the lack of centrosymmetry by the intrinsic puckered structure and confirm this observation by polarized Raman and second harmonic generation (SHG) spectroscopies.

13.4.2.5 Synthesis

The crystal growth and tape exfoliation methods lead to few-layer systems, whereas the liquid-phase exfoliation method allows the realization of bilayer SnS nanosheets through size-selective centrifugation [63]. The deposition of wafer-scale high-quality vertically standing SnS layers was possible by sulfur reduction from sputtered SnS_2 particles without postprocessing [70]. The phase purity was determined via Raman spectroscopy, and the morphology was extracted from HRTEM images. Analogous to related compounds, tin monosulfide monolayers are nontoxic and stable in ambient air with a large oxidation energy barrier of $E_A = 1.6$ eV [70]. A comprehensive review describes the enormous effort expended to find the best route for synthesizing and characterizing the optical and electronic properties of 2D tin-II sulfide (SnS) [71].

13.4.3 SnSe Monolayer

13.4.3.1 Chemistry and 2D Bonding

Since the number and type of valence electrons of SnSe, based on the electronic configurations of Sn atoms (Kr|$4d^{10}5s^2p^2$) and Se atoms (Ar|$3d^{10}4s^2p^4$), agree with those of SnS, the bonding behaviors are similar. However, the slightly larger size of the Se atoms compared with that of the S atoms reduces orbital overlap and thus the strength of covalent bonding. The difference in electronegativity between Sn (1.96) and Se (2.55) causes a Bader charge transfer of $\delta q = 0.855$ e from each Sn atom to each Se atom and leads to polar Sn–Se bonds with ionic character [1]. The published strongly deviating cohesive energies are $E_c = -3.28$ eV/atom [1] (PBE) and -4.57 eV/atom [20] (PBE) for the aw-SnSe monolayer, and $E_g = -3.22$ eV/atom [1] (PBE), -3.84 eV/atom [72], (PBE) and -4.56 eV/atom [20] (PBE) for the hb-SnSe monolayer. Thus, a marginally higher energetic stability is predicted for the aw-SnSe phase, which, however, is within the error involved in such calculations.

13.4.3.2 2D Structures

For the aw-SnSe phase, the lattice parameters are a = 0.426 nm, b = 0.445 nm, $d_1 = 0.289$ nm, $d_2 = 0.273$ nm [1], and $\Delta z = 0.275$ nm [22]. These lattice parameters agree with the results of independent calculations; for example, a = 0.425 nm, b = 0.435 nm, $d_1 = 0.271$ nm, $d_2 = 0.289$ nm, $\theta_1 = 94.53°$, $\theta_2 = 93.56°$, and $\theta_3 = 92.50°$ [21]. As the upper limit of monolayer thickness, the Se van der Waals diameter of $d_w = 0.380$ nm [2] and pucker effect of 0.259 nm [20] yield a thickness of $d_{w+\Delta z} = 0.639$ nm. As expected, this thickness is greater than the layer separation of $d_s = 0.589$ nm extracted from the bulk structure that involves interlayer interactions [25]. The calculated band gaps are $E_g = 1.14$ eV (PBE) and 1.67 eV (HSE) [20]. The space group of the puckered structure is $Pmn2_1$ [1].

In the aw-SnSe bilayers, the AC stacking order has an interlayer binding energy of $E_{int} = -40.7$ meV/Å^2, which is comparable to that of AB stacking with -40.5 meV/Å^2 [26]. The corresponding nearest layer distances in the bilayer are $d_{int} = 0.306$ nm and $d_{int} = 0.302$ nm, respectively. The binding energy of the bulk at $E_{int} = -146$ meV/atom confirms the strong interlayer charge transfer [73]. Accordingly, mechanical exfoliation may be difficult to achieve and may result in the preferred generation of few-layer assemblies. Calculations point to a stacking-dependent indirect-direct transition of the band gap of few-layer systems depending on the directionality of the interlayer interaction [74].

The structural data of hb-SnSe monolayers are a = 0.392 nm, d = 0.275 nm [1], and a buckling effect of $\Delta z = 0.153$ nm [20]. Recent DFT-GGA-PBE calculations yielded a = 0.389 nm, d = 0.299 nm, and $\theta = 80.55°$ [6]. The band gaps obtained by the different theoretical approaches are $E_g = 2.22$ eV (PBE) and 2.82 eV (HSE) [20].

13.4.3.3 Mechanics

The elastic moduli of the aw-SnSe monolayer are $E_{ac} = 14.0$ N/m (23.7 GPa) and $E_{zz} = 26.6$ N/m (45.1 GPa) for a thickness of $t = 0.589$ nm [23]. These values agree reasonably well with $E_{ac} = 11.0$ N/m and $E_{zz} = 24.9$ N/m [29]. Similar to aw GeS and aw SnS, a negative Poisson ratio of $\nu_{zx} = -0.210$ for stretching in the armchair and positive values of $\nu_{zy} = 0.352$, $\nu_{xy} = 0.851$ and $\nu_{yx} = 0.423$ have been reported [18]. The tensile strengths of aw-SnSe monolayers are further diminished compared to those of aw-SnS monolayers, with ultimate strength values at $\sigma_{ac} = 1.09$ N/m (1.85 GPa) and $\sigma_{zz} = 4.53$ N/m (7.69 GPa) [25]. Accordingly, the critical strains are strongly anisotropic at $\varepsilon_{ac} = 0.63$ and $\varepsilon_{zz} = 0.22$ [25].

13.4.3.4 Physics and Applications

The aw-SnSe monolayer is a semiconductor with an indirect band gap [21]. Similar to the behavior discussed for aw SnS, the application of uniaxial and biaxial strains induces indirect-to-direct transitions and semiconductor–metal transformations of bilayer SnSe, improving photovoltaic performance [75]. The authors observed an enhancement in the real and imaginary parts of the dielectric function of 14–30% with the effect of spectrally widening and improving absorption in the infrared and ultraviolet regions.

The group–IVA monochalcogenides GeS, GeSe, SnS, and SnSe have attracted enormous interest as potential materials for use in thermoelectric applications [76]. The thermoelectric performance, studied by DFT calculations combined with Boltzmann transport theory, reached as largest figure of merit, $ZT = 2.63$ for the aw phase of SnSe at 700 K, which was greater than the bulk value. Contrary to these results, a comparison of five distinct phases of SnSe (α, ß, γ, δ, and ε) based on similar calculations showed that the hb (ß) phase of SnSe monolayers has the best thermoelectric properties, with a ZT value of 2.06 at 300 °C [77].

13.4.3.5 Synthesis

The problems encountered in the preparation of SnSe monolayers are analogous to those described for SnS [20]. Several methods allow the growth of few-layer sheets, whereas synthesis of pure-phase single-layer SnSe remains a challenge due to the strong interlayer interaction. The growth of highly crystalline SnSe and $SnSe_2$ films is possible via vapor deposition by heating SnSe powder as the precursor [78]. Due to the different thermodynamic stabilities of SnSe and $SnSe_2$, phase control is possible by regulating the heating and substrate temperature. At a low substrate temperature (~210 °C) and proper heating temperature (650–700 °C), deposition of uniform $SnSe_2$ is possible. A substrate temperature of ~ 380 °C is sufficient to grow films consisting of SnSe.

13.4.4 SnTe Monolayer

13.4.4.1 Chemistry, 2D Bonding, and Structure

The chemical behavior of 2D SnTe follows from that of its homologues based on the related electronic configurations of Sn (Kr|4d^{10}5s^2p^2) and Te (Kr|4d^{10}5s^2p^4), supplying the same number and type of valence electrons responsible for covalent bonding. The Sn–Te bonds are only slightly polar owing to the small difference in the electronegativities of Sn (1.96) and Te (2.1), with minor charge transfer of $\delta q =$ 0.596 e from the Sn to the Te atoms. The cohesive energies of aw SnTe and hb SnTe are $E_c = -2.99$ eV/atom and -2.93 eV/atom (~3.00 eV/atom [79]), respectively, indicating weak bonding of these monolayers [1].

For aw-SnTe monolayers, the calculated lattice parameters are a = 0.454 nm, b = 0.458 nm, $d_1 = 0.293$, and $d_2 = 0.316$ nm [1] and the bonding angles are $\theta_1 = 92.1°$ and $\theta_2 = 95.5°$ [29]. The aw-SnTe monolayer is a semiconductor with an indirect band gap of $E_g = 0.666$ eV (PBE) [1]. The elastic moduli of the puckered lattice are $E_{ac} = 5.6$ N/m and $E_{zz} = 12.9$ N/m [29].

While hb SnTe has a smaller lattice constant than the aw structure at a = 0.415 nm the bond length of d = 0.295 nm is comparable. The buckled phase has much larger band gaps of $E_g = 1.62$ eV (PBE) and 2.44 eV (HSE06) [10]. First-principles calculation in the framework of DFT describe hb SnTe as a stable quasi-direct medium band gap semiconductor of $E_g \approx 2.00$ eV (PBE) and ~ 2.6 eV (HSE06) [79]. The study elucidates the strain sensitive electronic and optical properties with significant absorption and reflectivity in the UV region.

13.5 Plumbum Monochalcogenide Monoloayers (PbS, PbSe, and PbTe)

13.5.1 PbS Monolayer

13.5.1.1 Chemistry, 2D Bonding, Structure, and Mechanics

The electronic configurations of Pb atoms (Xe|5d^{10}6s^2p^2) and S atoms (Ne|3s^2p^4) determine 2D bonding and structure of stable PbS monolayers. In contrast to those of the other homologues, the most stable phase of lead monochalcogenide monolayers exhibits a square buckled structure that we name sb PbX (X = S, Se, Te) [80]. The structural data of this monolayer are a = 0.374 nm, d = 0.284 nm, $\Delta z = 0.197$ nm, and the buckling angle is $\theta = 21.6°$. The primitive cell in the square geometry contains two atoms. Bonding becomes partly ionic owing to the medium difference in the electronegativities of Pb (1.8) and S (2.58). The authors consider sb-PbS monolayers as a semiconductor with a large Rashba-like spin–orbit effect, controlled by out-of-plane buckling. There is a connection between relativistic spin–orbit interactions

(SOIs), quantum phenomena, effects such as the Rashba effect, Zeeman effect, spin Hall effect, and topological insulator (TI) states.

A comprehensive theoretical study discussed the changes in the geometrical parameters and electronic behavior of mono-, bi-, and trilayer assemblies of PbX (X = S, Se, Te) [81]. The postulated buckled PbX monolayer contains two sublayers along the (001) direction of bulk PbX with a small corrugation. The lattice parameters of this buckled PbS monolayer are a = 0.419 nm, d_1 = 0.269 nm (normal), and d_2 = 0.297 nm (in-plane), with an exceedingly small buckling effect of Δz = 0.0156 nm. In few-layer assemblies, the contribution of covalent-type interactions by lone electron pairs to van der Waals interactions increases the interlayer interaction energy and shortens the interlayer separation. Increasing the number of layers strongly decreases the band gap, e.g., from E_g = 1.65 eV for the monolayer to 0.98 eV for the trilayer. Remarkably, according to these DFT-PBE calculations, the bilayer of PbS has an extremely high electron carrier mobility of 252 000 $cm^2\ V^{-1}\ s^{-1}$ and a hole mobility of 16 000 $cm^2\ V^{-1}\ s^{-1}$ for mono- or trilayer PbS. The elastic modulus of E = 56.8 N/m [81] for sb PbS deviates strongly from the much smaller stiffness of E = 19.6 N/m of the hb-PbS monolayer [15].

For the honeycomb hb-PbS monolayer with the space group *P*3*m*1 theory predicts a lattice constant of a = 0.394 nm and a much larger buckling height of Δz = 0.147 nm [10]. The meta-stability of the buckled honeycomb structure was demonstrated by three independent methods for the twelve IVA–VIA monolayers MX (M = Si, Ge, Sn, and Pb; X = S, Se, and Te). Due to the broken inversion symmetry, these metastable structures possess an out-of-plane intrinsic electric polarization and piezoelectric response. The calculated electronic band gaps of hb PbS are E_g = 2.03 eV (PBE) and 2.85 eV (HSE). The published elastic constants yield a low Young's modulus of E = 19.6 N/m and for Poisson's ratio ν = 0.284. A coordination-preserving structural transformation of the buckled honeycomb phase of the PbS, PbSe, and PbTe monolayers with the space group *P*3*m*l leads to a lower-energy structure with the space group *Pmmn*. This symmetry is considered to present the global minimum of 2D PbS, PbSe, and PbTe [10].

13.5.2 PbSe Monolayer

13.5.2.1 Chemistry, 2D Bonding, Structure, Mechanics, and Synthesis

Analogous to the PbS monolayer, the most stable phase of the PbSe monolayer is the square buckled sb phase [80]. The set of geometrical parameters of sb PbSe is a = 0.382 nm, d = 0.296 nm, Δz = 0.231 nm, and a buckling angle of θ = 24.4° [80]. The Pb–Se bonds exhibit ionic character owing to a medium difference in the electronegativities of Pb (1.8) and Se (2.55). The sb-PbSe monolayer is an indirect semiconductor with a large Rashba-like spin–orbit effect. The ferroelectricity of the compound allows the switching of the electric polarization in the direction of buckling. As before, the interest in relativistic SOIs originates from their connection to several quantum phenomena.

As discussed for PbS, a low-buckled PbSe monolayer structure can be derived by cutting two sublayers along the (001) direction of bulk PbSe. The lattice parameters of the buckled PbSe monolayer are a = 0.436 nm, d_1 = 0.283 nm (normal), and d_2 = 0.309 nm (in-plane), with a marginal buckling effect of $\Delta z = 0.0034$ nm [81]. In few-layer assemblies, the contribution of covalent-type interactions by lone electron pairs to van der Waals interactions increases the interlayer interaction energy substantially. In addition, the formation of multiple layers reduces the band gap, for example, from $E_g = 1.26$ eV for the monolayer to 0.76 eV for the trilayer. This is consistent with the direct band gap values $E_g = 1.27$ eV (PBE) and 1.85 eV (HSE06) [82]. Furthermore, this work presents detailed information on the tunability of optoelectronic properties by tensile and compressive strains. The space group of the orthorhombic structure is *P4/nmm* [82]. The elastic modulus of 48.3 N/m [81] deviates strongly from the much smaller stiffness of 15.4 N/m of the hb-PbS monolayer [10].

For the metastable honeycomb buckled hb-PbSe monolayer with the space group *P*3*m*1, a lattice constant of a = 0.406 nm and a buckling height of $\Delta z = 0.159$ nm are predicted by theory [10]. The calculated band gaps are $E_g = 1.68$ eV (PBE) and 2.59 eV (HSE) and are smaller than those for hb-PbS monolayers. As mentioned already, the published elastic constants yield a small stiffness of E = 37 N/m and a Poisson ratio of $\nu = 0.31$ [10].

First-principles calculations predict for 2D PbSe topological insulator states protected by lattice symmetry instead of time-reversal symmetry [83]. The combination of strong spin–orbit interaction with quantum confinement effects leads to a topological phase transition with an even number of band inversion momentum space points. The PbSe nanostructure presents pairs of spin-polarized Dirac cones coming from the monolayer edges. Importantly, due to quantum confinement the gap is larger than that of the parent narrow band gap trivial insulator bulk PbSe, favoring a room-temperature 2D band gap with crystalline-protected Dirac states at the edges.

13.5.3 PbTe Monolayer

13.5.3.1 Chemistry and 2D Bonding

The chemical behavior of 2D PbTe follows that of its homologues based on the valence electrons of Pb ($6s^2p^2$) and Te atoms ($5s^2p_x{}^2p_y{}^2$), which have the same number and type of bonding electrons responsible for covalent bonding. The Pb–Te bonds are slightly polar owing to the small difference in the electronegativities of Pb (1.8) and Te (2.1). Unfortunately, detailed information on the cohesive energy and thus in the energetic stability of 2D-PbTe phases is still lacking.

13.5.3.2 2D Structures

According to DFT-PBE calculations, the most stable phase of the PbTe monolayer is the square buckled sb-phase of PbTe [80]. A set of geometrical constants of sb-PbTe is a = 0.401 nm, d = 0.316 nm, Δz = 0.265 nm, and the buckling angle is θ = 26.3° [14]. The sb-PbTe monolayer is an indirect semiconductor with a Rashba-like spin–orbit effect. According to a tight-binding analysis, the atomic composition and the buckling angle are the key parameters that control the Rashba effect. The atomic composition influences the mixing process of in-plane and out-of-plane orbitals via the SOI.

As discussed for PbS and PbSe, a low-buckled PbTe monolayer can be derived by cutting two sublayers along the (001) direction of bulk PbTe [81]. The lattice parameters of the low-buckled monolayer are a = 0.465 nm, d_1 = 0.303 nm (normal), and d_2 = 0.332 nm (in-plane), with a minor buckling effect of Δz = 0.0302 nm. In few-layer assemblies, the contribution of covalent-type interactions by lone electron pairs to van der Waals interactions increases the interlayer interaction energy. Increasing the number of layers reduces the band gap, for example, from E_g = 1.26 eV for the monolayer to 0.74 eV for the trilayer [81]. The elastic modulus of E = 32.9 N/m is comparable with that of phosphorene of 30 N/m.

For the honeycomb buckled monolayer hb PbTe with the space group $P3m1$, a lattice constant of a = 0.432 nm and a buckling height of Δz = 0.172 nm are predicted [10]. The calculated band gaps of this structure are E_g = 1.10 eV (PBE) and 2.12 eV (HSE). The elastic constants presented for hb-PbTe monolayers yield a small stiffness of only 16.5 N/m [10]. The Poisson ratio derived from the elastic constants is ν = 0.268 [10].

13.5.3.3 Physics and Applications

Several studies have investigated the thermoelectric properties of the three XTe (X = Ge, Sn, and Pb) monochalcogenides [50, 84, 85]. First-principles calculations combined with the Boltzmann transport theory revealed that the low lattice thermal conductivities of these monolayers result from resonant and weak bonding, as well as the avoidance of crossing between longitudinal acoustic phonons, and low-lying optical branches [50]. In this work, the GeTe monolayer was identified as the monolayer with the largest figure of merit of $ZT = 4.23$ [50].

Similar computations were used to study the room-temperature thermoelectric applications of the orthorhombic lead monochalcogenides PbX (X = S, Se, and Te). The zigzag direction of PbTe monolayers exhibits the largest electrical conductivity and a thermoelectric power factor of 10.80 W/m K^2 s at 0.66 eV [84]. A study concentrating on puckered PbTe monolayers reported a low lattice thermal conductivity of ~ 0.75 and ~ 0.79 W m^{-1} K^{-1} along the armchair and zigzag directions, respectively, and an optimal figure of merit of $ZT \sim 1.55$ along the zigzag direction at a carrier concentration of 3.94×10^{12} cm^{-2} at 900 K [85].

First-principles calculations and Boltzmann transport theory revealed enhanced thermoelectric performance in hexagonal monolayers of $(MX)_2$ (X = S, Se, and Te) with the space group P-3m1 [86]. The electronic conductivity, phononic thermal conductivity, Seebeck coefficient, power factor, and figure of merit were studied under biaxial tensile strain. The highest room temperature *ZT* values were found for p-type $(PbTe)_2$ monolayers at $ZT = 2.45$ in the zigzag direction and $ZT = 1.59$ in the armchair direction. As discussed before, the enormously reduced phonon thermal conductivity, ascribed by the authors to phonon scattering, is caused by phonon anharmonicity.

The investigation of the thermal transport of 2D PbTe and its use as a thermoelectric material is a topic of active research. A novel approach extended the scope of MD simulations by the development of an efficient machine-learned potential (MPL), based on a newly introduced MPL model called the neuro-evolution potential (NEP), suitable for PbTe monolayers [87]. The NEP has the potential of delivering a satisfactory accuracy, excellent training possibilities, and computational speed.

The development of 2D polar metals with polar structures and metallic behavior is highly desirable. First-principles calculations predict the formation of a polar metal by electrostatic doping of the buckled trigonal PbTe monolayer [88]. Owing to the noncentrosymmetric buckled geometry, a spontaneous out-of-plane electric polarization exists in this monolayer. The polar distortions can be preserved with carrier doping, enabling the doped material to function as a 2D polar metal. In fact, the PbTe monolayer has an intrinsic ferroelectricity whose ferroelectric distortions show compatibility with mobile carriers due to the special elastic-polar mode interaction that is important for the existence of a robust polar instability.

A recently published review describes the state-of-the-art knowledge of the ferroelectric and ferroelastic behavior of group IVA–VIA monolayers [89]. Relevant properties include switchable in-plane spontaneous polarization, soft elastic constants, structural degeneracy, thermally driven two-dimensional structural transformations, selective valley excitations, and valley Hall effects. The reviewed physical properties may be of interest for use in nonvolatile memories, valleytronics, spintronics, and optoelectronic devices.

13.5.3.4 Synthesis

As described above, PbS, PbSe, and PbTe have found considerable interest in the past years, for example, as potential thermoelectric materials. Early experiments achieved the deposition of thin PbSe and PbTe films by atomic layer deposition (ALD) using the compounds $Pb(C_{11}H_{19}O_2)$, $Se(Et_3Si)_2$, and $Te(Me_3Si)_2$ as ALD precursors for Pb, Se, and Te, respectively [90]. The process analysis revealed a Volmer-Weber-type growth of islands instead of the expected layer-by-layer ALD growth [90]. In the narrow temperature range of 170 °C to 210 °C the smoothest polycrystalline films were produced for both compounds [91].

The realization of nano-structures with stable ferroelectricity and high transition temperature is highly desired. Robust in-plane ferroelectricity has been discovered

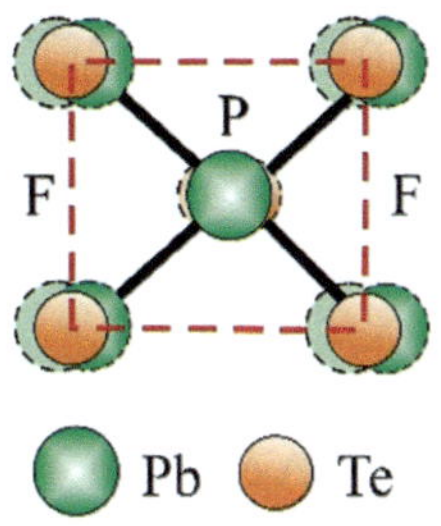

Fig. 13.9 Transitions between paraelectric P and ferroelectric F phases of 2D PbTe, showing the degenerate ferroelectric phases with positive and negative polarization (dashed). Adapted with permission from Ref. [93], Copyright 2023, American Physical Society

in atomic-thick SnTe and PbTe sheets [92]. Both compounds show a remarkably similar behavior, except that PbTe is paraelectric. Synthesis was achieved by state-of-the-art MBE on graphitized 6H-SiC(0001) substrates by relaxed van der Waals epitaxy. Characterization of the grown films by in situ reflection high-energy electron diffraction (RHEED) and scanning tunneling microscopy (STM) revealed randomly grown islands in size of 100 nm, which can be as large as ~ 1 μm for carefully optimized substrate temperature and controlled SnTe or PbTe flux. The deposited atomic-thick structures showed stable in-plane spontaneous polarization down to the 1-unit cell limit with a ferroelectric transition temperature of 270 K. Moreover, sheets with few unit cells thickness exhibit ferroelectricity at room temperature.

In addition to ab initio calculations demonstrating strain-induced ferroelectricity of 2D PbTe, MD simulations with ab initio-level accuracy, based on the deep potential framework, demonstrated robust ferroelectricity at room temperature in strained 2D PbTe [93]. The structural analysis indicates that the ferroelectric transition occurs by synchronized shifts in atomic positions (displacive), demonstrating the strong coupling between strain and polarization (see Fig. 13.9). Based on the absence of hysteresis the authors assume that the strain-induced and temperature-induced ferroelectric phase transitions are second-order transitions. The 2D ferroelectrics may find applications as ferroelectric field-effect transistors with low power dissipation, high-density nonvolatile memory, miniaturized sensor, and microactuator.

13.5.3.5 Synthesis

The growth of PbTe layers was achieved by the MBE method [94]. In this study, antiferromagnetic wide-gap EuTe layers and non-magnetic narrow-gap PbTe layers were used to fabricate monolayer short period $(EuTe)_n$ $/(PbTe)_{3n}$ superlattices. As observed for similar 2D materials, only in a narrow range of growth conditions strain-induced coherent islanding could be avoided, leading to surface roughening. Under suitable growth conditions STM analysis revealed an exceptionally smooth surface morphology.

13.6 Summary of Binary Group IVA–VIA Compounds

13.6.1 Binding–Structure Relationships

In the plot of cohesive energies versus bond lengths of binary IVA–VIA monolayers, we consider groups of chemically related compounds, defined by combining each group–IVA element with all group–VIA elements (see Fig. 13.10) [95]. The resulting five groups show the typical inverse dependence, where the carbon group separates from the silicon, germanium, and tin groups, which show dependencies that partly coincide. For most of the groups, the puckered structure has the highest energetic stability, closely followed by the buckled phase. As shown in Table 13.1, the bond lengths parallel to and between the planes of the puckered structure deviate only slightly in most cases. Thus, the application of mean bond lengths has no major influence on the conclusions. In addition to the four carbon monochalcogenides (C-group) with the largest cohesive energy and thus chemical binding strength, the other four related groups (Si-, Ge-, Sn-, and Pb-group) exhibit binding behaviors within a narrow range of cohesive energies between $E_c = -3$ eV/atom and -4 eV/atom.

The variable contributions of ionicity play a key role in the reduction of orbital overlap and, concurrently, the real strength of covalent-ionic bonding of group IVA–VIA monolayers. Since the electronegativities of C (2.55 eV), S (2.58 eV), and Se (2.55 eV) nearly coincide, a sizable electron charge moves only from the Te atoms (2.1 eV) to the C atoms in the C-group. In the other groups, charge transfer occurs to the chalcogenide atoms owing to the smaller electronegativities of the group–IVA elements Si (1.90 eV), Ge (2.01 eV), and Sn (1.96 eV). Since the difference

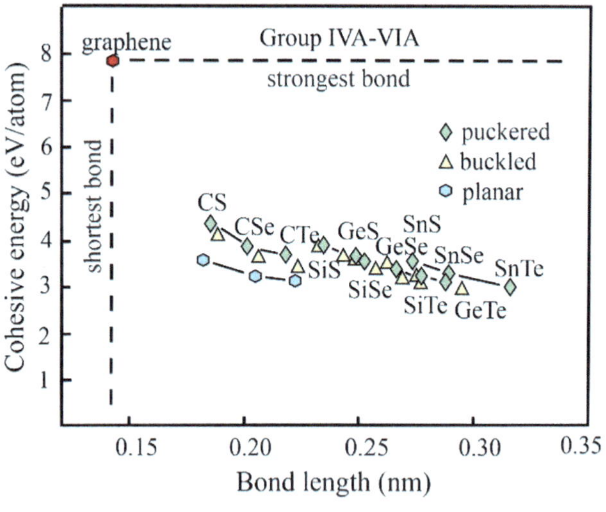

Fig. 13.10 Plot of the cohesive energies versus bond lengths of the hexagonal 2H-structures of group IVA–VIA monolayers. Adapted from Ref. [95], Copyright 2021, Royal Society of Chemistry

in electronegativity decreases within each group of related compounds going down the column of the periodic table, electrostatic interactions become less important with increasing atomic number, and the same is true for orbital overlap due to the increasing size of the chalcogen atoms. This is one reason for the characteristic inverse dependence of binding and cohesive energy. The decay of the cohesive energies decreases asymptotically, connected with a strongly increasing bond length. The inclusion of charge transfer, as determined by the Bader model, supports this view (see Table 13.1). The ionicity of bonds has a strong influence on the band gap of semiconductors. The minor changes in the atom size and electronegativity relative to carbon explain the striking similarity in bonding and the comparable mechanical properties considered in more detail below.

13.6.2 Mechanics–Structure Relationships

As expected from the reduced bonding strength, compared with those of planar honeycomb structures such as graphene, for example, the elastic moduli and intrinsic strengths exhibit a low stiffness and low ultimate strength of both the puckered and buckled phases. Figure 13.11 illustrates the enormous difference in mechanical stiffness and strength in comparison to the outstanding mechanical performance of covalently bonded graphene. [92]

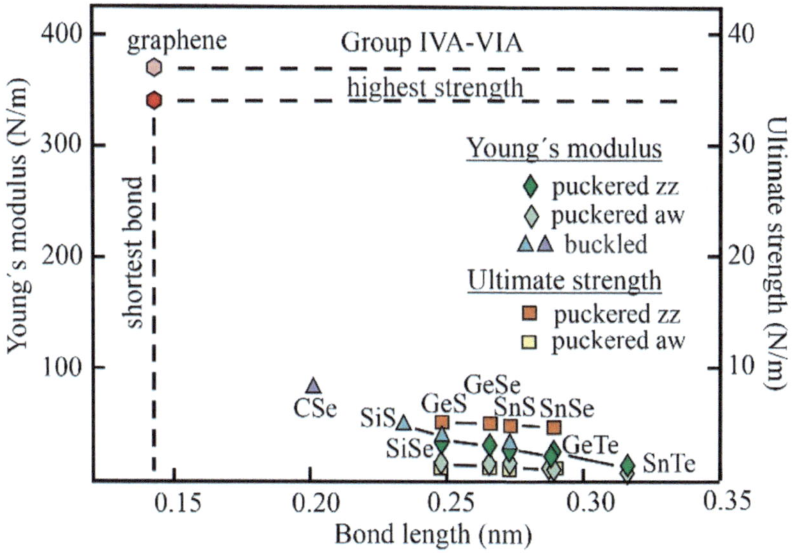

Fig. 13.11 Young's moduli and ultimate strengths versus bond lengths of the hexagonal 2H- structures of group IVA–VIA monolayers. Adapted from Ref. [95], Copyright 2021, Royal Society of Chemistry

Certainly, a low stiffness and/or strength can be critical issue for any practical application, especially observed for the heavier compounds. Interestingly, the stiffness and strength of Ge and Sn sulfides and selenides agree well with Griffith's rule in the armchair direction of the puckered phase but not in the stronger zigzag direction, where the calculated strengths are much greater than expected (see Fig. 13.11). Note that the large Young's modulus of the CSe monolayer is consistent with the shorter and stronger bonds in the C group and thus deviates from the behavior of the other four well-characterized groups of monolayers.

References

1. C. Kamal, A. Chakrabarti, M. Ezawa, Phys. Rev. B **93**, 125428 (2016)
2. S.S. Batsanov, Inorg. Mater. **37**, 871 (2001)
3. Q. Zhang, T. Xin, X. Lu, Y. Wang, Materials **11**, 431 (2018)
4. Q. Zhang, Y. Feng, X. Chen, W. Zhang, L. Wu et al., Nanomaterials **9**, 598 (2019)
5. A. Zaharo, A. Purqon, T. Winata, M. Saito, Jpn. J. Appl. Phys. **59**, 071006 (2020)
6. A. Rahim, W. Haider, A. Kahn, H. Kahn, H.U. Din et al., J. Comput. Electron. **21**, 725 (2022)
7. D. Rocca, A. Abboud, G. Vaitheeswaran, S. Lebègue, Beilstein J. Nanotech. **8**, 1338 (2017)
8. Z. Zhu, J. Guan, D. Liu, D. Tománek, ACS Nano **9**, 8284 (2015)
9. Y.-L. Zhu, J.-H. Juan, Y.-Q. Zong, S. Wang, K.-H. Xue et al., J. Mater. Sci. **54**, 11485 (2019)
10. S.P. Poudel, S. Barraza-Lopez, Phys. Rev. B **103**, 024107 (2021)
11. J.-W. Jiang, Y.-P. Zhou, *Handbook of Stillinger-Weber Potential Parameters for Two-Dimensional Atomic Crystals* (INtechOpen, 2017)
12. M.-Q. Le, Arch. Mech. **74**, 3 (2022)
13. B. Liu, M. Niu, J. Fu, Z. Xi, M. Lei et al., Phys. Rev. Mater. **3**, 054002 (2019)
14. C. Chowdhury, S. Karmakar, A. Datta, J. Phys. Chem. C **121**, 7615 (2017)
15. S. Karmakar, C. Chowdhury, A. Datta, J. Phys. Chem. C **120**, 14522 (2016)
16. S. Behzad, R. Chegel, J. Comput. Electron **22**, 971 (2023)
17. Y. Chen, Q. Sun, P. Jena, J. Mater. Chem. C **4**, 6353 (2016)
18. B. Sa, Z. Sun, B. Wu, Nanoscale **8**, 1169 (2016)
19. T.-N. Do, N.N. Hieu, N.A. Poklonski, N.T.T. Binh, C.Q. Nguyen et al., RSC Adv. **11**, 28381 (2021)
20. Z. Hu, Y. Ding, X. Hu, W. Zhou, X. Yu et al., Nanotechnology **30**, 252001 (2019)
21. Y. Xu, K. Xu, H. Zhang, Molecules **24**, 639 (2019)
22. R. Li, H. Cao, J. Dong, Phys. Lett. A **381**, 3747 (2017)
23. G. Qin, Z. Qin, W.-Z. Fang, L.-C. Zhang, S.-Y. Yue et al., Nanoscale **8**, 11306 (2016)
24. L.C. Gomez, A. Carvalho, Phys. Rev. B **92**, 085406 (2015)
25. Y. Gao, L. Zhang, G. Yao, H. Wang, J. Appl. Phys. **125**, 082519 (2019)
26. Z.-Y. Li, M.-Y. Liu, Y. Huang, Q.-Y. Chen, C. Cao et al., Phys. Chem. Chem. Phys. **20**, 214 (2018)
27. H. Wiedemeier, H.G. von Schnering, Z. Kristallographie **148**, 295 (1978)
28. H.-K. Choi, J. Cha, C.-G. Choi, J. Kim, S. Hong, Nanomaterials **11**, 2960 (2021)
29. L. Huang, F. Wu, J. Li, J. Chem. Phys. **144**, 114708 (2016)
30. X. Kong, J. Deng, L. Li, Y. Liu, X. Ding et al., Phys. Rev. B **98**, 184104 (2018)
31. L.C. Gomes, A. Carvalho, J. Appl. Phys. **128**, 121101 (2020)
32. S. Zhang, N. Wang, S. Liu, S. Huang, W. Zhou et al., Nanotechnology **27**, 274001 (2016)
33. R. Fei, W. Li, J. Li, L. Yang, Appl. Phys. Lett. **107**, 173104 (2015)
34. A. Priydarshi, Y.S. Chauhan, S. Bhowmick, A. Agarwal, J. Appl. Phys. **131**, 034101 (2022)
35. S. Song, Y. Zhang, J. Guan, S. Dong, Phys. Rev. B **103**, L140104 (2021)
36. M. Wan, Z. Zhang, S. Zhao, N. Zhou, Chinese Phys. B **31**, 096301 (2022)

37. P.D. Antunez, J.J. Buckley, R.L. Brutchey, Nanoscale **3**, 2399 (2021)
38. Y. Guo, S. Zhou, Y. Bai, J. Zhao, A.C.S. Appl, Mater. Interfaces **9**, 12013 (2017)
39. F. Li, X. Liu, Y. Wang, Y. Li, J. Mater. Chem. C **4**, 2155 (2016)
40. K. Chang, S.S.P. Parkin, J. Appl. Phys. **127**, 220902 (2020)
41. P. Ramasamy, D. Kwak, D.H. Lim, H.S. Ra, J.-S. Lee, J. Mater. Chem. C **4**, 479 (2016)
42. Y. Xu, H. Zhang, H. Shao, G. Ni, J. Li et al., Phys. Rev. B **96**, 245421 (2017)
43. Y. Ye, Q. Guo, X. Liu, C. Liu, J. Wang et al., Chem. Mater. **29**, 8361 (2017)
44. H. Shin, J.T. Krogel, K. Gasperich, P.R.C. Kent, A. Benali et al., Phys. Rev. Mater. **5**, 024002 (2021)
45. Y. Mao, C. Xu, J. Yuan, H. Zhao, Phys. Chem. Chem. Phys. **20**, 6929 (2018)
46. H.T.T. Nguyen, T.V. Vu, N.T.T. Binh, D.M. Hoat, N.V. Hieu et al., Chem. Phys. **529**, 110543 (2020)
47. C. Guo, C. Wang, T. Wang, Y. Liu, Nanotechnology **33**, 425701 (2022)
48. G. Qian, X. Hu, W. Chen, Q. Zhou, Chemosensors **10**, 246 (2022)
49. D.W. Boukhvalov, S. Nappini, M. Vorokhta, T.O. Menteş, L. Piliai et al., Adv. Funct. Mater. **31**, 2106228 (2021)
50. D. Zhang, Z. Zhou, H. Wang, Z. Yang, C. Liu, Nanoscale Res. Lett. **13**, 400 (2018)
51. W.V. Sinambela, S.A. Wella, F.S. Arsyad, N.T. Hung, A.R.T. Nugraha, Crystals **11**, 1290 (2021)
52. Q. Wang, L. Wu, A. Urban, H. Cao, P. Lu, Materials **14**, 4495 (2021)
53. D. Zhang, S. Hu, Y. Sun, X. Liu, H. Wang et al., ES Energy Environ. **10**, 59 (2020)
54. T. Zhang, J. Xu, M.-L. Li, J. Nanoelectron. Optoelectron. **15**, 1436 (2020)
55. M. Qiao, Y. Chen, Y. Wang, Y. Li, J. Mater. Chem. A **6**, 4119 (2018)
56. X. Zhang, F. Zhao, Y. Wang, X. Liang, Z. Zhang et al., ACS Appl. Mater. Interfaces **12**, 37108 (2020)
57. J. Singh, G. Singh, S.K. Tripathi, Mater. Today Proc. **45**, 5439 (2021)
58. W. Wan, C. Liu, W. Xiao, Y. Yao, Appl. Phys. Lett. **111**, 132904 (2017)
59. B. Xu, J. Deng, X. Ding, J. Sun, J.Z. Liu, NPJ Comput. Mater. **8**, 47 (2022)
60. P. Zhang, F. Zhao, P. Long, Y. Wang, Y. Yue et al., Nanoscale **10**, 15989 (2018)
61. A. Mubeen, A. Majid, M. Alkhedher, E.M. Tag-ElDin, N. Bulut, Materials **15**, 5578 (2022)
62. N. Higashitarumizu, H. Kawamoto, M. Nakamura, K. Shimamura, N. Ohashi et al., Nanoscale **10**, 22474 (2018)
63. J.R. Brent, D.J. Lewis, T. Lorenz, E.A. Lewis, N. Savjani et al., J. Am. Chem. Soc. **137**, 12689 (2015)
64. E. Akhoundi, M. Faghihnasiri, S. Memarzadeh, A.H. Firouzian, J. Phys. Chem. Solids **126**, 43 (2019)
65. Y. Zhang, B. Shang, L. Li, J. Lei, RSC Adv. **7**, 30327 (2017)
66. N. Higashitarumizu, H. Kawamoto, C.-J. Lee, B.-H. Lin, F.-H. Chu et al., Nat. Commun. **11**, 2428 (2020)
67. R. Gupta, B. Dongre, J. Carrete, C. Bera, J. Appl. Phys. **130**, 054301 (2021)
68. A. Pandit, B. Hamad, Appl. Surf. Sci. **538**, 147911 (2021)
69. J. Meng, J. Wang, J. Wang, Q. Li, J. Yang, J. Mater. Chem. A **10**, 3443 (2022)
70. M. Patel, H.-S. Kim, J. Kim, Nanoscale **9**, 15804 (2017)
71. K.J. Norton, F. Alam, D.J. Lewis, Appl. Sci. **11**, 2062 (2021)
72. X.-Q. Tian, J.-Y. Duan, M. Kiani, Y.-D. Wei, N. Feng et al., Nanoscale **12**, 13450 (2020)
73. H.-Y. Song, J.-T. Lü, Chem. Phys. Lett. **695**, 200 (2018)
74. H.I. Sirikumara, T. Jayasekera, J. Phys. Condens. Matter **29**, 425501 (2017)
75. T.V. Vu, H.D. Tong, T.K. Nguyen, C.V. Nguyen, A.A. Lavrentyev et al., Chem. Phys. **521**, 5 (2019)
76. A. Shafique Y.-H. Shin, Sci. Rep. **7**, 506 (2017)
77. Z.-Y. Hu, K.-Y. Li, Y. Lu, Y. Huang, X.-H. Shao, Nanoscale **9**, 16093 (2017)
78. D. Lu, C. Yue, S. Luo, Z. Li, W. Xue et al., Appl. Surf. Sci. **541**, 148615 (2021)
79. N. Fatahi, D.M. Hoat, A. Laref, S. Amirian, A.H. Reshak et al., Eur. Phys. J. B **93**, 32 (2020)
80. P.Z. Hanakata, A.S. Rodin, A. Carvalho, H.S. Park, D.K. Campbell et al., Phys. Rev. B **96**, 161401 (2017)

81. X.-B. Li, P. Guo, Y.-N. Zhang, R.-F. Peng, H. Zhang et al., J. Mater. Chem. C **3**, 6284 (2015)
82. Y. Mao, R. Wu, D. Ding, F. He, Comp. Mater. Sci. **202**, 110957 (2022)
83. E.O. Wrasse, T.M. Schmidt, Nano Lett. **14**, 5717 (2014)
84. B.U. Haq, S. AlFaify, T. Alshahrani, R. Ahmed, Q. Mahmood et al., Ceram. Int. **47**, 3380 (2021)
85. S. Tang, M. Wu, S. Bai, D. Luo, J. Zhang et al., J. Alloys Comp. **907**, 164439 (2022)
86. P.-Z. Jia, Z.-X. Xie, Y.-X. Deng, Y. Zhang, L.-M. Tang et al., Appl. Phys. Lett. **121**, 043901 (2022)
87. W. Sha, X. Dai, S. Chen, B. Yin, F. Guo, Mater. Today Phys. **34**, 101066 (2023)
88. T. Xu, J. Zhang, Y. Zhu, J. Wang, T. Shimada et al., Nanoscale Horiz. **5**, 1400 (2020)
89. S. Barraza-Lopez, B.M. Fregoso, J.W. Villanova, S.S.P. Parkin, K. Chang, Rev. Mod. Phys. **93**, 011001 (2021)
90. K. Zhang, A.D.R. Pillai, K. Bollenbach, D. Nminibapiel et al., ECS J. Solid State Sci. Technol. **3**, P207 (2014)
91. K. Zhang, A.D.R. Pillai, M. Tangirala, D. Nminibapiel, K. Bollenbach et al., Phys. Status Solidi (A) **211**, 1329 (2014)
92. K. Chang, J. Liu, H. Lin, N. Wang, K. Zhao et al., Science **353**, 274 (2016)
93. Z. Gong, J.Z. Liu, X. Ding, J. Sun, J. Deng, Phys. Rev. B **108**, 134112 (2023)
94. G. Springbok, J.J. Chen, N. Frank, Y. Ueta, G. Marschner et al., in *Narrow Gap Semiconductors* (CRC Press, Boca Raton, 1995)
95. P. Hess, Nanoscale Horiz. **6**, 856 (2021)

Chapter 14
Group–IVB Transition-Metal Dioxide and Dichalcogenide Monolayers

Abstract The chemical and physical properties of transition-metal dioxides (TMDOs), transition-metal dichalcogenides (TMDCs) and specifically group IVB–VIA monolayers depend on the number of *d* electrons or the *d*-electron count and thus on the group number in the periodic table, which controls possible chemical coordination in 2D space. The 2D compounds of Ti, Zr, and Hf and the group–VIA elements O, S, Se, and Te have found increasing interest. Group–IVB dichalcogenides have only sparsely filled *d* bands, and the Fermi level usually is situated between the filled σ band of the chalcogen and the empty *d* bands (d^0 compounds). The most stable monolayer structures consist of a hexagonal sublayer of metal atoms sandwiched between two chalcogen sublayers arranged as X–M–X with covalent-ionic M–X bonds. The two most important stable polymorphs, the trigonal prismatic structure (1T) and the octahedral structure (1H), realize such a three-sublayer arrangement. Owing to the different electronegativities, charge transfer occurs from the metal to the chalcogen atoms with greater electronegativity, causing the bonds to be partially ionic and in few cases strongly ionic. The thickness of the three sublayers that form the monolayer is typically 0.6–0.7 nm. Most TMDC monolayers are semimetals or semiconductors with modest band gaps, except for $TiSe_2$ and tellurides. With decreasing electronegativity, when going down the corresponding column in the periodic table, the band gap decreases, and the tellurides are semimetals because the chalcogen-derived valence band overlaps with the transition metal *d*-orbital-derived conduction band. Potential applications are based on the modest or large band gaps and large electron mobilities of TMDCs. Potential applications include catalysis, nanoelectronics, optoelectronics, energy conversion, and energy storage devices. Most of these compounds are nonmetallic and nonmagnetic. The Janus structures of group-IVB TMDCs, such as ZrSSe, ZrSeTe, ZrSTe, HfSSe, HfSeTe, and HfSTe, are of increasing interest due to the loss of inversion symmetry. In non-centrosymmetric structures with two different chalcogen sublayers piezoelectricity emerges along with robust electron mobility.

P. Hess, *Bonding, Structure, and Performance of Two-Dimensional Materials*, NanoScience and Technology, https://doi.org/10.1007/978-3-031-74161-6_14

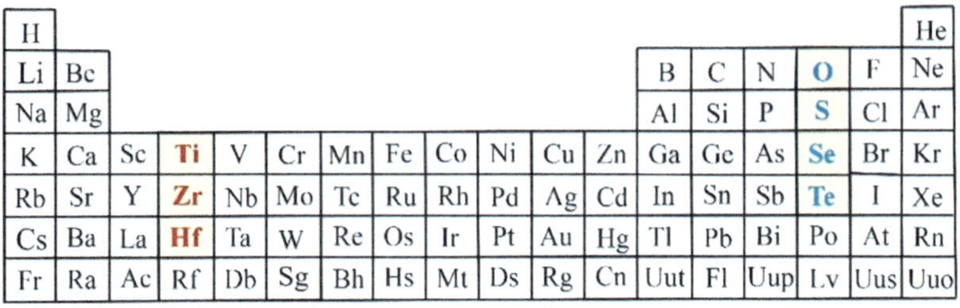

Fig. 14.1 Periodic table of the chemical elements highlighting group–IVB transition elements and group–VIA chalcogen elements involved in the group-IVB dichalcogenide monolayers

14.1 Titanium Dichalcogenide Monolayers (TiO_2, TiS_2, $TiSe_2$, and $TiTe_2$)

14.1.1 TiO_2 Monolayer

14.1.1.1 Chemistry and 2D Bonding

The scheme of the periodic table shown in Fig. 14.1 highlights the group–IVB elements and the chalcogenides considered in this chapter.

The electronic configuration of Ti atom (Ar|$3d^2 4s^2$) and O atoms ($1s^2 2s^2 p^4$) determine the type of 2D bonding in titanium dioxide monolayers. The oxidation state of the Ti atom (transition metal) is + 4, and that of the O atom (chalcogen) is − 2 [1]. Accordingly, the four valence electrons of Ti fill the valence shell of two O (chalcogen) atoms. The lone pair electrons of the chalcogen atoms terminate the outer sublayers of O atoms in the most stable 2D structures, consisting of a hexagonal sublayer of Ti atoms sandwiched between two sublayers of O atoms. In this triple-layer structure, the Ti atoms are sixfold coordinated, and each O atom binds to three Ti atoms. The nature of bonding is covalent-ionic owing to the enormous difference between the electronegativity of the Ti atom (1.54) and O atom (3.44), causing polarization of the covalent Ti–O bond by a substantial charge transfer from the $3d$ states of Ti to the $2p$ states of O. The TiO_2 monolayer with one of the highest ionic contributions in monolayers and a lattice with the largest separation of the O anions of the top and bottom sublayers has the highest stability due to geometrical minimization of electrostatic repulsion. Unfortunately, cohesive energies are not available for TiO_2 monolayers.

14.1.1.2 2D Structures

The best studied monolayers of 2D TiO_2 are the trigonal prismatic phase of 1T TiO_2 (T for trigonal) with space group P-3m1 and the octahedral 1H-TiO_2 structure (H for hexagonal) with space group P-6m2 (point group D_{3d}) (see Fig. 14.2) [1]. Note

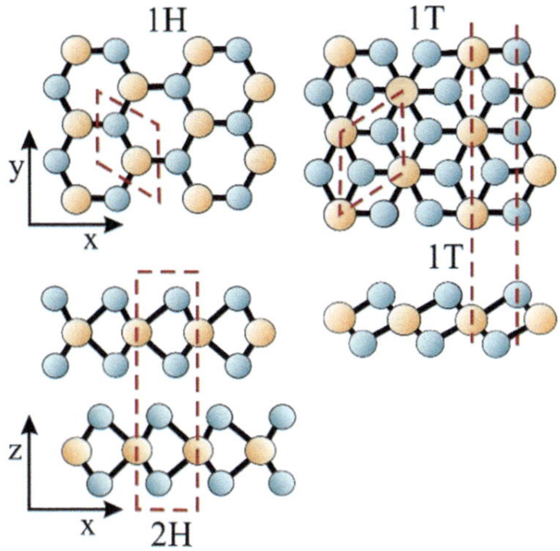

Fig. 14.2 Most stable structures of group IVB–VIA monolayers: Top and side view of hexagonal 1H- and 2H-structure, and trigonal prismatic 1T-structure; the dashed rhombus denotes the primitive cell

that for the H structure, often the structural parameters and properties of the 2H-TiO_2 bilayer structure are reported in the literature. Density functional theory (DFT) plays a major role in such calculations using the generalized gradient approximation (GGA) and Perdew–Burke–Ernzerhof (PBE) exchange–correlation functionals [2]. The 1T-TiO_2 monolayer seems to be more stable than the similar 1H structure, as indicated by its formation energy of $E_f = -3.10$ eV (FERE) compared with $E_f = -2.02$ eV (FERE) for the 2H phase [2]. In these calculations, the reference energies of the metal and oxygen atoms were based on the fitted elemental phase reference energies (FEREs). Note that the metastability of the H phase is still uncertain because calculations on TMDO monolayers are challenging.

The lattice parameters of 1T-TiO_2 monolayers based on DFT-PBE calculations are $a = 0.299$ nm, and a shorter value of $a = 0.288$ nm is obtained for the 2H phase [2]. However, the distance between the O atoms of $h_{OO} = 1.94$ nm is shorter for the 1T phase than for the 2H phase of $h_{OO} = 2.26$ nm [2]. The distance between the O atoms and their van der Waals diameter of $d_w = 0.310$ nm allow an estimate of the thickness of the 1T-TiO_2 phase of $d_{w+h} = 0.504$ nm (1T). Quite different bandgaps were obtained by the local density approximation (LDA) of $E_g = 1.10$ eV (LDA) and quasiparticle (QP) band structures including spin–orbit coupling (SOC) of 3.72 eV (G_0W_0) for the 2H phase and even larger gaps at $E_g = 2.65$ eV (LDA) and 5.74 eV (G_0W_0) for the 1T phase [2]. The band gaps based on the diffusion Monte Carlo (DMC) method of $E_g = 5.54$ eV for 2H-TiO_2 nanosheets and 5.69 eV for the 1T phase, derived for infinite periodic superlattices, are much larger than those estimated by DFT calculations [3]. An optical band gap of ~ 3.99 eV was found by taking into consideration the linear scaling relationship between the band gap and the exciton binding energy, which is consistent with the experimental value of ~ 3.84 eV [3, 4]. The space group, cohesive energy, lattice constant, bond length, atom distance, bond angle, and thickness of the TMDC monolayers are presented in Table 14.1.

Table 14.1 Space group, cohesive energy, lattice constant, bond length, X–X distance, bond angle, and corrugation corrected van der Waals diameter of group IVB–VIA monolayers

Monolayer	Space group	Cohesive energy, E_c (eV/atom)	Lattice constant a (nm)	Bond length d_{MX} (nm)	Distance h_{XX} (nm)	Bond angle θ_{X-M-X} (°)	Thickness d_{w+h} (nm)
TiO_2	*P*-3*m*1		0.299 [2]		0.194 [2]		0.504
TiS_2	*P*-3*m*1	− 6.12 [12], − 6.80 [11]	0.342 [11]	0.242 [11]	0.280 [11]	90.16 [11]	0.640
$TiSe_2$	*P*-3*m*1	− 5.64 [12], − 6.17 [11]	0.355 [11]	0.255 [11]	0.304 [11]	91.76 [11]	0.684
$TiTe_2$	*P*-3*m*1	− 5.03 [12], − 5.41 [11]	0.376 [11]	0.277 [11]	0.344 [11]	94.55 [11]	0.764
ZrO_2	*P*-3*m*1	− 7.71 [13]	0.328 [13]	0.212 [13]	0.193 [13]	79.05 [13]	0.503
ZrS_2	*P*-3*m*1	− 5.89 [13], − 7.35 [11]	0.369 [13]	0.257 [13]	0.290 [13]	88.59 [13]	0.650
$ZrSe_2$	*P*-3*m*1	− 5.35 [13], − 6.67 [11]	0.380 [13]	0.271 [13]	0.317 [13]	90.73 [13]	0.697
$ZrTe_2$	*P*-3*m*1	− 5.89 [11]	0.401 [11]	0.291 [11]	0.353 [11]	92.94 [11]	0.773
	P-6*m*2	− 4.62 [13]	0.393 [13]	0.294 [13]	0.373 [13]	78.83 [13]	0.793
HfO_2	*P*-3*m*1	− 7.90 [13]	0.324 [13]	0.211 [13]	0.193 [13]	79.38 [13]	0.503
HfS_2	*P*-3*m*1	− 6.00 [13], − 7.35 [11]	0.362 [11]	0.253 [11]	0.285 [11]	88.65 [11]	0.645
$HfSe_2$	*P*-3*m*1	− 5.42 [13], − 6.67 [11]	0.376 [13]	0.268 [13]	0.314 [13]	90.92 [13]	0.694
$HfTe_2$	*P*-3*m*1	− 5.80 [11]	0.398 [11]	0.288 [11]	0.347 [11]	92.58 [11]	0.767
	P-6*m*2	− 4.61 [13]	0.390 [13]	0.291 [13]	0.369 [13]	78.63 [13]	0.789

A global structural search using the constrained minima hopping method (MHM) together with the high-dimensional artificial neural network potential, based on the charge equilibrated neural network technique (CENT), revealed as a putative ground state a hexagonal structure that is energetically favored by 7 meV/atom compared to the lepidocrocite-type nanosheet and considered to be the most stable TiO_2 nanosheet [5]. In this structure, tetrahedral TiO_4 units form two layers stacked on top of each other by sharing corners and large pores. Note that the Ti atoms are fourfold coordinated, and the O atoms are twofold coordinated. The space group of this mechanically stable and chemically inert lattice is *P6/mmm.* The large band gap at $E_g = 5.05$ eV is tunable by strain engineering. This novel hexagonal sheet, which deviates from all known bulk structures of TiO_2, is a candidate for strain-tuned optical applications and photocatalytic water splitting. It is mechanically stable and chemically inert.

14.1.1.3 Physics and Applications

2D TiO_2 is a nontoxic low-cost transition-metal oxide with photocatalytic and electrocatalytic properties. A topic of present interest is to replace the thermodynamically stable {101} facet by high-energy facets such as {001} and {100} to enhance the catalytic activity of the material [6]. The authors present a fluorine-free method for performing the topochemical conversion of 1T-TiS_2 nanosheets to single-crystal anatase 2D TiO_2 that exposes {001} facets on the top and bottom and {100} facets on the sides of the nanosheet. Moreover, the anatase 2D-TiO_2 crystals exhibit an abundance of dangling bonds and surface defects. The goal of better understanding and optimizing the influence of facets, defects, and dopants on electrocatalytic performance is due to the enormous current interest in producing high-quality, well-defined TiO_2 monolayers.

DFT calculations elucidated the optical activity of 1H-TiO_2 monolayers from the visible to the ultraviolet region and revealed the possibility of tuning the electronic band gap at 3.48 eV by applying mechanical strain up to 10% [7]. Furthermore, calculations revealed the photocatalytic activity of the H-phase of TiO_2. Similarly, strain tuning of the novel hexagonal TiO_2 structure with the large band gap of 5.05 eV is possible, qualifying the structure for applications in optical devices and for photocatalytic water splitting due to the large surface area and the existing voids [5]. Additionally, hexagonal sheets may serve as building blocks for heterostructures with potential applications in energy conversion and storage devices.

Subnanometric 2D-TiO_2 networks are utilized in lamellar inorganic-organic supramolecular hybrid photocatalysts in the form of supramolecular 2D-TiO_2-surfactant heterostructures, where the TiO_2 network is stabilized by confinement between self-assembled monolayers of linear saturated fatty acids [8]. The optical properties and photocatalytic activity of composites with smaller band gaps show improved photoresponses due to band gap shifts and the extreme aspect ratio that is characteristic of 2D-TiO_2 confinement. The authors consider their results as proof-of-concept of the potential of these materials for creating custom-designed 2D-TiO_2-surfactant supramolecular photocatalysts.

Single-layer TiO_2 is an optional material for 2D gate oxides in electronic devices. The deposition of ~ 1.5 nm thick TiO_2 sheets on SiO_2/Si wafers from a colloidal suspension of immersed single-layer TiO_2 films allowed the study of the nature of the defects [9]. Atomic force microscopy (AFM) revealed edge thicknesses ranging from 1.3 to 1.75 nm, whereas the estimated crystallographic layer thickness is only ~ 0.7 nm. X-ray photoelectron spectroscopy (XPS) measurements confirmed that the binding energy of the Ti-2*p* peaks closely resembled the literature spectra but also demonstrated the existence of some oxygen vacancy defects. Analysis using spectroscopic ellipsometry (SE) by fitting a Lorentz oscillator model resulted in a similar layer thickness and indicated that the exact position of the defect states and energy levels depends on the estimated band gap and the charge state of the point defects.

14.1.1.4 Synthesis

Since the bulk TiO_2 crystal is a nonlayered material, the fabrication of TiO_2 monolayers is challenging. The fluorine-free topochemical conversion of 1T-TiS_2 nanosheets by desulfurization with hydrogen peroxide delivers ultrathin TiO_2 nanosheets of anatase with highly reactive exposed facets [6]. A solution-phase reaction of titanium tetrachloride and an oleylamine-sulfur complex under an inert atmosphere provided the 1T-TiS_2 samples. These samples showed a more efficient hydrogen evolution reaction (HER) and oxygen reduction reaction (ORR) than commercially available TiO_2 nanoparticles, without the incorporation of additional defects or doping.

Pure anatase inverse opal films of TiO_2 monolayers, consisting of a highly ordered hexagonal-patterned structure, were fabricated by using differently sized polystyrene spheres (PSs) as colloidal templates [10]. The $TiOSO_4$ solution employed as a TiO_2 precursor directly infiltrated the interstices of the 2D self-assembled PS opal template at an air/water interface, resulting in a $TiOSO_4$/PS opal composite film floating on the surface of water that could be deposited onto indium tin oxide (ITO) or silicon. Aqueous immersion after calcination afforded pure anatase 2D TiO_2 inverse opal ordered structures. The films showed structural changes from pale purple to blue to polychrome as the array period increases due to the redshift of the absorption wavelength. ITO substrates can be employed as active electrode materials in the fabrication of electrochromic devices.

14.1.2 TiS_2 Monolayer

14.1.2.1 Chemistry and 2D Bonding

The electronic configurations of Ti atoms (Ar|$3d^24s^2$) and S atoms (Ne|$3s^2p^4$) determine the type of 2D bonding in titanium disulfide monolayers. The oxidation state of Ti is + 4, and that of S is − 2, allowing various stable 2D bonding configurations. The 1T-TiS_2 monolayer, which consists of three sublayers with a titanium sublayer sandwiched between two sulfur sublayers, is the most stable monolayer structure. The Ti atom binds to six equivalent S atoms, forming edge-shared TiS_6 octahedrons, and the S atom binds to three equivalent Ti atoms, forming a trigonal pyramid structure. The calculated cohesive energies indicate strong covalent bonding that is enhanced by sizeable ionic interactions [11, 12]. The significant difference between the electronegativity of Ti atoms (1.54) and S atoms (2.58) causes bond polarization by a charge transfer of $\delta q = 1.49$ e from the 3*d* states of each Ti atom to the 3*p* states of two S atoms with $\delta q = 0.75$ e to each S atom. This transfer results in a very large fractional ionic character of 23.7% [13]. The cohesive energy of the 1T phase at $E_c = -6.80$ eV/atom [11] derived from pseudo potentials as implemented in the Vienna ab initio simulation package (VASP) code, modified to include the kinetic energy density required for meta-GGA (MGGA) calculations, is lower than $E_c = -6.12$ eV/atom obtained by the LDA approach [12]. The much lower binding strength of the metastable 2H phase is $E_c = -5.17$ eV/atom [13].

14.1.2.2 2D Structures

For the lattice parameters of 1T-TiS_2 monolayers, first-principles calculations using the strongly constrained and approximately normed (SCAN) MGGA approach give $a = 0.342$ nm as the lattice constant, $d_{Ti\text{-}S} = 0.242$ nm as the bond length, $h_{S\text{-}S} = 0.280$ nm as the distance between S atoms, and $\theta_{STiS} = 90.16°$ as the bond angle [11]. These values are partially in good agreement with other calculations presenting $a = 0.340$ nm, $d_{Ti\text{-}S} = 0.243$, $h_{SS} = 0.346$ nm, and $\theta_{STiS} = 89.08°$ [14]. The space group of 1T TiS_2 is *P*-3*m*1. Adding the distance between S atoms and their van der Waals diameter of $d_w = 0.360$ nm provides an estimate of the monolayer thickness of $d_{w+h} = 0.640$ nm. Interestingly, full hydrogenation on one side (1HC) or full hydrogenation on both sides (2HC) has little influence on the lattice constant or bond length. Hydrogenation increases the distance between the S atoms considerably and induces a transformation of the 1T into the 1H structure [15]. For comparison, the corresponding lattice parameters of the double hydrogenated structure are $a = 0.347$ nm, $d_{Ti\text{-}S} = 0.245$ nm, and $h_{SS} = 0.557$ nm. Despite insignificant changes in the lattice parameters, the ground state of TMDCs is tunable by hydrogenation from nonmagnetic to ferromagnetic and from semiconducting to half-metallic behavior, as discussed below.

The electronic nature of bulk and monolayer 1T-TiS_2 lattices is controversial. Recent calculations of the electronic band structure using the DFT-GGA-PBE and GGA-1/2 approaches indicated semiconducting behavior of the monolayer with an indirect band gap of $E_g = 0.78$ eV (GGA-1/2) [16]. In a former study, GGA calculations yielded metallic behavior, whereas the Heyd-Scuseria-Ernzerhof (HSE06) hybrid functional method delivered a band gap of $E_g = 0.62$ eV [17]. Such discrepancies are observed for small differences in the positive and negative indirect band gaps between the conduction and valence bands. Figure 14.3 displays the situation for small positive and negative band gaps.

The valence band (VB) derives from dispersing S-p orbitals with a valence band maximum (VBM) at the Γ point, and the conduction band (CB) derives from empty Ti-d orbitals with a conduction band minimum (CBM) at the M point, indicating an indirect semiconductor. The Fermi level is located between the filled S-derived σ band and the empty Ti-derived d band. Figure 14.4 presents a scheme of the typical band structure of group IVB–VIA monolayers with semiconducting properties [1, 18]. Figure 14.5 depicts the band structure in the case of semimetallic behavior, where the Fermi level coincides with one of the d bands.

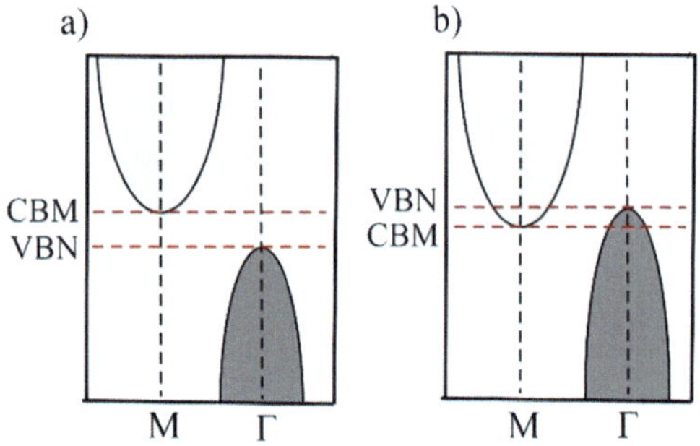

Fig. 14.3 Schematic diagram of **a** a small positive indirect band gap and **b** a small negative indirect band gap

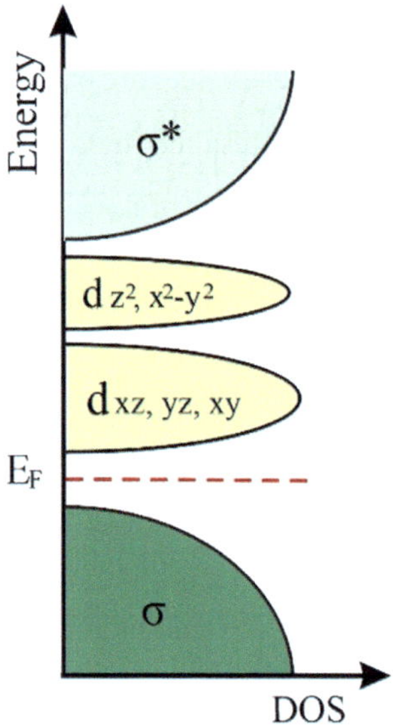

Fig. 14.4 Scheme of the band structure of group IVB–VIA semiconducting monolayers. The σ band of the VB derives from the p orbitals of S atoms, and the CB derives from the empty d orbitals of Ti atoms. The Fermi level is located between the filled σ band and the empty d bands. The filled and unfilled states are shaded with dark and light color, respectively. Reproduced with permission from Ref. [1], Copyright 2013, Springer Nature

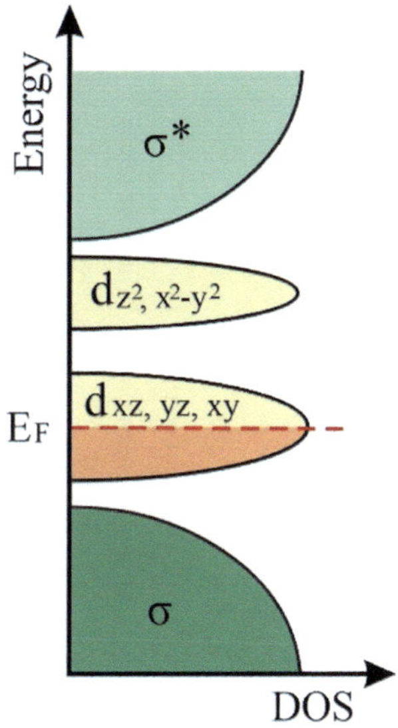

Fig. 14.5 Scheme of the band structure of group IVB–VIA semimetallic monolayers. The Fermi level is located within one of the Ti *d*-bands. The filled and unfilled states are shaded with dark and light color, respectively. Adapted with permission from Ref. [1], Copyright 2013, Springer Nature

14.1.2.3 Mechanics

As linear elastic properties of 1T-TiS_2 monolayers a Young's modulus of 85.2 N/m and in-plane and out-of-plane Poisson ratios of $\nu = 0.17$ and $\nu = 0.42$ were derived from first-principles calculations, which approximated the exchange–correlation energy by the SCAN-MGGA approach [11]. In addition, the authors attributed the bending stiffness of 1T TiS_2 a low value at 2.25 eV. The sparsely filled *d* bands of the d^0 compounds explain this low bending stiffness of 1T TiS_2 and of the other group–IVB compounds. An early study obtained a lower stiffness of only 76.3 N/m [12]. The symmetry, Young's modulus, Poisson's ratio, ultimate strength, and ultimate strain of group IVB–VIA monolayers are presented in Table 14.2.

14.1.2.4 Physics and Applications

A theoretical and experimental study combining photoemission spectroscopy (PES), Raman spectroscopy, and transport measurements investigated the effects of changes in chemical composition caused by Ti interstitials, Frenkel pairs, and oxidation on the corresponding transport properties of defective 1T-TiS_2 single crystals [19]. The analysis showed that bulk 1T TiS_2 is a self-doped semiconductor with a Fermi level close to the conduction band and not a semimetal. The low-temperature resistivity upturn behavior that is typical for doped semiconductors provides conclusive experimental evidence regarding the semiconducting nature of 1T-TiS_2 monolayers. The Ti interstitials and partial oxidation contribute to the high carrier concentration, whereas

Table 14.2 Symmetry, Young's modulus, Poisson's ratio, ultimate strength, and ultimate strain of group IVB–VIA monolayers

Monolayer	Symmetry	Young's modulus E_{ac}, E_{zz} (N/m)	Poisson 's ratio ν_{ac}, ν_{zz}, (ν_{out})	Ultimate strength σ_{ac}, σ_{zz} (N/m)	Ultimate strain ε_{ac}, ε_{zz}
TiS_2	1T phase	85.2 [11], 76.3 [12]	0.17 [11], (0.42) [11]		
$TiSe_2$	1T phase	59.7 [11], 63.9 [12]	0.23 [11], (0.43) [11]		
$TiTe_2$	1T phase	44.5 [11], 41 [12]	0.24 [11], (0.38) [11]		
ZrO_2	1T phase	209.6 [52]	0.24 [52]		
ZrS_2	1H phase	72, 68 [58]	0.41, 0.36 [58]	9.2, 6.6 [58]	0.19, 0.20 [58]
$ZrSe_2$	1H phase	62, 58 [58]	0.40, 0.34 [58]	8.3, 5.5 [58]	0.21, 0.22 [58]
$ZrTe_2$	1T phase	43.2 [11]	0.18, (0.44) [11]		
HfO_2		165.5 [90]	0.231 [90]		
HfS_2	1T phase	79.9 [96], 85.8 [11]	0.19 [96], 0.19, 0.52 [11]	8.3, 7.2 [96]	0.19 [96]
	2H phase	80, 76 [95]	0.40, 0.34 [95]	10.0, 7.6 [95]	0.21, 0.22 [95]
$HfSe_2$	1T Phase	77.8 [11]	0.21, (0.47) [11]		
	2H phase	69, 65 [95]	0.40, 0.35 [95]	9.0, 6.5 [95]	0.20, 0.21 [95]
$HfTe_2$	1T phase	46.8 [11]	0.15, (0.41) [11]		

Frenkel pairs acting as acceptors are responsible for reduced carrier concentrations and electrical conductivity.

One reason for the present interest in hydrogenated TMDCs is their magnetic behavior [15]. The DFT-GGA-PBE-based exchange energies predict that nonmagnetic TiS_2, $TiSe_2$, and $TiTe_2$ monolayers transform to ferromagnetic states and then back to nonmagnetic states with hydrogenation from 0 to 1HC and 2HC [15]. The partial density of states (PDOS) predicts that the nonmagnetic states are metallic or semiconducting, whereas ferromagnetic TiX_2-1HC (X = S, Se, and Te) monolayers are semiconductors. In these monolayers, the *d* electrons of the Ti atoms are spin polarized with magnetic moments of 0.46, 0.62, and 0.71 μ_B/Ti in 1T-TiS_2, 1T-$TiSe_2$, and 1T-$TiTe_2$ monolayers, respectively. The *p* electrons from the X atoms are only weakly spin polarized with smaller magnetic moments, which are antiparallel to the magnetic moments of the Ti atoms.

DFT calculations provided detailed information on the effects of functionalization and vacancy defects on the structural, electronic, and optical response of the TiS_2 monolayer [20]. This includes functionalization at the anion (S) side, such as substitution with P, N, or C dopants, and incorporation of single sulfur vacancy defects. In addition to the density of states (DOS), work-function, and optical absorption spectra, the study provides results on the hydrogen evolution reaction (HER) and the oxygen evolution reaction (OER) of functionalized and defective TiO_2 monolayers. The developed improved understanding of the catalytic activity is based on the reaction coordinates constructed from the adsorption free energies of the intermediates H*, O*, OH*, and OOH*.

A systematic study of 33 hydrogenated TMDC monolayers in the 2H phase revealed that these monolayers effectively catalyze the HER over a wide range of hydrogen coverages with a neutral thermal Gibbs free energy and a catalytic ability comparable to that of Pt [15]. The catalytic activity of 2H-TiS_2 monolayers is much better than that of the other hydrogenated MX_2 (M = Ti, Zr, and Hf; X = S, Se, and Te) compounds, which exhibit a catalytic activity only at certain hydrogen coverages, as revealed by first-principles calculations.

The search for abundant, cheap, and nontoxic ingredients for rechargeable metal-ion batteries inspired a theoretical study on the adsorption and diffusion properties of monovalent Li, Na, and K ions, as well as multivalent Be, Mg, Ca, and Al metal ions on 1T structured TiS_2 [17]. The total formation energy for ion adsorption and a Bader charge analysis revealed doubling of the charge storage density for divalent Mg and Ca ions compared with that for monovalent Li, Na, and K ions, whereas Be and Al ions form metallic clusters. The lowest average open-circuit voltage of 0.13 V emerged for the adsorption of Mg ions. The activation energy barriers for the diffusion of metal ions decrease from Li to K and from Be to Ca, and both K and Na exhibit a greater rate capability than Li. Replacement of Li by Mg leads to high energy density storage and high cell voltage, and in large stationary batteries replacement by Na is possible.

Simulations predict an enhanced catalytic performance of lithium-sulfur batteries by anchoring lithium polysulfides on TiS_2 monolayers to avoid degradation owing to dissolution of the lithium polysulfides in the electrolyte [21]. The TiS_2 monolayers provide appropriate adsorption energies between – 1.38 eV and – 3.54 eV for lithium polysulfides and ensure excellent battery activity during the charge and discharge processes due to their metallic behavior. Moreover, the surface of TiS_2 monolayers exhibits a lower Li_2S decomposition energy and Li-ion migration barrier.

First-principles calculations together with the Boltzmann transport theory elucidated the effect of biaxial strain on the phonon and electron thermal transport properties of 1T-TiS_2 monolayers [22]. With increasing tensile strain, the band gap transforms through indirect-to-direct-to-indirect transitions, increasing the Seebeck coefficient and power factor while decreasing the lattice thermal conductivity. Importantly, 8% tensile strain results in a much larger figure of merit of $ZT = 0.95$ for p-type doping than that obtained without applying strain.

The formation of charge-doping-induced charge density waves (CDWs) is an important phenomenon that has attracted increasing amounts of attention. First-principles calculations show that by proper electron doping, the (2×2) CDW instability can be avoided, and larger electron doping generates a $(\sqrt{7} \times \sqrt{3})$ CDW phase that involves spontaneous breaking of the hexagonal symmetry [23]. The stable doped 1T-TiS monolayer shows phonon-mediated superconductivity that can be further enhanced by strain. Combining doping with strain allows efficient modulation of phonon softening and electron–phonon coupling and leads to enhanced superconductivity.

When alkali atoms are vapor deposited on the surface of TMDCs, they function as strong electron donors and donate their valence electrons to the *d* band of the transition metal. Accordingly, alkali metal doping allows filling of the *d* band. Angle-resolved photoemission spectroscopy (ARPES) experiments revealed this process of forming an electron pocket around the M point [18]. The observed closing of the CDW gap between the σ-S band and the 3*d*-Ti states indicates the suppression of the CDW state by the filling process. Occupation of the 3*d*-Ti bands have also been associated with a transition to the superconducting state in bulk materials.

14.1.2.5 Synthesis

To date, molecular beam epitaxy (MBE) experiments are still lacking for TiS_2 monolayers compared with those performed for the other two Ti chalcogenides. This difference is likely related to the challenges in fabricating suitable and inexpensive sulfur sources for the MBE process [18].

The optimization of electrochemical lithium intercalation and exfoliation methods permitted the growth of single-layer TiS_2 (TaS_2) nanosheets [24]. Large-area AFM images of single-layer nanosheets deposited on a Si/SiO_2 substrate showed an average thickness of ~ 0.9 nm, clearly indicating the realization of monolayers. Notably, the growth of Pt and Au nanoparticles on these single-layer TiS_2 nanosheets provided good electrocatalytic activity in the HER process.

The growth of highly crystalline hexagonal TiS_2 nanosheets occurs at the gas/liquid interface between $TiCl_4$ vapor and Na_2S dissolved in equimolar NaCl–KCl melts at temperatures ranging from 700 °C to 850 °C [25]. The isolated TiS_2 sheets had a size of 10–20 μm and large thicknesses between 50 and 200 nm. Ultrasonic exfoliation enabled the extraction of few-layer nanosheets with sizes ranging from hundreds to thousands of nanometers and thickness values ranging from 3 to 5 nm.

The ease of environmental oxidation and the difficulties encountered in the determination of the exact number of layers in nanosheets limit the application of TiS_2 monolayers, which is the lightest TMDC. Layer-number-dependent Raman modes allow the identification of layer numbers between 1 and 5 for exfoliated few-layer assemblies [26]. Furthermore, a method is described that allows the production of oxidation-resistant inks of mono- and few-layered TiS_2 nanosheets at concentrations up to 1 g/ml that can be used to deposit thin films with controllable thickness over an area of square centimeters.

14.1.3 $TiSe_2$ Monolayer

14.1.3.1 Chemistry and 2D Bonding

The electronic configuration of Ti atoms (Ar|$3d^2 4s^2$) and the similarity of S with that of the Se atoms (Ar|$3d^{10} 4s^2 p^4$) suggests a similar 2D-bonding behavior of $TiSe_2$ and TiS_2 monolayers. The 1T-$TiSe_2$ monolayer, which consists of three sublayers with a titanium sublayer sandwiched between two Se sublayers, is the most stable 2D structure. The significant difference between the electronegativity of the Ti atom (1.54) and the Se atom (2.55) causes a polarization of the Ti–Se bonds by a charge transfer of $\delta q = 1.39$ e from the 3*d* states of each Ti atom to the 4*p* states of two Se atoms. This transfer process causes a significant fractional ionic character of 22.5% [13]. The valence and conduction bands derive predominantly from the Se-4*p* and Ti-3*d* states [18]. The relative positions of these bands determine the nature of the band gap of $TiSe_2$, which results in the generation of a semiconductor or semimetal. The cohesive energy of 1T-$TiSe_2$ monolayers at $E_c = -6.17$ eV/atom [11], based on first-principles calculations using the SCAN meta-GGA approach, indicates substantially stronger binding than the cohesive energy of − 5.64 eV/atom derived by the LDA approach [12]. For the metastable 2H-$TiSe_2$ phase, the binding energy resulting from $E_c = -4.68$ eV/atom is substantially smaller [13].

14.1.3.2 2D Structures

For the lattice parameters of 1T-$TiSe_2$ monolayers, the SCAN-MGGA calculations give a = 0.355 nm, $d_{\text{Ti-Se}} = 0.255$ nm, $h_{\text{SeSe}} = 0.304$ nm, and $\theta_{\text{SeTiSe}} = 91.76°$ [11]. The corresponding values based on the LDA approach are a = 0.343 nm, $d_{\text{Ti-Se}} = 0.251$ nm, $h_{\text{SeSe}} = 0.368$ nm, and $\theta_{\text{SeTiSe}} = 94.04°$ [12]. A monolayer thickness of $d_{w+h} = 0.684$ nm can be estimated by adding the distance between Se atoms and their van der Waals diameter of $d_w = 0.380$ nm. The 1T-$TiSe_2$ monolayer is a low-band gap semiconductor with little or no overlap of the bands (see Fig. 14.3). The electronic nature of bulk and monolayer $TiSe_2$ materials is controversial. Such problems appear when the difference in the positive and negative indirect band gaps is small.

As described already for the band gap of TiS_2, the gaps of $TiSe_2$ and $TiTe_2$ consist of a VB resulting from dispersing chalcogen *p* orbitals with a valence band maximum (VBM) at the Γ point, and the CB derives from the empty *d* orbitals of the transition metal with the CBM at the M point, forming a low-gap semiconductor or semimetal (see Figs. 14.4 and 14.5) [18]. In addition to the realized coordination, the electronic structure of TMDCs depends on the number of *d* electrons. From an orbital perspective, octahedrally coordinated transition metals exhibit degenerate d_{z^2}, $d_{x^2-y^2}$ and d_{xy}, d_{xz}, d_{yz} orbitals, and trigonal prismatic coordination causes splitting of the orbitals at the d_{z^2}, $d_{x^2-y^2}$ and d_{xy}, d_{xz}, d_{yz} energy levels [18].

Surprisingly, the normal (1×1) structure of $TiSe_2$ with three sublayers stacked in the order Se–Ti–Se and a bond length of d = 0.255 nm in octahedral prismatic coordination is stable only at higher temperatures. Measurements revealed that monolayers of titanium diselenide undergo a charge density wave (CDW) transition at the critical temperature of $T_{CDW} = 232$ K [27]. Angle-resolved photoemission spectroscopy (ARPES) measurements revealed a small absolute band gap at room temperature that increases below T_{CDW} in connection with the emergence of a (2×2) ordering of the CDW phase. Since HSE calculations indicate that the (2×2) phase is the ground state at T = 0 K, the authors interpret CDW ordering as a band structure effect based on energy minimization. The band gap defined by the top of the Se-4*p* band and the bottom of the 3*d*-Ti band is 98 meV at room temperature. The CDW ordering induces an opening of the band gap at the Fermi energy, which is 330 meV for the optimized structure at T = 0 K.

In a series of studies, several authors tried to identify the origin of the still poorly understood CDW ordering effect by attempting to explain the instability of the monolayer toward commensurate periodic lattice distortion (PLD) at low temperatures. Notably, in recent years, 1T-$TiSe_2$ monolayers have provided a platform for studying in detail the instabilities responsible for the transition to a symmetry-lowering CDW state, owing to the robust nature of ordering in this lattice.

Ab initio electronic structure calculations provided information on the PLDs and CDWs of the 1T-$TiSe_2$ monolayers [28]. Treating the exchange-correlation effects by the GGA approximation resulted in the best agreement with experimental data. As a normal phase, the calculations confirm the (1×1) structure of the three-layer Se–Ti–Se arrangement of the 1T-$TiSe_2$ phase. The Se-4*p*-based valence bands belong to the Γ point, and the Ti-3*d*-derived conduction bands are located at the M point [18]. Since they overlap, normal $TiSe_2$ monolayers are semimetals. The indirect gap, defined as the difference between the minimum of the conduction band and the maximum of the valence band, is $E_g = -0.446$ eV (GGA) and -0.242 eV (HSE). Below a certain threshold temperature, the soft mode has an imaginary frequency, and the normal structure is no longer stable. Thus, below this threshold temperature, the equilibrium configuration with lower energy is the distorted (2×2) superstructure. The phonon spectrum reflects the dynamical stability of this phase at T = 0 K. The results of first-principles calculations predict an insulating electronic state for the distorted phase with band gaps of 82 meV (GGA) and 325 meV (HSE). An interesting remaining mechanistic question is, to what extent is the CDW driven by electronic instability and by strong electron–phonon coupling?

Both bulk and monolayer 1T $TiSe_2$ are targets of intense research to clarify the nature of CDWs. Experimental evidence indicates the realization of a 2D-CDW phase without Fermi surface nesting in monolayer 1T $TiSe_2$ [29]. The ARPES analysis suggested the presence of an electron pocket at the Brillouin zone corner above the CDW transition temperature and an additional electron pocket and replica bands at the Brillouin zone center and corner below the T_{CDW} owing to back-folding of bands by the (2 × 2) superstructure potential. As a common driving force for CDW formation, the authors suggest exciton condensation, and a larger energy gap below the T_{CDW} enhances electron–hole coupling upon reducing dimensionality.

Analysis of the CDW state in 1T-$TiSe_2$ monolayers within the framework of DFT confirmed that the monolayer is unstable against PLD and CDW at low temperatures, while the undistorted and semimetallic states are stable only above a certain threshold temperature [28]. The CDW lattice distortions agree with the experimental data, including the unfolded and reconstructed band structures. This includes evidence for the role played by electron–electron interactions in the CDW instability. Below the transition temperature, a strong hybridization between the 2D plasmon and CDW excitations occurs [30]. This indicates the existence of a highly tunable plasmonic material with unconventional optical features in 2D semimetallic and metallic TMDCs, induced by charge-ordered states.

Temperature-dependent ARPES measurements of monolayer 1T-$TiSe_2$ samples, grown by MBE on graphene-terminated SiC substrates, revealed the characteristic signatures of (2 × 2) CDWs, such as strong coupling instabilities, where not all states coupled equally well to symmetry-breaking distortions with an electron pocket persisting to low temperatures as a nonbonding state [31]. At room temperature, the system has a band gap of 80 meV, which opens to 180 meV at low temperatures. A tight-binding model describes the low-temperature dispersion by including extra *d-p* hybridization, introduced by the lattice displacements of the ordering process. Interestingly, the introduction of sufficient Se vacancies suppresses the CDW ordering process. Complete suppression occurs at the critical doping level of 0.06 electrons per Ti atom. Suppression of CDW ordering in 1T $TiSe_2$ by intercalation, electron doping, or pressure leads to superconductivity as another fundamental quantum phenomenon. This work improves the understanding of the phenomenology, the nature and origin of this instability, however, remain controversial.

A recent theoretical approach used density functional perturbation theory (DFPT) to simulate phonon dynamics and to account for electron correlations, corrected with on-site Coulomb (Hubbard) interactions, to elucidate the controversial CDW ordering in 1T $TiSe_2$ [32]. To determine the accurate electronic structure, this work emphasizes the crucial role of electron correlations. The total phase diagram consists of commensurate and incommensurate CDW regions, where the latter coincide with the superconductive phase. In addition to the established roles of quantum lattice fluctuations and excitonic interactions, this work underlines the crucial role of momentum-dependent electron–phonon coupling and electron correlations.

The excitonic insulator transition model assumes that the exciton binding energy is related to the order parameter of the CDW phase transition and that the transition temperature is related to the exciton binding energy [33]. In TMDCs, the exciton binding energy increases markedly with the decreasing number of layers of a multilayer sheet. The increased exciton binding energy of the monolayer results in a larger CDW gap and larger transition temperature of the monolayer compared to those of the bulk. The exciton binding energy is tunable by the substrate material because it depends on the dielectric properties of the surrounding material. The authors tentatively attributed the pronounced peaks observed in the differential conductance of $TiSe_2$ monolayers to the formation of a coherent phase state in an excitonic condensate.

14.1.3.3 Mechanics

As linear elastic properties of $TiSe_2$, a Young's modulus of 59.7 N/m, and in-plane and out-of-plane Poisson's ratios of 0.23 and 0.43, respectively, were derived from first-principles calculations, where the exchange-correlation energy was approximated by the SCAN-MGGA approximation [11]. An early study reported a slightly greater stiffness of 63.9 N/m [12]. In addition, the authors determined as bending stiffness of the 1T-$TiSe_2$ monolayer 2.86 eV.

14.1.3.4 Physics and Applications

The combination of DFT calculations and Boltzmann transport theory revealed the effects of biaxial strain on the electronic and transport properties of 1T-$TiSe_2$ monolayers [34]. Different approaches using PBE, LDA + U, and HSE06 as exchange correlation functionals delivered a comparison between the bulk phase and the monolayer. Based on the more reliable HSE06 approach, the electronic band gap of the semimetallic material varies from zero to 120 meV and 390 meV by increasing the tensile strain from 0 to 2% and 4%, respectively. The strain-induced metal-to-insulator transition induces nonlinear changes in the electronic and thermoelectric properties. For example, every 2% of applied biaxial tensile strain decreased the electronic density by a factor of ten. This qualifies the material as a potential strain-controlled opto/electronic switch.

First-principles calculations elucidated the Li^+ storage properties of monolayer and bulk $TiSe_2$ in lithium-ion batteries [35]. In bulk $TiSe_2$, the intercalation mechanism results in a theoretical capacity of 260 mA h g^{-1} in the voltage window of 1.14–2.09 V, whereas the more efficient adsorption mechanism of the monolayer reaches a theoretical capacity of 780 mA h g^{-1} in a window of 0.18–1.43 V and fast diffusion kinetics. In both forms, Se functions as the main redox species in the initial stage, and then Ti and Se participate in the two-stage redox process. Furthermore, the materials exhibited low Li^+ diffusion barriers, e.g., 163 meV at high Li^+ concentrations and 39 meV at low Li^+ concentrations.

The properties of 2D materials are sensitive to defects and therefore defect engineering is an important strategy to improve catalytic activity. Experimental and theoretical results confirmed the existence of line defects in $TiSe_2$ monolayers induced by selenium vacancy defects [36]. The Gibbs free energy (ideally $\Delta G_{H^*} \approx 0$), which indicates the efficiency of H atom adsorption on the catalyst for the HER, were $\Delta G_{H^*} = -0.640$ eV for the pristine $TiSe_2$ monolayer on Au(111), $\Delta G_{H^*} = -0.107$ eV for the adsorbed $TiSe_2$ monolayer with a line defect, and $\Delta G_{H^*} = -0.073$ eV when the Se-deficient line defects were fully covered by Pt adatoms. For the Pt metal surface, the Gibbs free energy is $\Delta G_{H^*} = -0.090$ eV.

Additional chemical degrees of freedom are opened by extending the binary TMDCs to the ternary Janus systems TiXY (X, Y = S, Se, and Te; $X \neq Y$) [37]. A first-principles study determined the structural, mechanical, electronic, thermal, and optical properties of the common semimetallic 1T phase, which has a negative band gap, and the 2H phase, which is a semiconductor with an indirect band gap due to broken symmetry. For 1T TiSSe, a CDW transition occurs owing to Peierls instability, while for other metallic systems, such a CDW transition is missing. The reason for this behavior is that an increase in the elastic energy counteracts the decrease in energy by band gap opening. Note that the stiffness of 1T-TiSSe monolayers of 60.9 N/m nearly agrees with the stiffness of $TiSe_2$ at 59.7 N/m [37].

The growth of heterostructures of 1T-$TiTe_2$/1T-$TiSe_2$ with various twist angles is possible by MBE on bilayer-graphene SiC(0001) substrates [38]. The lattice mismatch produced by the long-wavelength periodic potential arising from the interlayer atomic registries introduces a Moiré pattern. Scanning tunneling microscopy (STM) allows the study of Moiré pattern-induced/enhanced CDW states and the resulting tessellation of alternate CDW and normal-state domains. Surprisingly, at twist angles near zero degrees, (2×2) CDW domains form in 1T $TiTe_2$, separated by normal (1×1) domains trapped in the Moiré pattern. Remarkably, these CDW states, ascribed to local strain variation and induced by atomic reconstructions, survived at room temperature despite the lower critical temperatures of the two constituents.

The condensation of excitons generated by electron-hole pairs with sufficiently reduced Coulomb screening relates to exotic physics such as superfluidity and insulating states [39]. Since direct detection of these effects is difficult, studying the interactions of excitons with electrons may replace the difficult direct detection of excitons. In fact, this was possibly due to their effect on the electron energy band of $TiSe_2$. Application of high-resolution scanning tunneling spectroscopy (HRSTS) and quasiparticle interference (QPI) revealed for the 1T-$TiSe_2$ bilayer, which is the 2D limit of the unconventional bulk $(2 \times 2 \times 2)$ CDW state, an extra state below the conduction band within the CDW gap region. Many-body calculations support the explanation that this extra state arises from the coupling of electrons with an excitonic mode.

14.1.3.5 Synthesis

Hexagonal $TiSe_2$ monolayers were accessible by MBE in an ultrahigh vacuum (UHV) chamber equipped with an integrated scanning tunneling microscopy (STM) system on a bilayer-graphene-terminated nitrogen-doped SiC(0001) wafer [40]. A Ta boat and a standard Knudsen diffusion cell evaporated the two high-purity source materials of Ti and Se. The step height of the layer-by-layer-grown sheets was 0.60 nm, and the in-plane lattice constant $a = 0.35$ nm, which agrees with that of a (0001)-oriented $TiSe_2$ structure. The concentration of Se vacancies and interstitial defects identified by STM measurements depended critically on the substrate temperature. The STM analysis confirmed the existence of CDWs in these monolayers as the correlation length gradually decreased with increasing density of defects.

In addition to the deposition of $TiSe_2$ monolayers by MBE under UHV conditions, monolayers and few-layer sheets could be grown by salt-assisted chemical vapor deposition (CVD) under atmospheric pressure employing a mixture of TiO_x and LiCl powder and Se powder as precursors at a growth temperature of ~ 600 °C [41]. AFM measurements performed under ambient conditions yielded a typical thickness of ~ 1.4 nm, which was assigned to monolayers contaminated by small, oxidized particles. High-angle annular dark-field scanning transmission electron microscopy (HAADF-STEM) and energy-dispersive X-ray spectrometry (EDX) analyses characterized the crystallographic structure and chemical composition of the as-grown samples. Low-temperature Raman spectroscopy revealed a thickness-dependent CDW phase transition at $T_{CDW} = 212.5$ K in few-layer $TiSe_2$ sheets, demonstrating the high quality of the as-grown crystals.

The synthesis of $TiSe_2$ monolayers and sheets by various methods is an active field of research, including the extension of the chemical vapor transport (CVT) technique from bulk to surface growth by introducing suitable growth substrates and by dramatically reducing the growth rate [42]. Remarkably, the deposited sub-10 nm flakes of 2D $TiSe_2$ were of comparable quality to the mechanically exfoliated flakes.

To achieve chemical exfoliation of $TiSe_2$ platelets with a thickness of 15–55 nm by the CVT method, the authors used ortho-dichlorobenzene functionalized with oleylamine (OLA) [43]. The addition of OLA supports the formation of a stable dispersion of large-area $TiSe_2$ sheets due to surface capping with OLA molecules. The X-ray diffraction (XRD) pattern confirmed the hexagonal structure and the $P\text{-}3m1$ space group of the $TiSe_2$ platelets.

14.1.4 $TiTe_2$ Monolayer

14.1.4.1 Chemistry and 2D Bonding

The electronic configuration of Ti atoms ($\text{Ar}|3d^2 4s^2$) and Te atoms ($\text{Kr}|4d^{10} 5s^2 p^4$) controls 2D covalent bonding in titanium ditelluride monolayers. The 1T-$TiTe_2$ monolayer consists, as discussed for 1T-$TiSe_2$, of three sublayers with a titanium sublayer sandwiched between two Te sublayers, as the most stable structure. The difference between the electronegativity of Ti (1.32) and Te (2.01) induces a charge transfer of $\delta q = 1.23$ e from the 3*d* states of each Ti atom to the 5*p* states of two Te atoms, introducing a fractional ionic character of 7.54% in 2H $TiTe_2$ [13]. For the 1T-$TiTe_2$ phase a similar ionicity can be expected. This charge transfer, and especially the larger 5*p*-Te orbital, are the origin of the greater overlap of the Ti-*d*-derived conduction band and the Te-*p*-derived valence band compared to that of the $TiSe_2$ monolayer. The calculated cohesive energies of $E_c = -5.41$ eV/atom [11] (SCAN-MGGA) and -5.03 eV/atom [12] (LDA) represent the weakest binding energies in this group of Ti compounds. However, the bonding strength is greater than that of the metastable H phase of $TiTe_2$ at $E_c = -4.92$ eV/atom [12] (LDA) and -4.12 eV/atom [13] (PBE).

14.1.4.2 2D Structures

For the lattice parameters of the 1T-$TiTe_2$ monolayers, the SCAN-MGGA calculations give a = 0.376 nm as the lattice constant, $d_{Ti\text{-}Te} = 0.277$ nm as the bond length, $h_{TeTe} = 0.344$ nm as the distance between Te atoms, and $\theta_{TeTiTe} = 94.55°$ as the bond angle [11]. The corresponding values based on the LDA approach are a = 0.374 nm, $d_{Ti\text{-}Te} = 0.278$ nm, $h_{TeTe} = 0.410$ nm, and $\theta_{TeTiTe} = 84.71°$ [14]. A monolayer thickness of $d_{w+h} = 0.764$ nm can be estimated by adding the distance between Te atoms and their van der Waals diameter of $d_w = 0.420$ nm. In the normal elevated temperature state, 1T $TiTe_2$ is a semimetal with the σ-Te band at the Γ point overlapping with the 3*d*-Ti band at the M point.

A theoretical study, investigating the $TiTe_2$ structure from the single-layer to the bulk limit, revealed the emergence of (2 × 2) CDW ordering with a transition temperature of $T_{CDW} = 92$ K only in the monolayer phase but not in the bilayer or multilayer phase according to ARPES and STM measurements, and scanning tunneling spectroscopy (STS) [44]. The authors exclude an excitonic mechanism in $TiTe_2$ as an explanation of this observation and favor the band-folding model, giving rise to gap opening at the intersection point between the conduction and folded valence band. Ideally, when gap opening occurs at the Fermi level, the valence states nearby will move to lower energies and thus reduce the total electronic energy.

It has been confirmed that 1T-$TiTe_2$ monolayers have a semimetallic character in the normal state [45]. DFT calculations of the lattice dynamics of few-layer and bulk 1T $TiTe_2$ indicate the absence of imaginary frequency modes and thus confirm dynamical stability at 0 K. Importantly, the monolayer undergoes a (2 × 2) lattice distortion at ≈ 110 K [45]. The lower critical temperature than that observed for 1T $TiSe_2$ indicates an intrinsically weaker instability. Temperature-dependent ARPES measurements and detailed model calculations predict a significant energy gain from orbital-selective band hybridization between the backfolded valence and conduction bands arising from a (2 × 2) lattice distortion. These factors suggest that electron–phonon coupling drives the CDW transition in the 1T-$TiTe_2$ phase. This model of low-energy electronic structural evolution explains why CDW ordering occurs only in the monolayer and is absent in bilayers and multilayers.

The stochastic self-consistent harmonic approximation (SSCHA) provides a mechanism for the different behaviors of suspended single-layer and bulk 1T $TiTe_2$ by analyzing the forces on supercells with atoms displaced from their equilibrium positions [46]. As demonstrated by nonperturbative anharmonic calculations, the CDW structure forms solely in single-layer $TiTe_2$ owing to an interplay of nonperturbative anharmonicity and an exchange enhancement of the electron–phonon interaction, generating more unstable harmonic phonon modes in the isolated monolayer. The single layer exhibits a larger instability at the harmonic level compared to that in the bulk.

14.1.4.3 Mechanics

For the linear elastic properties of 1T-$TiTe_2$ monolayers a Young's modulus at E = 44.5 N/m and in-plane and out-of-plane Poisson's ratios at $\nu = 0.24$ and $\nu = 0.38$, respectively, are available from first-principles calculations, where the exchange–correlation energy was approximated by the SCAN-MGGA approach [11]. The authors determined 3.29 eV as the bending stiffness of 1T-$TiTe_2$ monolayers.

Surprisingly, the stiffness of the Janus structure 1T TiSeTe with its Young's modulus of E = 42.5 N/m is even somewhat smaller than that of 1T $TiTe_2$, and the Poisson ratio is $\nu = 0.33$ [37]. The stiffness is in good agreement with the result of an independent calculation presenting E = 41.0 N/m [12].

14.1.4.4 Physics and Applications

Multilevel high-throughput calculations and machine learning techniques based on first-principles methods allowed the identification of metallic TMDCs with high performance as electrocatalysts in the HER process using criteria such as zero band gap, highest thermodynamic stability among available phases, low vacancy formation energy, and nearly zero hydrogen adsorption energy [47]. By a systematic high-throughput screening procedure, adsorption-centered neighboring atoms were identified as effective catalytic structures, and chalcogenide ion vacancies (X vacancies) in $TiTe_2$ or VSe_2 lead to a catalytic activity comparable to that of Pt(111). Importantly, the active electrons induced by lattice defects concentrate in the nearest transition metal atom and the nearest or next-nearest chalcogen atom rather than at a single-atom site.

14.1.4.5 Synthesis

The MBE technique allowed the deposition of $TiTe_2$ monolayers on bilayer-graphene-terminated SiC(0001) substrates [44]. In this case, the growth process is orientationally aligned with the substrate's in-plane crystallographic directions. In situ reflection high-energy electron diffraction (RHEED) measurements revealed a high-quality, well-ordered, and unstrained monolayer with a lattice constant of a = 0.378 nm, which agrees well with the bulk value of a = 0.3777 nm; however, this monolayer was incommensurate with the graphene lattice constant of a = 0.246 nm. Surprisingly, ARPES and STM/STS analyses revealed a (2 × 2) lattice distortion of the monolayer but not of the bilayer or multilayer assemblies. Consequently, $TiTe_2$ is a unique case in which the 2D limit is a singular point with unexpected physical properties that are to be explained.

MBE allows the growth of 1T-$TiTe_2$ monolayers on single-crystal van der Waals substrates such as MoS_2 and highly oriented pyrolytic graphite (HOPG) [48]. A mini electron-beam evaporator and a Knudsen effusion cell produced the atomic beams of Ti and Te, respectively. Analysis of the monolayer by STM and low-energy electron diffraction (LEED) confirmed the (1 × 1) structure and the (2 × 2) lattice distortion at 20 K. In contrast to the well-ordered monolayer, second-layer islands showed contrast variations on the atomic scale with some short-range order but no long-range periodicity. The STM data are consistent with the presence of intercalated Ti atoms in the multilayers. Computations indicate that such distortions by intercalation are much less common in more stable $TiTe_2$ than in V and Cr tellurides.

MBE deposition on bilayer graphene-terminated SiC wafers allowed the controlled growth of monolayers of $TiTe_{2x}Se_{2(1-x)}$ alloys over the entire compositional range [49]. Codeposition of Ti, Se, and Te induced the growth process, by applying a highly chalcogen rich flux introduced into the deposition chamber. The alloy composition was determined by XPS and its lattice constant by the shift of the RHEED pattern. Examination by ARPES elucidated the occurrence of the (2 × 2) instability through the semiconductor-semimetal transition. Closing the band gap of a semiconductor is a powerful route to control the microscopic mechanisms that drive collective phases like CDWs and excitonic insulator states by engineering to the normal-state band structure. Electronic structure calculations identified the relative strengths of excitonic and Peierls-like coupling processes.

14.2 Zirconium Dichalcogenide Monolayers (ZrO_2, ZrS_2, $ZrSe_2$, and $ZrTe_2$)

14.2.1 ZrO_2 Monolayer

14.2.1.1 Chemistry and 2D Bonding

The electronic configuration of Zr atoms (Kr|$4d^2 5s^2$) and O atoms ($1s^2 2s^2 p^4$) determine the type of 2D bonding in zirconium dioxide monolayers. Besides the valence electrons, which control coordination, the number of *d* electrons determines the electronic structure. The 1T-ZrO_2 monolayer, which consists of three sublayers with a zirconium sublayer sandwiched between two oxygen sublayers, is more stable than the 2H structure, as indicated by the formation energy of $E_f = -3.49$ eV (FERE) for the 1T phase compared with $E_f = -2.96$ eV (FERE) for the 2H phase [2]. Note that metastability of the 1T phase is still possible because calculations on TMDOs are challenging. The large difference in the electronegativity of Zr (1.22) and O (3.50) causes strong polarization of the predominantly *p-d* hybridized covalent Zr–O bonds by an extensive charge transfer of $\delta q = 2.54$ e from the 4*d* states of the Zr atom to the 2*p* states of two O atoms in the 1T-ZrO_2 structure [13]. Thus, in principle, a positively charged zirconium layer (Zr^{+4} cation) interacts with two negatively charged oxygen layers (O^{-2} anions). The strong ionic-covalent bonds have an extreme fractional ionic character of 67.1% and a cohesive energy of $E_c = -7.71$ eV/atom (PBE) that is comparable with that of graphene [13].

14.2.1.2 2D Structures

For the lattice parameters of 1T-ZrO_2 monolayers, DFT-PBE calculations give $a = 0.328$ nm as the lattice constant, $d_{Zr\text{-}O} = 0.212$ nm as the bond length, $h_{OO} = 0.193$ nm as the distance between O atoms, and as the bond angles $\theta_{OZrO} = 79.05$ and $\theta_{ZrOZr} = 100.95$ [13, 50]. The trilayer assembly O–Zr–O can be viewed as a positively charged hexagonal layer of Zr atoms sandwiched between two negatively charged hexagonal layers of O atoms. Adding the distance between O atoms and their van der Waals diameter of $d_w = 0.310$ nm gives an estimate of the monolayer thickness of $d_{w+h} = 0.503$ nm. The indirect gaps of the wide band gap insulator 1T ZrO_2 are $E_g = 4.51$ eV (GGA-PBE) and 7.51 eV (GW), as derived by different approaches [50, 51]. Owing to its large indirect band gap, dielectric permittivity, and mechanical stiffness, ZrO_2 is a promising 2D dielectric material that may be used in field-effect transistors (FETs) and heterojunctions [52]. The intense sharp absorption peaks in the UV spectral region qualify 2D ZrO_2 for the application as a UV detector.

14.2.1.3 Mechanics

The 1T-ZrO_2 monolayer, derived from the bulk structure that is known for its high chemical and thermal stability, has a large Young's modulus of $E = 209.6$ N/m and Poisson's ratio is $\nu = 0.24$ [52]. The elastic constants $c_{11} = 117.71$ GPa and $c_{12} = 28.32$ GPa satisfy Born-Huang's criteria of mechanical stability. Owing to its large elastic modulus and relatively small Poisson ratio, 1T-ZrO_2 monolayers have an enormous in-plane mechanical stability connected with a good out-of-plane flexibility.

14.2.1.4 Physics and Applications

Based on first-principles calculations, the physicochemical properties of the hexagonal 1T-ZrO_2 monolayers were determined. The standard treatment describes the interaction between the atomic core and valence electrons by projector augmented wave (PAW) potentials and the exchange and correlation of the valence electrons by the GGA-PBE approach [52]. These theoretical studies revealed the electronic structure, dielectric permittivity, phonon spectra, thermal properties, and optical response of the ZrO_2 monolayers. For the *p-d* hybridized ionic Zr–O bonds, the effective charges are 6.38 and − 3.19 for Zr and O atoms, respectively, which are larger than the nominal valences of + 4 and − 2, respectively. These values reflect the enormous charge transfer and ionicity mentioned above.

Zirconium dioxide or zirconia plays a significant role as nanomaterial in various applications such as catalyst, sensor, ceramics, and implant. Furthermore, in semiconductor devices, as coatings on cutting tools, in dentistry, and in drug delivery [53]. These wide applications are based on a variety of favorable properties, such as alkaline and acidic properties, chemical and mechanical stability, corrosion resistance, high thermal stability, luminescence, appropriate refractive indices, large dielectric constants, antibacterial, antioxidant, and antifungal properties. Current efforts concentrate on controlling the particle size, monodispersity, morphology, shape, stability, and the main problem of reducing the aggregation of particles. The review provides a critical survey of the methods of chemical synthesis and various applications of zirconia-based nanomaterials.

In a theoretical work, the potential use of ZrO_2 monolayers as cathode material for nonaqueous lithium-oxygen batteries was studied [54]. The authors suggest a peroxide-like adsorption geometry as the most stable configuration of LiO_2 that is different from the typical triangular O–Li–O geometry owing to the critical role played by substrate Zr atoms in stabilizing the LiO_2 cluster. Moreover, the system likely adopts a four-electron mechanism with lower overpotentials of 0.44 and 0.76 eV in the catalytic ORR and OER processes, respectively. Ab initio calculations indicate that the adsorption energy of the Li_2O_2 species is a good descriptor of the overall catalytic activity and that a small adsorption energy favors both a low overpotential and satisfactory catalytic performance.

14.2.1.5 Synthesis

In contrast to the enormous progress made in the synthesis of zirconia nanomaterials, the synthesis of high-quality monolayers is challenging. The fabrication of a 1T-ZrO_2 monolayer or bilayer may be possible by hard-bond cleavage from the (111) facet of nonlayered cubic ZrO_2 [50]. Interestingly, the cleavage to monolayer thickness leads to a reduction in the bond length from $d_{Zr\text{-}O} = 0.220$ nm to 0.212 nm, and of the Zr–O–Zr bond angle from 109.47° to 101.02°.

The bottom-up synthesis via the impregnation of graphene oxide (GO) in cyclohexane containing $Zr(OC_4H_9)_4$ as precursor, followed by a solvothermal reaction, resulted in the deposition of 3.6 nm thick ZrO_2 nanosheets. [55]. The solvothermal reaction transforms GO into reduced rGO, which can be removed by calcination of the composite in air to isolate free-standing ZrO_2 nanosheets. The selected-area electron diffraction (SAED) pattern revealed a polycrystalline nature of the nanosheets, which consisted of tetragonal ZrO_2 crystals.

The successful layer-by-layer growth of 2–10 monolayers thick ZrO_2 nanosheets of zirconia films, using a UHV-compatible sputter source and a Rh(111) substrate, was verified by STM and LEED [56]. While the underlying substrate influenced the structure of the 2–4 monolayer thick sheets, where the thicker sheets had a tetragonal (2 × 1) or distorted monoclinic (2 × 2) surface structure with respect to cubic ZrO_2(111). A transformation to the thermodynamically stable monoclinic phase occurs at 85 °C, where the film breaks, forming holes down to the substrate. The authors argue that oxygen deficiencies stabilize the tetragonal structure, whereas the monoclinic sheets are only weakly defective and exhibit band bending at defects and grain boundaries. The five monolayers thick tetragonal and monoclinic ZrO_2-Rh(111) based nanosheets possess the surface structure of either tetragonal or monoclinic zirconia. They exhibit a better homogeneity and stability than those of the previously known ZrO_2-Pt(111) system.

14.2.2 *ZrS_2 Monolayer*

14.2.2.1 Chemistry and 2D Bonding

The electronic configuration of Zr atoms (Kr|$4d^2 5s^2$) and S atoms (Ne|$3s^2 p^4$) determine the type of 2D bonding in zirconium disulfide monolayers. Strong hybridization of the S-3*p* orbitals with the Zr-4*d* orbitals is responsible for efficient *p*-*d* hybridization. The 1T-ZrS_2 monolayer, which consists of three sublayers with a zirconium sublayer sandwiched between two sulfur sublayers, seems to be more stable than the structurally quite similar H phase. The difference in the electronegativity of Zr (1.33) and S (2.58) causes a polarization of the *p*-*d* hybridized covalent Zr–S bonds by a significant charge transfer of $\delta q = 2.05$ e from the 4*d* states of Zr to the 3*p* states of two S atoms, i.e., $\delta q = 1.02$ e to each S atom [13]. Compared to the contribution of ionic forces in ZrO_2 monolayers, the fractional ionic character of ZrS_2 monolayers is substantially reduced to 32.3%, and the cohesive energy is only $E_c = -5.89$ eV/atom (PBE) for the 1T phase [13]. Note that a surprisingly low cohesive energy of − 7.35 eV/atom was obtained by SCAN-MGGA calculations [11], whereas the lattice parameters reported for the 1T-ZrS_2 monolayer agree well with those presented below.

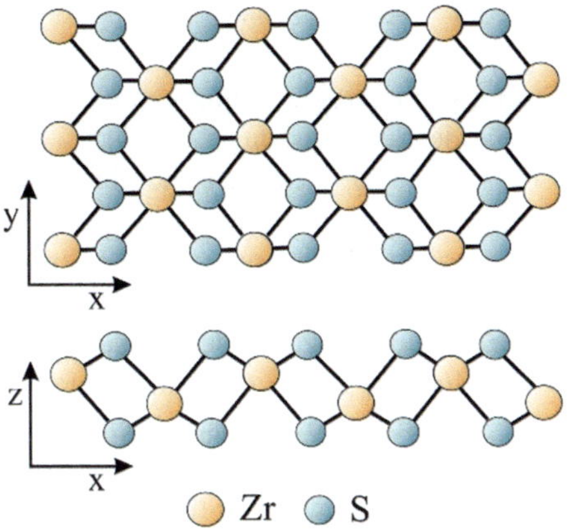

Fig. 14.6 Four-layer structure with the space group *Pmmn*. Starting with nominally three sublayers of the ZrS_2 monolayer, the central Zr atoms are no longer in the same plane but rather form a wavy layer. Adapted from Ref. [60]

14.2.2.2 2D Structures

For the lattice parameters of 1T-ZrS_2 monolayers, DFT-PBE calculations give a = 0.369 nm, $d_{Zr\text{-}S}$ = 0.257 nm, h_{SS} = 0.290 nm, and θ_{SZrS} = 88.59° [13]. These values are in good agreement with a = 0.367 nm, $d_{Zr\text{-}S}$ = 0.257 nm, h_{SS} = 0.287 nm, and θ_{SZrS} = 88.14° based on the SCAN-MGGA approximation [11]. The distance between S atoms and their van der Waals diameter of d_w = 0.360 nm provide an estimate of the monolayer thickness of d_{w+h} = 0.650 nm. The PBE band gap of the semiconducting monolayer is E_g = 1.19 eV (PBE), and the much larger quasiparticle gap is 2.56 eV (QP) [14]. A study of the impact of electron-electron and electron-hole interactions on the optoelectronic properties of ZrS_2 yielded a QP band gap of 2.81 eV (G_0W_0), while SOC induced splitting of the uppermost valence band did not alter the lowest conduction band [57]. Note that for the unstrained single-layer 1H-ZrS_2 lattice, much smaller gap values of E_g = 1.0 eV (PBE), 1.45 eV (HSE06), and 1.65 eV (G_0W_0) have been reported [58].

In addition to the well-established 1T and 1H phases, DFT-GGA-PBE calculations predicted the electronic, mechanical, and optical properties of a metastable orthorhombic structure. The symmetry of this ZrS_2 monolayer is *Pmmn* [59]. As shown in Fig. 14.6, in the monolayer starting with nominally three sublayers, the central Zr atoms are no longer in the same plane but rather form a wavy layer. Surprisingly, besides the bond angles, the lattice parameters of a = 0.367 nm, b = 0.649 nm, $d_{Zr\text{-}S}$ = 0.255 nm, h_{SS} = 0.289 nm, and θ_{SZrS} = 83.29° are nearly identical to those of the 1T structure [60]. The same is true for the cohesive energy of E_c = − 5.78 eV/atom [59]. This phase is a quasi-direct semiconductor with strain-tunable band gaps of E_g = 0.53 eV (PBE) and 1.53 eV (HSE06) [59]. The monolayer is dynamically stable, and ab initio molecular dynamics (MD) simulations indicate thermal stability up to ~ 500 K.

14.2.2.3 Mechanics

Strongly anisotropic Young's moduli of $E_{ac} = 83.6$ N/m and $E_{zz} = 63.6$ N/m and nearly identical Poisson's ratios of $\nu_{ac} = 0.09$ and $\nu_{zz} = 0.07$ have been reported for ZrS_2 monolayers with orthorhombic *Pmmn* symmetry [60]. These values agree only partly with the isotropic in-plane Young's modulus of E = 83.8 N/m, in-plane Poisson ratio of $\nu = 0.19$, and out-of-plane Poisson ratio of $\nu = 0.52$ for the 1T phase [11]. While the large stiffness values of the orthorhombic and the octahedral 1T phases agree, the Poisson ratios deviate substantially. Slightly anisotropic stiffnesses of $E_{ac} = 72$ N/m and $E_{zz} = 68$ N/m and Poisson's ratios of $\nu_{ac} = 0.41$ and $\nu_{zz} = 0.36$ were also obtained by first-principles calculations for the 1H monolayer [58]. As expected, the ultimate strength values of $\sigma_{ac} = 9.2$ N/m and $\sigma_{zz} = 6.6$ N/m show a large anisotropy, whereas the ultimate strains of $\varepsilon_{ac} = 0.19$ and $\varepsilon_{zz} = 0.20$ nearly coincide.

14.2.2.4 Physics and Applications

The electronic, dielectric, and optical properties of 2D ZrS_2 were analyzed within the framework of DFT calculations [61, 62]. For the 1T-ZrS_2 structure, a APW + l_0 analysis in the frame work of DFT indicated that the valence band consists predominantly of the *s*/*p*-states of the S atom and in a lower portion of the *s*/*p*/*d*-states of the Zr atom, while the (metallic) Zr-*d* states dominate the conduction band. The monolayer is extremely sensitive to light absorption due to the triple peaks at the Γ point of the VBM. Parallel band effects explain the intense light-matter interaction with an absorption of 1.06×10^6 cm^{-1} in the energy range of 5–10 eV, while the reflectivity reaches its largest values of 30–50% in the spectral range between 0 and 8 eV. The dielectric function has the highest peak at approximately 1.5–2.5 eV with remarkable anisotropy [62].

Thermoelectric conversion, for example, converting the waste energy of solid-state devices directly into electricity, is a topic of great relevance. Several studies have calculated thermoelectric parameters, such as electrical conductivity, Seebeck coefficient, and figure of merit [63, 64]. Due to the unfavorable connection of relevant properties, optimizing the figure of merit is a challenging endeavor. Tensile strain increases the power factor and decreases the lattice thermal conductivity, and therefore is enhancing the thermoelectric figure of merit. Furthermore, tensile strain widens the band gap, leading to a higher power factor and a decrease in the thermal conductivity due to stiffening of the out-of-plane phonon modes, which reduces the anharmonic scattering lifetime and thus increases the thermoelectric figure of merit. In fact, at a strain of 6%, a large value of $ZT = 2.4$ was found for p-type doped ZrS_2 monolayers at 300 K [63]. Similarly, recent calculations of the mode-dependent contributions to the thermal conductivity of strained ZrX_2 (X = S, Se, and Te) monolayers delivered figures of merit around two with the largest value for $ZrSe_2$ [64].

The development of gas sensors with high absorption rates, durability, gas recovery, and reusability for environmental air pollution control is important. Dispersion-corrected DFT calculations revealed the suitability of single S-vacancy-defected ZrS_2 monolayers for the detection of the toxic gases NO, NO_2, CO, and SO_2 [65]. Compared with perfect ZrS_2 monolayers, defective structures strongly enhance the adsorption energy and capacity, because efficient charge transfer intensifies the chemisorption process.

14.2.2.5 Synthesis

The gas-phase reaction of the evaporated precursors $ZrCl_4$ and sulfur powder under Ar/H_2 flow induced the growth of hexagonal ZrS_2 monolayers and few-layer sheets with a domain size of 1–3 μm for the deposited ZrS_2 hexagons grown on a hexagonal boron-nitride substrate [66]. The evaporation temperature of $ZrCl_4$ determines the number of ZrS_2 layers. Analysis by AFM, transmission electron microscopy (TEM), XPS, and Raman spectroscopy provided the characterization of the as-synthesized ZrS_2. The uniform thickness measured by AFM was 1.2 nm for the ZrS_2 hexagons, which is in reasonable agreement with the theoretical thickness value of $t = 0.6$–0.7 nm. A ZrS_2-based FET produced by photolithography and e-beam evaporation exhibited n-type transport behavior and an estimated mobility of 0.1–1.1 $cm^2\ V^{-1}\ s^{-1}$. This seems to be the only work claiming the synthesis of a single ZrS_2 monolayer.

The CVD method permits the epitaxial growth of high-quality and large-area ZrS_2 films on c-plane sapphire substrates with an atomically sharp interface [67]. The epitaxial films exhibited n-type semiconductor behavior and a mobility of 2.4 $cm^2\ V^{-1}\ s^{-1}$ at room temperature. Fabricated photodetectors showed excellent performance with a light on/off ratio of 10^6 and a detectivity of 2.6×10^{12} Jones, suggesting potential applications in optoelectronics.

Employing atomic layer deposition (ALD), which is a modified CVD method, individual pulses supply the precursors of each element onto the substrate to avoid gas phase reactions. For example, the precursors $ZrCl_4$ and H_2S permitted the deposition of uniform sheets of ZrS_2 ranging from a few monolayers to tens of nanometers thick films at ~ 400 °C [68]. At this temperature, the 8 nm thick sheets had a grain size of ~ 17 nm, as determined via TEM and scanning electron microscopy (SEM). The precursors and the growth temperature allow industrial upscaling of this deposition process. Owing to the environmental degradation by oxidation, it was necessary to deposit a protective $Al_xSi_yO_z$ layer on the ZrS_2 sheet without breaking the vacuum.

14.2.3 $ZrSe_2$ Monolayer

14.2.3.1 Chemistry and 2D Bonding

The electronic configuration of Zr atoms (Kr|$4d^2 5s^2$) and Se atoms (Ar|$3d^{10} 4s^2 p^4$) determine the type of 2D bonding in zirconium diselenide monolayers. Hybridization of the Se-4*p* orbitals and the Zr-4*d* orbitals is responsible for strong covalent bonding in the energetically stable 1T and 2H phase. The significant difference between the electronegativity of Zr (1.22) and Se (2.48) causes a polarization of the *p-d* hybridized Zr–Se bonds by a charge transfer of $\delta q = 1.87$ e from the Zr atom to two Se atoms in the 1T structure according to a Bader charge analysis [13]. The fractional ionic character of 31.1% nearly agrees for the 1T and 1H phases [13]. The 1T-$ZrSe_2$ monolayer, which consists of three sublayers with a zirconium sublayer sandwiched between two selenium sublayers, is energetically more stable than the 2H structure, as signified by the cohesive energies of $E_c = -5.35$ eV (PBE) for the 1T phase and − 5.21 eV (PBE) for the 2H phase [13]. Note that as mentioned for 1T-ZrS_2 monolayers a much larger binding energy of $E_c = -6.71$ eV/atom has been proposed by SCAN-MGGA calculations [11].

14.2.3.2 2D Structures

For the lattice parameters of 1T-$ZrSe_2$ monolayers, DFT-PBE calculations give a = 0.380 nm, $d_{Zr\text{-}Se} = 0.271$ nm, $h_{SeSe} = 0.317$ nm, $\theta_{SeZrSe} = 90.73°$ and $\theta_{ZrSeZr} = 89.27°$ [13]. These values are in very good agreement with a = 0.381 nm, $d_{Zr\text{-}Se} = 0.270$ nm, $h_{SeSe} = 0.312$ nm, and $\theta_{SeZrSe} = 90.14°$ resulting from the SCAN-MGGA approximation, despite the fact that this approach predicts much stronger bonding [11]. An estimate of the monolayer thickness of $d_{w+h} = 0.697$ nm is obtained by adding the van der Waals diameter of $d_w = 0.380$ nm and the distance between the Se atoms. The indirect band gaps calculated by various approaches are $E_g = 0.5$ eV (PBE) [14], 1.25 eV (HSE06) [58], 1.30 eV (G_0W_0) [58], and 1.54 eV (QP) [14].

The collective quantum phenomenon of CDWs is known as a wavelike modulation of the conduction electron density in metallic TMCDs. Dimensionality and Zr intercalation are possible reasons for the induction of CDWs. Surprisingly, 1T $ZrSe_2$ undergoes a CDW phase transition to a commensurate (2 × 2) CDW order below $T_{CDW} = 205$ K, following a semiconductor–metal transition, as revealed by an STM–STS analysis of epitaxially grown ultrathin sheets with thicknesses between one and ten monolayers [69]. This phenomenon is accompanied by a CDW energy gap of up to 22 meV around the Fermi level, leading to the simplest Fermi surface among all known CDW materials of TMDCs. The Fermi surface consists of only a Zr-4*d*-derived elliptical conduction band at the corners of the Brillouin zone, as revealed by Fourier transformed STM and ARPES data. Both Cu intercalation and pressure suppress the CDW order, and superconductivity develops, while CDW patches or incommensurate CDW orders can coexist with superconductivity. The results indicate the unique compatibility between the simple Fermi topology and the CDW phase in semiconductors and thus open new avenues to create and control many-body CDW phases in 2D materials.

14.2.3.3 Mechanics

1T-$ZrSe_2$ monolayers possess an in-plane Young's modulus of E $= 71.3$ N/m, and in-plane and out-of-plane Poisson's ratios of $\nu = 0.22$ and $\nu = 0.47$, respectively [11]. For 1H-$ZrSe_2$ monolayers, somewhat lower and anisotropic Young's moduli of $E_{ac} = 62$ N/m and $E_{zz} = 58$ N/m and Poisson's ratios of $\nu_{ac} = 0.40$ and $\nu_{zz} = 0.34$ were acquired by first-principles calculations [58]. A greater anisotropy was derived by these authors for the ultimate strengths of $\sigma_{ac} = 8.3$ N/m and $\sigma_{zz} = 5.5$ N/m, whereas the extracted ultimate strains of $\varepsilon_{ac} = 0.21$ and $\varepsilon_{zz} = 0.22$ nearly coincide.

14.2.3.4 Physics and Applications

A systematic theoretical study elucidated the thermoelectric properties of $ZrSe_2$ and $HfSe_2$ monolayers [70]. Both monolayers possess a thermoelectric figure of merit close to $ZT \approx 1$ at optimal doping. These values are due to the low lattice thermal conductivity, which is partly owing to the strong coupling of acoustic modes with low-frequency optical modes. The bilayers of $ZrSe_2$ and $HfSe_2$ are indirect semiconductors that exhibit degenerate conduction bands and stair-like density of states (DOSs) that provide a high-power factor for n-type doping [71]. Combination of these favorable properties with the low lattice thermal conductivity that originates from the low phonon frequency of acoustic modes, a maximum ZT value of 1.84 is found for $ZrSe_2$ bilayers and $ZT = 3.83$ for $HfSe_2$ bilayers at room temperature.

Vacancy defects and substitutional doping allow manipulation of the electronic structure and the generation of magnetism in nonmagnetic monolayers. For example, $ZrSe_2$ is a nonmagnetic semiconductor that transforms into a magnetic material by generating a single Zr vacancy or divacancy by one Zr and one Se vacancy with net moments of 1.756 μ_B and 1.295 μ_B, respectively, originating from the Se atoms surrounding the vacancies [72]. Moreover, the authors studied the replacement of one Zr atom in $ZrSe_2$ by 4*d* (Mo, Tc), 5*d* (W, Re), and 3*d* (Cr, Mn) transition metals, which can induce magnetism and metallic behavior. The monolayers doped with these atoms carry magnetic moments of 2.13 μ_B, 2.999 μ_B, 2.00 μ_B, 2.999 μ_B, 2.482 μ_B, and 3.00 μ_B, respectively. The magnetic moments are proportional to the number of unpaired electrons in the *d* orbitals. While the Cr- and Mo-doped monolayers acquire metallic character, the Mn-, Tc-, and Re-doped systems remain (magnetic) semiconductors, and the W-doped $ZrSe_2$ monolayer is a half-metal [72]. Interestingly, doping by one N or P atom or codoping by one N and P atom leads to a nonmagnetic metal [73].

Vertically stacked van der Waals heterostructures open the door to overcome lattice mismatch and enable the realization of new tailored physicochemical properties beyond those of the individual components [74]. A first-principles study investigated the possibility of further tailoring the electronic and optical properties of the ZrS_2/$ZrSe_2$ heterostructure by strain engineering. The unstrained heterostructure has a band structure with type-II alignment, and the small interface distance caused by enhanced interlayer interaction indicates a redistribution of charge with a built-in electric field at the interface. The band gap increases with tensile strain and decreases with compressive strain, and the band type converts from type-II alignment to type-I alignment by strain. Furthermore, under uniaxial and biaxial compressive strains, a redshift occurs, which indicates improved optical sensitivity.

According to first-principles DFPT calculations, the 1T Janus monolayers ZrSSe, ZrSeTe, and ZrSTe are dynamically, mechanically, and thermally stable [75]. Replacing one of the two chalcogen layers with a layer of different chalcogen atoms reduces the space group from $P\text{-}3m1$ (centrosymmetric) to $P3m1$ (noncentrosymmetric) owing to the absence of mirror planes perpendicular to the c axis. The resulting nanoscale piezoelectricity originates mainly from the ionic character of bonding that arises from decoupled band edges without hybridization between the chalcogen *p*-band and metal *d*-band edge states. The calculated piezoelectric coefficients of the 1T phase are $d_{22} = 5.58$ pm V^{-1} (ZrSSe), 8.67 pm V^{-1} (ZrSeTe), and 14.58 pm V^{-1} (ZrSTe). Applying a uniaxial tensile strength of 9% increases the piezoelectric coefficient of HfSSe to $d_{22} = 123.04$ pm V^{-1}, which is the level of state-of-the-art perovskites.

14.2.3.5 Synthesis

Mechanical exfoliation of $ZrSe_2$ sheets from the bulk to the monolayer limit permitted the investigation of the dependence of Raman spectra on dimensionality [76]. Under ambient conditions, Raman signals exist down to five monolayers without any shift in peak positions. Note that the detection limit of five layers in $ZrSe_2$ extends to bilayers in ZrS_2. The sensitivity of the $ZrSe_2$ sheets to oxidation under ambient conditions leads to partial oxidation of the films. Over the entire range of thicknesses studied, no photoluminescence appeared in accordance with the absence of indirect-to-direct band gap transitions.

The epitaxial growth of 1–10 monolayer thick 1T-$ZrSe_2$ sheets occurs on bilayer graphene-covered SiC(0001) wafers at 700 °C in a UHV chamber [69]. High-purity Zr and Se used as precursors were coevaporated by a single-pocket electron beam evaporator and a Knudsen diffusion cell, respectively. Analysis of the nanosheets followed by applying STM and STS in a UHV facility. The transfer of the as-prepared 1T-$ZrSe_2$ sheets into the ARPES chamber required the coating of several nanometer thick Se layers for protection.

The low-pressure CVD method involving epitaxial growth on c-plane sapphire substrates is a suitable method for the deposition of high-quality $ZrSe_2$ layers [77]. The $ZrCl_4$ powder and Se powder, used as precursors, are in the low-temperature

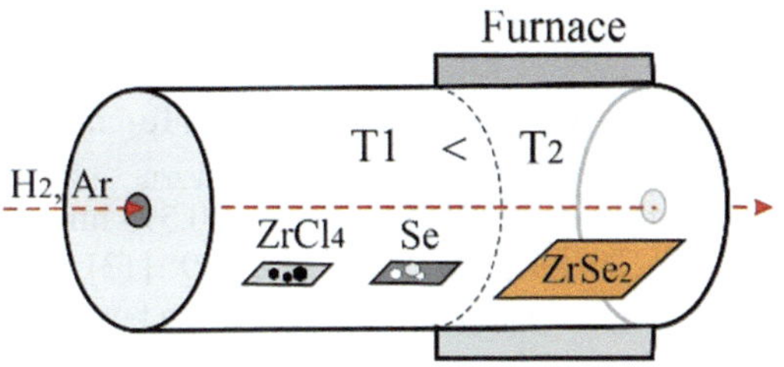

Fig. 14.7 Diagram of the CVD system for epitaxial growth of $ZrSe_2$ sheets on sapphire using $ZrCl_4$ and Se powder in a deposition chamber with a low temperature precursor zone and an elevated temperature around the substrate

zone, and the sapphire substrate is in the high-temperature zone (see Fig. 14.7). The deposited nanosheets showed an atomically sharp interface with the underlying substrate. A photodetector fabricated from this material showed a stable and appreciable photoresponse to visible light.

Band gap engineering, yielding a fully tunable band gap between 1.86 and 1.15 eV, was demonstrated for large-area 2D $ZrS_{2(1-x)}Se_{2x}$ (x = 0–1) semiconductors deposited on c-plane sapphire substrates with fully tunable compositions [78]. The alloys were prepared by a low-pressure single-step CVD process using $ZrCl_4$, S and Se powders as sources and a mixture of Ar and H_2 as carrier gases. The thickness of the deposited sheets varied between 10 and 20 nm. The fabricated photodetectors were sensitive to visible light with a fast response of ~ 100 μs and better performance for larger S contents.

14.2.4 *$ZrTe_2$ Monolayer*

14.2.4.1 Chemistry and 2D Bonding

The electronic configuration of Zr atoms (Ar|$4d^2 5s^2$) and Te atoms (Kr|$4d^{10} 5s^2 p^4$) determine the type of 2D bonding in zirconium ditelluride monolayers. Hybridization of the Te-5*p* orbitals with Zr-4*d* orbitals is responsible for the observed strong covalent bonding. In the trilayer arrangement a zirconium sublayer is sandwiched between two tellurium sublayers. The medium difference in the electronegativities between the values 1.22 (Zr) and 2.01 (Te) causes a reduced polarization of the *p*-*d* hybridized covalent bond compared to $ZrSe_2$. In the 2H-$ZrTe_2$ phase, a charge transfer of $\delta q = 1.58$ e from the 4*d* states of the Zr atom to the 5*p* states of two Te atoms generates a fractional ionic character of 13.8% [13]. Up to now, there is no agreement on whether the H or T structure is the most energetically and dynamically stable monolayer phase [11, 13]. The cohesive energy of $E_c = -5.89$ eV/atom [11] (SCAN-MGGA) obtained for the 1T phase indicates stronger binding for this phase than for the 2H phase with a predicted cohesion of only $E_c = -4.62$ eV/atom [13] (PBE). Note that the two approximations may give strongly deviating cohesive energies.

14.2.4.2 2D Structures

The lattice parameters of the 1T-$ZrTe_2$ monolayer are a = 0.401 nm, $d_{Zr\text{-}Te}$ = 0.291 nm, h_{TeTe} = 0.353 nm, and θ_{TeZrTe} = 92.94° [11]. For comparison, the lattice parameters of the 2H phase are a = 0.393 nm, $d_{Zr\text{-}Te}$ = 0.294 nm, h_{TeTe} = 0.373 nm, θ_{TeZrTe} = 78.83°, and θ_{ZrTeZr} = 83.99° [13]. By adding the van der Waals diameter of Te of d_w = 0.420 nm to the distance between Te atoms an estimate of monolayer thicknesses of d_{w+h} = 0.773 nm (1T) and 0.793 (2H) is found. Ultrathin sheets of 1T $ZrTe_2$ are described as semimetals in most publications; however, STM and STS measurements of deposited 1T-$ZrTe_2$ sheets, especially those grown epitaxially on bilayer graphene SiC(0001) substrates, show a semiconductor-like band structure with a small gap of ~ 0.4 eV [69]. The calculated band gaps of the 2H phase are E_g = 0.45 eV (PBE) [13], 0.18 eV (LDA), and 1.01 eV (GoWo@LDA) [2].

Compared to the behavior of 1T $ZrSe_2$, the formation of CDWs is weak, and no short-range CDW order occurs in the vicinity of native defects in 1T-$ZrTe_2$ monolayers, possibly due to the weak electron donor capacity [69]. Increasing the thickness of the sheets reduces the sensitivity to intercalated Zr defects, and the (2 × 2) order appears repeatedly. These observations support an identical CDW order in 1T $ZrSe_2$ and 1T $ZrTe_2$, albeit with a smaller CDW energy gap of ~ 11 meV around the Fermi level in 1T $ZrTe_2$.

Single- and few-layer $ZrTe_2$ sheets grown on InAs(111)/Si(111) substrates by van der Waals epitaxy exhibit different electronic structures [79]. Despite the connection by weak van der Waals forces, there is good registry between the $ZrTe_2$ epilayer and the substrate. Surprisingly, ARPES measurements indicate a crossing of valence and conduction bands at the Fermi level, revealing abrupt linear dispersion and thus massless Dirac fermions that exist down to the 2D limit, where the σ band exhibits a Dirac cone. Accordingly, $ZrTe_2$ monolayers, like graphene and $MoTe_2$, are potential candidates for topological Dirac semimetals. Figure 14.8 shows a typical band inversion and the opening of the band gap by strong and week SOC, forming a topological insulator [48]. The opening of the gap in the inverted bands due to

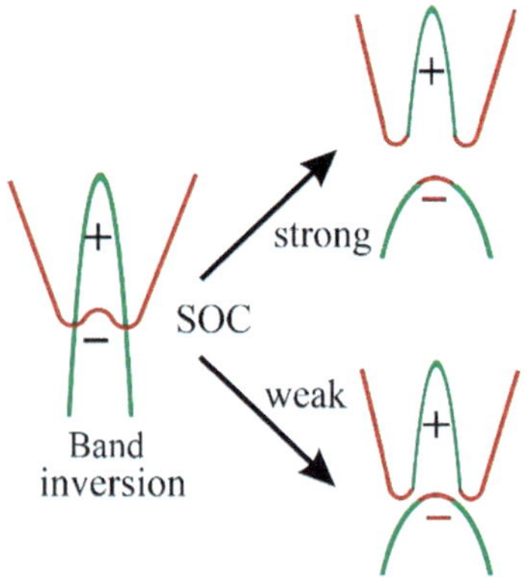

Fig. 14.8 Scheme of band inversion and opening of a gap by strong SOC interactions forming a topological insulator. Adapted with permission from Ref. [18], Copyright 2021, Elsevier

SOC can be detected by ARPES and STM. STS analysis identifies the conducting topologically protected edge states of the monolayer islands.

14.2.4.3 Mechanics

The 1T-$ZrTe_2$ monolayer has an in-plane Young's modulus of E = 43.2 N/m, and strongly deviating in-plane and out-of-plane Poisson's ratios of $\nu = 0.18$ and of $\nu = 0.44$, respectively [11]. This is the smallest stiffness and largest out-of-plane Poisson ratio among the group members of Zr-dichalcogenide monolayers. Note that the variation of the in-plane Poisson ratios within the group of Zr compounds is small.

14.2.4.4 Physics and Applications

A recent experimental and theoretical study investigated the interplay between topological properties and electron correlations in quantum materials that can induce exotic phenomena [80]. Characterization of epitaxially grown 1T-$ZrTe_2$ monolayers by STM and STS revealed not only the existence of CDWs but also a linear dispersion energy band near the Fermi energy, indicating the possible coexistence of topologically nontrivial behavior and electron correlation. Thus, this monolayer provides an interesting 2D platform for elucidating the many-body effects in topologically nontrivial materials.

The application of ARPES and STM measurements allowed the identification of a distinct correlated phase beyond the (2 × 2) CDW ground state in epitaxially grown monolayers of 1T $ZrTe_2$ [81]. The analysis identified a novel energy- and band-dependent folding process occurring in two steps, as evidenced by an exciton gas phase prior to condensation into the final CDW state. Figure 14.9 displays a schematic representation of the formation of the CDW state. Figure 14.10 schematically illustrates the periodic distortion of the lattice, and Fig. 14.11 illustrates the excitonic effect, which drives the band folding process, as verified by first-principles model calculations based on a microscopic Hamiltonian. Below $T_{CDW,}$ exciton condensation drives long-range CDW order formation, with PLD as a secondary effect. The recovery of the pristine band structure by carrier-density-dependent suppression of the CDW state corroborates the use of 1T-$ZrTe_2$ monolayers as a promising excitonic insulator (EI) material with a highly tunable electron structure.

Macroscopic quantum phenomena such as the Bose–Einstein condensation (BEC) and related phenomena such as the EI, which is a BEC of excitons bound by electron-hole interactions in a 2D material, are still characterized by an elusive state of realization [81]. A serious problem is to distinguish the EI state from CDWs, an already well-studied state in various monolayers. A hallmark for distinguishing EI from conventional CDWs is the detection of the precondensation of the exciton gas phase. As described above, ARPES and STM measurements revealed the existence of a distinct correlated phase beyond the (2 × 2) CDW ground state that emerges

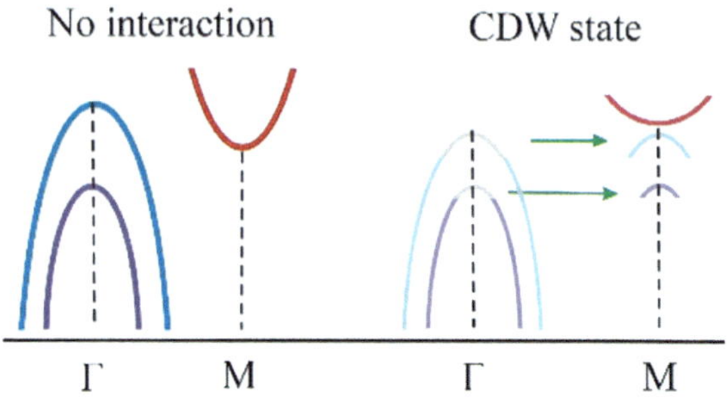

Fig. 14.9 Schematic band diagram illustrating the formation of the CDW state, which is the ground state at low temperature. Adapted from Ref. [81], Copyright 2023, Springer Nature

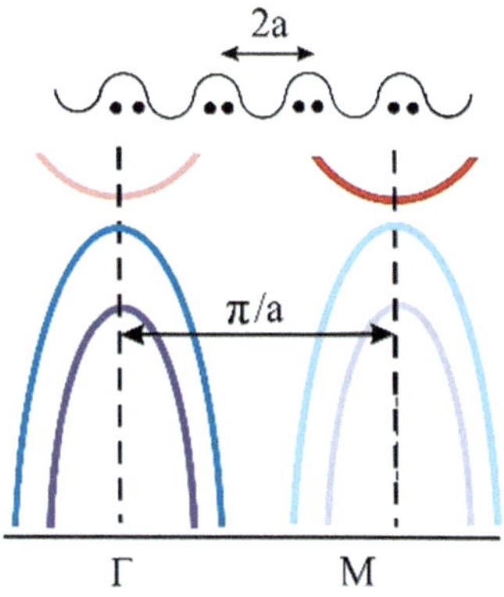

Fig. 14.10 Schematic diagram of the periodic lattice distortion (PLD) driving band folding. The first two VBs with similar orbital components exhibit strong spectral weight transfer and band folding. Adapted from Ref. [81], Copyright 2023, Springer Nature

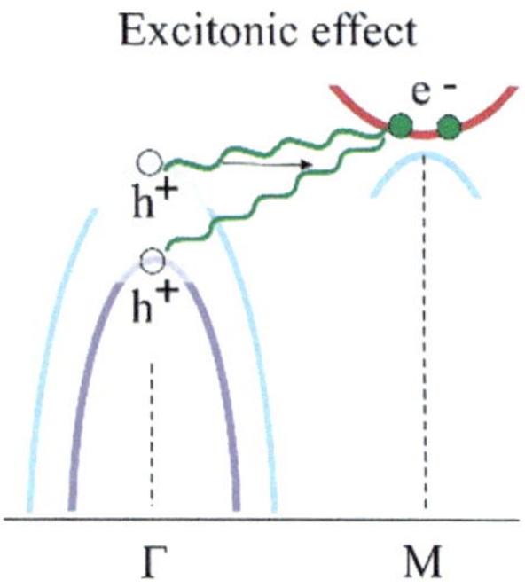

Fig. 14.11 Schematic diagram of the excitonic effect driving band folding. Adapted from Ref. [81], Copyright 2023, Springer Nature

in an epitaxially grown 1T-$ZrTe_2$ monolayer. These monolayers grown on bilayer graphene-terminated SiC substrates exhibit weak interfacial interactions.

The synthesis of single crystals of $ZrTe_{2(1-x)}Se_{2x}$ ($0 \leq x \leq 1$) was performed through the chemical vapor transport (CVT) method, which allowed the study of phase transitions induced by replacing Te atoms with Se atoms in hexagonal $ZrTe_2$ [82]. The study of the band structure by ARPES and first-principles DFT calculations revealed the semimetallic nature of $ZrTe_2$. The splitting of the valence bands at the Γ point due to the strong spin–orbit interaction of Te, decreased with Se substitution, and an indirect gap opened in the vicinity of the Fermi level with increasing Se concentration in $ZrTe_{2(1-x)}Se_{2x}$. Accordingly, the substituted Se atoms at the Te sites induce a change in the electronic structure from a semimetal to a semiconductor, as confirmed by DFT calculations.

Experimental results on the magneto-transport properties of $ZrTe_2$ indicate that these layered TMDCs could be topological 2D materials with exotic electrical transport properties qualifying them for quantum device applications [83]. The investigation of topological semimetals is an active field because they host low-energy particles such as type-I and type-II Dirac and Weyl fermions [84]. First-principles calculations predicted that $ZrTe_2$ is a type-II Dirac semimetal (DSM). In this case, the Dirac cone is overtilted, producing a finite Fermi surface consisting of needle-like electron and hole pockets that cross at the Dirac point (DP).

Measurements of the optical properties of $ZrTe_2$ monolayers suggest the coexistence of two types of charge carriers along with bound excitons in the infrared regime [85]. Hall measurements revealed the dominant electron-type carriers and their high mobility at low temperatures. The calculated thermal conductivity of ~ 5.4 W m^{-1} K^{-1} at ambient temperature qualifies $ZrTe_2$ for thermoelectric devices.

One possibility for tuning the band structure is the adsorption of molecules on the surface as nonbonded dopants [86]. Dispersion-corrected periodic DFT calculations provided information on the adsorption of NH_3 and NO_2 on Zr(Hf)X_2 (X = S, Se, and Te), elucidating the adsorption configuration, adsorption energy, and charge transfer. While adsorbed NH_3 donates electrons to the conduction band, NO_2 receives charge from the valence band. These changes in the conduction and valence bands affect the band gap. Furthermore, the adsorption energy and charge transfer are directly correlated. This allows the tuning of electronic properties, which will be detrimental, of course, in the case of unintentional adsorption of impurities.

In a recent comprehensive theoretical study, based on the VASP approach, structural, elastic, dielectric, and electronic data were determined for a large number of Zr compounds [51]. In addition, the elastic, dielectric, and dynamic properties of Zr-based monolayers (ZrX_2; X = O, S, Se, and Te) and Janus structures ZrXY (X, Y = O, S, Se, and Te; $X \neq Y$) were compared [51]. Note that only the usually more stable 1T phases were considered in these first-principles calculations.

14.2.4.5 Synthesis

Van der Waals epitaxy by MBE, as one of the best methods for the epitaxial growth of mono- and few-layer assemblies on bilayer graphene-terminated SiC is substrates, is currently used to synthesize high-quality monolayers [69, 79]. The quality of properties of the deposit depends critically on the selected substrate and surface preparation used to control the interface distance and interface interaction energy. UHV-based MBE permits the growth of high-quality single and few layers of $ZrTe_2$ semimetallic sheets on InAs(111)/Si(111) substrates available in larger wafer sizes [79]. The rotational commensurability, the van der Waals gap at the interface, and the Moiré pattern point to a good registry. A Moiré pattern indicates a stronger interphase interaction between the epilayer and substrate than expected from van der Waals forces. The valence and conduction bands show abrupt linear dispersions and cross at the Fermi level, as revealed by ARPES images. The resulting massless Dirac fermions can represent an electronic analog of the situation in graphene.

Single-layer and few-layer assemblies were grown by MBE in a UHV chamber allowing in situ RHEED and ARPES analysis [87]. High-purity Zr and Te samples were co-evaporated from an electron-beam evaporator and a Knudsen effusion cell and formed a nanosheet on a heated substrate. The ARPES measurements indicated a change in the band structure connected with a phase transition at ~ 180 K. Below this temperature a gap opened and an ultra-flat band top developed around the zone center. The experimental and theoretical analysis suggests the formation of an excitonic insulating ground state ('excitonic insulator') consisting of intrinsic bound electron-hole pairs in the $ZrTe_2$ monolayer.

14.3 Hafnium Dichalcogenide Monolayers (HfO_2, HfS_2, $HfSe_2$, and $HfTe_2$)

14.3.1 HfO_2 Monolayer

14.3.1.1 Chemistry and 2D Bonding

The valence electrons of the electronic configuration of Hf atoms (Xe|5$d^2$6s^2) and of O atoms (He|2s^2p^4) determine the covalent bonding behavior in hafnium dioxide monolayers. Based on the formal charge of Hf of + 4 and that of O of − 2, Hf would transfer four electrons to two O atoms to fill the valence shell according to the octet rule (d^0 configuration). The 1T-HfO_2 monolayer that consists of a hafnium sublayer sandwiched between two oxygen sublayers seems to be more stable than the structurally similar 2H structure, as indicated by the formation energies of $E_f = -$ 3.27 eV (PBE) for the 1T phase and $E_f = -$ 2.71 eV (PBE) for the 2H phase [2]. The extremely large difference between the electronegativity of Hf (1.23) and O (3.50) causes a significant polarization of the *p-d* hybridized bonds by a substantial charge transfer from the 5*d* states of Hf to the 2*p* states of O. According to a Bader analysis the Hf atom loses a charge of $\delta q = 2.32$ e, and each O atom obtains a charge of $\delta q = 1.16$ e [13]. The polar Hf–O bonds have an enormous fractional ionic character of 68.2% [13]. The DFT-PBE calculations suggest the realization of exceptionally strong ionic-covalent bonding with an extreme cohesive energy of $E_c = -$ 7.90 eV/atom [13] and – 6.73 eV/atom [88].

14.3.1.2 2D Structures

For the lattice parameters of 1T-HfO_2 monolayers, DFT-PBE calculations give a = 0.324 nm, $d_{Hf\text{-}O}$ = 0.211 nm, h_{OO} = 0.193 nm, θ_{OHfO} = 79.38° and θ_{HfOHf} = 100.6° as bond angles [13]. Density functional perturbation theory (DFPT) yields nearly identical lattice parameters: a = 0.323 nm, $d_{Hf\text{-}O}$ = 0.210 nm, and θ_{HfOHf} = 100.9° [88]. Adding the distance of O atoms and their van der Waals diameter of d_w = 0.310 nm yields a monolayer thickness of d_{w+h} = 0.503 nm. Formally, the (111) facet of the face-centered cubic phase of bulk HfO_2 allows a cleavage of the O–Hf–O trilayer unit of the 1T structure. A comparison of the cleavage energy of a free-standing monolayer per unit area from the nonlayered bulk of 3.54 J/m^2 with that of graphene of 0.37 J/m^2 indicates the difficulties encountered in the direct mechanical exfoliation of 1T-HfO_2 monolayers. For this monolayer indirect band gaps of E_g = 5.10 eV (GGA-PBE) and 6.73 eV (HSE06) have been calculated [88].

Replacement of one oxygen sublayer by a sulfur sublayer in HfO_2 yields the well-studied Janus 1T-HfSO monolayer [89]. The calculated lattice constant of HfSO is between those of the parent compounds HfS_2 and HfO_2, which is a behavior also observed for other properties. For example, the calculated bandgaps are E_g = 2.48 eV (HSE06) for HfS_2, 6.50 eV (HSE06) for HfO_2, and 2.98 eV (HSE06) for HfSO [89]. Such a behavior is also found for the related Janus structures HfSSe, HfSTe, and HfSeTe [75]. The different electronegativities of the bottom and top chalcogen atoms induce an intrinsic dipole moment that breaks the mirror symmetry of the Janus structures. These structures with broken inversion symmetry belong to the trigonal space group $P3m1$ and possess piezoelectric properties.

14.3.1.3 Mechanics

For HfO_2 monolayers, the evolutionary structure search (ESS) yielded a large Young's modulus of E = 165.5 N/m and small Poisson's ratio of ν = 0.231, as expected for the strong ionic-covalent bonding [90]. A smaller stiffness of E = 128.8 N/m and a similar Poisson ratio of ν = 0.236 were derived for the Janus monolayer HfSO [89]. These values follow the expectation that the linear elastic modulus of HfSO is smaller than that of HfO_2. For comparison, the stiffness of HfS_2 is E = 136.8 N/m, and the Poisson ratio is ν = 0.19.

14.3.1.4 Physics and Applications

The 1T-HfO_2 monolayer has in-plane and out-of-plane static dielectric constants of 27.35 and 4.80, respectively; therefore, it seems to be better suited for high-*k* gate dielectrics than Al_2O_3 [88]. The 1T phase, with its large band gap of E_g = 6.73 eV, is dynamically and thermally stable, whereas the 2H phase of HfO_2 is dynamically unstable.

The present interest in 2D FETs is based on their ultralow-power operation, high carrier mobility, and high current on/off ratio. Importantly, high-*k* HfO_2 is a promising alternative to SiO_2 as the gate dielectric that has an enormous influence on the performance of the transistor. This is the reason for the current search for methods allowing the growth of high-quality HfO_2 monolayers. A recent study verified the feasibility of scaling down the HfO_2 thickness to achieve a capacitance equivalent oxide thickness (CET) of ~ 1.5 nm with an array of top-gated metal-oxide-graphene field-effect transistors, using ALD for deposition of the HfO_2 layer [91]. The fabricated graphene heterojunction tunneling transistors achieved a record-low subthreshold swing value of < 60 mV/dec on an 8-inch glass wafer.

A possible way to improve the problems encountered by integrating high-*k* dielectrics with 2D semiconductors, caused by the high density of interface states and the damage to the monolayer 2D semiconductors, is through selective hydrogenation [92]. This strategy was successfully tested using the interface of monolayer MoS_2 and silicon nitride, as evidenced by the reduced inhomogeneous charge distribution, increased band offset, and untouched electronic properties of MoS_2. According to the authors, the application of this strategy to other interfaces than HfO_2/MoS_2, is possible and should be assessed.

The Janus 1T-HfSO monolayer has a suitable band gap for the activation of redox reactions. Therefore, this monolayer is a promising photocatalyst for initiating oxidation and reduction reactions to generate hydrogen and oxygen [89]. The extremely anisotropic mobility of both electrons and holes in the x–y plane is advantageous for electron–hole separation. A compressive strain of − 6% enlarges the inner charge polarization, charge carrier mobility, photovoltage, and the absorption in the visible range of 5×10^5 cm^{-1} by ~ 20%.

14.3.1.5 Synthesis

For the synthesis of dielectric HfO_2 sheets ALD was employed and CVD grown graphene monolayers served as substrate at 200 °C [91]. The e-beam evaporation of Hf seed layers induced the growth of high-quality pinhole-free HfO_2. The resulting uniform and conformal deposition of HfO_2 by the ALD process helped to scale down the thickness of the dielectric sheet.

DFT calculations delivered the atomistic mechanism for the growth of a high-quality interface by replacing the Se atoms of an interfacial $HfSe_2$ layer by O atoms [93]. This substitution reaction proceeds without energy barrier and allows layer-by-layer oxidation with an atomically sharp interface. In fact, a high-quality gate stack of native HfO_2 on 2D $HfSe_2$, fabricated via plasma oxidation, showed a suppressed interface trap density of $D_{it} \approx 5 \times 10^{10}$ cm^{-2} eV^{-1}. The dielectric constant of the chemically converted HfO_2 was $k \approx 23$ with a low leakage current of about 10^{-3} A cm^{-2} at an equivalent oxide thickness of ~ 0.5 nm. A fabricated gate stack of $HfO_2/HfSe_2$ incorporated in a FET showed an almost ideal subthreshold slope of ≈ 61 mV dec^{-1} at room temperature together with a high I_{on}/I_{off} ratio of ~ 10^8 and a small hysteresis of ~ 10 mV. The authors consider this as a major step forward to the realization of post-Si semiconducting devices for energy-efficient data-centric computing electronics.

14.3.2 HfS_2 Monolayer

14.3.2.1 Chemistry and 2D Bonding

Based on the electronic configurations of Hf atoms (Xe|$5d^2 6s^2$) and S atoms (Ne|$3s^2 p^4$), hybridization of the S-3*p* orbitals and Hf-5*d* orbitals is responsible for strong covalent bonding. The 1T-HfS_2 monolayer that consists of a hafnium sublayer sandwiched between two sulfur sublayers seems to be more stable than the 2H structure, as indicated by the slightly different formation energies of $E_f = -1.83$ eV (FERE) and -1.62 eV (FERE), respectively [2]. In these calculations, the reference energies of the metal and oxygen atoms were based on the fitted elemental phase reference energies (FEREs). The difference in the electronegativity of Hf atoms (1.23) and S atoms (2.44) causes a polarization of the covalent *p*-*d* hybridized bonds by a significant charge transfer of $\delta q = 1.91$ e from the 5*d* states of each Hf atom to the 3*p* states of two S atoms. This leads to a fractional ionic character of 33.6%, which is about half of the ionicity of HfO_2 monolayers [13]. The calculated strongly deviating cohesive energies of 1T-HfS_2 monolayers are $E_c = -4.73$ eV/atom [75], -6.00 eV/atom [13] (PBE), and -7.35 eV/atom [11] (SCAN-MGGA), partly indicating strong covalent-ionic bonding. For comparison, the cohesive energy of the 2H-HfS_2 phase is $E_c = -5.78$ eV/atom [13].

14.3.2.2 2D Structures

For the lattice parameters of 1T-HfS_2 monolayers, DFT-PBE calculations predicted $a = 0.364$ nm, $d_{Hf\text{-}S} = 0.255$ nm, and $h_{SS} = 0.289$ nm, and for the bond angles $\theta_{SHfS} = 88.90°$ and $\theta_{HfSHf} = 91.11°$ [11, 13]. The lattice constant is in good agreement with the experimental value of $a = 0.363$ nm [94]. With the distance of the S atoms and the van der Waals diameter of S atoms of $d_w = 0.360$ nm, a layer thickness of $d_{w+h} = 0.649$ nm can be estimated. Importantly, the reported band gaps of unstrained single-layer ZrS_2 with $E_g = 1.72$ eV (HSE06) and 1.90 eV (G_0W_0) are in good agreement with the experimental value of 1.8 eV [94, 95].

There are two well-studied Janus monolayers HfSSe and HfSTe, derived from the 1T-HfS_2 monolayer. First-principles calculations revealed the lattice parameters, dynamic stability, mechanical properties, and piezoelectric properties of these materials including those of the third Janus monolayer HfSeTe [74]. As discussed before for noncentrosymmetric Janus Zr structures, in the corresponding Hf-dichalcogenide structures the strong ionic character of interatomic bonding leads to high piezoelectricity together with robust electron mobility in the semiconducting phases.

14.3.2.3 Mechanics

The 1T-HfS_2 monolayer has an in-plane Young's modulus of E = 85.8 N/m, and anisotropic in-plane and out-of-plane Poisson's ratios of $\nu = 0.19$ and $\nu = 0.52$, respectively [1]. These values are in reasonable agreement with the values of E = 78 N/m and $\nu = 0.18$ calculated for the 1T phase [75]. First-principles calculations predicted slightly anisotropic stiffnesses of $E_{ac} = 80$ N/m and $E_{zz} = 76$ N/m and Poisson's ratios of $\nu_{ac} = 0.40$ and $\nu_{zz} = 0.34$ for the 2H phase [95]. While the stiffness of the trigonal phase agrees reasonably well with the larger value of the octahedral structure, the Poisson ratios deviate substantially. Previous calculations of the linear mechanical properties of 1T-HfS_2 monolayers predicted an isotropic Young's modulus of 79.9 N/m (136.8 GPa) based on a film thickness of t = 0.584 nm and a Poisson ratio of 0.19 [96].

Nonlinear properties, such as the ultimate strength, exhibit larger anisotropies at $\sigma_{ac} = 10.0$ N/m and $\sigma_{zz} = 7.6$ N/m, whereas the ultimate strains at $\varepsilon_{ac} = 0.21$ and $\varepsilon_{zz} = 0.22$ nearly coincide [95]. Previously published comparable values of the anisotropic ultimate strengths are $\sigma_{ac} = 8.3$ N/m (14.2 GPa) and $\sigma_{zz} = 7.2$ N/m (12.4 GPa) [96].

14.3.2.4 Physics and Applications

Calculations of the electronic, optical, and thermoelectric properties indicate a considerable influence of single-atom vacancies of Hf or S atoms and single-atom dopants in place of a S atom on the monolayer properties [97]. These energetically most favorable defects induce indirect-to-direct band gap transitions and semiconductor-to-metal transformations. Moreover, structural defects increase the light absorption of 1T HfS_2 in the infrared region and of 1H HfS_2 in the UV range. Interband transitions control light absorption between the VB-*p* states of S atoms and CB-*d* states of Hf atoms. The T and H phases have their maxima of light absorption and optical conductivity in the visible and UV regimes, respectively. An enhancement of the latter effect by structural defects is possible.

Calculations using the PBE and HSE06 functionals elucidated the thermoelectric properties of the HfS_2 monolayers [97]. The Seebeck coefficient of pristine 1T-HfS_2 monolayers at 2771 $\mu V\ K^{-1}$ increased to 2868 $\mu V\ K^{-1}$ by structural defects at 300 K. The figure of merit was $ZT \approx 0.96$ under these conditions, and Hf vacancies increased the value to $ZT \approx 1.23$. The corresponding values for the H phase are comparable. Manipulation of the band gap by introducing vacancies and dopants is possible, leading to an enhancement of thermoelectric performance. The electronic and optical properties qualify HfS_2 monolayers for practical applications as FETs, photodetectors, photocatalysts, and flexible optoelectronic devices.

In addition to the thermoelectric performance of intrinsic 1T-HfS_2 monolayers, a study investigated the improvement of these properties by strain, e.g., by biaxial strain engineering. Strain-induced band degeneracy improves electronic transport, and phonon dispersion may decrease thermal conductivity [98]. A 6% biaxial strain delivered for the lattice thermal conductivity 1.55 W m^{-1} K^{-1}, for the electronic thermal conductivity 1.44 W m^{-1} K^{-1}, for the power factor 22.9 mW mK^{-2}, and for the *ZT* value 2.29. For p-type HfS_2 monolayers at 7% strain, the corresponding values are 1.12 W m^{-1} K^{-1} for the lattice thermal conductivity, 1.53 W m^{-1} K^{-1} for the electronic thermal conductivity, 29.5 mW mK^{-2} for the greatly improved power factor, and 3.35 for the *ZT* value. The application of compressive vertical strains significantly changes the electronic and optical properties of the material [99].

The two Janus structures HfSSe and HfSTe have only slightly smaller piezoelectric coefficients than their Zr counterparts of $d_{22} = 4.68$ pm V^{-1} and 12.86 pm V^{-1}, respectively [75]. Remarkable is the enormous increase in the piezoelectric coefficient of HfSSe under the application of uniaxial tensile strain along the armchair direction to $d_{22} = 123.0$ pm V^{-1} at 9% strain.

14.3.2.5 Synthesis

A deposition method for high-quality HfS_2 monolayers, based on an improved CVD technique, used $HfCl_4$ and S powder for forming 2D-HfS_2 crystals on a thermally stable fluorophlogopite mica substrate under Ar and H_2 flow [100]. The migration energy barrier of Hf and S atoms on the inert mica surface is exceedingly small, resulting in a high migration rate and fast lateral growth. Thus, van der Waals epitaxy is possible on this substrate regardless of the lattice mismatch. An ultrasensitive near-infrared photodetector fabricated with these 2D HfS_2 crystals showed an ultrahigh responsivity exceeding 3.08×10^5 A W^{-1}. The realized ultrahigh photo gain exceeding 4.72×10^5 and the ultrahigh detectivity exceeding 4.01×10^{12} Jones compete with those of most phototransistors based on 2D materials.

The CVD method allowed layer-controlled epitaxial growth of high-quality HfS_2 sheets on c-plane sapphire substrates [101]. The flat 1T-HfS_2 layers showed an atomically sharp interface with the substrate, as demonstrated by XRD and TEM measurements. High-performance photodetectors fabricated from CVD-grown HfS_2 layers had an on/off ratio of > 10^3, a rise time of 130 μs, and a decay time of 155 μs.

14.3.3 $HfSe_2$ Monolayer

14.3.3.1 Chemistry and 2D Bonding

The electronic configurations of Hf atoms (Kr|$5d^2 6s^2$) and Se atoms (Ar|$3d^{10} 4s^2 p^4$) are responsible for the same type of 2D bonding in hafnium diselenide monolayers as described for HfS_2 and HfO_2. Hybridization of the Se-4*p* orbitals with the Hf-5*d* orbitals is responsible for strong covalent bonding. The 1T-$HfSe_2$ monolayer with a hafnium layer sandwiched between two selenide layers seems to be more stable than the similar 2H structure, as indicated by the formation energies of $E_f = -1.53$ eV (FERE) and -1.36 eV (FERE), respectively [2]. The difference in the electronegativity of Hf (1.23) and Se (2.48) causes a polarization of the *p-d* hybridized bonds by charge transfer of $\delta q = 1.76$ e from the 5*d* states of a Hf atom to the 4*p* states of two Se atoms [13]. This charge transfer causes a substantial fractional ionic character of the Hf–Se bonds of 32.3%. The published cohesive energies of the 1T phase are $E_c = -3.98$ eV/atom [75], – 5.42 eV/atom [13] (PBE), and – 6.67 eV/atom [11] (SCAN-MGGA) and the binding energy of the 2H phase is $E_c = -5.25$ eV/atom [13] (PBE), supporting a higher energetic stability of the 1T phase.

14.3.3.2 2D Structures

The lattice parameters of 1T-$HfSe_2$ monolayers determined via DFT-PBE calculations are a = 0.376 nm, $d_{Hf\text{-}Se} = 0.268$ nm, $h_{SeSe} = 0.314$ nm, $\theta_{SeHfSe} = 90.92°$, and $\theta_{HfSeHf} = 89.08°$ [11, 13]. These lattice parameters and the lattice constant are in good agreement with the experimental value of a = 0.374 nm [94] and recent calculations [14, 94]. By adding the distance between the Se atoms and their van der Waals diameter of $d_w = 0.380$ nm, a reasonable thickness value of the monolayer of $d_{w+h} = 0.694$ nm is obtained. The 1T-$HfSe_2$ monolayer is a semiconductor with an indirect band gap. Note that the band gaps reported for unstrained single-layer $HfSe_2$ at $E_g = 0.85$ eV (PBE), 1.25 eV (HSE06) and 1.30 eV (G_0W_0) are in reasonable agreement with the experimental value of $E_g = 1.13$ eV [94, 95].

14.3.3.3 Mechanics

The 1T-$HfSe_2$ monolayer has an isotropic in-plane Young's modulus of E = 77.8 N/m and anisotropic in-plane and out-of-plane Poisson's ratios of $\nu = 0.21$ and $\nu = 0.47$, respectively [13]. Moreover, smaller slightly anisotropic stiffnesses of $E_{ac} = 69$ N/m and $E_{zz} = 65$ N/m and Poisson's ratios of $\nu_{ac} = 0.40$ and $\nu_{zz} = 0.35$ were found by first-principles calculations for the 2H phase [95]. The ultimate strength values show the expected larger anisotropy of $\sigma_{ac} = 9.0$ N/m and $\sigma_{zz} = 6.5$ N/m, whereas the ultimate strains of $\varepsilon_{ac} = 0.20$ and $\varepsilon_{zz} = 0.21$ nearly coincide [95]. The lower stiffness values are in good agreement with the isotropic mechanical modulus of E = 65.4 N/m [102].

14.3.3.4 Physics and Applications

First-principles calculations based on DFT provided basic information on the influence of structural defects on the electronic structure and magnetic properties of $HfSe_2$ monolayers [103]. Removing a Hf or Se atom and introducing an S atom at the Se defect induces the transformation of $HfSe_2$ monolayers from a semiconductor to a metal. Moreover, defects can induce spin magnetic moments, where the total magnetic moment decreases as the number of Se defect atoms increases. The largest magnetic moment achieved is 2.985 μ_B. For S doping on Se defects, the $HfSe_2$ monolayer undergoes a transition from a metal to an indirect nonmagnetic semiconductor. Furthermore, the band gap is tunable by S doping to some extent.

Interestingly, theory predicts enhanced gas adsorption at Se vacancies of $HfSe_2$ monolayers, as revealed by studying the equilibrium distance of adsorbed molecules, their adsorption energy, the induced charge transfer, and the electron localization function (ELF) of pure and Se vacancy-defective $HfSe_2$ monolayers with adsorbed CO, NO, NO_2, and SO_2 molecules [104]. The resulting modification of the electronic structure leads to enhanced adsorption and charge transfer. Notably, the atom-projected DOS indicates hybridization between orbitals of the adsorbed molecule and the defective surface structure.

A first-principles study revealed that intrinsic and Pd-doped $HFSe_2$ monolayers exhibit increased adsorption energy and charge transfer, especially for toxic NO_2. Thus, they are better sensors with greater electrical responsivity for NO_2 than for SO_2 [105]. The favorable desorption behavior of the resistance-type NO_2 gas sensor that consists of Pd-doped $HfSe_2$ monolayers allows the recycling of the target molecule.

Despite the advantageous properties of silicon, such as a moderate band gap of 1.1 eV and the high-quality native insulator, SiO_2 is gradually replaced by high-k insulators with much larger dielectric constants of approximately 15–20 instead of ~ 4 for SiO_2 [106]. By spectroscopic analysis of the electronic band structure of air-stable transistors, fabricated with $HfSe_2$ sheets consisting of ≥ 3 layers in thickness and dielectric encapsulation, the suitability of these devices for practical applications was demonstrated. The high-k dielectrics with moderate band gap between 0.9 and 1.2 eV survive from multilayer sheets on the way down to monolayer thickness. ARPES measurements revealed a minimal change in the band gap with decreasing layer number, and TEM was employed to analyze the formation of amorphous native oxides. The native oxides exhibit promising performance by delivering an on/off ratio of $> 10^6$ at a current density of ~ 30 μA/μm. The work demonstrates the key attributes needed for technological integration in 2D nanomaterials, which are intrinsically scalable to the atomically thin limit.

A theoretical study investigated the structural, electronic, and optical properties of $HfSe_2$ and the Janus monolayer HfSSe, including their lateral heterostructures $(HfSe_2)_{4-n}$–$(HfSSe)_n$ (n = 1–3) [107]. The dynamically stable semiconductors exhibit indirect-to-direct band transitions when external strain is applied and semiconductor-to-metal transitions are induced at compressive strains of − 6%. Moreover, strong absorption of light from the visible to the ultraviolet regime qualifies the heterostructures for applications in optoelectronic devices.

14.3.3.5 Synthesis

Van der Waals MBE on HOPG and MoS_2 substrates allowed the growth of 1T-$HfSe_2$ sheets [108]. As diagnostic tools HAADF, STEM, TEM, XRD, RHEED, and Raman spectroscopy analyzed the properties of ~ 15 nm thick epitaxially grown crystals, consisting of approximately 24 layers. The band gap of ~ 1.1 eV measured by STS was much larger than the theoretically predicted value of $E_g = 0.5$ eV. The RHEED pattern suggested that the initial growth process was not pseudomorphic with the substrate lattice but proceeded unstrained with the 1T coordination.

One to six monolayers thick $HfSe_2$ sheets and $MoSe_2/HfSe_2$ van der Waals heterostructures were epitaxially grown on AlN(0001)/Si(111) substrates by MBE and characterized via in situ UHV-STM measurements [109]. The lattice constant a = 0.378 nm agrees with the calculated values given above and indicates the absence of strain, as expected for van der Waals epitaxy. ARPES and RHEED measurements revealed a high quality of the band structure of these $HfSe_2$ sheets. Instead of the theoretically predicted type III $MoSe_2/HfSe_2$ heterointerface, the measurements indicated a tendency to form a weak type II interface with the prospect of applications in optoelectronics.

14.3.4 $HfTe_2$ Monolayer

14.3.4.1 Chemistry and 2D Bonding

The valence electrons of the electronic configurations of the Hf atom (Kr|$5d^2 6s^2$) and the Te atom (Kr|$4d^{10} 5s^2 p^4$) determine the type of 2D bonding in hafnium ditelluride monolayers. The hybridization of the 5*p* orbitals of Te with the 5*d* orbitals of Hf is responsible for the strong covalent bonding in $HfTe_2$ monolayers. Like in the other Hf-dichalcogenides, the monolayer consists of three sublayers, with the hafnium sublayer sandwiched between two Te sublayers. It seems that the 1T structure of $HfTe_2$ forms the ground state, as observed for most group IVB–VIA monolayers with the d^0 configuration [11]. The difference in the electronegativity of Hf (1.23) and Te (2.01) causes a polarization of the *p-d* hybridized covalent Hf–Te bonds. This is connected with a charge transfer from the Hf-5*d* states to the Te-5*p* states of $\delta q = 1.48$ e that occurs from one Hf atom to two Te atoms in the similar 2H phase [13]. The values available for the cohesive energy of the 1T phase deviate substantially from $E_c = -3.98$ eV [75] to $E_c = -5.80$ eV [11] (SCAN-MGGA). The reported cohesive energy of the 2H phase is situated in between these extreme values at $E_c = -4.61$ eV/atom (PBE) [13].

14.3.4.2 2D Structures

SCAN-MGGA calculations for the 1T-$HfTe_2$ monolayer structure delivered the lattice parameters a = 0.398 nm, $d_{Hf\text{-}Te} = 0.288$ nm, $h_{TeTe} = 0.347$ nm, and $\theta_{TeHfTe} = 92.58°$ [11]. For comparison, the lattice parameters of 2H-$HfTe_2$ monolayers, determined via the DFT-PBE approach, are a = 0.390 nm, $d_{Hf\text{-}Te} = 0.291$ nm, $h_{TeTe} = 0.369$ nm, $\theta_{TeHfTe} = 78.63°$ and $\theta_{HfTeHf} = 84.14°$ [13]. The distance between Te atoms and their van der Waals diameter of $d_w = 0.420$ nm yield the monolayer thicknesses $d_{w+h} = 0.767$ nm (1T) and $d_{w+h} = 0.789$ nm (2H). While the lattice parameters of the two phases are similar, their bonding angles deviate substantially. Note that $HfTe_2$ is a gapless halfmetal with the same band gap of $E_g = 0.02$ eV for the 1T and 2H phases [94].

14.3.4.3 Mechanics

Investigations of the mechanical properties of $HfTe_2$ monolayers are widely missing, and only few results are available for the linear mechanical properties. The 1T-$HfTe_2$ monolayer has an in-plane Young's modulus of $E = 46.8$ N/m, which is the lowest of the group IVB–VIA compounds, and extremely anisotropic in-plane and out-of-plane Poisson's ratios of $\nu = 0.15$ and $\nu = 0.41$, respectively [11].

14.3.4.4 Physics and Applications

Using DFT and nonequilibrium Green's function (NEGF)-based methods, the sensing capabilities of 1T-$HfTe_2$ monolayers for hazardous gases such as CO, CO_2, NO, NO_2, NH_3, and N_2O and for the environmental gases O_2, N_2, H_2, and H_2O were studied [110]. Only in the case of NO a significant, and for NO_2 and O_2 a moderate charge transfer between the adsorbed molecule and the 1T-$HfTe_2$ surface was detected, and a semimetal-to-metal transition was induced. Simulations of the *I–V* characteristics of the pristine and covered 1T-$HfTe_2$ surfaces revealed the resistance change due to adsorption and thus determined the output and performance of the resistance sensor. These calculations indicate a significant increase in current for NO, an insignificant increase for NO_2 and O_2, and almost no increase for the rest of the examined gases given above.

In a theoretical work, investigating sensing materials with high response and selectivity, the adsorption behaviors of pristine $HfTe_2$ monolayers and that of a surface with a dispersed single Pt atom adsorbed at the hollow site between the three Te atoms of the $HfTe_2$ monolayer were compared for the hazardous gases NO_2, NO, CO, SO_2, H_2S, and NH_3 [111]. For the pristine surface, long-range interactions dominated for most gases except NO, but the introduction of Pt increased the adsorption energies of all six molecules. The short-range chemical interactions induced between the adsorbed molecule and the Pt atom explain the observed enhancement effects.

Hafnium ditelluride nanosheets exhibit an extraordinary surface-enhanced Raman scattering (SERS) sensitivity, uniformity, and reproducibility [112]. The nanosheets were prepared by liquid exfoliation combined with a hydrothermal reaction and did not suffer from background interference. The Raman enhancement factor of 2.32×10^6 achieved for rhodamine 6G through efficient charge transfer, demonstrates the ultrasensitivity of this well-studied monolayer. Furthermore, a reliable and label-free SERS fingerprint analysis of uric acid molecules revealed arrested SERS signals with detection limits as low as 0.1 mmol/L. The enormous sensitivity originates from charge transfer from the valence band of $HfTe_2$ to the lowest unoccupied molecular orbital (LUMO) of the dye molecule.

Contrary to the situation in bulk $HfTe_2$, the 1T-$HfTe_2$ monolayers show a stable (2 × 2) CDW order caused by an enhanced electron–phonon coupling (EPC) [113]. The accompanying lattice distortions induce a metal-to-semiconductor transition. By suppressing the CDW by electron doping, Ising superconductivity occurs with a significantly enhanced in-plane critical field in the centrosymmetric monolayer because spin-orbital locking protects Ising pairing. This result extends the possibilities for designing and controlling novel quantum states in 2D materials.

14.3.4.5 Synthesis

The UHV-based MBE method allowed the growth of $HfTe_2$ monolayers on the three substrates Si(111)/InAs(111), SiC/graphene, and sapphire/MoS_2, which served as examples to study the influence of the substrate on the electronic properties of the epilayer [114]. A line scan suggested that the thickness of the epilayer was t ~ 0.66 nm. ARPES measurements excluded the possibility of strong chemical interaction between the epilayer and the substrate owing to a change in hybridization. This indicates that the band gap features remained identical in all three cases, allowing only a shift in the band energy. Electron doping is a plausible reason for the energy shift induced by substrate interactions. In principle, the energy shift could move the Dirac point closer to the Fermi level to benefit from Dirac fermions.

14.4 Summary of Group IVB–VIA Monolayers

14.4.1 Binding–Structure Relationships

Figure 14.12 shows the cohesive energy versus bond length of the M–X bonds for the MX_2 (M: Ti, Zr, and Hf; X: O, S, Se, and Te) monolayers. Note that the X–X distances, describing the separation of the two outer chalcogen sublayers, are much greater than the lengths of the chemical bonds of the TMDO and TMDC compounds. The cohesive energies displayed in the plot mostly originate from calculations using the exchange-correlation energy based on the DFT-GGA-PBE approach, which mostly delivers the lower or medium binding energy, whereas the LDA and SCAN-MGGA approximations predict much larger binding energies for the stable trilayer systems. Altogether, all cohesive energies of the three-sublayer assemblies are exceptionally large, and in few cases, the estimated binding energies deviate substantially. It is important to note the enormous contribution of ionic forces to bonding, especially for compounds containing a second- or third-row element.

The cohesive and binding energies of all compounds follow the established inverse relationship between the orbital overlap (covalent bonding strength) and bond length proposed by tight-binding theory despite strong ionic contributions to bonding. The increasing atom size and change in the electronic structure of the

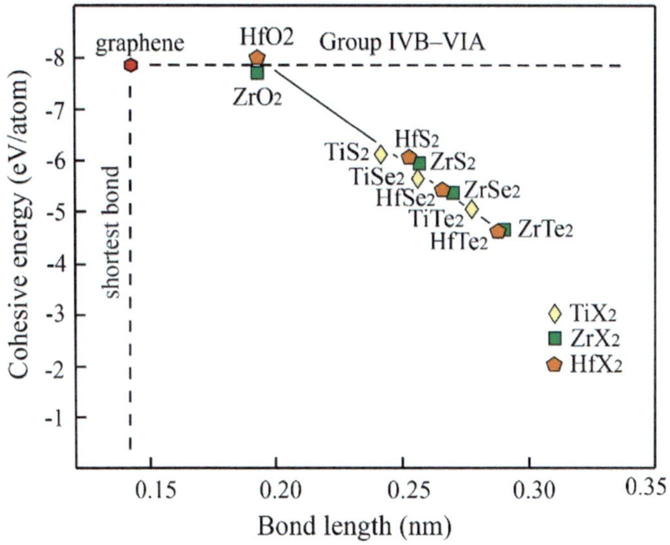

Fig. 14.12 Plot of the cohesive energies versus bond lengths of the hexagonal 1T structures of group IVB–VIA monolayers

chalcogens are responsible for the parallel decrease in the binding strength of the related groups with increasing bond length. From the significantly deviating cohesive energies, Fig. 14.12 preferably presents the lower binding values. Interestingly, the Hf compounds partly exhibit stronger binding and somewhat shorter bonds as the corresponding Zr compounds. This follows from the smaller van der Waals radius of Hf atoms compared with Zr atoms even though Hf is the heavier element in this column. Note the overall striking similarity of binding energies and bond lengths in the whole group of IVB–VIA monolayers.

14.4.2 Mechanics–Structure Relationships

Figure 14.13 shows calculated Young's moduli and few ultimate strength values as a function of the bond lengths. Note that contrary to the behavior of the cohesive energies, the low stiffnesses and ultimate strengths fit into the range of mechanical properties expected from comparable layered monolayers. In the cases of anisotropic behavior, the usually larger stiffness in the armchair direction shows a greater deviation of the ultimate strength from Griffith's rule. In the weaker zigzag direction, the simple rule predicting a general relation between stiffness and strength of perfect lattices often gives a good estimate.

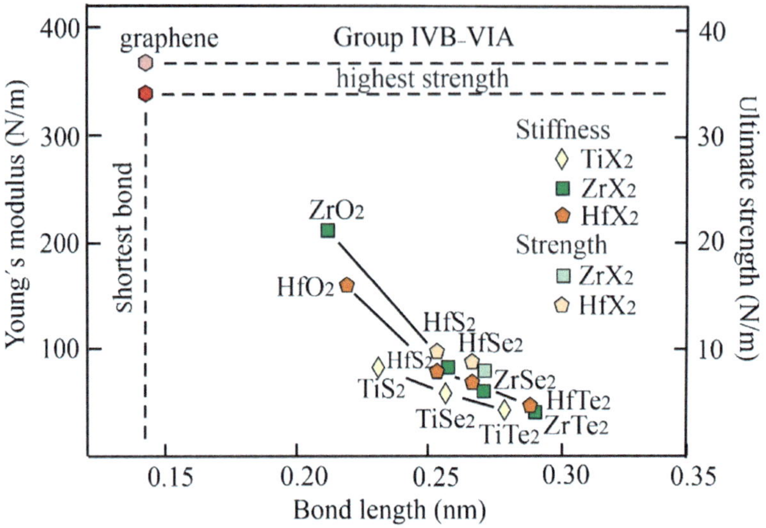

Fig. 14.13 Young's moduli and ultimate strengths versus bond lengths of the hexagonal structures of group IVB–VIA monolayers

14.4.3 Reviews Covering Group IVB–VIB Monolayers

Several comprehensive review articles systematically treating all or a larger number of group–IVB dichalcogenides, discuss in more detail selected aspects from methods used for synthesis, tools employed to characterize the geometric and electronic structure, specific physical properties, and already existing and potential applications in devices [115–117]. These studies provide a general picture of our present knowledge and understanding of group IVB–VIA monolayers.

References

1. M. Chhowalla, H.S. Shin, G. Eda, L.-J. Li, K.P. Loh et al., Nat. Chem. **5**, 263 (2013)
2. F.A. Rasmussen, K.S. Thygesen, J. Phys. Chem. C **119**, 13169 (2015)
3. X. Huang, H. Zhang, X.-L. Cheng, Int. J. Quant. Chem. **121**, e26643 (2021)
4. N. Sakai, Y. Ebina, K. Takada, T. Sasaki, J. Am. Chem. Soc. **126**, 5851 (2004)
5. H.A. Eivari, S.A. Ghasemi, H. Tahmasbi, S. Rostami, S. Faraji et al., Chem. Mater. **29**, 8594 (2017)
6. M. Zarattini, C. Dun, L.H. Isherwood, A. Felten, J. Filippi et al., J. Mater. Chem. A **10**, 13884 (2022)
7. V.R. Patel, Y. Sonvane, P.B. Thakor, Mater. Today Proc. **47**, 563 (2021)
8. H. Lozano, S. Devis, J. Aliaga, M. Alegria, H. Guzmán, Int. J. Mol. Sci. **23**, 4006 (2022)
9. X.F. Song, L.F. Hu, D.H. Li, L. Chen, Q.Q. Sun et al., Sci. Rep. **5**, 15989 (2015)
10. H. Li, J. Robichaud, Y. Djaoued, RCS Adv. **11**, 8065 (2021)
11. N.K. Nepal, L. Yu, Q. Yan, A. Ruzsinszky, Phys. Rev. Mater. **3**, 073601 (2019)

12. C. Ataca, H. Şahin, S. Ciraci, J. Phys. Chem. C **116**, 8983 (2012)
13. G. Özbal, R.T. Senger, C. Sevik, H. Sevinçli, Phys. Rev. B **100**, 085415 (2019)
14. H.L. Zhuang, R.G. Hennig, J. Phys. Chem. C **117**, 13169 (2013)
15. Y. Qu, H. Pan, C.T. Kwok, Sci. Rep. **6**, 34186 (2016)
16. M. Berwanger, R. Ahuja, P.C. Piquini, Catalysts **10**, 94 (2020)
17. A. Samad, A. Shafique, Y.-H. Shin, Nanotechnology **28**, 175401 (2017)
18. K. Lasek, J. Li, S. Kolekar, P.M. Coelho, L. Guo et al., Surf. Sci. Rep. **76**, 100523 (2021)
19. K. Chen, M. Song, Y.-Y. Sun, H. Xu, D.-C. Qi et al., Appl. Phys. Lett. **116**, 121901 (2020)
20. T. Das, S. Chakraborty, R. Ahuja, G.P. Das, ChemPhysChem **20**, 608 (2018)
21. W. Zhao, L.-C. Xu, R. Li, Y. Guo, Z. Yang et al., Mater. Today Commun. **30**, 103196 (2022)
22. G. Li, K. Yao, G. Gao, Nanotechnology **29**, 015204 (2018)
23. J.-H. Liao, Y.-C. Zhao, Y.-J. Zhao, X.-B. Yang, Y. Chen, J. Appl. Phys. **127**, 044301 (2020)
24. Z. Zeng, C. Tan, X. Huang, S. Bao, H. Zhang, Energy Environ. Sci. **7**, 797 (2014)
25. M. Tan, Z. Wang, J. Peng, X. Jin, J. Am. Ceram. Soc. **98**, 1423 (2015)
26. P.C. Sherrell, K. Sharda, C. Grotta, J. Ranalli, M.S. Sokolikova, ACS Omega **3**, 8655 (2018)
27. P. Chen, Y.-H. Chan, X.-Y. Fang, Y. Zhang, M.Y. Chou et al., Nat. Commun. **6**, 8943 (2015)
28. B. Singh, C.-H. Hsu, W.-F. Tsai, V.M. Pereira, H. Lin, Phys. Rev. B **95**, 245136 (2017)
29. K. Sugawara, Y. Nakata, R. Shimizu, P. Han, T. Hitosugi et al., ACS Nano **10**, 1341 (2016)
30. Z. Torbatian, D. Novko, arXiv: 2307.11370v1 (2023)
31. M.D. Watson, A. Rajan, T. Antonelli, K. Underwood, I. Marković, 2D Mater. **8**, 015004 (2021)
32. D. Novko, Z. Torbatian, I. Lončarić, Phys. Rev. B **106**, 245108 (2022)
33. S. Kolekar, M. Bonilla, Y. Ma, H.C. Diaz, M. Batzill, 2D Mater. **5**, 015006 (2018)
34. S.N. Sadeghi, M. Zebarjadi, K. Esfarjani, J. Mater. Chem. C **7**, 7308 (2019)
35. C. Zhao, D. Wang, R. Lian, D. Kan, Y. Dou et al., Appl. Surf. Sci. **540**, 148314 (2021)
36. Z. Song, J. Yi, J. Qi, Q. Zheng, Z. Zhu et al., Nano Res. **15**, 4687 (2022)
37. A. Mogulkoc, Y. Mogulkoc, S. Jahangirov, E. Durgun, J. Phys. Chem. C **123**, 29922 (2019)
38. W.-M. Zhao, L. Zhu, Z. Nie, Q.-Y. Li, Q.-W. Wang et al., Nat. Mater. **21**, 284 (2022)
39. L. Zhu, W.-M. Zhao, Z.-Y. Jia, H. Li, X. Xie et al., Chin. Phys. Lett. **40**, 057101 (2023)
40. J.-P. Peng, J.-Q. Guan, H.-M. Zhang, C.-L. Song, L. Wang et al., Phys. Rev. B **91**, 121113 (2015)
41. H. Wang, Y. Chen, M. Duchamp, Q. Zeng, X. Wang et al., Adv. Mater. **30**, 1704382 (2018)
42. J. Wang, H. Zheng, G. Xu, L. Sun, D. Hu et al., J. Am. Chem. Soc. **138**, 16216 (2016)
43. A. Kumar, R. Sharma, S. Yadav, S.K. Swami, R. Kumari et al., Dalton Trans. **50**, 3894 (2021)
44. P. Chen, W.W. Pai, Y.-H. Chan, A. Takayama, C.-Z. Xu et al., Nat. Commun. **8**, 516 (2017)
45. T. Antonelli, W. Rahim, M.D. Watson, A. Rajan, O.J. Clark et al., NPJ Quant. Mater. **7**, 98 (2022)
46. J.S. Zhuo, R. Bianco, L. Monacelli, I. Errea, F. Mauri et al., 2D Mater. **7**, 045032 (2020)
47. N. Ran, B. Sun, W. Qiu, E. Song, T. Chen et al., J. Phys. Chem. Lett. **12**, 2102 (2021)
48. K. Lasek, P.M. Coelho, K. Zberecki, Y. Xin, S.K. Kolekar et al., ACS Nano **14**, 8473 (2020)
49. T. Antonelli, A. Rajan, M.D. Watson, S. Soltani, J. Houghton et al., Nano Lett. **24**, 215 (2024)
50. J. Weng, S.-P. Gao, RSC Adv. **9**, 32984 (2019)
51. Y. Zhang, Physica E **134**, 114855 (2021)
52. Y. Zhang, H.-X. Chen, L. Duan, J.-B. Fan, Thin Solid Films **721**, 138549 (2021)
53. N. Kumari, S. Sareen, M. Verma, S. Sharma, A. Sharma et al., Nanoscale Adv. **4**, 4210 (2022)
54. S. Lu, K. Zhu, X. Hu, A.C.S. Appl, Mater. Interfaces **14**, 13410 (2022)
55. S. Takenaka, S. Uwai, S. Ida, H. Matsune, M. Kishida, Chem. Lett. **42**, 1188 (2013)
56. P. Lackner, Z. Zou, S. Mayr, J.-I.J. Choi, U. Diebold et al., Surf. Sci. **679**, 180 (2019)
57. K.W. Lau, C. Cocchi, C. Draxl, Energy Environ. Sci. **9**, 841 (2016)
58. M. Salavati, Front. Struct. Civil Engineer **13**, 486 (2019)
59. M.M. Abutalib, Res. Phys. **12**, 903 (2019)
60. Q. Song, X. Liu, H. Wang, X. Wang, X. Ni, et al. https://doi.org/10.21203/rs.3.rs-810582/v1
61. Q. Zhao, Y. Guo, K. Si, Z. Ren, J. Bai et al., Phys. Status Solidi (b) **254**, 1700033 (2017)
62. T.V. Vu, A.A. Lavrentyev, D.V. Thuan, C.V. Nguyen, O.Y. Khyzhun et al., Superlatt. Microstruct. **125**, 205 (2019)

63. H.Y. Lv, W.J. Lu, D.F. Shao, H.Y. Lu, Y.P. Sun, J. Mater. Sci. C **4**, 4538 (2016)
64. R. D'Souza, S. Mukherjee, S. Ahmad, J. Appl. Phys. **126**, 214302 (2019)
65. H.T.T. Nguyen, D.-Q. Hoang, T.P. Dao, C.V. Nguyen, H.V. Phuc et al., Superlatt. Microstruct. **140**, 106454 (2020)
66. M. Zhang, Y. Zhu, X. Wang, Q. Feng, S. Qiao et al., J. Am. Chem. Soc. **137**, 7051 (2015)
67. Y. Tian, Y. Cheng, J. Huang, S. Zhang, H. Dong et al., Nano Res. **15**, 6628 (2022)
68. M. Mattinen, G. Popov, M. Vehkamäki, P.J. King, K. Mizohata, Chem. Mater. **31**, 5713 (2019)
69. M.-Q. Ren, S. Han, J.-Q. Fan, L. Wang, P. Wang et al., Nano Lett. **22**, 476 (2022)
70. G. Ding, G.Y. Gao, Z. Huang, W. Zhang, K. Yao, Nanotechnology **27**, 375703 (2016)
71. P. Yan, G.-Y. Gao, G.-Q. Ding, D. Qin, RSC Adv. **9**, 12394 (2019)
72. X.-Y. Gao, J.-M, Zhang, A. Ali, X.-M. Wei, Y.-H. Huang, Thin Solid Films **732**, 138790 (2021)
73. X. Zhao, H. Zhang, T. Chen, Y. Gao, H. Wang et al., Superlatt. Microstruct. **120**, 659 (2018)
74. J.-M. Zhang, X.-Y. Gao, X.-M. Wei, Y.-H. Huang, A. Ali et al., Thin Solid Films **755**, 139332 (2022)
75. D. Dimple, N. Jena, A. Rawat, R. Ahammed, M.K. Mohanta et al., J. Mater. Chem. A **6**, 24885 (2018)
76. S. Mañas-Valero, V. Garcia-Lópes, A. Cantarero, M. Galbiati, Appl. Sci. **6**, 264 (2016)
77. Y. Tian, M. Zheng, Y. Cheng, Z. Yin, J. Jiang et al., J. Mater. Chem. C **9**, 13954 (2021)
78. J. Huang, Y. Tian, Y. Cheng, X. Li, S. Zhang et al., Sci. China Mater. **66**, 1870 (2023)
79. P. Tsipas, D. Tsoutsou, S. Fragkos, R. Sant, C. Alvarez et al., ACS Nano **12**, 1696 (2018)
80. L.-N. Yang, Y.-J Xu, Q.-Y. Li, Y.-X. Meng, Y.-F. Zhao et al., Appl. Phys. Lett. **120**, 073105 (2022)
81. Y. Song, C. Jia, H. Xiong, B. Wang, Z. Jiang et al., Nat. Commun. **14**, 1116 (2023)
82. Z. Muhammad, B. Zhang, H. Lv, H. Shan, Z.U. Rehman et al., ACS Nano **14**, 835 (2020)
83. H. Wang, C.H. Chan, C.H. Suen, S.P. Lau, J.-Y. Dai, ACS Nano **13**, 6008 (2019)
84. S. Fragkos, P. Tsipas, E. Xenogiannopoulou, Y. Panayiotatos, A. Dimoulas, J. Appl. Phys. **129**, 075104 (2021)
85. Z. Muhammad, M. Usman, S. Ullah, B. Zhang, Q. Lu et al., Inorg. Chem. Front. **8**, 3885 (2021)
86. S.S. Raya, A.S. Ansari, B. Shong, Nanomaterials **10**, 1215 (2020)
87. Q. Gao, Y.-H. Chan, Y. Wang, H. Zhang, P. Jinxu et al., Nat. Commun. **14**, 994 (2023)
88. J. Weng, S.-P. Gao, Phys. Chem. Chem. Phys. **20**, 26453 (2018)
89. V.D. Dat, T.V. Vu, New J. Chem. **46**, 1557 (2022)
90. Y. Wang, J. Ren, Phys. Chem. Chem. Phys. **22**, 4481 (2020)
91. S.-J. Jeong, Y. Gu, J. Heo, J. Yang, C.-S. Lee et al., Sci. Rep. **6**, 20907 (2016)
92. Y. Yang, T. Yang, T. Song, J. Zhou, J. Chai et al., Nano Res. **15**, 4646 (2022)
93. T. Kang, J. Park, H. Jung, H. Choi, S.-M. Lee et al., Adv. Mater. **2312747** (2024)
94. T.M.D. Huynh, D.K. Nguyen, T.D.H. Nguyen, V.K. Dien, H.D. Pham et al., Front. Mater. **7**, 569756 (2021)
95. M. Salavati, Front. Struct. Civil Engin. **13**, 486 (2019)
96. J. Kang, H. Sahin, F.M. Peeters, Phys. Chem. Chem. Phys. **17**, 27742 (2015)
97. D. Singh, R. Ahuja, A.C.S. Appl, Energy Mater. **2**, 6891 (2019)
98. H. Wang, Y.-S. Lan, B. Dai, X.-W. Zhang, Z.-G. Wang et al., ACS Omega **6**, 29820 (2021)
99. D.M. Hoat, R. Ponce-Pérez, T.V. Vu, J.F. Rivas-Silva, G.H. Cocoletzi, Optik **225**, 165718 (2021)
100. L. Fu, F. Wang, B. Wu, N. Wu, W. Huang et al., Adv. Mater. **29**, 1700439 (2017)
101. D. Wang, X. Zhang, H. Liu, J. Meng, J. Xia et al., 2D Mater. **4**, 031012 (2017)
102. S. Li, M. Zhou, X. Wang, F. Zheng, X. Shao et al., Phys. Lett. A **384**, 126534 (2020)
103. X. Zhao, Y. Gao, H. Zhang, H. Wang, T. Wang et al., J. Magnet. Magn. Mater. **479**, 192 (2019)
104. G. Yang, P. Yan, C. Zhu, Y. Gu, N. Lu et al., Adv. Theory Simul. **2**, 1900052 (2019)
105. H. Cui, P. Jia, X. Peng, Appl. Surf. Sci. **513**, 145863 (2020)
106. M.J. Mieczko, C. Zhang, H.R. Lee, H.-H. Kuo, B. Magyari-Köpe et al., Sci. Adv. **3**, e1700481 (2017)

107. V.V. On, J. Guerrero-Sanchez, R. Ponce-Pérez, J.F. Rivas-Silva, G.H. Cocoletzi et al., Int. J. Quant. Chem. **122**, e26857 (2021)
108. R. Yue, A.T. Barton, H. Zhu, A. Azcati, L.F. Pena et al., ACS Nano **9**, 474 (2015)
109. K.E. Aretouli, P. Tsipas, D. Tsoutsou, J. Marquez-Velasco, E. Xenogiannopoulou et al., Appl. Phys. Lett. **106**, 143105 (2015)
110. D. Chakraborty, P. Johari, A.C.S. Appl, Nano Mater. **3**, 5160 (2020)
111. J. Hu, S. Zhai, Q. Zhang, H. Cui, X. Jiang, Appl. Surf. Sci. **605**, 154572 (2022)
112. Y. Li, H. Chen, Y. Guo, K. Wang, Y. Zhang et al., Photon. Res. **9**, 1039 (2021)
113. D. Wang, J. Wu, C. Si, Nano Res. **16**, 11521 (2023)
114. P. Tsipas, P. Pappas, E. Symeonidou, S. Fragkos, C. Zacharaki et al., APL Mater. **9**, 101103 (2021)
115. C. Yan, C. Gong, P. Wangyang, J. Chu, K. Hu et al., Adv. Funct. Mater. **28**, 1803305 (2018)
116. C.M.O. Bastos, R. Besse, J.L.F. Da Silva, G.M. Sipahi, Phys. Rev. Mater. **3**, 044002 (2019)
117. L.-Y. Feng, R.A.B. Villaos, H.N. Cruzado, Z.-Q. Huang, C.-H. Hsu et al., Chin. J. Phys. **66**, 15 (2020)

Chapter 15
Group–VB Transition-Metal Dioxide and Dichalcogenide Monolayers

Abstract The group–VB elements V, Nb, and Ta and the chalcogen elements O, S, Se, and Te are involved in the group VB–VIA monolayers discussed in this chapter. Most transition-metal dioxide (TMDO) and transition-metal dichalcogenide (TMDC) monolayers are nonmagnetic in their pristine form. This is different for group–VB dichalcogenide materials, which exhibit ferromagnetic properties. They possess interesting intrinsic features, such as a *d* electron, which is responsible for the specific chemical behavior and metallic character (d^1 group). Short- and long-range Coulomb interactions (electron-electron correlations) and sizable electron-phonon interactions give rise to collective ordering effects such as charge density waves (CDWs), Mott insulating phases, and superconductive phases. Charge density waves consist of a periodically distorted atomic lattice, and the charge density acquires a broken-symmetry state. The 2D-CDW phases have attracted particular attention as powerful resources for novel quantum phenomena. The competition of charge, spin, Mott insulating, and superconducting states produces a rich phase diagram of V-based, Nb-based, and Ta-based dichalcogenides. In particular, CDWs, which consist of spontaneous reorganization of electrons into periodic charge modulations, have attracted increasing amounts of attention as a new tool for realizing nanosheets for quantum-based applications. While the term CDW is widely used, the expression 'periodic lattice distortion' (PLD) describes the observed lattice conversion much better. A related hot topic is the investigation of magnetic 2D materials. Due to the rapid development of synthesis strategies for high-quality monolayers, the discovery of interesting novel structures is progressing quite fast. Since techniques for accurately measuring the magnetism of monolayers are still lacking, our knowledge of their magnetic properties is still in its infancy. Except for $TaTe_2$, molecular beam epitaxy (MBE) allowed the growth of the compound monolayers on various substrates.

P. Hess, *Bonding, Structure, and Performance of Two-Dimensional Materials*, NanoScience and Technology, https://doi.org/10.1007/978-3-031-74161-6_15

15.1 Vanadium Dichalcogenide Monolayers (VO_2, VS_2, VSe_2, and VTe_2)

15.1.1 VO_2 Monolayers

15.1.1.1 Chemistry and 2D Bonding

The scheme of the periodic table shown in Fig. 15.1 displays the group-VB elements V, Nb, and Ta and the chalcogen elements O, S, Se, and Te involved in the VB–VIA monolayers.

The electronic configurations of V atoms (Ar|$3d^3 4s^2$) and O atoms ($1s^2 2s^2 p^4$) determine the type of 2D bonding in vanadium dioxide monolayers. The *d-p* hybridization of the outermost 3*d* orbitals of V atoms and the outermost 2*p* orbitals of O atoms forms strong covalent bonds with extensive electron charge accumulation between the V and O atoms. In the best studied structures, the valence electrons form covalently bonded triple-layer structures with a sublayer of V atoms in the center sandwiched between two O sublayers. Each O atom forms three covalent bonds with V atoms in a bridging position, and each V atom is bound to six O atoms, as described for the group IVB–VIA compounds. The minor structural differences in the trilayer structure of the 1H and 1T phase can be seen in Fig. 15.2.

Moreover, in Fig. 15.3 the structures of the 1H monolayer and the 2H phase realized in bilayers and the bulk phase are displayed. The enormous difference in the

H																	He
Li	Be											B	C	N	O	F	Ne
Na	Mg											Al	Si	P	S	Cl	Ar
K	Ca	Sc	Ti	V	Cr	Mn	Fe	Co	Ni	Cu	Zn	Ga	Ge	As	Se	Br	Kr
Rb	Sr	Y	Zr	Nb	Mo	Tc	Ru	Rh	Pd	Ag	Cd	In	Sn	Sb	Te	I	Xe
Cs	Ba	La	Hf	Ta	W	Re	Os	Ir	Pt	Au	Hg	Tl	Pb	Bi	Po	At	Rn
Fr	Ra	Ac	Rf	Db	Sg	Bh	Hs	Mt	Ds	Rg	Cn	Uut	Fl	Uup	Lv	Uus	Uuo

Fig. 15.1 Periodic table highlighting the transition metals of group VB and chalcogens of group VIA involved in group VB–VIA monolayers

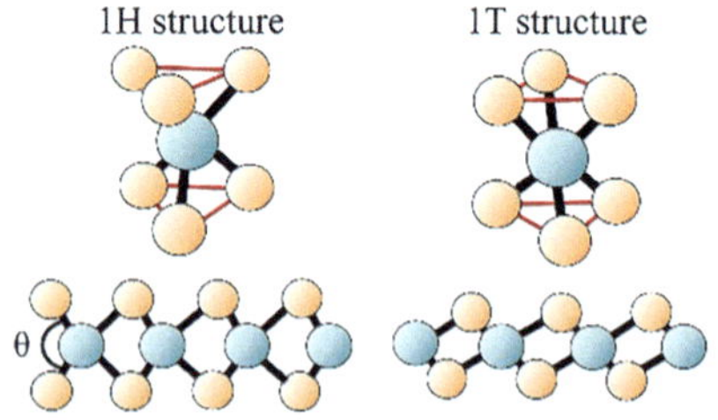

Fig. 15.2 Perspective view and side view of the atomic configuration of the honeycomb H and the centered honeycomb T monolayer structure of group-VB transition-metal dichalcogenides

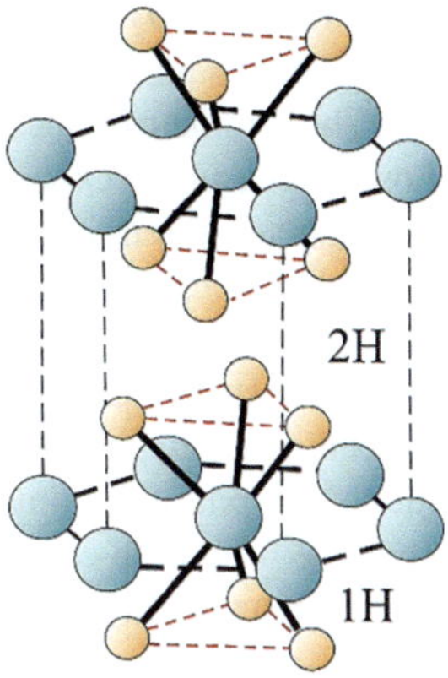

Fig. 15.3 Perspective view of the 1H structure of the monolayer and the 2H structure realized in bilayers and the bulk material

electronegativities of V (1.63) and O (3.44) leads to a substantial ionic contribution to bonding and thus to strongly polar V–O bonds. A Bader charge transfer analysis of the 1H phase yields a charge transfer of $\delta q = 1.79$ e from each V atom to two O atoms [1]. The charge transfer induces a positively charged inner V sublayer and negatively charged outer O sublayers. The strong covalent-ionic bonding produces a cohesive energy of $E_c = -7.21$ eV/atom, which was calculated with the exchange correlation potential of the local density approximation (LDA) for the H phase [1]. This large binding energy has not been confirmed by density functional theory (DFT) calculations, which approximate the exchange correlation potential by the generalized gradient approximation (GGA) using Perdew Burke Ernzerhof (PBE) functionals. According to this commonly used approach, the cohesive energy is only $E_c = -5.26$ eV/atom [2].

15.1.1.2 2D Structures

The 1H structure of VO_2 monolayers has a lattice constant of a = 0.270 nm, a bond length of $d_{V\text{-}O} = 0.192$ nm, a distance between O atoms of $h_{O\text{-}O} = 0.224$ nm, and a bond angle of $\theta_{OVO} = 71.34°$ (LDA) [1]. These values agree well with those calculated by the DFT-GGA-PBE-D2 approach of a = 0.274 nm, $d_{V\text{-}O} = 0.195$ nm, $h_{O\text{-}O} = 0.227$ nm, and $\theta_{OVO} = 71.29°$ [2]. By adding the van der Waals diameter of O atoms of $d_w = 0.310$ nm and the distance between O atoms an estimate of the layer thickness of $d_{w+h} = 0.534$ nm is obtained. The VO_2 monolayer has metallic and magnetic properties and preserves the electronic and magnetic properties of the bulk material because only weak van der Waals forces with a low exfoliation energy of 16.23 meV/Å^2 must be overcome when removing a VO_2 monolayer (graphene: 21.6 meV/Å^2). The space group of the H monolayer is *P*-6*m*2. The states with spin up and spin down cross the Fermi level with a nonsymmetrical band structure of spin up and spin down electrons. This generates a total magnetic moment of 0.22 μ_B/atom, originating from hybridization and polarization processes between V-*d* and O-*p* orbitals [2]. Notably, both the 1H-VO_2 monolayer and the 1T-VO_2 monolayer are energetically stable, however, dynamically unstable at room temperature owing to imaginary frequencies [3]. The symmetry, cohesive energy, lattice constant, bond length, chalcogen distance, bond angle, and thickness of group VB–VIA monolayers are presented in Table 15.1.

A square-pyramidal 2D-VO_2 structure is thermodynamically and dynamically stable [3]. In this square-pyramidal S-VO_2 monolayer, a V atom binds to five instead of six neighboring O atoms, resulting in the space group *P4/nmm-6m*. The optimized lattice constant of a = 0.375 nm is much larger than those of the hexagonal structures. The authors calculated an even lower cohesive energy of $E_c = -7.38$ eV/atom for the S-VO_2 phase and thus stronger bonding than for the 1H-VO_2 phase. Based on the negative formation energy of $E_f = -0.271$ eV they concluded that this structure is thermodynamically stable at room temperature [3]. This study predicts as cohesive energy of 1H-VO_2 monolayers $E_c = -7.11$ eV/atom, comparable with the value presented above. As the ground state of the semiconducting S-VO_2 phase exhibits no magnetism due to the antiferromagnetic coupling between the two nearest V atoms, a biaxial compressive strain of 8% results in ferromagnetism. The direct band gap of the semiconductor is $E_g = 1.25$ eV (PBE). Moreover, hydrogenation and fluorination of the monolayer induce large magnetic moments. Functionalization allows tuning from a nonmagnetic semiconductor to a magnetic semiconductor and to a half-metal by changing the structure by functionalization from the S-VO_2-2F monolayer to the S-VO_2-2H monolayer and to the S-VO_2-HF monolayer, respectively.

15.1.1.3 Mechanics

Early LDA calculations delivered a large Young's modulus of 172 N/m for the H phase of vanadium dioxide monolayers [1]. This value may be considered as an upper limit of the in-plane stiffness because it is based on the LDA approximation. Comprehensive work based on the atomic finite element approach with a valence force field method provided Young's moduli and Poisson's ratios for 34 TMDCs. For the VO_2 monolayers, these calculations yielded much lower and slightly anisotropic Young's moduli at $E_{ac} = 134.6$ N/m and $E_{zz} = 137.2$ N/m, and similar Poisson's ratios at $\nu_{ac} = 0.17$ and $\nu_{zz} = 0.16$ [4]. The space group, Young´s modulus and Poisson´s ratio of group VB–VIA monolayers are presented in Table 15.2.

15.1.1.4 Physics and Applications

VO_2 monolayers were deposited on glass slides by reactive RF sputtering, which led to an oriented deposit with needle-shaped nanostructures, allowing the study of the high visible transmittance of these monolayers and the thermochromic thin films [5]. For monolayers, the observed transmittance of 67% in the visible region should advance their practical application in smart windows. The fabrication of a blue-green colored 2D-VO_2 monolayer with inverse opal behavior for thermochromic applications was possible by an improved dynamic hard-template infiltration strategy

Table 15.1 Symmetry, cohesive energy, lattice constant, bond length, X–X distance, bond angle, and corrugation corrected van der Waals diameter (thickness) of group VB-VIA monolayers

Monolayer	Symmetry	Cohesive energy, E_c (eV/atom)	Lattice constant a (nm)	Bond length d_{MX} (nm)	Distance h_{XX} (nm)	Bond angle θ_{XMX} (°)	Thickness d_{w+h} (nm)
VO_2	1H	– 7.21 [1], – 5.26 [2]	0.270 [1]	0.192 [1]	0.224 [1]	71.34 [1]	0.534
VS_2	1H	– 5.82 [1], – 4.97 [11]	0.309 [1]	0.231 [1]	0.295 [1]	79.14 [1]	0.655
VSe_2	1H	– 5.32 [1], – 4.51 [11]	0.324 [1]	0.245 [1]	0.317 [1]	80.49 [1]	0.701
VTe_2	1T	– 4.75 [1], – 4.01 [11]	0.346 [1]	0.264 [1]	0.400 [1]	98.35 [1]	0.820
NbS_2	1T, (H)	– 6.55 [1], (– 5.44 [42])	0.330 [1]	0.245 [1]	0.362 [1]	95.25 [1]	0.722
$NbSe_2$	1H	– 6.08 [1], – 4.92 [42]	0.340 [1]	0.257 [1]	0.333 [1]	80.68 [1]	0.713
$NbTe_2$	1T, (H)	– 5.46 [1], (– 3.34 [42])	0.356 [1]	0.277 [1]	0.424 [1]	100.05 [1]	0.844
TaS_2	1H	– 5.92, – 6.87 [42]	0.334 [42]	0.248 [42]	0.313 [42]	78.11 [42]	0.673
$TaSe_2$	1H	– 5.34, – 6.31 [42]	0.347 [13]	0.261 [13]	0.335 [13]	79.87 [13]	0.715
$TaTe_2$	1H	– 4.67, – 5.62 [42]	0.370 [13]	0.281 [13]	0.366 [13]	81.19 [13]	0.786

Table 15.2 Space group, Young's modulus and Poisson's ratio of group VB–VIA monolayers

Monolayer	Space group	Young's modulus E_{ac}, E_{zz} (N/m)	Poisson's ratio ν_{ac}, ν_{zz}
VO_2	*P*-6*m*2	134.6, 137.2 [4], 172.0 [1]	0.17, 016 [4]
VS_2	*P*-6*m*2	85.7, 88.0 [4], 93 [15]	0.27, 0.25 [4], 0.305 [15]
VSe_2	*P*-6*m*2	81.4, 83.5 [4], 75.7 [15]	0.23, 0.21 [4], 0.317 [15]
VTe_2	*P*-3*m*1	67.3, 69.3 [4], 54.5 [1]	0.27, 0.25 [4]
NbS_2	*P*-3*m*1	96.6 [1], 87.5, 89.9 [4]	0.26, 0.24 [4]
$NbSe_2$	*P*-6*m*2	87.2 [1], 80.2, 82.6 [4]	0.28, 0.26 [4]
$NbTe_2$	*P*-3*m*1	64.1 [1]	
TaS_2	*P*-6*m*2	52.3, 57.8 [65], 86.8, 89.2 [4]	0.27 [65], 0.26, 0.25 [4]
$TaSe_2$		80.7, 83.1 [4]	0.28, 0.26 [4]
$TaTe_2$	*P*-3*m*1	55, 55.7 [65]	0.10 [65]

that uses a sacrificial polystyrene sphere (PS) photonic crystal as the template [6]. Laser stimuli and temperature-dependent Raman spectroscopy helped to identify a reversible metal-insulator transition (MIT) phase change. The films showed high optical contrast, low change in visible transmittance, and a mean MIT temperature of 57.9 °C, demonstrating their potential for meeting the color requirements of industrial windows.

First-principles calculations suggest the suitability of metallic VO_2 monolayers as anode material in battery applications with high capacity, high electrical conduction, and excellent ion mobility [7]. Metal adatoms such as Li, Na, K, Mg, and Ca preferably occupy the center of the hexagonal V–O rings and function as charge donors in fast charging/discharging processes. The calculations revealed excellent mobilities with low diffusion barriers of 0.043 eV for K, 0.119 eV for Li, 0.098 eV for Na, 0.517 eV for Mg, and 0.306 eV for Ca and specific capacities of up to 968, 613, and 815 mA h g^{-1} for Li, Na, and Mg, respectively; these values are significantly greater than those of graphene.

A DFT-based study investigated the adsorption of CH_4, CO, and H_2O on sandwich 2D VO_2 tablets [8]. The adsorption energies of – 229.5 meV (CH_4), – 239.1 meV (CO), and – 388.3 meV (H_2O) indicate a strong preference for H_2O adsorption, which is difficult to desorb and therefore hinders the application as a gas sensor. In addition to the change in the electronic properties resulting from significant charge transfer, a redshift of the dielectric function of the VO_2 sheets occurs, indicating a notable change in the optical properties. This finding qualifies VO_2 nanosheets for potential application as highly selective optical gas sensors.

15.1.1.5 Synthesis

The MIT observed in 2D VO_2 near room temperature with a change in the electrical conductivity of up to six orders of magnitude offers switchable electronic states of interest for flexible low-power electronic and photonic applications [9, 10]. To study the interplay between electrons and the lattice in this symmetry-lowering structural phase transition, ultrathin free-standing membranes with pronounced MITs are of interest. Available methods allow the fabrication of VO_2-based free-standing membranes with a lateral size of millimeters and thicknesses down to 5 nm [9]. A dissolvable $Sr_3Al_2O_6$ sacrificial layer between the VO_2 film and the c-Al_2O_3 substrate allowed the transfer of the material to any other surface. Furthermore, an intermediate Al_2O_3 buffer layer detaching together with the 2D-VO_2 film tremendously enhanced the MIT process.

15.1.2 VS_2 Monolayer

15.1.2.1 Chemistry and 2D Bonding

The valence electrons of the electronic configurations of the V atoms (Ar|$3d^3 4s^2$) and S atoms (Ne|$3s^2 p^4$) determine the possible 2D bonding arrangements in vanadium disulfide monolayers. Owing to the only partially filled *d* band with a single *d* electron, bonding of the monolayer has a metallic character. An analysis of the electronic density of states (DOS) indicates that hybridization of the 3*d* orbitals of the V atoms and the 3*p* orbitals of the S atoms are responsible for strong 2D covalent bonding. This results in the formation of triple-layer structures with a V sublayer in the center sandwiched between two S sublayers, where each S atom forms three covalent bonds with V atoms in a bridging position and each V atom is bound to six S atoms. Extensive electron charge accumulation between the V and S atoms confirms the formation of strong covalent V–S bonds. The reduced difference in the electronegativities of V (1.63) and S (2.58) leads to a smaller ionic contribution to bonding than observed for the V–O bonds. A Bader charge transfer analysis of the stable hexagonal 1H phase yields a charge transfer of $\delta q = 1.29$ e from the V atom to two S atoms and thus to a transfer of $\delta q = 0.65$ e to each S atom [1]. The covalent-ionic bonds produce a cohesive energy of the 1H phase of $E_c = -5.82$ eV/atom and an identical value for the 1T phase, calculated by using the LDA approximation [1]. Spin-polarized DFT calculations of electronic and magnetic properties indicate that the 1H-VS_2 monolayer with a cohesive energy of $E_c = -4.97$ eV/atom is -26 meV more stable than the 1T-VS_2 monolayer [11]. Because the same effect exists for the bilayers, these authors concluded that for VS_2 and VSe_2, which exhibit similar behaviors, the 1H phase is energetically more stable, despite the large errors involved in such calculations [11].

15.1.2.2 2D Structures

The lattice constant of 1H-VS_2 monolayers is a = 0.309 nm, and the other lattice parameters are $d_{V\text{–}S}$ = 0.231 nm, $h_{S\text{–}S}$ = 0.295 nm, and θ_{SVS} = 79.14° (LDA) [1]. These values deviate slightly from those calculated by the DFT-GGA-PBE approach for the 1H structure of a = 0.317 nm, $d_{V\text{–}S}$ = 0.236 nm, and $h_{S\text{–}S}$ = 0.298 nm [12]. The distance between S atoms and their diameter yield a thickness of d_{w+h} = 0.655 nm. According to the projections of the DOS, there are no states at the Fermi level, indicating the semimetallic nature of this phase. The magnetic moment of the 1H phase is 1.01 μ_B. While the 2H-VS_2 bilayer (bulk) is metallic, the monolayer becomes semimetallic due to the enhanced quantum confinement of electrons in two dimensions [10]. The space group of this phase is *P*-6*m*2 [11]. Note that there is no final agreement on the nature of the ground state of the VS_2 monolayer, which is also described as an indirect semiconductor with intrinsic ferromagnetism and an in-plane magnetic anisotropy, based on DFT-GGA calculations [11, 12].

The lattice parameters of the 1T-VS_2 monolayer are a = 0.310 nm, $d_{V\text{–}S}$ = 0.231 nm, $h_{S\text{–}S}$ = 0.343 nm, and θ_{SVS} = 95.44° (LDA) [1]. A systematic theoretical study on the suitability of TMDCs for photocatalytic water splitting presented similar geometrical data [13]. As mentioned above, the metallic 1T phase is ~ 26 meV less stable than the 1H-VS_2 monolayer. This is evident from the presence of delocalized states near the Fermi level, originating predominantly from the 3*d* orbitals of the V atoms (see Fig. 15.4) [11]. Generally, 1H-VX_2 monolayers have much larger magnetic moments than the 1T-VX_2 monolayers, e.g., for the 1T phase of VS_2, the magnetic moment is 0.51 μ_B, and for the 1H phase it is 1.01 μ_B [11].

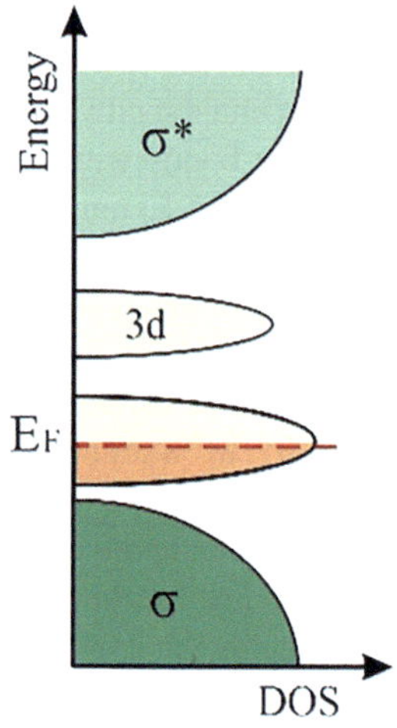

Fig. 15.4 Schematic representation of the band structure (DOS) with 3*d* bands of metallic 1T-VS_2 monolayers. The filled and unfilled states are shaded with dark and light green color, respectively. Adapted with permission from Ref. [1], Copyright 2017, American Chemical Society

15.1.2.3 Mechanics

To date, only Young's moduli of various phases of vanadium disulfide monolayers have been calculated, with slightly different values for the 1H phase at E = 106.0 N/m and the 1T phase at E = 104.3 N/m [1]. The linear mechanical properties extracted from the calculated elastic constants of H-type VS_2 predict a lower stiffness of E = 93 N/m and a Poisson ratio of $\nu = 0.305$ [15]. The results derived by finite element calculations are $E_{ac} = 85.7$ N/m and $E_{zz} = 88.0$ N/m for Young's moduli and $\nu_{ac} = 0.27$ and $\nu_{zz} = 0.25$ for Poisson's ratios [4].

15.1.2.4 Physics and Applications

Nanometer thick metallic VS_2 nanosheets have been suggested as a platform for multifunctional electrode applications [15]. Progress has been made in the fabrication of such layered materials that can be used as building blocks to construct next generation electronic and energy-storage devices. Tuning of the electronic and optical properties by changing the structure is of particular interest if several stable crystal structures exist for the same 2D material. This is the case for VS_2 monolayers with a stable semiconducting 1H-VS_2 phase and an energetically slightly metastable, however, dynamically stable metallic 1T-VS_2 phase. Charge doping allows the initiation of transitions between these phases [16]. For VS_2 monolayers, the transition between the semiconducting and the metallic phase generates a large electronic contrast, whereas other group–VB TMDCs exhibit a metal-to-metal transition. The H phase of VS_2 is a bipolar magnetic semiconductor whose valence and conduction states near the Fermi level carry opposite spin polarizations. Therefore, a small amount of charge doping (electron or hole) induces an electronic phase transition from the semiconductor to a half metal [16]. When the hole concentration reaches a certain value, a transition from the H to the T phase occurs, which is associated with an electronic phase transition from the half metal to a normal metal.

The present information on the coexistence of piezoelectricity and magnetism in magnetic H-type VS_2, VSe_2 monolayers, and Janus-type VSSe semiconductors originates from DFT-GGA-PBE and density functional perturbation theory (DFPT) [15]. According to phonon frequency analysis, the three monolayers are dynamically stable. While VS_2 and VSe_2 exhibit high in-plane piezoelectricity, the Janus-VSSe monolayer is a magnetic semiconductor with in-plane and vertical piezoelectric properties owing to the broken symmetry in the z direction. The corresponding d_{11} values (e.g., $d_{11} = 2.34$ pm V^{-1} for VS_2) are larger than those of traditional 3D piezoelectric materials such as α quartz ($d_{11} = 2.27$ pm V^{-1}). According to theory, the H structure is the energetically preferred free-standing phase, however, synthesis in the adsorbed state yields the T structure.

A transition from the bulk to the monolayers of VS_2 and VSe_2 changes the structure from a metal to an insulator [17]. The enhanced Coulomb interaction, occurring with decreasing film thickness, introduces insulating behavior, and the layer becomes ferromagnetic with possible spin currents through thermal gradients across the ferromagnetic layer. With increasing thickness, the few-layer sheets become metallic with a large anomalous Hall conductivity (AHC) of ~ 100 S/m [17].

Layered TMDCs, including VS_2, have attracted great interest as alternative anode materials in ion batteries, which are among the most important rechargeable energy storage devices [18]. The adsorption energy and diffusion coefficients of lithium and sodium ions in H-type monolayers and T-type bulk VS_2 electrodes were studied by DFT calculations, which revealed a low adsorption energy and high ion diffusion rate. The capacity can reach up to 466 mA h g^{-1}, and the output voltages are 0.72 and 0.48 V for lithium and sodium ion batteries in monolayer VS_2 and 0.88 and 0.6 V in bulk VS_2, respectively. Note that the theoretically estimated Li and Na diffusion coefficients in bulk VS_2 are five and seven orders of magnitude greater than those reported, for example, for bulk MoS_2.

In future applications of 2D VS_2 in Li-ion batteries, owing to its good cycling stability and large capacity, in addition to its ionic conductivity, the electronic conductivity will play a significant role. First-principles calculations indicate that W, Re, and Co are the most effective dopants for improving the electronic conductivity of all transition elements, where a narrow band gap and shorter V–S bond length contribute to an increase in the electronic conductivity [19].

A proposed facile chemical vapor deposition (CVD) route for the direct growth of metallic vanadium disulfide nanosheets allows the production of the building blocks needed for multifunctional electrode applications [14]. These nanosheets feature a sub-10 nm thickness, a domain size of tens of micrometers, and excellent electrical conductivity. The materials enable various applications, such as contact electrodes that reduce the contact resistance and supercapacitor electrodes in aqueous electrolytes. The nanosheets enable contact electrodes for monolayer MoS_2 with contact resistances of ~1/4 to that of Ni/Au metals.

Electrical-driven water electrolysis is of foremost importance because of its high efficiency and eco friendliness. The fabrication of highly active, stable, and inexpensive electrocatalysts is a critical issue in the development of green energy technology. The attractive features of TMDC catalysts include a high surface area, active edge states, and high mechanical flexibility. Theoretical work elucidated the effect of tension on the hydrogen evolution reaction (HER) activity for VS_2 monolayers used as electrocatalysts [20]. Tuning the applied tension provides access to the optimal condition, namely, near-zero Gibbs free energy. The tension-induced improvement in the catalytic activity is due to an increase in the free carrier density around the Fermi level.

Engineering the basal plane of VS_2 monolayers by doping with 3*d* transition metal atoms (Ti–Ni) further improved the efficiency of the HER catalyst [21]. The reason for the better performance of doped surfaces in comparison to pristine surfaces is charge modulation at S sites. The three transition metals Ti, Mn, and Co are perfect catalysts for the HER, with $\Delta G_H \approx 0$ eV, and are even superior to Pt. Notably, besides the type of transition metal, the doping concentration clearly plays a crucial role.

First-principles computations assisted in the development of efficient bifunctional electrocatalysts to boost the oxygen reduction reaction (ORR) and the oxygen evolution reaction (OER) [22]. The authors provide a systematic investigation of the catalytic activity of a series of single transition-metal atoms anchored at the VS_2 monolayer for ORR/OER. The results revealed that Ni@VS_2 provides low overpotentials for both the ORR (0.45 V) and the OER (0.31 V). The effect originates from the moderate interaction with oxygenated intermediates according to the established scaling relationship and volcano plot.

The CVD method allowed the growth of stable single-crystal sheets of 1T-VS_2 multilayers. X-ray diffraction (XRD), transmission electron microscopy (TEM), Raman spectroscopy, and atomic force microscopy (AFM) were employed for the analysis of the deposits [23]. The suitability of 1T-VS_2 few-layer sheets as electrocatalysts for the HER was demonstrated by the electrochemical reduction of water. The polarization curve (j–E plot) indicated a negligible overpotential for initializing the HER, like that of the reference Pt electrode. A current density of 10 mA cm^{-2} caused a slightly larger overpotential of − 68 mV than that observed for Pt, which was only − 43 mV. The authors attributed the excellent performance to the high activity of the basal plane and edge sites, which both approach a H adsorption free energy approximating $\Delta G_H \approx 0$ eV. The study identified electrochemical desorption as the rate-limiting step of the reaction process.

15.1.2.5 Synthesis

Monolayers of 1T VS_2 were prepared on bilayer graphene deposited on 4H-SiC(001) substrates by epitaxial growth [24]. The 2D-VS_2 deposit showed octahedral symmetry and a similar in-plane lattice parameter as its bulk material and it was epitaxially aligned to graphene with small tensile strain. The experiments did not confirm the theoretically predicted spin-polarized band structures. Angle-resolved photoemission spectroscopy (ARPES) indicated gap opening at low temperatures. Furthermore, a temperature-dependent evolution of a CDW gap occurred with a transition temperature of $T_{CDW} \approx 400$ K.

For the growth of VS_2 monolayers, a two-step MBE method allowed the deposition of sulfur-based TMDCs on an inert graphene-covered Ir(111) substrate [25]. DFT and DFPT calculations of the acoustic phonon dispersion of the undistorted 1T-VS_2 monolayer confirmed structural instability and a tendency toward CDW formation, involving nonlinear coupling of transverse and longitudinal phonons. Both scanning tunneling microscopy (STM) and scanning tunneling spectroscopy (STS) measurements, as well as ab initio calculations, confirmed the realization of CDWs, however, different from the standard model, the gap resides in the unoccupied states of VS_2 above the Fermi level, accompanied by a topological metal–metal transition at the Fermi level.

A topical review addresses various routes for synthesizing high-quality metallic TMDCs, including chemical vapor transport, mechanical/chemical exfoliation, MBE, and scalable CVD methods employing transition metal chlorides and transition metal oxides mixed with alkali halides [26]. The study focuses on the controllable growth of VS_2, VSe_2, NbS_2, $NbSe_2$, TaS_2, and $TaSe_2$ sheets.

15.1.3 VSe_2 Monolayer

15.1.3.1 Chemistry and 2D Bonding

The valence electrons of the electronic configurations of the V atoms (Ar|$3d^34s^2$) and Se atoms (Ar|$3d^{10}4s^2p^4$) determine the type of 2D bonding in vanadium diselenide monolayers. An analysis of the electronic density of states (DOS) indicates that hybridization of the 3*d* orbitals of the V atoms and 4*p* orbitals of the Se atoms form covalently bonded triple-layer structures with a V sublayer in the center sandwiched between two Se sublayers, as described for the VO_2 and VS_2 monolayers. The difference in the electronegativities of V (1.45) and Se (2.48) leads to an ionic contribution to bonding comparable to that in the V–S bonds. A Bader charge transfer analysis for the most stable hexagonal 1H phase yields a charge transfer of $\delta q = 1.05$ e from one V atom to two Se atoms [1]. Covalent-ionic bonding produces a cohesive energy for the 1H phase of $E_c = -5.32$ eV/atom (LDA) [1]. This should be compared with the much smaller binding energy obtained by the PBE approach with a cohesive energy of – 4.51 eV/atom[11] and the two independently calculated nearly coinciding values at $E_c = -5.16$ eV/atom (LDA) and – 5.13 eV/atom (GGA) [27].

15.1.3.2 2D Structures

The lattice parameters of the 1H-VSe_2 monolayer are a = 0.324 nm, $d_{V\text{–}Se}$ = 0.245 nm, h_{SeSe} = 0.317 nm, and θ_{SeVSe} = 80.49° (LDA) [1]. These values deviate only slightly from those calculated by the DFT-PBE approach for the 1H phase of a = 0.333 nm, $d_{V\text{–}Se}$ = 0.250 nm, $h_{Se\text{–}Se}$ = 0.321 nm. Adding the diameter of Se atoms and their separation yields a layer thickness of d_{w+h} = 0.380 nm + 0.321 nm = 0.701 nm. Spin-polarized DFT calculations of the electronic and magnetic properties show that the difference in the energetic stabilities between the 1H and the 1T phases of VSe_2 monolayers is further diminished compared to that of the VS_2 monolayer with $E_c = -4.50$ eV [11] (PBE) and – 5.32 eV/atom [1] (LDA) for 1T VSe_2] [11]. According to the projections of the DOS, there are no states at the Fermi level, indicating a narrow indirect band gap semiconductor with a band gap of ~ 0.24 eV. The magnetic moment of the 1H phase is 1.01 μ_B. Remarkably, the 2H-VSe_2 bilayer is metallic, whereas the monolayer transforms to a semiconductor due to the enhanced quantum confinement of electrons in two dimensions [11]. The space group of the H structure is *P*-6*m*2.

For comparison, the lattice parameters of the 1T-VSe_2 monolayer consist of a similar lattice constant of a = 0.324 nm and a similar bond length of $d_{V\text{–}Se}$ = 0.244 nm;

however, there is a larger thickness due to the larger distance between the Se atoms of $h_{Se-Se} = 0.366$ nm and a much larger bond angle of $\theta_{SeVSe} = 97.04°$ (LDA) [1].

15.1.3.3 Mechanics

Calculations of the elastic moduli of the two stable phases of VSe_2 monolayers yielded similar values for the 1H phase at E = 82.9 N/m and the 1T phase at E = 80.2 N/m [1]. The linear mechanical properties extracted from published elastic constants of 1H-VSe_2 monolayers are E = 75.7 N/m and $\nu = 0.317$ [14]. In addition, finite element calculations with a valence force field model yielded comparable slightly anisotropic stiffnesses of $E_{ac} = 81.4$ N/m and $E_{zz} = 83.5$ N/m and Poisson's ratios of $\nu_{ac} = 0.23$ and $\nu_{zz} = 0.21$ [4].

15.1.3.4 Physics and Applications

A comprehensive study describes the use of in situ microscopic and spectroscopic techniques such as STM/STS, synchrotron-based photoelectron spectroscopy (PES), ARPES, and X-ray absorption spectroscopy (XAS) to characterize VSe_2 monolayers grown by MBE on highly oriented pyrolytic graphite (HOPG) [28]. The measurements revealed a thickness of the metallic 1T-VSe_2 monolayer at 0.7 nm. Element-specific X-ray magnetic circular dichroism (XMCD), complemented by magnetic susceptibility measurements, characterized the VSe_2 monolayer as a frustrated magnet, with its spins exhibiting subtle correlations, albeit in the absence of long-range magnetic order. The XMCD measurements were performed in a magnetic field of up to 7 T and at temperatures down to 2 K. The observed results are attributed to the relative stability of the ferromagnetic and antiferromagnetic ground states, arising from the specific structural features. Notably, at this point, it is not possible to exclude unintended magnetic contributions from other sources, such as contaminations and defects. According to ARPES measurements, the slightly greater stability of the 1H phase is incorrect. The inclusion of Hubbard on-site repulsion in the DFT calculations stabilizes the 1T structure and antiferromagnetic ordering. The presence of structural features, such as misaligned domains and vacancies, may lift the small energy difference to ferromagnetic ordering.

Unfortunately, contradictory experimental and theoretical results are still being reported. An important step forward was achieved by combining first-principles methods with dynamical mean-field theory (DMFT) to study the magnetic properties of VSe_2 by applying the charge self-consistent LDA-DMFT approach [29]. This theory still produces a ferromagnetic ground state, as consistently obtained by DFT-GGA calculations, however, the paramagnetic state of the bulk is correctly described by the extended theory. The calculated small magnetic moment of 0.6 μ_B per V ion is in reasonable agreement with the recently measured strongly reduced magnetic moment of ~ 0.31 μ_B and thus resolves the inconsistency of the measured moments that are too large [29].

15.1.3.5 Synthesis

The exfoliation energy of 18.9 meV/Å^2 is comparable to that of graphene (21.0 meV/Å^2) and in principle suggests easy preparation of VSe_2 monolayers [27]. Nevertheless, exfoliation of monolayers from layered bulk VSe_2 is challenging. Electrochemical exfoliation with organic tetrapropyl-ammonium cations as intercalants separated the 1T-VSe_2 monolayer flakes in high yield (see Fig. 15.5) [30]. Passivation of the exfoliated flakes is possible by the addition of a thiol group to protect the flake for subsequent characterization. The methods used to detect strong ferromagnetism include superconducting quantum interference devices (SQUIDs), XMCD, and magnetic force microscopy (MFM) at room temperature. Ferromagnetic behavior and charge density waves (CDWs) exist in 1T-VSe_2 nanosheets with 4–8 monolayers [31]. In these nanosheets, the CDW transition temperature increases to $T_{CDW} = 135$ K.

A one-pot colloidal route was used in which Se powder was employed as a precursor to replace expensive, air-sensitive, and toxic compounds. The more reactive vanadyl acetylacetonate ($VO(acac)_2$) served as a precursor to prepare single-crystalline single-layer VSe_2 samples. Tapping-mode AFM measurements resulted in a monolayer thickness of ~ 0.4 nm [32]. Oxygen plasma exposure (OPE) increases the number of active sites and thus the catalytic activity of single-layer VSe_2. Moreover, OPE treatments delivered an overpotential of 0.206 V for a drive current density of 10 mA cm^{-2} and a Tafel slope of 88 mV per decade.

The growth of high-quality VSe_2 monolayers is possible by MBE in an ultrahigh vacuum (UHV) chamber using ultrapure vanadium and selenium samples for the deposition on a HOPG substrate [33]. Contrary to the findings of other reports, the monolayers were surprisingly stable in air. Furthermore, STS measurements revealed characteristic magnetism-related peaks in the electronic DOS of the as-grown VSe_2 monolayers. Note that magnetic properties are difficult to uncover owing to the lack of high-quality samples and the presently available measurement techniques.

Strong ferromagnetic ordering emerges in VSe_2 monolayers and van der Waals heterostructures grown on graphite (HOPG) or MoS_2 substrates by MBE. Encapsulation of the chemically unstable monolayer has been achieved by a Se capping layer [34]. Notably, the ferromagnetic ordering with a large magnetic moment did not disappear above room temperature, suggesting that VSe_2 is useful for applications

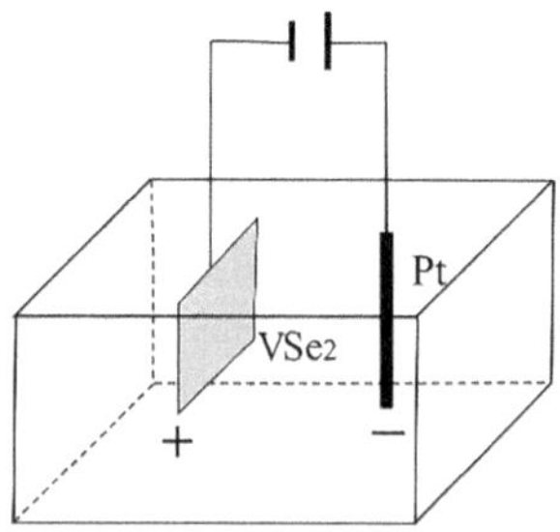

Fig. 15.5 Scheme of the electrochemical setup employed for cathode exfoliation of VSe_2 sheets with tetrapropylammonium cations as the intercalant

as a material for van der Waals spintronics that uses the spin degree of freedom of electrons rather than their charge. Compared with those of the multilayer samples, in single layers the magnetic moment was significantly enhanced.

15.1.4 VTe_2 Monolayer

15.1.4.1 Chemistry and 2D Bonding

The valence electrons of the electronic configurations of the V atoms (Ar|$3d^3 4s^2$) and the Te atoms (Kr|$4d^{10} 5s^2 p^4$) determine the type of 2D bonding in vanadium ditelluride monolayers. Hybridization of the V-3*d* orbitals and the Te-5*p* orbitals results in the formation of stable phases of covalently bonded triple-layer 1H and 1T structures with a vanadium sublayer in the center, sandwiched between two Te sublayers, as described for the other vanadium dichalcogenide monolayers. The relatively small difference in the electronegativity of V atoms (1.45) and Te atoms (2.01) introduces a modest ionic contribution to bonding. A Bader charge transfer analysis of the most stable hexagonal 1H phase yields a charge transfer of $\delta q = 0.80$ e from each V atom to two Te atoms [1]. The polar covalent-ionic bonds produce a cohesive energy of the 1H phase at $E_c = -4.72$ eV/atom (LDA) and -4.75 eV/atom (LDA) for the 1T phase [1]. Note that such small differences between cohesive energies are within the errors involved in such calculations and may not allow a decision to be made, which phase is energetically more stable. These values should be compared with the cohesive energies obtained by the PBE approach of $E_c = -3.99$ eV/atom (1H) and -4.01 eV/atom (1T) indicating much smaller binding energies [11]. A lower binding energy of 1H-VTe_2 monolayers is supported by recent calculations of the cohesive energy at $E_c = -3.92$ eV/atom [35]. The cohesive energies obtained in this work for the more stable 1H-VSe_2 phase and 1H-VS_2 monolayers are $E_c = -4.67$ eV/atom and -5.20 eV/atom, respectively.

15.1.4.2 2D Structures

The lattice parameters of 1H-VTe_2 monolayers are $a = 0.348$ nm, $d_{V\text{–}Te} = 0.266$ nm, $h_{TeTe} = 0.348$ nm, and $\theta_{TeVTe} = 81.90°$ (LDA) [1]. These values deviate slightly from those calculated by the GGA-PBE approach for the 2H phase of $a = 0.359$ nm, $d_{V\text{–}Te} = 0.272$ nm, and $h_{TeTe} = 0.351$ nm [12].

The corresponding lattice parameters of the 1T-VTe_2 monolayers are $a = 0.346$ nm, $d_{V\text{–}Te} = 0.264$ nm, $h_{TeTe} = 0.400$ nm, and $\theta_{TeVTe} = 98.35°$ (LDA) [1]. Adding the van der Waals diameter of Te atoms of $d_w = 0.420$ nm to the distance between Te atoms yields a layer thickness if $d_{w+h} = 0.820$ nm. The stable phase is metallic, as evidenced by the presence of delocalized states near the Fermi level, originating predominantly from the $3d^1$ electrons of the V atoms, which are also

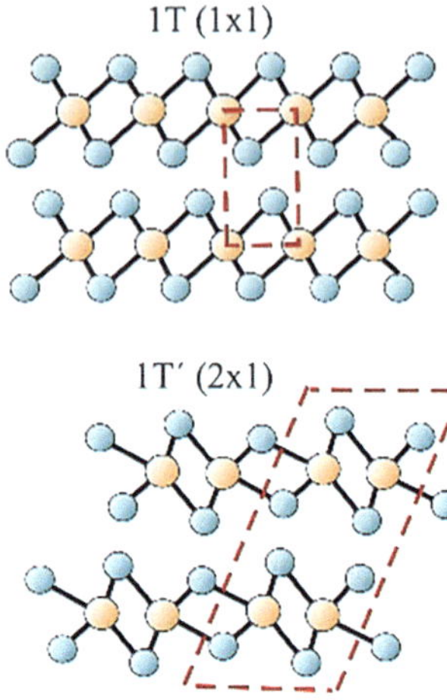

Fig. 15.6 Comparison of the stable hexagonal 1T structure with the distorted 1T′ structure. Adapted with permission from Ref. [37], Copyright 2019, American Chemical Society

involved in CDW order and magnetism (see Fig. 15.4) [11]. According to first-principles calculations, the 1T-VTe_2 monolayers are antiferromagnetic metals with the space group P-3m1 [36]. The calculations identified three metastable phases, two antiferromagnetic semiconductors with space groups $P$21/m and $C$2/m and a ferromagnetic half-metal phase with space group P3m1. Electron doping transfers the $C$2/m phase to the P3m1 phase, and hole doping transfers the C2/m structure to the P-3m1 phase [36].

Theory predicts that due to the loss of interlayer interactions, the stable distorted 1T' structure of the bulk and multilayer sheets transforms to the hexagonal 1T structure of the monolayer (see Fig. 15.6) [37]. X-ray photoelectron spectroscopy (XPS) analysis confirmed electron transfer from the d bands of V atoms to the Te atoms in the monolayer. This charge transfer obviously reduces in-plane 3d orbital hybridization and thus favors the undistorted 1T structure. Imaginary phonon modes cause an instability of the CDW phase that exists as a (4×4) periodic lattice distortion (PLD) and is detectable by low-temperature scanning tunneling microscopy (STM).

Furthermore, first-principles calculations propose a puckered pentagonal VTe_2 (PP-VTe_2) structure that exhibits strong intrinsic ferromagnetism with semiconductive behavior [38]. The magnetic semiconductor has an indirect narrow band gap of $E_g = 0.33$ eV. In the pentagonal motifs, the distance between neighboring V and Te atoms, located in two separate sublayers, is shorter than in all other bonds inside one sublayer, indicating strong interlayer bonding. The structure has the space group $Pca2_1$ and is dynamically and mechanically stable. Interestingly, multiferroic coupling occurs between 2D ferroelasticity and in-plane magnetization, allowing unprecedented control by external strains and electric fields. The estimated Curie temperature of the PP-VTe_2 phase is 110 K.

15.1.4.3 Mechanics

Finite element calculations with a valence force field method revealed slightly anisotropic stiffnesses for the 1H-VTe_2 monolayer at $E_{ac} = 67.3$ N/m and $E_{zz} = 69.3$ N/m and Poisson's ratios at $\nu_{ac} = 0.27$ and $\nu_{zz} = 0.25$ [4]. Surprisingly, these elastic moduli are substantially greater than those calculated by the LDA approximation at E = 49.7 N/m (H) and E = 54.5 N/m (T) [1].

15.1.4.4 Physics and Applications

The monolayers of VSe_2 and VTe_2 show novel physical properties not found in the bulk, such as ferromagnetism, quantum phases such as CDWs, quantum Hall effects, and superconductivity. However, fundamental questions concerning the structure of the ground state and various metastable phases, including their intrinsic properties, are still unresolved. For example, controversial experimental reports on long-range magnetism in the 1T-VTe_2 monolayer still exist.

Employing the first-principles pseudopotential plane-wave method, the spin-polarized electronic structures of the Janus vanadium dichalcogenides VXY (X, Y = S, Se, and Te; X $\neq$ Y) were elucidated [35]. Atomic population charges and differences in electron density produce a high electrostatic potential gradient perpendicular to the layer plane in VSTe and VSeTe monolayers. Spin polarization of the 3*d* orbital component of V atoms introduces pronounced energetic spin splitting of electronic states near the Fermi level, which is responsible for a semimetal band structure and an increasing optoelectronic band gap. The strong SOC of V and chalcogen atoms and a high intrinsic electric field introduce Rashba spin splitting, generating the largest net spin in VSeTe of 1.64 ($\hbar/2$) per unit cell.

15.1.4.5 Synthesis

The prediction of the intrinsic ferromagnetism of VTe_2 monolayers at room temperature initiated the synthesis of high-quality monolayers in a UHV-MBE chamber using high-purity V and Te samples for evaporation on an insulating $Al_2O_3(001)$ substrate with hexagonal symmetry, a smooth surface, and the absence of dangling bonds [39]. Capping all the samples in situ with 15 nm Te provided sufficient protection. Irrespective of the sheet thickness and temperature, no magnetic field hysteresis loop appeared, clearly indicating the absence of intrinsic long-range ferromagnetic order. Confirmation of this result for in-plane oriented applied magnetic fields excludes the existence of an anomalous Hall effect (AHE) for any of the investigated VTe_2 samples.

Synthesis by MBE leads to the formation of monolayers with the 1T-VTe_2 structure as deposit on bilayer graphene substrates [40]. Low-energy electron diffraction (LEED) measurements identified a (4×4) CDW ordering below the CDW transition temperature of 186 K. Moreover, it was possible to identify anisotropic CDW gaps, arising from the 3*d* orbitals of V. The anisotropic CDW gap found in the calculated phonon spectrum is due to imperfect nesting of the CDW wavevector. First-principles calculations indicate phonon softening at the same vector.

As mentioned before, monolayers of the 1T-VTe_2 phase grow on bilayer graphene/SiC by MBE via codeposition of V and Te atoms on the graphene surface [41]. The good agreement in the Fermi-surface topology between ARPES measurements and first-principles band calculations confirms the existence of the octahedral 1T-VTe structure. At low temperatures, the 1T-VTe_2 phase assumes a metallic state. Contrary to the results described above, the observed large triangular Fermi surface at the K point satisfied an almost perfect nesting condition and suppressed the formation of the CDW at low temperature. This behavior deviates from the behavior of VSe_2 monolayers with a well-defined CDW and gapped insulating state.

15.2 Niobium Dichalcogenides (NbS_2, $NbSe_2$, and $NbTe_2$)

15.2.1 NbS_2 Monolayer

15.2.1.1 Chemistry and 2D Bonding

The valence electrons of the electronic configurations of Nb atoms (Kr|$4d^3 5s^2$) and S atoms (Ne|$3s^2 p^4$) determine the type of 2D bonding in niobium disulfide monolayers. The odd number of valence electrons in Nb is responsible for the metallic character of 2D bonding. In the stable 1T and 1H phase, in-plane covalent bonds form the chemically bonded S–Nb–S triple-layer structure, where each S atom participates in three covalent bonds with Nb atoms in a bridging position and each Nb atom is bound to six S atoms, as discussed before for the vanadium compounds. The difference in the electronegativities of Nb (1.6) and S (2.58) leads to an ionic contribution to bonding. According to a Bader analysis, the transfer of electrons from each Nb to two S atoms induces a depletion of charges on Nb atoms of $\delta q = 1.52$ e and excess charges of $\delta q = 0.76$ e on each S atom. The covalent-ionic bonding produces a large cohesive energy of the 1T phase of $E_c = -6.55$ eV/atom, calculated with the exchange correlation potential of the LDA approximation [1]. As is well known, the LDA approximation often overestimates the energy of materials. Other LDA calculations obtained a similar cohesive energy of – 6.45 eV/atom for the 1H phase and, as expected, a much smaller value of only – 5.44 eV/atom by the PBE approach [42].

15.2.1.2 2D Structures

For the lattice parameters of 1T-NbS_2 monolayers, LDA calculations delivered a lattice constant of $a = 0.330$ nm, a bond length of $d_{Nb–S} = 0.245$ nm, a separation of

S atoms of $h_{SS} = 0.362$ nm, and a bond angle of $\theta_{SNbS} = 95.25°$ [1]. The diameter of S atoms and their distance in the 1T structure give a thickness of $d_{w+h} = 0.722$ nm. The 1T-NbS_2 monolayer is a metal. The geometrical parameters calculated with the PBE functional for the 2H-NbS_2 structure of a = 0.336 nm, $d_{Nb–S} = 0.249$ nm, $h_{SS} = 0.314$ nm, and $\theta_{SNbS} = 77.97°$ partly deviate strongly from those of the 1T phase, which has a much larger thickness and bond angle [13]. For example, the thickness of the 1T-NbS_2 monolayer is only $d_{w+h} = 0.674$ nm.

15.2.1.3 Mechanics

The strong covalent-ionic bonding of 1T-NbS_2 monolayers results in a medium stiffness for the three-layer arrangement of E = 96.6 N/m [1]. The application of the finite element method with the valence force field model predicts smaller slightly anisotropic Young's moduli of $E_{ac} = 87.5$ N/m and $E_{zz} = 89.9$ N/m and Poisson's ratios of $\nu_{ac} = 0.26$ and $\nu_{zz} = 0.24$ [4]. Results for the nonlinear mechanical properties are still lacking.

15.2.1.4 Physics and Applications

The origin of phonon-mediated superconductivity in 2H NbS_2 with a critical temperature of $T_c = 5.7$ K is of special interest because this monolayer is known as the only phonon-mediated superconductor without a CDW phase [43]. To elucidate this specific finding, the nature of the superconducting gap and pairing mechanism were studied by the fully anisotropic ab initio Migdal–Eliashberg theory, including Coulomb interactions. The theory describes electron-electron and electron-phonon interactions without any adjustable parameters. Hot spots on the Fermi surface with unusually strong electron-phonon interactions are responsible for the superconductivity and a large contribution to electron pairing. Low-energy anharmonic phonons dominate the electron-lattice coupling, which places the system on the verge of CDW instability. A possible extension of this mechanism to monolayers with CDWs remains an open question.

A study based on ab initio calculations allowed a simultaneous treatment of the fundamental short- and long-range Coulomb interactions controlling electron-electron correlations and electron-phonon interactions [44]. The electronic correlations resulting from this combination of interactions deviate substantially from that of the separate interaction terms. The combination of the lattice structure, spin–orbit coupling (SOC), and enhanced spin and charge susceptibilities suggest that the competing ordering tendencies are subject to frustration effects. Owing to the complex competition between magnetic and charge instabilities, an understanding of the requirements for realizing superconductivity needs a complete analysis of all related aspects. Owing to an unexpected coexistence of strong charge and spin fluctuations NbS_2 is close to spin and charge order.

Contradictory experimental results indicate that an Au(111)-supported 1H-NbS_2 monolayer exhibits no CDW, whereas a 1H-NbS_2 monolayer, supported by a

SiC(0001) substrate, undergoes (3 × 3) reconstruction, as derived from first-principles anharmonic quantum calculations [45]. Calculations indicate that free-standing 1H-NbS_2 monolayers undergo (3 × 3) reconstruction, in agreement with the findings for SiC(0001)-supported monolayers. Note that due to the extreme sensitivity of a CDW to external influences, a small compressive strain of $\varepsilon = 0.5\%$ removes the CDW order.

2D-VbS_2 nanosheets synthesized by thermolysis of an $NbCl_5$ precursor employing the CVD method, allowed the selective and reversible detection of NO_2 at room temperature [46]. Different NbS_2 edge configurations controlled the sensing abilities, which varied with the conditions used during synthesis. DFT calculations revealed the sensing mechanism and elucidated the effects of modified edge configurations.

15.2.1.5 Synthesis

While it is difficult to control the thickness of NbS_2 flakes exfoliated from bulk crystals due to metallic interlayer interactions, the controllable growth of the metallic 3R phase of few-layer NbS_2 sheets was demonstrated by CVD on atomically flat h-BN [47]. The 2H-NbS_2 and 3R-NbS_2 bilayers have the same monolayer structure, whereas the bilayers and multilayers exhibit different stacking orders (see Fig. 15.7) [48]. Structural characterization by AFM, optical microscopy, and electron microscopy revealed single-domain sizes in the micrometer range, and electrical transport measurements confirmed the metallic character down to a thickness of few layers.

The CVD method was applied using $NbCl_5$ and S powder as precursors and h-BN as a substrate to produce large trigonal prismatic NbS_2 monolayers and 3R-polytype few-layer assemblies [48]. Micro-Raman spectroscopy, atomic resolution scanning transmission electron microscopy (STEM), and electrical measurements approved the structure. This procedure can be generalized to other group–VB dichalcogenides, which are of interest owing to the intriguing structure-dependent CDW and superconductivity.

Furthermore, the deposition of 3R-NbS_2 nanosheets is possible using a Nb-HAD complex and H_2S gas via a solid-state reaction without a substrate or catalyst [50]. Scanning electron microscopy (SEM), TEM, XRD, XPS, and Raman spectroscopy characterized the morphology, structure, and chemical composition of the samples. Temperature-dependent electrical resistivity measurements and tests of the electrocatalytic activity in the HER provided additional insight into the specific properties of the NbS_2 nanosheets.

The deposition of high-quality sheets of NbS_2 by CVD on a sapphire substrate allowed layer-controlled epitaxially growth of oriented few-layer assemblies with 1.5–10 nm thickness and grain sizes of up to 500 μm [49]. Raman spectroscopy,

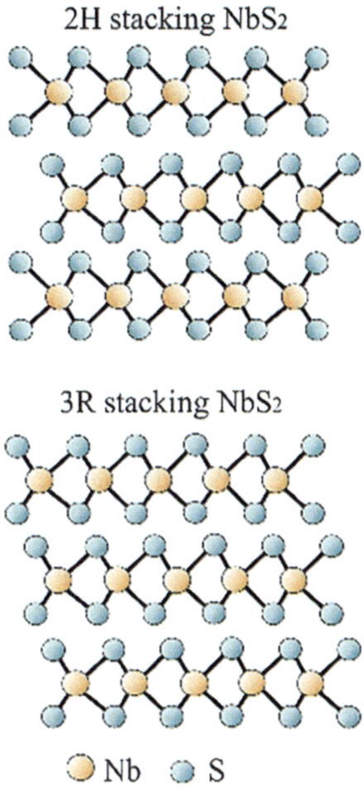

Fig. 15.7 Bilayer and multilayer stacking order of monolayers with a 2H structure and 3R structure. Adapted with permission from Ref. [49], Copyright 2021 American Chemical Society

aberration-corrected scanning transmission electron microscopy (STEM), and electrical transport measurements revealed a transition from the metallic 3R polytype to the superconducting 2H polytype with increasing layer thickness. The superconductivity of the layered 2H-NbS_2 sheets appears below 3 K.

15.2.2 *$NbSe_2$ Monolayer*

15.2.2.1 Chemistry and 2D Bonding

The valence electrons of the electronic configurations of Nb atoms (Kr|$4d^3 5s^2$ or $4d^4 5s^1$) and Se atoms (Ar|$3d^{10} 4s^2 p^4$) determine the type of 2D bonding in niobium diselenide monolayers. In the more stable 1H structure and the metastable 1T-phase, in-plane covalent bonding forms a chemically bonded triple-layer with a Nb sublayer in the center sandwiched between two Se sublayers. The difference in the electronegativities of Nb (1.6) and Se (2.55) induces an ionic contribution to bonding. According to Bader analysis, a transfer of electrons occurs from Nb atoms to Se atoms, leading to a depletion of charges on each Nb atom of $\delta q = 1.23$ e and an excess of charges on each Se atom of $\delta q = 0.62$ e. The covalent-ionic bonding produces a sizeable cohesive energy of the 1H structure of $E_c = -6.08$ eV/atom, calculated by using the exchange correlation potential of the local density approximation (LDA), which differs only marginally from the energetic stability of the 1T phase of -6.04 eV/atom (LDA) [1]. A comparison of these values with those of the PBE approach for the 1H phase at $E_c = -5.96$ eV/atom and 1T phase at -4.92 eV/atom are in line with the previously discussed experience of an overestimate of binding energies by the LDA approximation [42].

15.2.2.2 2D Structures

For the lattice parameters of the 1H-$NbSe_2$ monolayer, LDA calculations delivered the lattice constant $a = 0.340$ nm, the bond length $d_{Nb–Se} = 0.257$ nm, the Se–Se distance $h_{SeSe} = 0.333$ nm, and the bond angle $\theta_{SeNbSe} = 80.68°$ [1]. Adding the van der Waals diameter of Se and the separation of Se atoms in the 1H structure yields an estimate of the layer thickness of $d_{w+h} = 0.713$ nm. This structure is of particular interest because 3D $NbSe_2$ is a superconductor. Furthermore, the lattice parameters agree quite well with $a = 0.348$ nm, $d_{Nb–Se} = 0.262$ nm, $h_{SeSe} = 0.337$ nm, and $\theta_{SeNbSe} = 79.90°$, calculated by the PBE approach for the 1H structure [13].

15.2.2.3 Mechanics

The large covalent-ionic binding energy of the 1H-$NbSe_2$ monolayer results in a stiffness of the three-layer system of $E = 87.2$ N/m that is substantially greater than the 70.5 N/m of the 1T phase [1]. The finite element method with the valence force field model yielded slightly lower anisotropic Young's moduli at $E_{ac} = 80.2$ N/m and $E_{zz} = 82.6$ N/m and Poisson's ratios at $\nu_{ac} = 0.28$ and $\nu_{zz} = 26$ [4].

15.2.2.4 Physics and Applications

Niobium diselenide is one of the best studied superconducting materials because the Ising superconductivity exists down to the monolayer limit. The 2H-$NbSe_2$ sheets are d^3 materials with a variety of symmetry broken states, CDWs, superconductivity, and magnetism. Conventional electron-phonon coupling is responsible for the superconducting state, and the emergence of charge density states depends on soft-phonon modes, whereas Coulomb repulsion plays only a marginal role [51]. Experimental control of the strength of the interactions occurring via quantum confinement indicates the proximity of $NbSe_2$ to a correlated insulating state [51]. The confinement-controlled interactions indicate the coexistence of competing interactions, which induce a transition from the superconducting to an insulating quantum correlated state. Variation in the island sizes and their relative separations in submonolayer deposits of $NbSe_2$ on a graphite substrate provided relevant experimental information. In fact, the analysis revealed competing interactions allowing a transition from the superconducting to an insulating state. The electron-phonon interactions, which are independent of system size, give rise to the superconducting ground state with increasing system size, whereas Coulomb interactions, which depend strongly on system size, drive the system into a Coulomb-gaped correlated state. Thus, in addition to conventional electron-phonon interactions, electronic correlations may have a crucial impact on the emergence of both CDW orders and superconductivity.

An alternative method to describe full momentum-dependent spin susceptibility that is important for describing magnetism and superconductivity combines standard fixed-spin-moment calculations with the direct determination of the energy of spin spirals stabilized by an artificial Hubbard interaction [52]. This concept is applicable to Ising superconductivity in $NbSe_2$ monolayers. While previous estimates suggested proximity to ferromagnetism, this analysis yields a more complicated picture of spin fluctuations with a sharply peaked fluctuation spectrum that would change the interband pairing interaction process and thus affect the superconducting state.

By alloying the 1H-$NbSe_2$ monolayer with sulfur (S), the transition temperature for superconductivity shows a nonmonotonic behavior [53]. According to first-principles calculations, this effect depends on the presence of selenium vacancies in $NbSe_2$; these vacancies are magnetic pair-breaking defects present at considerable concentrations in as-grown $NbSe_2$ monolayers instead of the suggested fractal superconductivity. Both S and Si can occupy the Se sides and reduce the pair-breaking effect. Moreover, the DOS at the Fermi level and the proximity to magnetism in the alloy are both reduced.

High-quality submonolayer $NbSe_2$ deposits grown on bilayer graphene on 6H-SiC(0001) substrates served as samples for the experimental characterization of the collective ground states [54]. For single-layer $NbSe_2$, a (3 × 3) CDW uniform superlattice develops. At 5 K, STS measurements indicate the formation of a small CDW gap at the Fermi energy. At $T_c = 1.9$ K, the resistance starts to decrease, indicating the onset of superconducting fluctuations, whereas the superconducting transition midpoint occurs at 0.65 K, and the zero-resistance point at 0.46 K. These measurements document the existence of superconductivity at the single-layer limit; however, there is a substantial reduction in the onset from the bulk value at $T_c = 7.2$ K.

The effect of quantum confinement on the electronic and dielectric properties was studied by first-principles calculations using a pseudopotential and a numerical pseudoatomic orbital basis set for mono-, bi-, four-, and six-layer sheets and bulk compounds of NbS_2, $NbSe_2$, and $NbTe_2$ [55]. A metallic character of the electronic band structure was observed, irrespective of quantum confinement. Transitions from chalcogen *p* bands to Nb *d* bands are the source of most interband transitions. For in-plane and out-of-plane polarizations, the redshift of the π and $\pi + \sigma$ plasmon peaks was determined by changing the layer thickness from the bulk to the monolayer. The interband transition energies did not change; however, an additional interband transition occurred at 3.2 eV for $NbTe_2$.

15.2.2.5 Synthesis

The growth of triangular or truncated triangular crystals of $NbSe_2$ on an h-BN substrate is possible employing MBE for synthesis and by supplying niobium and selenium atoms with an electron-beam evaporator and a low-temperature Knudsen cell, respectively [56]. A precisely controlled supply rate and an ultraflat surface of h-BN play decisive roles in realizing high-quality deposits. This procedure leads to much larger lateral sizes—up to several hundred nanometers—of metallic monolayers compatible with device fabrication. Note that the realization of metallic layers by top-down methods is challenging due to strong interlayer interactions.

The controlled growth of high-quality and single-phase 1T $NbSe_2$ and 2H $NbSe_2$, as well as their mixed phases, was achieved by MBE on bilayer graphene/SiC and graphite substrates [57]. STS analysis provided clear evidence for an exchange interaction in a 1T-$NbSe_2$/2H-$NbSe_2$ heterostructure consisting of localized magnetic moments in 1T $NbSe_2$, which is a charge transfer insulator, and the electron/Cooper pairs of the metallic/superconducting 2H-$NbSe_2$ phase. Of course, this is not a definite proof of localized magnetic moments in an isolated monolayer; however, the realization of the pure phases of 1T $NbSe_2$ and 2H $NbSe_2$ provides an appealing platform for investigating the interplay of charge, spin, and lattice degrees of freedom in two dimensions.

15.2.3 $NbTe_2$ Monolayer

15.2.3.1 Chemistry and 2D Bonding

The electronic configurations of Nb (Kr|$4d^3 5s^2$) and Te (Kr|$4d^{10} 5s^2 p^4$) atoms determine the type of 2D bonding in niobium ditelluride monolayers. In the most stable 1T structure, in-plane covalent bonding results in the formation of the chemically bonded triple-layer structure with a Nb sublayer in the center, sandwiched between two Te sublayers. The difference in the electronegativities of Nb (1.6) and Te (2.1) leads to an ionic contribution to bonding. According to a Bader analysis, a transfer of electrons occurs from the Nb to the Te atoms, leading to a depletion of charges on Nb atoms of $\delta q = 0.90$ e and an excess of charges on each Te atom of $\delta q = 0.45$ e. The covalent-ionic bonding produces a cohesive energy of the 1T structure at $E_c = -5.46$ eV/atom, calculated by using the exchange correlation potential of the LDA approach [1]. As discussed for $NbSe_2$, the LDA approach delivers a comparable cohesion of $E_c = -5.35$ eV/atom for the H structure, while the PBE approach yields -4.34 eV/atom [42]. Owing to the similarity of the bonding arrangements in the 1T and 1H structures, their cohesive energies should be comparable.

15.2.3.2 2D Structures

For the lattice parameters of the more stable 1T-$NbTe_2$ monolayers, LDA calculations yielded the lattice constant a = 0.356 nm, the bond length $d_{Nb–Te} = 0.277$ nm, the Te–Te distance $h_{Te–Te} = 0.424$ nm, and the bond angle $\theta_{TeNbTe} = 100.05°$ [1]. Adding the van der Waals diameter of Te and the separation of Te atoms in the 1T structure yields as thickness of $d_{w+h} = 0.844$ nm. The 1H-$NbTe_2$ phase is energetically less favorable than the 1T-$NbTe_2$ phase by approximately 52 meV. The published lattice parameters of the 1H phase based on the PBE approach are a = 0.370 nm, $d_{Nb–Te}$ = 0.282 nm, and $h_{Te–Te}$ = 0.369 nm, predicting a larger lattice constant and bond length. However, the distance between the Te atoms is much shorter, and thus, the thickness of the H phase monolayer is reduced [42]. The 1T-$NbTe_2$ structure is a metal with a Young's modulus of 64.1 N/m [1].

15.2.3.3 Physics and Applications

The existence of a nontrivial topology, electronic correlations, and CDWs in $NbTe_2$ qualify the material for the exploration of new quantum states. Theory predicts that high-symmetry 1T-$NbTe_2$ monolayers, stabilized by anharmonicity at room temperature, have a nontrivial topology that changes from the quantum spin Hall (QSH) state to a quantum anomalous Hall (QAH) state, when the Hubbard potential exceeds a critical value due to the overall sensitivity to electronic interactions [58]. Interestingly, at low temperature, the nontrivial topology coexists with a (4×4) CDW order, and the CDW structure exhibits an insulating nontrivial gap in either the QSH or the QAH phase, demonstrating the possibility of modulating the topological state by CDWs.

Electron correlation and related quantum effects in two dimensions have attracted widespread attention. The application of STM to characterize the lattice structure and of ARPES to elucidate the overall electronic band structure allowed the investigation of monolayers of $NbTe_2$ that were grown by van der Waals epitaxy on bilayer graphene deposited on a SiC substrate [59]. The analysis revealed the formation of 1T-$NbTe_2$ monolayers with a metallic CDW state and the periodicity $(\sqrt{19} \times \sqrt{19})$ coexisting with a partially occupied star of David (SOD) lattice (see Figs. 15.8 and 15.9) [60]. The coexistence of CDW order and SOD lattice distortions is described in more detail for 1T-$TaSe_2$. In the DS arrangement, the corner atoms are displaced from the original position toward the central atom, and the half-filling condition to realize the Mott-insulating phase with a half-filled metallic band is satisfied. However, the existence of complex energy bands in the CDW phase has prohibited a final understanding of the mechanism leading to the formation of the CDW-Mott phase. The uncovered metallic $(\sqrt{19} \times \sqrt{19})$ CDW state that coexists with an unusual DS lattice distortion in 1T $NbTe_2$ deviates from the gapped Mott-insulating state of 1T-$NbSe_2$ monolayers.

In an experimental study of CDW phases in addition to the triaxial compression (4×4) order, uniaxial (4×1) order, and large-scale $(\sqrt{19} \times \sqrt{19})$ CDW order, an additional $(\sqrt{28} \times \sqrt{28})$ large-scale order was detected via STM characterization [61]. Contrary to prior predictions, not the (4×4) order but the $(\sqrt{19} \times \sqrt{19})$ order with large periodicity is suggested as the energetically stable CDW system at high growth or annealing temperatures of > 400 °C. The systematic development of epitaxial growth procedures for $NbTe_2$ monolayers on a bilayer graphene/SiC substrate and STM/STS characterization opened the door to this enriched detection of CDW phases.

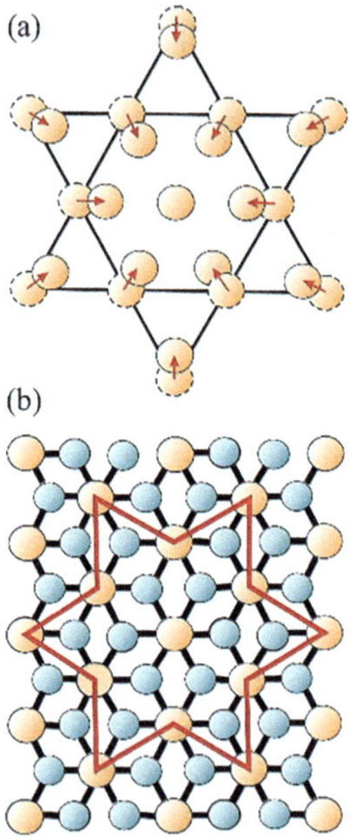

Fig. 15.8 **a** Scheme of the star-of-David (SOD) indicating the displacements of the group-VB atoms. **b** Scheme showing the placement of the SOD in the structure of the monolayer. Partially adapted from Ref. [60], Copyright 2021, Springer Nature

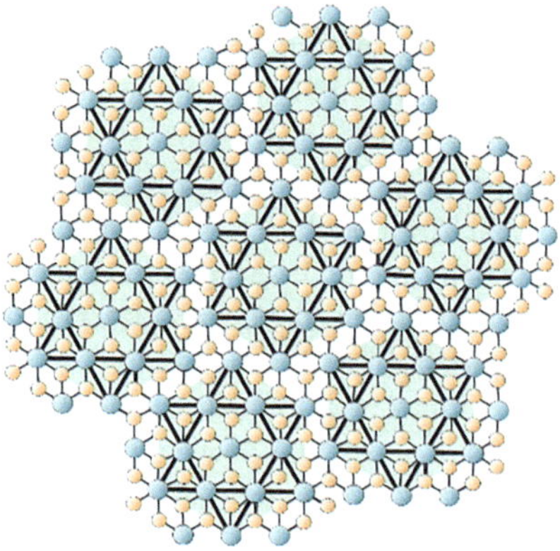

Fig. 15.9 Scheme of the 1T-$TaSe_2$ structure of the monolayer with SOD clusters and ($\sqrt{13} \times \sqrt{13}$) periodicity. Reproduced from Ref. [60], Copyright 2021, Springer Nature

Ultrathin crystalline nanosheets of $NbTe_2$ were introduced as low-cost metal-free 2D surface-enhanced Raman scattering (SERS) materials for ultrasensitive detection of aromatic molecules and antibiotics. The detection is based on fluorescence quenching and charge transfer enhancement, contributing to the Raman effect of the analytes [62]. The reproducibility was satisfactory, and the ultralow detection limit of 10^{-9} M was comparable to that of metal-based SERS substrates. For example, the detection limits of ciprofloxacin and enrofloxacin were 35.1 and 35.9 ppb, respectively.

15.2.3.4 Synthesis

As discussed above for other compounds, van der Waals epitaxial deposition, using coevaporation of Nb and Te atoms under UHV conditions, allowed the growth of high-quality monolayers of 1T-$NbTe_2$ [59]. The thickness of the islands grown on the bilayer graphene/SiC substrate was ~ 1.0 nm, as measured via STM, and the band structure was determined via in situ analysis using ARPES. There is a good correspondence between the experimental band structure and the theoretical prediction. The observation of the star-of-David (SOD) lattice distortion coexisting with the ($\sqrt{19} \times \sqrt{19}$) periodicity characterizes a new class of TMDC materials with a unique correlated metallic ground state with no signature of a Mott gap in contrast to the gapped Mott-insulating state of 1T-$NbSe_2$ monolayers [59].

Furthermore, a two-step method permitted the deposition of single-layer and few-layer assemblies of $NbTe_2$ [62]. The first step delivered multilayer sheets from bulk $NbTe_2$ crystals by the liquid-phase method, which subsequently were further delaminated in a second step to few-layer systems by the hydrothermal method. Various analytical techniques, such as TEM, scanning electron microscopy (SEM), and AFM, characterized the structure, morphology, and size of the $NbTe_2$ deposit.

15.3 Tantalum Dichalcogenide Monolayers (TaS_2, $TaSe_2$, and $TaTe_2$)

15.3.1 TaS_2 Monolayer

15.3.1.1 Chemistry and 2D Bonding

The valence electrons of the electronic configurations of Ta atoms (Xe|$5d^3 6s^2$) and S atoms (Ne|$3s^2p^4$) determine the type of 2D bonding in tantalum disulfide monolayers. In the most stable phases of TaS_2 with trigonal prismatic (2H) and octahedral (1T) coordination, in-plane covalent bonding results in the formation of the chemically bonded S-Ta-S triple-layer structure, formed by the 1H and 1T phases described previously. The difference in the electronegativities of Ta (1.5) and S (2.58) induces an ionic contribution to bonding by a charge transfer of $\delta q = 1.1$ e from each Ta atom to two S atoms [42]. The strongly polar Ta–S bonds generate a low cohesive energy for the 1H-TaS_2 monolayers of $E_c = -6.87$ eV/atom (LDA) and -5.92 eV/atom (PBE) [42].

15.3.1.2 2D Structures

The 2D phases of 2H-TaS_2 and 1T-TaS_2 are among the few theoretically and experimentally well-studied compounds. The 2H structure with a slightly smaller lattice

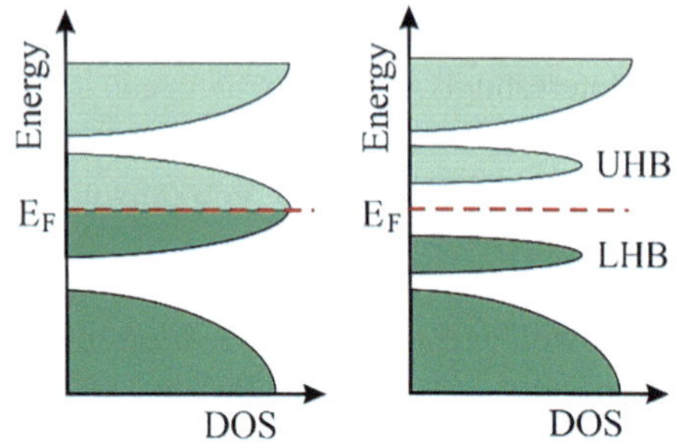

Fig. 15.10 Splitting of the band by Coulomb interaction into a peak below the Fermi level (LHB) and one above the Fermi level (UHB) located at the center of the DS

constant of a = 0.332 nm is energetically more stable than the 1T monolayer with a = 0.335 nm, based on the PBE and SCAN functionals with van der Waals corrections [63]. The H structure that forms the ground state, is a nonmagnetic metal. The PBE-based lattice parameters of the 1H phase are a = 0.334 nm, $d_{Ta\text{–}S} = 0.248$ nm, $h_{SS} = 0.313$ nm, and $\theta_{STaS} = 78.11°$ [13, 42]. The corresponding lattice parameters obtained by the LDA approximation are slightly smaller, consistent with the greater binding energy [13, 42]. Adding the van der Waals diameter of the S atom and the separation of the S atoms of the 1H structure, yields an estimate of the layer thickness of $d_{w+h} = 0.673$ nm.

In particular, the 1T-TaS_2 monolayer has attracted enormous attention because it exists in several metastable phases. It undergoes first-order phase transitions from the high-symmetry 1T phase to the low-temperature $(\sqrt{13} \times \sqrt{13})$ phase. At liquid helium temperature, the 1T phase shows a $(\sqrt{13}\times\sqrt{13})$ commensurate CDW pattern [64]. The STS spectra revealed an energy gap of $E_g = 0.45$ eV at the Fermi level and few peaks. Additional information on the electronic structure comes from DFT calculations. Based on the space symmetry of the lattice, the Ta atoms in a David star form three groups: the central atom, the six nearest atoms, and the six next-nearest atoms (see Fig. 15.8). With a valence state of + 4 for Ta and − 2 for S, the Ta atom should contribute one unpaired 5*d* electron to the band near the Fermi energy, while chemical bonds connect the nearest and next-nearest Ta atoms, isolating the central atom [64]. The nearest and next-nearest Ta atoms contribute extended bands fully occupied by 5*d* electrons, whereas the central atom contributes a half-filled band near the Fermi level. Coulomb interactions split the band into two sharp peaks, one above and one below the Fermi level. The two states located at 0.202 eV and − 0.311 eV at the center of the DS represent the upper Hubbard and lower Hubbard bands, respectively (see Fig. 15.10). The good agreement between theory and experiment suggests that the monolayer is a Mott insulator with a band gap of $E_g = 0.45$ eV.

15.3.1.3 Mechanics

Linear and nonlinear mechanical properties of 2H- and 1T-TaS_2 are available for multilayer sheets in the thickness range of 3.5–12.6 nm, as determined by microscopy-based nanoindentation measurements [65]. The measured Young's modulus of 2H-TaS_2 sheets of 85.9 GPa (57.8 N/m) was independent of the thickness and is in satisfactory agreement with the corresponding calculated value of 77.7 GPa (52.3 N/m) (PBE) [65]. The elastic moduli given in 2D units (N/m) are based on the estimated layer thickness of $d_{w+h} = 0.673$ nm. The calculated Poisson's ratio is 0.27 (PBE). For comparison, the finite element method with the valence force field model yielded much larger slightly anisotropic Young's moduli of $E_{ac} = 86.8$ N/m and $E_{zz} = 89.2$ N/m and similar Poisson's ratios of $\nu_{ac} = 0.26$ and $\nu_{zz} = 0.25$ [4].

The measured breaking strength of 5.01 GPa (3.4 N/m) deviates strongly from the calculated value of 15.8 GPa (10.6 N/m), obviously due to incorporated defects in the measured sample [65]. The results show that metallic sheets can sustain ultimate strains like semiconducting sheets and can serve as electrical contact materials in flexible electronics.

15.3.1.4 Physics and Applications

The development of materials allowing a controllable switching between different properties is an active field of research. An interesting approach is based on the manipulation of CDWs by charge carrier doping, as studied in 1T-TaS_2 monolayers. While hole doping can suppress commensurate CDWs (CCDWs), the structure remains stable upon electron doping [66]. The application of light or a positive perpendicular electric field can excite an electron from the lower Hubbard band (LHB) to the upper Hubbard band (UHB) and generate a hole in the LHB. The hole is located at the center of the David star, leaving a polaron with an excess charge (see Fig. 15.11). Doping more holes into the David stars leads to a destruction of the CCDW phase that ultimately transforms to a nearly commensurate or metastable phase. In this way, switching between the CCDW phase, Mott insulating states, metallic states, superconducting states, and topological states is possible by tuning the carrier density in monolayer and bulk samples.

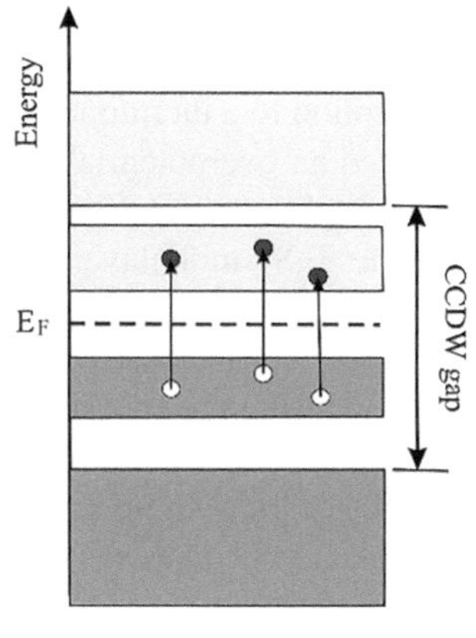

Fig. 15.11 Excitation of electrons from the LHB to the UHB generates holes at the center of a David star, leading to destruction of the commensurate CDW phase

An in-depth analysis of the hysteretic electronic phase transitions in 1T-TaS_2 monolayers, including the CDW phases, was performed based on the interplay of temperature-dependent electronic interactions [67]. The electronic states are associated with delicate temperature-dependent competition and/or cooperation among stacking-dependent interlayer interactions, intralayer electron-electron correlations, and electron-phonon coupling leading to the existence of a commensurate CDW (CCDW), triclinic CDW (TCDW), a spotty-CDW state, a network-CDW state (NCDW), a belt-TCDW state, and a mosaic-CDW state.

Combined STM, STS, and ARPES measurements together with DFT calculations provide information on the influence of structural defects on the CDWs of 1T-TaS_2 sheets [68]. The analysis indicated that the band gap depends on the interplay between layer stacking, structural defects, concentrating on sulfur vacancies and sulfur-oxygen substitution. Oxygen substitution leads to semiconducting behavior with an open band gap, a spin-polarized flat band, magnetization or metallic behavior with a closed dispersive band at the Fermi level, and quenched magnetic moments. The sulfur vacancy has an insignificant impact on the electronic properties and affects only one CDW site, contrary to defects working as dopants.

Notably, atomically thin 1T-TaS_2 sheets may find application as memristors, which are materials with different states of electrical resistance that can be switched by changing the voltage or current [69]. Transport studies have identified intermediate resistance states nearly commensurate with the commensurate CDW transition in 1T TaS_2. A vertical heterostructure geometry allowed simultaneous electrical driving and scanning photocurrent imaging. The authors assumed a mostly uniform change in true metastable CDW states and demonstrated repeatable and bidirectional switching across multiple CDW states without changing the temperature.

The catalytic activity of group-VB chalcogenides is not only based on unsaturated edges but also on highly active surface basal planes, e.g., in the hydrogen evolution reaction (HER) [70]. Liquid-phase exfoliation of 2H-TaS_2 crystals using 2-propanol delivers high-performance electrocatalysts by producing single/few-layer flakes deposited as catalytic films. Thermal treatment improves the activity by increasing the porosity, which promotes ion access to the basal planes. Furthermore, hybridization of 1H TaS_2 and 1H $TaSe_2$ tunes the Gibbs free energy of hydrogen adsorption to a thermoneutral value. With 0.5 M H_2SO_4, the heterogeneous catalyst showed an overpotential at a cathodic current density of 10 mA cm^{-2} of 120 meV and a high mass activity of 314 mA g^{-1} at an overpotential of 200 meV.

The TaS_2 monolayer is Pauli-paramagnetic up to a strain of 6% [71]. At 6% strain, the material becomes ferromagnetic with a magnetic moment of 0.24 μ_B per Ta atom that increases further with strain, reaching 0.48 μ_B per Ta atom at 10% strain. Lifting the spin degeneracy generates the magnetic moment that arises from the Ta 5*d* states near the Fermi level. By hydrogenation of the monolayer on one side to $HTaS_2$, the spin up and spin down states disappear from the Fermi level, and the monolayer becomes an indirect semiconductor with a band gap of ~ 0.75 eV because there are no states at the Fermi level. The transition from paramagnetic metallicity to semiconducting behavior depends on the electron transfer from hydrogen filling the empty Ta 5*d* states above the Fermi level.

A comprehensive review covers recent progress achieved in the investigation of 2D tantalum dichalcogenides and their future applications [72]. This includes a discussion of important 2D structures (1T, 1T′, and 2H) and their electronic, magnetic, optical, and layer-dependent properties. Moreover, the study addresses recent advances made in the development of potential applications in several areas of research, such as field emission, energy storage, fuel cells, photocatalysis, and solar cells. The review provides guidance for improving the performance of sensing devices, energy storage and energy conversion devices, as well as field emission devices. Finally, the work discusses trends of future developments and provides a perspective on the required advancements needed to further develop the fascinating field.

15.3.1.5 Synthesis

Monolayers of 1T TaS_2 were prepared by MBE in a UHV chamber, which provided in situ STM analysis at 4.2 K [64]. FeS served as a sulfur source and graphitized 6H-SiC(0001) crystals with an epitaxial bilayer of graphene were used as substrates. STM allowed local topographic imaging experiments, and STS measurements elucidated the electronic structure without disrupting the vacuum.

The ambient pressure CVD method allowed the deposition of high-quality 1T-TaS_2 nanosheets on mica with control of the layer number by AFM measurements [73]. The mica substrate has excellent epitaxial growth characteristics and may be advantageous compared to Au, SiO_2/Si, h-BN, and sapphire substrates owing to charge transfer. Sulfur powder and $TaCl_5$ served as source materials, and a mixture of H_2/Ar (5% H_2) functioned as the carrier gas. Temperature-dependent resistance measurements and Raman spectroscopy were used to monitor the thickness-dependent nearly commensurate CDW (NCCDW)/CCDW phase transition of the as-grown sample. Below 180 K, the 1T-TaS_2 crystal contains ‘star of David’ (SOD) clusters, and at 300 K, the 1T-TaS_2 crystal can transform from the CCDW to the partially filled NCCDW phase with satellite domains.

An experimental study investigated the stability of 2D-TaS_2 sheets toward O_2 exposure for large-area 2D-TaS_2 crystals grown on single-crystal Cu(111) films via a multistep PVD process [74]. For analysis, Auger electron spectroscopy, LEED, and PES were employed to characterize the composition, crystallinity, and electronic structure. As revealed by LEED and simulations, the (3×3) unit cells of TaS_2 matched the (4×4) unit cells of the Cu(111) substrate. Interestingly, exposure to O_2 leads to a reduction in the film thickness. Temperature-programmed desorption spectroscopy confirmed the assumption that O_2 caused the reduction of thickness by the detection of SO_2 and its spontaneous desorption between 100 and 500 K.

15.3.2 TaSe Monolayer

15.3.2.1 Chemistry and 2D Bonding

The electronic configurations of Ta atoms (Xe|$5d^3 6s^2$) and Se atoms (Ar|$3d^{10} 4s^2 p^4$) determine the type of 2D bonding in tantalum diselenide monolayers. In the most stable 1H phase of 2D $TaSe_2$ with trigonal prismatic coordination, in-plane covalent bonding results in the formation of the chemically bonded Se–Ta–Se triple-layer structure. The difference in the electronegativities of Ta atoms (1.5) and Se atoms (2.55) leads to polar bonds with sizable ionic contribution induced by a charge transfer of $\delta q = 0.92$ e (PBE) from each Ta atom to two Se atoms [42]. The cohesive energies calculated by different approaches are $E_c = -5.34$ eV/atom (PBE) and -6.31 eV/atom (LDA) for the 1H phase [42].

15.3.2.2 2D Structures

The 2D phases of 1T-$TaSe_2$ and 1H-$TaSe_2$ are well-studied monolayer structures, theoretically and experimentally. The 2H structure with a lattice constant a = 0.345 nm is more stable than the 1T monolayer with a nearly identical value of a = 0.347 nm [63]. A complete set of lattice parameters of 1H-$TaSe_2$ monolayers of a = 0.347 nm, $d_{Ta\text{–}Se} = 0.261$ nm, $h_{SeSe} = 0.335$ nm, and $\theta_{SeTaSe} = 79.87°$ was obtained by PBE calculations [13]. These lattice parameters agree quite well with those of another study [42]. The thickness of the monolayer can be estimated by adding the van der Waals diameter of Se atoms and the separation of the Se atoms in the 1H structure, yielding $d_{w+h} = 0.715$ nm.

The 1T-$TaSe_2$ phase is isostructural and isoelectronic to 1T-TaS_2 and therefore exhibits the already discussed $\left(\sqrt{13} \times \sqrt{13}\right)$ CDW pattern [64]. The in-plane lattice constant of a = 0.344 nm is comparable with the values given above, and the lattice constant of the CCDW is a = 0.122 nm. In contrast to other predictions for 1T TaS_2, the band structure calculations suggest metallic behavior of the 1T-$TaSe_2$ monolayer, contrary to the spatially resolved STS measurements, which suggest an energy gap of $E_g = 0.2$ eV. These contradictory results indicate that $TaSe_2$ is not a Mott insulator, however, this issue requires further work.

15.3.2.3 Mechanics

The finite element method with the valence force field model yielded slightly anisotropic Young's moduli of $E_{ac} = 80.7$ N/m and $E_{zz} = 83.1$ N/m and Poisson's ratios of $\nu_{ac} = 0.28$ and $\nu_{zz} = 26$ [4]. Surprisingly, the stiffness of the $TaSe_2$ monolayer is only slightly smaller and the Poisson's ratios are only slightly larger than those of the TaS_2 monolayer.

15.3.2.4 Physics and Applications

Metastable 1T-TaX_2 (X = S, Se) monolayers are receiving increasing attention as correlated materials for accessing many-body quantum phenomena. There is consensus on the nature of the ground state as a peculiar star of David (SOD) CDW superlattice (Fig. 15.8). The unit cell of the SOD contains 13 5*d* conduction electrons originating from the 13 Ta atoms that form the cluster [75]. However, a controversy still exists concerning the nature of the insulating states. One explanation is that the metal-to-Mott insulator transition is due to the reduction in dimensions from the bulk to the monolayer [75]. Below a critical thickness of approximately seven layers, the reduced average coordination number quenches the kinetic energy of the initially itinerant electrons and drives the material into a Mott insulation state. This explanation of a dimensionality-driven metal–insulator transition explains the long-standing contradiction between the metallic bulk and insulating surface of 1T-$TaSe_2$ monolayers.

An experimental study confirmed the existence of a state above the Fermi level with an overall band gap of ~ 0.7 eV at $\overline{\Gamma}$, ascribed to the upper Hubbard band, using combined time-resolved and angle-resolved photoemission spectroscopy (trARPES) [76]. The origin of this state was attributed to band structure modifications, which are induced by the CDW phase without the need for a Mott correlation effect. For charge migration from the tips toward the center of the star, a charge-transfer gap instead of a Mott gap would appear. These results suggest that electron-phonon coupling rather than electronic instability dominates the formation of the CDW. The authors agreed with the statement that the 1T-$TaSe_2$ monolayer is a Mott system; however, the results confute the Mott nature of this monolayer.

A combined experimental and theoretical study revealed the existence of a robust 2D-CDW-Mott phase with a transition temperature onset of ~ 530 K in monolayers of 1T $TaSe_2$ stabilized by band engineering [60]. ARPES analysis of epitaxially grown 1T-$TaSe_2$ and 1T-$NbSe_2$ monolayers provided the necessary fundamental information. The lower Hubbard band survives under external perturbations such as carrier doping and photoexcitation, in contrast to its bulk counterpart. The enhanced Mott–Hubbard and CDW gaps originating from the lattice distortion suggest the stabilization of a nonmagnetic CDW–Mott insulator phase well above room temperature.

Interestingly, first-principles calculations indicate the transition of a Mott insulating phase into a charge-transfer insulating phase and then into a normal metallic phase by increasing the compressive biaxial strain [77]. The CDW order with SOD lattice distortions induces the formation of two types of orbitals, namely localized and extended *d* orbitals in Ta. The Mott insulating state originates from electron correlation in the localized orbital, which is separated from the extended orbitals in energy. The metallization of the CDW phase occurs due to immersion of the localized *d* orbital of Ta in the extended-orbital-spanned Fermi field and by quenching of Hubbard splitting by orbital hybridization.

The $TaSe_2$ monolayer is paramagnetic up to a strain of 6% [71]. At 6% strain, the material becomes ferromagnetic with a magnetic moment of 0.34 μ_B per Ta atom, and the effect increases further with strain, reaching 0.50 μ_B per Ta atom at 10% strain. The magnetic moment arises from the 5*d* states of Ta near the Fermi level and is induced by lifting spin degeneracy. By hydrogenation of the monolayer on one side of the monolayer ($HTaSe_2$), the spin up and spin down states disappear from the Fermi level, and the monolayer becomes an indirect semiconductor with a band gap of ~ 0.67 eV, because there are no states at the Fermi level. Electron donation from hydrogen and filling of the empty Ta 5*d* states above the Fermi level are responsible for the transition from paramagnetic metallicity to semiconducting behavior.

Modulating the CDW order and superconductivity by high pressure tunes the properties of bulk van der Waals heterostructures 4*Hb*-$TaSe_2$, which consist of alternately stacked 1T-$TaSe_2$ and 1H-$TaSe_2$ monolayers [78]. Remarkably, superconductivity competes with the intralayer and adjacent-layer CDW order, leading to substantially and continually boosted superconductivity under compression. This interplay between CDW and superconductivity opens new possibilities for designing materials with tailored properties. Note that low temperature STM and STS revealed the roles of the narrow electronic band in bulk 1T-TaS_2 and 4*Hb*-TaS_2, where the latter is a superconducting compound with alternating 1T- and 1H-TaS_2 layers, and the 1H-TaS_2 layer has a weak CDW pattern and reduces the CDW coupling between the adjacent layers [79].

Good van der Waals contacts in 2D metal-semiconductor heterostructures and metal electrodes are a challenge due to Schottky barriers, which severely limit electrical performance. Whether a Schottky barrier or Ohmic contact is realized at the interface, depends on the work functions. For an n-type semiconductor with a smaller work function than that of the metal (electrode) the Fermi levels tend to reach the same level and the band of the semiconductor bends upward, forming a Schottky barrier, where the height and width of the barrier determine the depletion of electrons (see Fig. 15.12). If the work function of the semiconductor is greater the band bends downward inducing electron accumulation in the interface region and an Ohmic contact. In heterostructure stacking plays a crucial role, as studied for the metal $TaSe_2$ and the semiconductor WSe_2. The deposition of $TaSe_2$ on presynthesized 1H-WSe_2 monolayers, employing a two-step CVD method, yields 2D layered $TaSe_2$/WSe_2 heterostructures on a sapphire substrate [80]. TEM measurements revealed a high-quality and clean interface. Moreover, surface potential and electrical transport measurements confirmed a high field-effect mobility and on/off ratio. Thus, these observations indicate potential improvements in the mobility of field-effect transistors (FETs) and their on/off ratios, and the performance of 2D electronic devices.

15.3.2.5 Synthesis

The selective synthesis of 1H-$TaSe_2$ monolayers with octahedral structures and 1T-$TaSe_2$ monolayers with trigonal-prismatic structures was first achieved on bilayer graphene by the MBE method [81]. The 1H-$TaSe_2$ monolayer has a metallic electronic structure with a dispersive band across the Fermi level, whereas the 1T-$TaSe_2$

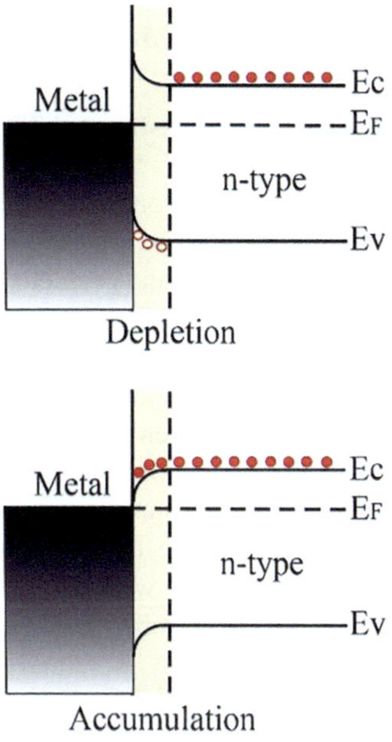

Fig. 15.12 Scheme of the formation of **a** a Schottky contact, where E_c is the conduction band, E_F the Fermi level, and E_v the valence band. **b** An Ohmic contact with an accumulation of electrons in the interface region. Adapted from Ref. [80], Copyright 2023, Royal Society of Chemistry

phase shows a sizable gap at the Fermi level, indicating insulating behavior contrary to the metallic nature of the bulk 1T-$TaSe_2$ crystal, as confirmed by ARPES.

Epitaxy yields large-area 1T-$TaSe_2$ monolayers grown on atomically smooth Se-treated GaP(111)B-(1 $\times$ 1) surfaces, employing doped and undoped substrates [82]. Reflection high-energy electron diffraction (RHEED) measurements controlled the Se passivation step and the growth process. STM analysis identified a superstructure resulting from the combination of CDW reconstruction in both chiral directions and a Moiré pattern with the substrate. Moreover, STS experiments revealed a fully open Mott insulating gap. The possibility of manipulating the Mott gap by STS together with long-range electrical transport processes may lead to the development of large-scale 2D Mott materials.

The phase-controlled epitaxial growth allowed the study of interfacial effects in heterostructures. Contrary to previous findings stating a quasi-freestanding nature of monolayers grown on bilayer graphene substrates, strong interactions between 1T- and 1H-$TaSe_2$ monolayers and bilayer graphene substrates were detected [83]. This was evidenced by kinked band structures and significant charge transfer to the $TaSe_2$ monolayer. As expected, in 1H-$TaSe_2$ monolayers the effective interaction leads to a reduced interface distance and strongly reduced work function. On the other hand, 1T-$TaSe_2$ monolayers interact only weakly with bilayer graphene substrates without charge transfer and band hybridization.

15.3.3 $TaTe_2$ Monolayer

15.3.3.1 Chemistry and 2D Bonding

The valence electrons of the electronic configurations of Ta (Xe|5$d^3$6s^2) and Te atoms (Kr|4d^{10}5$s^2$$p^4$) determine the type of 2D bonding in tantalum ditelluride monolayers. The coordination and geometry and the realized structures are related to the d-electron count of Ta. In $TaTe_2$ monolayers, the distorted octahedral 1T' phase is 25 meV more stable than the 1H phase per formula unit [63]. In-plane covalent bonding forms the standard chemically bonded Te–Ta–Te triple-layer structures as described above. The difference in the electronegativities of Ta (1.5) and Te (2.1) introduces a medium ionic contribution to bonding. The charge transfer from the metal to the two chalcogen atoms amounts to $\delta q = 0.91$ e (PBE). The cohesive energies calculated for the metastable 1H phase at $E_c = -4.67$ eV/atom (PBE) and -5.62 eV/atom (LDA) signify the weakest bonding in this group of VB-dichalcogenide monolayers but still indicate robust bonding [42].

15.3.3.2 2D Structures

The lattice constants of the stable 1T'-$TaTe_2$ monolayer are a = 0.657 nm and b = 0.338 nm [63]. For comparison, the lattice constant of the 2H structure is a = 0.367 nm, and that of the 1T structure a = 0.365 nm, where both values are based on the PBE-van der Waals approach. The lattice parameters of the 2H phase, calculated with the PBE functionals, are a = 0.370 nm, $d_{Ta-Te} = 0.281$ nm, $h_{TeTe} = 0.366$ nm, and $\theta_{TeTATe} = 81.19°$ [13]. The monolayer thickness can be estimated by adding the van der Waals diameter of the Te atoms and their known distance in the 2H structure, yielding $d_{w+h} = 0.786$ nm.

A first-principles study investigated besides the structural and electronic properties the magnetic behavior of the H phase of 2D TaX_2 (X = S, Se, and Te) and the corresponding T phase [63]. While the normal phases of TaS_2 and $TaSe_2$ are nonmagnetic metals with the H structure as the ground state, the H phase of $TaTe_2$ is a magnetic metal and the T′ structure is a nonmagnetic metal as the ground state. The synthesis of monolayers and the detailed elucidation of realizable structures are just at the beginning. However, an extensive study of layered 1T′-$TaTe_2$ crystals using ARPES measurements to analyze bulk single crystals, compared the experimental results with calculations employing the DFT-LDA approach [84]. As expected, the Fermi surface topology was quite complicated, necessitating an analysis of monolayers without additional distortions caused by interlayer interactions. Remarkably, the surface electronic structure of layered 1T′-$TaTe_2$ sheets resembled that of 2H-$TaTe_2$, and the bulk electronic structure had more similarities with that of the hypothetical 1T-$TaTe_2$ sheet. Importantly, up to 180 K, the Fermi surface topology was temperature independent. This confirms the independence of the CDW order from surface nesting and the stability of the surface of the 2H phase up to this temperature.

15.3.3.3 Mechanics

Linear and nonlinear mechanical properties of $TaTe_2$ are available for multilayer sheets in the thickness range from 3.5 to 12.6 nm, as determined by microscopy-based nanoindentation measurements [65]. The measured Young's modulus of 2H-$TaTe_2$ sheets of 70 GPa (55 N/m) was independent of the thickness and agreed well with the calculated value of 70.9 GPa (55.7 N/m) (PBE). The results given in 2D units (N/m) are based on the estimated thickness of $d_{w+h} = 0.786$ nm. The calculated Poisson's ratio is extremely small at 0.10 (PBE). Remarkably, the measured breaking strength of 7.18 GPa (5.6 N/m) nearly matches the calculated value of 6.2 GPa (4.9 N/m) and fulfills Griffith's rule surprisingly well [65].

15.3.3.4 Physics and Applications

As discussed before, quasi 2D 1T-TaX_2 (X = S, Se, and Te) sheets incorporate CDWs at low temperatures as manifested in periodic lattice distortions (PLDs) [85]. To study the effect of CDWs on the Fermi surface, the authors calculated Fermi surface cross sections in various planes and determined the nesting vectors. The full potential of the linearized augmented plane-wave method reproduces the lattice distortions caused by the CDWs, using the total energy and force calculations performed within the supercells. The results indicate instability of these compounds toward the CDW state. This confirms the presence of gaping in the Fermi surface and supports the crucial role played by electron-phonon coupling in the stabilization of CDWs.

The magnetic moments of 1H-$TaTe_2$ monolayers are antiferromagnetic below 2% strain and ferromagnetic above 2% strain [71]. At a strain of 4%, the ferromagnetic moment is 0.28 μ_B per Ta atom, and at 10%, the value increases to 0.57 μ_B per Ta atom. Hydrogenation on one side ($HTaTe_2$) induces a transformation of the antiferromagnetic state into a direct semiconducting state with a band gap of $E_g = 0.62$ eV. The donation of an electron from the hydrogen atom to the empty Ta-5*d* state induces this transition.

Interestingly, 2D-$TaTe_2$ sheets promote the ambient electrochemical N_2-reduction reaction (NRR) for ammonia production [86]. The abundance of Ta vacancies and surface hydrophobicity of dispersed metallic few-layer $TaTe_2$ nanosheets can tune this process. The main active sites are the edges and exposed Ta atom sites, which dramatically enhance the binding of N and reduce the free energy change at the potential-determining step. This synergistic strategy boosts the reduction of strongly bonded N_2 molecules, leading to a remarkable NH_3 faradic efficiency of 32.2% at a mild overpotential.

15.3.3.5 Synthesis

The spontaneous formation of electronic orders was first demonstrated by photoemission and scanning probe measurements on few-layers thick 1T-$TaTe_2$ nanosheets deposited by MBE [87]. The analysis revealed the realization of a variety of metastable CDW orders, including the $(\sqrt{19} \times \sqrt{19})$ superstructure, which can be stabilized by controlling the postgrowth annealing temperature, and exists in films thicker than a monolayer, of up to 8 layers.

The versatility of MBE for growing metastable phases was demonstrated by the deposition of metastable polymorphic phases, such as hexagonal or octahedral $TaTe_2$ monolayers, which were analyzed via in situ STM and first-principles calculations [88]. The applied substrate of graphene/Ir(111) is known for its minimal interaction with TMDCs and the possibility of incorporating low defect structures. Careful adjustment of the deposition parameters leads to the coexistence of the 1H and 1T phases. Furthermore, 1T $TaTe_2$ exhibits several super periodic structures in coexistence with CDWs.

Control of the deposition parameters such as the substrate temperature allowed the selective deposition of 1H- and 1T-$TaTe_2$ monolayers with a higher energy than the 1T′ ground state using MBE [89]. Moreover, post-growth annealing of the 1H-$TaTe_2$ monolayer induces a transition of the $(\sqrt{19} \times \sqrt{19})$ superstructure to a new (2×2) superstructure with two gaps in the electronic structure. The different electronic structures were identified by ARPES and first-principles calculations. Theory revealed the role played by phonon instabilities in the formation of superstructures and the effect of local atomic distortions on the modification of the electronic structure.

Two extensive review articles, systematically covering several transition-metal dichalcogenide groups, include group-VB dichalcogenides and all aspects of synthesis, characterization, electronic structure, and performance [90, 91]. The first review concentrates on van der Waals metallic transition-metal dichalcogenides [90]. This includes synthesis of high-quality wafer-scale nanosheets grown by conventional methods (e.g., CVD and CVT), but also small area samples modified by Li intercalation, electron beam irradiation, light illumination, pressure, and strains. The theoretical and experimental results on the band structures of metallic TMDCs and their dependence on the number of layers are reviewed. In addition, distinct physical properties such as PLD, magnetoresistance, superconductivity, topological insulators, and Weyl semimetals are elucidated. The second review concentrates on the recent progress made in synthesis and characterization of 2D TMDCs from a vacuum surface science perspective [91]. Accordingly, the focus of this work are deposition methods such as advanced MBE with the possibility of characterization under UHV conditions, preferably as single-layer material, for example, by STM and ARPES. Other topics are compositional modifications, phase switching, and control between 1H and 1T (1T′) phases. This includes phase control by tuning growth conditions or post-growth modification, e.g., by electron doping. New phases not existing in the bulk but achievable as monolayers by MBE such as Mott insulators, CDW materials, and topological insulators are of special interest in this comprehensive overview covering besides VB-TMDCs also IVB-, VIB-, VIIB-, and VIIIB-TMDCs. Specifically, the group of vanadium dichalgogenides has found significant attention, based on promising applications in electronics, spintronics, and energy-related fields. In a recent review the band structure, exchange parameters, and magnetic anisotropy constants were calculated to determine the Curie temperatures, hysteresis curves, and energy of spin-wave excitations [92]. In another review on this subject, the structures, fabrication methods, key properties, and potential applications of the vanadium dichalcogenides MX_2 (X = S, Se, and Te) and their heterostructures are discussed [93].

15.4 Summary of Group VB–VIA Monolayers

15.4.1 Binding–Structure Relationships

Figure 15.13 shows the cohesive energy versus bond length of the M–X bonds of MX_2 (M = V, Nb, and Ta; X = O, S, Se, and Te) monolayers. Note that the separation of chalcogen atoms in the triple layer is much longer than the mixed covalent-ionic metal–chalcogen bonds. The plot displays the smaller values of the published binding energies preferably calculated by the DFT-GGA-PBE approach, which in some cases deviate significantly from the corresponding LDA binding energies that often overestimate the energy of the system (see Table 15.1).

The plot documents the inverse behavior between the cohesive energy and bond length of (metallic) bonding, following tight-binding theory. Note the obvious parallel decrease in the stiffness of the individual VB-dichalcogenide groups with increasing bond length that is due to the increasing atom size and change in the electronic structure of the chalcogens.

The diagram shows the large deviation between published cohesive energies of VO_2, indicating the enormous problems encountered in the determination of accurate

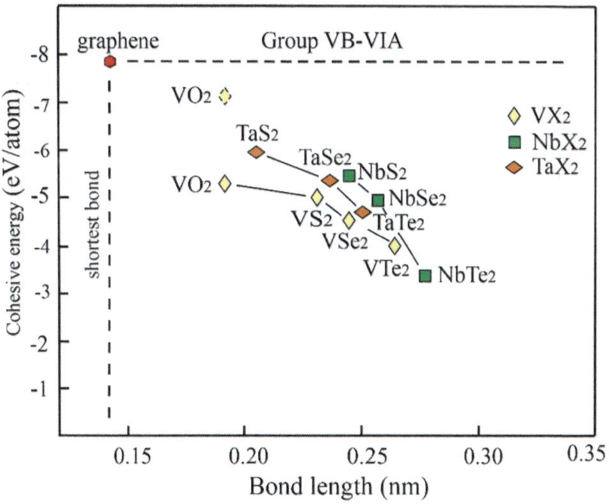

Fig. 15.13 Diagram of the cohesive energy versus bond length of group-VB dichalcogenides

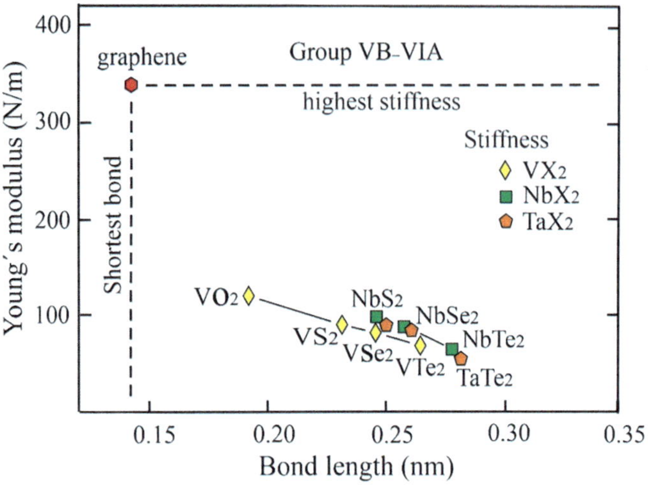

Fig. 15.14 Diagram of the Young's modulus versus bond length of group-VB dicalcogenides

binding energies specifically for the oxides of transition metals with substantial ionic contributions. As a consequence of these difficulties, the cohesive energies of the oxides, NbO_2 and TaO_2, are still not available.

15.4.2 Mechanics–Structure Relationships

Table 15.2 presents the presently available linear elastic properties, such as Young's moduli and Poisson's ratios. Figure 15.14 displays the dependence of selected stiffness values on the bond length of the investigated group–VB dichalcogenides.

A comparison of the elastic modulus of graphene with those of the group VB–VIA monolayers demonstrates a low mechanical stability in comparison to their cohesive energies and that of the extremely stiff graphene. The inverse dependence is less steep than the dependence of the binding energies owing to the significant reduction in stiffness. The three-layer structure may increase the overall cohesion and binding energy; however, the ultimate strength may decrease because it is determined by the weakest bonds.

References

1. C. Ataca, H. Şahin, S. Ciraci, J. Phys. Chem. C **116**, 8983 (2012)
2. A.H. Tobar, J.F. Murillo, C.O. López, J.A.R. Martínez, M.J. Espitia, Comput. Condens. Matter **25**, e00498 (2020)
3. Z.-K. Tang, X.-B. Li, D.-Y. Zhang, Y.-N. Zhang, L.-M. Liu, J. Mater. Chem. C **3**, 3189 (2015)
4. D.-T. Nguyen, in Lecture Notes in Mechanical Engineering, The AUN/SEED-Net Joint Regional Conference in Transportation, Energy, and Mechanical Manufacturing Engineering. Springer, Berlin, pp. 687–700 (2021)
5. Y. Zhang, X. Wu, R. Wang, Z. Qiu, X. Guo, Mater. Focus **2**, 118 (2013)
6. H. Li, H. Djaoued, J. Robichaud, Y. Djaoued, J. Mater. Chem. C **8**, 11572 (2020)
7. Y. Wang, N. Song, X. Song, T. Zhang, Q. Zhang et al., RSC Adv. **8**, 10848 (2018)
8. L. Lin, R. Chen, J. Huang, P. Wang, H. Tao et al., J. Phys. Condens. Matter **33**, 055502 (2021)
9. K. Han, L. Wu, Y. Cao, H. Wang, C. Ye et al., ACS Appl. Mater. Interfaces **13**, 16688 (2021)
10. D. Lee, B. Chung, Y. Shi, G.-Y. Kim, N. Campbell et al., Science **362**, 1037 (2018)
11. A.H.M.A. Wasey, S. Chakrabarty, G.P. Das, J. Appl. Phys. **117**, 064313 (2015)
12. H.-R. Fuh, C.-R. Chang, Y.-K. Wang, R.F.L. Evans, R.W. Cantrell et al., Sci. Rep. **6**, 32625 (2016)
13. H.L. Zhuang, R.G. Hennig, J. Phys. Chem. C **117**, 13169 (2013)
14. J. Yang, A. Wang, S. Zhang, J. Liu, Z. Zhong et al., Phys. Chem. Chem. Phys. **21**, 132 (2019)
15. Q. Ji, C. Li, J. Wang, J. Niu, Y. Gong et al., Nano Lett. **17**, 4908 (2017)
16. N. Luo, C. Si, W. Duan, Phys. Rev. B **95**, 205432 (2017)
17. H.-R. Fuh, B. Yan, S.-C. Wu, C. Felser, C.-R. Chang, New J. Phys. **18**, 113038 (2016)
18. W. Wang, Z. Sun, W. Zhang, Q. Fan, Q. Sun et al., RSC Adv. **6**, 54874 (2016)
19. Y. Cui, W. Fan, X. Liu, J. Ren, Y. Gao, Comp. Mater. Sci. **200**, 110767 (2021)
20. H. Pan, Nanoscale Res. Lett. **11**, 113 (2016)
21. L. Wang, W. Gao, X. Chen, Y. Liu, A.H. Qureshi et al., J. Phys. Chem. C **125**, 12004 (2021)
22. Z. Qin, Z. Wang, J. Zhao, Nanoscale **14**, 6902 (2022)
23. J. Yuan, J. Wu, W.J. Hardy, P. Loya, M. Lou et al., Adv. Mater. **27**, 5605 (2015)
24. H.J. Kim, B.K. Choi, I.H. Lee, M.J. Kim, S.-H. Chun et al., Curr. Appl. Phys. **30**, 8 (2021)
25. C. van Efferen, J. Berges, J. Hall, E. van Loon, S. Kraus et al., Nat. Commun. **12**, 6837 (2021)
26. Z. Zhang, P. Yang, M. Hong, S. Jiang, G. Zhao, Nanotechnology **30**, 182002 (2019)
27. S. Daguer, N. De La Espriella, C.O. López, J. Phys. Conf. Series **2046**, 012037 (2021)
28. P.K.J. Wong, W. Zhang, F. Bussolotti, X. Yin, T.S. Herng et al., Adv. Mater. **31**, 1901185 (2019)
29. T.J. Kim, S. Ryee, M.J. Han, S. Choi, 2D Mater. **7**, 035023 (2020)
30. W. Yu, J. Li, T.S. Herng, Z. Wang, X. Zhao et al., Adv. Mater. **31**, 1903779 (2019)
31. K. Xu, P. Chen, X. Li, C. Wu, Y. Guo, et al., Angew. Chem., Int. Ed. **52**, 10477 (2013)
32. W. Zhao, B. Dong, Z. Guo, G. Su, R. Gao, ChemComm. **52**, 9228 (2016)
33. Z.-L. Liu, X. Wu, Y. Shao, J. Qi, Y. Cao et al., Sci. Bull. **63**, 419 (2018)
34. M. Bonilla, S. Kolekar, Y. Ma, H.C. Diaz, V. Kalappattil et al., Nat. Nanotech. **13**, 289 (2018)
35. M.-H. Lv, C.-M. Li, W.-F. Sun, Nanomaterials **12**, 382 (2022)
36. B.-D. Guo, J.-H. Liao, W.-Q. Xie, Y.-H. Peng, Y.-J. Zhao, Phys. Lett. A **446**, 128260 (2022)
37. P.M. Coelho, K. Lasek, K.N. Cong, J. Li, W. Niu et al., J. Phys. Chem. Lett. **10**, 4987 (2019)
38. X. Li, Z. Zhu, Q. Yang, Z. Cao, Y. Wang, et al., Nano Res. ISSN 1998-0124 (2021)
39. O. Concepción, L. Mulder, D.H. Wielens, A. Brinkman, Solids **3**, 500 (2022)
40. Y. Wang, J. Ren, J. Li, Y. Wang, H. Peng et al., Phys. Rev. B **100**, 241404(R) (2019)
41. K. Sugawara, Y. Nakata, K. Fujii, K. Nakayama et al., Phys. Rev. B **99**, 241404(R) (2019)
42. Y. Ding, Y. Wang, J. Ni, L. Shi, S. Shi et al., Physica B **406**, 2254 (2011)
43. C. Heil, S. Poncé, H. Lambert, M. Schlipf, E.R. Margine et al., Phys. Rev. Lett. **119**, 047003 (2017)
44. E.G.C.P. van Loon, M. Rösner, G. Schönhoff, M.I. Katsnelson, T.O. Wehling, NPJ Quantum Mater. **3**, 32 (2018)
45. R. Bianco, I. Errea, L. Monacelli, M. Calandra, F. Mauri, Nano Lett. **19**, 3098 (2019)

46. Y. Kim, K.C. Kwon, S. Kang, C. Kim, T.H. Kim et al., ACS Sens. **4**, 2395 (2019)
47. S. Zhao, T. Hotta, T. Koretsune, K. Watanabe, T. Taniguchi, et al., 2D Mater. **3**, 025027 (2016)
48. X. Wang, J. Lin, Y. Zhu, C. Luo, K. Suenaga et al., Nanoscale **9**, 16607 (2017)
49. Z. Wang, C.-Y. Cheon, M. Tripathi, G.M. Marega, Y. Zhao et al., ACS Nano **15**, 18403 (2021)
50. P. Chithaiah, S. Mallik, D.C. Binwal, A. Bhui, C.N.R. Rao, New J. Chem. **47**, 17098 (2023)
51. S.C. Ganguli, V. Vaňo, S. Kezilebieke, J.L. Lado, P. Liljeroth, Nano Lett. **22**, 1845 (2022)
52. S. Das, I.I. Mazin, Comput. Mater. Sci. **200**, 110758 (2021)
53. D. Wickramaratne, I.I. Mazin, Nat. Commun. **13**, 2376 (2022)
54. M.M. Ugeda, A.J. Bradley, Y. Zhang, S. Onishi, Y. Chen, Nat. Phys. **12**, 92 (2016)
55. A. Kumar. P.K. Ahluwalia, J. Alloys Comp. **550**, 283 (2013)
56. T. Hotta, T. Tokuda, S. Zhao, K. Watanabe, T. Taniguchi et al., Appl. Phys. Lett. **109**, 133101 (2016)
57. M. Liu, J. Leveillee, S. Lu, J. Yu, H. Kim, et al., Sci. Adv. **7**, eabi6339 (2021)
58. K. Zhang, N. Zou, Y. Ren, J. Wu, C. Si et al., Adv. Funct. Mater. **32**, 2111675 (2022)
59. T. Taguchi, K. Sugawara, H. Oka, T. Kawakami, Y. Saruta et al., Phys. Rev. B **107**, L041105 (2023)
60. Y. Nakata, K. Sugawara, A. Chainani, H. Oka, C. Bao et al., Nat. Commun. **12**, 5873 (2021)
61. Y. Bai, T. Jian, Z. Pan, J. Deng, X. Lin et al., Nano Lett. **23**, 2107 (2023)
62. K. Wang, Z. Guo, Y. Li, Y. Guo, H. Liu et al., ACS Appl. Nano Mater. **3**, 11363 (2020)
63. T. Jiang, T. Hu, G. Zhao, Y. Li, S. Xu et al., Phys. Rev. B **104**, 075147 (2021)
64. H. Lin, W. Huang, K. Zhao, S. Qiao, Z. Liu et al., Nano Res. **13**, 133 (2020)
65. A. Sheraz, N. Mehmood, M.M. Çiçek, I. Ergün, H.R. Rasouli et al., Nanoscale Adv. **3**, 3894 (2021)
66. D.F. Shao, R.C. Xiao, W.J. Lu, H.Y. Lv, J.Y. Li et al., Phys. Rev. B **94**, 125126 (2016)
67. Y. Geng, L. Lei, H. Dong, J. Guo, S. Mi et al., Phys. Rev. B **107**, 195401 (2023)
68. I. Lutsyk, K. Szalowski, P. Krukowski, P. Dabrowski, M. Rogala et al., Nano Res. **16**, 11528 (2023)
69. T. Patel, J. Okamoto, T. Dekker, B. Yang, J. Gao et al., Nano Lett. **20**, 7200 (2020)
70. L. Najafi, S. Bellani, R. Oropesa-Nuñez, B. Martín-Garcia, M. Prato et al., ACS Catal. **10**, 3313 (2020)
71. P. Manchanda, V. Sharma, H. Yu, D.J. Sellmyer, R. Skomski, Appl. Phys. Lett. **107**, 032402 (2015)
72. S. Kumar, S. Pratap, N. Joshi, R. Trivedi, C.S. Rout et al., Micro Nanostruct. **181**, 207627 (2023)
73. X. Pan, T. Yang, H. Bai, J. Peng, L. Li et al., Nanomaterials **13**, 1806 (2023)
74. A.K. Pathan, A. Gupta, M.E. Vaida, Nanotechnology **32**, 505605 (2021)
75. N. Tian, Z. Huang, B.G. Jang, S. Guo, Y.-J. Yan, et al., Nat. Sci. Rev. **11**, nwad144 (2024)
76. C.J. Sayers, G. Cerullo, Y. Zhang, C.E. Sanders, R.T. Chapman et al., Phys. Rev. Lett. **130**, 156401 (2023)
77. K. Zhang, C. Si, C.-S. Lian, J. Zhou, Z. Sun, J. Mater. Chem. C **8**, 9742 (2020)
78. L. Yan, C. Ding, M. Li, R. Tang, W. Chen et al., Nano Lett. **23**, 2121 (2023)
79. C. Wen, J. Gao, Y. Xie, Q. Zhang, P. Kong, at al., Phys. Rev. Lett. 126, 256402 (2021)
80. P. Qiao, J. Xia, X. Li, Y. Li, J. Cao et al., Nanoscale **15**, 17036 (2023)
81. Y. Nakata, T. Yoshizawa, K. Sugawara, Y. Umemoto, T. Takahashi et al., ACS Appl. Nano Mater. **1**, 1456 (2018)
82. H. Koussir, Y. Chernukha, C. Sthioul, E. Haber, N. Peric et al., Nano Lett. **23**, 9413 (2023)
83. H. Lee, H. Im, B.K. Choi, K. Park, Y. Chen et al., Nano Convergence **11**, 14 (2024)
84. I. Kar, K. Dolui, L. Harnagea, Y. Kushnirenko, G. Shipunov et al., J. Phys. Chem. C **125**, 1150 (2021)
85. S. Sharma, L. Nordström, B. Johansson, Phys. Rev. B **66**, 195101 (2002)
86. Z. Zhao, J. Park, C. Choi, S. Hong, X. Hui et al., Innovation **3**, 100190 (2022)
87. J. Hwang, Y. Jin, C. Zhang, T. Zhu, K. Kim et al., Adv. Mater. **34**, 2204579 (2022)
88. I.D. Bernardo, J. Ripoll-Sau, J.A. Silva-Guillén, F. Calleja, C.G. Ayani et al., Small **19**, 2300262 (2023)

89. R. Feng, W. Wang, C. Bao, Z. Zhang, F. Wang et al., Adv. Mater. **36**, 2302297 (2024)
90. G.H. Han, D.L. Duong, D.H. Keum, S.J. Yun, Y.H. Lee, Chem. Rev. **118**, 6297 (2018)
91. K. Lasek, J. Li, S. Kolekar, P.M. Coelho, L. Guo et al., Surf. Sci. Rep. **76**, 100523 (2021)
92. M. Jafari, W. Rudzinski, J. Barnas, A. Dyrdal, Sci. Rep. **13**, 20947 (2023)
93. M. Hossain, H. Zhang, Y. Huangfu, M.Z. Saeed, B. Qin, et al., Mater. Today Adv. **21**, 100451 (2024)

Chapter 16
Group–VIB Transition-Metal Dioxide and Dichalcogenide Monolayers

Abstract Transition-metal dioxide (TMDO) and transition-metal dichalcogenide (TMDC) monolayers with MX_2 stoichiometry consist of a central sublayer of transition-metal atoms, M = Cr, Mo, and W, sandwiched between two sublayers of chalcogenide atoms, X = O, S, Se, and Te. These compounds present an ideal platform for exploring 2D chemistry and physics owing to their various combinations of elements and many existing polytypes. A characteristic feature controlling the physicochemical properties of this group of compounds is the half-filled *d* band (d^2 group) of the transition metals. An early comprehensive study of monolayers with MX_2 stoichiometry presented the structural, chemical, and physical data of 88 compounds based on first-principles calculations. The monolayers consist of three sublayers bonded by varying covalent and ionic forces, and the few-layer assemblies usually are bound to each other by much weaker van der Waals attraction. The fabrication of monolayers and few-layer sheets is possible by micromechanical cleavage or liquid-phase exfoliation if layered bulk crystals exist. Chemical vapor deposition (CVD) and metal organic chemical vapor deposition (MOCVD) allow the bottom-up synthesis of any stable compound. The controlled growth of high-quality single monolayers is possible by molecular-beam epitaxy (MBE), which uses ultra-high vacuum (UHV) and therefore allows in situ detection of the growth process, e.g., by reflection high-energy electron diffraction (RHEED) and thus the analysis of intrinsic properties. Indentation measurements, providing Young's moduli by atomic force microscopy (AFM), reveal the strongly varying stiffness and ultimate strength of monolayers and few-layer systems. For example, Young's modulus of MoS_2 is 176 N/m, and the maximum breaking strain is 0.10, to be compared with 340 N/m and 0.25 for graphene, respectively. Some of the best studied monolayers (MoS_2, WS_2, $MoSe_2$, WSe_2, and $MoTe_2$) possess a direct band gap and qualify for applications in electronics and optoelectronics. Monolayer structures, which have no inversion centers, qualify for potential applications in the new field of valleytronics. For heavy atoms with *d* orbitals, strong spin–orbit coupling (SOC) removes spin degeneracy in the valence and conduction bands. The resulting energy splitting between the spin-up and spin-down states opens access to novel devices in spintronics.

P. Hess, *Bonding, Structure, and Performance of Two-Dimensional Materials*,
NanoScience and Technology, https://doi.org/10.1007/978-3-031-74161-6_16

16.1 Chromium Dichalcogenide Monolayers (CrO_2, CrS_2, $CrSe_2$, and $CrTe_2$)

16.1.1 CrO_2 Monolayer

16.1.1.1 Chemistry and 2D Bonding

The scheme of the periodic table shown in Fig. 16.1 displays the group–VIB transition-metal elements and group–IVA chalcogen elements considered in this chapter.

The electronic configurations of Cr (Ar|$3d^5 4s$) and O atoms ($1s^2 2s^2 p^4$) determine the type of bonding in 2D chromium dioxide. The *d-p* hybridization of the outermost $3d$ orbitals of Cr atoms and the outermost $2p$ orbitals of O atoms form strong covalent bonds (see Fig. 16.2). This bonding process leads to a covalently bonded three-sublayer structure with a Cr sublayer in the center sandwiched between two O sublayers.

Each O atom forms three covalent bonds with Cr atoms at a bridge position, and each Cr atom is bound to six O atoms (see Fig. 16.3). The extensive electron charge accumulation between the Cr and O atoms is responsible for the observed strong covalent Cr–O bonds. In addition, the significant difference in the electronegativities of Cr (1.66) and O (3.44) leads to a substantial ionic contribution to bonding and consequently to strongly polar Cr–O bonds. A Bader charge transfer analysis of the most stable hexagonal 1H phase yields a charge transfer of $\delta q = 1.54$ e [1]

H																	He
Li	Be											B	C	N	O	F	Ne
Na	Mg											Al	Si	P	S	Cl	Ar
K	Ca	Sc	Ti	V	Cr	Mn	Fe	Co	Ni	Cu	Zn	Ga	Ge	As	Se	Br	Kr
Rb	Sr	Y	Zr	Nb	Mo	Tc	Ru	Rh	Pd	Ag	Cd	In	Sn	Sb	Te	I	Xe
Cs	Ba	La	Hf	Ta	W	Re	Os	Ir	Pt	Au	Hg	Tl	Pb	Bi	Po	At	Rn
Fr	Ra	Ac	Rf	Db	Sg	Bh	Hs	Mt	Ds	Rg	Cn	Uut	Fl	Uup	Lv	Uus	Uuo

Fig. 16.1 Periodic table highlighting the transition metals of group VIB and the chalcogen atoms of group VIA, which form the VIB–VIA monolayers

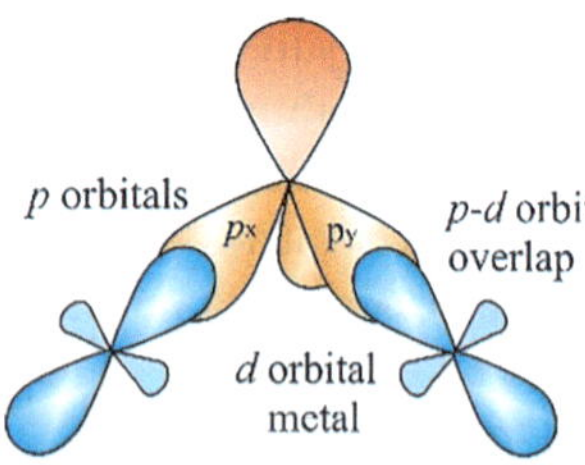

Fig. 16.2 Schematic representation of bonding arrangements based on *d–p* hybridization in MX_2 monolayers involving *d* orbitals of the metal atom and *p* orbitals of the chalcogen atoms

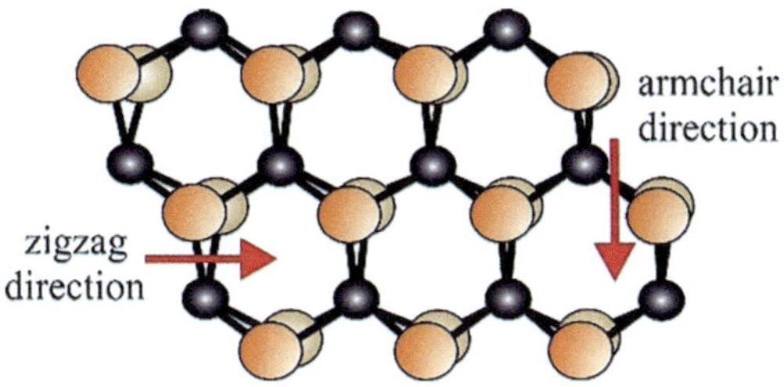

Fig. 16.3 Scheme of the three-layer structure of MX_2 monolayers with the metal atoms in the center sandwiched between two layers of chalcogen atoms. Each chalcogen atom X forms three covalent bonds with metal atoms at a bridge position, and each metal atom is bound to six X atoms

(1.48 e) [2] from each Cr atom to two O atoms and thus to the transfer of $\delta q =$ 0.77 e (0.74 e) [2] to each O atom. This charge transfer generates an extremely large fractional ionic character of 54.5% [2]. The strong ionic-covalent bonding results in low cohesive energies for CrO_2 monolayers. For the cohesive energy of the most stable 1H phase the reported strongly deviating results are $E_c = -6.52$ eV/atom [1], which was calculated by using the exchange correlation potential of the local density approximation (LDA), and $E_c = -5.34$ eV/atom [2], which was obtained by using density functional theory (DFT), approximating the exchange correlation potential by the generalized gradient approximation (GGA) using Perdew Burke Ernzerhof (PBE) functionals. Note that the LDA approximation often overestimates the energy of materials. Recent GGA-PBE calculations delivered cohesive energies of the 1H phase of $E_c = -5.79$ eV/atom [3] and of the metastable 1T phase of − 5.69 eV/atom and therefore they support a lower binding energy of the latter phase [4].

16.1.1.2 2D Structures

The distinct phases 1H, 1T, and 1T′ have similar three-sublayer structures (see Fig. 16.4). The 1H phase, which has hexagonal symmetry and trigonal prismatic coordination, is a semiconductor with the space group *P*-6*m*2 and the most stable phase of chromium dichalcogenide monolayers [5]. Furthermore, the metallic 1T phase with space group *P*-3*m*1 is metastable for most TMDO and TMDC compounds, except for the heavy compounds such as $MoTe_2$ and WTe_2. The third phase is the distorted 1T′ phase, in which pairs of metal atoms move closer to each other, as indicated schematically in Fig. 16.4. This distortion arises from the Peierls instability of the lattice based on electron-phonon coupling. The space group of this latter 1T′ phase belongs to $P2_1/m$ [5].

The 1H structure and 2H structure have different layer stacking in bilayer and multilayer arrangements, as displayed in Fig. 16.5. The 1H structure consists of layers stacked on top of each other such that the unit cell contains only one formula unit, belonging to space group *P*-6*m*2. For the shifted stacking of the 2H bilayer and

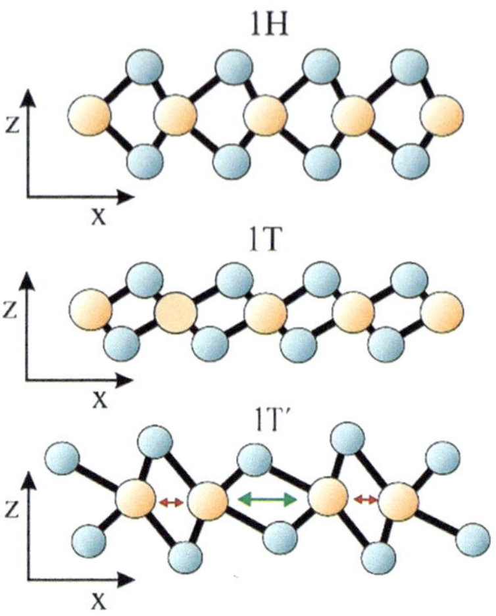

Fig. 16.4 Side view of the semiconducting hexagonal 1H phase, the metallic octahedral 1T phase, and the disturbed 1T′ phase derived from the 1T structure by shifting the metal atoms

bulk structure, the unit cell contains two formula units, and the space group changes to $P6_3/mmc$.

The lattice constant of the 1H structure of CrO_2 monolayers is a = 0.258 nm, the bond length is $d_{Cr–O} = 0.188$ nm, the distance between O atoms is $h_{OO} = 0.229$ nm, and the bond angle is $\theta_{OCrO} = 75.21°$ (LDA) [1]. These lattice parameters deviate substantially from a = 0.284 nm, $d_{Cr–O} = 0.205$ nm, $h_{OO} = 0.247$ nm, and the bond angle $\theta_{OCrO} = 74.05°$ (GGA-PBE-D2) [3]. By adding the distance between O atoms $h_{OO} = 0.229$ nm and the van der Waals diameter of O atoms of $d_w = 0.310$ nm a monolayer thickness of $d_{w+h} = 0.539$ nm is obtained. Importantly, the 1H-CrO_2 monolayer and 2H few-layer assemblies are thermodynamically and dynamically stable structures [1, 2].

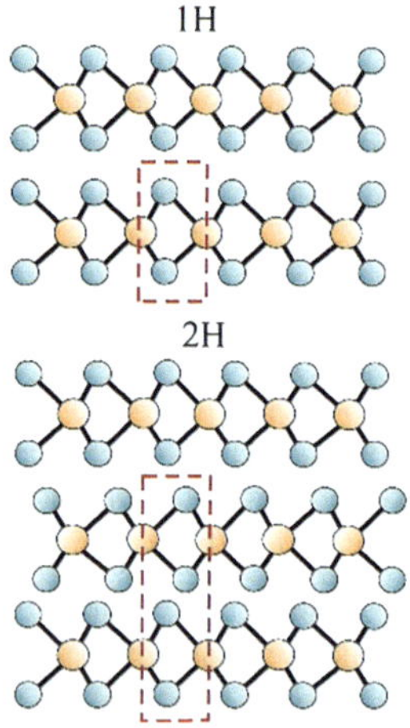

Fig. 16.5 Side view of the 1H multilayer structure with one formula unit per unit cell and the 2H multilayer structure with two formula units per unit cell

The semiconducting 1H-CrO_2 phase has a band gap of $E_g = 0.5$ eV and a much larger GW_0-corrected gap of 1.80 eV [1]. These values can be compared with the LDA band gap of $E_g = 0.43$ eV and the G_0W_0 gap of 1.64 eV provided for 2H CrO_2 [6]. Note that group–VIB transition-metal dioxides are indirect semiconductors, while group–VIB transition-metal dichalcogenides usually are direct semiconductors.

The lattice constant of the 1T phase is a = 0.289 nm and thus much larger than that of the 1H phase, whereas the bond length of $d_{Cr–O} = 0.192$ nm is comparable [4]. The metastable 1T-CrO_2 monolayer is thermodynamically and dynamically stable and belongs to the space group *P*-3*m*1 [4]. Unexpectedly large formation energies of $E_f = -1.78$ eV/atom [4] (PBE) and − 2.14 eV/atom [3] (PBE) were reported in the literature for the 1T-CrO_2 phase. The symmetry, cohesive energy, bond length, X–X distance, bond angle, and thickness of the group of VIB–VIA monolayers are collected in Table 16.1.

16.1.1.3 Mechanics

For the chromium dichalcogenide monolayers, only linear elastic properties, such as the elastic constants, Young's moduli, and Poisson's ratios, are currently available. Calculations by the DFT-GGA-PBE approach consider lattice extensions, where elastic deformations respond linearly and reversibly to strain. From the calculated elastic constants of the 1H-CrO_2 phase, a large Young's modulus of 193.3 N/m and Poisson's ratio of 0.34 were derived [7]. However, a much larger stiffness of 220.9 N/m can be found in the early literature [1]. Young's moduli and Poisson's ratios derived by an efficient automated toolkit that is based on DFT and ab initio molecular dynamics (MD) calculations, are much lower than the presented values [8]. For example, for the stiffness of 1H-CrO_2 monolayers, these automated calculations give E = 133.5 N/m and for Poisson's ratio $\nu = 0.25$; for the 1T-CrO_2 phase a stiffness of only E = 63.0 N/m, and auxetic behavior with a negative Poisson ratio of $\nu = -0.07$ have been predicted [8]. Note that the elastic constants of chromium dioxide monolayers fulfill the Born-Huang criteria for mechanical stability [7, 8]. Importantly, a comprehensive work covers the elastic mechanical properties of 34 TMDO and TMDC monolayers, including the group VIB–VIA compounds [9]. This work predicts large slightly anisotropic stiffnesses of $E_{ac} = 212.4$ N/m and $E_{zz} = 216.1$ N/m, and small Poisson's ratios of $\nu_{ac} = 0.13$ and $\nu_{zz} = 0.12$ for 1H-CrO_2 monolayers [9]. According to this study, for the same transition metal M, the Young's modulus and shear modulus of MO_2 monolayers have maximal values, and for the MTe_2 monolayer, they have minimal values. The space group, Young´s modulus, Poisson´s ratio, ultimate strength, and ultimate strain of group VIB–VIA monolayers are presented in Table 16.2.

Table 16.1 Symmetry, cohesive energy, lattice constant, bond length, X–X distance, bond angle, and corrugation corrected van der Waals diameter (thickness) of group VIB–VIA monolayers with stoichiometry MX_2

Monolayer	Symmetry	Cohesive energy E_c (eV/atom)	Lattice constant a (nm)	Bond length d_{M-X} (nm)	X–X distance h_{XX} (nm)	Bond angle θ_{XMX} (°)	Thickness d_{w+h} (nm)
CrO_2	H phase	– 5.34 [2], – 5.79 [3]	0.258 [1]	0.188 [1]	0.229 [1]	75.21 [1]	0.539
CrS_2	H phase	– 4.17 [2], – 4.10 [19]	0.297 [1]	0.225 [1]	0.292 [1]	80.86 [1]	0.652
$CrSe_2$	H phase	– 3.64 [2], – 3.59 [19]	0.313 [1]	0.238 [1]	0.311 [1]	81.54 [1]	0.691
$CrTe_2$	H phase	– 3.09 [2]	0.339 [1]	0.258 [1]	0.338 [1]	81.56 [1]	0.758
MoO_2	H phase	– 6.24 [2], – 6.20 [3]	0.283 [2]	0.205 [2]	0.248 [2]	74.18 [2]	0.558
MoS_2	H phase	– 5.14 [2], – 5.07 [62]	0.318 [2]	0.241 [2]	0.313 [2]	80.77 [2]	0.673
$MoSe_2$	H phase	– 4.60 [2], – 4.53 [62]	0.332 [2]	0.254 [2]	0.334 [2]	82.12 [2]	0.714
$MoTe_2$	H phase	– 4.04 [2], – 3.97 [62]	0.355 [2]	0.273 [2]	0.361 [2]	82.74 [2]	0.781
WO_2	H phase	– 7.02 [2], – 8.17 [3]	0.283 [2]	0.205 [2]	0.248 [2]	74.34 [2]	0.558
WS_2	H phase	– 5.80 [2], – 5.71 [62]	0.319 [2]	0.242 [2]	0.314 [2]	80.87 [2]	0.674
WSe_2	H phase	– 5.19 [2], – 5.10 [62]	0.332 [2]	0.255 [2]	0.335 [2]	82.44 [2]	0.715
WTe_2	H phase $1T_d$	– 4.54 [2], – 4.46 [62]	0.355 [2] 0.350	0.274 [2] 0.273	0.362 [2]	82.89 [2]	0.782

Table 16.2 Space group, Young's modulus, Poisson's ratio, ultimate strength, and ultimate strain of group VIB–VIA monolayers with stoichiometry MX_2

Monolayer	Space group	Young's modulus E_{ac}, E_{zz} (N/m)	Poisson's ratio ν_{ac}, ν_{zz}	Ultimate strength σ_{ac}, σ_{zz} (N/m)	Ultimate strain ε_{ac}, ε_{zz}
CrO_2	*P*-6*m*2	193.3 [7], 212.4, 216.1 [8]	0.34 [7], 0.13, 0.12 [9]		
CrS_2	*P*-6*m*2	112.0 [7], 110.4 [19]	0.27 [19], 0.26 [19]	12.9 (u), 12.2 (b) [19]	0.24 (u), 0.18 (b) [19]
$CrSe_2$	*P*-6*m*2	88.0 [7], 85.6 [19]	0.30 [7], 0.31 [19]	10.7 (u), 9.76 (b) [19]	0.26 (u), 0.16 (b) [19]
$CrTe_2$	*P*-6*m*2	63.9 [7], 63.9 [41], 71.4 [8]	0.35 [7], 0.30 [41], 0.30 [48]		
MoO_2	*P*-6*m*2	199.2 [7], 198.3 [51]	0.37 [7], 0.367 [51]		
MoS_2	*P*-6*m*2	150, 148 [68], 136 [69]	0.25 [69]	16.8 [68], 15.8, 10.8 [69]	0.28, 0.19 [68]
$MoSe_2$	*P*-6*m*2	128, 126 [68], 105.7 [106]	0.28 [106]	16.2, 9.2 [68], 15.0,12.9 [106]	0.29. 0.16 [68], 0.41. 0.27 [106]
$MoTe_2$	*P*-6*m*2	98.3, 96.5 [68], 88.9 [8]	0.21 [8], 0.25 [104]	13.4, 6.2 [68]	0.32, 0.14 [68]
WO_2	*P*-6*m*2	231.7 [7], 234.5 [8], 250 [1]	0.34 [7], 0.32 [8]		
WS_2	*P*-6*m*2	165, 162 [68], 177 [70]		20.2, 13.4 [68]	0.28, 0.18 [68]
WSe_2	*P*-6*m*2	141, 139 [68], 127.7 [8]	0.17 [8]	17.7, 11 [68]	0.30, 0.17 [68]
WTe_2	$1T_d$ *P*-6*m*2	71, 105 [152] 107, 106 [68]	0.26, 0.38 [152]	 14.6, 7.3 [68]	 0.32, 0.15 [68]

16.1.1.4 Physics and Applications

The 1T-CrO_2 monolayer is a high-temperature Dirac half-metal with an intrinsic 3*d* ferromagnetic ground state and a Curie temperature of up to 507 K [4]. A half-metal behaves as a metal in one spin channel and as an insulator or semiconductor in the opposite spin channel. The 3*d* states of Cr and the high Curie temperature explain the robust ferromagnetism. Materials such as 1T CrO_2 with complete spin polarization and high-speed conduction electrons are promising candidates for high-efficiency spintronics applications, which use the spin degree of freedom in electronic devices.

Combined first-principles calculations and Monte Carlo simulations predicted the existence of a novel 2D CrO_2 structure, wherein the electrons around the Fermi level are in an intrinsic completely spin-polarized state with an ultrawide energy range [10]. The space group of this orthorhombic structure is *Pmmn*, which differs from the previously discussed phases. The magnetically ordered spin state of the dynamically and thermodynamically stable material is stable up to 280 K and further stabilizes under biaxial tensile strain or hole doping. Such highly spin-polarized materials behaving as a magnetic semimetal are of increasing interest as magnetic storage materials or spin injection sources.

A theoretical study suggested the deposition of an epitaxial CrO_2 monolayer on ferroelectric $PbTiO_3$ as a potential system for ferroelectric polarization-based catalysis with switchable surface chemistry [11]. The surface of CrO_2–$PbTiO_3$ allows direct NO_x decomposition and CO oxidation. In these reactions ferroelectric polarizations, cycling between positive and negative polarization states, drive the desired reaction in a cyclic manner. Negative polarization is the preferred state at low temperatures and raising the temperature to values near the Curie point transforms the system into the paraelectric state or positive polarization. Accordingly, the surface cycles between strong binding (initiating dissociation of molecules) and weak binding (easy desorption) (see Fig. 16.6). The rate-determining step of the periodic reaction process is NO dissociation on the paraelectric surface.

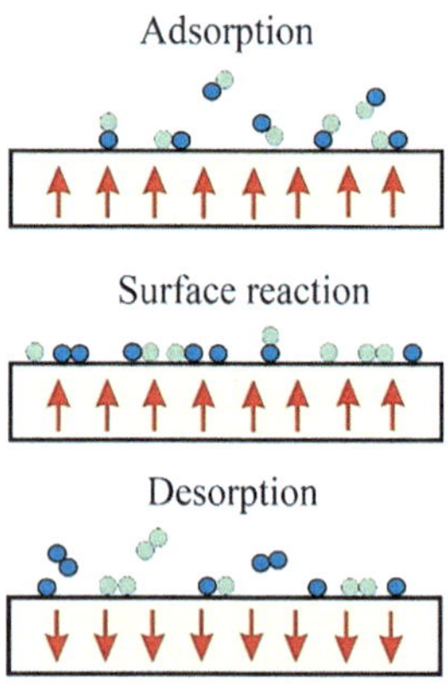

Fig. 16.6 Scheme of catalytic transformation of 2NO into N_2 and O_2 molecules. This includes chemisorption of NO, dissociation of NO by the positively poled surface, and efficient desorption of the reaction products N_2 and O_2 by the negatively poled surface

First-principles calculations suggest that the magnetic properties of CrO_2 monolayers can be controlled by hydrogenation [12]. In hydrogenated monolayers H_xCrO_2 ($0 \leq x \leq 2$) the behavior transforms from a ferromagnetic half-metal ($x = 0$) to a small-gap insulator for hydrogenation at $x = 0.75$. At $x = 1$ and $x = 1.25$, the monolayer behaves as a bipolar antiferromagnetic insulator and becomes an antiferromagnetic insulator at $x = 2$. The authors consider this finding as a first step toward realizing tunable 2D magnetic materials.

16.1.1.5 Synthesis

The development of diverse techniques for growing CrO_2 sheets has improved their quality in recent years [13–15]. To date, the synthesis of perfect 1T-CrO_2 monolayers has not been successful. A standard CVD method using chromyl chloride (CrO_2Cl_2) as a precursor allowed the growth of highly spin-polarized metallic CrO_2 films on TiO_2(100) [13]. The (100) oriented films, grown in registry with the substrate, showed a spin polarization of 81%. The Curie temperature of the ferromagnetic films was 395 K, and the coercivity ~ 100 Oe at 298 K. Importantly, the high Curie temperature of CrO_2 makes studies on spin-polarized features accessible at room temperature.

Pulsed laser ablation allowed the deposition of nanostructured single-phase CrO_2 films on MgO(100) using a metallic Cr target in an O_2 environment [14]. Scanning electron microscopy (SEM) and AFM revealed the synthesis of orthogonally arranged strained needlelike and plate-like structures. X-ray photoemission (XRP) confirmed the existence of (110) or (200) CrO_2 crystallites with an epitaxial relationship to the substrate and the protection of the 30–50 nm thick CrO_2 films by a thin CrO_3 overlayer without any residual metallic Cr atoms.

The setup used for epitaxial growth of CrO_2 films on single-crystal TiO_2, employing Cr_8O_{21} as a precursor in flowing oxygen, allowed transfer of the film to the substrate under UHV conditions [15]. In multilayer sheets, CrO_2 films with lattice-matched (001) and (110) surfaces could be stabilized up to ~ 99 °C, and were stable under UHV conditions up to ~ 150 °C. The presence of oxygen stabilizes the ordered structures, which become metastable without sufficient oxygen (~1 atm) and well-ordered CrO_2 surfaces can be recovered by annealing disordered phases in oxygen. Optimized pressure–temperature conditions will minimize interfacial spin relaxation in multilayer devices.

The investigation of the atomic structure of chromium dioxide catalysts involved the analysis of well-defined CrO_2 bilayer islands and Cr_2O_7 dinuclear clusters by scanning tunneling microscopy (STM) and theory. Under different redox treatments, chromium dioxide nanoislands featuring a CrO_2 bilayer structure and clusters of similar thickness and Cr_2O_7 stoichiometry can reversibly be converted into each other if stabilized by a metal surface such as Au(111) and a specific environment [16]. Oxidation by O_3 can disperse layered CrO_2 nanoislands into Cr_2O_7 dinuclear clusters that can be converted back by annealing.

In a theoretical work the stability of the (100), (001), (110), and (011) surfaces of CrO_2 have been studied, which is the prototype of a half-metal [17]. This study investigated the interplay between work function and stability and concluded that oxygen-terminated CrO_2 surfaces exhibit the greatest stability. The valency of the surface atoms controls the surface stability, where the oxygen atoms, with their lower valency, allow the Cr atoms to retain their high coordination, as in the bulk. Note that the situation is different for intermetallic compounds.

16.1.2 CrS_2 Monolayer

16.1.2.1 Chemistry and 2D Bonding

A smaller orbital overlap, due to the increasing size of S atoms with a van der Waals diameter of d_w = 0.360 nm [18] (O: 0.310 nm), controls bonding in chromium disulfide monolayers with the valence electrons provided by the electronic configuration of S atoms (Ne|$3s^2p^4$) and Cr atoms (Ar|$3d^54s$). Compared with CrO_2, charge shielding by inner electrons and the smaller difference in the electronegativities of S atoms (2.58) and O atoms (3.44) reduce the ionic interaction and binding energy. Nonetheless, the previously discussed three-sublayer structures with a positively charged central Cr sublayer and negatively charged outer S sublayers retain the most stable structural arrangements. The calculated Bader charge transfer amounts to δq = 0.92 e [1] (1.00 e [2]) from each Cr atom to two S atoms. Owing to the reduced charge transfer, the fractional ionic character is only 19.1% [2]. Due to the strongly diminished orbital overlap, the calculated cohesive energy of the most stable hexagonal 1H phase has the values E_c = – 5.30 eV/atm [1] (LDA) and – 4.17 eV/atom (PBE) [2], where the latter value is in good agreement with – 4.10 eV/atom (PBE) [19].

16.1.2.2 2D Structures

The hexagonal 1H phase of CrS_2 is the most stable structure of chromium disulfide monolayers. Owing to the larger size of the S atom, the lattice constant increases to a = 0.297 nm, the bond length to $d_{Cr–S}$ = 0.225 nm, the distance between S atoms to h_{SS} = 0.292 nm, and the bond angle to θ_{SCrS} = 80.86° in comparison to CrO_2 [1]. Succeeding publications reported similar values, e.g., a = 0.304 nm, $d_{Cr–S}$ = 0.229 nm, h_{SS} = 0.294 nm, θ_{SCrS} = 79.82°, and θ_{CrSCr} = 83.26° [2, 19]. The van der Waals diameter of S atoms and their separation in the 1H phase yield as monolayer thickness d_{w+h} = 0.652 nm with the van der Waals diameter d_w = 0.360 nm [18]. The formation energies are E_f = – 1.70 eV [19] (PBE) for the 1H-CrS_2 phase and E_f = – 0.662 eV [6] (PBE) for the 2H-CrS_2 structure.

The antiferromagnetic metallic 1T-CrS_2 phase has the structural parameters a = 0.328 nm, b = 0.540 nm, and $d_{Cr–S}$ = 0.316 nm [20]. The distorted 1T′-CrS_2 phase exhibits a further extended lattice with a = 0.332 nm, b = 0.562 nm, a short distance between Cr atoms d_1 = 0.299 nm, and a long distance between Cr atoms

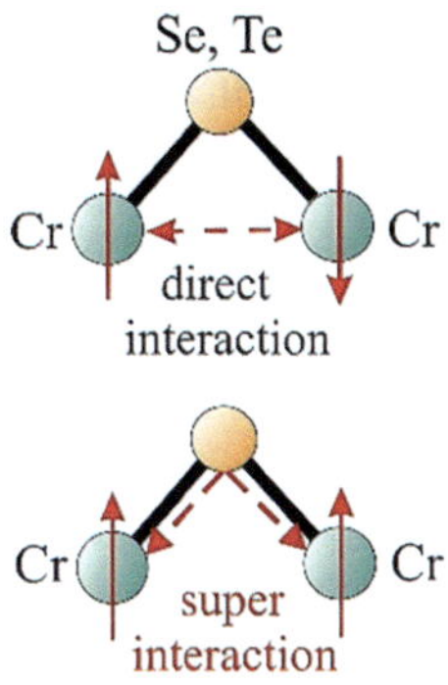

Fig. 16.7 Schematic representation of Cr–Cr direct exchange and the superexchange interaction via chalcogen atoms (Cr–X–Cr). Spin switching processes induce the transformation between ferroelectric and antiferroelectric states. Adapted with permission from Ref. [21], Copyright 2015, American Physical Society

$d_2 = 0.355$ nm [20]. According to DFT calculations, the 1T′ phase is a ferromagnetic semiconductor with a Curie temperature of ~ 1000 K. The enhanced magnetic moments of the S atoms and their opposite signs to those of the Cr atoms suggest the super exchange via Cr–S–Cr interactions in the 1T′ phase (see Fig. 16.7). As a unique property of T′-CrS_2 monolayers, lattice deformations by uniaxial tensile or compressive strain in the zigzag direction convert the T′ phase into a spin-up or a spin-down half-metal state. The tunability under tensile (compressive) strain promotes selectively the spin-up (spin down) orbital interaction at the valence band bottom. The authors present as an example a prototype spin-valve device based on a 2D CrS_2 heterostructure that could significantly reduce energy consumption in a miniaturized spintronic device.

16.1.2.3 Mechanics

Weaker bonding leads to a drastic decrease of the elastic modulus of the 1H-CrS_2 monolayer to E = 112.0 N/m and a Poisson ratio of $\nu = 0.27$ [7]. These values agree well with those of a recent publication, reporting similar mechanical parameters of E = 110.4 N/m and $\nu = 0.26$ [19]. Polar plots demonstrate the isotropic behavior of both stiffness and Poisson ratio. A somewhat larger stiffness of E = 121.2 N/m and $\nu = 0.25$ were obtained for the 2H-CrS_2 phase by screening a large configuration space in high-throughput calculations (ElasTool toolkit) [8]. Note the comparable stiffness of 1H CrS_2 and that of the 1T-CrS_2 monolayer at E = 108.7 N/m, however, a negative Poisson ratio at $\nu = -0.01$ [8]. The greatest Young's modulus of 129.0 N/m has been obtained by the LDA approach [1].

The ultimate strength that describes the stress the material can withstand without damage is $\sigma_{un} = 12.9$ N/m for uniaxial stress and $\sigma_{bi} = 12.2$ N/m for biaxial stress [19]. These values are in good agreement with Griffith's rule. The corresponding ultimate strain values are $\varepsilon_{un} = 0.24$ (uniaxial) and $\varepsilon_{bi} = 0.18$ (biaxial). A larger strain indicates a higher ductility of the material. Note that these nonlinear mechanical properties are only slightly greater than those previously predicted by a simple reference model [22].

16.1.2.4 Physics and Applications

As mentioned above, chromium disulfide monolayers exhibit diverse structural, electronic, and magnetic properties, including a nonmagnetic direct semiconductor 1H phase, a metallic antiferromagnetic 1T phase, and a ferromagnetic semiconductor 1T′ phase. [20]. The band gaps reported for the direct semiconductor are 1.07 eV (LDA) and 1.84 eV (GW_0-corrected gap) [1] in comparison with 0.90 eV (LDA) and 1.54 eV (G_0W_0) [6]. At the Heyd-Scuseria-Ernzerhof (HSE06) level of theory, the CrS_2 phase with the direct band gap and piezoelectric properties is tunable between 1.18 and 1.86 eV (HSE06) by applying tensile strain, which also induces a direct-to-indirect band gap transition at a tensile strain > 2% [23].

A recent comprehensive work addresses the structural, electronic, magnetic, and catalytic properties and the hydrogen-controlled phases of 33 TMDCs functionalized by hydrogen on one or both sides. This study includes the chromium compounds CrS_2, $CrSe_2$, and $CrTe_2$ [24]. The comprehensive work covers the lattice parameters of single- and double-hydrogenated compounds and discusses the switching between semiconducting, metallic, and half-metallic states by phase transitions and the accompanying change in the magnetic ground state between nonmagnetic, ferromagnetic, paramagnetic, and antiferromagnetic states. For example, CrS_2 and $CrSe_2$ switch with increasing hydrogenation from the nonmagnetic state to the antiferromagnetic and finally to the ferromagnetic state. Correspondingly, the electronic properties switch from semiconducting to metallic or half-metallic by hydrogenation. Furthermore, $CrTe_2$ and $HCrTe_2$ are antiferromagnetic; however, $2HCrTe_2$ is ferromagnetic.

As suggested by the authors, the layered Janus sheets of CrSSe may serve as a suitable material for lithium-ion batteries, enabling lithium-free cathodes to be accessible in practical applications [25]. They estimated an intercalative capacity of 190 mA h g^{-1} for the layered material with all the charges compensated by anionic S and Se. The unique magnetoelectric properties and the ferroelasticity stimulated a more detailed investigation of the intrinsic properties [26]. In addition to the lattice parameters of tetrahedral 2D CrSSe, the space group of the suggested structure of *Pmm*2 was determined. Furthermore, the calculations indicated a high critical temperature of $T_c = 790$ K, a large magnetic anisotropy of 0.622 meV/Cr, and a robust half-metallicity.

The Rashba spin splitting (RSS) and piezoelectricity of various Janus chromium dichalcogenide monolayers CrXY (X, Y = S, Se, and Te; $X \neq Y$) were studied by first-principles calculations [27]. Owing to the mirror asymmetry of the Janus structure and strong SOC, the strain-free CrXY monolayer has large Rashba parameters and superior intrinsic piezoelectric responses, which are orders of magnitude greater than those of MoXY monolayers. In addition, the study covers strain engineering of the band gap and reveals in detail the modulation of the band structure, RSS, and piezoelectric properties under biaxial strain. The application of strain induces semiconductor-to-metal and indirect-direct phase transitions in these compounds. By varying the chemical composition of the compounds $MX_{2(1-x)}Y_{2x}$, $x = 0–1$, (M = Cr, Mo, and W; $X \neq Y$ = S, Se, and Te), extensive fine tuning of the structural, electronic, and piezoelectric properties is possible [28]. This additional chemical degree of freedom may be a useful strategy for the design of nanoscale devices.

For the dynamically and thermally stable Janus monolayer CrSCl, which is an out-of-plane ferromagnetic semiconductor, theory predicts a large vertical piezoelectric response with a high Curie temperature [29]. The calculated out-of-plane piezoelectric strain coefficient of $d_{31} = -1.58$ pm/V is greater than that of most 2D materials. Tensile strain is favorable for further extending the Curie temperature, the magnetic anisotropy energy, and the d_{31} coefficient.

DFT calculations combined with the semiempirical Boltzmann transport theory allowed the evaluation of the phonon thermal transport of 1H-CrS_2 and 1H-$CrSe_2$ monolayers, while ab initio MD simulations predicted a thermal stability of up to 300 K [30]. The high thermal conductivities of 131.7 W m^{-1} K^{-1} (CrS_2) and 88.6 W m^{-1} K^{-1} ($CrSe_2$) can be ascribed to strong bonding, large Young's moduli, a relatively small mass, high phonon group velocities, and long phonon relaxation times. The authors considered phonon boundary scattering as the dominant scattering process.

16.1.2.5 Synthesis

Experimental studies on chromium TMDCs, including their synthesis, have not received much attention. The CVD method allows the growth of few-layer flakes of CrS_2 with thicknesses down to the monolayer [31]. Interestingly, the three phases 1H, 1T, and 1T′ exist in the monolayer ('pudding phenomenon'). The 1T′ phase is derived from the 1T phase by the dimerization of metal atoms, as verified by MD simulations. Interestingly, the 2H phase is a nonmagnetic semiconductor with a direct band gap, the 1T phase is metallic with a ferromagnetic spin configuration and a magnetic moment of 0.531 μ_B/Cr, and the 1T′ phase is semimetallic with a gap of approximately 10 meV and a magnetic moment of 2.206 μ_B/Cr. The polymorphic 2D material of chromium disulfide may be of interest as a multifunctional material in nanoelectronics, valley electronics, and spintronics.

An air-stable magnetic 1T-CrS_2 phase has been successfully grown on a mica substrate by van der Waals epitaxial deposition via a CVD process [32]. AFM, STM, and energy-dispersive X-ray (EDX) spectroscopy were used to verify the crystallinity, stoichiometry, and purity of the layers. The crystallites exhibited room-temperature ferromagnetism, a moderate conductivity of 1.78×10^2 S m^{-1}, and a coercive field of 28.7 Oe.

16.1.3 $CrSe_2$ Monolayer

16.1.3.1 Chemistry and 2D Bonding

The valence electrons of the electronic configuration of Cr atoms (Ar|$3d^5 4s$) and Se atoms (Ar|$3d^{10} 4s^2 p^4$) determine the type of 2D bonding in chromium diselenide monolayers. Like in the related compounds, the hexagonal 1H phase with a chemically bonded three-sublayer assembly is the most stable 2D structure. In $CrSe_2$ monolayers, the van der Waals diameter of S atoms of $d_w = 0.360$ nm increases slightly to 0.380 nm in Se [18], which weakens covalent bonding by reducing orbital overlap [33]. Also other factors such as the increasing charge shielding by inner electrons weakens the strength of Cr–Se bonds. Interestingly, the electronegativities of Se atoms (2.55) and S atoms (2.58) atoms are nearly identical. Remarkably, chromium disulfide and diselenide monolayers have a similar surprisingly large fractional ionic character of approximately 19.1% and 18.0%, respectively [2]. The ionicity is based on the transfer of an electron charge of $\delta q = 0.77$ e (0.81 e) [2] from the inner Cr atom to two outer Se atoms [1]. Especially the larger atom size and greater shielding reduce the chemical binding energy and thus the calculated cohesive energies are $E_c = -4.77$ eV/atom [1] (LDA), – 3.64 eV/atom [2] (PBE), and – 3.59 eV/atom [19] (PBE), depending on the applied approach.

16.1.3.2 2D Structures

The hexagonal 1H phase of $CrSe_2$ monolayers has the highest stability, as found for 2D CrS_2. Owing to the further increased size of the chalcogen atom, the lattice constant extends to a = 0.313 nm, the bond length to $d_{Cr-Se} = 0.238$ nm, the distance between Se atoms to $h_{SeSe} = 0.311$ nm, and the bond angle to $\theta_{SeCrSe} = 81.54°$ (LDA) [1]. An estimate of the thickness of the monolayer yields $d_{w+h} = 0.691$ nm. Calculations based on the PBE approach resulted in similar lattice parameters of a = 0.321 nm, $d_{Cr-Se} = 0.243$ nm, $h_{SeSe} = 0.314$ nm, $\theta_{SeCrSe} = 80.6°$, and $\theta_{CrSeCr} = 82.59°$ (PBE) [2, 17]. The formation energy attained by the fitted elemental phase reference energy (FERE) is $E_f = -0.65$ eV/atom, and that obtained by the PBE approximation is – 0.474 eV/atom [6]. The band gaps of the 1H-$CrSe_2$ phase obtained by different approaches are 0.86 eV [1] (LDA), 1.51 eV [1] (GW_0-corrected gap), 0.70 eV [6] (LDA), and 1.21 eV [6] ($G_0 W_0$).

A comprehensive study based on DFT calculations clarified the roles of epitaxial strain in tuning intra- and inter-plane magnetism in epitaxial $CrSe_2$ and $CrTe_2$ mono- and bilayers [34]. Mismatched lattice constants at a van der Waals epitaxial interface may introduce in-plane strain in the epitaxial layer ('epitaxy strain'). The in-plane strain varies the intra- and interplane magnetic couplings in $CrSe_2$ ($CrTe_2$) monolayers and bilayers. The intralayer magnetic ground state is tunable between ferromagnetic and three antiferromagnetic orders within 2.5% (4.5%) in-plane strain. The tunability primarily concerns change of the Cr–Se–Cr and Se–Cr–Se angles and a

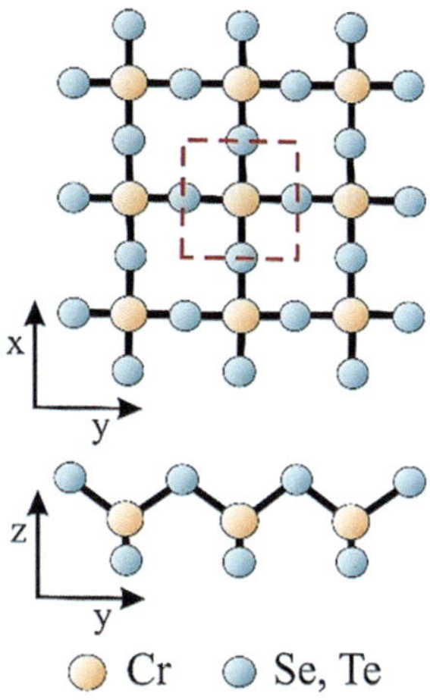

Fig. 16.8 Top and side view of the square lattice consisting of three sublayers. The central metal atoms bind only four chalcogen atoms. Adapted with permission from Ref. [35], Copyright 2023, Elsevier

control of the strength and type of intralayer spin-exchange couplings by competing intralayer Cr–Cr interactions and interlayer magnetism in bilayers.

In addition, for $CrSe_2$ and $CrTe_2$ monolayers, first-principles calculations predict the existence of a square lattice [35]. In this three-layer structure, with a sublayer of Cr atoms sandwiched between two layers of Se (Te) atoms, the Cr atoms bind to only four chalcogen atoms (see Fig. 16.8). This structure belongs to the space group *P*-4*m*2. Due to the different electronegativities, each Cr atom loses two electrons, and each Se (Te) atom gains one electron. Accordingly, the Cr atom with a valence electron configuration ($3d^5 4s^1$) transforms to the ($3d^4$) configuration with Cr^{2+}. These monolayers are ferromagnetic half-metals with a large spin, large half-metal gap, and high Curie temperature. The lattice constant of a = 0.379 nm is larger than the value given above for 1H $CrSe_2$, and the cohesive energy at $E_c = -3.87$ eV/atom (PBE) indicates comparable bonding strength. The Heisenberg model implies that the Curie temperature is well above room temperature and characteristic volt-ampere curves at room temperature ensure fully spin-polarized transport properties of the spintronic device. The study predicts a significant improvement in ferromagnetism by applying biaxial tensile strain.

16.1.3.3 Mechanics

The further decrease in the stiffness of the H phase of 2D $CrSe_2$ within the group of VIB–VIA monolayers is smaller than before and yields a value of 88.0 N/m for Young's modulus and a slight increase in Poisson's ratio to $\nu = 0.30$ [7]. These values agree well with E = 85.6 N/m and $\nu = 0.31$ [19]. Other available mechanical properties are E = 98.0 N/m with $\nu = 0.28$ [8], which agree better with the greater stiffness of E = 104.6 N/m calculated by LDA [1].

The values of the ultimate strengths for uniaxial and biaxial stresses are $\sigma_{un} = 10.7$ N/m and $\sigma_{bi} = 9.8$ N/m, respectively [19]. Remarkably, these values are in good agreement with Griffith's rule. The corresponding ultimate strain values of $\varepsilon_{un} = 0.26$ (uniaxial) and $\varepsilon_{bi} = 0.16$ (biaxial) indicate an increasing anisotropy of the nonlinear mechanical behavior that is also evident in the ultimate strength values.

16.1.3.4 Physics and Applications

As discussed above, the arrangement of parallel electronic spins that create an intrinsic dipole moment seems to be possible in low-dimensional 2D materials without an external magnetic field. Current interest concentrates on magnetism in 2D van der Waals materials and the possibility of strain engineering as an effective route to manipulate magnetism in low-dimensional compounds.

Like their bulk counterpart, metastable 1T-$CrSe_2$ monolayers can be ferromagnetic and antiferromagnetic [21]. Importantly, applying a biaxial tensile strain of > 2% to the $CrSe_2$ monolayer spin ordering induces transformation to the ferromagnetic state. This switching process is due to the competition between the ferromagnetic Cr–Se superexchange interaction and the antiferromagnetic direct Cr–Cr exchange interaction (see Fig. 16.7). At a small tensile strain of 4%, the Curie temperature moves from $T_c = 63.7$ K to above room temperature to $T_c = 363.1$ K. At a strain of 10%, the Cr-3*d* and Se-4*p* states become completely spin polarized, and the $CrSe_2$ monolayer is approximately half metallic, a feature highly desirable for applications in spintronic devices. Similarly, 1T-$CrTe_2$ monolayers are ferromagnetic with a Curie temperature of 310 K. In this latter case, a critical compressive strain of − 1% allows the switch between the ferroelectric state and the antiferroelectric state.

As mentioned above, the modulation of magnetism in $CrSe_2$ monolayers is a topic of intense research [36]. First-principles calculations clarified the impact of hydrogenation on magnetic and electronic properties. At low hydrogenation of $CrSe_2$ monolayers, the magnetic ground state of the dynamically and thermally stable monolayers transforms from the antiferromagnetic to the ferromagnetic state. Moreover, in addition to the Curie temperature, the magnetic anisotropy energy increases significantly from 0.32 meV/Cr to 0.68 meV/Cr after hydrogenation. This difference is attributed to interatomic charge redistribution induced by hydrogenation. This opens the possibility of realizing magnetoelectric nanodevices in advanced spintronic applications.

The development of 2D magnetic materials with air and water stability is critical since several known 2D materials are not stable under ambient conditions. As predicted by first-principles calculations, monolayers of $CrSe_2$ exhibit exceptional ambient stability for several months because the dissociative chemisorption of O_2 is endothermic with an energy barrier of 0.53 eV [37]. However, under biaxial tensile strain, the reaction barrier decreases, and the dissociation process becomes exothermic. The authors attributed this behavior to enhanced charge transfer to the adsorbed oxygen molecule that activates the O–O bond by filling antibonding states, which facilitates bond rupture. This is an important result because during fabrication processes lattice mismatch may introduce interfacial strain that reaches a tensile strain of ~ 1%, where the reaction process becomes exothermic.

16.1.3.5 Synthesis

The deposition of $CrSe_2$ and Cr_2O_3 monolayers allows the investigation of the role played by reduced dimensions in low-dimensional magnetic order [38]. In epitaxially grown layers, metallic 1T-$CrSe_2$ with a lattice constant of a = 0.33 nm coexisted with the distorted 1T″ phase with (2 × 2) surface periodicity and a lattice constant of a = 0.36 nm. The MBE-grown monolayer of $CrSe_2$ was antiferromagnetic, whereas the Cr_2O_3 layer was ferromagnetic with a Curie temperature of ~ 200 K, as revealed by X-ray magnetic circular dichroism (XMCD) combined with DFT calculations. Annealing at 300 °C transforms the $CrSe_2$ monolayer into Cr_2O_3 (see Fig. 16.9).

While existing 2D van der Waals ferromagnets obtained by mechanical exfoliation from layered bulk materials are often unstable under ambient conditions, CVD of $CrSe_2$ on dangling-bond-free WSe_2 substrates allows the deposition of stable $CrSe_2$ nanosheets with tunable thickness, down to the monolayer limit [39]. The deposited $CrSe_2$/WSe_2 heterostructure with a van der Waals bonded interface exhibited ferromagnetic behavior, and magnetic properties were still observable after months of environmental exposure to air. Calculations indicate that interlayer coupling plays a critical role in the magnetic order of few-layer $CrSe_2$ nanosheets by charge transfer from the WSe_2 substrate and interlayer coupling within the $CrSe_2$ sheets. This work opens the door to robust 2D magnets of interest in spintronic and magnetoelectronic devices.

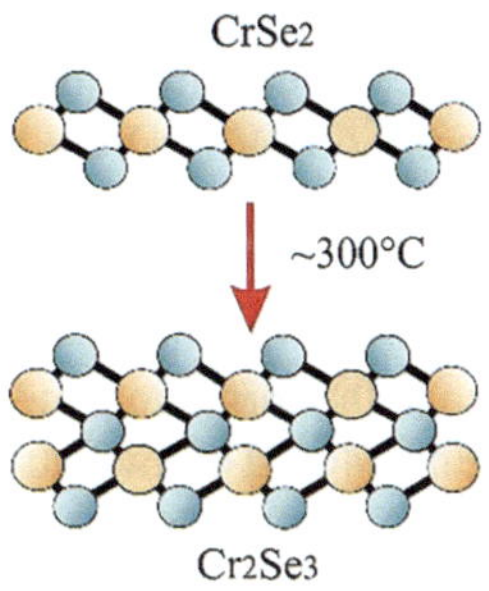

Fig. 16.9 Transformation of the 1T-$CrSe_2$ monolayer into a Cr_2Se_3 monolayer, consisting of five-sublayers, by annealing at 300 °C. Adapted with permission from Ref. [38], Copyright 2021, American Chemical Society

16.1.4 $CrTe_2$ Monolayer

16.1.4.1 Chemistry and 2D Bonding

In this group of chromium dichalcogenides with an electronic configuration of Cr atoms of (Ar|$3d^5 4s$) and Te atoms of (Kr|$4d^{10}5s^2p^4$) and energetically and dynamically stable 1H and 1T structures, Te has the largest van der Waals diameter of d_w = 0.420 nm [18] and the smallest electronegativity of 2.10. Accordingly, the Bader charge transfer reaches only δq = 0.46 e (0.56 e) [2] that is transferred from each Cr atom to two Te atoms [1]. Consequently, the fractional ionic character of the bonding decreases from approximately 54.5% in CrO_2 to the lowest value of 4.73% in $CrTe_2$ [2]. Moreover, according to different theoretical approximations, 1H-$CrTe_2$ monolayers have the smallest binding strength among these compounds with a cohesive energy at E_c = – 4.17 eV/atom [1] (LDA), or even weaker binding, with a cohesion at E_c = – 3.09 eV/atom [2] (PBE). In general, the reduced bonding strength has an enormous influence on the mechanical performance especially of the heaviest 2D compounds.

16.1.4.2 2D Structures

The hexagonal 1H phase of $CrTe_2$ has the largest lattice constant and bond length among the chromium dichalcogenide monolayers. The lattice constant increases to a = 0.339 nm, the bond length to $d_{Cr–Te}$ = 0.258 nm, the Te–Te distance to h_{TeTe} = 0.338 nm, and the bond angle to θ_{TeCrTe} = 81.56° (LDA) [1]. As monolayer thickness, the van der Waals diameter and the separation of Te atoms yield d_{w+h} = 0.758 nm. Independent calculations obtained similar lattice parameters at a = 0.348 nm, $d_{Cr–Te}$ = 0.264 nm, h_{TeTe} = 0.341 nm, θ_{TeCrTe} = 80.63°, and θ_{CrTeCr} = 82.66° (PBE) [2]. The calculated band gaps of the 1H-$CrTe_2$ phase obtained by different approaches are 0.60 eV (LDA) [1], 1.12 eV [1] (GW_0), 0.45 eV [6] (LDA), and 0.77 eV [6] (G_0W_0).

According to recent work, the 1T-$CrTe_2$ structure represents the ground state with a formation energy of E_f = – 10.09 eV per formula unit, which is slightly lower than the formation energy of the 1H-$CrTe_2$ structure of E_f = – 9.75 eV per formula unit [40]. The two structures are separated by an energetic transition barrier of ~ 0.5 eV [40]. The lattice parameters of the fully relaxed 1T-$CrTe_2$ monolayer are a = 0.369 nm, $d_{Cr–Te}$ = 0.270, and θ_{CrTeCr} = 86.3°, as calculated by an optimized PBE approach.

Note that the calculation of electronic and magnetic properties requires the self-consistent Hubbard correction parameter U (PBE + U) [40]. Using the value U = 5.8 eV, a magnetic moment per formula unit of 3.05 μ_B is found for the 1T-$CrTe_2$ monolayer and 3.08 μ_B for the bulk. A unique property is the decrease in the Curie temperature with increasing film thickness. The discovery of a Curie temperature of ~ 310 K for bulk 1T $CrTe_2$ initiated several studies. Remarkably, the calculated magnetic moment of the monolayer is out-of-plane, whereas it lies in the plane for two or more layers [40]. Tensile biaxial strain induces a transition of the magnetic moment of the monolayer from out-of-plane to in-plane and compressive bilayer strain causes the bilayer magnetic moment to switch from in-plane to out-of-plane. Long range magnetic order at finite temperatures in monolayers is the subject of the Mermin-Wagner theorem. According to this theorem, magnetic anisotropy prevents the distortion of magnetic order by thermal fluctuations. In 2D, the large SOC of the Te atoms is responsible for this magnetic anisotropy together with the superexchange coupling between the Cr-3*d* and Te-5*p* orbitals [40].

16.1.4.3 Mechanics

The H structure of $CrTe_2$ monolayers has the lowest Young's modulus at $E = 63.9$ N/m and largest Poisson's ratio at $\nu = 0.35$ of the group members [7]. Larger values are $E = 71.4$ N/m and $\nu = 0.45$ [8]. As expected, the largest stiffness of $E = 77.4$ N/m results from LDA calculations [1]. An extended investigation of the 2H phase of 25 semiconductor TMDO and TMDC monolayers included Young's moduli, Poisson's ratios, piezoelectric stresses, and piezoelectric strain coefficients [41]. The relaxed ion elastic moduli presented in this work for CrS_2, $CrSe_2$, and $CrTe_2$ are $E = 111.9$ N/m, 87.9 N/m, and 63.9 N/m, with Poisson's ratios of $\nu = 0.27$, 0.30, and 0.30, respectively. These stiffnesses and Poisson ratios are in good agreement with the values given by other studies.

16.1.4.4 Physics and Applications

While previous calculations and experiments on piezoelectric properties have focused mainly on Mo- and W-based TMDCs, first-principles calculations on an extended number of compounds, MX_2 (M = Cr, Mo, W, Ti, Zr, Hf, and Sn; X = O, S, Se, and Te), revealed that the SnS_2 and $CrTe_2$ monolayers have the largest piezoelectric stress and piezoelectric strain coefficients, which are comparable or even larger than those of well-known conventional piezoelectric bulk materials [41]. These piezoelectric coefficients were derived from the slope of the polarization change induced by applying uniform strain. The work suggested that 2H $CrTe_2$ could be used in future 2D piezoelectric applications, such as transducers, ultrasensitive sensors, nanoscale electromechanical systems, and energy harvesting devices owing to its reduced size, weight, and energy consumption.

1T-$CrTe_2$ van der Waals sheets have been increasingly studied as 2D materials with long-range ferromagnetic order in the single-layer limit at room temperature. Several studies have treated this problem theoretically and shown that in metallic 2D 1T-$CrTe_2$ crystals, intrinsic long-range spin ordering should occur above room temperature and down to the ultrathin limit [42]. The Mermin–Wagner theorem postulates that at finite temperatures, a 2D isotropic spin–S Heisenberg model with finite-range exchange interactions can be neither ferromagnetic nor antiferromagnetic [43]. Accordingly, the symmetry of a 2D system must be lowered, e.g., by an external magnetic field or strain or intrinsically by strong magnetic anisotropy driven by SOC. First-principles calculations show that in addition to tensile strain, the charge density wave (CDW) phase can also induce the localization of the *d* orbitals and generate magnetization pointing out of the plane [44].

16.1.4.5 Synthesis

Early synthesis of the (metastable) layered 1T-$CrTe_2$ structure involved complete structural characterization and measurement of the magnetization and electrical resistivity between 4 and 300 K [45]. The electrical resistance exhibited metallic behavior below room temperature, and the magnetization and specific heat curves indicated a transition to a ferromagnetic state at $T_c = 310$ K with magnetic moments ordered parallel to the basal plane. From their measurements and DFT calculations, the authors concluded that ferromagnetism is of itinerant nature.

MBE in an integrated MBE/STM ultrahigh vacuum system allowed the deposition of single- and few-layer 1T-$CrTe_2$ van der Waals crystals on a bilayer graphene/SiC substrate [46]. STM, XRD, and transmission electron microscopy (TEM) analyses confirmed the single-crystalline nature of the 1T phase with octahedral symmetry. The nanosheets showed intrinsic ferromagnetism, with a Curie temperature of up to $T_c = 300$ K, preserved to the monolayer thickness, large magnetic moments, and a magnetic spin-split band structure. Owing to the possibility of extending MBE to wafer sizes, these nanosheets may be of interest in the development of 2D magnet-based spintronic technologies. Similarly, for 1T $CrTe_2$ grown on a GaAs(111)B substrate, films with thicknesses between 4 and 35 monolayers showed robust out-of-plane ferromagnetism, with a Curie temperature of 191 K [47].

The combination of spin-polarized STM experiments and first-principles calculations provided new information on the magnetism of 1T-$CrTe_2$ van der Waals crystals grown on a SiC-supported bilayer graphene substrate by MBE [48]. While the bulk material is ferromagnetic, antiferromagnetic order can be observed at the atomic scale, which shows an intriguing noncollinear spin reorientation under a magnetic field. This complex antiferromagnetic spin texture is consistent with moderate magnetic anisotropy and provides new insight into layer-dependent magnetic order transitions.

An MBE study of $CrTe_x$ films on Si(111) substrates elucidated a thickness-dependent structural phase change with self-intercalation during growth [49]. In situ RHEED, STM, X-ray photoemission spectroscopy (XPS), and ex situ XRD provided a detailed analysis of the growth process. After deposition of the first $CrTe_2$ layer, which served as a buffer layer, chromium atoms started to intercalate into the layers of $CrTe_2$, and the Cr_3Te_4 phase dominated the subsequent growth to the multiple layer $Cr_3Te_4/CrTe_2$ ($CrTe_x$). Subsequent superconducting quantum interference device (SQUID) measurements verified the ferromagnetism of single- and few-layer chromium tellurides.

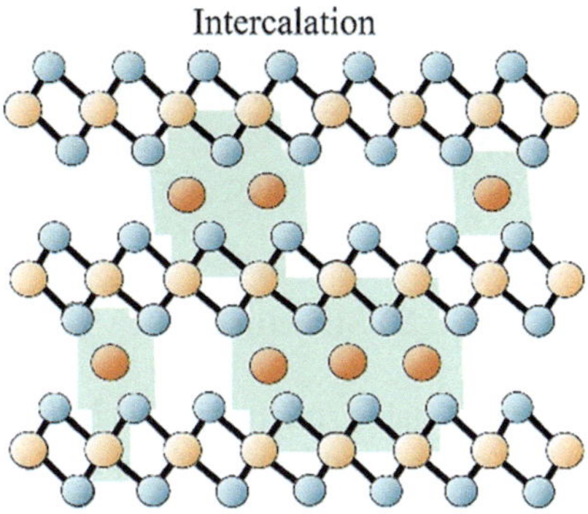

Fig. 16.10 Scheme of three-monolayer TMDC sheet intercalated by the same or a different metal atom. The green areas schematically display the induced interlayer interactions, which are responsible for ferroelectric order by spin alignment

A similar approach to 2D ferroelectric materials is the synthesis of $Cr_{(1+\delta)}Te_2$ monolayers [50]. The δ symbol represents the amount of intercalated Cr in comparison to that in the $CrTe_2$ layer and may be tuned by postgrowth annealing between 0.5 and 1. Such a nanosheet consists of at least two $CrTe_2$ units covalently linked together by additional Cr atoms that are bound to six adjacent neighbors and thus form a stable nano system (see Fig. 16.10). Under optimized conditions, these materials also grow by van der Waals epitaxy on dissimilar van der Waals substrates. Note that metastable 1T-$CrTe_2$ monolayers disproportionate into elemental Te and the stable $Cr_{(1+\delta)}Te_2$ intercalation compound upon annealing. LEED and STM analyses provided the characterization of the MBE-deposited $Cr_{(1-\delta)}Te_2$ monolayers [50]. A robust ferromagnetic material grows on van der Waals heterostructures by varying the amount of intercalated Cr atoms or by alloying with other transition metals, where tuning is possible.

16.2 Molybdenum Dichalcogenide Monolayers (MoO_2, MoS_2, $MoSe_2$, and $MoTe_2$)

16.2.1 MoO_2 Monolayer

16.2.1.1 Chemistry and 2D Bonding

The valence electrons of the electronic configuration of Mo atoms (Kr|4$d^5$5s) and O atoms (He|2s^2p^4) determine the type of 2D bonding in molybdenum dioxide monolayers. By *d*-*p* hybridization the outermost 4*d* orbitals of the Mo atoms and the outermost 2*p* orbitals of the O atoms form a covalently bonded three-sublayer structure with a Mo sublayer in the center, sandwiched between two O sublayers. Each O atom forms covalent bonds with three Mo atoms in a bridging position and each Mo atom is bound to six O atoms. The large difference in the electronegativities between Mo atoms (2.16) and O atoms (3.44) induces a substantial charge transfer of $\delta q =$

1.67 e from each Mo atom to two O atoms, generating strong covalent-ionic bonds with a sizable fractional ionic character of 33.6% [2]. For the most stable H structure, cohesive energies of $E_c = -6.24$ eV/atom [2] and -6.20 eV/atom [3] (GGA-PBE) and a much lower value of $E_c = -7.55$ eV/atom [3] (LDA) have been calculated.

16.2.1.2 2D Structures

The most stable structure of 2D MoO_2 is the 2H phase with a lattice constant of a = 0.283 nm, a bond length of $d_{Mo–O} = 0.205$ nm, a distance between O atoms of $h_{OO} = 0.248$ nm, and the bond angles $\theta_{OMoO} = 74.18°$ and $\theta_{MoOMo} = 87.39°$ (PBE) [2] in good agreement with a = 0.278 nm, $d_{Mo–O} = 0.200$ nm, $h_{OO} = 0.243$ nm, and $\theta_{O–Mo–O} = 73.9°$ (LDA) [3]. An estimate of the monolayer thickness of $d_{w+h} = 0.558$ nm is obtained by adding the van der Waals diameter of O atoms of $d_w = 0.310$ nm and their distance of $h_{OO} = 0.248$ nm. The calculated negative formation energies based on different approaches are -1.73 eV (PBE) [6], -1.50 eV (PBE) [3] and -1.94 eV (FERE) [6]. The MoO_2 monolayer is a nonmagnetic semiconductor that is easily cleavable from the bulk material with a low exfoliation energy of 19.2 meV/Å^2 [3].

The 1H-MoO_2 monolayer is an indirect semiconductor with estimated band gaps of 0.99 eV (PBE), 2.04 eV (GW_0), and 2.42 eV (LDA + GW_0) [51]. The change in bonding induced by tensile or compressive strain implies that the electronic band structure can be tuned and that the energy band gap decreases with increasing biaxial tensile strain up to 4%, while a larger strain induces a semiconductor-to-semimetal transition. However, no transition occurs when a uniaxial tensile strain is applied [51].

16.2.1.3 Mechanics

The linear mechanical properties presently available for 2H MoO_2 include elastic constants, which meet the Born–Huang criteria for mechanical stability, predicting a very large stiffness of E = 199.2 N/m and a Poisson ratio of $\nu = 0.37$ [7]. Both, Young's modulus and Poisson's ratio are in good agreement with other calculations, e.g., E = 198.3 N/m and $\nu = 0.367$, obtained by studying the elastic deformation in the harmonic regime [51]. The ElasTool toolkit calculations predict quite similar values of E = 199.4 N/m and $\nu = 0.35$ [8]. Since the bond lengths increase under tensile strain and decrease under compressive strain, phonon softening and hardening occur. Theoretically, for 1T-MoO_2 monolayers, a surprisingly small stiffness of only E = 78.3 N/m and an extremely large Poisson ratio of $\nu = 0.53$ are reported [8]. Unfortunately, a detailed study of the nonlinear mechanical properties of MoO_2 monolayers is still lacking.

16.2.1.4 Physics and Applications

A theoretically suggested route to realizing MoO_2 monolayers is based on the oxidation of already synthesized defective MoS_2 monolayers, based on the better thermodynamic stability of the oxide [52]. However, the large energy barrier of dissociative chemisorption of molecular oxygen kinetically limits the oxidation process at low temperatures. Generally, the excellent oxidation resistance and resulting environmental stability are important properties of 2D TMDOs. Importantly, this situation changes with the presence of defects.

The large family of 2D oxides consists of oxygen and a metallic or quasimetallic element and contains 2D metals, semiconductors, and insulators [53]. The addition of oxygen provides excellent air stability. The MoO_2 nanosheets have metallic characteristics with low electrical resistivity and excellent electronic conductivity, as revealed by the calculated electronic structure and density of states (DOS). Furthermore, 2D MoO_2 has high chemical stability and electrochemical activity. For example, surface plasmon resonances can be excited owing to the high density of free electrons. Interestingly, 2D MoO_3 nanosheets exhibit several types of conductivity and even insulating properties. The use of the latter oxide in p-type 2D oxide field-effect transistors (FETs) complements some deficiencies of other common 2D materials in p-type FETs [53].

The planarity and stability endow chemical-vapor-deposited MoO_2 nanosheets with superior uniformity of the Raman signal; therefore, these nanosheets can serve as novel metal oxide substrates for surface-enhanced Raman spectroscopy (SERS) [54]. The concentration limit of 4×10^{-8} M and the enhancement factor of 2.1×10^5 are comparable to the detection limits of traditional noble metals and superior to those of other 2D materials. The enhancement effect originates not from charge transfer but from the surface plasmon resonance (SPR) effect and becomes more pronounced by decreasing the thickness of the nanosheet below 6 nm. The superior performance of the material is based on the sensitivity, stability, uniformity, and reusability of MoO_2 nanosheets for SERS analysis.

16.2.1.5 Synthesis

In principle, we can discriminate between van der Waals oxides with a layered structure that allow top-down separation of nanosheets, and 2D non van der Waals oxides where bottom-up syntheses such as CVD is necessary [55]. Only a limited number of TMDOs have a layered structure, such as MoO_3, which has a stable orthogonal phase. However, structural degradation and damage during the stripping process are severe problems. Most TMDOs exhibit strong ionic interactions with 3D chemical bonding, which results in reactive compounds with dangling bonds at the surface. The bottom-up preparation and separation of synthesized nanosheets with defined atomic thickness is difficult; however, these nanosheets exhibit adjustable optical, electrical, magnetic, and catalytic properties.

Recent synthesis of 2D MoO_2 concentrated on nanosheets because monolayers are difficult to realize. In contrast to monolayers, few-layer nanosheets already exist for several TMDOs, including MoO_2. The MoO_2 nanosheets were grown by CVD using MoO_3 powder and sublimated sulfur as precursors. The current thickness limit of CVD-deposited nanosheets with high crystallinity is in the range of 5.5–11.5 nm [56]. Structural analysis was performed by SEM and Raman spectroscopy. The as-prepared nanosheets had a high electrical conductivity of 3600 S/cm. It is worth noting that theory predicts dynamic and thermal stability up to ~ 200 °C. Above 250 °C, the metallic MoO_2 sheets oxidize in air into insulating MoO_3 flakes.

In a subsequent study electron-beam lithography (EBL) was employed for the deposition of approximately 5–10 nm thick MoO_2 nanosheets on Si/SiO_2 substrates by CVD with good crystalline quality using molybdenum trioxide as the precursor [57]. Temperature-dependent Raman spectra revealed the thermal stability of the wide-bandgap semiconductor up to 503 K. The structure was determined via XRD and TEM. The as-fabricated nanosheets possessed a metallic conductivity of ~ 200–475 S/cm under ambient conditions, which resulted from their unique structure.

The deposition of ~ 25 nm thick sheets of a hexagonal phase and of ~ 30 nm thick films of a rectangular phase using two-zone CVD and MoO_3 powder as a precursor, confirmed the previous results [58]. Importantly, Raman spectroscopy permitted the identification of mixed phases as main matrix. Analysis by AFM, SEM, optical photography, and cathodoluminescence (CL) provided additional information on the nanosheets. The measured I–V curves of all the fabricated devices revealed linear metallic behavior owing to Ohmic contact between sheet and electrode. The measurements confirmed the high melting point, good thermal stability, and excellent electrical conductivity of the material.

A recent work achieved the fabrication of MoO_2 nanosheets by annealing MoO_3 powder in an Ar-H_2 (10% H_2) atmosphere at 750 °C, balancing the sublimation of solid MoO_3 with the reduction of gaseous MoO_3, and spin-coating of the obtained MoO_2 nanosheets on a quartz substrate [59]. The high quality of the nanosheets was demonstrated by extensive characterization via Raman spectroscopy, SEM, TEM, XRD, and AFM. The MoO_2 nanosheets exhibited a high damage threshold, high chemical stability, ultrafast optical response in optical switches, and optical modulation for generating high-repetition-rate devices such as compact Q-switched lasers and stable mode-locked lasers working as saturable absorbers.

A recent investigation studied the controlled synthesis of metallic layered MoO_2 flakes via CVD, using MoO_3 powder as the precursor and SiO_2/Si as substrate [60]. XRD, STM, and TEM analyses confirmed the high quality of the deposited monoclinic multilayer crystals, which had thicknesses ranging from 3.5 to 106 nm. The ultrathin flakes showed a conductivity exceeding 10^6 S/m, comparable to that of graphene. The incorporation of MoO_2 flakes as van der Waals contacts in a designed MoS_2 FET confirmed the anticipated high carrier mobility of 33.6 $cm^2\ V^{-1}\ s^{-1}$ and an optimized Schottky barrier height.

16.2.2 MoS_2 Monolayer

16.2.2.1 Chemistry and 2D Bonding

The similarity of the electronic configurations of Mo atoms (Kr|$4d^5 5s$) and S atoms (Ne|$3s^2p^4$) with those of MoO_2, leads to the same 2D bonding types for MoS_2 monolayers. The hybridization between $4d$ molybdenum electrons and $3p$ chalcogen electrons determines the electronic structure and physicochemical properties of the 2D MoS_2 structures. An extensive electron charge accumulation between the Mo and S atoms produces strong covalent Mo–S bonds. Moreover, the different electronegativities of Mo (2.16) and S atoms (2.58) lead to an ionic contribution to bonding and thus to polar Mo–S bonds. A Bader charge transfer analysis of the most stable hexagonal 2H phase predicts a charge transfer of $\delta q = 1.07$ e from each Mo atom to two S atoms, causing only a small fractional ionic character of 4.3% in comparison to that of MoO_2 [2]. For the covalent-ionic bonding of the 1H structure, nearly identical cohesive energies of $E_c = -5.14$ eV/atom [2], -5.19 eV/atom [61], and -5.07 eV/atom [62] were derived by various authors, using the GGA-PAW and DFT-GGA-PBE approach, respectively. Not surprising, the LDA approximation delivers substantially smaller cohesive energies of -6.35 eV/atom [1] and -6.25 eV/atom [62]. The smallest cohesive energy reported for MoS_2 monolayers at $E_c = -7.86$ eV/atom implies extraordinarily strong bonding [63].

16.2.2.2 2D Structures

For 2D MoS_2 with the most stable 2H symmetry, the lattice constant is $a = 0.318$ nm, the bond length is $d_{Mo-S} = 0.241$ nm, the distance between S atoms is $h_{SS} = 0.313$ nm, and the bond angles are $\theta_{SMoS} = 80.77°$ and $\theta_{MoSMo} = 82.55°$ (PBE) [2]. The lattice parameters of the H structure published by others are similar, e.g., $a = 0.311$ nm, $d_{Mo-S} = 0.237$ nm, $h_{SS} = 0.311$ nm, and $\theta_{SMoS} = 81.62°$ [1, 63–65]. An extensive study presents the lattice constants, bond lengths, and elastic constants of molybdenum and tungsten chalcogenides. A comparison of the results obtained by the PBE, LDA, and optB88 approaches indicates good agreement between the three approximations [66]. By adding the distance between S atoms and the van der Waals diameter of S atoms of $d_w = 0.360$ nm, a monolayer thickness of $d_{w+h} = 0.673$ nm is found. The H structure is thermodynamically stable, with calculated formation energies of $E_f = -0.83$ eV/atom [1] and -1.42 eV/atom [62]. Furthermore, the phonon dispersion curves indicate dynamic stability of the H structure for all molybdenum dichalcogenides [2].

The lattice parameters of the 1T structure at $a = 0.319$ nm, $b = 0.554$ nm (rectangular unit cell) and $d_{Mo-S} = 0.243$ nm are only slightly larger than those of the H structure and increase further for the 1T′ phase to $a = 0.321$ nm, $b = 0.571$ nm (rectangular unit cell), and $d_{Mo-S} = 0.240$–0.254 nm [65]. The synthesis conditions under which specific phases are formed and the possible pathways for transitions among different phases were investigated in detail [64]. Notably, by gliding one S plane, the 1H and 1T phases can be transformed into each other. These calculations suggested the existence of another metastable monolayer structure, namely, a distorted octahedral z-MoS_2 phase with zigzag Mo–Mo chains and lattice parameters of a = 0.635 nm, b = 0.655 nm, d_{Mo-S} = 0.245 nm, and d_{Mo-Mo} = 0.329 nm [64].

Besides the electronic structures and elastic properties of MX_2 (M = Mo, W; X = O, S, Se, and Te) monolayers, their bilayers were studied by the LDA, PBE, and van der Waals (vdW)-DF exchange-correlation functionals in the form of optB88-vdW [66]. Five bilayer structures with indirect bandgaps were investigated by the optB88-vdW approach and the exchange-correlation functional of the local density approximation (LDA) [66]. As expected, the stacking configuration has only a minimal influence on the in-plane monolayer structure, whereas the interlayer distances of metal atoms in different layers vary strongly from 0.52 nm to 0.78 nm depending on the stacking order of the bilayer. For a particular bilayer structure, the LDA approach proposes much smaller lattice parameters and interlayer distances than does the optB88-vdW method.

Interestingly, the polymorphic 1H and 1T structures of MoS_2 can form a heterostructure by coherently connecting the two phases, apparently with a strain-free phase boundary. It was impossible to resolve the lattice mismatch between the two different phase regions of the two polymorphs [67]. Notably, only high-resolution scanning transmission electron microscopy (STEM) verified the coexistence of a semiconducting and metallic phase within the chemically homogeneous combined 2D-MoS_2 nanosheets.

16.2.2.3 Mechanics

The hexagonal 1H structure of MoS_2 has two high-symmetry directions, with an orientation along the M–X bonds (armchair direction) and perpendicular to the M–X bonds (zigzag direction), when projected on the central plane. Determination of Young's modulus from the linear part of the slope of the stress–strain curve indicates a small anisotropy in the stiffness. Taking as thickness of the monolayer of 0.673 nm results in the stiffnesses of E_{ac} = 150 N/m and E_{zz} = 147.7 N/m in 2D units [68]. This finding is not in good agreement with the stiffness of E = 136 N/m and Poisson's ratio of ν = 0.25 [69]. Different theoretical approaches delivered E = 127.3 N/m (PBE), 141.2 N/M (LDA), and 123.5 N/m (optB88) [66]. For comparison, the linear elastic properties of the ElasTool toolkit are E = 135.9 N/m and ν = 0.22 for 1H MoS_2, and a lower stiffness of E = 111.2 N/m and a negative Poisson ratio of ν = –0.07 for the 1T-MoS_2 monolayer [8].

Surprisingly, indentation measurements of CVD-based monolayers by AFM revealed a much greater stiffness of 171 ± 11 N/m [70]. A larger in-plane stiffness at 180 ± 60 N/m was confirmed by separate AFM measurements of single and bilayer MoS_2 samples exfoliated from bulk crystals [71]. Interestingly, AFM indentation measurements of 5–25 monolayers thick freely suspended MoS_2 nanosheets yielded a much larger stiffness of 202 N/m (300 GPa) [72].

A recent study on the negative Poisson ratio of single-layer MoS_2 confirmed the negative in-plane values of $\nu_{xy} = -0.0736$ and $\nu_{yx} = -0.0750$ for a serrated structure with the space group *R*-3*m* [73]. The unique zigzag geometry and strong bonding between the Mo-4*d* orbitals and the S-3*p* orbitals are the main reasons for the auxetic effect that results in lateral expansion when the monolayer is subjected to uniaxial tensile stress.

The ultimate strengths exhibit a large anisotropy of $\sigma_{ac} = 18.4$ N/m (27.4 GPa) and $\sigma_{zz} = 11.4$ N/m (16.9 GPa) [68]. As a consequence, this induces a large anisotropy in the ultimate strains of $\varepsilon_{ac} = 0.28$ and $\varepsilon_{zz} = 0.19$, which signifies the critical stress response. Other calculations have shown similar anisotropies for the ultimate strengths, e.g., $\sigma_{ac} = 15.8$ N/m (23.5 GPa) and $\sigma_{zz} = 10.8$ N/m (16.0 GPa) [69]. The AFM measurements discussed above imply an average breaking strength of $\sigma = 15$ N/m, which is in satisfactory agreement with the other strength values [71].

First-principles calculations of the tensile stress-strain relations predicted ultimate tensile strengths for uniaxial and biaxial tension along the armchair and zigzag directions in excellent agreement with the experimental results [74]. As a failure mechanism, the authors identified out-of-plane soft-mode phonon instability under uniaxial strain along the armchair direction and under biaxial tension. The single MoS_2 layer becomes unstable when the bond angle $\theta_{SMoS} = 80.69°$ reaches the critical value of ~ 64° under tension at the critical strain. This is due to the depletion of electron density at the Mo–S bond. Furthermore, the study showed that the direct semiconductor transforms to an indirect semiconductor at small tensile strains (< 2%) and transforms into a metal before reaching the ultimate strength.

Ultimate strength values, which are ~ 10% of the measured Young's modulus, suggest the strength limit of a perfect defect-free crystalline material according to Griffith's rule. The reason for the large deviation between the experimental and theoretical stiffness values requires additional investigations. In a study combining theory and AFM indentation measurements, a low Young's modulus of 130 N/m and a large intrinsic strength of 16.5 N/m were determined for MoS_2 [75]. The ultimate strength or critical stress of 2D materials is an important mechanical property to judge the performance during the fabrication processes and practical use, which may be connected with the occurrence of strain.

Moreover, the elastic bending modulus that describes the weakest deformation mode is available for the well-studied MoS_2 monolayer. The interest derives from the coupling of mechanical and electrical properties in flexible and stretchable electronics made of 2D materials. One approach used the derivation of an analytical formula for the bending modulus from an empirical interaction potential [76]. The advantage of this approach is its independence from the thickness of the layer. The reason for this behavior is its derivation from the surface and not from the volume energy density. The value obtained for MoS_2 monolayers of D = 9.61 eV is in the upper range of calculated bending moduli. Owing to the chemically bonded three-layer system, D = 9.61 eV is much larger than, for example, the bending modulus of graphene (1.4 eV).

In another approach, the energetics of single-wall nanotubes determined by first principles are fitted to the Helfrich Hamiltonian to determine the configurational energy of the membrane [77]. This parameter-free approach does not require thickness and specifies direction-dependent bending rigidities of $D_{ac} = 9.10$ eV and $D_{zz} = 9.6$ eV, in agreement with the previously discussed result, derived from an analytical formula that is based on an empirical potential. This study reveals the bending rigidities of single monolayers of MX_2 (M = Mo, W; X = S, Se).

With respect to approaches such as targeted molecular mechanics (MM), four-point bending, and the nanotube method applied to determine the bending stiffness, according to continuum theory the four-point bending method provided the most accurate results [78]. As additional successful approaches with interaction potentials, the application of the revised Stillinger–Weber and reactive empirical bond-order (REBO) potentials give reasonable bending stiffnesses of 8.7–13.4 eV. In addition to the single-layer bending stiffness, these approaches supply the effective deformed conformation.

16.2.2.4 Physics and Applications

The 1H-MoS_2 monolayer is a semiconductor with a direct band gap, where the valence band maximum (VBM) and conduction band minimum (CBM) are located at the K point of the first Brillouin zone. The transformation of the indirect gap of the bulk down to the bilayer and into a direct gap at the monolayer limit is due to the absence of interactions with adjacent layers and quantum confinement. The calculated band gaps vary substantially with the applied approach, predicting gaps at 1.58 eV (LDA) and 2.48 eV (G_0W_0) [6]. An independent first-principles study of the band alignments and band gaps of hexagonal MX_2 monolayers yielded similar gaps of 1.59 eV (PBE) and 2.02 eV (HSE06) for MoS_2 [79]. A comparison with available experimental results indicates an underestimation by the PBE approach and a slight overestimation by the HSE06 approach [79]. This study showed that the VBM and CBM originate from the repulsion between the $d_z{}^2$ orbital of the Mo atom and the $p_x + p_y$ orbitals of the S atom. A simple model explains the chemical trends for MoS_2, $MoSe_2$, and $MoTe_2$. Some of the monolayers and their heterostructures show appropriate band alignments for spontaneous water splitting, photovoltaics, and optoelectronics. The band alignment of MoS_2 monolayers is suitable for spontaneous photosplitting of water.

A recent review describes both the progress and remaining challenges in the potential use of TMDC monolayers for thermoelectric applications, and self-powered, flexible, and wearable devices [80]. This includes the modulation of electron and transport properties by tuning the band structure, i.e., by chemical functionalization. State-of-the-art theoretical and available experimental data for the MX_2 (M = Mo, W, and Ti; X = S, Se, and Te) compounds provided the information needed for an analysis of the main factors controlling the thermoelectric properties. This includes the manipulation of electrical and thermal transport by thickness variation, structural defects, and applied mechanical strain. While the thermal conductivity is much greater, the thermoelectric performance is still lower than that of conventional bulk materials, and the measured Seebeck coefficient is much lower than theoretically suggested. The main challenges remaining are the reliable fabrication of high-quality 2D-TMDCs and the general lack of sufficient experimental data.

The application of 2D TMDCs as cost-effective electrocatalytic materials for the generation of hydrogen in a renewable energy conversion process is an active field of research. The progress made in the use of TMDCs and their composites as electrocatalysts for acidic and alkaline electrolysis and as photocatalysts for the formation of hydrogen is reviewed in a comprehensive study [81]. An important aspect that requires further investigation is the activation of the basal plane by chalcogenide and metal vacancies, point defects, and grain boundaries to enable fast electron transfer because in perfect monolayers only the edge sites are electrocatalytically active. Most of the published work on the hydrogen evolution reaction (HER)and oxygen evolution reaction (OER) has concentrated on MoS_2 and $MoSe_2$, while the performance of most members of the large family of TMDCs has not been fully elucidated.

An important development is the search for asymmetric TMDCs with reduced structural symmetry, such as XMY, which is usually described as a Janus structure, e.g., S–Mo–Se. Another emerging strategy is the formation of heterostructures by stacking different TMDC monolayers together by weak van der Waals forces, e.g., MoS_2/WS_2. Unfortunately, an established set of activity parameters for a consistent judgment of the efficiency of the considerable number of feasible electrocatalysts is still lacking.

The fabrication of the Janus SMoSe monolayer is possible by selective sulfurization of the top Se sublayer of $MoSe_2$ monolayers [82]. Raman spectroscopy, TEM, X-ray photoelectron spectroscopy (XPS), and time-of-flight secondary ion mass spectrometry were used to analyze the geometrical and electronic structures of the material. DFT calculations explain the high activity of basal planes in the HER by the synergistic effect of defects and the structural strain inherent in the Janus structure.

The fabrication of a field-effect transistor (FET) employing single-layer MoS_2 obtained by scotch tape-based exfoliation, covered with a 270 nm thick SiO_2 nm film, and transferred to doped silicon substrates started in 2011 [83]. The MoS_2 monolayer served as a conductive channel, and HfO_2 served as a gate insulator. This transistor had a current on/off ratio of 1×10^8 at room temperature and a mobility of ~ 200 $cm^2\ V^{-1}\ s^{-1}$. The advantages of single-layer MoS_2 devices include a nanometer-thick planar geometry, high band gap, flexibility, transparency, and chemical and mechanical stability. Progress in the field of single-layer electronic MoS_2 devices, together with the realization of digital circuits and transistors operating in the gigahertz range, has made MoS_2 monolayers with their direct band gap interesting for the investigation of phototransistors with high sensitivity and low noise, as well as light-emitting diodes and solar cells [84].

The environmental stability of TMDC monolayers is a critical issue, especially in the case of high carrier mobility. First-principles calculations of MoS_2, $MoSe_2$, $MoTe_2$, WS_2, WSe_2, and WTe_2 monolayers indicate that due to the weak physical adsorption of O_2 on perfect 2D TMDC, the surfaces remain intact under normal conditions even though oxidation is thermodynamically favored [85]. However, if chalcogen vacancy defects exist, which are the most common defects in real TMDC monolayers, chemisorption and dissociation occur at the vacancy sites. Importantly, the dissociation of O_2 on MoS_2 surfaces is difficult owing to the large energy barrier, whereas O_2 dissociates exothermically by overcoming the energy barrier on oxidized $MoTe_2$ and WTe_2 monolayers. Adsorption of oxygen molecules on the chalcogen vacancy induces a flat band at the top of the VB that partly disappears after dissociation of O_2 at the vacancy site. At the VB maximum, a very sharp peak appears corresponding to the flat band in the partial density of states (PDOS). Consequently, oxidation may be detrimental to the electronic properties of TMDCs.

Notably, aging of a CVD-based MoS_2 sheet was observed [86]. According to this study, dramatic aging occurred during several months of exposure to the environment, including extensive cracking, changes in morphology, and severe quenching of direct gap photoluminescence. As revealed by XPS and Auger electron spectroscopy (AES), the observed effects are due to gradual oxidation along the grain boundaries and the adsorption of organic contaminants. As a solution the authors suggest, encapsulating the monolayer with a 10–20 nm thick optically transparent parylene C polymer.

16.2.2.5 Synthesis

Contrary to the statements found in most publications the MoS_2 monolayer was synthesized before graphene. This is not surprising because the exfoliation energies of these two monolayers are comparable. In 1986, a monolayer of MoS_2 was prepared by exfoliation employing intercalation with lithium followed by reaction with water [87]. XRD analysis confirmed the presence of a one-molecule-thick layer. Diverse exfoliation methods, including mechanical, solvent-assisted, and chemical exfoliation, still play key roles in the synthesis of TMDC monolayers.

The synthesis of TMDC monolayers and nanosheets has attracted enormous attention in recent years, and 2D MoS_2 seems to be one of the best studied 2D compounds. Two reviews provide insight into the recent advances achieved in the synthesis and characterization of 2D TMDCs [88, 89]. These treatments describe in detail the growth of monolayers and few-layer nanosheets by vapor-phase deposition methods. The authors elucidate the underlying mechanism such as evaporation, thermal decomposition of precursors, transport and migration of reactants, as well as nucleation and growth of 2D crystals. A better insight into the individual mechanistic processes allows important aspects such as layer thickness, phase, composition, and number of defects in the deposit to be controlled.

Micromechanical exfoliation of MoS_2 is a widespread tool for preparing nanosheets and single layers of MoS_2 with superior quality, as initially discovered for the isolation of graphene [90]. However, this method is not suitable for nonlayered bulk materials or large-scale production owing to its low throughput. In the direct sonication process, the layered bulk material incorporated in an organic solvent or water-ethanol mixture disperses the van der Waals bonds and exfoliates into monolayers and few-layer sheets. This simple exfoliation technique can provide a high-yield production of single-layer MoS_2 [90].

For exfoliation by Li intercalation, the bulk material is incorporated as a cathode, and lithium foil is employed as an anode in a Li-ion battery setup [91]. The Li_nMoS_2 (n = number of Li atoms) compound obtained by the discharge process is sonicated in water or ethanol, where the generated H_2 gas further separates the MoS_2 layers, leading to well-dispersed nanosheets. Lithium intercalation is the best chemical exfoliation method and involves functionalization of MoS_2 and sonication to disperse the nanosheets. In contrast to those samples produced by mechanical exfoliation, nanosheets fabricated by chemical Li intercalation and exfoliation contain 2H and 1T phases; however, the 1T phase is not stable at approximately 300 °C [90].

A method of practical interest is CVD, owing to the possibility of large area growth and the possibility of controlling the thickness of the deposited layer. The deposition of a defined number of layers still is a severe problem, e.g., the growth of a perfect bilayer. Uncontrolled nucleation and local growth of multilayers usually prohibit the realization of a perfect bilayer. Various forms of vapor-phase processes have been employed to produce well-defined MoS_2 layers, for example, to fabricate CVD-based MoS_2 FETs [88]. Despite an enormous effort to generate large-scale and uniform atomic layers, the quality of these materials has not yet reached that of good mechanically exfoliated samples. Three reviews discuss the crucial improvements needed to go beyond prototypes to practical applications. These include single-layer control, defect-free growth, and controlled doping and alloying [92–94].

Controlling the nucleation density of Mo during CVD at the surface of a SiO_2/Si or c-plane sapphire substrate yielded uniform 90 μm long MoS_2 monolayer crystals and 695 × 395 μm^2 monolayered deposits [95]. The application of such a MoS_2 monolayer as a sensor material allowed ultrasensitive NO_2 detection with a detection limit of 1.4 ppb. First-principles DFT calculations revealed that the adsorption of NO_2 induces a reduction of the electronic band gap of MoS_2 monolayers of 0.7 eV by molecular orbital hybridization.

Further improvements of the CVD method allowed the growth of uniform 12-inch MoS_2 monolayers [96]. This deposition process required the controlled release of the precursors and the use of substrates covered with predeposited amorphous Al_2O_3 to ensure uniform growth of MoS_2. To confirm the high quality and suitability of the produced devices for integrated circuit manufacture, transistor arrays were fabricated using this method.

The epitaxial growth of monolayers provides films with better quality owing to the use of a UHV setup that reduces contamination (see Fig. 16.11). A comprehensive study of the optical and structural properties of a single-layer MoS_2 film grown by MBE on a sapphire substrate with 6 μm crystallites supports this conclusion [97]. Diagnostic tools such as AFM, SEM, XPS, Raman spectroscopy, and optical microscopy were applied to assess the structural quality. Spectroscopic ellipsometry in the spectral range from 250 to 1700 nm under variable incident angles provided accurate optical constants. The superior quality of the MBE-grown MoS_2 monolayers follows from the almost two times larger quantum yield of photoluminescence observed experimentally.

Recent work demonstrated the highly reproducible epitaxial growth of wafer-scale single-crystal monolayers of MoS_2 deposited on industry-compatible sapphire

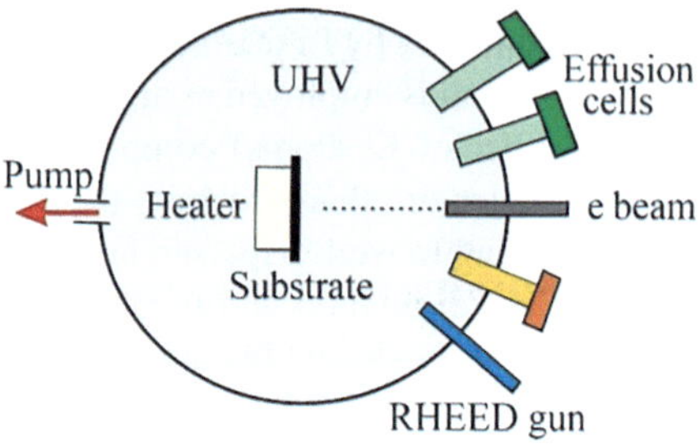

Fig. 16.11 Scheme of UHV setup for MBE and in situ analysis of deposits under ultrahigh vacuum conditions

substrates [98]. The key factor for the epitaxial growth of high-quality samples was the special precursor supply route involving precise optimization of the chalcogenide to metal precursor ratio. The performance of produced electronic devices confirmed the high quality of the 2D MoS_2 crystals. Altogether, this study demonstrated a reliable and highly efficient route for batch production of wafer-scale 2D materials in nanoelectronics and optoelectronics.

The technique of gas source MBE allows the realization of high-quality large-scale growth [99]. A high growth temperature and ultralow precursor flux suppressed the formation of anti-aligned grains. These conditions allowed the growth of a continuous, uniform, and highly crystalline monolayer of MoS_2 by complete Mo sulfurization during MoS_2 co-deposition, as confirmed by Raman mapping. Commercially available 200 mm wafer size templates of SiO_2 and Al_2O_3 served as substrates for growing large-scale fiber textured MoS_2 monolayers and multilayers at temperatures between 300 °C and 550 °C. The authors carefully proved by Raman spectroscopy, photoluminescence, and XPS that complete sulfurization of Mo occurred at the chosen low growth rates.

The selection of the optimal substrate plays a significant role in the fabrication of high-quality monolayers [100]. Therefore, this is the focus of a review concentrating on the epitaxial growth of large-scale films with large domain sizes and high domain alignment. The absence of covalent bonding between nanosheet and substrate in van der Waals epitaxy reduces the requirement for lattice matching. Thus, weak interlayer interactions guarantee minimal changes in monolayer properties, while heterostructures can also exhibit strong interlayer coupling. Note that the often-reported epitaxial deposition by a CVD-type process is a low-cost version without UHV conditions.

With respect to phase engineering, the development of a general method for bottom-up synthesis of semimetallic 1T′-TMDCs under mild conditions is highly desirable. A scalable colloidal synthesis method was employed to fabricate 1T′-MoS_2, 1T′-$MoSe_2$, 1T′-WS_2, and 1T′-WSe_2 with high phase purity [101]. The tested electrocatalytic activity of 1T′-MoS_2 provided a proof-of-concept. In fact, the polymorphs showed excellent performance in hydrogen production with abundant in-plane active sites, impressive endurance in electroactivity, and high conductivity coupled with the contribution of lattice strain.

A general article on epitaxial growth of various phases addresses the compelling class of TMDC materials [102]. This work discusses the fundamental aspects of the epitaxial growth of van der Waals bonded monolayers and heterostructures, including nondestructive characterization techniques. The study comprises source materials, deposition methods such as vapor phase deposition, substrate properties, and film-substrate orientation and bonding. Moreover, this work proposes future directions for the development of optimized substrates, in situ and full wafer characterization methods, and the growth of bilayer heterostructures (Table 16.1).

16.2.3 $MoSe_2$ Monolayer

16.2.3.1 Chemistry and 2D Bonding

Compared with MoS_2, bonding in molybdenum diselenide monolayers is determined by the slightly increased size of Se atoms with the electronic configuration ($Ar|3d^{10}4s^2p^4$) and a similar electronegativity of Se (2.55) that is greater than that of Mo (2.16). The most stable structure is the three-layer arrangement with a positively charged central Mo sublayer and negatively charged outer Se sublayers. The calculated Bader charge transfer amounts to $\delta q = 0.83$ e from each Mo atom to two Se atoms, inducing a fractional ionic character of only 3.7% [2]. Owing to the reduced orbital overlap, the calculated cohesive energies are $E_c = -4.60$ eV/atom [2] and -4.53 eV/atm [62] (PBE), whereas the LDA values at -5.82 eV/atom [1] and -5.73 eV/atom [62] indicate stronger bonding. Much stronger bonding in $MoSe_2$ is proposed by an extreme cohesive energy of $E_c = -7.22$ eV/atom [63]. The decreasing binding strength when going down the column of chalcogens is due to the increase in the atomic size from O atoms ($d_w = 0.31$ nm) to S atoms ($d_w =$ 0.36 nm), Se atoms ($d_w = 0.38$ nm), and Te atoms ($d_w = 0.42$ nm) as denoted by the van der Waals diameters and increased charge shielding by inner electrons [18].

16.2.3.2 2D Structures

Most calculations consider the semiconducting H structure of 2D $MoSe_2$ to be the most stable phase. For this phase with the hexagonal sandwich-like structure, DFT-GGA-PBE calculations yielded a lattice constant of $a = 0.332$ nm, a bond length of $d_{Mo–Se} = 0.254$ nm, a distance between Se atoms of $h_{SeSe} = 0.334$ nm, and the bond angles $\theta_{SeMoSe} = 82.12°$ and $\theta_{MoSeMo} = 81.54°$ (PBE) [2]. These lattice parameters do not deviate much from those calculated by the LDA approach: $a =$ 0.324 nm, $d_{Mo–Se} = 0.250$ nm, $h_{SeSe} = 0.332$ nm, and $\theta_{SeMoSe} = 83.05°$ [1]. The calculated band gaps of the H structure are 1.44 eV [2] (PBE), 1.62 eV [1] (LDA), and 2.31 eV [1] (GW_0). Adding the van der Waals diameter of Se atoms of $dw =$ 0.380 nm and their distance h_{SeSe}, yields an estimate of the monolayer thickness of $d_{w+h} = 0.714$ nm. There are two well studied polymorphs of $MoSe_2$, the H phase and 1T phase. Recent studies concentrate on the structural stabilities of these two polymorphs [103]. A recent publication describes the metallic 1T phase of $MoSe_2$ as the thermodynamically more stable $MoSe_2$ monolayer [104].

Compared with MoS_2, $MoSe_2$ has a larger interlayer spacing, narrower band gap, and more pronounced metallic properties. A comparison of the lattice constants of a rectangular unit cell of the 1T phase of a = 0.329 nm and b = 0.570 nm with those of the 1H phase of a = 0.332 nm and b = 0.575 nm and of the bond lengths of $d_{Mo–Se}$ = 0.257 nm (1T) and $d_{Mo–Se}$ = 0.255 nm (1H) indicates quite similar lattices [65]. This work also presents the lattice parameters and bond lengths of the distorted 1T′ structure at a = 0.326 nm, b = 0.604 nm, and $d_{Mo–Se}$ = 0.253–0.270 nm. Lattice distortions in multilayer 1T $MoSe_2$ can induce the formation of chemical bonds between the Mo atoms and thus a transition from the 1T to the 1T′ phase. The controversy on the most stable phase is not surprising in view of the similarity of the lattice parameters and especially the quasi-identical bond lengths.

16.2.3.3 Mechanics

The integration of 2D TMDCs into devices exposes the 2D material to strains owing to a mismatch of lattice parameters or thermal expansion. Therefore, knowledge of the mechanical behavior, such as stiffness, critical strain, and ultimate fracture strength is important for the design of 2D devices [104]. As mentioned above, going down the column of chalcogens from MoS_2 to $MoSe_2$ and $MoTe_2$, the bonding energy decreases with increasing atom size due to reduced orbital overlap and stronger shielding of the interaction of valence electrons with the nuclei due to the increasing number of electrons in the inner shells. These effects determine the linear and nonlinear mechanical properties and performance.

The given cohesive energy leads to Young's moduli of $MoSe_2$ monolayers of E_{ac} = 128 N/m (178.8 GPa) and E_{zz} = 126 N/m (176.0 GPa) [68]. To calculate the stiffness values in 2D units, a thickness of d_{w+h} = 0.714 nm was assumed, in reasonable agreement with the experimental result of t = 0.70 nm [105]. In contrast to the situation described for MoS_2 monolayers, the elastic modulus was measured with a testing device that enables uniform loading of a free-standing monolayer or bilayer membrane suspended over a perforated substrate yielding 127 N/m (177.2 GPa) [105]. This stiffness confirms the results given above but deviates from the DFT-based stiffness of E = 105.7 N/m and Poisson's ratio of ν = 0.28 [106]. For comparison, the ElasTool toolkit delivers E = 115.4 N/m and ν = 0.21 for the 2H-$MoSe_2$ phase and E = 112.9 N/m and ν = − 0.02 for 1T-$MoSe_2$ monolayers [8].

A distinct anisotropy is found for the calculated ultimate strength values of σ_{ac} = 16.2 N/m (22.7 GPa) and σ_{zz} = 9.2 N/m (12.9 GPa) and also for the ultimate strains of ε_{ac} = 0.29 and ε_{zz} = 0.16 [68]. Independent calculations yielded ultimate tensile stresses with much smaller anisotropy of σ_{ac} = 15 N/m and σ_{zz} = 12.9 N/m and substantially larger ultimate strains of ε_{ac} = 0.41 and ε_{zz} = 0.27 [106]. Importantly, the measured fracture strength in CVD-grown nanosheets was only σ = 3.4 N/m (4.8

GPa), possibly due to flaws [105]. According to Griffith's fracture theory, a fracture inducing crack of ~ 33 nm in length would explain such a low fracture strength [105].

16.2.3.4 Physics and Applications

The 2D-$MoSe_2$ sheets are among the best studied TMDCs. As mentioned above, the reasons for the current interest are the stability of the 1T phase, the large interlayer spacing due to weak van der Waals interactions, the direct band gap of ~ 1.5 eV, and the more pronounced metallic character than, for example, that observed for 2D MoS_2 [103]. Potential applications of 2D $MoSe_2$ include the fields of electrocatalysis and photocatalysis. Various synthesized 2D-$MoSe_2$ materials exhibited high electrocatalytic activity due to their high conductivity and abundant active centers, such as chalcogenide vacancies, which have low formation energies or oxygen substitution defects. Furthermore, owing to its narrow band gap, $MoSe_2$ absorbs light in the near-infrared region and has a high photo-corrosion resistance. The incorporation of suitable absorbers in heterojunctions yields photocatalysts with an almost full spectrum of absorption. In addition, the study reviews the published work on the application of $MoSe_2$ nanosheets in the field of energy storage [103]. This includes lithium-ion batteries, sodium-ion cells, potassium-ion batteries, and lithium-sulfur batteries, which need further improvement in their stability and capacity. The large interlayer distance of ~ 0.70 nm, which allows faster migration of charge carriers, is of particular interest for future developments.

The conversion of light into an electrical signal in optoelectronics employing transition-metal diselenide is an intense field of research [104]. Here, the gradual transformation of the indirect band gap of multilayers into a direct gap with the reduction of thickness to the monolayer stage that is accompanied by enhanced photoluminescence, has boosted enormous research efforts in the improvement of electroluminescence devices, photodetectors, solar cells, catalysts, FETs, and light emission diodes (LEDs).

In addition, the use of TMDCs as gas or biosensor has attracted widespread attention. Good detection limits are achieved owing to the inherent large specific area, the active sites, the fast carrier mobility, and the rapid electrical response [103]. For example, a gas sensor realized by a $MoSe_2$-FET device delivered a high sensitivity for NH_3, NO, and especially for NO_2 [107]. The reduced conductance is consistent with the large adsorption energy found for these compounds in comparison to the adsorption of CO and CO_2 molecules. An external electric field enhances the adsorption process significantly by modulating charge transfer and the magnetic moments of NO and NO_2.

Defect and vacancy engineering is a novel strategy for improving the interaction of gases and biomolecules with 2D materials in sensing applications [108]. Recent experiments realized this plan by a single-step laser-assisted bottom-up growth process of $MoSe_2$ monolayers with a tunable vacancy concentration of 1–20%. The adjustable power density of the CO_2 laser radiation controlled the vaporization dynamics of the stoichiometric $MoSe_2$ powder, which was used as a single precursor.

The time-controlled heating process varied between a few tens of milliseconds and seconds, and the background pressure regulated the confinement of the vapor. This method allowed the growth of $MoSe_{2-x}$ crystals with tunable uniform vacancy concentrations and thus tunable stoichiometries. Analysis of growth and structure by Raman and photoluminescence spectroscopy, STEM, AFM, optical microscopy, and time-correlated single-photon counting (TCSPC) allowed the quantification of the effects of laser power, irradiation time, and background pressure.

16.2.3.5 Synthesis

The synthesis of 2D molybdenum dichalcogenides is a field of intense research. Based on the unique band structure with a narrow direct band gap of ~ 1.5 eV, high optical absorbance, and large spin splitting energy, current interest has focused on 2D $MoSe_2$. The low chemical reactivity of Se impedes the realization of a versatile bottom-up growth technique for the preparation of monolayers suitable for electrical devices by the CVD method.

A key step forward was the observation that NaCl enhances the growth rate and reduces the growth temperature by the formation of intermediate Na-containing species, which reduce the activation energy [109]. NaCl-assisted CVD is a straightforward method for synthesizing large-scale single-crystal $MoSe_2$ monolayers with a size of up to 250 μm. The fabricated monolayers had a mobility of up to 38 $cm^2\ V^{-1}\ s^{-1}$ (average value: 12 $cm^2\ V^{-1}\ s^{-1}$). The optical responsivity of the monolayer was 42 $mA\ W^{-1}$, and the external quantum efficiency 8.22%, like that of MoS_2 monolayers.

A large-scale growth technique allows batch fabrication of atomically thin electronic and optoelectronic devices and circuits without film transfer [110]. This rapid deposition method is based on spin coating of the Mo precursor, assisted by NaCl, where sodium ions increase the reaction rate. This improvement allows the growth of uniform monolayers of up to a few inches in size and with grain sizes up to 120 μm within 5 min. The fabricated FETs had a high carrier mobility of approximately 27.6 $cm^2\ V^{-1}\ s^{-1}$ and an on/off ratio of ~ 10^5. Importantly, this CVD method is also applicable to related TMDCs, such as MoS_2, WS_2, and WSe_2.

A facile growth technique for monolayers is ambient pressure chemical vapor deposition (APCVD). The inevitable use of hydrogen during deposition and the formation of cracks could be circumvented by synthesizing large-area, continuous, and highly crystalline monolayers and nanosheets of up to 3 nm thickness on SiO_2/Si substrates at higher reaction temperatures of Se [111]. To account for the formation mechanism in more detail, the authors proposed the application of the wing model, boundary layer theory, and diffusion theory.

A broad range of morphologies, ranging from hexagons to triangles, sawtooth hexagons, dendrites, and fractals, were found by carefully controlling the vapor composition and growth rate during CVD synthesis of $MoSe_2$ monolayers and few-layer flakes on SiO_2/Si substrates using MoO_3 and Se powder as the solid precursors [112]. The authors revealed that the diffusion process on the substrate and the growing film were responsible for the observed differences in shape. The variable size and shape strongly affected the photoluminescence response of the deposited $MoSe_2$ flakes. Optical microscopy, SEM, AFM, and Raman spectroscopy confirmed the deposition of a monolayer regardless of the morphology and analyzed the quality of the material. Micro-Raman spectroscopy provided a structural fingerprint for the identification of the diverse $MoSe_2$ phases and supported the AFM thickness measurements.

The more advanced MBE uses an elaborate ultrahigh vacuum environment to avoid further contamination of high-purity source materials. In addition, in situ analysis is possible and allows better control of the growth process and the detection of the intrinsic crystal structure. Thus, MBE has important advantages as a high-quality technique, however, at the expense of a costly stainless-steel apparatus and sophisticated detection equipment. CVD is supposed to provide a cost-effective method for fabricating monolayers and heterostructures with high yield and sufficient quality for future applications, while MBE is an ideal research tool for fundamental studies, providing new insight into the atomic mechanism of deposition and growth processes.

Van der Waals MBE on an $Al_2O_3(0001)$ substrate yielded nearly perfect singly oriented $MoSe_2$ monolayers [113]. The results of in situ and real-time monitoring of the evolution of optical anisotropy during the growth process indicated two phases: the growth of small islands and subsequent coalescence. The $MoSe_2$ nuclei and domains align in their zigzag direction along the $Al_2O_3(0001)$ direction and maintain this preferential orientation during coalescence, forming a singly oriented monolayer. An increase in the optical anisotropy during island growth indicates a dominating interplay with the substrate, while a decrease in optical anisotropy during coalescence signifies progressing lateral interactions between islands during monolayer formation.

The polymorphism of atomically thin $MoSe_2$ layers and the stabilization of the metastable phase of 1T $MoSe_2$ depend on the coverage and deposition temperature during epitaxial growth [114]. For example, the 1T phase and the 1H phase of $MoSe_2$ can be grown in comparable quantities by MBE, and the 1T phase gradually converts into the more stable 1H phase before the first monolayer is completed. This kind of information is relevant for the development of phase-selective synthesis.

While the deposition of large-area $MoSe_2$ monolayers seems to be difficult by conventional growth methods, a simple postselenization route delivered deposit sizes greater than 3.5 inch, with excellent homogeneity [115]. Moreover, the direct deposition of n-$MoSe_2$/p-Si heterojunctions on a patterned p-Si substrate permits the growth of high-quality van der Waals devices. These devices showed a self-driven broadband photoresponse from the ultraviolet to the infrared region with an impressive responsivity of 720.5 mA W^{-1}, a specific detectivity of 10^{13} Jones, and a response time following the pulsed nanosecond optical signal.

A detailed up-to-date perspective considers the standard methods used for synthesis and recent advancements in the research field of van der Waals MBE. It concentrates on the realization of transition-metal sulfide, selenide, and telluride monolayers, few-layer nanosheets, and their heterostructures [116]. These layered

TMDCs are promising building blocks for the next generation of electronics and optoelectronics.

16.2.4 $MoTe_2$ Monolayer

16.2.4.1 Chemistry and 2D Bonding

Bonding in molybdenum ditelluride monolayers depends on the *d* electrons of Mo atoms with the electronic configuration (Kr|4$d^5$5s) and the *p* valence electrons of Te atoms with the configuration (Kr|4d^{10}5s^2p^4). The increase in the van der Waals diameter of Te atoms to d_w = 0.42 nm [18] and nearly identical electronegativities of Te (2.10) and Mo (2.16) (Pauling scale) are responsible for the weakest covalent-ionic bonding of $MoTe_2$ monolayers in the molybdenum dichalcogenide group of monolayers. Note that according to a Bader charge transfer analysis a charge of δq = 0.52 e is transferred from each Mo atom to two Te atoms [2]. This amounts to a marginal fractional ionic character of only 0.09%. These physicochemical properties of $MoTe_2$ lead to the calculated cohesive energies of E_c = − 3.97 eV/atom [62] and − 4.04 eV/atom [2] with weakest bonding in this group according to the PBE approach and to E_c = − 5.13 eV/atom [62] and − 5.22 eV/atom [1] based on the LDA approximation. The strongest cohesion proposed in the literature is E_c = − 6.54 eV/atom for the 1H-$MoTe_2$ monolayer [63].

16.2.4.2 2D Structures

The best-characterized phases of 2D-$MoTe_2$ are the semiconducting 1H and 2H structures. Early DFT-GGA-PBE calculations delivered a lattice constant of a = 0.355 nm, a bond length of $d_{Mo–Te}$ = 0.273 nm, a distance between Te atoms of h_{TeTe} = 0.361 nm, and the bond angles θ_{TeMoTe} = 82.74° and θ_{MoTeMo} = 81.07° [2]. These values do not deviate much from the lattice parameters of the H phase (a = 0.346 nm, $d_{Mo–Te}$ = 0.269 nm, $d_{Te–Te}$ = 0.359 nm, and θ_{TeMoTe} = 83.88°) based on LDA calculations [1]. Adding the van der Waals diameter of Te atoms of d_w = 0.420 nm and their distance in the monolayer, yields an estimate of the layer thickness of d_{w+h} = 0.781 nm. The bulk and few-layer systems possess an indirect band gap, whereas the monolayer becomes a semiconductor with a direct band gap [117]. The space group of the hexagonal 2H-$MoTe_2$ phase is *P*6$_3$/*mmc* [117].

While the 1T phase of $MoTe_2$ has not found much attention, this is the case for the monoclinic 1T′ phase that is a distorted 1T phase with in-plane Mo–Mo bonds forming zigzag metal chains. An additional structurally comparable lattice besides the 1T′ phase is the T_d structure with distorted octahedral coordination. The 1T and 1T′ phase differ in the few-layer arrangement of the monolayers (see Fig. 16.12) [117, 118]. Note that the atoms in different layers are in good alignment in the T_d-$MoTe_2$ structure, however, the layers slide a short distance and form a tilt angle of

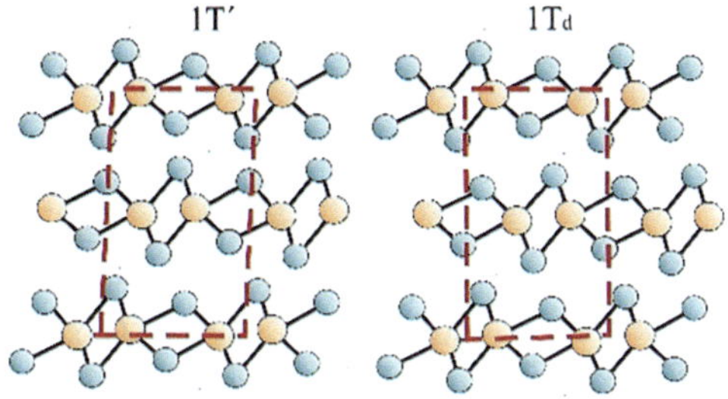

Fig. 16.12 Structures of layered topological 1T′-$MoTe_2$ and $1T_d$-$MoTe_2$ sheets. Note that the atoms in different layers are in good alignment in the T_d-$MoTe_2$ structure, whereas the layers slide a short distance and form a tilt angle of 0.94° in 1T′ $MoTe_2$. Reproduced from Ref. [117], Copyright 2022, MDPI

0.94° in 1T′ $MoTe_2$. The metallic 1T′-$MoTe_2$ phase has a monolithic structure and belongs to the space group $P2_1/m$ [117]. Interestingly, the energy difference between the 2H phase and 1T′ phase is only 35 meV/unit, where the unit cell contains 12 atoms [118]. This makes selective fabrication of the pure phases extremely difficult despite the slightly higher stability of the 2H-$MoTe_2$ phase; therefore, both phases may coexist in a deposited sample. Furthermore, phase transitions between these structures occur easily. Optimizing the kinetic rate of nucleation resulted in single-crystalline 2H-$MoTe_2$ deposits with a domain size of several hundred micrometers [119].

The T_d-$MoTe_2$ phase has an orthogonal structure and belongs to the space group $Pmn2_1$ [117]. The structures of the T_d and the T′ phases are remarkably similar; therefore, distinguishing the two structures is nearly impossible. However, due to its broken symmetry, the T_d phase exhibits interesting properties, such as superconductivity, which is pressure- and thickness-dependent and possibly a topologically nontrivial state.

16.2.4.3 Mechanics

The mechanical properties of $MoTe_2$ monolayers with the smallest stiffness and lowest strength of the considered Mo dichalcogenide monolayers has not yet received the same attention as the MoS_2 and $MoSe_2$ monolayers. Thus, only limited information is available on the mechanical behavior of these compounds. The calculated stiffness values are $E_{ac} = 98.3$ N/m (125.9 GPa) and $E_{zz} = 96.5$ N/m (123.5 GPa); the extremely anisotropic ultimate strength values are $\sigma_{ac} = 13.4$ N/m (17.1 GPa) and $\sigma_{zz} = 6.2$ N/m (7.88 GPa); and the ultimate strains are $\varepsilon_{ac} = 0.32$ and $\varepsilon_{zz} = 0.14$ [68]. The mechanical properties in 2D units were estimated by assuming a layer thickness of $d_{w+h} = 0.781$ nm, which is in reasonable agreement with DFT calculations [120]. Young's modulus and Poisson's ratio of previous DFT calculations are smaller at E = 72.5 N/m and $\nu = 0.25$ [104]. The ElasTool toolkit values are in between with the values E = 88.9 N/m and $\nu = 0.21$ for 2H $MoTe_2$ and E = 89.5 N/m and $\nu = 0.02$

for 1T-$MoTe_2$ monolayers [8]. Calculations of the elastic moduli, based on uniaxial strain, are available for the three polymorphs, namely for the isotropic hexagonal 2H phase (117.8 GPa), the anisotropic distorted octahedral 1T′ phase (E_{ac} = 124.5 GPa and E_{zz} = 96.0 GPa), and the octahedral T_d phase (E_{ac} = 124.9 GPa and E_{zz} = 97.6 GPa) [120].

Measured elastic properties, based on the application of biaxial strain, are available for the three polymorphs 2H, 1T′, and T_d [120]. These experiments yielded Young's moduli at 110 GPa for 2H-$MoTe_2$, 99 GPa for 1T′-$MoTe_2$, and 102 GPa for T_d-$MoTe_2$ monolayers. The corresponding ultimate strengths are 5.6 GPa, 2.1 GPa, and 2.5 GPa, respectively. The layer thicknesses given for these phases show so enormous differences that a transformation to 2D units seems not to be reasonable. The measurements indicate that the polymorphs have a comparable stiffness and that the results for the 2H phase are consistent with the other results. The experimental strength values of the two anisotropic phases 1T′ and Td are approximately one half of the strength of the isotropic 2H phase. This behavior occurs because the in-plane elastic modulus is less sensitive to the specific orientation of bonds and defects in a complex arrangement of bonds than the breaking strength.

16.2.4.4 Physics and Applications

The current interest in 2D $MoTe_2$ originates from its Janus derivatives, its rich polymorphism including the T′, and T_d phases, and extensive stacking polytypes [106, 121]. The diverse phases are related to a range of quantum states, such as quantum spin Hall (QSH) states, superconductivity, anisotropic SOC, and Weyl semimetals (WS) [107, 122]. Owing to the small energy barrier needed for the transformation of 2H $MoTe_2$ to the 1T′ phase, the material becomes a desired object of phase engineering. In particular, the temperature-induced transformation of the 1T′ phase to the T_d phase that breaks centrosymmetry offers a platform for studying emerging quantum states.

As an example of interfacial engineering, the thermally induced phase transition from Mo_6Te_6 nanowires, grown on a $SrTiO_3$(001) surface by MBE, to 1T′-$MoTe_2$ monolayers in a superconductive state was demonstrated [123]. A combination of STM and scanning tunneling spectroscopy (STS) measurements, as well as ex situ electrical transport and magnetization experiments, revealed a critical temperature of $T_c \approx 30$ K, approximately two orders of magnitude greater than that of the 3D crystal.

A review discusses the topological phase transition between the metallic type 1T′ phase and the topological T_d phase that is associated with the crystal structure, charge transport, and electronic band structure [124]. Various stimuli that affect structural transitions, such as reduced dimensionality, pressure, charge doping, and chemical substitution are discussed in detail, and correlations between apparently unrelated structural and superconducting phase transitions are revealed in this elaborate treatment.

For the direct synthesis of hydrogen peroxide by a two-electron oxygen reduction reaction (2e-ORR) in acidic media, usually employing expensive noble-metal-based electrocatalysts, 1T′-$MoTe_2$ monolayers can function as efficient electrocatalysts [125]. The electrocatalytic reduction of O_2 to H_2O_2 is favored by the binding energy of − 4.24 eV of OOH* at the surface-exposed Te sites with high basal plane activity toward the 2e-ORR process instead of the competing four-electron ORR step.

The development of 2D-$MoTe_2$ heterophase homojunctions, which involve the combination of a semiconducting phase with a metallic phase and an extremely small transition energy barrier, is an interesting research field [117]. Future applications include the demand for low contact resistance and clean interfaces in high-performance FETs, which compromise a semiconducting phase as the channel and a metallic phase as the electrode. Despite various challenges 2D-$MoTe_2$ heterophase homojunctions are expected to be a first step to the super-large-scale integrated circuits (IC) with extremely high integrity based on pure 2D materials.

16.2.4.5 Synthesis

Despite enormous effort in the development of new growth strategies, the deposition of high-quality wafer-scale materials under complete phase control remains a challenge. A mini review covers the various synthesis strategies for the realization of large-area 2D topological semimetal and transition-metal tellurides [126]. In particular, this study covers the progress made in the realization of 1T′-$MoTe_2$ nanosheets and their potential applications in electronics, optoelectronics, terahertz and spintronic devices.

The synthesis of 1T′-$MoTe_2$ monolayers, deposited by CVD with a lateral size of 500 μm, has attracted increasing amounts of attention as a promising possibility for the development of innovative devices such as topological transistors, as well as quantum and spintronic devices [127]. Encapsulation of the CVD-grown samples with h-BN membranes after synthesis delays degradation that is otherwise occurring immediately due to the large instability of the monolayer in air. The protection layer extended the lifetime to time scales sufficient for characterization and device fabrication.

To synthesize high-quality and large-scale monolayers and few-layer assemblies of 1T′ $MoTe_2$ with crystal sizes of up to 1 mm in length by ambient pressure CVD, KCl and NaCl salts were employed as growth promotors together with $(NH_4)_6Mo_7O_{24} \times 4H_2O$ as the Mo source [128]. The achieved quality, attained stoichiometry, and atomic structure of the as-grown 1T′ $MoTe_2$ sheets were determined by AFM, XPS, Raman spectroscopy, and TEM analysis. Variable-temperature electric transport measurements confirmed the semimetallic nature of the material. These results pave the way to nanophotonic research in the mid- and far-infrared spectral regions.

MBE experiments allowed the first demonstration of the coexistence of the 2H and 1T′ phases in 2D-$MoTe_2$ nanosheets owing to the small difference in the formation energy and demonstrated the degree of tunability by changing the deposition conditions (‘phase-change electronics’) [129]. STM and STS analyses revealed the atomic structures and intrinsic electronic properties of the two phases. This included quantum confinement and quantum interference effects in the 2H- and 1T′-$MoTe_2$ domains, respectively, as effected by the domain boundaries in the deposited samples.

The in-situ analysis by RHEED measurements provided the structure and morphology of the growing $MoTe_2$ monolayer during deposition employing MBE on a Se-treated GaAs(111)B substrate [130]. The RHEED patterns of the as-grown monolayers showed the coexistence of the 2H phase and the 1T′ phase. By annealing the sample at 600 °C under Te flux, the RHEED pattern signified the transformation of the 1T′ phase to the 2H phase. Note that under Te-deficient conditions, no transition occurred from the 2H to the 1T′ structure. Instead, the 2D-$MoTe_2$ phase started to evolve into nanowire-like structures at the edges of holes due to the desorption of Te. This work offers a strategy for the large-scale synthesis of monolayers with precisely controlled phases, because, in principle, it is possible to extend the growth of $MoTe_2$ films over the entire 2-in GaAs wafer.

A comprehensive review treats 2D $MoTe_2$, 2D WTe_2, and their alloys, which are of special interest, owing to their unique properties, arising from their polymorphic structure [118]. This includes preparation methods such as exfoliation, CVD, MOCVD, and MBE used for synthesis, as well as physical properties such as large spin–orbit interaction, nontrivial band structure, Weyl semimetal states, superconductive states, and ferromagnetism. In addition, the review introduces as potential device applications FETs, memory devices, spin-to-charge conversion devices, and solar cells.

An already mentioned fundamental problem is the rapid degradation of as grown samples in the environment, which impedes accurate characterization of the intrinsic properties of the deposit. Due to weak bonding in 2D $MoTe_2$ materials, the surface oxidizes easily to MoO_3 and TeO_2. Therefore, in situ protective capping layers of AlN or SiO_2 help to avoid changes of intrinsic properties before analysis. Note that the 1T′-$MoTe_2$ structure is more vulnerable to oxidation in air than 2H $MoTe_2$ (Table 16.2).

16.3 Tungsten Dichalcogenide Monolayers (WO_2, WS_2, WSe_2, and WTe_2)

16.3.1 WO_2 Monolayer

16.3.1.1 Chemistry and 2D Bonding

The valence electrons of the electronic configuration of W atoms (Xe|5$d^4$6s^2) and O atoms (He|2s^2p^4) determine the strong 2D bonding in tungsten dioxide monolayers. The *d-p* hybridization of the outermost W-5*d* orbitals and the outermost O-2*p* orbitals form stable covalently bonded three-layer structures with a W core sublayer sandwiched between two O sublayers, where each O atom forms three covalent bonds with W atoms in a bridging position and each W atom is bound to six O atoms. The difference in the electronegativities of W atoms (2.36) and O atoms (3.44) induces a substantial charge transfer of $\delta q = 1.83$ e from each W atom to two O atoms, forming strong covalent-ionic bonds with a large fractional ionic character of 25.3% [2]. The most stable H structure has a cohesive energy of $E_c = -7.02$ eV/atom that indicates stronger bonding than in CrO_2 (− 5.34 eV/atom) and MoO_2 monolayers (− 6.24 eV/atom) (GGA-PBE) [2]. Accurate calculations including SOC, by using fully relativistic pseudopotentials, based on the GGA approach, reported strongly deviating cohesive energies at $E_c = -8.16$ eV/atom (CrO_2), − 8.59 eV/atom (MoO_2), and − 8.43 eV/atom (WO_2) [131]. Other studies support a similar strong cohesion at $E_c = -8.19$ eV/atom [1] and − 8.17 eV/atom [3], calculated by the LDA approach. Note that these values imply the strongest bonds found in the families of TMDOs and TMDCs.

16.3.1.2 2D Structures

The most stable structure of 2D WO_2 is the 2H phase with a lattice constant of $a = 0.283$ nm, a bond length of $d_{W\text{–}O} = 0.205$ nm, a distance between O atoms of $h_{OO} = 0.248$ nm, and the two bond angles $\theta_{OWO} = 74.34°$ and $\theta_{WOW} = 87.27°$ (GGA-PBE) [2], in good agreement with the lattice parameters $a = 0.280$ nm, $d_{W\text{–}O} = 0.203$ nm, $h_{OO} = 0.245$ nm, and $\theta_{OWO} = 74.12°$, as calculated by the LDA approach [1, 3]. Adding the van der Waals diameter of O atoms $d_w = 0.310$ nm and their monolayer distance h_{OO}, yields an estimate of the monolayer thickness of $d_{w+h} = 0.558$ nm. The large formation energy of the 1H structure of $E_f = -2.24$ eV supports an enormous thermodynamic stability of the material [1]. The 1H-WO_2 monolayer is an indirect semiconductor with the bandgaps $E_g = 1.37$ eV [1] (LDA), 2.87 eV [1] (GW_0), and 2.16 eV [132] (HSE06) obtained by different theoretical approximations.

16.3.1.3 Mechanics

Only linear mechanical properties, such as the elastic constants, which fulfill the mechanical Born–Huang stability criteria, are presently available for WO_2 monolayers [7]. Young's modulus at $E = 231.7$ N/m and Poisson's ratio at $\nu = 0.34$, were derived from these constants [7]. Larger stiffness values at $E = 250$ N/m [1] and the anisotropic values $E_{ac} = 284.4$ N/m and $E_{zz} = 289.6$ N/m have been obtained by other studies [132]. As expected from the large binding energies, these stiffnesses belong to the largest values observed for TMDOs. For comparison, the ElasTool toolkit yields as linear mechanical properties $E = 234.5$ N/m and $\nu = 0.32$ for 2H-WO_2 monolayers, which are in remarkably good agreement with the values derived from the elastic constants [8].

16.3.1.4 Physics and Applications

There is a general interest in 2D metal oxides with natural abundance, chemical inertness, notable absorption in the UV region of > 10^5 cm^{-1}, and high carrier mobility, qualifying for possible applications in advanced electronics and opto-electronics [132]. A comprehensive study using systematic screening and unbiased particle swarm optimization by first-principles calculations, investigated eighteen metal-oxide monolayers, including 1H WO_2, owing to their wide band gaps of E_g = 1.22–6.48 eV, their superior oxidation resistance in an open-air environment, and ultrahigh carrier mobilities of up to 8540 cm^2 V^{-1} s^{-1} (InO monolayer) All 18 metal oxides exhibit negative formation energies in the range between $E_f = -0.47$ and -4.27 eV per atom, indicating the possibility of synthesizing these compounds in the laboratory, e.g., nine by exfoliation from a layered bulk structure.

16.3.1.5 Synthesis

The cohesive energy and the formation energy indicate energetic and thermodynamic stability of 2H-WO_2 sheets. While 2D CrO_2 and 2D MoO_2 have exfoliation energies comparable to that of graphene, the exfoliation energy of bulk WO_2 is 62.55 meV/$Å^2$ and thus roughly three times greater than that of graphene. Therefore, more effort is needed to cleave layered WO_2, especially in the case of a separation of single monolayers [3].

Interestingly, to date, the only reported successful synthesis involving the growth of a stoichiometric and epitaxially ordered 2D-WO_3 bilayer has been achieved by MBE on an Ag(111) substrate [133]. Microscopic and spectroscopic measurements combined with theoretical calculations revealed the geometry and the electronic and vibrational properties of the 2D material. The bilayer WO_3 phase consists of a staggered arrangement of WO_6 octahedra linked together by corner and edge sharing in a structure that resembles the bilayer of the layered bulk lattice of α-MoO_3. The basically defect-free and large-area layer coupled only weakly with the Ag(111) substrate, which is consistent with the prediction by DFT and suggests the stability of the free-standing WO_3 nanosheets. The WO_3 bilayer has a band gap of $E_g = 2.35$ eV and anisotropic elastic moduli of E_{ac} =197.8 N/m and E_{zz} = 198.1 N/m [133].

16.3.2 *WS_2 Monolayer*

16.3.2.1 Chemistry and 2D Bonding

The valence electrons of the electronic configuration of S atoms (Ne|$3s^2p^4$) and W atoms (Xe|$5d^4 6s^2$) are responsible for the similarity of 2D bonding in WS_2 and WO_2 monolayers. The extensive electron charge accumulation between the W and S atoms produces strong covalent W–S bonds. Moreover, the small difference in the electronegativities of W (2.36) and S (2.58) induces an ionic contribution to bonding, generating weakly polar W–S bonds. A Bader charge transfer analysis of the most stable triple layer hexagonal 1H phase predicts a surprisingly large charge transfer of $\delta q = 1.21$ e from each W atom to two S atoms, resulting in a fractional ionic character of only 1.2% [2]. For the H structure the reported cohesive energies are $E_c = -6.94$ eV/atom [1] (LDA) and -5.80 eV/atom [2] (PBE). These binding energies agree quite well with those of another study proposing $E_c = -6.90$ eV/atom (LDA) and -5.71 eV/atom (PBE) [62]. Moreover, an extreme cohesive energy of $E_c = -7.91$ eV/atom postulates stronger bonding for the WS_2 monolayers [63]. In general, the cohesive energies decrease with the atom size and electronic configuration of the chalcogen atoms going down the corresponding column in the periodic table.

16.3.2.2 2D Structures

Important related 2D-WS_2 crystal structures with a S–W–S three-layer arrangement are the stable hexagonal 2H phase, the metastable 1T phase with octahedral coordination, and the 1T′ phase, which is a distorted 1T phase [63, 65]. First-principles calculations, using the strongly constrained and appropriately normed (SCAN) metaGGA approach, provided the phase stability of the Mo- and W-TMDC monolayers [63]. There is a minor stabilization of the H phase over the T phase at the monolayer level, however, further stabilization of the adsorbed T structure, e.g., by interfacial strain exerted by the substrate, may change this situation.

Note that for small zigzag nanoribbons, the dimerized 1T′ phase is the actual ground state of WS_2 monolayers [65]. This is explained by the competition between edge energy, which favors the nonpolar 1T′ edge over the polar 1H edge, and the energy of atoms in the center of the ribbon, which favor the 1H ground state of infinite monolayers.

For WS_2 monolayers with the most stable H symmetry, the lattice constant is a = 0.319 nm, the bond length is $d_{W\text{-}S} = 0.242$ nm, the distance between S atoms is $h_{SS} = 0.314$ nm, and the bond angles are $\theta_{SWS} = 80.87°$ and $\theta_{WSW} = 82.47°$ (PBE) [2]. Adding the van der Waals diameter of S atoms of $d_w = 0.360$ nm and their distance h_{SS}, yields an estimate of the monolayer thickness of $d_{w+h} = 0.674$ nm. Note that the presented lattice parameters are essentially identical to those of the MoS_2 monolayer and also agree well with the lattice parameters a = 0.313 nm, $d_{W\text{-}S} = 0.239$ nm, $h_{SS} = 0.313$ nm, and $\theta_{SWS} = 81.74°$ calculated by the LDA approximation [1, 62]. An extensive comparison of the structural parameters of molybdenum and tungsten chalcogenides calculated by the PBE, LDA, and optB88 approaches yielded nearly identical results [66]. The 2H structure is thermodynamically stable, with suggested formation energies of $E_f = -1.37$ eV/atom (PBE) and -1.65 eV/atom (LDA) [62]. Furthermore, the phonon dispersion curves indicate dynamic stability of the 2H structure for all tungsten dichalcogenides [2, 6].

The available lattice parameters of the 1T structure of WS_2 monolayers are a = 0.322 nm, b = 5.58 nm (based on a rectangular unit cell), and $d_{W\text{-}S} = 0.244$ nm, and agree well with those of the H structure [65]. For the 1T′ phase, the corresponding values are a = 0.320 nm, b = 0.576 nm (based on a rectangular unit cell), and $d_{W\text{-}S}$ = 0.243–0.257 nm [65].

16.3.2.3 Mechanics

The hexagonal 1H structure of WS_2 has two high-symmetry directions, one oriented along the M–X bonds (armchair direction) and the other oriented perpendicular to the M–X bonds (zigzag direction), when projected on the central plane. Determination of Young's modulus from the linear part of the slope of the stress–strain curve indicates a small anisotropy of the stiffness. Taking as monolayer thickness $d_{w+h} = 0.674$ nm, the stiffnesses in 2D units are $E_{ac} = 165$ N/m and $E_{zz} = 162$ N/m [68]. This finding is in reasonable agreement with the previously published stiffness of E = 151.5 N/m [1]. Indentation measurements on CVD monolayers by atomic force microscopy (AFM) revealed a greater stiffness of 177 ± 12 N/m, as already discussed for MoS_2 [70]. For comparison, the smaller ElasTool toolkit values are E = 151.5 N/m and $\nu = 0.19$ for 2H WS_2 and E = 110.4 N/m and $\nu = -0.06$ for 1T-WS_2 monolayers [8].

The ultimate strength exhibits a larger anisotropy of $\sigma_{ac} = 20.2$ N/m and $\sigma_{zz} = 13.4$ N/m [68]. As expected from these strength values, the ultimate strains at $\varepsilon_{ac} = 0.28$ and $\varepsilon_{zz} = 0.18$ also display a large anisotropy as an indicator of the irreversible stress response. The transition-metal sulfides WS_2 and MoS_2 are well known for their favorable tribological properties. The microscopic reason for the low friction coefficient is the easy sliding behavior between weakly interacting adjacent monolayers upon application of weak shear forces.

16.3.2.4 Physics and Applications

The figure of merit $ZT = S^2\sigma T/\kappa$ characterizes the efficiency of thermoelectrical materials, where S is the Seebeck coefficient, σ is the electrical conductivity, and κ is the contribution of electrons and phonons to thermal conductance. Hydrothermally synthesized metallic 1T-WS_2 nanosheets exhibit a measured electrical conductivity of 45 S/cm and a large Seebeck coefficient of 30 μV/K that originates from an increased electrical conductivity and a lower thermal conductivity of the 1T phase in comparison with those of the 2H-WS_2 nanosheet [134].

A recent detailed discussion of the most important properties of the two similar monolayers of WS_2 and MoS_2 includes tribological, piezoelectric and magnetic properties, superconductivity, and the oxidation behavior [135]. Moreover, the review covers various emerging applications such as their use in electronics, optoelectronics, memory devices, lubrication, catalytic, environmental, and bio-medical applications, and as supercapacitor and sensor.

As mentioned above, of special interest are the tribological properties due the weak interlayer interaction of these monolayers. This leads to ultra-low friction and reduced wear of the lubricating oil containing WS_2 and MoS_2 nanosheets as additives, due to the effective penetration of the nanosheets into the crevices and pits, preventing the moving parts from direct contact [135]. Furthermore, the optical properties are of great interest. WS_2 monolayers with their direct band gap are expected to strongly absorb photons with an energy that is larger than the gap, which influences the optical performance of the material including photoluminescence and photoconductivity.

16.3.2.5 Synthesis

Remarkably, shortly after the synthesis of MoS_2 monolayers, 2D WS_2 nanosheets were realized in 1988 using mechanical exfoliation assisted by Li-ion intercalation [136]. After the discovery of graphene, several groups developed improved methods that are suitable for the preparation of TMDCs, especially for WS_2 and MoS_2 monolayers [135]. Various improvements in the exfoliation process, such as ultrasonication and centrifugation, have allowed the preparation of high-quality monolayers and few-layer assemblies. However, these samples do not qualify for device applications owing to the still small size of the deposits and the lack of sufficient control over the number of layers.

Therefore, bottom-up methods, including chemical vapor deposition (CVD) and physical vapor deposition (PVD) play vital roles. Single-step growth by sulfurization of WO_3 powders yields large-area triangular single-crystalline WS_2 monolayers with a size of up to hundreds of micrometers [137]. The high yield of light emission and valley-selective circular dichroism confirmed the high optical quality of these monolayers. The investigation of the influence of various substrates, such as SiO_2, graphene/SiO_2, and graphene on irreversible oxidative degradation provided new insight into the behavior of CVD-grown TMCDs in air [138]. Interestingly, on the SiO_2 substrate, oxidation begins within weeks, whereas for the use of graphene, no sign of oxidation could be observed, and for the graphene/SiO_2 substrate, oxidation was suppressed significantly. The authors attribute the enhanced oxidation resistance to the difference in the screening of the surface electric field caused by graphene and the substrate and a reduction of defects in the actual WS_2 monolayer [135].

Metalorganic CVD (MOCVD), employing a multistep growth process, allows wafer-scale epitaxial growth of fully coalesced unidirectional WS_2 monolayers on 2-inch diameter c-plane sapphire with a low in-plane rotational twist of 0.09° [139]. TEM analysis confirmed the absence of widespread inversion domain boundaries. Instead, translational boundaries form when WS_2 domains merge with a slight offset lattice. The high nucleation density along the sapphire surface steps, providing growth conditions that promote surface diffusion, lateral domain growth, and coalescence by preserving the aligned domain structure, can explain the pronounced unidirectional growth of the monolayer.

Pulsed direct current sputtering from a WS_2 target on SiO_2/Si, Si, and Al_2O_3 substrates is another successful approach to deposit WS_2 mono- and few-layer sheets [140]. Controlling the number of layers was possible by TEM, Raman spectroscopy, and photoluminescence measurements. XRD analysis revealed the (001) texture, AFM the atomic smoothness and uniformity of the layers, and XPS the chemical composition. The authors concluded that sputtering opens the door for batch production of high-uniformity and high-quality 2D materials. Importantly, postdeposition sulfurization using H_2S in the chamber during cooling of the substrate leads to stoichiometric films.

Of particular interest is the synthesis of the 1T-WS_2 phase due to its potential applications in electronic devices such as supercapacitors, electrocatalyst for hydrogen evolution, and phase-transition memories [141]. Small flakes produced by ion intercalation in combination with solution-based exfoliation contained 1T patches embedded in a 2H matrix. A refined CVD method produced monolayer 1T-WS_2 crystals with a size of up to 80 μm [141]. Using an iron oxide and sodium chloride catalyst allowed nucleation of 1T WS_2 at the infant stage of the growth process, forming a butterfly-like single crystal with the 1T phase in one wing and the 2H phase in the other. Diverse types of phase boundaries form at the 1T–2H interface, and the 1T phase persists at temperatures above 800 °C. Notably, atomic images taken from a 1T WS_2–MoS_2 heterojunction show a coherent and defectless interface with a sharp atomic transition.

Pulsed-laser deposition (PLD) allows the growth of wafer-scale WS_2 monolayers on insulating $SrTiO_3$ and Ni substrates, which are of interest for spintronics applications [142]. The deposition process was initiated by Nd:YAG laser radiation (355 nm) using a commercial WS_2 sample in a UHV chamber filled with high-purity Ar (10^{-5}–10^{-1} mbar) (see Fig. 16.13). XPS and Raman spectroscopy provided the phase control and thickness measurements. The growth process preserved the interface integrity of the deposited semiconductor. The Ni substrate showed a dramatically enhanced growth rate, and the WS_2/Ni electrode was resistant to oxidation under extended exposure to ambient conditions.

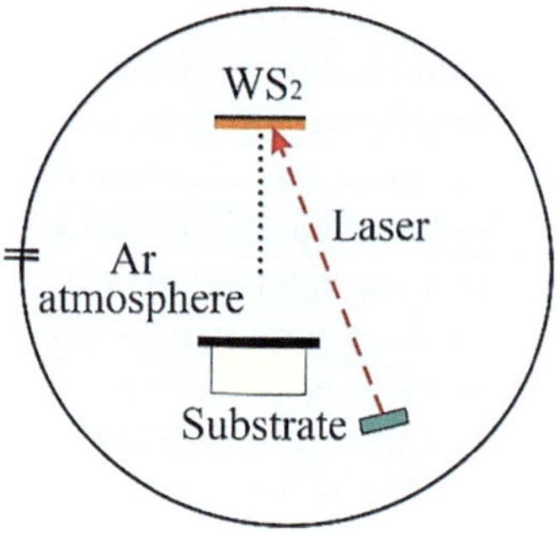

Fig. 16.13 Scheme of the UHV setup used for pulsed-laser deposition of WS_2 sheets from a WS_2 target on a $SrTiO_3$ or Ni substrate

16.3.3 WSe_2 Monolayer

16.3.3.1 Chemistry and 2D Bonding

Besides those of 2D WS_2, the monolayers of WSe_2 are well-studied TMDCs. The valence electrons of the electronic configuration of Se atoms (Ar|$3d^{10}4s^2p^4$) and W atoms (Xe|$5d^4 6s^2$) are responsible for the similarity of 2D bonding with that of WS_2. The extensive electron charge accumulation between the W and Se atoms produces strong covalent W–Se bonds. Moreover, the slightly different electronegativities of W atoms (2.36) and Se atoms (2.55) lead to a small ionic contribution to bonding and thus to slightly polar W–Se bonds. A Bader charge transfer analysis of the most stable hexagonal 2H phase yields a transfer of $\delta q = 0.92$ e from each W atom to two Se atoms, resulting in a fractional ionic character of only 0.9% [2]. Especially the covalent contribution to bonding of the H structure accounts for the cohesive energies of $E_c = -6.36$ eV/atom [1] (LDA) and -5.19 eV/atom [2] (PBE). These cohesive energies agree quite well with $E_c = -6.31$ eV/atom (LDA) and -5.10 eV/atom (PBE) [62]. The also published extreme cohesive energy of WSe_2 at $E_c = -7.20$ eV/atom (WS_2: -7.91 eV/atom) is nearly identical with that of $MoSe_2$ of -7.22 eV/atom [63]. Importantly, the cohesive energies predicting lower binding energies, which are favored here, are much smaller than those reported for the strongly ionic WO_2 monolayers.

16.3.3.2 2D Structures

The most stable well characterized polymorphic form of the presently known 2D-WSe_2 structures is the hexagonal 1H phase [1, 2, 7]. Additional polymorphic forms of 2D WS_2 crystals are the metastable 1T phase with octahedral coordination, and the distorted 1T′ phase [63, 65]. First-principles calculations, using the SCAN metaGGA approach, provided the phase stabilities of the Mo- and W-TMDC monolayers [63]. In addition to a minor stabilization of the H phase over the T phase at the monolayer level, further stabilization of the adsorbed T structure is possible, e.g., by interfacial interaction on a substrate.

For 2D WSe_2 with the most stable 1H symmetry, the lattice parameters are $a = 0.332$ nm, $d_{W\text{–}Se} = 0.255$ nm, $h_{SeSe} = 0.335$ nm, $\theta_{SeWSe} = 82.44°$ and $\theta_{WSeW} = 81.30°$ (PBE) [2]. Note that these lattice parameters are essentially identical with those of the $MoSe_2$ monolayer, due to the identical van der Waals diameters of Mo and W atoms. The LDA approach delivers the lattice parameters $a = 0.325$ nm, $d_{W\text{–}Se} = 0.251$ nm, $h_{SeSe} = 0.334$ nm, and $\theta_{SeWSe} = 83.24$ °C [1]. Adding the van der Waals diameter of Se atoms of $d_w = 0.380$ nm and their distance h_{SeSe}, yields an estimate of the monolayer thickness of $d_{w+h} = 0.715$ nm. An extensive theoretical comparison of the structural parameters of molybdenum and tungsten dichalcogenides calculated by the PBE, LDA, and optB88 approximations yields nearly identical values, confirming the accuracy of calculated lattice parameters [66]. The 2H structure is also the thermodynamically most stable form of few-layer sheets with calculated formation energies of $E_f = -1.09$ eV/atom (PBE) and -1.38 eV/atom (LDA) [62]. The phonon dispersion curves indicate dynamic stability of the 2H structure of all tungsten dichalcogenides [2, 6].

16.3.3.3 Mechanics

The two high-symmetry directions of the hexagonal 1H structure of WSe_2 extend along the M–X bonds (armchair direction) and perpendicular to the M–X bonds (zigzag direction). Determination of the elastic modulus from the linear part of the slope of the stress-strain curve indicates a small anisotropy in the stiffness. Taking the estimated thickness of WSe_2 monolayers of $d_{w+h} = 0.715$ nm, results in the stiffness values $E_{ac} = 141$ N/m and $E_{zz} = 139$ N/m, in 2D units [68]. This is in reasonable agreement with the stiffness of $E = 130$ N/m [1]. For suspended multilayers of WSe_2, AFM indentation experiments delivered as Young's modulus 167.3 GPa, which amounts to a lower stiffness of 120 N/m [143]. Since the measured stiffness was independent of the number of layers, a reduction in stiffness by interlayer sliding can be excluded. Thus, the lower stiffness may be due to defects in the multilayer assembly. For comparison, the ElasTool toolkit values are $E = 127.7$ N/m and $\nu = 0.17$ for 2H WSe_2 and $E = 113.5$ N/m and $\nu = -0.04$ for 1T WSe_2 [8].

The ultimate strength exhibits a large anisotropy of $\sigma_{ac} = 17.7$ N/m and $\sigma_{zz} = 10.8$ N/m [68]. Correspondingly, the ultimate strains exhibit a large anisotropy of $\varepsilon_{ac} = 0.30$ and $\varepsilon_{zz} = 0.17$, which are the indicators of the structural stress response. As expected for a defective WSe_2 multilayer, the measured ultimate strength and ultimate strain values are lower than their ideal values, which is indeed the case for $\sigma = \sim 8.9$ N/m and $\varepsilon = 0.073$ [143].

A prerequisite for the application of TMDCs is their mechanical stability under various loading conditions. Extensive nanomechanical measurements and simulations revealed a large scatter in the fracture strength and fatigue lifetime of 2D WSe_2 (2D MoS_2), resembling common engineering ceramics such as SiN and SiC [144]. Thermal fluctuations and the presence of diverse types of defects explain the observed low reliability of these 2D materials.

16.3.3.4 Physics and Applications

As discussed above, TMDC monolayers form several stable phases such as 1H, 1T, 1T′, $1T_d$ with similar stabilities. Moreover, the formation of CDWs is observed, which attract intense interest in fundamental studies and practical applications. Potential applications include piezoelectricity, energy storage, catalysis, electronics, optoelectronics, spintronics, and energy harvesting [145]. Controlling phase transitions by strain engineering, electron doping, alloying, electric fields, and chemical adsorption provides another degree of freedom for novel properties and special applications.

Both, first-principles calculations and experimental studies provided atomistic insight into the catalytic function of in-plane intrinsic defects and of reconstructed edge atoms of 2H-WSe_2 sheets [146]. The calculations revealed that in the process of photocatalytic reduction of CO_2 to CH_4, reconstructed and imperfect edge configurations can adsorb CO_2 in linear and bent forms more efficiently than in-plane point defects. The experiments revealed that the internal quantum efficiency (IQE) of the CO_2-to-CH_4 transformation was a reciprocal function of the perimeter of monolayer flakes. The consumed electron rate per edge atom was two orders of magnitude greater than that of in-plane defects, as confirmed by nanoscale redox mapping at the WSe_2 monolayer-liquid interface. These results prove that the edge is the most preferred location for charge transfer.

Oriented adsorption of NO_2 occurs on perfect and defective WSe_2 monolayers, as revealed by a DFT study by large-scale sampling. The study disclosed atomic details of the adsorption behavior on a perfect monolayer and on typical defects such as selenium vacancies, tungsten vacancies, and antisite defects of selenium on a tungsten site [147]. No modification of the anisotropic adsorption behavior by typical defects could be observed. The out-of-plane polarization is believed to play a key role in the vertical and oriented adsorption along the two directions of the W–Se bond and the Se–W–Se trough. As an estimate of detection sensitivity, MD simulations provided moderate adsorption energies of 0.35, 0.42, 0.95, and 0.78 eV for NO_2 on perfect monolayers, for selenium vacancies, for tungsten vacancies, and selenium on tungsten sites, and for the specific charge transfer processes the values $\delta q = 0.205$ e, 0.208 e, 0.164 e, and 0.160 e, respectively.

16.3.3.5 Synthesis

To ultimately apply CVD to grow high-quality and large-area monolayers, the deposition temperature must be compatible with current silicon technology. As demonstrated for WSe_2, the use of moisture-assisted defective tungsten oxide (WO_3) powder as a precursor provides access to low-temperature growth of TMDCs (~550 °C) [149]. The reason for the decrease in the growth temperature was the formation of defects on the precursor crystal by moisture, as indicated by DFT calculations.

Layer-by-layer growth of scalable WSe_2 10-layer sheets is possible by MBE on insulating $Al_2O_3(001)$ (sapphire) substrates, characterized by weak van der Waals forces between neighboring layers [150]. Recording the intensity of RHEED signals during growth signifies layer-by-layer growth without island formation. A film thickness of 6.4 nm, calculated from the Laue oscillation period, is consistent with a thickness estimate of 10 monolayers. A fabricated top-gate electric double-layer transistor (EDLT), constructed using an ionic liquid, was employed to evaluate the transport properties of the sheet.

Synthesis of 1T′-WSe_2 monolayers via a heating process using as the only precursor tetraethylammonium tetraselenotungstate [$(Et_4N)_2WSe_4$], allowed the fabrication of metallic 1T′-WSe_2 monolayers [148]. The triethylammonium produced after the reaction is an electron donor that yields negatively charged WSe_2.

Stabilization by triethylammonium cations as intercalants between the layers induces the formation of metastable 1T′-WSe_2 monolayers. Interestingly, oxygen-containing crystalline substrates with a lower lattice mismatch produced higher purity 1T′-WSe_2 sheets than amorphous substrates owing to uniform triethylammonium intercalation.

16.3.4 WTe_2 Monolayer

16.3.4.1 Chemistry and 2D Bonding

The electronic configurations of W atoms (Xe|5$d^4$6s^2) and Te atoms (Te|4d^{10}5s^2p^4) with the typical valence electrons of this group are responsible for the formation of stable monolayers with three sublayers, as discussed for the other tungsten dichalcogenides. The electron charge accumulation between the W and Te atoms generates predominantly covalent W–Te bonds. Note that in this case, the Pauling scale gives a slightly larger electronegativity for W (2.36) than for Te (2.10). Nevertheless, a Bader charge transfer analysis of the most stable hexagonal 2H phase yields a substantial charge transfer of $\delta q = 0.58$ e from each W atom to two Te atoms, causing a small ionic contribution to bonding of 1.68% [2]. The primarily covalent bonding of the H structure generates cohesive energies of $E_c = -5.68$ eV/atom [1] (LDA) and − 4.54 eV/atom [2] (PBE). These binding energies agree quite well with previous results of $E_c = -5.63$ eV/atom (LDA) and − 4.46 eV/atom (PBE) [62]. An extreme cohesive energy of $E_c = -6.49$ eV/atom has been reported for the 1T′-WTe_2 monolayer [63]. Owing to weaker bonds in 2D WTe_2, WTe_2 is the monolayer with the lowest mechanical performance of the W chalcogenides. Note that the cohesive energies calculated for the metastable H phase may also be useful estimates for the most stable $1T_d$ phase because the differences between these phases are expected to be small.

16.3.4.2 2D Structures

From the studied polymorphic 2D-WTe_2 crystal structures, the $1T_d$ phase with octahedral coordination is the most stable phase with a Te–W–Te triple-sublayer arrangement. At the monolayer level, the $1T_d$ phase is only slightly favored over the 1H phase by 99 meV per formula unit; therefore, the 1H structure may be easily stabilized, e.g., by the strain exerted by a substrate [151]. Note that the $1T_d$-WTe_2 monolayer is a semimetal as the bulk phase, where the valence band and conduction band overlap at the center of the hexagonal phase in the Brillouin zone [151].

Due to the inherent instability of the centrosymmetric 1T-WSe_2 structure as a free-standing monolayer, the structure transforms to the distorted $1T_d$ polytype with strongly distorted octahedral coordination around the metal atom [151]. The W atoms move toward each other and form a slightly buckled W–W chain along the armchair axis. The available lattice constants of the $1T_d$ phase are $a = 0.350$ nm, $b = 0.630$ nm,

$d_{W\text{–}Te} = 0.273$, and $d_{W\text{–}W} = 0.285$ nm. Note that the latter is the bond length of the tungsten atoms in the zigzag chain in the distorted $1T_d$ phase. The T_d structure is a distorted version of the centrosmmetric 1T structure with inversion symmetry.

The 1T′ phase of WTe_2 derives from the instability of the 1T phase under free-standing conditions and is a distorted version of the 1T structure [152]. The three-layer structure consists of buckled sublayers of Te atoms with central W atoms residing at the distorted octahedral sites. The lattice parameters of the 1T′-WTe_2 monolayer are a = 0.350 nm, b = 0.633 nm, and $d_{W\text{–}W} = 0.288$ nm, which are essentially identical to those of the T_d phase. This resemblance is not surprising considering the tight similarity of the bond arrangements of the two structures. Since the interlayer binding energy of a single layer to the bulk phase is ~28 meV/A^2, which is comparable to that of MoS_2 (26 meV/A^2), mechanical exfoliation should be possible.

For the metastable 2H structure, the reported lattice parameters are a = 0.355 nm, $d_{W\text{–}Te} = 0.274$ nm, $h_{TeTe} = 0.362$ nm, and $\theta_{TeWTe} = 82.89$ °C (PBE) [2], which are in very good agreement with the results of other publications [68]. Adding the van der Waals diameter of the Te atom of $d_w = 0.420$ nm and their distance in the H phase of h_{TeTe}, yields an estimate of the monolayer thickness of $d_{w+h} = 0.782$ nm. The similarity of the structural parameters of molybdenum and tungsten dichalcogenides disclosed by comparing PBE, LDA, and optB88 calculations can be explained by the nearly identical sizes of the atoms, e.g., van der Waals diameters and nearly identical electronegativities [66].

16.3.4.3 Mechanics

Determination of the elastic modulus of the hexagonal 1H structure of WTe_2 along the two high-symmetry directions from the linear part of the slope of the stress–strain curve indicates a small anisotropy in the stiffness. Taking as the thickness of the monolayer $d_{w+h} = 0.782$ nm (layer separation in the bulk t = 0.70 nm [151]), results in the stiffness values $E_{ac} = 107$ N/m and $E_{zz} = 106$ N/m in 2D units [68]. Other calculations delivered a larger anisotropy with stiffnesses of $E_{ac} = 71$ N/m and $E_{zz} = 105$ N/m and Poisson's ratios of $\nu_{ac} = 0.26$ and $\nu_{zz} = 0.38$ for the $1T_d$-WTe_2 phase. For the strongly anisotropic 1T′ phase of WTe_2 monolayers, the anisotropic elastic moduli are $E_{ac} = 75.8$ N/m and $E_{zz} = 106.6$ N/m, and the Poisson ratios are $\nu_{ac} = 0.24$ and $\nu_{zz} = 0.34$ [152]. The corresponding values for the linear mechanical properties of the 1T′ phase are E = 86.8 N/m, $\sigma_{in} = 0.35$ and $\sigma_{out} = 0.60$ [63]. For comparison, the ElasTool toolkit values are E = 97.7 N/m and $\nu = 0.14$ for 2H WTe_2 and E = 42.2 N/m and $\nu = -0.11$ for 1T WTe_2 [8].

The ultimate strength exhibits a large anisotropy of $\sigma_{ac} = 14.6$ N/m and $\sigma_{zz} = 7.3$ N/m [68]. Correspondingly, a large anisotropy exists for the ultimate strains of $\varepsilon_{ac} = 0.32$ and $\varepsilon_{zz} = 0.15$. As mentioned above, these medium stiffness and strength values are a consequence of weaker covalent and ionic bonding.

16.3.4.4 Physics and Applications

The most stable 1T′-WTe_2 monolayer is one of the first 2D crystals with an experimentally verified quantum spin Hall phase, as theoretically predicted for the 1T′ phase of TMDCs, which form the distorted (2×1) superstructure with a doubled period [153]. The monolayer has a robust bond inversion of approximately 1 eV and an indirect band gap above 300 K, whereas the signatures of a quantum-spin Hall insulator (QSHI) could be detected only at 100 K. Calculations based on nonperturbative methods predict a weak influence of thermal expansion that tends to increase the indirect band gap, and a slight reduction of band inversion with increasing temperature. Accordingly, the topological phase and finite band gap should exist above room temperature.

Interest in the 1T′-WTe monolayers is also rapidly growing due to predictions of possible band structure manipulations. As mentioned previously, the 1T′ phase is a semimetal with QSH edge states entangled with the bulk states. However, first-principles DFT calculations indicate that a transition from the semimetallic 1T′ state to a stable QSHI can be induced by uniaxial tensile (compressive) strains along the a and b axes of the 1T′ phase if the strain is not reversing the band order between the W-*d* orbitals and the Te-*p* orbitals [152]. Such a transition opens the door to topologically nontrivial monolayers and practical applications in strain-tunable topological quantum electronics. Moreover, it is the first material with extremely large and nonsaturating magnetoresistance at low temperatures [152].

A first-principles study investigated the phonon transport properties in distinct phases of 2D WTe_2 (2D $MoTe_2$) [155]. In addition to the lattice thermal conductivity of the H and T′ phase the phonon dispersion, phonon heat capacity, phonon group velocity, phonon lifetime, and phonon mean free paths were calculated. The highly anisotropic maximal lattice thermal conductivity of T′-WTe_2 monolayers is reduced by ~ 78% at 300 K in comparison to the isotropic value of the H phase. This reduction, which is of interest for thermoelectric applications, has been attributed to increased phonon scattering and loss of phonon band gap.

The semimetallic character of $1T_d$-WTe_2 sheets depends on the SOC, and the structural stability originates from the metallic 1T structure [151]. Electrical transport measurements indicate a metal-to-insulator transition in electrical conductivity and the change in the sign of the Seebeck coefficient from negative (n-type) to positive (p-type) at approximately 373 K are remarkable. An analysis of the Raman signals revealed a stiffening of most Raman bands with decreasing thickness and a softening of the bands of bulk and few-layer systems with increasing temperature.

The anisotropic $1T_d$-WTe_2 monolayers possess thermal conductivities along the principal lattice directions of $\kappa_{ac} = 14.1$ W m^{-1} K^{-1} and $\kappa_{zz} = 7.55$ W m^{-1} K^{-1} at room temperature. The anisotropy originates from the in-plane linear acoustic phonon branches [154]. Such a low value reduces the figure-of-merit and thus qualifies the material for thermoelectric applications. A first-principles study confirmed the enormous anisotropy of the thermal conductivity of WTe_2 monolayers at $\kappa_{ac} =$ 20 W m^{-1} K^{-1} and $\kappa_{zz} = 9$ W m^{-1} K^{-1} [155]. The size effect can persist up to 10 μm, and the anisotropy decreases with decreasing sample size owing to the suppression of low-frequency anisotropic phonons.

16.3.4.5 Synthesis

Synthesis by chemical vapor transport employing polycrystalline WTe_2 powder and $TeCl_4$ as a transport agent, results in the formation of single crystals of $1T_d$ WTe_2 [151]. For the preparation of samples used for the AFM and TEM analysis, the simplest and most common method of cleaving few-layer $1T_d$-WTe_2 flakes by mechanical exfoliation from the layered bulk crystal was employed, using the scotch tape separation method. Usually, a substrate provides additional stabilization of mechanically exfoliated few-layer sheets, e.g., the Si/SiO_2 (300 nm) wafer employed in these experiments [151].

Large-scale WTe_2 ($MoTe_2$) monolayers and few-layer sheets with a lateral size > 300 μm were directly synthesized by CVD at large scale. A mixed compound consisting of compound metal oxide, metal chloride, and Te with weight ratio 1:1:1 was used as source of W (Mo) to induce the reaction of Te with the metal source [156]. Raman spectroscopy provided information on the quality of the deposit, and atom-resolved STEM provided the atomic structure and allowed the identification of quasi-one-dimensional tungsten-tellurium zigzag chains along the armchair direction. Electric-transport measurements at low temperatures revealed a novel semimetal-to-insulator transition in WTe_2 layers and an enhanced superconductivity in few-layer $MoTe_2$ sheets.

Reproducible CVD experiments delivered large-area single-crystal $1T'$-WTe_2 monolayer flakes with a surface coverage of ~ 20% [157]. Without a CVD-grown graphene protection layer, the flakes degraded within minutes under ambient conditions. The chemical composition was determined by XPS, the monolayer height was measured by AFM, the vibrational modes were analyzed by Raman spectroscopy, and the structural phase was clarified by aberration-corrected scanning transmission electron microscopy (ACSTEM). Measurements of the electrical properties revealed metallic conduction at low temperatures and a weak antilocalization effect, indicating strong SOC.

A recent report demonstrated probing during CVD growth with atomic resolution under a controlled inert atmosphere, transfer to STEM imaging, and characterization of the large-scale intrinsic structure of $1T'$-WTe_2 monolayers [158]. Large-scale mapping revealed the presence of collective lattice distortions such as anisotropic rippling. This effect modulates and contributes to the intrinsic defect distribution of Te vacancies, as demonstrated by DFT calculations.

16.4 Summary of Group–VB Dichalcogenide Monolayers

16.4.1 Binding–Structure Relationships

Figure 16.14 shows the cohesive energy versus bond length of the short M–X bonds for the MX_2 (M: Cr, Mo, and W; X: O. S, Se, and Te) VB–VIA monolayers. Note that the M–M separation is much greater than that of the metal-chalcogen bonds. The plot shows the binding energies calculated by the DFT-GGA-PBE approach that are significantly lower than the LDA energies, as described in the text.

The plot documents the inverse behavior between the cohesive energy, which is a measure of the binding strength, and bond length of the covalent-ionic M–X bonds, as predicted by tight-binding theory. The theoretical results presently available for the three transition metal–chalcogen groups follow a regular inverse dependence and allow the identification of strongly deviating results. Note that the nearly parallel decrease in the cohesive energy of the three groups with increasing bond length is caused by an increase in the atom size and changes in the electronic structure of the chalcogens. The diagram demonstrates the surprisingly large cohesion and critical strengths of the heavy tellurides, which are of particular interest for future applications. The metal-oxide monolayers with a substantial ionic contribution to bonding and good mechanical performance are not yet systematically studied. For these monolayers with covalent-ionic bonding the LDA approximation suggests cohesion energies comparable and greater than that of graphene.

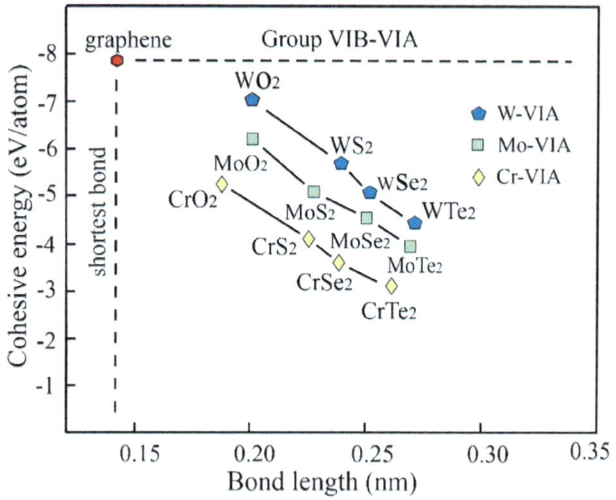

Fig. 16.14 Plot of the cohesive energies versus bond lengths of the M–X bonds of group VIB–VIA monolayers

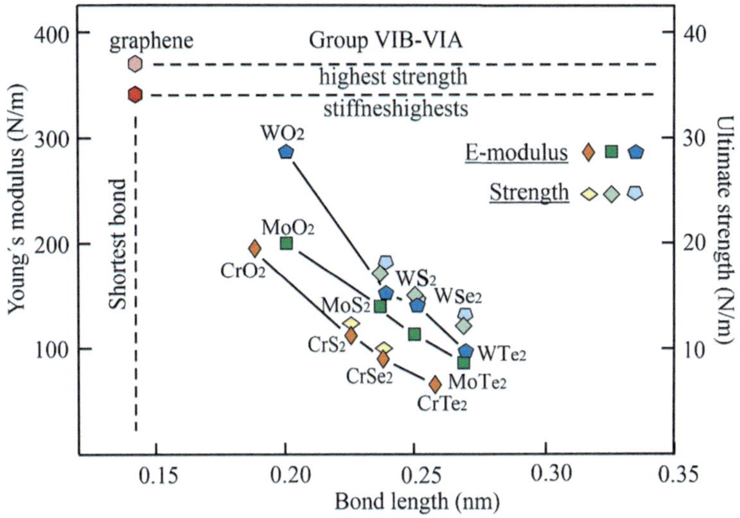

Fig. 16.15 Young's moduli and ultimate strengths versus M–X bond lengths of the group VIB–VIA monolayers

16.4.2 Mechanics–Structure Relationships

Figure 16.15 shows the calculated Young's moduli and the ultimate strengths as a function of the M–X bond length. Note that the three-layer assemblies with their strong covalent-ionic bonding show a sensitive dependence of the mechanical stiffness and strength on bond length, as previously described for single-layer structures.

The few calculations of ultimate strengths presently available follow only for the chromium group Griffith's rule, whereas the strengths are much greater than one-tenth of the elastic modulus for the molybdenum and tungsten groups. Assuming a higher accuracy for the calculated Young's moduli as for nonlinear mechanical properties, the ultimate strengths seem to be greater than expected for a perfect structure. Of course, it cannot be excluded that the simple rule is not applicable.

References

1. C. Ataca, H. Şahin, S. Ciraci, J. Phys. Chem. C **116**, 8983 (2012)
2. G. Özbal, R.T. Senger, C. Sevik, H. Sevinçli, Phys. Rev. B **100**, 085415 (2019)
3. A.H. Tobar, J.F. Murillo, C.O. López, J.A.R. Martínez, M.J. Espitia, Comput. Condens. Matter **25**, e00498 (2020)
4. S. He, P. Zhou, Y. Yang, W. Wang, L.Z. Sun, Nanoscale Adv. **3**, 3093 (2021)
5. M. Fukuda, J. Zhang, Y.-T. Lee, T. Ozaki, Mater. Adv. **2**, 4392 (2021)
6. F.A. Rasmussen, K.S. Thygesen, J. Phys. Chem. C **119**, 13169 (2015)
7. D. Çakir, F.M. Peeters, C. Sevik, Appl. Phys. Lett. **104**, 203110 (2014)

8. S.M. Kastuar, C.E. Ekuma, Z.-L. Liu, Sci. Rep. **12**, 3776 (2022)
9. D.-T. Nguyen, Elastic Mechanical Properties of Transition Metal Dichalcogenides Monolayer Using Atomic Finite Element Method, in The AUN/SEED-Net Joint Regional Conference in Transportation, Energy, and Mechanical Manufacturing Engineering, pp. 687–700 (2022)
10. X. Deng, Z. Li, Phys. Chem. Chem. Phys. **22**, 17038 (2020)
11. A. Kakekhani, S. Ismail-Beigi, ACS Catal. **5**, 4537 (2015)
12. W. Ouyang, B. Shi, T. Su, X. Cheng, H. Gao et al., J. Phys. Condens. Matter **35**, 305001 (2023)
13. W.J. DeSisto, P.R. Broussard, T.F. Ambrose, B.E. Nadgorny, M.S. Osofsky, Appl. Phys. Lett. **76**, 3789 (2000)
14. N.F. Heinig, H. Jalili, K.T. Leung, Appl. Phys. Lett. **91**, 253102 (2007)
15. P.G. Ivanov, K.M. Bussmann, J. Appl. Phys. **105**, 07B107 (2009)
16. Z. Yi, L. Lin, Y. Chang, X. Bao, PNAS **119**, e2120716119 (2022)
17. J.J. Attema, M.A. Uijttewaal, G.A. de Wijs, R.A. de Groot, Phys. Rev. B **77**, 165109 (2008)
18. S.S. Batsanov, Inorg. Mater. **37**, 871 (2001)
19. S.B. Sharma, R. Paudel, R. Adhikari, G.C. Kaphle, D. Paudyal, Physica **146**, 115517 (2023)
20. K. Chen, J. Deng, Y. Yan, Q. Shi, T. Chang et al., NPJ Comput. Mater. **7**, 79 (2021)
21. H.Y. Lv, W.J. Lu, D.F. Shao, Y. Liu, Y.P. Sun, Phys. Rev. B **92**, 214419 (2015)
22. P. Hess, Phys. Chem. Chem. Phys. **20**, 7604 (2018)
23. S.-B. Chen, Z.-Y. Zeng, X.-R. Chen, X.-X. Yao, J. Appl. Phys. **128**, 12511 (2020)
24. Y. Qu, H. Pan, C.T. Kwok, Sci. Rep. **6**, 34186 (2016)
25. S.-Y. Yang, D.-R. Shi, T. Wang, X.-Y. Yue, L. Zheng et al., J. Mater. Chem. A **8**, 25739 (2020)
26. F. Han, X. Yan, A. Bergara, W. Li, H. Yu et al., Phys. Chem. Chem. Phys. **25**, 29672 (2023)
27. S. Chen, Z. Zeng, B. Lv, S. Guo, X. Chen et al., Phys. Rev. B **106**, 115307 (2022)
28. C. Yu, X. Chen, C. Wang, Z. Wang, J. Phys. Chem. Solids **135**, 109081 (2019)
29. S.-D. Guo, X.-S. Guo, Y.-T. Zhu, Y.-S. Ang, Appl. Phys. Lett. **121**, 062403 (2022)
30. S. Tang, D. Wan, S. Bai, S. Fu, X. Wang et al., Phys. Chem. Chem. Phys. **25**, 22401 (2023)
31. M.R. Habib, S. Wang, W. Wang, H. Xiao, S.M. Obaidula et al., Nanoscale **11**, 20123 (2019)
32. H. Xiao, W. Zhuang, L. Loh, T. Liang, A. Gayen et al., Adv. Mater. Interfaces **9**, 2201353 (2022)
33. P. Hess, ChemPhysChem **23**, e202100880 (2022)
34. L. Wu, L. Zhou, X. Zhou, C. Wang, W. Ji, Phys. Rev. B **106**, L081401 (2022)
35. Z. Wang, M. Zhang, Y. Ge, W. Wan, Y. Liu, Results Phys. **51**, 106687 (2023)
36. M. Alsubaie, C. Tang, D. Wijethunge, D. Qi, A. Du, A.C.S. Appl, Electron. Mater. **4**, 3240 (2022)
37. J. Chen, L. Zhou, N. Liu, J. Qiao, X. Zhou et al. arXiv:2204.01400 (2022)
38. M. Liu, Y.L. Huang, J. Gou, Q. Liang, R. Chua et al., J. Phys. Chem. Lett. **12**, 7752 (2021)
39. B. Li, Z. Wan, C. Wang, P. Chen, B. Huang et al., Nat. Mater. **20**, 818 (2021)
40. Y. Liu, S. Kwon, G.J. de Coster, R.K. Lake, M.R. Neupane, Phys. Rev. Mater. **6**, 084004 (2022)
41. M.M. Alyörük, Y. Aierken, D. Çakir, F.M. Peeters, C. Sevik, J. Phys. Chem. C **119**, 23231 (2015)
42. X. Sun, W. Li, X. Wang, Q. Sui, T. Zhang, Nano Res. **13**, 3358 (2020)
43. N.D. Mermin, H. Wagner, Phys. Rev. Lett. **17**, 1133 (1966)
44. A.O. Fumega, J. Phillips, V. Pardo, J. Phys. Chem. C **124**, 21047 (2020)
45. D.C. Freitas, R. Weht, A. Sulpice, G. Remenyi, P. Strobel et al., J. Phys. Condens. Matter **27**, 176002 (2015)
46. X. Zhang, Q. Lu, W. Liu, W. Niu, J. Sun et al., Nat. Commun. **12**, 2492 (2021)
47. Y. Sun, P. Yan, J. Ning, X. Zhang, Y. Zhao et al., AIP Adv. **11**, 035138 (2021)
48. J.-J. Xian, C. Wang, J.-H. Nie, R. Li, M. Han et al., Nat. Commun. **13**, 257 (2022)
49. F. Yu, Y. Yin, G. Liu, Q. Tian, Q. Meng et al., App. Phys. Lett. **120**, 261602 (2022)
50. K. Lasek, P.M. Coelho, P. Gargiani, M. Valvidares, K. Mohseni et al., Appl. Phys. Rev. **9**, 011409 (2022)
51. F. Ersan, S. Sarikurt, Phys. Chem. Chem. Phys. **21**, 19904 (2019)

52. K.C. Santosh, R.L. Roberto, M.W. Robert, C. Kyeongjae, J. Appl. Phys. **117**, 135301 (2015)
53. X. Hu, K. Liu, Y. Cai, S.-Q. Zang, T. Zhai, Small Sci. **2**, 2200008 (2022)
54. H. Wu, X. Zhou, J. Li, X. Li, B. Li et al., Small 1802276 (2018)
55. M. Yang, Z. Ye, M.A. Iqbal, H. Liang, Y.-J. Zeng, Nanoscale **14**, 9576 (2022)
56. J. Li, J. Yin, X. Li, J. Zhou, W. Guo, Mater. Lett. **174**, 188 (2016)
57. E. Pu, D. Liu, P. Ren, W. Zhou, D. Tang et al., AIP Adv. **7**, 025015 (2017)
58. N. Wazir, C. Ding, X. Wang, X. Ye, X. Lingling et al., Nanoscale Res. Lett. **15**, 156 (2020)
59. X. Feng, D. Xin, Z. Zhang, J. Liu, T. Ning et al., Opt. Laser Technol. **157**, 108632 (2023)
60. L. Fang, H. Liu, W. Guan, B. Zheng, J. Liang et al., Sci. China Mater. **66**, 1504 (2023)
61. C. Ataca, E. Aktürk, H. Şahin, S. Ciraci, J. Phys. Chem. C **115**, 3934 (2011)
62. Y. Ding, Y. Wang, J. Ni, L. Shi, S. Shi et al., Physica B **406**, 2254 (2011)
63. N.K. Nepal, L. Yu, Q. Yan, A. Ruzsinszky, Phys. Rev. Mater. **3**, 073601 (2019)
64. M. Kan, J.Y. Wang, X.W. Li, S.H. Zhang, Y.W. Li et al., J. Phys. Chem. C **118**, 1515 (2014)
65. F. Güller, A.M. Llois, J. Goniakowski, C. Noguera, Phys. Rev. B **91**, 075407 (2015)
66. Z. Fan, Z. Wei-Bing, T. Bi-Yu, Chin. Phys. B **24**, 097103 (2015)
67. G. Eda, T. Fujita, H. Yamaguchi, D. Voiry, M. Chen et al., ACS Nano **6**, 7311 (2012)
68. J. Li, N.V. Medhekar, V.B. Shenoy, J. Phys. Chem. C **117**, 15842 (2013)
69. J. Kang, H. Sahin, F.M. Peeters, Phys. Chem. Chem. Phys. **17**, 27742 (2015)
70. K. Liu, Q. Yan, M. Chen, W. Fan, Y. Sun et al., Nano Lett. **14**, 5097 (2014)
71. S. Bertolazzi, J. Brivio, A. Kis, ACS Nano **5**, 9703 (2011)
72. A. Castellanos-Gomez, M. Poot, G.A. Steele, H.S.J. van der Zant, N. Agrait et al., Nanoscale Res. Lett. **7**, 233 (2012)
73. Y. Zhu, X. Cao, Y. Tan, Y. Wang, J. Hu et al., Coatings **13**, 283 (2023)
74. T. Li, Phys. Rev. B **85**, 235407 (2012)
75. R.C. Cooper, C. Lee, C.A. Marianetti, X. Wei, J. Hone et al., Phys. Rev. B **87**, 035423 (2013)
76. J.-W. Jiang, Z. Qi, H.S. Park, T. Rabczuk, Nanotechnology **24**, 435705 (2013)
77. K. Lai, W.-B. Zhang, F. Zhou, F. Zeng, B.-Y. Tang, J. Phys. D Appl. Phys. **49**, 185301 (2016)
78. S. Xiong, G. Cao, Nanotechnology **27**, 105701 (2016)
79. J. Kang, S. Tongay, J. Zhou, J. Li, J. Wu, Appl. Phys. Lett. **102**, 012111 (2013)
80. W. Zhou, H. Gong, X. Jin, Y. Chen, H. Li et al., Front. Phys. **10**, 842789 (2022)
81. R. Sukanya, D.C. da Silva Alves, C.B. Breslin, J. Electrochem. Soc. **169**, 064504 (2022)
82. J. Zhang, S. Jia, I. Kholmanov, L. Dong, D. Er et al., ACS Nano **11**, 8192 (2017)
83. B. Radisavljevic, A. Radenovic, J. Brivio, V. Giacometti, A. Kis, Nat. Nanotech. **6**, 147 (2011)
84. D. Lembke, S. Bertolazzi, A. Kis, Acc. Chem. Res. **48**, 100 (2015)
85. H. Liu, N. Han, J. Zhao, CSC Adv. **5**, 17572 (2015)
86. J. Gao, B. Li, J. Tan, P. Chow, T.-M. Lu et al., ACS Nano **10**, 2628 (2016)
87. P. Joensen, R.F. Frindt, S.R. Morrison, Mat. Res. Bull. **21**, 457 (1986)
88. W. Choi, N. Choudhary, G.H. Han, J. Park, D. Akinwande et al., Mater. Today **20**, 116 (2017)
89. S. Zhao, L. Wang, L. Fu, iScience **20**, 527 (2019)
90. X. Huang, Z. Zeng, H. Zhang, Chem. Soc. Rev. **42**, 1934 (2013)
91. R. Yang, L. Mei, Q. Zhang, Y. Fan, H.S. Shin et al., Nat. Protoc. **17**, 358 (2022)
92. J. You, D. Hossain, Z. Luo, Nano Convergence **5**, 26 (2018)
93. J. Wang, T. Li, Q. Wang, R. Shi, N. Wang et al., Mater. Today Adv. **8**, 100098 (2020)
94. T. Kang, T.W. Tang, B. Pan, H. Liu, K. Zhang et al., ACS Mater. Au **2**, 665 (2022)
95. C. Patel, R. Singh, M. Dubey, S.K. Pandey, S.N. Upadhyay, A.C.S. Appl, Nano Mater. **5**, 9415 (2022)
96. Y. Xia, X. Chen, J. Wei, S. Wang, S. Chen et al., Nat. Mater. **22**, 1324 (2023)
97. G.A. Ermolaev, M.A. El-Sayed, D.I. Jakubovsky, K.V. Voronin, R.I. Romanov et al., Nanomaterials **11**, 1411 (2021)
98. P. Yang, F. Liu, X. Li, J. Hu, F. Zhou et al., Small Methods **7**, 2300165 (2023)
99. S. El Kazzi, W. Mortelmans, T. Nuytten, J. Meersschaut, P. Carolan et al., J. Appl. Phys. **123**, 135702 (2018)
100. Z. Wei, Q. Wang, L. Li, R. Yang, G. Zhang, Nano Res. **14**, 1598 (2021)
101. Z. Liu, K. Nie, X. Qu, X. Li, B. Li et al., J. Am. Chem. Soc. **144**, 4863 (2022)

102. T.H. Chaudhury, X. Zhang, Z.Y. Al Balushi, M. Chubarov, J.M. Redwing, Ann. Rev. Mater. Sci. **50**, 155 (2020)
103. L. Banu, V. Veerapandy, P. Vajeeston, Mater. Today: Proceed. **64**, 1808 (2022)
104. Y. Zhao, Y. Yan, J.-M. Lee, Nanoscale **14**, 1075 (2022)
105. D.M. Guzman, A. Strachan, J. Appl. Phys. **115**, 243701 (2014)
106. Y. Yang, X. Li, M. Wen, E. Hacopian, W. Chen et al., Adv. Mater. **29**, 1604201 (2017)
107. H. Sun, P. Agrawal, C.V. Singh, Mater. Adv. **2**, 6631 (2021)
108. W. Ai, L. Kou, X. Hu, Y. Wang, A.V. Krasheninnikov et al., J. Phys. Condens. Matter **31**, 445301 (2019)
109. N. Azam, M.G. Boebinger, S. Jaiswal, R.R. Unocic, P. Fathi-Hafshejani et al., ACS Appl. Nano Mater. **5**, 9129 (2022)
110. J. Li, W. Yan, Y. Lv, J. Leng, D. Zhang, RSC Adv. **10**, 1580 (2020)
111. D. Zhang, C. Wen, J.B. Mcclimon, P.M. Das, Q. Zhang et al., Adv. Electron. Mater. **7**, 2001219 (2021)
112. H. Yan, T. Yu, H. Li, Z. Li, H. Tang et al., Nanoscale **13**, 8922 (2021)
113. C. Gonzáles, J.P.B. Silva, A.S. Viana, K. Gwozdz, O. Conde, Appl. Surf. Sci. **605**, 154742 (2022)
114. Y. Huang, Y. Wei, Y. Li, C. Hu, W. Shen et al., Appl. Surf. Sci. **628**, 157352 (2023)
115. Y. Wei, C. Hu, Y. Li, X. Hu, K. Yu et al., Nanotechnology **31**, 315710 (2020)
116. Y. Wu, S.-E. Wu, J. Hei, L. Zeng, P. Lin et al., Nano Res. **16**, 11422 (2023)
117. D.K. Singh, G. Gupta, Mater. Adv. **3**, 6142 (2022)
118. J. Guo, K. Liu, Nanomaterials **12**, 110 (2022)
119. R. Zhou, J. Wu, Y. Chen, L. Xie, Chin. J. Chem. **40**, 989 (2022)
120. X. Xu, S. Chen, S. Liu, X. Chen, W. Xu, J. Am. Chem. Soc. **141**, 2128 (2019)
121. Y. Sun, J. Pan, Z. Zhang, K. Zhang, J. Liang et al., Nano Lett. **19**, 761 (2019)
122. Y. Deng, X. Zhao, C. Zhu, P. Li, R. Duan et al., ACS Nano **15**, 12465 (2021)
123. Y. Qi, P.G. Naumov, M.N. Ali, C.R. Rajamathi, W. Schnelle et al., Nat. Commun. **7**, 11038 (2016)
124. B. Yang, C. Zhao, B. Xia, H. Ma, H. Chen et al., Quantum Front. **2**, 9 (2023)
125. S. Paul, S. Talukdar, R.S. Singh, S. Saha, Phys. Status Solidi **17**, 2200420 (2023)
126. X. Sun, X. Zhu, Y. Wang, Y. Li, Chin. J. Catal. **43**, 1520 (2022)
127. W. Zhuang, Z. Chen, X. Wang, Adv. Phys. **7**, 2034529 (2022)
128. S. Pace, L. Martini, D. Convertino, D.H. Keum, S. Forti et al., ACS Nano **15**, 4213 (2021)
129. K. Chen, Z. Chen, X. Wan, Z. Zheng, F. Xie et al., Adv. Mater. 1700704 (2017)
130. J. Chen, G. Wang, Y. Tang, H. Tian, J. Xu et al., ACS Nano **11**, 3282 (2017)
131. A. Ohtake, X. Yang, J. Nara, ANPJ 2D Mater. Appl. **6**, 35 (2022)
132. J.A. Reyes-Retana, F. Cervantes-Sodi, Sci. Rep. **6**, 24093 (2016)
133. Y. Guo, L. Ma, K. Mao, M. Ju, Y. Bai et al., Nanoscale Horiz. **4**, 592 (2019)
134. F.R. Negreiros, Th. Obermüller, M. Blatnik, M. Mohammadi, A. Fortunelli et al., J. Phys. Chem. C **123**, 27584 (2019)
135. M. Piao, J. Chu, X. Wang, Y. Chi, H. Zhang et al., Nanotechnology **29**, 025705 (2018)
136. A.H. Kumar, A.K. Pandey, J. Joardar, in Handbook of Advanced Ceramics and Composites, Eds. Y. Mahajan, R. Johnson, Springer Nature, pp. 1–47 (2021)
137. B.K. Miremadi, S.R. Morrison, J. Appl. Phys. **63**, 4970 (1988)
138. C. Cong, J. Shang, X. Wu, B. Cao, N. Peimyoo et al., Adv. Optical Mater. **2**, 131 (2014)
139. K. Kang, K. Godin, Y.D. Kim, S. Fu, W. Cha et al., Adv. Mater. 1603898 (2017)
140. M. Chubarov, T.H. Choudhury, D.R. Hickey, S. Bachu, T. Zhang et al., ACS Nano **15**, 2532 (2021)
141. M.M.S. Villamayor, S. Husain, R. Oropesa-Nuñez, F.O.L. Johansson, R. Lindblad et al., Nanoscale **14**, 6331 (2022)
142. Y.-C. Lin, C.-H. Yeh, H.-C. Lin, M.-D. Siao, Z. Liu et al., ACS Nano **12**, 12080 (2018)
143. F. Godel, V. Zatko, C. Carrétéro, A. Sander, M. Galbiati et al., ACS Appl. Nano Mater. **3**, 7908 (2020)
144. R. Zhang, V. Koutsos, R. Cheung, Appl. Phys. Lett. **108**, 042104 (2016)

145. T. Cui, S. Mukherjee, M. Onodera, G. Wang, B. Kumral et al., Matter **5**, 2975 (2022)
146. H. Chen, J. Zhang, D. Kan, J. He, M. Song et al., Crystals **12**, 1381 (2022)
147. M. Qorbani, A. Sabbah, Y.-R. Lai, S. Kholimatussadiah, S. Quadir et al., Nat. Commun. **13**, 1256 (2022)
148. L.M. Sassi, A. Krishnamoorthy, J.A. Hachtel, S. Susarla, A. Apte et al., 2D Mater. **9**, 045026 (2022)
149. F.H. Tian, Y. Zhang, J. Dong, L. Sun, X. Xue, Mater. Today Chem. **24**, 100850 (2022)
150. M. Nakano, Y. Wang, Y. Kashiwabara, H. Matsuoka, Y. Iwasa, Nano Lett. **17**, 5595 (2017)
151. Y.J. Park, H.-S. So, H. Hwang, D.S. Jeong, H.J. Lee et al., ACS Nano **16**, 11059 (2022)
152. M.K. Jana, A. Singh, D.J. Late, C.R. Rajamathi, K. Biswas et al., J. Phys. Condens. Matter **27**, 285401 (2015)
153. H. Xiang, B. Xu, J. Liu, Y. Xia, H. Lu et al., AIP Adv. **6**, 095005 (2016)
154. A. Marrazzo, Phys. Rev. Mater. **7**, L021201 (2023)
155. J. Shen, D. Han, B. Zhang, R. Cao, Y. Liu et al., Physica E **145**, 115509 (2023)
156. J. Ma, Y. Chen, Z. Han, W. Li, 2D Mater. **3**, 045010 (2016)
157. J. Zhou, F. Liu, J. Lin, X. Huang, J. Xia et al., Adv. Mater. **29**, 1603471 (2017)
158. C.H. Naylor, W.M. Parkin, Z. Gao, H. Kang, M. Noyan et al., 2D Mater. **4**, 021008 (2017)
159. K. Niu, M. Weng, S. Li, Z. Guo, G. Wang et al., Adv. Sci. **8**, 2101563 (2021)

Chapter 17
Group–XB Transition-Metal Dioxide and XB Dichalcogenide Monolayers

Abstract The group of XB transition-metal dioxide (TMDO) and XB transition-metal dichalcogenide (TMDC) monolayers with MX_2 stoichiometry consist of a central layer of transition-metal atoms, M = Ni, Pd, and Pt sandwiched between two layers of chalcogenide atoms, X = O, S, Se, and Te. Among the dynamically, thermally, and mechanically stable structures, the octahedral T phase usually forms the most stable structure. In addition to the less important metastable H phase, a pentagonal orthorhombic phase with lower symmetry and strongly anisotropic properties has attracted increasing amounts of attention from researchers. Nickel, an inexpensive group-XB transition metal with a similar *d*-electron configuration as the noble metals Pd and Pt, is of interest as an alternative material. Nickel oxide and Ni dichalcogenides are relevant as electrode materials for electrochemical energy storage. Noble-type TMDCs with almost full occupation of *d* orbitals exhibit attractive properties, such as largely tunable electronic structures. Therefore, these materials may be useful for electronics, optoelectronics, catalysis, and sensors. Pentagonal $PdSe_2$ and hexagonal $PdTe_2$ monolayers and their few-layer sheets are receiving increasing attention. Owing to the low symmetry of pentagonal $PdSe_2$, new properties are possible, and for hexagonal $PdTe_2$ the combination of superconductivity with topical nontrivial states has attracted the interest of researchers. Several TMDCs and their Janus compounds are promising thermoelectric materials that can serve as alternatives to conventional devices, which are based on the Seebeck effect, using a temperature gradient to obtain clean energy.

P. Hess, *Bonding, Structure, and Performance of Two-Dimensional Materials*, NanoScience and Technology, https://doi.org/10.1007/978-3-031-74161-6_17

H																	He
Li	Be											B	C	N	**O**	F	Ne
Na	Mg											Al	Si	P	**S**	Cl	Ar
K	Ca	Sc	Ti	V	Cr	Mn	Fe	Co	**Ni**	Cu	Zn	Ga	Ge	As	**Se**	Br	Kr
Rb	Sr	Y	Zr	Nb	Mo	Tc	Ru	Rh	**Pd**	Ag	Cd	In	Sn	Sb	**Te**	I	Xe
Cs	Ba	La	Hf	Ta	W	Re	Os	Ir	**Pt**	Au	Hg	Tl	Pb	Bi	Po	At	Rn
Fr	Ra	Ac	Rf	Db	Sg	Bh	Hs	Mt	Ds	Rg	Cn	Uut	Fl	Uup	Lv	Uus	Uuo

Fig. 17.1 Scheme of the periodic table highlighting the group–XB elements Ni, Pd, and Pt and the main group–VIA elements O, S, Se, and Te

17.1 Nickel Dichalcogenide Monolayers (NiO_2, NiS_2, $NiSe_2$, and $NiTe_2$)

17.1.1 NiO_2 Monolayer

17.1.1.1 Chemistry and 2D Bonding

The scheme of the periodic table shown in Fig. 17.1 displays the group–XB elements Ni, Pd, and Pt and the main group–VIA elements O, S, Se, and Te.

The valence electrons of the electronic configuration of Ni atoms (Ar|$3d^8 4s^2$) and O atoms ($1s^2 2s^2 p^4$) determine the type of 2D bonding in 2D nickel dioxide lattices. The *d-p* hybrids of the outermost 3*d* orbitals of Ni and the outermost 2*p* orbitals of O form strong covalent bonds. The hybridization leads to covalently bonded three-layer structures, such as the 1T and 1H phase, with a Ni sublayer in the center sandwiched between two O sublayers (see Fig. 17.2). Each O atom forms three covalent bonds with Ni atoms at a bridge position, and each Ni atom is bound to six O atoms. An extensive electron charge accumulation between the Ni and O atoms is responsible for the strong covalent Ni–O bonds. The significant difference in the electronegativities of Ni (1.91) and O (3.44) induces a sizable ionic contribution to bonding and consequently to strongly polar Ni–O bonds. A charge transfer analysis of the most stable hexagonal 1T phase yields a charge transfer of $\delta q = 1.32$ e from each Ni to two O atoms [1]. The published cohesive energies range between $E_c = -5.59$ eV/atom [1] (Coulomb + LDA +U) and − 6.35 eV/atom [2] (DFT). A recently published value of − 5.78 eV/atom supports the lower binding energy [3].

17.1.1.2 2D Structures

Figure 17.2 shows the most stable octahedral 1T phase of NiO_2 monolayers with top and side views of the chemically bonded three-layer assembly. Previous calculations of the lattice parameters of the 1T-NiO_2 monolayer presented a = 0.277 nm for the lattice constant, $d_{Ni\text{–}O} = 0.184$ nm for the bond length, $h_{OO} = 0.244$ nm for the distance between the O atoms, and $\theta_{ONiO} = 82.82°$ for the bond angle [1]. These values are partly in good agreement with those of a recent study that proposed the

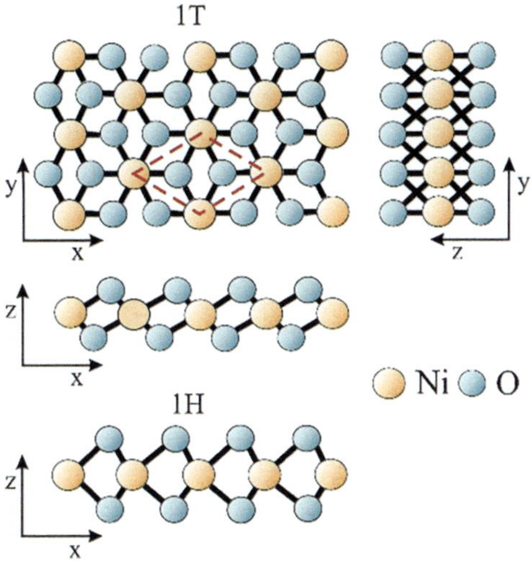

Fig. 17.2 Top view and side views of the hexagonal 1T triple-sublayer structure with space group *P*-3*m*1 and side view of the 1H structure with the space group *P*-6*m*2

lattice parameters a = 0.282 nm, $d_{Ni\text{–}O}$ = 0.188 nm, h_{OO} = 0.190 nm, θ_{ONiO1} = 96.84° and $\theta_{O\text{–}Ni\text{–}O2}$ = 83.16° [3]. Adding the van der Waals diameter of O atoms of d_w = 0.310 nm and the distance between O atoms h_{OO} = 0.244 nm, yields an estimate of the monolayer thickness of d_{w+h} = 0.554 nm. For the indirect semiconductor, band gaps of E_g = 1.39 eV, calculated by the Perdew Burke Ernzerhof (PBE) functionals, and 2.95 eV, based on the Heyd-Scuseria-Ernzerhof (HSE06) approach, were estimated [3]. The space group, cohesive energy, bond length, X–X distance, bond angle, and thickness of group XB–VIA monolayers are presented in Table 17.1.

17.1.1.3 Mechanics

First-principles calculations provided elastic constants for the mechanically stable 1T-NiO_2 monolayer, which fulfill the Born–Huang stability criteria [3]. This study reports for the in-plane stiffness or Young's modulus E = 124.7 N/m, and for Poisson's ratio ν = 0.25. The calculated nonlinear mechanical properties are an ultimate strength of σ = 13.6 N/m and an ultimate strain of ε = 0.15 [3]. Previous calculations based on the local density approximation (LDA) delivered a larger Young's modulus of E = 146.6 N/m for the 1T phase of NiO_2 monolayers, based on the harmonic range of elastic deformation, where the structure responds linearly and reversibly to strain [1]. The larger stiffness is supported by the slightly anisotropic values E_{ac} = 141.8 N/m and E_{zz} = 144.3 N/m derived by the PBE approach [4]. The most stable 1T structure was determined by the unbiased particle swarm optimization (PSO) method, as implemented in the crystal structure analysis by the particle swarm optimization (CALYPSO) code. The space group, Young´s modulus, Poisson´s ratio, ultimate strength, and ultimate strain of group XB–VIA monolayers are presented in Table 17.2.

Table 17.1 Space group, cohesive energy, lattice constant, bond length, distance of outer X atoms, bond angle, and corrugation corrected van der Waals diameter (thickness) of group XB–TMDO and XB–TMDC monolayers

Monolayer	Space group	Cohesive energy E_c (eV/atom)	Lattice constant a, b (nm)	Bond length $d_{M\text{-}X1}$, $d_{M\text{-}X2}$ (nm)	Distance d_{XX}, h_{XX} (nm)	Bond angle θ_{XMX} (°)	Thickness d_{w+h} (nm)
NiO_2	*P*-3*m*1	– 5.59 [1], – 5.78 [3]	0.277 [1]	0.184 [1]	0.244 [1]	82.82 [1]	0.554
NiS_2 O-NiS_2 P-NiS_2	*P*-3*m*1 *P*2$_1$/*c* *P*-42$_1$*m*	– 4.97 [1], – 5.21 [2] – 4.26 [13], – 4.82 [12] – 4.87 [12]	0.335 [11] 0.521, 0.533 [13] 0.528 [12]	0.226 [11] 0.217, 0.218 [13] 0.214 [12]	0.233 [11] 0.114 [13] 0.214 [12]	84.46 [11]	0.594 0.474 0.496
$NiSe_2$	*P*-3*m*1 *P*2$_1$/*c*	– 4.66 [1], – 4.68 [2] – 3.85 [13]	0.346 [1] 0.548, 0.568 [13]	0.234 [1] 0.2306, 0.2314 [13]	0.315 [1] 0.139 [13]	84.59 [1]	0.695 0.519
$NiTe_2$	*P*-3*m*1 *P*2$_1$/*c*	– 4.40 [1] – 3.48 [13]	0.364 [1] 0.595, 0.625 [13]	0.252 [1] 0.2489, 0.2495 [13]	0.347 [1] 0.153 [13]	87.33 [1]	0.735 0.573
PdO_2	*P*-3*m*1	– 4.93 [3]	0.307 [3]	0.203 [3]	0.196 [3]	81.5 [3]	
PdS_2	*P*-3*m*1 *P*2$_1$/*c*	– 4.10 [14], – 4.71 [36]	0.353 [34] 0.549, 0.559 [34]	0.240 [34] 0.234, 0.235 [34]	0.249 [31] 0.210 [34]		0.609
$PdSe_2$	*P*2$_1$/*c*	– 4.05 [35], – 4.30 [36],	0.571, 0.590 [44]	0.245, 0.246 [44]	0.242 [44]	93.6, 86.4 [44]	
$PdTe_2$	*P*-3*m*1 *P*2$_1$/*c*	– 4.07 [53] – 3.61 [35]	0.396 [53] 0.614, 0.644 [55]	0.267 [53] 0.262, 0.263 [55]	0.273 [53] 0.280 [55]	83.91 [53]	0.693
PtO_2	*P*-3*m*1	– 4.37 [62], – 5.77 [3]	0.313 [3]	0.205 [3]	0.193 [3]	80.34 [3]	0.503
PtS_2	*P*-3*m*1	– 4.68 [70], – 5.73 [53]	0.352 [53]	0.237 [53]	0.245 [53]	84.25 [53]	0.605
$PtSe_2$	*P*-3*m*1	– 4.42 [70], – 5.32 [53]	0.368 [53]	0.249 [53]	0.260 [53]	84.83 [53]	
$PtTe_2$	*P*-3*m*1	– 4.28 [70], – 5.07 [53]	0.395 [53]	0.266 [53]	0.274 [53]	84.15 [53]	

Table 17.2 Space group, Young's modulus, Poisson's ratio, ultimate strength, and ultimate strain of group XB–TMDO and XB–TMDC monolayers

Monolayer	Space group	Young's modulus E_{ac}, E_{zz} (N/m)	Poisson's ratio ν_{ac}, ν_{zz}; ν_{in}, ν_{out}	Ultimate strength σ_{ac}, σ_{zz} (N/m)	Ultimate strain ε_{ac}, ε_{zz}
NiO_2	*P*-3*m*1	124.7 [3], 146.6 [1]	0.25 [3]	13.6 [3]	0.15 [3]
NiS_2	*P*-3*m*1 *P*-2$_1$/*c*	50.7 [16] 48.9 [13]	0.38 [16] 0.212 [13]	8.0, 6.45 [16] 3.63, 3.87 [13]	0.18, 0.14 [16] 0.14 [13]
$NiSe_2$	*P*-3*m*1 *P*-2$_1$/*c*	64.8 [22] 30.1 [13]	0.27 [22] 0.52 [13]	~ 3.30, ~ 3.35 [13]	0.20, 0.18 [13]
$NiTe_2$	*P*-3*m*1 *P*-2$_1$/*c*	41.6, 41.2 [22] 22.9, 61.7 [13]	-0.044, -0.120 [13]	4.69, 4.52 [22] ~ 2.62, ~ 2.72 [13]	0.22, 02.55 [22] 0.22, 0.18 [13]
PdO_2	*P*-3*m*1	103.3 [3]	0.24 [3]	11.9 [3]	0.15 [3]
PdS_2	*P*2$_1$/c	42.4, 74.1 [36]	0.038, 0.053 [15]	1.9 [36]	0.16, 0.18 [36]
$PdSe_2$	*P*2$_1$/c	28.6, 65.3 [36]	0.025, 0.040 [15]	2.0, 2.1 [36]	0.19, 0.17 [36]
$PdTe_2$	*P*-3*m*1 *P*2$_1$/c	61.8 [53] 23.2, 51.0 [15]	0.32, 0.64 [53] -0.001, -0.002 [15]		
1T-PtO_2	*P*-3*m*1	116.5 [3]	0.175 [3]	13.4 [3]	~ 0.15 [3]
1T-PtS_2	*P*-3*m*1	82.1 [71], 99 [70]	0.268 [71], 0.29 [70]		
1T-$PtSe_2$	*P*-3*m*1	67.5 [71], 88 [70]	0.271 [71], 0.31 [70]		
1T-$PtTe_2$	*P*-3*m*1	61.2 [71], 73 [70]	0.274 [71], 0.31 [70]		

17.1.1.4 Physics and Applications

To determine the accurate electronic structure and spin properties of transition-metal oxide and transition-metal dichalcogenide monolayers by density functional theory (DFT)-based calculations, the spin-orbit interaction was calculated by using fully relativistic pseudopotentials (FRUP) compared with scalar relativistic pseudopotentials (SRUP) [2]. This work confirms the requirement of using a fully relativistic pseudopotential approach to accurately predict the properties of most monolayers containing transition metals. For NiO_2 monolayers, the two approaches delivered an identical band gap of $E_g = 1.26$ eV, confirming the small gap predicted by the PBE approach.

By employing DFT and the semiclassical Boltzmann transport equations, the structural, electronic, and transport properties of NiO_2 monolayers were derived at 300, 500, and 700 K [5]. The authors determined the exact values of the relaxation time and mobility of the charge carriers at different temperatures. The resulting maximum Seebeck coefficient of -2674 μV/K, the electrical conductivity of 9.92×10^5 S/m, and the peak electronic thermal conductivity of 7.22 W m^{-1} K^{-2} yielded a figure of merit of $ZT = 0.506$ [5].

Solar water splitting is one of the promising techniques to store energy in chemicals. Partially hydrogenated 2D-NiO_2 sheets, containing nickel oxyhydroxide (NiOOH), provide an efficient catalyst for the oxygen evolution reaction (OER) that is significantly more reactive than bulk NiOOH [6]. Several OER mechanisms examined by the DFT study could not explain the outstanding reactivity of this catalyst. Therefore, the authors postulated reaction centers at the surface of NiO_2 and not only the usually assumed defect sites at edge locations. Furthermore, the work provides evidence for an inhibiting effect of hydrogen ad-atoms on the NiO_2 surface. The work delivers added information leading to an improved understanding of the reaction mechanism and can be considered as a key to further progress in the development of tailored 2D catalysts.

A principal issue is the replacement of noble-metal materials by 2D Ni-based materials such as oxides, sulfides, phosphides, selenides, and hydroxides. For example, 2D Ni-based materials with short and open channels, efficient mass diffusion capability, and abundant accessible active sites are suitable for electrochemical energy storage (EES) applications [7]. A detailed review describes the relevant properties, growth methods, and potential applications of these materials. The design strategies employed for improving the electrochemical performance include, for example, vacancy generation and doping by heteroatoms. Moreover, the work covers the formation of 3D-NiO_2 nanostructures and their application as electrode material for EESs, including supercapacitors, alkali (Li, Na, K)-ion batteries, and multivalent metal (Zn, Mg, Ca)-ion batteries.

The transition-metal dioxide monolayers MO_2 (M = Mn, Co, Ni) are of interest as cathode materials for Li- and Na-ion batteries [8]. Due to their long bonding distance, Li and Na atoms move easily on the monolayer surface with a low diffusion barrier. The species Li_2MO_2 and Na_2MO_2 are stable with a negative intercalation energy. MnO_2 has the lowest diffusion barrier, highest specific capacity, and smallest lattice expansion under Li intercalation; however, it has the lowest cell voltage. NiO_2 has the largest cell voltage and the largest volume expansion in Li_2NiO_2.

17.1.1.5 Synthesis

Infinite layers of NiO_2 exist in epitaxial thin films of $LaNiO_2$, which is an oxygen-deficient perovskite with infinite layers of $Ni^{1+}O_2$. These films were prepared by low-temperature reduction of $LaNiO_3$ single-crystal films on a $NdGaO_3$ substrate [9]. Analysis via high-angle annular dark-field and bright-field scanning transmission electron microscopy (STEM) revealed a layer stacking sequence of NiO_2/$LaNiO_2$. Resistivity measurements revealed a T^2 dependence between 400 and 150 K and a negative Hall coefficient.

In the field of complex oxide heterostructures, the determination of properties at the film-substrate interface requires careful analysis due to the substantial restructuring of the interface. Using $LaNiO_3$ as a model system, the initial stages of the growth process on perovskite ($LaAlO_3$)(Sr_2AlTaO_6)(001) substrates were analyzed via in situ synchrotron X-ray scattering during molecular beam epitaxy (MBE) [10]. During the deposition process, the surface layers of the substrate rearrange such that a NiO_2 monolayer is weakly bound, whereas a LaO monolayer bonds strongly forming a La(Al, Ta)O_3 ultrathin film by incorporating (Al, Ta)O_2 from the surface.

17.1.2 *NiS_2 Monolayer*

17.1.2.1 Chemistry and 2D Bonding

The valence electrons of Ni atoms ($3d^8 4s^2$) and S atoms ($3s^2 p^4$) determine the type of 2D bonding in 2D nickel disulfide. The *d-p* hybridization of the outermost 3*d* orbitals of Ni and the outermost 3*p* orbitals of the S atoms form strong covalent bonds with a tendency toward d^2sp^3 hybridization. The most stable types of hybridization lead to the previously discussed covalently bonded 1T or 1H three-layer structures with a Ni sublayer in the center, sandwiched between two S sublayers. As described for the NiO_2 monolayers, each S atom forms three covalent bonds with Ni atoms at a bridging position, and each Ni atom is bound to six S atoms (see Fig. 17.2). The extensive electron charge accumulation between the Ni and S atoms is responsible for the formation of the strong covalent Ni–S bonds. The difference in the electronegativities of Ni (1.91) and S (2.58) leads to an ionic contribution to bonding and consequently to polar Ni–S bonds. A charge transfer analysis of the most stable hexagonal 1T phase predicts a small charge transfer of $\delta q = 0.42$ e from each Ni to two S atoms [1]. The cohesive energy of the 1H phase at $E_c = -4.78$ eV/atom designates a slightly lower stability than that reported for the 1T phase at −4.97 eV/atom (LDA) [1]. A study based on the generalized gradient approximation (GGA) indicates a somewhat larger binding energy for the most stable 1T phase of E_c = –5.21 eV/atom [2].

17.1.2.2 2D Structures

Early calculations of the lattice parameters of 1T-NiS_2 monolayers with a = 0.335 nm, d_{Ni-S} = 0.226 nm, h_{SS} = 0.233 nm, θ_{SNiS} = 84.22° and θ_{NiSNi} = 95.81° [11] are only partly in agreement with the lattice constant and bond length of a recent study at a = 0.335 nm, d_{Ni-S} = 0.226 nm, d_{S-S} = 0.303 nm, and h_{SS} = 0.234 nm [12]. Adding the van der Waals diameter of S atoms of d_w = 0.360 nm and the distance between S atoms of h_{SS} = 0.233 nm yields an estimate of the monolayer thickness of d_{w+h} = 0.594 nm. The space group of the 1T phase is *P*-3*m*1, and the calculated band gaps of the indirect semiconductor are E_g = 0.51 eV [1] (LDA) and 1.11 eV [12] (HSE06). The 1H-NiS_2 phase with the space group *P*-6*m*2 is a metal [12].

A first-principles study determined the structural, mechanical, and electronic properties of the 2D pentagonal NiX_2 (X = S, Se, and Te) structures [13]. The orthorhombic pentagonal structure derives from bulk pyrite-type NiS_2 with the space group *P*2$_1$/*c* (see Fig. 17.3). The unit cell contains two Ni atoms and four X atoms. In this orthorhombic structure, the Ni atoms coordinate with four chalcogen atoms in the pentagons to form NiX_4 planar squares; however, the pentagons are puckered or crumpled like in pentagonal graphene. The vertical height h refers to the vertical distance between the uppermost layer and the bottom layer. The lattice parameter increases with increasing atomic size and decreasing electronegativity in the order of NiS_2 < $NiSe_2$ < $NiTe_2$. All three structures are thermally, thermodynamically, and mechanically stable. The lattice parameters of the NiS_2 monolayer are a = 0.522 nm, b = 0.533 nm, d_{Ni-S1} = 0.217 nm, d_{Ni-S2} = 0.218 nm, d_{S-S} = 0.212 nm,

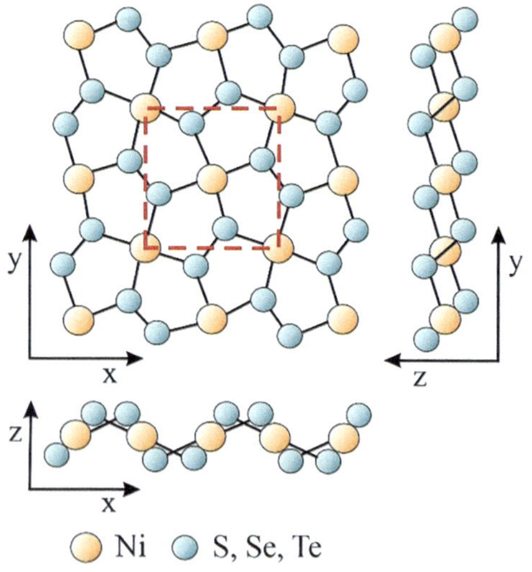

Fig. 17.3 Top view and side views of the orthorhombic pentagonal three-sublayer structure with the space group *P*2$_1$/*c*. Adapted with permission from Ref. [13], Copyright 2023, Elsevier

$h_{SS} = 0.114$ nm, and the cohesive energy is $E_c = -4.26$ eV/atom [13]. Similar lattice parameters have been presented by others, however, with a much lower cohesive energy of $E_c = -4.82$ eV/atom [12]. Adding the van der Waals diameter of S atoms of $d_w = 0.360$ nm and the distance between S atoms $h_{SS} = 0.114$ nm yields an estimate of the monolayer thickness of $d_{w+h} = 0.474$ nm. The orthorhombic phases partly possess a large in-plane anisotropy due to their reduced lattice symmetry. The physicochemical properties and the mechanical behavior reflect the structural anisotropy.

Pentagonal structures with $P2_1/c$ symmetry, dynamic stability, and strong cohesion have been studied for the three group-XB sulfides NiS_2, PdS_2, and PtS_2. [14]. An analysis of bonding by the electron localization function (ELF) indicated charge accumulation around the S atoms and charge depletion around the metal atoms. As cohesive energies the calculation delivered $E_c = -4.70$, -4.10, and -5.01 eV/atom, and the presented indirect band gaps of the three semiconductors are $E_g = 0.87$ eV, 1.08 eV, and 2.01 eV, respectively [14]. Biaxial strain induces a semiconductor-to-metal transition.

A first-principles study determined the structural, mechanical, electronic, and optical properties of the pentagonal orthorhombic 2D-MX_2 and 2D-M_2X_3 (M = Ni, Pd; X = S, Se, and Te) structures [15]. The lattice parameters agree quite well with those given above for this phase, and the cohesive energy of $E_c = -4.82$ eV/atom is comparable with that of the 1T phase. The calculated band gaps vary between $E_g = 0.82$ eV (PBE) and 2.40 eV (HSE06) with and without inclusion of spin–orbit coupling (SOC) [15]. The compounds are hyperbolic materials operating in a broad range of the electromagnetic spectrum, from the infrared to the ultraviolet region. For example, the hyperbolic regions of 2D NiS_2 appear in the visible range (1.70–1.93 eV) and the ultraviolet range (3.06–3.87 eV).

DFT-based ab initio calculations revealed the existence of another dynamically, thermally, and energetically stable unique pentagonal-structured NiS_2 monolayer with the symmetry $P\text{-}42_1\,m$ (see Fig. 17.4) [12]. This pentagonal monolayer derives from a marcasite-type P-NiS_2 bulk network with a tetragonal lattice. Here, each Ni atom binds with four S atoms, and each threefold coordinated S atom binds with two Ni atoms and one S atom. The lattice parameters of these pentagonal NiS_2 monolayers are $a = 0.528$ nm, $d_{Ni\text{–}S} = 0.214$ nm, $d_{S\text{–}S} = 0.214$ nm and $h_{SS} = 0.136$ nm [12]. Note the similarity with the lattice parameters of the pentagonal NiS_2 structure with symmetry $P2_1/c$ [12]. Adding the van der Waals diameter of S atoms and the distance between S atoms yields an estimate of the monolayer thickness of $d_{w+h} = 0.496$ nm. Notably, the tetragonal NiS_2 monolayer with a cohesive energy of $E_c = -4.87$ eV/atom has been considered as the most stable structure from both the kinetic and energetic point of view. A small strain induces a transition of the indirect band gap of $E_g = 1.94$ eV into a direct band gap [12].

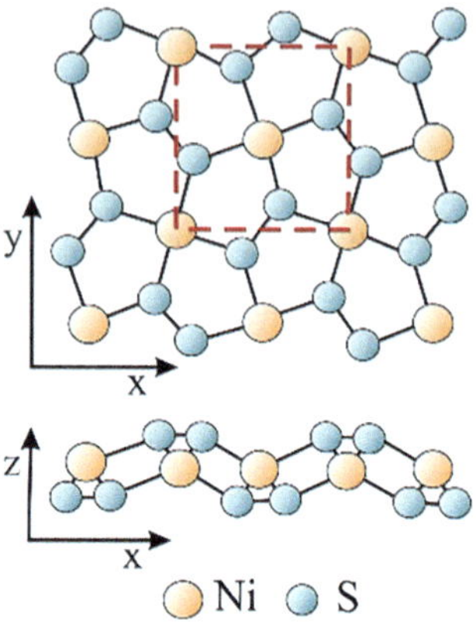

Fig. 17.4 Top view and side views of the pentagonal-structured tetragonal phase with the space group $P\text{-}42_1\,m$. Adapted with permission from ref. [12], Copyright 2020, Royal Society of Chemistry

17.1.2.3 Mechanics

The calculated Young´s modulus at E = 86.2 N/m, derived by the LDA approach for the more stable 1T-NiS_2 monolayers, is substantially smaller than the stiffness of the 1T-NiO_2 monolayer [1]. More detailed calculations of the mechanical properties based on the GGA-PBE approach yielded only E = 50.7 N/m as elastic modulus and the result for the Poisson ratio was $\nu = 0.38$ [16]. The elastic constants confirmed the mechanical stability of the 1T-NiS_2 phase according to the Born–Huang criteria. The calculated ultimate strength in the armchair direction is $\sigma_{ac} = 8.0$ N/m at an ultimate strain of $\varepsilon_{ac} = 0.18$, whereas it is only $\sigma_{zz} = 6.45$ N/m in the zigzag direction with a lower ultimate strain of $\varepsilon_{zz} = 0.14$ [16].

For the pentagonal NiS_2 monolayer with space group $P2_1/c$, both Young's modulus and Poisson's ratio show isotropic behavior, with values at E = 48.9 N/m and $\nu = 0.212$ [13]. Similarly, the nonlinear properties such as the critical strengths at $\sigma_{ac} = 3.63$ N/m and $\sigma_{zz} = 3.87$ N/m and critical strains at $\varepsilon_{ac} = 0.14$ and $\varepsilon_z = 0.14$ exhibit only low anisotropy. Contrary to these results, a study presented strongly anisotropic in-plane Young's moduli at $E_{ac} = 58.9$ N/m and $E_{zz} = 92.5$ N/m and anisotropic Poisson's ratios at $\nu_{ac} = 0.062$ and $\nu_{zz} = 0.098$ for pentagonal NiS_2 with $P2_1/c$ symmetry [15].

17.1.2.4 Physics and Applications

Strain can change the location of electron bands and thus the band gap of 2D structures. Under tensile strain, the band gap decreases linearly in the xy plane, e.g., from $E_g = 0.49$ eV at $\varepsilon_{xx} = 0$ to $E_g = 0.08$ eV at $\varepsilon_{yy} = 0.14$ [16]. In particular, at a tensile strain of $\varepsilon = 0.14$ in the y direction, the NiS_2 structure shows metallic character with a band gap of 0 eV. Furthermore, mechanical strain is an important versatile tool for controlling the electronic properties, optical conductivity, effective mass of charged particles, and transport properties [16]. For example, the mobility of charge carriers increases under mechanical strain of $\varepsilon = 0.14$ by 61.6% in the x direction and 71.7% in the y direction. The optical conductivity increases significantly under strain, and therefore NiS_2 is a potential candidate for fiber optic applications.

Tuning of the band structure by biaxial strain varies the thermoelectric parameters and may distort phonon dispersion, leading to a decrease in the lattice thermal conductivity [11]. The figure of merit of unstrained NiS_2 is $ZT = 0.73$ for n-type carriers and $ZT = 0.28$ for p-type carriers. Specifically, compressive strain enhances the figure to $ZT = 0.82$ for n-type doping. On the other hand, tensile strain improves the ZT value for p-type doping up to $ZT = 0.89$.

The nonmagnetic semiconductor of a perfect 1T-NiS_2 monolayer can be transformed into a magnetic semiconductor (Mn-doped), half-metal (V- and Fe-doped), magnetic (Cr-doped), or nonmagnetic (Co-doped) metal by substituting 3*d* transition-metal atoms in the monolayer, as revealed by DFT-based calculations [17]. Accordingly, the magnetism of NiS_2 monolayers can be controlled. The doped monolayers can be used in nanoelectronic and spintronic devices, where V, Cr, Mn, and Fe atoms induce total magnetic moments of 1.056 μ_B, 2.651 μ_B, 2.921 μ_B, and 1.735 μ_B, respectively.

DFT-based calculations revealed the thermal, dynamic, and mechanical stability of monolayer NiX_2 (X = S, Se, and Te) compounds, their electronic structure, and potential use [18]. The NiS_2, $NiSe_2$, and $NiTe_2$ monolayers possess high storage capacities of 874, 496, and 342 mA h g^{-1}, respectively. In addition to their low diffusion energy barriers of 0.23, 0.24, and 0.27 eV as well as 0.12, 0.13, and 0.12 eV, when used as anodes in lithium-ion batteries or sodium-ion batteries, respectively, these materials are promising anode materials. While the sulfide and selenide monolayers are semiconductors, the $NiTe_2$ monolayer is an intrinsic superconductor with a transition temperature of $T_c = 2.24$ K that increases to 5.48 K by an in-plane compressive strain of 6%.

DFT calculations and semiclassical Boltzmann transport theory were used to predict the electronic, optical, and thermoelectric properties of the Janus monolayers SNiX (X = O, Se, and Te) [19]. While SNiO and SNiSe monolayers are semiconductors with indirect band gaps, SNiTe has metallic character. The *n*-type SNiO and SNiSe monolayers have high Seebeck coefficients of 650 and 550 μV/K, large power factors of 20.83×10^{10} and 25.16×10^{10} W m^{-1} s^{-1} K^{-2}, and a figure of merit of $ZT \approx 1$.

According to a previous GGA-PBE study of the three Janus compounds SNiSe, SNiTe, and SeNiTe with asymmetric chemical bonding, the SNiSe monolayer has a semimetallic band structure [20]. There is an asymmetric distribution of valence electrons between the S and Se atoms, as verified by ELF. The latter method yields a large amount of electron gas-like features around the more electronegative chalcogen atoms, which are relevant for the conduction properties. As discussed in more detail below, the random phase approximation (RPA) predicts a broad optical response of SNiSe monolayers with significant peaks in the infrared, visible, and ultraviolet regions of the spectrum.

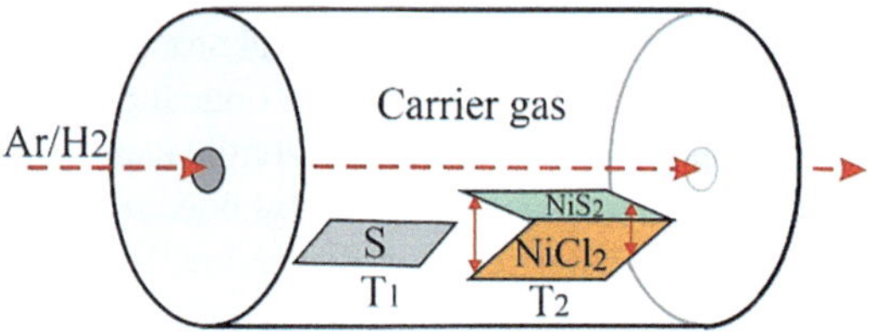

Fig. 17.5 Scheme of the setup for the CVD of NiS_2 using sulfur powder and $NiCl_2$ as precursors

17.1.2.5 Synthesis

While the growth of monolayers has not been reported, ultrathin NiS and NiS_2 nanoplates with multiple stoichiometric ratios and layer thicknesses of 2.0 and 7.0 nm, respectively, could be deposited by chemical vapor deposition (CVD) under an Ar/H_2 stream by tuning the growth temperature and the amount of sulfur powder [21]. The SiO_2/Si substrate was tilted above the surface with the $NiCl_2$ sample, used as the Ni source (see Fig. 17.5). X-ray diffraction (XRD) and transmission electron microscopy (TEM) were used to identify the hexagonal phase of the NiS nanoplates and the cubic phase of the NiS_2 single-crystal nanoplates. Electrical transport studies showed that the 2D NiS and NiS_2 nanoplates are metallic with excellent conductivities of 4.6×10^5 and 6.3×10^5 S/m, respectively. These values qualify the nanoplates for use as electrodes in 2D electronics.

17.1.3 $NiSe_2$ monolayer

17.1.3.1 Chemistry and 2D Bonding

The valence electrons of the electronic configurations of Ni (Ar|$3d^84s^2$) and Se atoms (Ar|$3d^{10}4s^2p^4$) determine the type of 2D bonding in 2D nickel diselenide. The *d-p* hybrids of the outermost 3*d* orbitals of Ni and the outermost 4*p* orbitals of Se form strong covalent Ni–Se bonds. The most stable types of hybridization lead to the previously discussed covalently bonded 1T and 1H three-layer structures with a Ni sublayer in the center sandwiched between two Se sublayers. The difference in the electronegativities of Ni (1.91) and Se (2.55) introduces an ionic contribution to bonding forming polar Ni–Se bonds. A charge transfer analysis of the most stable hexagonal 1T phase yields a transfer of $\delta q = 0.27$ e from each Ni to two Se atoms [1]. The nearly identical cohesive energies of $E_c = -4.66$ eV/atom [1] (LDA) and -4.68 eV/atom [2] (GGA), resulting from these different theoretical approaches, are in good agreement with the slightly larger value of $E_c = -4.39$ eV/atom [22] (PBE), published recently. The total energy of the T phase is 0.47 eV/unit cell lower than that of the H phase [22].

17.1.3.2 2D Structures

Early calculations of the lattice parameters of 1T-$NiSe_2$ monolayers resulted in the values a = 0.346 nm, $d_{Ni–Se}$ = 0.234 nm, h_{SeSe} = 0.315 nm, and $\theta_{Se–Ni–Se}$ = 84.59° [1]. These results deviate slightly from more recent calculations, giving a = 0.354 nm, $d_{Ni–Se}$ = 0.239 nm, and $h_{Se–Se}$ = 0.321 nm [22]. Adding the van der Waals diameter of Se atoms and the distance between Se atoms yields an estimate of the monolayer thickness of d_{w+h} = 0.695 nm. The 1T-$NiSe_2$ monolayer is a semiconductor with a small band gap of E_g = 0.19 eV (PBE) and 0.54 eV (HSE06) [22]. Small band gap materials are suitable anode materials owing to their better electronic conductivity due to adsorbed metal ions.

DFT-based ab initio calculations predict a dynamically, thermally, and energetically stable pentagonal $NiSe_2$ monolayer with the symmetry $P2_1/c$ [13]. Note that in this orthorhombic three-layer structure, each Ni atom binds only to four Se atoms instead of six Se atoms. The lattice constants of a = 0.548 nm and b = 0.568 nm are much larger than those of the T and H phases; the two Ni–Se bonds $d_{Ni–Se1}$ = 0.2306 nm and $d_{Ni–Se2}$ = 0.2314 nm are nearly identical, the X–X bonds are $d_{X–X}$ = 0.243 nm; and the vertical height is only h_{SeSe} = 0.139 nm [13]. Adding the van der Waals diameter of Se atoms and the distance between Se atoms yields an estimate of the monolayer thickness of d_{w+h} = 0.519 nm. The cohesive energy is E_c = − 3.85 eV/atom, indicating a gradual decrease in the binding energy from NiS_2 to $NiSe_2$. The calculated band gaps of the indirect pentagonal $NiSe_2$ semiconductor are E_g = 1.28 eV (PBE) and 2.31 eV (HSE06), without consideration of the negligible spin–orbit coupling (SOC).

17.1.3.3 Mechanics

The calculated elastic constants demonstrate the mechanical stability of the hexagonal 1T-$NiSe_2$ phase according to the Born–Huang criteria, with an elastic modulus at E = 64.8 N/m and a Poisson ratio at ν = 0.27 [22].

The mechanical properties of the $NiSe_2$ monolayers with pentagonal $P2_1/c$ symmetry exhibit negligible anisotropy, with a low Young's modulus of E = 30.1 N/m and a large Poisson ratio of ν = 0.52 [13]. The values of the slightly anisotropic critical strengths are σ_{ac} = ~ 3.30 N/m and σ_{zz} = ~ 3.35 N/m, and those of the critical strains ε_{ac} = 0.20 and ε_{zz} = 0.18.

Different to these results, a large anisotropy is proposed for the pentagonal lattice of O-$NiSe_2$ with the space group $P2_1/c$, which can be observed in the enormous direction dependence of Young's moduli of E_{ac} = 37.4 N/m and E_{zz} = 72.8 N/m, as well as the negative Poisson's ratios of ν_{ac} = − 0.018 and ν_{zz} = − 0.036 [15]. Note that such auxetic materials, which expand in one direction if stretched in another direction, are highly desirable for tissue engineering.

17.1.3.4 Physics and Applications

First-principles calculations indicate that T-$NiSe_2$ sheets, which have a small band gap, are promising anode materials for potassium-ion batteries because of their low redox potential for K atoms and abundant available resources [22]. The calculated performance predicts a large capacity of 247 mA h g^{-1}, a small diffusion barrier of K atoms on T $NiSe_2$ of 0.05 eV, a low average voltage of 0.49 V in the energy range of ideal anode materials, and a superior carrier mobility of 1685 cm^2 V^{-1} s^{-1}, implying fast transport of electrons. These features suggest fast transport of electrons in T $NiSe_2$.

As mentioned above, first-principles calculations predict the stability of the Janus materials SNiSe, SNiTe, and NiSeTe with the hexagonal H structure of the monolayer, and the largest energetic and dynamic stability for the SNiSe monolayer with a semimetallic character, owing to several bands crossing the Fermi level [20]. The Ni–Se interaction plays a dominant role in the band profile near the Fermi energy, and ELF plots reveal different carrier distributions among the chalcogen atoms. According to the optical properties calculated with the random phase approximation (RPA), the SNiSe monolayers exhibit a broad optical response with significant peaks in the infrared, visible, and ultraviolet regions [20]. The cohesive energy of SNiSe is only $E_c = -1.68$ eV/atom and decreases further to -1.48 eV/atom (SNiTe) and -1.13 eV/atom (SeNiTe) with increasing size. The chemical asymmetry between the top and bottom sublayers changes the electric dipole moment, which may positively influence the catalytic behavior. The wide range of potential applications includes batteries, supercapacitors, electrocatalysis, and electronic/spintronic devices.

17.1.3.5 Synthesis

High-quality 2D $NiSe_2$ films were grown epitaxially by a single-step direct selenization process on a Ni(111) single crystal that served simultaneously as the source material and substrate (see Fig. 17.6) [23]. The formation of a ($\sqrt{3} \times \sqrt{3}$) superlattice structure of $NiSe_2$ films on the ($\sqrt{7} \times \sqrt{7}$) superlattice of the Ni(111) substrate was confirmed by X-ray photoemission spectroscopy (XPS), which is used for monitoring the whole fabrication process, besides low-energy electron diffraction (LEED), and scanning tunneling spectroscopy (STS) combined with DFT calculations. This fabrication process opens the door to potential applications of $NiSe_2$ nanosheets, especially in electrocatalysis and as supercapacitors in energy storage devices.

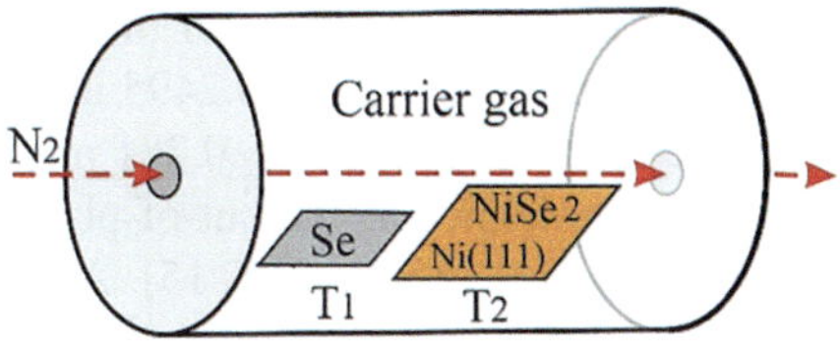

Fig. 17.6 Scheme of the setup for the growth of $NiSe_2$ by CVD. In the direct single-step selenization process, a single crystal Ni(111) surface served as the source material and substrate

17.1.4 $NiTe_2$ Monolayer

17.1.4.1 Chemistry and 2D Bonding

The valence electrons of the electronic configurations of Ni (Ar|$3d^8 4s^2$) and Te atoms (Kr|$4d^{10} 5s^2 p^4$) control the type of 2D bonding in 2D nickel ditelluride. The *d-p* hybrids of the outermost 3*d* orbitals of Ni and the outermost 5*p* orbitals of the Te atoms form covalent bonds with medium strength. The most stable hybrids form the already discussed 1T and metastable 1H three-layer structure with a Ni sublayer in the center, sandwiched between two Te sublayers. The electron charge accumulation between the Ni and Te atoms is responsible for the strength of the covalent Ni–Te bonds. The small difference in the electronegativities of the Ni atoms (1.91) and Te atoms (2.10) induces a minor ionic contribution to bonding responsible for slightly polar Ni–Te bonds. According to the Pauling scale, charge transfer would occur from the Ni atoms to the Te atoms. Analysis of the most stable hexagonal 1T phase, however, suggests a minor loss of charge of $\delta q = 0.06$ e of each Te atom [1]. The cohesive energies of the 1T-$NiTe_2$ and 1H-$NiTe_2$ monolayers at $E_c = -4.40$ eV/atom and -4.31 eV/atom, respectively, were obtained by the LDA approach and therefore should be considered as upper limits, because this method overestimates the energy [1].

17.1.4.2 2D Structures

Early calculations of the lattice parameters of 1T-$NiTe_2$ monolayers yielded the values: a = 0.364 nm, $d_{Ni-Te} = 0.252$ nm, $h_{TeTe} = 0.347$ nm, and $\theta_{Te-Ni-Te} = 87.33°$ [1]. Adding the van der Waals diameter of Te atoms and the distance between Te atoms, yields an estimate of the monolayer thickness of $d_{w+h} = 0.735$ nm. The 1H structure has a comparable lattice constant of a = 0.359 nm, a comparable bond length of $d_{Ni-Te} = 0.254$ nm, however, a smaller Te–Te separation of $h_{TeTe} = 0.293$ nm, and a much smaller bond angle of $\theta_{Te-Ni-Te} = 70.55°$ [1]. The 1T and 1H phase are metals.

DFT-based ab initio calculations predict a dynamically, thermally, and energetically stable pentagonal-structured $NiTe_2$ monolayer with the symmetry $P2_1/c$, where each Ni atom binds only to four Te atoms instead of six Te atoms in the T and H phases [13]. The lattice constants of a = 0.595 nm and b = 0.625 nm are no longer identical and are much larger than those obtained for the T and H phases. The two bond lengths $d_{Ni-Te1} = 0.2489$ nm and $d_{Ni-Te2} = 0.2495$ nm differ only marginally, and the bond length of the Te–Te bond is $d_{Te-Te} = 0.281$ nm. Note that this bond is much longer than the layer width defined as the out-of-plane distance between the two Te sublayers, which is $h_{TeTe} = 0.155$ nm [13, 15]. Adding the van der Waals diameter of Te atoms and the distance between Te atoms, yields a monolayer thickness of only $d_{w+h} = 0.573$ nm. The cohesive energy at $E_c = -3.48$ eV/atom (PBE) designates the weakest bonding of this group [13]. The band gap of the indirect semiconductor increases to $E_g = 0.95$ eV (PBE) and 1.89 eV (HSE06) [15].

17.1.4.3 Mechanics

Extensive classical molecular dynamics (MD) simulations elucidated the linear and nonlinear mechanical properties of the 1T-$NiTe_2$ and 1H-$NiTe_2$ monolayers and reveal the fracture mechanism [24]. This approach gives for the 1T phase similar Young's moduli of $E_{ac} = \sim 41.6$ N/m and $E_{zz} = \sim 41.2$ N/m. For the ultimate tensile stress, the authors calculated $\sigma_{ac} = 4.69$ N/m and $\sigma_{zz} = 4.52$ N/m, and for the fracture strains $\varepsilon_{ac} = 0.22$ and $\varepsilon_{zz} = 0.255$. Interestingly, the reported stiffness and material strength of the 1T phase are smaller than those reported for the 1H phase [24]. Furthermore, the authors studied the effects of temperature, defect concentration, and strain rate on the brittle failure of 2D $NiTe_2$. An early study of the elastic modulus proposed a comparable isotropic stiffness of $E = 43.7$ N/m for the 1T structure and a somewhat smaller stiffness of $E = 41.0$ N/m for the H phase [1].

The pentagonal structure with $P2_1/c$ symmetry shows an extreme anisotropy for the $NiTe_2$ monolayers [13]. Actually, the elastic modulus varies between $E_{ac} = 22.9$ N/m and $E_{zz} = 61.7$ N/m, and the negative Poisson ratios alter between $\nu_{ac} = -0.044$ and $\nu_{zz} = -0.120$ [13]. The values of the only marginally anisotropic critical strength are $\sigma_{ac} = \sim 2.62$ N/m and $\sigma_{zz} = \sim 2.79$ N/m, and those of the corresponding critical strains are $\varepsilon_{ac} = 0.22$ and $\varepsilon_{zz} = 0.18$.

Interestingly, the strong anisotropy has been proposed by previous calculations, which reported Young's moduli at $E_{ac} = 22.6$ N/m and $E_{zz} = 60.6$ N/m and Poisson's ratios at $\nu_{ac} = -0.037$ and $\nu_{zz} = -0.10$ [15]. Note that the negative Poisson ratios originate from the specific geometric structure and bonding characteristics, causing a distinct auxetic effect in $NiTe_2$ monolayers. Importantly, the two studies on the mechanical behavior of pentagonal $NiTe_2$ monolayers agree quite well, contrary to the situation observed for pentagonal NiS_2 and $NiSe_2$ monolayers.

17.1.4.4 Physics and Applications

The Seebeck coefficient, power factor, and thermoelectric *ZT* value of low-cost pentagonal NiX_2 (X = S, Se, and Te) monolayers were determined between 300

K and 600 K using semiclassical Boltzmann transport theory and deformation potential theory [25]. Low lattice thermal conductivities are predicted at 600 K for pentagonal $NiTe_2$ monolayers, leading to large *ZT* values of 1.81 and 1.58 in the armchair and zigzag directions, respectively, indicating high performance of the anisotropic pentagonal monolayer. The large anisotropy of the *ZT* values is due to the direction dependence of the electrical and thermal transport processes.

A comprehensive work on 33 2D TMDCs investigated the changes in structural, electronic, magnetic, and catalytic properties caused by the hydrogenation of monolayers, using first-principles calculations [26]. In principle, hydrogenation can switch the magnetic state between non, ferro-, para-, and antiferro-magnetic states and the electronic state can be changed between that of a semiconductor, metal, and half-metal. Upon hydrogenation the stable phases of TMDC monolayers can be transformed from the 1T to the 1H phase or vice versa. Specifically, hydrogenation is a tool for realizing magnetic and electronic switching in group XB-TMDC monolayers. While NiS_2 and $NiSe_2$ switch to a half-metal upon single hydrogenation and to a metallic state upon double hydrogenation, the nonmagnetic $NiTe_2$ monolayer maintains its metallic state regardless of hydrogenation.

Based on its broadband optical response, ultrafast carrier recovery time, strong light-matter interaction, and robust chemical stability, 2D $NiTe_2$ is a suitable material for ultrafast pulse generation via the passive mode-locking technique [27]. Few-layer nanosheets (< 5 nm) prepared by combining conventional chemical synthesis of $NiTe_2$ powder and liquid-phase exfoliation were used to fabricate a saturable absorber by depositing samples on a tapered fiber. An Er-doped mode-locked fiber laser with a central wavelength at 1567.3 nm and a pulse width of 1.07 ps was realized, using 2D $NiTe_2$ samples as saturable absorber.

17.1.4.5 Synthesis

CVD allowed the synthesis of large-area single crystals of 2D $NiTe_2$ with highly uniform thickness [28]. The setup provided separate heating of the Te and the $NiCl_2$ source and deposition occurred on a tilted SiO_2/Si substrate located above the $NiCl_2$ powder reservoir (see Fig. 17.6). Controlling the growth temperature allowed the realization of domain sizes of up to ~ 440 μm and precise tuning of the layer thickness of the nanosheets from ~ 0.9 nm thick monolayers to multilayer sheets of the hexagonal 1T-$NiTe_2$ phase. XRD, TEM, and high-resolution scanning transmission electron microscopy (HRSTEM) were employed for analysis. The 2D-$NiTe_2$ sheets exhibited strong thickness-tunable electrical transport properties with a conductivity of up to 7.8×10^5 S/m.

The CVD method also allowed the deposition of thickness-controlled large domains (~1.5 mm) of metallic 1T-$NiTe_2$ nanosheets on atomically flat mica substrates by varying the growth temperature [29]. AFM and Raman spectroscopy were employed to control the layer thickness (> 20.5 nm). The CVD-synthesized deposit showed an ultrahigh conductivity of ~ 1.15×10^6 S/m, a high electrocatalytic activity in the pH-universal HER, and an ultrahigh environmental stability for more than one year even after harsh treatments.

17.2 Palladium Dioxide and Dichalcogenide Monolayers (PdO_2, PdS_2, $PdSe_2$, and $PdTe_2$)

17.2.1 PdO_2 Monolayer

17.2.1.1 Chemistry and 2D Bonding

The scheme of the periodic table presented in Fig. 17.1 displays the elements palladium and oxygen with the valence electrons in the outer valence shell of the electronic configuration of Pd (Kr|4d^{10}) and O ($1s^2 2s^2 p^4$), which determine the type of 2D bonding in 2D palladium dioxide. The *d-p* hybrids of the outermost 4*d* orbitals of Pd and the outermost 2*p* orbitals of the O atoms form a three-layer structure with strong covalent bonds. In the most stable 1T configuration of the three-layer structure, a Pd sublayer is in the center sandwiched between two O sublayers. Each O atom forms three covalent bonds with Pd atoms at the bridge position, and each Pd atom is bound to six O atoms (see Fig. 17.2). The large difference in the electronegativities between Pd (2.2) and O (3.44) leads to a strong ionic contribution to bonding, with a charge transfer of $\delta q = 1.28$ e from each Pd atom to two O atoms ($\delta q = 0.64$ e to each O), forming strongly polar Pd–O bonds [3]. The binding strength resulting from the cohesive energy of $E_c = -4.93$ eV/atom, calculated for 1T-PdO_2 monolayers, is substantially smaller than that of the 1T-NiO_2 monolayer at $E_c = -5.78$ eV/atom [3].

17.2.1.2 2D Structures

Recent calculations of the lattice parameters of 1T-PdO_2 monolayers with a = 0.307 nm, $d_{Pd\text{–}O} = 0.203$ nm, $h_{OO} = 0.196$ nm, $\theta_{O\text{–}Pd\text{–}O1} = 98.54°$, and $\theta_{O\text{–}Pd\text{–}O2} = 81.46°$ [3] are in good agreement with those of a similar study delivering a = 0.309 nm, $d_{Pd\text{–}O} = 0.204$ nm, $\theta_{O\text{–}Pd\text{–}O} = 81.50°$ and $\theta_{PdOPd} = 98.48°$ [30]. The estimated layer thickness of t = 0.381 nm [30] is smaller than $d_{w+h} = 0.506$ nm, obtained by adding the distance between O atoms and the van der Waals diameter of O atoms. For the indirect band gap of the semiconductor, energy gaps of $E_g = 1.50$ eV (PBE) and 3.00 eV (HSE06) have been reported [3], in reasonable agreement with a study of the absolute band edge positions and corresponding band gaps of $E_g = 1.30$ (LDA) and 2.84 G_0W_0 [31].

17.2.1.3 Mechanics

First-principles calculations for the mechanically stable 1T-PdO_2 monolayer yielded as Young's modulus $E = 103.3$ N/m and as Poisson's ratio $\nu = 0.24$ [3]. The corresponding isotropic ultimate fracture strength is $\sigma = 11.9$ N/m, and the ultimate strain $\varepsilon \approx 0.15$. The elastic constants of several TMDOs, calculated by density functional perturbation theory (DEPT), allowed the identification and characterization of several 2D high-κ auxetic oxides, including their negative Poisson ratios [32]. The elastic constants presented for 2D $PdO_2(110)$, deliver as slightly anisotropic elastic moduli $E_{ac} = 130.6$ N/m and $E_{zz} = 128.0$ N/m and as Poisson's ratios $\nu_{ac} = 0.379$ and $\nu_{zz} = 0.377$.

17.2.1.4 Physics and Applications

The lattice thermal conductivity of the PdO_2 monolayer is 19.06 W m^{-1} K^{-1} at 300 K, according to first-principles calculations [3]. The carrier mobility, calculated by employing the deformation potential theory, exhibits anisotropic behavior of the effective mass and deformation potential in the armchair and the zigzag directions. The highest electron and hole mobilities are 1288.1 and 19.18 cm^2 V^{-1} s^{-1}, respectively, and the lattice thermal conductivity is 19.06 W m^{-1} K^{-1} [3]. A similar thermal conductivity of 15.85 W m^{-1} K^{-1} was obtained at room temperature, leading to a *ZT* value of 0.87 for n-type doped PdO_2 monolayers [30].

17.2.1.5 Synthesis

A remote inductively coupled O_2/Ar plasma, removes in an atomic layer etching-like (ALE) process layer-by-layer from CVD-grown $PdSe_2$ flakes [33]. The oxygen plasma oxidizes the top layer of $PdSe_2$, forming volatile PdO_2 and SeO_2. AFM measurements indicate a clear trend between the oxygen concentration and the number of layers etched per cycle. XPS measurements detected no traces of PdO_2 or SeO_2 after in situ inert atmosphere annealing at low temperature, suggesting complete desorption of all byproducts. In addition to the controlled thinning process to produce $PdSe_2$ nanosheets, the sublimation of PdO_2 may be used to deposit thin PdO_2 films on a suitable surface.

17.2.2 PdS_2 Monolayer

17.2.2.1 Chemistry and 2D Bonding

The valence electrons of the outer shell of the electronic configuration of Pd (Kr|4d^{10}) and S (Ne|3s^2p^4) determine the type of 2D bonding in palladium disulfide monolayers. The *d-p* hybrids of the outermost 4*d* orbitals of Pd atoms and the outermost 3*p* orbitals of S atoms form stable triple-layer structures. There is an ongoing controversial discussion on the most stable phase of 2D PdS_2. Different from the 1T phase, a pentagonal three-layer structure with Pd layers located between S layers (S–Pd–S) has been suggested. According to a global search employing the PSO method, in the most stable pentagonal structure the Pd atom is coordinated with four S atoms, and these four Pd–S bonds are situated in the same plane, forming planar squares as building blocks [34]. Each S atom forms three bonds, two Pd–S bonds and one S–S bond with a neighboring S atom. The covalent S–S bonds may contribute to the greater stability of the orthorhombic PdS_2 structure. In this network, two Pd atoms and three S atoms form a wrinkled pentagon motif. The difference in the electronegativities of Pd (2.2) and S (2.58) induces to a small ionic contribution to bonding, which leads to slightly polar Pd–S bonds. According to a Bader charge transfer analysis, each Pd atom transfers a charge of $\delta q = 0.15$ e to the outer layers that is shared by two S atoms [35]. The calculated cohesive energies are $E_c = -4.10$ eV/atom [14] (PBE), − 4.28 eV/atom [35] (PBE), and − 4.71 eV/atom (PBE) [36].

17.2.2.2 2D Structures

Palladium dichalcogenides exhibit phases with comparable total energies, which makes the determination of the stable phase difficult [34]. For instance, the pentagonal structure of PdS_2 monolayers is only slightly more stable than the 1T phase. The lattice parameters of the pentagonal structure are a = 0.549 nm, b = 0.559 nm, $d_{Pd\text{–}S1} = 0.235$ nm, $d_{Pd\text{–}S2} = 0.234$ nm, and $d_{S\text{–}S} = 0.210$ nm [34], in agreement with a subsequent publication [14]. The orthorhombic structure belongs to the space group $P2_1/c$. The indirect band gap of $E_g = 1.60$ eV (HSE06), which decreases with increasing thickness, can also be effectively reduced by applying uniaxial or biaxial tensile strain [34].

The available lattice parameters of the metastable 1T-PdS_2 monolayers are a = 0.353 nm [34], $d_{Pd\text{–}S} = 0.240$ nm [34], and $h_{SS} = 0.249$ nm [31] with a much smaller lattice constant. The thickness of the 1T phase at $d_{w+h} = 0.609$ nm can be estimated by adding the separation and the van der Waals diameter of the S atoms. The quasiparticle (QP) band structure with SOC in the G_0W_0 approximation is $E_g = 2.48$ eV, to be compared with 1.11 eV (LDA) [31].

17.2.2.3 Mechanics

The pentagonal structure of PdS_2 monolayers results in strongly anisotropic linear mechanical properties with stiffnesses of $E_{ac} = 58$ N/m and $E_{zz} = 82$ N/m [34]. These stiffness values are in good agreement with $E_{ac} = 56.5$ N/m and $E_{zz} = 78.3$ N/m and the calculated Poisson ratios are $\nu_{ac} = 0.038$ and $\nu_{zz} = 0.053$ [15]. The distinctive anisotropy effect is in excellent agreement with $E_{ac} = 57.8$ N/m and $E_{zz} = 80.1$ N/m [35]. Another study reported smaller stiffness values of $E_{ac} = 42.4$ N/m and $E_{zz} = 74.1$ N/m, however, with the typical anisotropy of the orthorhombic phase [36]. Penta-PdS_2 monolayers can withstand an ultimate strain of $\varepsilon_{ac} = 0.16$ and a fracture stress of $\sigma_{ac} = 3.44$ GPa in the armchair direction and $\varepsilon_{zz} = 0.18$ and $\sigma_{zz} = 3.43$ GPa in the zigzag direction. The low ultimate strains indicate flexibility up to the critical strain. A proposed isotropic in-plane Young's modulus of the hexagonal 1T-PdS_2 structure is $E = 74.5$ N/m [37].

17.2.2.4 Physics and Applications

A comprehensive review discusses the atomic structure, methods of synthesis, defects and dopants, layer-dependent band structures, electronic properties, electrochemical activities, and broadband optoelectronic properties of 2D noble-metal (Pd, Pt) dichalcogenides (S, Se) [38]. Furthermore, this work concentrates on a comparison of special features of group XB TMDCs with those of group VIB TMDCs to determine the similarities and differences originating from the different nature of the metals involved.

A study using a combination of the PSO method with first-principles calculations predicted the existence of two polymorphs of dynamically, thermally, and mechanically stable puckered Janus PdSSe monolayers with pentagonal structure [39]. The two polymorphs exhibit anisotropic Young's moduli and Poisson's ratios, band gaps of $E_g = 2.10$ and 2.06 eV, and band edges, which perfectly straddle the redox potential of water. In addition, they exhibits excellent light absorption in the visible and ultraviolet regions with absorption coefficients of ~ 10^6 cm^{-1}. Moreover, the large anisotropic carrier mobilities can promote the separation of photogenerated electrons and holes and thus reduce the probability of recombination during the photocatalytic water splitting process.

Angle-resolved polarized Raman spectroscopy revealed the anisotropy of the optical properties of few-layer PdS_2 sheets in the temperature region of 12–300 K [40]. Furthermore, a fabricated field-effect transistor (FET) based on PdS_2 sheets, revealed a tunable ambipolar transport mechanism with field-effect mobilities of up to ~ 388 cm^2 V^{-1} s^{-1} and an on/off ratio of ~ 800. These results are expected to open the door to high-performance optoelectronic devices.

A theoretical study identified PdS_2 monolayers as highly sensitive, selective, and reusable gas sensors for the toxic and environmentally hazardous molecules NO_2,

NO, and SO_2 [41]. In addition to the usual analysis of the adsorption energy, separation of molecules from the surface, charge transfer, electronic structure, work function, and recovery time, this work suggests a novel strategy for screening suitable sensing materials by comparing the frontier molecular orbitals of the gas molecule with the band edges of the adsorption material.

17.2.2.5 Synthesis

While the realization of PdS_2 monolayers has not been achieved, the centimeter-scale synthesis of few-layer PdS_2 sheets was demonstrated by combining physical vapor deposition (PVD) and plasma-enhanced CVD (PE-CVD) [40]. A thin layer of Pd was sputtered on a SiO_2/Si substrate, and sulfurization was achieved by heating sulfur powder and using pure N_2 as the plasma feed gas and carrier gas. Measurements of the height profile by atomic force microscopy (AFM) resulted in a thickness of $t = 4.4$ nm for the deposited sheets. Raman spectroscopy, TEM, AFM, XPS, and energy-dispersive X-ray spectroscopy (EDS) analyses provided the experimental data for the deposit, and DFT-based first-principles calculations helped to interpret the experimental information.

Sulfurization of thin palladium films using an Ar + H_2 mixture as carrier gas, sulfur powder at 250 °C, and a growth temperature of the PdS_2 film around 450–550 °C produced PdS_2 nanosheets [42]. These conditions are compatible with the back-end-of-line processes used in semiconductor industry. The quality of the synthesized n-type semiconducting films with high mobility of 10.4 $cm^2\ V^{-1}\ s^{-1}$ was demonstrated by a fabricated PdS_2 photodetector with a broadband photoresponse from 450 to 1550 nm.

17.2.3 *$PdSe_2$ Monolayer*

17.2.3.1 Chemistry and 2D Bonding

The electronic configurations of Pd atoms (Kr|4d^{10}) and Se atoms (Ar|3d^{10}4s^2p^4) with similar outer shell valence electrons as S atoms, determine the type of 2D bonding in palladium diselenide monolayers. The fully occupied 4d^{10} orbital leads to a 4d^8 electron configuration with Pd^{2+} ions. The hybridization between the 4$d_z{}^2$ orbitals of the Pd atom and 4p/3$d_z{}^2$ orbitals of the Se atom is responsible for the low-symmetry three-layer structure with Pd layers located between two Se layers [43]. As already discussed for PdS_2, in the most stable pentagonal structure the Pd atom is coordinated with four Se atoms, and these four Pd–Se bonds are positioned in the same plane, forming a planar square as a building block [34]. Each Se atom is involved in two Pd–Se bonds and one Se–Se bond with a neighboring Se atom, where the covalent Se–Se bonds increase the stability of the orthorhombic structure. Two Pd atoms and three Se atoms form a wrinkled pentagon. The structure is sensitive to defects such as Se vacancies, which break the symmetry and induce large structural distortions [38]. The small difference in the electronegativities of the Pd atoms (2.2) and Se atoms (2.55) leads to a small ionic contribution to bonding and consequently to slightly polar Pd–Se bonds. According to a Bader charge transfer analysis, each Pd atom transfers a charge of only $\delta q = 0.045$ e to two neighboring Se atoms in the outer layers [35]. The so far reported cohesive energies of penta-$PdSe_2$ monolayers are $E_c = -4.05$ eV/atom [35] (PBE) and a smaller cohesive energy of $E_c = -4.30$ eV/atom (PBE), indicating slightly stronger bonding for pentagonal PdS_2 than the other study [36].

17.2.3.2 2D Structures

The lattice parameters of the pentagonal $PdSe_2$ structure are a = 0.571 nm, b = 0.590 nm, $d_{Pd\text{–}Se1} = 0.245$ nm, $d_{Pd\text{–}Se2} = 0.246$ nm, $d_{Se1\text{–}Se2} = 0.242$ nm, and the bond angles are $\theta_{Se1\text{–}Pd\text{–}Se1} = 93.6°$ and $\theta_{Se1\text{–}Pd\text{–}Se2} = 86.4°$ [44]. These parameters are in very good agreement with a = 0.594 nm, b = 0.568 nm, $d_{Pd\text{–}Se1} = 0.245$ nm, $d_{Pd\text{–}Se2} = 0.245$ nm, and $d_{Se1\text{–}Se2} = 0.242$ nm [36]. Orthorhombic $PdSe_2$ sheets with an odd number of layers belong to the space group $P2_1/c$, while those with an even number of layers are allocated to the space group $Pca2_1$ [43]. The puckered pentagonal orthorhombic structure of the $PdSe_2$ monolayer is a semiconducting crystal with an indirect band gap of 1.33 eV (PBE) that decreases with increasing thickness and finally reaches zero in the bulk phase with semimetallic features.

17.2.3.3 Mechanics

Theory provided the first information on the strongly anisotropic mechanical properties of pentagonal $PdSe_2$ monolayers, predicting as Young's moduli $E_{ac} = 37.7$ N/m and $E_{zz} = 61.5$ N/m and as anisotropic Poisson ratios $\nu_{ac} = 0.025$ and $\nu_{zz} = 0.040$ [15]. The anisotropic stiffnesses agree well with $E_{ac} = 38.3$ N/m and $E_{zz} = 60.6$ N/m [35]. The stiffnesses of $E_{ac} = 28.6$ N/m and $E_{zz} = 65.3$ N/m with an even larger anisotropy were reported in another study [36]. Penta-$PdSe_2$ monolayers can withstand an ultimate strain of $\varepsilon_{ac} = 0.19$ and a fracture stress of $\sigma_{ac} = 3.40$ GPa in the armchair direction and $\varepsilon_{zz} = 0.17$ and $\sigma_{zz} = 3.46$ GPa in the zigzag direction. The low ultimate strains indicate flexibility up to the critical strain. Due to the decreasing binding energy, both stiffness and strength of the pentagonal monolayer decrease from PdS_2 to $PdSe_2$ and further to $PdTe_2$ [36].

17.2.3.4 Physics and Applications

The thermoelectric properties of penta-PdX_2 (X = S, Se, and Te) monolayers were determined by first-principles calculations combined with the Boltzmann transport theory. Phonon dispersion and ab initio molecular dynamics (AIMD) simulations confirmed the dynamic and thermal stability of the 2D materials [45]. The calculations revealed strong anisotropy of the thermal conductivity, power factor, and thermoelectric figure of merit. The largest ZT values evaluated for p-type monolayers were $ZT = 0.85$ (PdS_2), 1.18 ($PdSe_2$), and 2.42 ($PdTe_2$).

The information currently available on the short- and long-term environmental stability of CVD-grown few-layer $PdSe_2$ sheets is based on aging studies [46]. Short-term (<160 h) ambient aging of a pristine n-type $PdSe_2$ device induced a decrease in n-type conduction and an increase in p-type conduction, which has been attributed to activated chemisorption of O_2 at Se vacancy sites. The created defect complex eliminates the Se vacancy gap states near the conduction band, consistent with the observed increase in p-type conduction. Long-term aging (>430 h) leads to complete suppression of n-type conduction and a further increase in p-type conduction owing to PdO_2 formation, as confirmed by XPS analysis. The PdO_2 deposit is volatile at low temperatures, and annealing partly restores n-type conduction; however, this process is not fully reversible.

Owing to the demand for photocatalysts with wide-ranging absorption of solar radiation and efficient carrier separation, the anisotropic in-plane electronic and optical properties of $PdSe_2$ monolayers were studied via DFT [47]. These properties are realized due to the low symmetry of the pentagonal lattice. The suitable band gap and band edge positions lead to a high solar-to-hydrogen energy conversion efficiency of up to 12.59%. According to nonadiabatic molecular dynamics (NA-MD) simulations combined with time-dependent DFT (TD-DFT) calculations, $PdSe_2$ monolayers show an extremely slow electron–hole recombination process of approximately 1.9 ns.

2D $PdSe_2$ sheets provide an ideal platform for studying defect-induced phenomena, e.g., by controlling electronic and magnetic properties via defect engineering. First-principles calculations indicate that owing to the special pentagonal morphology Se vacancies V_{Se} form easily and migrate across the $PdSe_2$ layer [48]. They are responsible for in-gap states and trapping of carriers. Complex vacancies such as V_{Pd+4Se}, which are thermodynamically stable, induce spin-polarized states with a total magnetic moment of $2\mu_B$ per defect, allowing the introduction of magnetization. The low defect formation energies explain the enormous number of defects after mechanical exfoliation. The production of various defect patterns with controllable electron-beam energy provides access to additional remarkable properties that are absent in defect-free monolayers.

A comprehensive review concentrates not only on the stable structures, 2D properties, and deposition techniques but also on various potential applications of 2D $PdSe_2$, including heterostructures such as FETs, Schottky junction transistors, polarized light detectors, p–n junction-based rectifiers, sensors, thermoelectric devices, and system integration into infrared image sensors [43].

17.2.3.5 Synthesis

Mechanical exfoliation is a widely used method for producing highly crystalline flakes for the fabrication of individual devices [49]. Noteworthy, large-area pentagonal $PdSe_2$ samples with different thicknesses obtained by gold-assisted exfoliation allowed the first observation of the unique linear dichroism conversion phenomenon in pentagonal 2D $PdSe_2$, using polarization-resolved absorption spectroscopy. Unfortunately, this technique is difficult to apply in industrial production due to the lack of large-area samples, lack of uniformity, and necessity of controlling the number of layers.

The established method of postchalcogenation through sulfurization or selenization of Mo, W, and Pt to fabricate the corresponding TMDCs was extended to the realization of 2D-$PdSe_2$ sheets [50]. In fact, the growth of wafer-scale high-quality $PdSe_2$ films on 2-inch Pd-covered SiO_2/Si wafers with sheet thicknesses ranging from 1.2 to 20 nm was demonstrated by selenization of predeposited Pd layers. Optical absorption and ultraviolet photoemission spectroscopy (UPS) measurements confirmed the theoretically postulated transition from a semiconducting monolayer to the semimetallic bulk phase. Raman spectroscopy, AFM, and XPS analyses revealed the thickness and chemical composition of the samples. An extremely sensitive, fast, and broadband photodetector exhibited a responsivity of 300.2 mA/W and a specific detectivity of $\approx 10^{13}$ Jones. The comprehensive review describes the currently available deposition strategies, including the geometric structure, layer thickness, domain size, and quality of the deposited 2D-$PdSe_2$ sheets [50]. Realization of a versatile growth method for high-quality single-crystal monolayers of puckered pentagonal $PdSe_2$ suitable for mass production is still lacking.

The synthesis of pentagonal $PdSe_2$ and Pd_2Se_3 monolayers and few-layer sheets of $PdSe_2$ with inherently asymmetric structures is possible on the honeycomb lattice of graphene and 2D-MoS_2 substrates by direct selenization of a predeposited molecular Pd precursor in a low-temperature one-step selenization process [51]. The thickness of the grown layer depends on the amount of deposited precursor. Annular dark-field scanning transmission electron microscopy (ADF-STEM) was employed to analyze the structural data, phase variation, and mechanism of the heteroepitaxial growth process at the atomic level. In principle, the selenization strategy enables wafer-scale growth of pentagonal $PdSe_2$.

Interestingly, the growth of 2D $PdSe_2$ through a two-step selenizing process of predeposited Pd yields the theoretically predicted octahedral phase that is one type of usually stable hexagonal phases. Surprisingly, octahedral $PdSe_2$ occurs in the form of octahedral air-stable nanotubes [52]. Preliminary experiments with the hexagonal phase related to photoelectrochemical water splitting indicated potential applications in catalysis.

17.2.4 PdTe Monolayer

17.2.4.1 Chemistry and 2D Bonding

The electronic configuration of Pd atoms (Kr|4d^{10}) with fully occupied 4d^{10} orbitals transforms to the 4d^8 configuration of Pd^{2+} by hybridization of the 4$d_z{}^2$ orbitals of Pd atoms and the 5p orbitals of Te atoms with the electronic configuration Te (Kr|4d^{10}5$s^2$$p^4$). Contrary to the situation in PdS_2 and $PdSe_2$, the 1T-$PdTe_2$ phase of the three-layer structures is assumed to represent the most stable phase of $PdTe_2$ monolayers [53]. The electronic character of the 1T monolayer is semimetallic, as verified by its band structure, in which the negative system band gap decreases as the number of layers approaches the monolayer regime. The small difference in the electronegativities of Pd (2.2) and Te (2.1) (Pauling scale) leads to a minor ionic contribution to bonding with slightly polar Pd–Te bonds. Note that according to a Bader charge transfer analysis, a charge of $\delta q = 0.105$ e is transferred from each Pd atom to two neighboring Te atoms [35]. The use of the strongly constrained and appropriately normed (SCAN) meta-GGA (MGGA) approach to describe the exchange-correlation energy resulted in a cohesive energy of $E_c = -4.07$ eV/atom, which specifies the weakest bonding for 1T-$PdTe_2$ monolayers in the group of Pd compounds [53]. Calculations based on the PBE approach propose for pentagonal $PdTe_2$ $E_c = -3.61$ eV/atom, predicting for the pentagonal phase a further reduced binding strength [35].

17.2.4.2 2D Structures

The 1T structure of the $PdTe_2$ monolayer has the lattice parameters a = 0.396 nm, $d_{Pd\text{–}Te} = 0.267$ nm, $h_{TeTe} = 0.273$ nm, and a bond angle of $\theta_{Te\text{–}Pd\text{–}Te} = 83.91°$ [53]. Adding the van der Waals diameter of Te atoms of $d_w = 0.420$ nm and the distance between Te atoms h_{TeTe}, yields an estimate of the monolayer thickness of $d_{w+h} = 0.693$ nm. Quite similar lattice parameters were calculated for the 1H structure of the $PdTe_2$ monolayer [15]. The hexagonal $PdTe_2$ monolayers have a small indirect band gap of $E_g = 0.17$ eV in comparison to the pentagonal structure [54].

First-principles calculations and AIMD simulations confirmed the dynamic and thermal stability of the metastable pentagonal $PdTe_2$ monolayer in addition to that of the hexagonal phase [55]. The calculated lattice parameters are a = 0.614 nm, b = 0.644 nm, $d_{Pd\text{–}Te1} = 0.262$ nm, $d_{Pd\text{–}Te2} = 0.263$ nm, and $d_{Te\text{–}Te} = 0.280$ nm. The calculated indirect band gaps of the pentagonal semiconductor are $E_g = 1.24$ eV (PBE) and 1.84 eV (HSE) [55].

17.2.4.3 Mechanics

The 1T-$PdTe_2$ monolayer has a small elastic modulus of E = 61.8 N/m and large in-plane and out-of-plane Poisson ratios of $\nu_{in} = 0.32$ and $\nu_{out} = 0.64$ [53]. The slightly anisotropic stiffnesses presented for the 1H phase are only $E_{ac} = 37.07$ N/m and $E_{zz} =$ 37.27 N/m with similar Poisson ratios of $\nu_{ac} = 0.311$ and $\nu_{zz} = 0.312$ [15]. A study of the mechanical properties of pentagonal $PdTe_2$ revealed a large anisotropic behavior with stiffnesses at $E_{ac} = 23.2$ N/m and $E_{zz} = 51.0$ N/m and negative Poisson ratios at $\nu_{ac} = -0.001$ and $\nu_{zz} = -0.002$ [15]. The recently published strongly anisotropic stiffnesses of $E_{ac} = 23.5$ N/m and $E_{zz} = 51.8$ N/m agree quite well with the former results [35].

17.2.4.4 Physics and Applications

The stable hexagonal structure of 1T $PdTe_2$ undergoes a transition from the narrow-gap semiconducting phase of the monolayer to the metallic phase in multilayer sheets [56]. These multilayer sheets show robust superconductivity with a higher transition temperature than expected from the empirical inverse thickness dependence in the ultrathin limit. Moreover, first-principles calculations revealed a thickness-tuned band topology that is caused by a substantial orbital-dependent interlayer coupling. This coupling is the reason for the experimentally observed thickness dependence of the atomic and electronic properties. Combined with the predicted topologically nontrivial states, introduced by the Te-*p* orbitals of the honeycomb lattice, 2D $PdTe_2$ allows a detailed exploration of the interplay between superconductivity and topology in the 2D limit.

The dependence of the thermal conductivity of hexagonal 1T-$PdTe_2$ monolayers and its Janus structures on dimensional parameters such as length, the number of layers, and in addition on temperature was studied based on nonequilibrium molecular dynamics (NE-MD) simulations [57]. The thermal conductivity increases with increasing length (10–200 nm) and decreases with thickness of the sample due to the suppression of out-of-plane phonon modes already by interlayer van der Waals interactions. The lowest thermal conductivity, found for the Janus structures, is caused by the large phonon scattering from mass disorders and the breakdown of the inversion symmetry in comparison with the pristine sample.

A theoretical study compared the thermoelectric performance of hexagonal and pentagonal $PdTe_2$ monolayers [54]. Using the semiempirical Wiedemann–Franz law to calculate the electronic thermal conductivity, yielded for both structures ZT values above 3. Moreover, employing the Boltzmann transport equation to calculate the electronic thermal conductivity, the pentagonal phase showed better thermoelectric performance; however, both structures had the same figure of merit, $ZT \approx 0.8$.

First-principles calculations and the Boltzmann transport theory provided the thermoelectric properties of monolayers of the Janus compounds XMY (M: Pd, Pt; X, Y = S, Se, and Te; X $\neq$ Y), such as the thermal conductivity, Seebeck effect, power factor, and figure of merit for various carrier concentrations [58]. The lattice thermal conductivities decreased in the order of SPdSe, SPdTe, and SePdTe with *ZT* values close to one at 300 K. The maximum value of $ZT = 2.54$ was predicted for a monolayer of SePdTe at 900 K.

According to SCAN + SOC-based calculations of monolayer and bulk structures, the monolayer of PdS_2 has an orthorhombic pyrite structure, and that of $PdSe_2$ a hexagonal 1T structure with band gaps of $E_g = 1.399$ eV and 0.778 eV, respectively [59]. The indirect band gaps disappear with increasing thickness, forming stable pyrite bulk phases with semimetallic features. In contrast to these changes in electronic behavior with thickness, both the bulk $PdTe_2$ phase and the stable 1T-$PdTe_2$ monolayer possess semimetallic character. Importantly, 2D $PdTe_2$ and 2D $PtTe_2$ are Dirac semimetals that are receiving increasing attention owing to potential intrinsic quantum phenomena such as topological superconductivity (TSC).

17.2.4.5 Synthesis

The deposition of noble TMDCs is an active field of research. Despite strong interlayer coupling and orbital hybridization, it was possible to grow $PdTe_2$ monolayers and few-layer sheets. Millimeter-sized high-quality $PdTe_2$ monolayers were epitaxially grown under UHV conditions at a temperature of 470 °C by direct single-step tellurization of a Pd(111) surface, based on the low reaction barrier between Pd and Te [60]. XPS, STM, LEED, and scanning tunneling microscopy (STM) analyses confirmed the hexagonal structure and high crystallinity of the as-grown monolayers over a large area.

Hexagonal monolayers and 1T-$PdTe_2$ sheets (1–20 ML) were prepared by a bottom-up approach via MBE on a $SrTiO_3$(001) substrate [56]. This method allows controlled deposition by van der Waals bonded few-layer sheets of high-quality $PdTe_2$ with thickness-dependent atomic, electronic, and superconducting properties. The diagnostic methods used for analysis were STM/STS, ex situ transport measurements, combined with first-principles calculations.

Layer-by-layer growth of hexagonal $PdTe_2$ sheets by MBE on bilayer graphene (BLG)/SiC substrates at 210 °C yielded high-quality few-layer sheets with a thickness down to three monolayers [61]. The quality of the films was determined by in situ STM and angle-resolved photoemission spectroscopy (ARPES). The three monolayers thick films showed isolated spin-polarized surface resonance states separated from the bulk bands due to quantum size effects. Below a critical thickness of six monolayers, a hybridization gap appeared and increased gradually with further decreasing thickness. A topological surface Dirac cone opens this gap in the band structure. The realization of gapped topological surface states and isolated surface resonances is of interest in the field of quantum computing.

17.3 Platinum Dichalcogenide Monolayers (PtO_2, PtS_2, $PtSe_2$, and $PtTe_2$)

17.3.1 PtO_2 Monolayer

17.3.1.1 Chemistry and 2D Bonding

The valence electrons of the electronic configuration of Pt atoms (Xe|$4f^{14}5d^{8}6s^{2}$) and O atoms (He|$2s^{2}p^{4}$) determine the type of 2D bonding in 2D platinum dioxide. The *d–p* hybridization of the outermost 5*d* orbitals of Pt atoms and the outermost 2*p* orbitals of O atoms form strong covalent bonds. Hybridization prefers a covalently bonded three-layer structure with a Pt sublayer in the center sandwiched between two O sublayers. The significant difference in the electronegativities of Pt (2.2) and O (3.44) leads to a strong ionic contribution to bonding with strongly polarized Pt–O bonds. The different electronegativities induce a charge transfer of $\delta q = 1.37$ e from each Pt atom to two O atoms, i.e., a charge of $\delta q = 0.68$ e to each O atom. This charge transfer generates a positively charged inner Pt sublayer and negatively charged outer O sublayers [3]. The oxidation states of O and Pt approach − 2 and + 4, respectively, in the case of PtO_2. The calculated cohesive energy of the stable 1T-PtO_2 monolayer is $E_c = -5.77$ eV/atom and thus nearly identical with the cohesive energy of NiO_2 monolayers (− 5.78 eV/atom) [3]. Independent PBE-based calculations, however, delivered a cohesive energy of $E_c = -4.37$ eV/atom, predicting notably weaker bonding [62].

17.3.1.2 2D Structures

The calculated lattice parameters of the hexagonal 1T-PtO_2 monolayer are a = 0.313 nm, $d_{Pt\text{–}O} = 0.205$ nm, $h_{OO} = 0.193$ nm, and $\theta_{OPtO} = 80.34°$ [3]. Adding the van der Waals diameter of O atoms and the distance between O atoms, yields an estimate of the monolayer thickness of $d_{w+h} = 0.503$ nm. For the indirect semiconductor, large band gaps of $E_g = 1.83$ eV (PBE) and 3.34 eV (HSE06) are presented, as expected for bonds with large ionic contributions [3]. The lattice constant agrees well with other calculations [62]. The dynamically and mechanically stable three-layer structure has the space group *P*-3*m*1 [62].

Inspired by a study of *Pmmn*-ZrS_2 monolayers, first-principles calculations, MD simulations, and the phonon spectrum, revealed the thermodynamic, dynamic, energetic, thermal, and mechanical stability of orthorhombic PtO_2 monolayers with the space group *Pmmn* [63]. The optimized lattice parameters of the direct semiconductor are a = 0.305 nm, b = 0.576 nm, $d_{Pt\text{–}O1} = 0.195$ nm, $d_{Pt\text{–}O2} = 0.267$ nm, $\theta_{PtOPt1} = 103.11°$, and $\theta_{PtOPt2} = 110.24°$. Without spin-orbit coupling (SOC), the *Pmmn* PtO_2 monolayer is a nodal line semimetal protected by the glide mirror symmetry. This phase undergoes a nontrivial topological phase transition to a topological insulator (TI) under the SOC effect.

17.3.1.3 Mechanics

First-principles calculations for the mechanically stable 1T-PtO_2 monolayer yielded a Young's modulus of E = 116.5 N/m and a Poisson ratio of $\nu = 0.175$ [3]. As isotropic nonlinear mechanical properties, the authors present an ultimate strength of $\sigma = 13.4$ N/m and an ultimate strain of $\varepsilon \approx 0.15$ [3]. Furthermore, this publication presents slightly anisotropic elastic moduli of $E_{ac} = 122.6$ N/m and $E_{zz} = 123.3$ N/m.

17.3.1.4 Physics and Applications

Recent first-principles calculations derived missing information on the elastic constants, electronic structure, and thermal conductivities of the 1T phase of the group XB dioxides NiO_2, PdO_2, and PtO_2 [3]. A large lattice thermal conductivity of PtO_2 monolayers at 300 K of $\kappa = 17.43$ W m^{-1} K^{-1} was revealed by these calculations [3]. The largest electron and hole mobilities of the anisotropic PtO_2 monolayer are 404.71 and 270.60 cm^2 V^{-1} s^{-1}, respectively, are based on the application of deformation potential theory. The corresponding value of the lattice thermal conductivity obtained for NiO_2 monolayers is $\kappa = 53.55$ W m^{-1} K^{-1} and the largest electron and hole mobilities are 13,707.96 and 53.25 cm^2 V^{-1} s^{-1}, respectively.

Calculations based on the DFT approach and Boltzmann's transport equations delivered a lattice thermal conductivity of $\kappa = 12.41$ W m^{-1} K^{-1} at 300 K for PtO_2 monolayers [30]. Owing to the larger value of the power factor and the smaller value of the thermal conductivity in comparison to those of PdO_2, the *ZT* value for n-type doped PtO_2 monolayers of 1.04 is greater than that of PdO_2 monolayers at $ZT = 0.87$.

The thermoelectric properties of the four PtX_2 (X = O, S, Se, and Te) monolayers were compared by first-principles calculations within the ballistic transport regime at 300 K [64]. In these compounds, the high valley degeneracy in the electronic bands helps to improve the Seebeck coefficient and power factor, and in addition, the low dimension of the 2D structures reduces the lattice thermal conductivity. These factors lead to the following figures of merit $ZT = 1.65$ ($PtTe_2$), 1.42 (PtS_2), 0.94 (PtO_2), and 0.84 ($PtSe_2$), which is in good agreement with the value given above for PtO_2 at room temperature.

First-principles DFT calculations suggest the application of the van der Waals hetero-bilayer PtO_2/GaN for photocatalytic water splitting [65]. The bilayer has an indirect band gap with an absorption coefficient of ~ 2×10^5 cm^{-1} in the visible spectrum. Furthermore, the band gaps have a type-II band orientation, which is favorable for efficient spatial carrier separation in photocatalytic water splitting. The band edges with type-II band alignment possess sufficient kinetic overpotential for the HER and the OER. Notably, homogeneous compressive and tensile biaxial strains induce a change in the band gap, shift the absorption coefficient, and tune the band edges.

The already experimentally synthesized Janus monolayers OPtS and OPtSe exhibit additional beneficial properties for photocatalytic water splitting [66]. In addition to the high electron mobility and large optical absorption coefficient, the large disparity between the electron and hole mobilities and the intrinsic dipole-induced built-in electric field improve the efficient separation of the photogenerated electrons and holes. Furthermore, under the potentials of photoexcited electrons and holes, S and Se vacancies can effectively lower the free energy difference of the HER and induce a spontaneous catalytic reaction.

Spherical PtO_2 nanoparticles decorated by a chemical reduction process on a MoS_2 nanostructured gas sensor enhance the response of the pristine MoS_2 gas sensor to NH_3 [67]. The gas response of the as-obtained decorated sensor was almost 20 times greater than that of the pristine MoS_2 gas sensor under optimal working conditions. The catalytic spillover effect of PtO_2 nanoparticles and the formation of *p-p* heterojunctions on the MoS_2 surface have been invoked to explain the enormous improvement.

17.3.1.5 Synthesis

To date, there have been no reports on the realization of extended free-standing or surface-stabilized PtO_2 monolayers. A sharp LEED pattern appears by oxidation of the Pt(111) surface at 1250–1350 K and a partial oxygen pressure of 5×10^{-5} mbar, induced by self-limited growth of well-ordered PtO_2 nanoclusters with a triangular shape and an O–Pt–O trilayer structure with a thickness of $t = 0.32$ nm [68]. The experiments were performed in an ultra-high vacuum (UHV) chamber equipped with STM, LEED, and Auger electron spectroscopy (AES). The LEED data of the nano-structured PtO_2/Pt(111) surface showed a characteristic (2×2) pattern in which the tensile strain was reduced to ~ 2%. The PtO_2 nanoclusters provide ordered and uniformly sized catalytic centers.

First-principles calculations based on GGA-PBE functionals and the projector-augmented wave (PAW) method, describing the electron-electron and electron-core interactions, suggest that PtO_2 monolayers can be exfoliated from layered bulk PtO_2 crystals and may survive even at high temperatures [69]. Moreover, the PtO_2 monolayer may form a stable van der Waals heterostructure with MoS_2 that closely matches the lattice parameters. Furthermore, according to theory, the PtO_2/MoS_2 heterostructure has a type-II band structure with electron and hole carriers distributed in different spatial layers of the heterostructure. Together with the excellent light absorption, the PtO_2/MoS_2 heterostructure qualifies for wide application in photovoltaics and nanoelectronics.

17.3.2 PtS_2 Monolayer

17.3.2.1 Chemistry and 2D Bonding

The valence electrons of the outer shells of the electronic configuration of Pt atoms (Xe|$4f^{14}5d^{8}6s^{2}$) and S atoms (Ne|$3s^{2}p^{4}$) determine the type of 2D bonding in 2D platinum disulfide. The *d-p* hybrids of the outermost Pt-5*d* orbitals and the outermost S-3*p* orbitals form strong covalent bonds. The hybridization process prefers the covalently bonded three-layer 1T structure with a Pt sublayer in the center sandwiched between two S sublayers. The difference in the electronegativities of Pt (2.2) and S (2.58) leads to charge donation of $\delta q = 0.4$ e from an inner Pt atom to two outer S atoms and thus to an ionic contribution to bonding [70]. The published cohesive energies are $E_c = -4.68$ eV/atom [70] (PBE) and -5.73 eV/atom [53] (meta-GGA) for the 1T-PtS_2 monolayer. The cohesive energy of pentagonal PtS_2 monolayers is in the same range at $E_c = -5.01$ eV/atom [14].

17.3.2.2 2D Structures

The lattice parameters of the 1T-PtS_2 monolayer are a = 0.352 nm, $d_{Pt\text{–}S} = 0.237$ nm, $h_{SS} = 0.245$ nm, and $\theta_{S\text{–}Pt\text{–}S} = 84.25°$ [53], in good agreement with the results of succeeding studies [70, 71]. Adding the van der Waals diameter of S atoms and the distance between S atoms, yields an estimate of the monolayer thickness of $d_{w+h} = 0.605$ nm. A larger difference can be found for the separation of S atoms at $h_{SS} = 0.324$ nm, implying a greater monolayer thickness [72]. The values of the indirect band gaps presented in the literature for the semiconductor are $E_g = 1.81$ eV (PBE) and 2.58 eV (HSE06) [71]. Of interest are calculations of the band gap for thicknesses between 1 and 10 layers [73]. The authors tested several pseudopotentials combined with different van der Waals corrections and found the best results for the GGA-PW91 approach combined with the optB88-vdW functional for van der Waals corrections, in view of the best agreement of the calculated lattice parameters with available experimental values. The calculations revealed a significant decrease of the monolayer band gap of the indirect semiconductor from $E_g = 1.68$ eV to 0.30 eV for a sheet with 10 layers, whereas the formation energy was almost constant [73]. Note that the gap of the ten-layers sheet already approaches the gap of the bulk at $E_g = 0.25$ eV, indicating that the major changes occur within the first ten layers [74].

The lattice parameters of the pentagonal PtS_2 monolayer are a = 0.548 nm, b = 0.557 nm, $d_{Pt\text{–}S1} = 0.234$ nm, $d_{Pt\text{–}S2} = 0.233$ nm, and $d_{S\text{–}S} = 0.215$ nm [14]. The indirect band gap of the semiconducting monolayer is $E_g = 2.01$ eV (PBE). The large binding strength of the low-symmetry phase with a cohesive energy of $E_c = -5.01$ eV/atom originates from strong polar Pt–S bonds and pure covalent S–S bonds.

17.3.2.3 Mechanics

First-principles calculations revealed a Young's modulus of E = 105.8 N/m for the mechanically stable 1T-PtS_2 monolayer and $\nu_{in} = 0.29$ and $\nu_{out} = 0.58$ for the in-plane and out-of-plane Poisson's ratios [53]. Other calculations obtained a substantially smaller elastic modulus of = 82.1 N/m, however, a similar Poisson ratio of ν = 0.268 [71]. While the lower stiffness is in good agreement with E = 81 N/m[75], a larger stiffness of E = 99 N/m and a similar Poisson ratio of ν = 0.29 can also be found in the literature [70]. As the atomic number of the chalcogen atom increases, the stiffness decreases from 99 N/m (PtS_2) to 88 N/m ($PtSe_2$) and 71 N/m ($PtTe_2$). Note that other studies presented systematically lower stiffness values at 82.10 N/m (PtS_2), 67.5 N/m ($PtSe_2$), and 61.2 N/m ($PtTe_2$) [71], as well as 81 N/m (PtS_2), 64 N/m ($PtSe_2$), and the smallest value of 40 N/m ($PtTe_2$) [75].

The literature gives examples in which the enlargement of an existing defect induced counterintuitive mechanical behavior concerning the expected mechanical degradation. Classical MD simulations elucidated the mechanical properties of PtS_2 monolayers with rectangular defects [76]. Extension of the defect in the zigzag direction and application of uniaxial tension in the same direction increased the Young's modulus only slightly, while both the fracture strength and strain showed greater effects on increasing the defect length. On the other hand, uniaxial tension in the perpendicular armchair direction significantly decreased strain and critical stress values and thus the stiffness as the defect enlarged.

17.3.2.4 Physics and Applications

First-principles calculations (DFT-PBE) were used to determine the transport and thermoelectric properties of PtS_2 monolayers [77]. Since the conversion of thermal energy into electrical energy depends on the motion of carriers (electrons and holes), 2D materials with unique electron and phonon transport characteristics are currently receiving increased attention. Calculations of the thermal and electronic transport properties provided the *ZT* values of the PtS_2 monolayer. The largest *ZT* values for p-type and n-type pentagonal PtS_2 monolayers were 0.29 and 0.56, respectively. Under 1% strain, the figure of merit increases to $ZT = 0.7$.

The replacement of the upper chalcogen atoms in hexagonal 1T-PtS_2 and 1T-$PtSe_2$ monolayers (P-3m1) by oxygen atoms in a selective oxidation process yields OPtS and OPtSe with the lower symmetry of the space group $P3m1$. The structural, electronic, optical, and thermoelectric properties of these dynamically stable Janus structures were determined via first-principles calculations [78]. The indirect semiconductors have energy gaps of E_g = 1.35 eV [Wu-Cohen (WC)] and 2.44 eV (HSE06) for OPtS, as well as 0.978 eV (WC) and 1.98 eV (HSE06) for OPtSe. Moreover, the thermoelectric parameters, including the electrical and thermal conductivity, Seebeck coefficient, power factor, and the figure of merit at $ZT \approx 1$, were determined.

The easiest way to obtain Janus-structured SPtSe is to selectively convert the upper Se sublayer of a $PtSe_2$ monolayer that is strongly bonded on a Pt(111) surface,

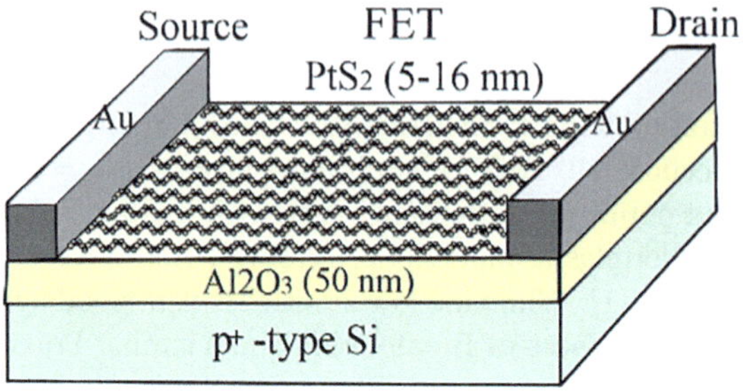

Fig. 17.7 Scheme of a FET device prepared from few-layer PtS_2 sheets on an Al_2O_3/Si substrate as the basis. Adapted with permission from Ref. [80], Copyright 2019, John Wiley and Sons

to an SPtSe monolayer by sulfurization under a H_2S atmosphere [79]. Observation of the transformation process is possible in situ and operando by grazing incidence synchrotron X-ray diffraction (GIXRD) measurements. While the crystalline long-range order of the as-grown monolayer disappears during thermal annealing, the presence of a sulfur source yields highly ordered SPtSe crystals. As indicated by angle-resolved XPS, the selective substitution of the topmost layer is due to different kinetic barriers for the substitution of top and bottom Se atoms. Interestingly, longer exposure to H_2S at higher temperatures allows the formation of PtS_2 pinned to the Pt(111) surface. Note that the change in chemical composition breaks the inversion symmetry and introduces spontaneous electrical polarization and a piezoelectric character to the monolayer.

A FET device fabricated with few-layer PtS_2 sheets exhibited an excellent electronic mobility of > 62.2 cm^2 V^{-1} s^{-1} and a high on/off ratio of $> 10^6$ at room temperature (see Fig. 17.7) [80]. The transfer of the mechanically exfoliated PtS_2 flakes to an Al_2O_3/Si substrate was the initial step. Electron beam lithography provided the source and drain of the FET, and sequential thermal evaporation completed the deposition of the metals (chromium and gold). Since 2D-PtS_2 photodetectors possess absorption from the visible to mid-infrared region with a fast photoresponse time of 175 μs at 830 nm, PtS_2 sheets are candidates for high-performance electronic and mid-infrared optoelectronic devices.

17.3.2.5 Synthesis

The main reasons for the difficulties encountered in the synthesis of high-quality PtS_2 monolayers are the mobility of Pt atoms and the formation of the more stable PtS phase [81]. Therefore, careful control of the critical deposition parameters such as the amount of supplied sulfur, the amount (thickness) of the predeposited Pt layer, and the gas flow rate is necessary during the thermally induced atmospheric CVD process. Fine tuning of the growth parameters allows a controlled transition from PtS to PtS_2 deposits, as demonstrated by Raman spectroscopy and XPS analysis. Importantly, various substrates, such as SiO_2/Si, sapphire, p-type Si, and mica deliver appropriate surfaces under optimized deposition conditions.

Nevertheless, the synthesis of high-quality extended PtS_2 monolayers is still challenging. However, large-area few-layer PtS_2 sheets were deposited on an inch-scale Si wafer with a 300 nm thick SiO_2 layer [82]. A Pt layer was sputtered by PVD on a SiO_2/Si substrate, and direct sulfurization was achieved by heating the sulfur powder and using nitrogen as the carrier gas in a plasma-enhanced CVD furnace. Characterization of the PtS_2 sheets by PL measurements, Raman spectroscopy, scanning electron microscopy (SEM), high-resolution TEM (HR-TEM), AFM, and XPS confirmed the quality and uniformity of the approximately 5 nm thick sheets corresponding to 9–10 monolayers.

Importantly, the deposition of uniform bilayer 1T-PtS_2 single crystals on a mica substrate is possible by CVD [83]. Depending on the growth temperature, the morphology changes from triangular to hexagonal. Interestingly, the stable bilayer flakes showed promising properties for the HER, with an overpotential of 184 mV at 10 mA/cm, a Tafel slope of 81 mV/dec, an exchange current density of 43 μA/cm^2, and an outstanding stability.

17.3.3 *$PtSe_2$ Monolayer*

17.3.3.1 Chemistry and 2D Bonding

The valence electrons of the outer shell of the electronic configuration of Pt atoms (Xe|$4f^{14}5d^8 6s^2$) and Se atoms (Ar|$3d^{10}4s^2p^4$) determine the type of 2D bonding in 2D platinum diselenide. The *d-p* hybridization of the outermost 5*d* orbitals of Pt and the outermost 4*p* orbitals of Se form strong covalent bonds. Hybridization prefers the covalently bonded three-layer 1T structure with a Pt sublayer in the center sandwiched between two Se sublayers. Based on the slightly greater electronegativity of Se (2.55) than that of Pt (2.2), a minor charge transfer from Pt to Se atoms is expected. Surprisingly, a Bader charge analysis indicates that no donation of charge occurs between Pt and Se [70]. The published cohesive energies of 1T-$PtSe_2$ monolayers range from − 4.42 eV/atom [70] (PBE) to − 5.32 eV/atom [53] (meta-GGA).

17.3.3.2 2D Structures

The lattice parameters of the 1T-$PtSe_2$ monolayer are a = 0.368 nm, $d_{Pt\text{–}Se}$ = 0.249 nm, h_{SeSe} = 0.260 nm, and $\theta_{Se\text{–}Pt\text{–}Se}$ = 84.83° [53], in good agreement with the results of other studies [70, 71]. A larger difference can be found for the separation of the Se atoms with h_{SeSe} = 0.339 nm, which leads to a much thicker monolayer [72]. Adding the van der Waals diameter of Se atoms of d_w = 0.380 nm and the distance between Se atoms h_{SeSe} = 0.260 nm yields an estimate of the monolayer thickness of d_{w+h} = 0.640 nm. For this semiconductor with indirect band gap, the calculated energy gaps are 1.40 eV (PBE) and 1.94 eV (HSE06) [71].

Of general interest are calculations of the band gap for thicknesses between 1 and 10 layers of the octahedral 1T phase [73]. The authors assessed several pseudopotentials combined with different van der Waals correction functionals and found the best results for the GGA-PW91 approach combined with the optB88-vdW functional for van der Waals corrections. The selection was based on the best agreement of the calculated lattice parameters with the experimental values. The calculations reveal a large decrease in the band gap from $E_g = 1.18$ eV (monolayer) to $-$ 0.71 eV (10 layers), where a negative value indicates a semimetal [73].

Note that $PtSe_2$ transforms from a type-II Dirac semimetal in the bulk to an indirect semiconductor in the monolayer [84]. Remarkably, the complete transition from the semimetal to the semiconductor occurs between the trilayer and monolayer, with the disappearance of interlayer interaction. Such an atomically thin 1T-$PtSe_2$ monolayer is an excellent electrocatalyst with a comparable HER efficiency as a traditional Pt catalyst.

Based on DFT calculations, using the full-potential of the linearized augmented plane wave (FP-LAPW) method and the modified Becke–Johnson potential (mBJ), electronic properties such as the density of states (DOS), the energy band structure, and optical properties of monolayer and bulk $PtSe_2$ were compared [85]. These calculations confirmed that the bulk material indeed is a semimetal, whereas nanosheets of $PtSe_2$ were identified as semiconductors with an energy gap of 1.5 eV.

17.3.3.3 Mechanics

First-principles calculations predict as Young's modulus $E = 87.0$ N/m for the mechanically stable 1T-$PtSe_2$ monolayer and $\nu_{in} = 0.26$ and $\nu_{out} = 0.59$ as in-plane and out-of-plane Poisson's ratios [53]. This stiffness is in good agreement with 88 N/m calculated by the DFT-PBE approach yielding as Poisson's ratio $\nu = 0.31$ [70]. Another study obtained a smaller Young's modulus of 67.5 N/m but a comparable in-plane Poisson's ratio of $\nu = 0.271$ [71]. Assuming a monolayer thickness of $d_{w+h} = 0.640$ nm for $PtSe_2$ to transform the slightly anisotropic Young's moduli of $E_a = 115.9$ GPa and $E_b = 117.7$ GPa from 3D to 2D units, yields medium stiffnesses of $E_a = 74.2$ N/m and $E_b = 75.3$ N/m [86].

17.3.3.4 Physics and Applications

Regarding charge transfer and electrochemical activation, all Pt dichalcogenides are suitable electrocatalysts owing to the fast heterogeneous electron transfer (HET) rates observed in electrochemically reduced Pt dichalcogenides [87]. Their HER performance follows the trend $PtTe_2 > PtSe_2 > PtS_2$. According to DFT calculations, this correlation relates to the change in the electronic structure from a semiconductor (PtS_2) to a semimetal ($PtSe_2$) and a metal ($PtTe_2$). Large-area and edge-rich multilayer $PtSe_2$ sheets synthesized via selenization of Pt, employing magnetron sputtering, improved the HER activity of layered $PtSe_2$ by increasing the number of active edge sites [88]. Tuning the sheet thickness allowed a control of the density of edge sites at the atomic scale. As predicted by DFT calculations, the HER activity increases significantly with the density of active edge sites, which can be as high as 81% with a maximum cathodic current density of 227 mA/cm. Figure 17.8 compares the simulated free energies for HER on a bare surface, an antisite defect with Se replaced by Pt (Pt_{Se}), a Se vacancy defect (V_{Se}), and the 1T edge at low coverage ($\leq$1/4 monolayer). In good agreement with the simulations, a linear relationship between the edge density and the HER activity is observed.

Traditional gas sensors based on semiconducting oxide materials show excellent sensitivity, however, suffer from poor recovery and elevated temperature operation, which leads to significant power consumption and safety concerns. Metal-selenide materials, such as WSe_2, $MoSe_2$ or $PtSe_2$, allowing low-power operation, have received increasing attention because of their layered structure, high surface-to-volume ratio, reactive surface area, and low-power sensing [89]. In these selenide materials sensing is based on pronounced charge transfer, electrocatalytic activity, and tremendous absorption capabilities. They qualify for innovative sensing applications, e.g., as flexible and wearable gas sensors.

Few-layer $PtSe_2$ nanosheets (7–12 nm) coated on a glass substrate exhibit antibacterial activity against *Escherichia coli* under dark and light irradiation conditions according to the ISO standard procedure for semiconducting photocatalytic materials [90]. Selenization by the thermally assisted conversion (TAC) process provided the 2D-$PtSe_2$ layers employed in this study. Electron paramagnetic resonance (EPR) analysis confirmed the production of reactive oxygen species such as *OH under light irradiation. In the dark, the antibacterial activity increased with the Pt deposition time and the viability of the bacteria was reduced to 30% (Pt deposition 3 s) and to 15% (Pt deposition 10 s) after treatment for 6 h. Under light irradiation, the activity of $PtSe_2$ was like that in the dark for Pt deposition of 3 s; however, for Pt deposition of 10 s, the reaction was highly effective, with a reduction in bacterial viability to 1.2% after 6 h. In conclusion, the effect of reducing the viability of bacteria is related to the constructive interaction of several parameters, such as increasing film thickness, film roughness, and surface coverage.

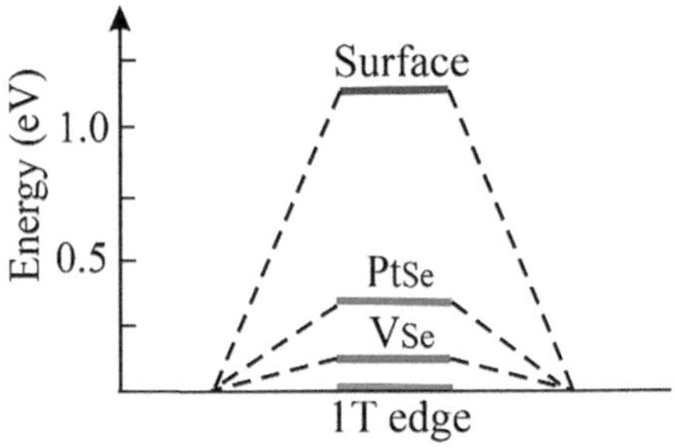

Fig. 17.8 Free energy diagram for HER on a bare surface, an antisite defect with Se replaced by Pt (Pt_{Se}), a Se vacany defect (V_{Se}), and the 1T edge at low coverage ($\leq$1/4 monolayer). Adapted with permission from Ref. [88], Copyright 2017, Elsevier

17.3.3.5 Synthesis

The deposition of high-quality single-crystal $PtSe_2$ monolayers was achieved by epitaxial growth in a UHV chamber employing direct selenization of a Pt(111) substrate by evaporation of Se from a Knudsen cell via a precursor-free method [91]. The growth process was monitored in situ by XPS, the atomic structure of the epitaxial monolayers was determined by STM and LEED, the thickness was obtained by high-angle ADF-STEM measurements, and the electronic structure was determined by ARPES, which confirmed the semiconducting nature of the deposited monolayer.

An alternative and simpler chemical deposition method for growing large-scale $PtSe_2$ monolayers is based on the immersion of a Pt(111) substrate into a Na_2Se aqueous solution [92]. Annealing of the sample under UHV conditions induced significant modifications in the structure and intrinsic electronic properties. Synchrotron-based high-resolution XPS, low-temperature STM, and LEED measurements revealed various temperature-dependent phase transitions. For example, the LEED analysis revealed a complex transition from a (4×4) structure to a (2×2) superstructure.

Top-down liquid-phase exfoliation (LPE), together with centrifugation, produces widely basal plane defect-free $PtSe_2$ nanosheets of varying sizes and thicknesses [93]. The authors linked quantitative data on the lateral sheet dimensions obtained by AFM measurements and the information on dimensions contained in the optical spectra. This procedure established a connection between the lateral size and layer number of the LPE nanosheets based on optical absorbance and extinction spectroscopy. Such a connection between sheet properties and Raman data allows quality control via optical spectroscopy and would be extremely useful as a characterization technique in high-throughput production processes.

The synthesis of large-scale ($\approx$150 μm in length) high-quality 1T-$PtSe_2$ nanoribbons is possible by atmospheric pressure CVD (APCVD) using a $(NH_4)_2PtCl_4$ solution and Se powder as precursors for deposition on an annealed Au(111) foil [94]. In particular, the monolayer samples showed strong catalytic activity for the HER process owing to the abundant effective catalytic sites. Theoretical and experimental analyses of the overall activity of the ribbon edges, base planes, and defects of the $PtSe_2$ nanoribbons revealed that the best performance was achieved at the monolayer level.

As mentioned above, exfoliation allows the fabrication of 2D crystals with the highest quality. Such monocrystalline $PtSe_2$ sheets with layer-controlled thicknesses between 1 and 10 layers were exfoliated on fused silica substrates. The samples were used to establish criteria for film quality using characteristic Raman signatures as a reference [95]. As demonstrated by the authors, the crystalline quality can be assessed from the width and height of the E_g Raman mode, with a linewidth narrowing down to 4.2 cm^{-1} for the highest quality monolayer and 2.5 cm^{-1} for thicker exfoliated samples (7 monolayer to bulk). Instead of using the commonly employed shift in the E_g peak for layer count identification or the intensity ratio of the A_{1g} and E_g peaks, the authors recommend a specific peak pattern of the A_{1g} and *LO* peaks as quality criteria of grown films.

A comprehensive review of 2D-$PtSe_2$ monolayers and few-layer sheets concentrates on diverse deposition methods, such as mechanical exfoliation, CVD, thermally assisted conversion (TAC), MBE, and chemical vapor transport (CVT). In addition to a description of recent achievements in fundamental science such as structural, electronic, and optical properties, various potential applications of these materials are discussed. The presentation includes photodetectors, mode-locked lasers, FETs, photocatalysts, gas sensors, and HER [96].

17.3.4 PtTe₂ monolayer

17.3.4.1 Chemistry and 2D Bonding

The electronic configurations of Pt atoms (Xe|$4f^{14}5d^{8}6s^{2}$) and Te atoms (Kr|$4d^{10}5s^{2}p^{4}$) determine the type of 2D bonding in 2D platinum ditelluride. The valence electrons of the outermost 5*d* orbitals of Pt and the outermost 5 *s* and 5*p* orbitals of the Te atoms form covalent bonds by d–sp hybridization that is restricted to two dimensions. This d^2sp^3-type hybridization leads to the three-sublayer 1T-$PtTe_2$ structure, as discussed for the other Pt dichalcogenides, with a Pt sublayer in the center sandwiched between two Te sublayers. According to the Pauling scale, the electronegativity of Pt atoms (2.2) is slightly greater than that of Te atoms (2.1). Accordingly, a charge transfer of $\delta q = 0.3$ e occurs from the Te atom to the Pt atom, indicating a change of polarization in $PtTe_2$ [70]. The calculated cohesive energies of $PtTe_2$ monolayers are $E_c = -4.28$ eV/atom [70] (PBE) and -5.07 eV/atom [53] (meta-GGA).

17.3.4.2 2D Structures

The lattice parameters of 1T-$PtTe_2$ monolayers are a = 0.395 nm, $d_{Pt–Te} = 0.266$ nm, $h_{TeTe} = 0.274$ nm, and $\theta_{TePtTe} = 84.15°$ [53], in good agreement with the results of other studies [70, 71]. Adding the van der Waals diameter of Te atoms and the distance between Te atoms, yields an estimate of the monolayer thickness of $d_{w+h} = 0.694$ nm. For the semiconductor with indirect band gap, the calculated energy gaps are $E_g = 0.76$ eV (PBE) and 1.13 eV (HSE06) [71]. Of general interest are calculations of the band gap for thicknesses between 1 and 10 monolayers [73]. As described above, the authors used the GGA-PW91 approach combined with the optB88-vdW correction. The calculations revealed a large decrease in the band gap from 0.4 eV (monolayer) to -1.19 eV (10 layers), where the negative value indicates a semimetal [73]. Note that the semimetallic $PtTe_2$ bulk phase transforms to a semiconductor in the monolayer. Remarkably, the complete transition from the semimetal to the semiconductor occurs between bilayer and monolayer [73]. The reason for the postulation of a metallic character of bulk $PtTe_2$ seems to be the neglect of SOC [97].

The optimized lattice parameters of the metastable pentagonal $PtTe_2$ structure are $a = 0.611$ nm, $b = 0.642$ nm, $d_{Pt–Te1} = 0.263$ nm, $d_{Pt–Te2} = 0.262$ nm, and $d_{Te–Te} = 0.282$ nm [98]. The monolayer with the space group $P2_1/c$ is an indirect semiconductor with calculated band gaps of $E_g = 1.37$ eV (PBE) and 2.07 eV (HSE06). Determination of the phonon spectra and ab initio MD simulations confirmed the dynamic, thermal, and thermodynamic stability of the pentagonal phase of 2D $PtTe_2$. As discussed below, $PtTe_2$ monolayers exhibit the smallest lattice thermal conductivities and the greatest figures of merit in the group of Pt dichalcogenides [98].

17.3.4.3 Mechanics

First-principles calculations for the mechanically stable 1T-$PtTe_2$ monolayer yielded as Young's modulus $E = 81.4$ N/m and for the in-plane and out-of-plane Poisson's ratios $\nu_{in} = 0.26$ and $\nu_{out} = 0.57$ [53]. This stiffness is in reasonable agreement with the value of $E = 73$ N/m presented together with an in-plane Poisson ratio of $\nu = 0.31$ [70]. Further calculations obtained an even smaller elastic modulus of only $E = 61.2$ N/m and a comparable Poisson ratio of $\nu = 0.274$ [71]. In addition, this work presents both Young's moduli and Poisson's ratios for the Janus PtXY ($X \neq Y =$ S, Se, and Te) monolayers with a mechanical performance in between that of the corresponding compounds containing only one chalcogen element.

17.3.4.4 Physics and Applications

In a systematic theoretical study, the formation energies and band structures of the dichalcogenides of Nb, Mo, Ta, W, Ti, V, Zr, Hf, and Pt were calculated [99]. The band structure calculations performed for single-layer Pt dichalcogenides by the many-body G_0W_0 approximation identified only PtS_2 and $PtSe_2$ as potential photocatalysts for water splitting by comparing the band edge positions with the redox potentials of water [99]. In terms of charge transfer, all Pt chalcogenides exhibit efficient electrochemical activation, as is evident in fast heterogeneous electron transfer rates [87]. In addition, the large solvation enthalpies of these candidates suggest their stability in aqueous solution.

An active field of research is the enhancement of catalytic activity by increasing the density of active sites on TMDC catalysts, e.g., by defect engineering. First-principles calculations revealed the possibility of defect engineering of the basal planes of $PtSe_2$ and $PtTe_2$ [100]. Defects such as double vacancies and B-doping can introduce quasi-localized states in the band gap, and these states interact with H atoms, improving the HER activity. The activity of tellurides is superior in terms of the free energy ΔG_{H*}, and B doping was more effective than vacancy doping in triggering the HER.

The application of the Vienna ab initio simulation package (VASP) and BoltzTraP codes allowed the determination of the thermoelectric properties of the anisotropic pentagonal PtX_2 (X = S, Se, and Te) monolayers [98]. Penta-$PtTe_2$ monolayers exhibit the smallest lattice thermal conductivities of $\kappa = 1.77$ and 5.17 W m^{-1} K^{-1} at 300 K, which result in *ZT* values of 2.60 and 1.14 in the x and the y directions, respectively. At 600 K, the value of p-type pentagonal $PtTe_2$ increases to $ZT = 5.03$ in the x direction. Note that for monolayers with the same structure, a large lattice constant and mass and a low vibration frequency result in a low lattice thermal conductivity. Calculations confirmed the low phonon thermal conductivities of the Pt dichalcogenides that can be enhanced by applying biaxial tensile strain. This behavior was explained by the enhancement of the phonon scattering rate due to strong phonon anharmonicity [101].

First-principles calculations and deformation potential theory revealed the role of strain engineering on the mechanical behavior, phonon dispersion, electronic structure, and electronic transport properties of pentagonal $PtTe_2$ monolayers [102]. Thermal stability is confirmed by ab initio MD simulations up to 600 K. Specifically, the large anisotropy of the hole mobility is demonstrated in this work.

The controllable wafer-scale production of type-II Dirac heterostructures of $PtTe_2$/graphene with tunable thickness, variable size, exact stoichiometry, and complementary metal-oxide-semiconductor (CMOS) compatibility, allowed the realization of self-driven, broadband photodetectors [103]. Due to the built-in electric field at the junction area, the photodetector is self-driven with broadband photodetection between 405 and 1850 nm. Moreover, the photodetector had a response time of ~ 8.4 μs, and a responsivity without bias of ~ 0.52 A/W.

The enormous progress made in recent years in the investigation of 2D noble transition-metal dichalcogenide (NTMDC) MX_2 (M = Pd, Pt; X = S, Se, and Te) materials has been reviewed extensively [104]. Increasing interest is based on the unique properties of these compounds, such as nearly full occupation of *d* orbitals with abundant *d* electrons, largely tunable electronic structures, and strongly changing properties with the number of layers. In addition, to introducing the structure and characterization of FETs, this review concentrates on the production of FETs and photodetectors and applications in catalysis and as sensors. NTMDCs are emerging materials for the application in optical and mechanical devices, and as gas sensor[105]. The relevant chemical and physical features of these materials include besides abundant valence electrons environmental stability and high carrier mobility.

17.3.4.5 Synthesis

CVD allows the production of nanosheets of $PtTe_2$ with hexagonal and triangular structures, lateral dimensions of up to 80 μm, and thicknesses of 1.8–20 nm [106]. The growth temperature, flow rate, and nature of the carrier gas influence the thickness of the deposit and therefore allow its tuning. Obviously, the strong interlayer interaction and orbital hybridization hinder the growth of extended monolayers. The XRD, TEM, Raman, and electron diffraction analyses demonstrated the high quality of the single crystals. Electrical transport measurements revealed the excellent conductivity of the nanosheets, with values up to 2.5×10^6 S/m.

As described above for $PdTe_2$, the tellurization of Pt(111), taking advantage of the direct surface reaction of Te atoms, permits the growth of patterned $PtTe_2$ monolayers [60]. The deposit showed a well-ordered (2 × 2) pattern realized by arrays of Te vacancies. STM, XPS, and LEED analyses confirmed the large area and high quality of the monolayers.

The effect of growth temperature on the quality of $PtTe_2$ sheets deposited by MBE on highly oriented pyrolytic graphite (HOPG) was investigated via XPS, Raman spectroscopy, and STM/STS combined with DFT calculations. In addition, the layer-dependent electrical properties were determined [107]. The stack of a monolayer and bilayer $PtTe_2$ deposit grows below 250 °C, whereas at approximately 300 °C, a large-area continuous bilayer is formed. Theoretical calculations support the pronounced stability of bilayers over other assemblies above a critical size of ~ 100 nm^2. DFT calculations confirmed the indirect band gap of the semiconducting monolayer with a gap of 0.80 eV (PBE).

17.4 Summary of Group–XB Transition-Metal Dioxide and Dichalcogenide Monolayers

17.4.1 *Binding–Structure Relationships*

Figure 17.9 shows the cohesive energy versus bond length of the short M–X bonds for the MX_2 (M: Ni, Pd, and Pt; X: O, S, Se, and Te) monolayers. Note that in the stable three-layer structures, the separation of M atoms is larger than the mixed metal–chalcogen bonds. The plot shows the lower values of the often-deviating cohesive energies calculated by different theoretical approaches. The results document the inverse relationship between cohesive energy and bond length of the polar M–X bonds with variable ionicity, as predicted by tight-binding theory.

Note the tight coincidence of the dependencies of the cohesive (binding) energy versus bond length plots for the related groups of monolayers with diverse transition metals. Interestingly, graphene fits into this correlation as a single atomic layer, while the other monolayers consist of covalent-ionic three-layer assemblies. The difference in the cohesive energies of separate phases, such as the 1T and 1H triple-layer structures is small. This is due to the similar hybridization between the *d* orbitals of the transition atoms and the *p* orbitals of the chalcogen atoms. The related bonding types are also responsible for the approximately identical decay of the three plots, where the size of the atoms plays a dominant role. In contrast to the situation observed for the main group elements, the van der Waals diameters of Ni (0.40 nm), Pd (0.410 nm), and Pt (0.410 nm) and the related electronic structure imply that the orbitals and thus the binding energies of each transition metal with that of a particular chalcogenide atom are comparable. Note that most transition metals have similar van

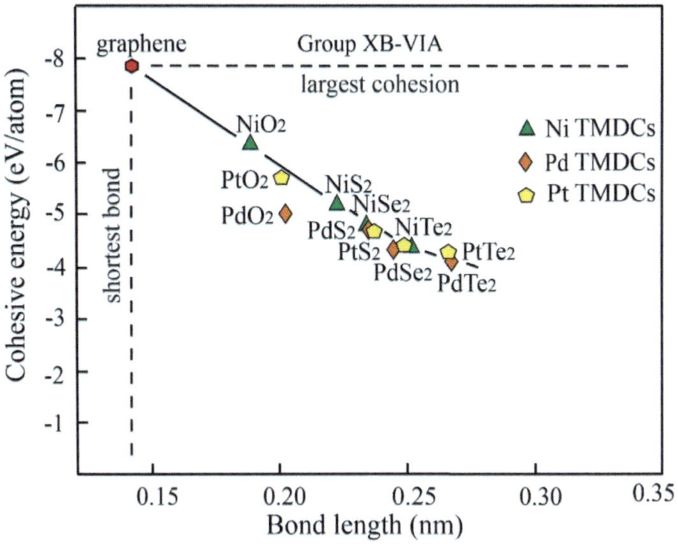

Fig. 17.9 Plot of the cohesive energies versus M–X bond lengths of group XB–VIA monolayers with MX_2 stoichiometry

der Waals diameters in the range of 0.42 ± 0.02 nm, e.g., the elements from group IVB to group XIIB.

17.4.2 Mechanics–Structure Relationships

Figure 17.10 shows calculated Young's moduli and the few available ultimate strength values as a function of the M–X bond lengths. Note that both, stiffness and strength of chemically bonded three-layer assemblies do not depend sensitively on the bond length, as observed for single-layer structures. Furthermore, the reported in-plane stiffnesses are surprisingly small for covalently bonded three-layer networks with respect to graphene, which is not correlated with the stiffness-bond length plots. The reason is that for a certain bond length the cohesive energy is much smaller than that found for strongly bonded single-atomic monolayers. In fact, the total binding energies of the three-layer assemblies are surprisingly small. This explains the greater flexibility of the three-layer networks.

Most calculated ultimate strengths presently available, follow Griffith's rule reasonably well. A larger deviation is visible for the 1T-PtO_2 monolayer, where the fracture strength is noticeably greater than one-tenth of Young's modulus. This could be due to the extreme ionicity of bonding. The finding suggests that also in three-layer networks the critical fracture strength is connected with the elastic modulus. The similarity of the mechanical behavior of the considered compounds is

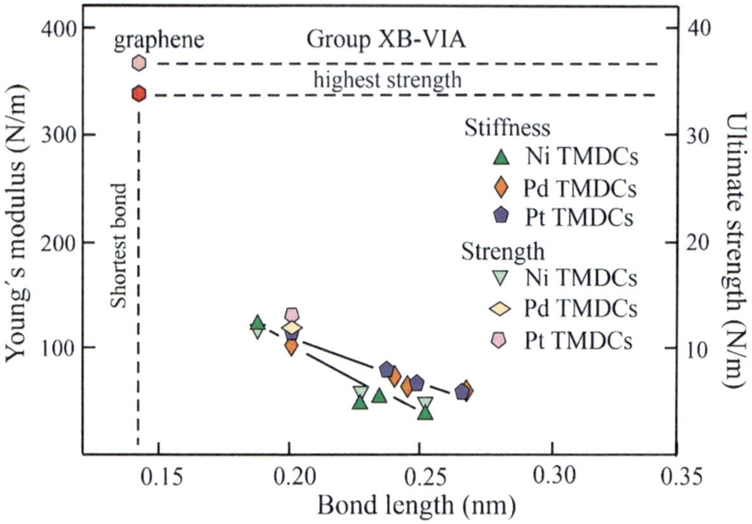

Fig. 17.10 Plot of Young's moduli and ultimate strengths versus the MX bond lengths of group XB–VIA monolayers with MX_2 stoichiometry

a consequence of the dominant formation of analogously bonded 2D structures with the same MX_2 stoichiometry.

References

1. C. Ataca, H. Şahin, S. Ciraci, J. Phys. Chem. **116**, 8983 (2012)
2. J.A. Reyes-Retana, F. Cervantes-Sodi, Sci. Rep. **6**, 24093 (2016)
3. W.-Y. Fang, Y. Chen, P. Ye, H.-R. Wei, X.-L. Xiao et al., Acta Phys. Sin. **70**, 246301 (2021)
4. Y. Guo, L. Ma, K. Mao, M. Ju, Y. Bai et al., Nanoscale Horiz. **4**, 592 (2019)
5. A.F. Wani, B. Rani, U.B. Sharopov, S. Dhiman, K. Kaur, Int. J. Energy Res. **46**, 8527 (2022)
6. J. Zaffran, M.C. Toroker, ChemElectroChem **4**, 2764 (2017)
7. W. Xu, X. Zhao, F. Zhan, Q. He, H. Wang et al., Energy Storage Mater. **53**, 79 (2022)
8. C.C. Leong, H. Pan, S.K. Ho, PhysChemChem Phys. **18**, 7527 (2016)
9. A. Ikeda, Y. Krockenberger, H. Irie, M. Naito, H. Yamamoto, Appl. Phys. Express **9**, 061101 (2016)
10. Y. Li, F. Wrobel, X. Yan, A. Bhattacharya, J. Sun et al., Appl. Phys. Lett. **118**, 061601 (2021)
11. R. Anisha, S. Kumar, K. Srivastava, Tankeshwar. Mater. Today Commun. **34**, 105169 (2023)
12. C.-T. Wang, S. Du, Phys. Chem. Chem. Phys. **22**, 7483 (2020)
13. Y. Mao, J. Zhang, S. Bai, M. Wu, D. Luo et al., Comput. Mater. Sci. **216**, 111873 (2016)
14. H. Yang, Y. Li, Z. Yang, X. Shi, Z. Lin et al., Vacuum **174**, 109176 (2020)
15. W. Xiong, K. Huang, S. Yuan, J. Mater. Chem. C **7**, 13518 (2019)
16. N.H. Linh, T.T. Quang, N.M. Son, T.T. Thang, V.V. Thanh et al., JST Eng. Technol. Sustain. Dev. **33**, 34 (2023)
17. H. Khalatbari, S.I. Vishkayi, M. Oskouian, H.R. Soleimani, Sci. Rep. **11**, 5779 (2021)
18. R. Ku, L. Yan, K. Xue, J. Zhang, K. Pang et al., J. Phys. Chem. C **126**, 6925 (2022)
19. S. Ahmadi, A.S.S. Molla, Z. Moradi, L. Eslami, Appl. Surf. Sci. **616**, 156560 (2023)

20. A. Sengupta, Comput. Mater. Sci. **206**, 111278 (2022)
21. C. Dai, B. Li, J. Li, B. Zhao, R. Wu et al., Nano Res. **13**, 2506 (2020)
22. Y. Shen, J. Liu, X. Li, Q. Wang, A.C.S. Appl, Mater. Interfaces **11**, 35661 (2019)
23. Y. Shao, S. Song, X. Wu, J. Qi, H. Lu et al., Appl. Phys. Lett. **111**, 113107 (2017)
24. F. Jamil, R.A.S.I. Subad, T.S. Akash, P. Bose, Comput. Condens. Matter **26**, e00522 (2021)
25. S. Tang, S. Bai, M. Wu, D. Luo, J. Zhang et al., Phys. Chem. Chem. Phys. **24**, 5185 (2022)
26. Y. Qu, H. Pan, C.T. Kwok, Sci. Rep. **6**, 34186 (2016)
27. J. He, H. Lu, L. Tao, Y. Zhao, Z. Zheng et al., Infrared Phys. Technol. **123**, 104195 (2022)
28. B. Zhao, W. Dang, Y. Liu, B. Li, J. Li et al., J. Am. Chem. Soc. **140**, 14217 (2018)
29. J. Shi, Y. Huan, M. Xiao, M. Hong, X. Zhao et al., ACS Nano **14**, 9011 (2020)
30. A.F. Wani, L. Patra, M. Srinivasan, J. Singh, S.A.M. Abdelmohsen et al., Adv. Theory Simul. 2300158 (2023)
31. F.A. Rasmussen, K.S. Thygesen, J. Phys. Chem. C **119**, 13169 (2015)
32. Y. Hu, J. Jiang, P. Zhang, Z. Ma, F. Guan et al., Nat. Commun. **14**, 6555 (2023)
33. A.N. Hoffman, Y. Gu, J. Tokash, J. Woodward, K. Xiao et al., ACS Appl. Mater. Interfaces **12**, 7345 (2020)
34. Y. Wang, Y. Li, Z. Chen, J. Mater. Chem. C **3**, 9603 (2015)
35. M. Xie, X. Xia, Y. Tai, X. Guo, J. Jang et al., Vacuum **212**, 111982 (2023)
36. D. Raval, S.K. Gupta, P.N. Gajjar, R. Ahuja, Sci. Rep. **12**, 2964 (2022)
37. E.A. Moujaes, W.A. Diery, J. Phys. Condens. Matter **31**, 455502 (2019)
38. E. Chen, W. Xu, J. Chen, J.H. Warner, Mater. Today Adv. **7**, 100076 (2020)
39. F. Xiao, W. Lei, W. Wang, L. Xu, S. Zhang et al., J. Mater. Chem. C **9**, 7753 (2021)
40. X. Zhang, G. Su, J. Lu, W. Yang, W. Zhuang et al., ACS Appl. Mater. Interfaces **13**, 43063 (2021)
41. Y. Ma, X. Gong, F. Xiao, Y. Liu, X. Ming, A.C.S. Appl, Nano Mater. **6**, 12470 (2023)
42. K. He, W. Xu, J. Tang, Y. Lu, C. Yi et al., Small **19**, 2206915 (2023)
43. Y. Wang, J. Pang, Q. Cheng, L. Han, Y. Li et al., Nano-Micro Lett. **13**, 143 (2021)
44. M. Sun, J.-P. Chou, L. Shi, J. Gao, A. Hu et al., ACS Omega **3**, 5971 (2018)
45. Y.-S. Lan, X.-R. Chen, C.-E. Hu, Y. Chen, Q.F. Chen, J. Mater. Chem. A **7**, 11134 (2019)
46. A.N. Hoffman, Y. Gu, L. Liang, J.D. Fowlkes, K. Xiao et al., **3**, 50 (2019)
47. C. Long, Y. Liang, H. Jin, B. Huang, Y. Dai, A.C.S. Appl, Energy Mater. **2**, 513 (2019)
48. A.V. Kuklin, L.V. Begunovich, L. Gao, H. Zhang, H. Ågren, Phys. Rev. B **104**, 134109 (2021)
49. J. Yu, X. Kuang, Y. Gao, Y. Wang, K. Chen et al., Nano Lett. **20**, 1172 (2020)
50. L.-H. Zeng, D. Wu, S.-H. Lin, C. Xie, H.-Y. Yuan et al., Adv. Funct. Mater. **29**, 1806878 (2019)
51. K.-L. Tai, J. Chen, Y. Wen, H. Park, Q. Zhang et al., ACS Nano **14**, 11677 (2020)
52. Z. Wu, L. Lu, X. Liang, C. Dun, S. Yan et al., J. Phys. Chem. C **124**, 10935 (2020)
53. N.K. Nepal, L. Yu, Q. Yan, A. Ruzsinszky, Phys. Rev. Mater. **3**, 073601 (2019)
54. B. Marfoua, J. Hong, A.C.S. Appl, Mater. Interfaces **11**, 38819 (2019)
55. Y.-S. Lan, X.-R. Chen, C.-E. Hu, Y. Cheng, Q.-F. Chen, J. Mater. Chem. A **7**, 11134 (2019)
56. C. Liu, C.-S. Lian, M.-H. Liao, Y. Wang, Y. Zhong et al., Phys. Rev. Mater. **2**, 094001 (2018)
57. T. Miao, M. Xiang, D. Chen, M. An, W. Ma, Int. J. Heat Mass Transfer **183**, 122099 (2022)
58. W.-L. Tao, J.-Q. Lan, C.-E. Hu, Y. Cheng, J. Zhu et al., J. Appl. Phys. **127**, 035101 (2020)
59. L.-Y. Feng, R.A.B. Villaos, Z.-Q. Huang, C.-H. Hsu, F.-C. Chuang, New J. Phys. **22**, 053010 (2020)
60. L. Liu, D. Zemlyanov, Y.P. Chen, 2D Mater. **8**, 045033 (2021)
61. J. Cook, S. Mardanya, Q. Lu, C. Conner, M. Snyder et al., Nano Lett. **23**, 1752 (2023)
62. H.A.H. Mohammed, G.M. Dongho-Nguimdo, D.P. Joubert, Physica E **127**, 114514 (2021)
63. B. Yin, L. Yang, S. He, P. Zhou, J. Zhang et al., Mater. Today Commun. **37**, 107484 (2023)
64. J. Zhang, Y. Xie, Y. Hu, H. Shao, Appl. Surf. Sci. **532**, 147387 (2020)
65. S.H. Khan, M.S. Islam, R. Islam, A. Iskanderani, I.M. Mehedi et al., IEEE Access **9**, 109510 (2021)
66. H. Shen, Y. Zhang, G. Wang, W. Ji, X. Xue et al., Phys. Chem. Chem. Phys. **23**, 21825 (2021)

67. T.T.H. Duong, H.H. Hau, L.T. Hong, L.A. Vu, C.M. Hung et al., Mater. Sci. Semicond. Process. **151**, 106990 (2022)
68. S.A. Krasnikov, S. Murphy, N. Berdunov, A.P. McCoy, K. Radican et al., Nanotechnology **21**, 335301 (2010)
69. J. Li, S.-Y. Xie, Phys. Lett. A **384**, 126286 (2020)
70. Z. Kahraman, A. Kandemir, M. Yagmurcukardes, H. Sahin, J. Phys. Chem. C **123**, 4549 (2019)
71. X. Ge, X. Zhou, D. Sun, X. Chen, ACS Omega **8**, 5715 (2023)
72. R.A. Ghazi, D.A. Sabur, R.S. Al-Hasnawy, H.O. Muhsen, B.B. Kadhim et al., Key Eng. Mater. **886**, 48 (2021)
73. R.A.B. Villaos, C.P. Crisostomo, Z.-Q. Huang, S.-M. Huang, A.A.B. Padama et al., NPJ 2D Mater. Appl. **3**, 2 (2019)
74. Y. Zhao, J. Qiao, P. Yu, Z. Hu, Z. Lin et al., Adv. Mater. **28**, 2399 (2016)
75. J. Du, P. Song, L. Fang, T. Wang, Z. Wei et al., Appl. Surf. Sci. **435**, 476 (2018)
76. X. Chang, Y. Ji, H. Li, Comput. Mater. Sci. **196**, 110552 (2021)
77. X.-C. Liang, X.-J. He, Y.-C. Ding, Y.-J. Hao, J. Zhu, Physica E **132**, 114744 (2021)
78. D.K. Nguyen, D.M. Hoat, A. Bafekry, V.V. On, J.F. Rivas-Silva et al., Physica E **131**, 114732 (2021)
79. R. Sant, M. Gay, A. Marty, S. Lisi, R. Harrabi et al., NPJ 2D Mater. Appl. **4**, 41 (2020)
80. Z. Wang, P. Wang, F. Wang, J. Ye, T. He et al., Adv. Funct. Mater. **30**, 1907945 (2020)
81. G. Bassi, R. Wadhwa, S. Deswal, P. Kumar, M. Kumar, J. Alloys Comp. **955**, 170233 (2023)
82. J. Lu, X. Zhang, G. Su, W. Yang, K. Han et al., Mater. Today Phys. **18**, 100376 (2021)
83. Y.-Q. Zhao, L. Zhu, B. Zhou, S. Jiang, Phys. Chem. Chem. Phys. **25**, 11311 (2023)
84. J. Shi, Y. Huan, M. Hong, R. Xu, P. Yang et al., ACS Nano **13**, 8442 (2019)
85. F. Ghasemi, R. Taghavimendi, A. Bakhshayeshi, Opt. Quant. Electron. **52**, 492 (2020)
86. S. Deng, L. Li, Y. Zhang, A.C.S. Appl, Nano Mater. **1**, 1932 (2018)
87. X. Chia, A. Adriano, P. Lazar, Z. Sofer, J. Luxa et al., Adv. Funct. Mater. **26**, 4306 (2016)
88. S. Lin, Y. Liu, Z. Hu, W. Lu, C.H. Mak et al., Nano Energy **42**, 26 (2017)
89. A. Sharma, G. Gupta, Mater. Sci. Eng. B **290**, 116333 (2023)
90. N. Todorova, N. Minev, V. Marinova, K. Buchkov, V. Videva et al., Appl. Surf. Sci. **611**, 155534 (2023)
91. Y.-L. Wang, L. Li, W. Yao, S. Song, J.T. Sun et al., Nano Lett. **15**, 4013 (2015)
92. Y. Tong, M. Bouaziz, H. Oughaddou, H. Enriquez, K. Chaouchi et al., RSC Adv. **10**, 30934 (2020)
93. B.M. Szydlowska, O. Hartwig, B. Tywoniuk, T. Hartman, T. Stimpel-Lindner, 2D Mater. **7**, 045027 (2020)
94. Z. Li, M. Huang, J. Li, H. Zhu, Adv. Funct. Mater. **33**, 2300376 (2023)
95. M. Tharrault, E. Desgué, D. Carisetti, B. Plaçais, C. Voisin et al., arXiv:2307.15520v4 (2023)
96. Y. Gong, Z. Lin, Y.-X. Chen, Q. Khan, C. Wang et al., Nano-Micro Lett. **12**, 174 (2020)
97. D. Dai, H.-J. Koo, M.-H. Whangbo, C. Soulard, X. Rocquefelte et al., J. Solid State Chem. **173**, 114 (2003)
98. W.-L. Tao, Y.-Q. Zhao, Z.-Y. Zeng, X.-R. Chen, H.-Y. Geng, A.C.S. Appl, Mater. Interfaces **13**, 8700 (2021)
99. H.L. Zhuang, R.G. Hennig, J. Phys. Chem. C **117**, 20440 (2013)
100. H. Huang, X. Fan, D.J. Singh, W. Zheng, ACS Omega **3**, 10058 (2018)
101. P.-Z. Jia, Z.-X. Xie, Y.-X. Deng, Y. Zhang, L.-M. Tang et al., Appl. Phys. Lett. **121**, 043901 (2022)
102. D. Wan, S. Bai, M. Wu, J. Zhang, D. Luo et al., Vacuum **215**, 112343 (2023)
103. W. Yu, Z. Dong, H. Mu, G. Ren, X. He et al., ACS Nano **16**, 12922 (2022)
104. L. Pi, L. Li, K. Liu, Q. Zhang, H. Li et al., Adv. Funct. Mater. **29**, 1904932 (2019)
105. Y. Hu, W. Zheng, S. Fan, J. Zhang, X. Liu, Appl. Phys. Rev. **10**, 031306 (2023)
106. H. Ma, P. Chen, B. Li, J. Li, R. Ai et al., Nano Lett. **18**, 3523 (2018)
107. L. Zhang, T. Yang, Arramel, Y.P. Feng, A.T.S. Wee et al., Nanoscale **14**, 7650 (2022)

Chapter 18
Summary and Outlook

Abstract The extension of materials science to two dimensions is an extremely active and successful field of research. The treatment of the enormous amount of information already available on 2D materials can be best characterized by the fundamental chemical principles governing the periodic table. The 'aufbau' principle not only helps to understand the related features of monolayers that allows to catagorize the impressive results, it also induced the prediction of new stable 2D structures as already demonstrated in several cases. Since monolayers and few-layer assemblies appear as insulators, semiconductors, and metals, it is possible to construct nanoscale devices. The summary highlights the enormous progress made in realizing physical properties such as topological, superconducting, ferroelectric, piezoelectric, thermoelectric, and magnetic properties in 2D materials by presenting state-of-the-art examples. In addition, the first successful steps in the construction of nanoscale devices in catalysis, photovoltaics, field-effect transisitors, batteries, and sensors are summerized. An outlook to future developments is presented, whenever appropriate.

18.1 Materials in Two Dimensions

18.1.1 Basics

Going from the bulk to the ultimate single-layer limit, usually drastic changes of the physicochemical properties occur at a thickness below approximately 10 monolayers. The gradual decrease of interlayer interactions and especially the absence of these interactions in an isolated monolayer modify the electronic structure. Furthermore, the change from three dimensions to two dimensions generates quantum phenomena by spatial confinement to two dimensions. Consequently, drastic changes of the properties of two-dimensional materials and the manifestation of novel phenomena occur, originating from the nature of the primarily covalently bonded atomic structures and

P. Hess, *Bonding, Structure, and Performance of Two-Dimensional Materials*,
NanoScience and Technology, https://doi.org/10.1007/978-3-031-74161-6_18

the presence of a drastically enlarged surface. The existence of nontrivial topologies, electronic correlations, and charge density waves qualify the material for the exploration of new quantum states.

Due to the enormous number of surface atoms especially in single-atomic layers, defects such as vacancies, substitutions, and dangling bonds may increase the reactivity significantly and consequently environmental deterioration occurs, e.g., by oxidation or chemisorption. However, it also opens the door to enhanced surface chemistry in heterogeneous catalysis. Unfortunately, up to now this topic has found only limited attention. First promising results are available, e.g., for several transition-metal dichalcogenides (TMDCs) including not only monolayers and nanosheets but also more complex heterostructures with tailored properties.

There is an ongoing effort to convert elements of the periodic table and their stable stoichiometric combinations into covalently bonded 2D materials, sometimes called modern 'alchemy'. Theory already approached this goal successfully, whereas experimental verification is far behind the enormous achievements already made by theory. Theoretical predictions helped to synthesize hitherto unknown monolayers and calculations and simulations are always needed to interpret experimental results. The enormous amount of theoretical information presently available gives a clear picture of 2D bonding, realizable 2D structures, and the mechanical stability of monolayers, embracing the crucial elements of the periodic table, including their atomic, binary, and complex heteroatomic compounds.

Theory clearly dominates our current picture of the 2D world, which has demonstrated the ability to accurately predict novel structures and to predict their physicochemical properties. However, the further development of the field needs combined work on both fronts because theoretical predictions always require experimental verification and experiments need guidance by theory to interpret the measurements. To close the large gap existing between already existing theoretical predictions and experimental verification is one of the critical issues to advance practical applications in this field. In fact, many potential applications have been suggested by theory, waiting for experimental confirmation and practical realization.

Carbon atoms with their small size, ideal electronic configuration forming energetically preferred two-center two-electron bonds, and outstanding physical properties, have an exceptional position in the periodic table. In the columns on the left and right side of carbon the missing or additional valence electrons change the ideal bonding situation and this relates to new types of bonding and a variety of novel 2D structures often with lower symmetry. This situation is responsible for a principal issue in the development of 2D materials, an enormous structural diversity. The inherent multivalency of elements such as B (+ 3 and − 5) or Se (+ 6 and − 2) is the reason for the formation of various metastable hexagonal, tetragonal, rhombic, and monoclinic 2D structures. Specifically, the group of chalcogenides with the largest difference in oxidation states plays a dominant role in the diversity of stable and metastable 2D structures.

Going down the columns of the periodic table from the second period with the smallest atoms and strong covalent bonding the atom size increases and this reduces the covalent bonding strength due to decreasing orbital overlap. In addition, electron screening weakens bonding in the heavier compounds. Consequently, in the fifth and sixth period bonding and mechanical performance are enormously reduced. The loss of mechanical stability and strength may drastically increase the reactivity of monolayers, which are now more susceptible to adsorption, chemisorption, and substitution reactions. Nevertheless, interest in monolayers consisting of heavy elements is steadily increasing owing to unique quantum properties such as topological features, based on the existence of distinct low-symmetry structures.

An important degree of freedom in the determination and improvement of 2D materials properties is modification of the chemical composition. Chemical diversity is a valuable tool that opens the possibility of tuning specific properties, however, at the prize of rising complexity introduced by an increasing number of different elements. A well-studied important method for fine-tuning physicochemical properties is the introduction of Janus monolayers based on partial substitution of an element by a different (related) element to create a 2D material with tailored chemical composition. The underlying principle is the reduction of symmetry that is of interest due to emerging quantum features originating from structural asymmetry. The name Janus is based on the Roman god of beginnings and endings with two faces.

Even though high-quality large-scale elemental monolayers are expected to be the first 2D materials reaching the level of widespread industrial applications. Nevertheless, more complex stoichiometries [1], ternary compounds [2], heterostructures [3, 4], and superlattices [5] will play a significant role, owing to additional options to tailor relevant chemical and physical properties and to realize novel behavior. All these exciting new possibilities and phenomena drive the explosive extension of research of the 2D world, incorporating more elements of the periodic table and sophisticated combinations to heteroelemental monolayers, few-layer sheets, and complex heterostructures.

18.1.2 Realization of 2D Materials

The realization of high-quality free-standing monolayers, allowing the direct measurement of their intrinsic properties is an exception. It has been achieved only in very few selected cases with exceptional strong bonding such as graphene and molybdenum disulfide. In the current situation, size-limited exfoliation ('scotch tape') often delivers better quality samples than most other deposition methods. Quasi-free-standing mono- and few-layer networks of 2D materials, also prepared by top-down or bottom-up methods and stabilized by van der Waals interaction on a surface, are a main source of experimental information. Since these generally applicable methods need no layered bulk material for preparation they will play a dominant role in the further development of the field to practical applications. Moreover, the deposition of van der Waals-bonded monolayers will make fundamental contributions to the mechanisms of growth processes by in situ analysis of intrinsic physicochemical properties. Notably, van der Waals epitaxy without ultra-high vacuum (UHV) conditions will play a key role in the growth of high-quality quasi-free-standing monolayers, few-layer assemblies, heterostructures, and superlattices.

Nevertheless, UHV van der Waals epitaxy using various in situ diagnostic techniques is one of the most valuable methods for in situ characterization during or directly after growth, especially of chemically unstable deposits. Of course, the massive technical effort needed in comparison to exfoliation techniques and vapor-flow deposition will hamper its widespread application. Diagnostic tools such as angle-resolved photoemission spectroscopy (ARPES) and scanning tunneling microscopy (STM) allow the investigation of the band gap and electronic states, e.g., the identification of inverted gap states due to spin–orbit coupling (SOC). Scanning tunneling spectroscopy (STS) allows the identification of conducting topologically protected edge states in topological insulators (TIs). These diagnostic tools are a requisite for optimizing growth processes to the level needed to reach the quality crucial for specific practical applications. The last step of fabricating 2D materials by low-cost methods, suitable for large scale production, in a standard process still is a challenge. Often, the produced devices show a much lower performance than expected by theory, as a result of the inferior quality of the integrated 2D materials.

18.1.3 Theory of 2D Materials

The calculation of lattice parameters by various approaches usually delivers good agreement. Larger deviations are regularly observed for the band gap of semiconductors calculated by different approximations. The larger gaps obtained by the Heyd-Scuseria-Ernzerhof (HSE) approach often are in better agreement with experiments than, for example, the density functional theory (DFT)-generalized gradient approximation (GGA)-Perdew-Burke-Ernzerhof (PBE) computations. Unfortunately, authors seldom discuss the large deviations found for the cohesive energy, calculated by various approaches, and therefore the reasons for deviations extending up to a factor of two are unknown in most cases. Owing to this uncertainty, sometimes the determination of the energetically most stable phase of several existing polymorphs is not possible. The problem of contradictory results on the stability of polymorphs is difficult to solve experimentally because even in the case of weak substrate interaction the influence of the substrate may change relevant properties of the deposit. Polymorphs with similar binding energies are usually grown as mixed deposit and the realization of the pure phases is difficult. It is necessary that authors compare their results with already existing literature values and discuss possible reasons for the observed deviations. Theoretical advancement in the discovery and identification of novel stable and metastable 2D materials has considerably outpaced the experimental realization of already identified promising candidates. Presently, most encouraging are chemical methods for expanding synthesis techniques to still unknown 2D materials such as vapor deposition under flow conditions and van der Waals epitaxy on suitable substrates.

18.2 Bonding Concepts of 2D Materials

18.2.1 Semiconductors and Metals

Semiconducting and insulating 2D materials with strong covalent-ionic bonding that produces a bandgap can be synthesized under ambient conditions. Since the more reactive 2D metals easily oxidize under ambient conditions they may require vacuum equipment during synthesis, characterization, and device fabrication, or a protection layer. Owing to this special effort needed to avoid oxidation and environmental contamination the investigation of metals, hold together by weaker metallic bonding, is far behind the development of monolayers and few-layer assemblies of semiconductors and insulators.

The band gaps of 2D semiconductors range from mid-infrared in bilayer graphene and black phosphorus, to the visible region in transition-metal dichalcogenides (TMDCs), and the ultraviolet range in h-BN [6]. The gaps are highly tunable by control of the layer number in nanosheets, strain engineering, chemical doping, alloying, formation of heterostructures, intercalation, an external electric field, and substrate engineering. In addition, the available work covers band gap closing, metal-to-insulator transitions in TMDCs mediated by structural phase transitions. As mentioned above, the band gaps calculated by different approaches may deviate substantially and experimental values are rare.

In general, metallic monolayers exhibit smaller cohesive energies. This causes serious problems in the synthesis of monolayers, whereas it is possible to fabricate nanosheets, e.g., for Ga. For group–IIA monolayers with two valence electrons first-principles calculations delivered the cohesive energies and lattice parameters for several familiar structures. To determine the most stable structure especially in the case of metallic phases, theory must be further developed in tight connection with experiments. For example, theory provided the properties for various metallic boron-based monolayers and their homologues. With three valence electrons in the outer shell these atoms cannot form a closed-shell structure by conventional bonding, however, they can form extended covalent 2D networks by electron-deficient bonds. The missing valence electron reduces the bonding strength and makes these networks reactive. Nevertheless, the delocalization of binding electrons between three or more atoms yields reasonable cohesive energies at least in the case of borophenes that allow the isolation of monolayers. The complexity of bonding and structural diversity of the already known metallic borophenes open new opportunities to discover novel exciting boron nanostructures even semiconducting ones.

Recent advances achieved in wet-chemical synthesis of 2D metallic nanomaterials (e.g. noble metals, non-noble metals, and bimetallic alloys) triggered their application in a variety of electrochemical reactions, despite the thermodynamically unfavorable large surface energy of specific facets [7]. Organic ligands and polymeric ligands allow the diminution of surface energy during synthesis and the final stabilization of the 2D structure. With their unique bonding and specific properties these metallic materials provide interesting perspectives for a series of potential technological applications. The flexibility in bonding, mechanical flexibility, in-plane elasticity, thermal stability, thermal conductivity, optical transparency, and intrinsic strength make these 2D monolayers and sheets are of great interest in materials science.

18.2.2 Insulators

With the strong ionic bonding contribution, chemical and mechanical stability, electronic band gap of $E_g = 6.8$ eV, low dielectric constant, and excellent chemical inertness, planar h-BN monolayers are an ideal insulator for replacing SiO_2 in silicon technology [8] (Chap. 77, [26]). In addition to being an ultimately thin insulator, it is particularly attractive as a barrier or encapsulant of reactive monolayers.

However, 1T-HfO_2 monolayers with an even larger difference in the electronegativities of Hf and O exhibit a similar band gap of $E_g = 6.73$ eV together with large dielectric constants. Their synthesis is possible by plasma enhanced atomic layer deposition [9, 10]. It seems to be better suited for high-k gate dielectrics than Al_2O_3. DFT calculations elucidated the atomistic mechanism for the growth of a high-quality interface by replacing the Se atoms of an interfacial $HfSe_2$ layer by O atoms. This substitution reaction proceeds without energy barrier and allows layer-by-layer oxidation with an atomically sharp interface. In fact, a high-quality gate stack of native HfO_2 on 2D $HfSe_2$ results from plasma oxidation.

The chemically converted HfO_2 sheets showed a high dielectric constant and a low leakage current. A fabricated gate stack of HfO_2/$HfSe_2$ incorporated in a field-effect transistor (FET) showed an almost ideal subthreshold slope. This work presents a major step forward to the realization of post-Si semiconducting devices for energy-efficient computing electronics [10] (Chap. 14, [93]).

DFT-PBE calculations delivered an indirect wide band gap of 1T ZrO_2 insulators of $E_g = 4.51$ eV [11] (Chap. 14, [52]). Owing to the large indirect band gap, dielectric permittivity, and mechanical stiffness, ZrO_2 is another promising dielectric 2D material that is suitable for FETs and heterojunctions.

18.2.3 2D Materials with d Electrons

A promising class of 2D materials are transition-metal dioxides and dichalcogenides with a varying number of d electrons (d-electron count), such as a partly-filled d band (group IVB), a half-filled d band (group VIB), and a completely-filled d band (group XB), which control the allowed chemical coordination in 2D space and thus the intrinsic physicochemical properties. The bonding strength and consequently the mechanical strength depend on the d-band filling of the valence electrons. The filling process of d bands increases from group IV (~sparsely filled) to group VI (~ half filled) to group X (~ completely filled) [12]. The stable covalently bonded two-

and three-layer assemblies cover a broad range of symmetries. Theoretical predictions indicate further progress in the synthesis and application of low-symmetry monolayers with exotic properties. Current work on piezoelectric behavior, transport and conductivity properties to advance thermoelectric devices, and electrochemical catalytic processes in 2D materials delivered just the first contributions to this bonding type. In few cases the performance surpasses that of the 3D counterparts. As for other bonding types, the bottle neck is the experimental verification of fundamental predictions made by theory.

18.2.4 Hypercoordinated 2D Materials

The currently not well studied class of hypercoordinated monolayers has a high potential for the realization of novel structures with exotic symmetries. Meanwhile, the theoretical treatment of 2D hypercoordinate chemistry is a well developing field, however, even for the most stable hypercoordinated monolayers experimental realization still is a challenge [13, 14] (Chap. 4, [1, 45]). At first glance, according to the valence-shell electron-pair repulsion (VSEPR) rule, hypercoordinated moieties seem to be energetically unfavorable owing to repulsion of the bonding electrons in a high-density arrangement. Nevertheless, quasi-planar tetracoordinate, pentacoordinate, and hexacoordinate carbon moieties are energetically, dynamically, and thermally stable in two dimensions.

Besides the coordination numbers found in 2D materials of the main group elements, theory predicts planar motifs with coordination numbers of eight or even nine. The corresponding bonding-structure plots demonstrate surprisingly efficient bonding of these quasi-planar hypercoordinate structures with their unconventional bonding types. As discussed in the corresponding chapter for IIA–IVA and IIIA–IVA monolayers, effective charge transfer mechanisms stabilize these rule-breaking quasi-planar hypercoordinate structures with multicenter bonding.

The remaining task is the development of strategies for stabilizing the hypercoordinated networks during the fabrication process. A severe problem originates from the extensive variation of bond strengths in complex structures. This is because the weakest bonds in the whole network control the ultimate fracture strength. Consequently, the fabrication process is sensitive to additional energy input during growth. An experimental breakthrough, managing the obviously critical grow conditions on the basis of a molecular-level understanding of the grow process would provide a strong impetus to the field of hyper-coordinated chemistry with versatile bonding capabilities. Since metals prefer high-coordination bonding, progress in experimental realization of hypercoordinated networks would strongly influence the future development of 2D metallic materials.

18.3 Distinctive Properties of 2D Materials

18.3.1 Mechanics of 2D Materials

The various chapters in this book provide a systematic collection of linear mechanical parameters such as lattice constants, Young's moduli, and Poisson's ratios that are available for most known monolayers of the periodic table. Sometimes, relevant information is lacking for the nonlinear properties such the ultimate strength, fracture toughness, and critical strain. The tables and plots displayed in each chapter for related compounds demonstrate the variation of these properties and the degree of scatter in the published results. Moreover, an enormous lack of experimental data and especially of nonlinear properties is clearly visible. For the ultimate strengths and ultimate strains, and especially for fracture toughness, even theoretical information is lacking. In view of the large deviations obtained for ultimate strength and strain values, it is necessary to introduce standardized test methods to allow a valuable comparison of independent measurements. Furthermore, a detailed discussion of the differences obtained by various theoretical approaches is urgently needed.

Two up-to-date articles discuss the mechanical characterization and relevance of 2D mechanical properties [15, 16]. In the first comprehensive article the progress made in the characterization of the mechanics of free-standing inorganic 2D materials is summarized [15]. The complete experimental characterization of the linear and nonlinear mechanical properties of free-standing 2D materials is limited to a very small number of monolayers such as graphene, h-BN, MoS_2, etc. This comprehensive work covers all aspects of experimental techniques including atomic force microscopy (AFM) based nanoindentation, in situ tensile and fracture testing, bulge testing, Raman spectroscopy, Brillouin light scattering, and buckling-based metrologies to determine the elastic constants, fracture behavior, as well as bending and interfacial properties. In addition, the study treats a variety of mechanical applications such as resonators, microphones, and various nanoelectromechanical sensors.

The second comprehensive overview on recent advances in mechanics of 2D materials covers many already known but also some novel 2D materials, new characterization techniques, and computational methods [16]. Emphasis is placed on the role played by mechanical properties in materials design, manufacturing, integration, and performance, based on a better understanding of deformation and failure processes. The presentation of the enormous progress made in understanding the mechanical behavior includes sections with the topics: elastic properties, in-plane failure, fatigue, interfacial shear and friction, and adhesion. These sections discuss recent experimental and computational methods, experimental and theory-based in-plane and out-of-plane moduli, measured and simulated failure properties, fatigue processes, as well as frictional and interfacial shear processes between 2D materials and substrates.

18.3.2 2D Topological Insulators

A unique feature of 2D materials is quantum behavior at the thickness limit of monolayers. For example, 2D topological insulators realize an intriguing new state of quantum matter [17]. This state exhibits an electronic structure with a gaped interior and gapless edge states incorporating spin-orbit interaction and spin states that counter-propagate at the boundaries with opposite spins. This results in a nonzero quantum spin Hall (QSH) conductance. With the discovery of the mechanisms governing topological phase states in 2D materials such as opening a gap in the electronic band structure by SOC in 2D Dirac semimetals and band inversion in narrow-band semiconductors the investigation of 2D TIs has found increasing attention.

Examples for systems with nontrivial topological behavior suggested by theory are besides graphene the low buckled silicene, germanene, and stanene monolayers with stronger SOC effects than graphene. For epitaxially grown low-buckled germanene, experimental evidence proved for the first time that single-layer germanene is a QSH insulator with a bulk gap and robust metallic edges [18] (Chap. V, [70]). Applying a perpendicular critical electric field closes the topological gap and generates a Dirac semimetal. Heavier elements find increasing attention as potential topological materials for room temperature applications, as for example, graphene-like stanene and plumbene with a planar structure, induced by strong substrate interaction. Besides group–IVA elements, elemental graphene-like Xenes of group VA and group VIA are of interest as potential QSH materials [19].

The band gap opened by SOC of ~ 0.1 eV endows square selenene and tellurene moieties with TI properties, hosting nontrivial edge states. The authors consider the low-symmetry square structure as a new template for studying highly anisotropic band dispersions. While previous models of TIs involved only *s*-state electrons, the low-symmetry square lattice involves *p*-state hybridizations around the Fermi level. This gives access to novel Dirac or semi-Dirac semimetals and TIs. Thus, the square phase provides an interesting platform for exploring novel electronic and spintronic applications [20] (Chap. 11, [4]).

Notably, there is experimental evidence that elemental tellurene monolayers prepared in the helical (α) phase by an anisotropic ultrasonic treatment behave as 2D TIs [21] (Chap. 11, [36]). Parallel chains, weakly bonded by van der Waals forces, possess an extremely high carrier mobility and a tunable band gap. As observed by STS, the inner part of the tellurene monolayer showed a conduction band energy different from that of the valence band, whereas gapless states with a Dirac point appeared along the edges. The transition from the gapless edge state to the bulk gap region extended more than 3 nm from the edge to the interior of the nanoflake. Built-in strains explain the appearance of the 2D TI phase.

Owing to strong SOC, buckled bismuthene is the topological insulator with the largest nontrivial band gap in its natural state of the group-VA elements. Importantly, π-type bonding of supported bismuthene with the SiC(0001) substrate yielded a honeycomb superstructure with an energetically more favorable planar configuration

[22] (Chap. 8, [87]). This bonding structure leads to huge on-site SOC of p_x and p_y orbitals and breaks the inversion symmetry. However, there is an Achilles heel of the VA elements, i.e., their environmental degradation that decreases from P to Bi. This problem can at least partially be circumvented by oxidation of bismuthene on one side, yielding a nontrivial topological phase with broken inversion symmetry, p_x–p_y band inversion at the Γ point, metallic edge states, and a band gap of 0.26 eV at the Γ point opened by SOC [23] (Chap. 8, [91]).

Elemental systems such as 2D bismuthene and 2D tellurene have the advantage in providing simple chemical structures as single-element monolayers. These systems may deliver the basis for the upcoming advancements made in the development of spintronics. First-principles calculations identified TIs, topological semimetals, trivial semimetals, and trivial insulators, e.g., in double layers of bismuthene and antimonene [24] (Chap. 8, [92]). For example, double-layer bismuthene in AB stacking order with strong atomic SOC transforms into the TI phase through band inversion according to first-principles calculations.

Binary monolayers opened the door to a variety of distinct TIs. For example, IIIA elements combined with bismuth in buckled hexagonal structures form a new class of 2D TIs [25] (Chap. 7, [107]). Band inversion occurs in pristine GaBi, InBi, and TlBi bilayers. It can be expected that buckled structures may play a role as a versatile platform for hosting nontrivial topological states and spin-polarized Dirac fermions.

An interesting theoretical work discussed the use of Janus monolayers as a novel platform for the integration of topological physics, namely, spintronics and valleytronics. The Janus monolayer $CSb_{1.5}Bi_{1.5}$ represents a spin-valley-coupled quantum spin Hall (QSH) insulator with topological Rashba-splitting edge states due to missing horizontal mirror symmetry [26] (Chap. 10, [26]). The valence band maximum (VBM) and conduction band minimum (CBM) are at the valley points. The inequivalent valleys have opposite Berry curvatures and spin moments and therefore can generate a nontrivial spin-valley-coupled quantum spin-valley Hall insulator that is robust against biaxial strain.

Theory already has predicted a variety of 2D topological materials comprising elemental monolayers, bilayers, Janus systems, and heteroelemental compounds. Moreover, theory already covers systems with large gap by designing structures with a high concentration of heavy metal atoms and a large SOC effect to realize high-temperature topological states. Successful experiments would induce a breakthrough in the verification of theoretically predicted stable systems. Furthermore, this would guide theory to provide more realistic computationally designed systems [25].

The 2024 roadmap on 2D topological insulators with an extensive collection of papers presents a status update of the field [27]. This comprehensive work, consisting of 13 contributions to specific topics presents reviews of recent advances, and discussion of the remaining challenges in the theoretical understanding, material synthesis, physical characterization, and device perspectives.

18.3.3 Superconducting 2D Materials

The family of TMDCs consists of materials with a variety of symmetry broken phases and quantum ground states. The development of 2D superconductors still is in the initial stages. According to the Bardeen-Cooper-Schrieffer (BCS) theory the phonon mediated attractive electron–electron interaction yields electron pairs with opposite spins and momenta.

A well-defined stoichiometric 2D material is the honeycomb borophene oxide B_2O monolayer, which exhibits intrinsic superconducting features that are tunable by strain [28] (Chap. 12, [5]). Within the framework of BCS theory, metallic B_2O monolayers exhibit intrinsic superconducting behavior with a transition temperature of $T_c \approx 10.3$ K. The superconductivity is tunable by tensile strain of 1% to an optimal critical temperature of $T_c \approx 14.7$ K. The reason for the enhancement effect is the appearance of out-of-plane soft-mode vibrations involving the light O atoms. According to first-principles calculations metallic δ-BS nanosheets with an orthorhombic lattice are intrinsic BCS-type superconductors with large electron-phonon coupling and $T_c \approx 21.56$ K [29] (Chap. 12, [11]).

The origin of phonon-mediated superconductivity in 2H NbS_2 with a critical temperature of $T_c = 5.7$ K is of special interest because this monolayer is known as the only phonon-mediated superconductor without a charge density wave (CDW) phase [30] (Chap. 15, [43]). The specific nature of the superconducting gap and pairing mechanism were studied by the fully anisotropic ab initio Migdal–Eliashberg theory, including Coulomb interactions. Hot spots on the Fermi surface with unusually strong electron-phonon interactions are responsible for the superconductivity and large contribution to electron pairing. Low-energy anharmonic phonons dominate the electron-lattice coupling, which places the system on the verge of CDW instability.

One of the best studied superconducting materials is $NbSe_2$ because Ising superconductivity exists down to the monolayer limit. Interestingly, recent results could prove the coexistence of attractive electron-phonon interactions driving superconductivity and repulsive Coulomb interactions driving correlated insulating behavior by slightly increasing Coulomb interactions [31] (Chap. 15, [51]). These results clearly demonstrate the present possibilities for elucidating quantum states using high-quality epitaxial samples and the available diagnostic tools. The chemical vapor deposition (CVD) method allows the growth of high-quality superconducting $NbSe_2$ monolayers without the need of UHV conditions [32]. Atomic-resolution scanning transmission electron microscopy confirmed the low concentration of point defects and grain boundaries and allowed the realization of 2D superconductivity. Protected with a graphene capping layer the transition temperature of as-grown $NbSe_2$ layers increased from 1.0 K in the monolayer to 4.56 K in a 10-layer nanosheet.

High-throughput density functional perturbation theory (DFPT) was applied to elucidate electron–phonon coupling and to predict the critical temperature of 165 materials, screened from the JARVIS-DFT database, including 34 dynamically stable structures [33]. This study includes layered 2D materials such as NbS_2, $NbSe_2$, and

NbO_2, with the highest transition temperature of $T_c = 17.5$ K for NbO_2. A comparison of experimental values with theory yields for the critical temperature of 2H NbS_2 an experimental value at $T_c = 5.7$ K [30] and a theorical value at $T_c = 9.3$ K [33]; for the as grown 10-layer $NbSe_2$ sheet $T_c = 4.56$ K [32] and a value at Tc ≈ 1.0 K [32] for the $NbSe_2$ monolayer. The high-throughput DFT data can guide future computational and experimental studies.

18.3.4 Ferroelectric 2D Materials

Ferroelectricity is a quantum order occurring by spontaneous alignment of electronic dipoles. The structural analysis indicates that the ferroelectric transition occurs by a synchronized ('displacive') shift in atomic positions, demonstrating the strong coupling between strain and polarization. The displacement of atoms in the polarization direction disrupts the centrosymmetric geometry and simultaneously lowers the energy of the system. Symmetry reduction favors the existence of ferroelectricity. Therefore, low dimensional 2D materials may prefer ferroelectric behavior. Reversal polarization occurs through a phase transition between two symmetry-equivalent ferroelectric states with opposite polarizations. Even though that ferroelectricity usually disappears below a critical thickness due to a depolarizing electric field or electron screening, polar states remain stable in hyperferroelectric materials of ultrathin atomic layers [34] (Chap. 9, [13]).

Binary heteroatomic VA–VA monolayers were the first 2D materials recognized of extending ferroelectricity to two dimensions. For example, the aw-SbN and aw-BiN monolayers, as phosphorene-like lattices with ridged structures and lone-pair electrons, exhibits ferroelectric behavior [35, 36] (Chap. 9, [11, 14]). The spontaneous electric polarization is switchable by an electric field. It is possible to switch the spontaneous electric polarization from the (100) to the (010) direction by applying a strain of approximately 2%. Notably, aw SbN has one of the largest spontaneous polarizations of 7.8×10^{-10} C/m that can retain until melting at ~ 1700 K and the ferroelectric polarization of aw BiN can reach a value of 5.8×10^{-10} C/m, being comparable with that of 3D $BiFeO_3$.

From the four well-studied group–IVA monochalcogenides (GeS, GeSe, SnS, and SnSe) the aw-GeS phase exhibits the optimal ferroelectric parameters, with a spontaneous polarization of 4.9×10^{-10} C/m, on par with the largest polarizations reported [37] (Chap. 13, [34]). Applying uniaxial tensile strain further improves polarization. The induction of ferroelectricity by phonon frustration, i.e., the competition between ferroelectric and antiferroelectric modes, allows for the possibility of tuning polarity. First-principles calculations revealed an in-plane spontaneous polarization of aw GeTe [38] (Chap. 13, [58]). Displacement of the Ge atoms in the x direction with respect to the Te atoms breaks the centrosymmetry in an orthorhombic structure. For a monolayer thickness of ~ 1 nm, the intrinsic polarization is 3.28×10^{-10} C/m. The stability of ferroelectricity is determined by the Curie temperature, which is 570 K for the aw-GeS phase. The in-plane coercive field for reversing polarization is tunable by a vertical field. Remarkably, for the bilayer, the ferroelectric state still is the ground state.

A novel concept for energy harvesting by a tribological nanogenerator was studied by the van der Waals force-induced interlayer ferroelectric-to-antiferroelectric transition via interlayer sliding of the top layer in bilayer group–IVA monochalcogenides MX (M: Ge, Sn; X = S, Se) [39] (Chap. 13, [59]). In these bilayers, AA and AC stacking prefer a ferroelectric state with parallel polarization, while AB and AD stacking favor an antiferroelectric state with an antiparallel arrangement of polarization. Tribological interlayer sliding permits reversible and hysteretic transitions between the two states. Consequently, a change in the stacking order allows a change in the electric polarization with high magnitude and rate. This process has advantages for energy harvesting over existing piezoelectric and triboelectric nanogenerators. Note that the interlayer sliding induced polarization change generates an open-circuit voltage two orders of magnitude higher than that of MoS_2-based nanogenerators.

Of particular interest in future developments are stacking orders, e.g., of aw-SbN sheets that realize parallel polarization for all monolayers introducing strong ferroelectricity in an ordered multilayer. Research on reversible spontaneous electronic polarization will continue in the direction of non-volatile compact memory devices and field-effect transistors. Besides the extension of the ferroelectric family to the 2D limit, the co-existence of ferroelectricity and the QSH effect in the same material leads to novel possibilities in the development of topological materials with multiple quantum properties [34].

18.3.5 *Piezoelectric 2D Materials*

The piezoelectric effect refers to the induction of macroscopic polarization by the application of mechanical stress. Because it is possible to generate large strains in most 2D materials, electromechanical effects play a significant role in strain tuned processes such as mechanical sensors and actuators and strain engineering in 2D materials. Low symmetry, as the absence of mirror planes, combined with weak chemical bonding introduce large piezoelectric defects.

We first consider a noncentrosymmetric structure with the monoclinic space group that can be obtained by exfoliation from bulk Te [40] (Chap 11, [32]). For the monoclinic phase of tellurene large thermal expansion and large in-plane piezoelectric coefficients were estimated that are a factor of 23 and 12 greater than those of the widely used MoS_2 monolayer, respectively. Theory predicts giant in-plane polarization with stress and strain coefficients of $e_{22} = -83.9 \times 10^{-10}$ C/m and $d_{22} = -333.1$ pm/V, as well as $e_{22} = -42.6 \times 10^{-10}$ C/m and $d_{22} = -302.2$ pm/V in the zigzag direction under biaxial and uniaxial strain conditions, respectively.

The novel phases of δ-AsN, δ-SbN, and δ-BiN monolayers show robust ferroelectricity and extraordinary piezoelectricity with a relaxed-ion piezoelectric stress coefficient of $e_{11} = 12.4 \times 10^{-10}$ C/m and a strain coefficient of $d_{11} = 43.87$ pm/V for δ-BiN [41]. The spontaneous polarization, ferroelectric switching energy barrier, and piezoelectric properties can be modulated effectively by the application of external strain.

Multiple valence and conduction band valleys characterize the band structures of the group–IVA monochalcogenides (GeS, GeSe, SnS, and SnSe), which make these monolayers suitable for a variety of applications such as piezotronics and ferroelectricity [42] (Chap. 13, [33]). The polar character is introduced by the formation of monolayers because in the bulk, the layers have alternative electrical dipoles. According to first-principles calculations, the puckered GeS, GeSe, SnS, and SnSe monolayers possess piezoelectric coefficients approximately one to two orders of magnitude greater than those of other 2D materials. The large piezoelectric effect results from the unique puckered symmetry and weak chemical bonds, which are more sensitive to applied stresses. These materials will allow the fabrication of ultrasensitive mechanical detectors and efficient energy conversion devices. For example, the aw-SnSe monolayers possess the greatest relaxed-ion piezoelectric coefficients of $e_{11} = 34.9 \times 10^{-10}$ C/m and $d_{11} = 250.6$ pm/V.

While previous calculations and experiments of piezoelectric properties have focused mainly on few selected transition-metal dioxides (TMDOs) and TMDCs, first-principles calculations on an extended number of MX_2 (M = Cr, Mo, W, Ti, Zr, Hf, and Sn; X = O, S, Se, and Te) compounds, revealed that the SnS_2 and $CrTe_2$ monolayers have the largest piezoelectric stress and piezoelectric strain coefficients, which are comparable to well-known conventional piezoelectric bulk materials [43] (Chap. 16, [41]). These piezoelectric coefficients were derived from the slope of the polarization change induced by applying uniform strain. In the group of investigated monolayers 2H $CrTe_2$ monolayers showed the greatest relaxed-ion piezoelectric strain coefficient of $d_{11} = 17.1$ pm/V ($e_{11} = 8.06 \times 10^{-10}$ C/m) together with 2D SnS_2.

Several of the presented monolayers have the potential to be used in future 2D piezoelectric applications, such as transducers, ultrasensitive sensors, nanoscale electromechanical systems, low power electronics, and energy harvesting devices owing to their reduced size, weight, and energy consumption. The main obstacle is the preparation of suitable samples. In addition, theory forecasts the use of piezoelectric nanomaterials in future drug delivery systems and therapeutic approaches [44]. The outstanding piezoelectric effect of noncentrosymmetric 2D structures efficiently generates positive and negative charges under mechanical forces. The effect allows precise drug targeting and ensures high responsiveness by the stimulus, thereby controlling cellular destinies. The abundant active sites facilitate catalytic processes through piezoelectric coupling, offering multimodal synergistic therapeutic strategies.

18.3.6 Thermoelectric 2D Materials

Thermoelectric properties of 2D materials are a topic of enormous interest since monolayers have per se a lower lattice thermal conductivity than bulk materials, favoring thermoelectric performance. The thermoelectric figure of merit *ZT* is the standard parameter to quantify thermoelectric performance. However, optimizing the *ZT* value is a challenging task owing to the tight coupling of the relevant properties such as the Seebeck coefficient, electron conductivity, and electronic and phonon (lattice) thermal conductivity. As a rule, a good thermoelectric material should be a good electric conductor, however, a poor thermal conductor [45] (Chap. 10, [73]).

Several publications have used DFT calculations and Boltzmann transport theory to study the thermoelectric properties of group–IVA monochalcogenides. The calculations revealed that the low lattice thermal conductivities of these monolayers result from resonant and weak bonding, the avoidance of crossing between longitudinal acoustic phonons, and low-lying optical branches [46] (Chap. 13, [51]). In this work, the GeTe monolayer has been identified as the monolayer with one of the largest figures of merit of $ZT = 4.23$, and therefore GeTe monolayers are good candidates for use in room-temperature thermoelectric devices.

Few-layer systems offer new possibilities to optimize thermoelectric properties. For example, $ZrSe_2$ and $HfSe_2$ bilayers are indirect semiconductors that exhibit degenerate conduction bands and stair-like shaped density of states (DOSs) that provide a high power factor for n-type doping [47] (Chap. 14, [71]). A combination of these favorable properties with the low lattice thermal conductivity that originates from the low phonon frequency of acoustic modes, yields a *ZT* value of 1.84 for n-doped 1T-$ZrSe_2$ bilayers and a maximum value of $ZT = 3.83$ for 1T-$HfSe_2$ bilayers at 300 K, indicating superior performance in comparison to their bulk counterparts.

To describe the production of electrical energy from heat by using thermoelectric materials the Seebeck coefficient S and the electrical conductivity σ deliver the thermoelectric power factor ($S^2\sigma$) that together with the thermal conductivity is the main factor behind the thermoelectric performance. Compared with other 2D materials the 2D compounds GeN, SiN, and CAs exhibit exceptionally high-power factors of 3190, 1473, and 774 W m^{-1} K^{-2} at 300 K, indicating an enormous potential for thermoelectric device applications [48] (Chap. 10, [2]). Furthermore, the calculated optical parameters indicate a high-power conversion efficiency of 27.25% for SnBi.

18.3.7 Magnetic 2D Materials

The observation of magnetic properties in 2D materials needs more attention to clarify the origin of the effect of long-range magnetic ordering. Present explanations reach from the induction by intrinsic defects or impurities to the coupling between magnetic dopants and charge carriers.

Like their bulk counterpart, metastable 1T-$CrSe_2$ monolayers can be ferromagnetic and antiferromagnetic [49] (Chap. 16, [35]). Applying a biaxial tensile strain of $> 2\%$ to the $CrSe_2$ monolayer, spin ordering induces transformation from the antiferromagnetic to the ferromagnetic state. This switching process is due to the competition between the ferromagnetic Cr–Se superexchange interaction and antiferromagnetic Cr–Cr direct exchange interaction. At a tensile strain of 4%, the Curie temperature moves from $T_c = 63.7$ K to $T_c = 363.1$ K. At a strain of 10%, the Cr-3*d* and Se-4*p* states become completely spin polarized, and the $CrSe_2$ monolayer is approximately half metallic, a feature highly desirable for applications in spintronics.

The 1T-$CrTe_2$ monolayer has a magnetic moment of 3.05 μ_B per formula unit nearly identical to 3.08 μ_B obtained for the bulk phase. Remarkably, the calculated magnetic moment of the monolayer is out-of-plane, whereas it lies in the plane for two or more layers [50] (Chap. 16, [40]). Tensile biaxial strain induces a transition of the magnetic moment of the monolayer from out-of-plane to in-plane and compressive bilayer strain causes the bilayer magnetic moment to switch from in-plane to out-of-plane. A critical compressive strain of –1% in a bilayer allows the switch between the ferroelectric and antiferroelectric states. The large SOC of Te atoms is responsible for this magnetic anisotropy together with the superexchange coupling between the Cr-3*d* and Te-5*p* orbitals. The 1T-$CrTe_2$ monolayer is ferromagnetic with a Curie temperature of 310 K. Long range magnetic order at finite temperatures in monolayers is the subject of the Mermin-Wagner theorem. This theorem states that magnetic anisotropy prevents the distortion of magnetic orders by thermal fluctuations.

Strong ferromagnetic ordering emerges in VSe_2 monolayers and van der Waals heterostructures grown on highly oriented pyrolytic graphite (HOPG) or MoS_2 substrates by molecular beam epitaxy (MBE), contrary to the paramagnetic behavior of the bulk. Encapsulation of the chemically unstable few-layer sheets has been achieved by a Se capping layer [51] (Chap. 15, [34]). Notably, the ferromagnetic ordering with a large magnetic moment did not disappear above room temperature, suggesting that VSe_2 is useful for applications in van der Waals spintronics, using the spin degree of freedom of electrons rather than their charge.

Both, strain and H functionalization induce in Ta-dichalcogenide monolayers TaX_2 (X = S, Se, and Te) magnetic phase transitions involving paramagnetic, ferromagnetic, and modulated antiferromagnetic states. The TaS_2 ($TaSe_2$) monolayer is paramagnetic up to a strain of 6% [52] (Chap. 15, [70]). At 6% strain, the material becomes ferromagnetic with a magnetic moment of 0.24 μ_B (0.34 μ_B) per Ta atom that increases further by strain, reaching 0.48 μ_B (0.50 μ_B) per Ta atom at 10% strain. Lifting the spin degeneracy generates a magnetic moment that arises from the Ta-5*d* states near the Fermi level. By hydrogenation of the monolayer on one side to TaS_2–1H, the spin up and spin down states disappear from the Fermi level, and the monolayer becomes an indirect semiconductor. The transition from paramagnetic metallicity to semiconducting behavior depends on the electron transfer from hydrogen, filling the empty Ta-5*d* states above the Fermi level.

First-principles calculations provided basic information on the influence of structural defects on the electronic structure and magnetic properties of $HfSe_2$ monolayers [53] (Chap. 14, [103]). Removing a Hf or Se atom and introducing an S atom at the Se defect side induces the transformation of $HfSe_2$ from a semiconductor to a metal. Moreover, defects can induce spin magnetic moments, where the total magnetic moment decreases as the number of Se defect atoms increases. The largest magnetic moment achieved was 2.985 μ_B.

Vacancy defects and substitutional doping allow the generation of magnetism in nonmagnetic monolayers of $ZrSe_2$ that is a nonmagnetic semiconductor that transforms into a magnetic material by generating a single Zr vacancy or a divacancy by a Zr and Se vacancy. The formation of these vacancies is connected with net moments of 1.756 μ_B and 1.295 μ_B, respectively, originating from the Se atoms surrounding the vacancies [54] (Chap. 14, [72]). Moreover, the authors studied the replacement of one Zr atom in $ZrSe_2$ by 3*d* (Cr, Mn), 4*d* (Mo, Tc), and 5*d* (W, Re), transition metals, which can induce magnetism and metallic behavior. The monolayers doped with these atoms carry magnetic moments of 2.482 μ_B, 3.00 μ_B, 2.13 μ_B, 2.999 μ_B, 2.00 μ_B, and 2.999 μ_B, respectively. The magnetic moments are proportional to the number of unpaired electrons in the *d* orbital. While the Cr- and Mo-doped monolayers accept metallic character, the Mn-, Tc-, and Re-doped systems remain (magnetic) semiconductors, and the W-doped $ZrSe_2$ monolayer is a half-metal.

Vacancy defects and doping transform non-magnetic 1T-PdO_2 monolayers to spintronic 2D materials [55]. A single Pd vacancy induces half-metallicity with a large magnetic moment of 3.95 μ_B, originating from the O atoms located around the vacancy site. Interestingly, a single O vacancy preserved the non-magnetic nature and doping by B, N, and F atoms induced either halfmetallic PdO_2 monolayers or diluted magnetic semiconductors.

The Heisenberg model with finite-range exchange interactions predicts a competition between ferromagnetism and dimensionality. Accordingly, it is difficult for 2D materials to maintain magnetism. The introduction of anisotropy enables long-range ferromagnetic order in atomically layered materials and thus reduces the difficulty to maintain magnetism [56]. Nevertheless, problems remain to overcome weak ferromagnetic coupling and low Curie temperatures. This includes besides theory the development of techniques to improve coupling such as layer engineering, strain engineering, and interface engineering.

18.4 2D Devices

18.4.1 Catalysis

The development of 2D catalysts is an important field of research. This includes the development of specific properties of 2D materials and their incorporation to enhance the catalytic activity of electrocatalysts and photocatalysts. With the realization of efficient 2D catalysts that can compete with traditional Pt catalysts this field has found enormous attention. Typical features are high surface area, active edge states, and high mechanical flexibility. The already theoretically proposed and experimentally realized devices are based on a wide range of chemical compositions and 2D bonding arrangements and cover important chemical reaction processes previously studied under 3D conditions.

Electrical-driven water electrolysis is of foremost importance because of its high efficiency and eco friendliness. The fabrication of highly active, stable, and inexpensive electrocatalysts is a critical issue in the development of green energy technology. First-principles calculations elucidated the effect of tension on the hydrogen evolution reaction (HER) for VS_2 monolayers used as electrocatalysts [57] (Chap. 15, [20]). Tuning the applied tension provides access to the optimal reaction conditions, namely, near-zero Gibbs free energy. The tension-induced improvement in the catalytic activity is due to an increase in the free carrier density around the Fermi level.

Engineering the basal plane of the VS_2 monolayer structure by doping with 3*d* transition metal atoms (Ti–Ni) further improved the efficiency of the HER catalyst [58] (Chap. 15, [21]). The reason for the better performance of doped surfaces in comparison to pristine surfaces is charge modulation at S sites. The three transition metals Ti, Mn, and Co are perfect dopants for enhancing the HER process, with $\Delta G_{H^*} \approx 0$ eV, and are even superior to Pt. Besides the type of transition metal, the doping concentration clearly plays a crucial role.

The favorable effects of adsorption of As and Bi atoms on the Sb sites of metallic GeSb on the electrocatalytic performance in pristine and modified structures of graphene-like GeSb monolayers and those modified by substitutional doping were studied by the DFT approach [59] (Chap. 10, [62]). A moderate adsorption energy and Gibbs free energy is achieved by physisorption of hydrogen over the Sb-top site and the lowest Gibbs free energy of $\Delta G_{H^*} = 0.31$ eV is found under substitutional As doping, compared with hydrogen adsorption over the Sb-top site, on pristine GeSb monolayers. A large amount of charge accumulates at the hydrogen atom owing to its significant electronegative nature. The electron density modulation in GeSb produces a HER activity of 72% under As-doped conditions.

Interestingly, $PtSe_2$ transforms from a semimetal in the bulk to an indirect semiconductor in the monolayer [60] (Chap. 17, [85]). Remarkably, the complete phase transition occurs between the trilayer and monolayer, with the disappearance of interlayer interactions as verified by large-scale CVD-based samples. Such an atomically thin 1T-$PtSe_2$ monolayer is an excellent electrocatalyst with a record HER efficiency comparable to traditional Pt catalysts, indicating an enormous enhancement in catalytic efficiency of TMDCs.

Electrochemical hydrogen evolution via water splitting is a highly promising technique for fossil-fuel substitution in the future. A comprehensive review summarizes recent advances of transition-metal sulfides (TMSs) for the HER, oxygen reduction reaction (OER), and overall water splitting [61]. The work covers the strategies and concepts of synthesis, characterization, and catalytic mechanisms. Moreover, it elucidates existing deficiencies, challenges, and future development directions of TMSs in HER, OER, and overall water splitting processes. The authors propose that the OER is the major obstacle because of the sluggish kinetics of the complex reaction process. They suggest the development of highly efficient, low cost, and earth-abundant electrocatalysts to overcome the slow kinetics. Among the various non-noble-metal-based electrocatalysts, the authors consider 2D-TMS materials as potential candidates.

Fixation of nitrogen is a key reaction achieved by the Haber–Bosch process under intense energy consumption. Remarkably, 2D-$TaTe_2$ sheets prepared by exfoliation from the bulk phase promote the ambient electrochemical N_2-reduction reaction (NRR) for ammonia production [62] (Chap. 15, [85]). The main active sites are the edges and exposed Ta atom sites, which dramatically enhance the binding of N atoms and reduce the free energy change of the rate-determining step. This synergistic strategy boosts the reduction of the triple N≡N bonds, which belong to the strongest chemical bonds, leading to a remarkable NH_3 faradic efficiency of 32.2% at a mild overpotential.

Another promising approach is the application of 2D binary compounds, which allow the construction of tailored heterostructures. Boron-phosphide monolayers, for example, satisfy the conditions for water splitting, and biaxial tensile strain further improves the photocatalytic performance. First-principles calculations propose 2D boron phosphide–silicon carbide (BP/SiC) van der Waals heterostructures with their direct band gap and enhanced optical absorption, compared to those of the constituents with indirect gaps, as promising candidates for water splitting at pH = 1–3 [63] (Chap. 7, [94]). The type-II band alignment, with the VBM and CBM belonging to different monolayers, potentially separates the free electrons and holes at the interface.

Photocatalysis by Janus compounds with suitable band edges for catalytic processes and enhanced absorption of visible light based on a direct band gap, is a versatile chemical concept for catalysis. Importantly, the dipole moment generated by the intrinsic electric field along the z direction helps to separate the photoinduced charges into different surfaces. Moreover, the Gibbs free energy decreases under an external electric field, which may enhance the relevant reaction steps. Compared with the parental Al_2S_2 monolayer, the Janus compound Al_2OS exhibits suitable band edges for the HER process and enhanced absorption of visible light based on a direct band gap [64] (Chap. 12, [20]). In addition, the Gibbs free energy decreases from $\Delta G_{H*} = 0.97$ eV to 0.69 eV under an external electric field, which enhances the hydrogen adsorption energy.

Solar water splitting is one of the promising techniques to store energy in chemicals. Partially hydrogenated NiO_2 monolayers, containing nickel oxyhydroxide (NiOOH), provide an efficient catalyst for the OER that is significantly more reactive than bulk NiOOH [65] (Chap. 17, [6]). Various OER mechanisms examined by a DFT study could not explain the outstanding reactivity of this catalyst. Therefore, the authors postulated reaction centers at the intrinsic NiO_2 surface itself and not as usually assumed, only at defect sites and edge locations. Furthermore, the work provides evidence for an inhibiting effect of hydrogen ad-atoms on the NiO_2 monolayer surface.

An interesting theoretical study suggests epitaxial CrO_2 monolayers on ferroelectric $PbTiO_3$ as a system for ferroelectric polarization-based catalysis with switchable surface chemistry [66] (Chap. 6, [11]). The CrO_2–$PbTiO_3$ surface allows direct NO_x decomposition. The ferroelectric polarizations, cycling between positive and negative polarization states, drive the decomposition reaction in a cyclic manner. Negative polarization is the preferred state at low temperatures and raising the temperature to values near the Curie point causes transformation into the paraelectric state or positive polarization. Thus, the surface cycles between positive polarization with strong bonding and rapid dissociation of molecules and negative polarization with weak bonding and easy desorption of reaction products. The concept is an elegant technique to realize the Sabatier principle of adsorbate-surface interaction; strong enough to drive the reaction forward and weak enough to permit the desorption of reaction products.

Few-layer $PtSe_2$ nanosheets (7–12 nm) coated on a glass substrate by selenization of a pre-deposited Pt film exhibit antibacterial activity against *Escherichia coli* under dark and light irradiation conditions according to the established standard procedure for semiconducting photocatalytic materials [67] (Chap. 17, [91]). Under light irradiation the antibacterial activity increased substantially with increasing film thickness, roughness, and surface coverage. Moreover, reactive oxygen species form at the surface, which consists mainly of hydroxyl radicals.

A review describes the development of TMDO and TMDC photocatalysts [68]. The work pays particular attention to emerging strategies improving the ability of light harvesting and photoinduced charge transfer and charge separation to enhance photocatalytic performance. This includes elemental doping, surface functionalization, and heterojunctions of semiconducting and conductive 2D materials with thermodynamically favorable redox reactions. Further improvements are possible by morphological control of the exposed facets during the deposition process. Another strategy is tuning the stoichiometry of the catalyst, e.g., by the implementation of sub-stoichiometric phases. Finally, the authors suggest the use of 2D/2D heterojunctions owing to greater stability and better coupling of the interfaces.

In addition, the review considers distinct kinds of 2D-TMDO and TMDC photocatalysts for HER, organic pollutant degradation, carbon reduction, and microbial disinfection [68]. The work critically discusses several improvement strategies such as elemental doping, surface functionalization, sub-stoichiometry phases, and heterojunction formation (2D/2D). The Janus compound CrSTe provided the highest efficiency.

A recent review summarizes the recent literature on the use of TMDCs as a platform for catalytic CO_2 and N_2 electroreduction in aqueous electrolytes [69]. For the CO_2 reduction reaction (CO_2R) and the nitrogen reduction reaction (NRR) the review describes the progress made in unveiling the reaction mechanisms at different sites and the best strategies for fabricating the most effective catalysts by exploiting the structural and electronic requirements. Due to favorable CO_2 adsorption on the metal edge sites TMDCS outperform state-of-the-art catalysts for CO_2 to CO reduction in ionic liquids. The efficiency of the NRR is based on the resemblance of TMDCs to the active centers of nitrogenase enzymes. Importantly, the same strategy employed to optimize the adsorption of hydrogen (HER) is used to enhance the interaction with CO_2 and N_2. Since the electrochemical nitrogen reduction to NH_3 suffers from slow kinetics it competes with HER, proton adsorption limits CO_2R and NRR and makes it difficult to steer the reaction process toward the desired reaction.

The fabrication of highly active, stable, and inexpensive electrocatalysts are the critical issues in the development of the water splitting technology. Besides the already mentioned TMS compounds, the TMDC catalysts show similar attractive features and may contribute to the realization of suitable 2D materials. Further progress depends on the effort needed for the synthesis of the first catalyst with the necessary quality and performance.

The 2D world allows the construction of a large variety of chemical compositions such as Janus structures and van der Waals nanocomposites to systematically optimize the relevant reaction parameters and tailor the desired properties. Importantly, the realization of the concept requires easy-to-manage stable chemical systems that allow the synthesis of high-quality tailored arrangements of monolayers. The already investigated catalysts demonstrate a wide field of potential applications.

18.4.2 Photovoltaics

A thermophotovoltaic (TPV) cell converts the thermal emission from a high-temperature emitter directly into electric power. As advantage, this method is not restricted to solar energy as photovoltaics but works also with other heat sources such as combustion, nuclear power, or waste heat. The AC-stacking order realized in buckled grey antimony, exhibits a high conversion efficiency in TPV cells [70] (Chap. 8, [70]). The AC stacked bilayer has a narrow band gap in the optimum range for TPV devices of 0.35–0.75 eV. Furthermore, the van der Waals bonded bilayer has a large interlayer distance as a consequence of a much smaller interaction energy than, for example, for Sb AB stacking. Remarkably, the TPV device shows an energy conversion efficiency of $\eta = 31\%$ that is comparable to the efficiencies of other cells. This concept, based on van der Waals bonded few-layer sheets of antimony, seems to be realizable experimentally.

A comprehensive theoretical study investigated the optical and solar energy conversion capabilities of group IVA–VA monolayers [48] (Chap. 10, [2]). The study evaluated as solar parameters the optical band gap, short-circuit current density, open-circuit voltage, and fill factor of 17 stable IVA–VA compounds. The tabulated values revealed an outstanding solar power conversion efficiency of $\eta = 27.25\%$ for binary SnBi monolayers that comes from an unusually high short-circuit current density and low open-circuit voltage. This value seems to be one of the highest conversion efficiencies predicted for 2D compounds.

According to theory shift current photovoltaic devices are potential candidates for cheap, sustainable, and efficient electricity generation [71]. The name derives from the 'shift' of intracell coordinates of an excited electron due to its asymmetrical wave function that drives the current. Importantly, shift current photovoltaic devices may generate photovoltages above the band gap limitation not possible for traditional p-n junction cells according to the Shockley-Queisser efficiency limit. Calculations delivered the solar-generated shift current and efficiencies for 326 different 2D materials. The out-of-plane shift currents of multilayer Cr-based compounds with direct band gaps around ~ 1 eV show great promise for future applications.

The investigation of Janus heterostructures is finding increasing interest, because theory and experiments indicate a potential use as photocatalysts for water splitting and in photovoltaic cells. For example, according to first principles calculations, the SePtTe/InS van der Waals heterojunction has a direct band gap of 1.08 eV (HSE), efficient optical absorption, high carrier mobility, and a staggered type-II band alignment [72]. The optoelectronic properties are tunable by an external electric field and vertical strain. Remarkably, the power conversion efficiency is tunable to $\eta = 24.91\%$ at a flux of absorbed phonons beyond 4.41 mA/cm^2.

18.4.3 *Field-Effect Transistor*

The interest in 2D FETs is based on their ultralow-power operation, high carrier mobility, and high current on/off ratio. These features qualify them as an alternative channel material as Si-based FETs have reached their scaling limit. Despite the high-quality of the native insulator SiO_2 with moderate band gap of (1.1 eV), high-k dielectrics with much larger dielectric constants are gradually replacing SiO_2. Important advantages are the uniform thickness of monolayers and few-layer sheets and their tunable band gap.

Following the fabrication of air-stable selenium nanosheet FETs with a gate length of 5 nm, ab initio quantum transport simulations were employed to simulate a sub-5 nm gate-length double-gate monolayer selenene FET [73] (Chap. 11, [15]). This study showed that selenene has the potential to deliver a channel material that can scale Moore's law down to a gate length of only 3 nm. These devices can meet the 2013 standards for high-performance applications in the armchair and zigzag directions in the 2028 horizon project.

Similarly, in view of the progress made in the synthesis of 2D tellurene sheets, besides their application in FETs, the potential use in ultrascaled complementary metal-oxide semiconductor (CMOS) circuits is discussed owing to the one-dimensional chiral structure and electrical advantages [74] (Chap. 11, [34]). The problem of the low band gap is solved by using few-layer sheets or the 1D nanowire limit with $E_g > 1$ eV. Exact ab initio quantum transport simulations predict performance limits of sub-5 nm for tellurene metal-oxide-semiconductor field-effect transistors (MOSFETs). The authors concluded that tellurene monolayers may offer the possibility of continuing Moore's law to a limit of 3 nm.

Currently, SiO_2 is gradually replaced by high-k insulators such as $HfSe_2$ and $ZrSe_2$ with much larger dielectric constants of approximately $k \approx 15$–20 instead of ~ 4 for SiO_2 [75] (Chap. 14, [106]). Spectroscopic analysis of the electronic band structure of air-stable transistors, fabricated with $HfSe_2$ nanosheets consisting of ≥ 3 monolayers and dielectric encapsulation, substantiated the suitability of these devices for practical applications. The high-k dielectrics show moderate band gaps of 0.9–1.2 eV from multilayer sheets to monolayer thickness. Transmission electron microscopy (TEM) analysis indicated the formation of amorphous native oxides. The native oxide HfO_2 exhibits promising performance by delivering an on/off ratio of

$> 10^6$. The work demonstrates the key attribute needed for technological integration into 2D materials, which requires the property to be intrinsically scalable to the atomically thin limit. The 1T-HfO_2 monolayer with a large band gap of 6.73 eV has in-plane and out-of-plane static dielectric constants of 27.35 and 4.80, respectively. Therefore, it seems to be better suited as high-k gate dielectrics than Al_2O_3 and it is a promising alternative to SiO_2. The study verified the feasibility of scaling down the HfO_2 thickness to achieve a capacitance equivalent oxide thickness (CET) of ~ 1.5 nm from an array of top-gated metal-oxide-graphene FETs using atomic layer deposition (ALD) for deposition of the HfO_2 layer.

Metal-semiconductor contacts play a vital role in the performance of electronic devices such as FETs as crucial components especially in ultracompact electronic devices. First-principles calculations predict a Schottky contact with the Schottky barrier varying between 0.36 and 0.51 eV, depending on the stacking order for the heterostructure NbS_2/BSe, consisting of a metallic NbS_2 monolayer and a semiconducting BSe monolayer [76]. Both, the tunneling probability and the contact tunneling-specific resistivity are low. This finding suggests the applicability of such nano-heterostructures in emerging nanoelectronics such as Schottky diodes, photodetectors, and FETs. High-quality van der Waals contacts were previously described for the 2D metal-semiconductor heterostructure $TaSe_2$/WSe_2 (Chap. 15, [79]). Importantly, TEM measurements revealed a high-quality and clean interface. Moreover, surface potential and electrical transport measurements confirmed a high field-effect mobility and on/off ratio.

An up-to-date review describes the role played by 2D electronics in the post-Moore area [77]. The in-depth roadmap discusses the progress made in digital circuits, analog circuits, heterogeneous integration, sensing circuits, artificial intelligence chips, and quantum chips. Moreover, the extensive review speculates on the future development of 2D electronics and the most accessible and promising pathways for further improvements in the field of advanced electronics.

18.4.4 Batteries

The search for abundant, cheap, and nontoxic ingredients for rechargeable metal-ion batteries inspired theoretical studies on the adsorption and diffusion properties of monovalent Li, Na, and K ions, as well as multivalent Be, Mg, Ca, and Al metal ions on semimetallic 2D 1T-TiS_2 nanosheets [78] (Chap. 14, [17]). The total formation energy for ion adsorption and a Bader charge analysis revealed doubling of the charge storage density for divalent Mg and Ca ions compared with that for monovalent Li, Na, and K ions, whereas Be and Al ions form metallic clusters. The lowest average open-circuit voltage emerged for the adsorption of Mg ions. The activation energy barriers for the diffusion of metal ions decrease from Li to K and from Be to Ca. Divalent Mg and Ca can store a higher power density than Li, and both K and Na exhibit a greater rate capability than Li.

Layered TMDCs, including VS_2, have attracted great interest as alternative anode materials in ion batteries, which are among the most important rechargeable energy storage devices [79] (Chap. 15, [17]). The adsorption energy and diffusion coefficients of lithium and sodium ions for H-type monolayer and T-type bulk VS_2 electrodes were studied by DFT calculations, which revealed a low adsorption energy and high ion diffusion rate. The capacity and the output voltages reach high values for lithium and sodium ion batteries in monolayer VS_2. Note that the theoretically estimated Li and Na diffusion coefficients in bulk VS_2 are five and seven orders of magnitude greater than those reported, for example, for bulk MoS_2.

In future applications of 2D-VS_2 sheets in Li-ion batteries the electronic conductivity will play a significant role, combined with the good cycling stability, large capacity, and ionic conductivity. First-principles calculations indicate that W, Re, and Co are the most effective dopants for further improving the electronic conductivity of transition elements, where a narrow band gap and short V–S bond length contribute to this increase in the electronic conductivity [80] (Chap. 15, [18]).

DFT calculations predict enhanced performance of pre-lithiated MS_2 (M = Mo, W, and V) monolayers used as the anode material [81]. While the adsorption energy of additional Li atoms is not significantly changed, favoring multilayer adsorption, the Li diffusion barrier on pre-lithiated MS_2 is negligible small due to charge accumulation between the diffusing Li ions and the pre-lithiated Li layer. The authors recommend this finding as a general procedure to improve storage capacities and the charge-discharge performance in Li-ion batteries.

A first-principles study investigated the possibility of enhancing the electrochemical performance of lithium-sulfur batteries with hb-PN monolayers [82] (Chap. 9, [8]). Due to the chemical interactions between electrodes and intermediates produced during the charging and discharging process, 2D electrode materials can improve the electrochemical performance by overcoming the shuttle effect and by increasing diffusion rates. During the discharge process sulfur (S_8) and its intermediate lithium polysulfide products (Li_2S_n, n = 1, 2, 4, 6, and 8) with low conductivity and long-chains dissipate in the electrolyte (shuttle effect). The hb-PN monolayer can suppress the shuttle effect and the extremely small diffusion barriers for Li ions and, for example, Li_2S_8 allow a higher battery charge and discharge rate. The chemisorption of the Li atoms in lithium polysulfides at the N atoms of the hb-PN surface effectively suppresses the shuttle effect.

18.4.5 Sensors

Ultrathin crystalline nanosheets of $NbTe_2$, synthesized via liquid-phase exfoliation, provide a low-cost metal-free 2D surface-enhanced Raman scattering (SERS) material for ultrasensitive detection of aromatic molecules and antibiotics. The detection is based on fluorescence quenching and enhancement of charge transfer, contributing to the Raman effect of the analytes [83] (Chap. 15, [61]). The reproducibility was satisfactory, and the ultralow detection limit of 10^{-9} M was comparable to that of metal-based SERS substrates. For example, the detection limits for ciprofloxacin and enrofloxacin were 35.1 and 35.9 ppb, respectively.

The planarity and stability endow chemical vapor-deposited MoO_2 nanosheets with superior uniformity of the Raman signal in the whole area. These nanosheets can serve as novel metal oxide substrates for SERS [84] (Chap. 16, [54]). The concentration limit of 4×10^{-8} M and the enhancement factor of 2.1×10^5 are comparable to the detection limits of traditional noble metals. The enhancement effect mainly originates from the surface plasmon resonance (SPR) effect and becomes more pronounced with decreasing thickness of the nanosheet below 8 nm, which was the thickness used in the experiments. Another reason for the superior performance is the reusability of MoO_2 nanosheets.

Using DFT and nonequilibrium Green's function-based methods, the sensing capabilities of 1T-$HfTe_2$ monolayers for hazardous gases such as CO, CO_2, NO, NO_2, NH_3, and N_2O and for the environmental gases O_2, N_2, H_2, and H_2O were studied [85] (Chap. 14, [110]). Only in the case of NO a significant, and for NO_2 and O_2 a moderate charge transfer between the adsorbed molecule and the 1T-$HfTe_2$ surface was detected that affects other surface properties such as, for example, a semimetal-to-metal transition. Simulations of the *I–V* characteristics of the pristine and covered 1T-$HfTe_2$ surfaces revealed a resistance change due to adsorption. This characterizes the output and performance of the resistance gas sensor. The calculations demonstrated a significant increase in current for NO, an insignificant increase for NO_2 and O_2, and almost no effect for the rest of the examined gases.

A theoretical study compared the response and selectivity of the adsorption behaviors of pristine $HfTe_2$ monolayers and that of a surface with a dispersed single Pt atom adsorbed at the hollow site between the three Te atoms, for the hazardous gases NO_2, NO, CO, SO_2, H_2S, and NH_3 [86] (Chap. 14, [111]). For the pristine surface, long-range interactions dominated for most of the gases except for NO. The introduction of Pt increased the adsorption energies for all six molecules. The contribution of short-range chemical interactions between the adsorbed molecule and the surface is enhanced by the Pt atom.

A strategic review discusses the enhancement of chemiresistive gas sensing for 2D WS_2 and 2D WSe_2 sheets [87]. The fabrication of these sensors with high stability, high sensitivity, and distinct electronic properties is possible for a low prize. Decoration with noble-metal nanoparticles yields heterojunctions or composites, which induce enhanced and faster response, shorter recovery time, reduced operating temperatures, and extremely low detection limits.

A comprehensive article reviews recent developments of sensing gas molecules and volatile organic compounds by TMDCs, MXenes, and black phosphorus [88]. Most 2D sensors consist of a chemiresistive gas sensor, FET, quartz-crystal microbalance, or optical fiber. Their working principle is based on physisorption, chemisorption with charge transfer, surface reaction, mass loading, and change of the refractive index. The overview covers both experimental and theoretical aspects and provides perspectives for the future development of gas sensors with a reliability and durability comparable with commercially available metal-oxide gas sensors.

References

1. Z. Lu, G.P. Neupane, G. Jia, H. Zhao, D. Qi et al., Adv. Funct. Mater. **30**, 2001127 (2020)
2. W. Gao, Z. Zheng, P. Wen, N. Huo, J. Li, Nanophotonics **9**, 2147 (2020)
3. K. Khan, A.K. Tareen, M. Aslam, R. Wang, Y. Zhang et al., J. Mater. Chem. C **8**, 387 (2020)
4. A.D. Bartolomeo, Nanomaterials **10**, 579 (2020)
5. P. Chen, Z. Zhang, X. Duan I, X. Duan II, Chem. Soc. Rev. **47**, 3129 (2018)
6. A. Chaves, J.G. Azadani, H. Alsalman, D.R. da Costa, R. Frisenda et al., NPJ 2D Mater. Appl. **4**, 29 (2020)
7. Z. Li, L. Zhai, Y. Ge, Z. Huang, Z. Shi et al., Mater. Sci. Rev. **0**, nwab142 (2022)
8. R.J.P. Román, F.J.R.C. Costa, A. Zobelli, C. Elias, P. Valvin et al., 2D Mater. **8**, 044001 (2021)
9. K.M. Price, S. Najmaei, C.E. Ekuma, R.A. Burke, M. Dubey et al. **2**, 4085 (2019)
10. T. Kang, J. Park, H. Jung, H. Choi, S.-M. Lee et al., Adv. Mater. **2312747** (2024)
11. Y. Zhang, H.-X. Chen, L. Duan, J.-B. Fan, Thin Solid Films **721**, 138549 (2021)
12. N.K. Nepal, L. Yu, Q. Yan, A. Ruzsinszky, Phys. Rev. Mater. **3**, 073601 (2019)
13. Y. Wang, Y. Li, Z. Chen, Acc. Chem. Res. **53**, 887 (2020)
14. B. Song, L.-M. Yang, WIREs Comp. Molec. Sci. **14**, e1699 (2024)
15. X. Zhang, A. Beyer, Nanoscale **13**, 1443 (2021)
16. G. Wang, H. Hou, Y. Yan, R. Yagatramka, A. Shirsalimian et al., Int. J. Extrem Manuf. **5**, 032002 (2023)
17. L. Kou, Y. Ma, Z. Sun, T. Heine, C. Chen, J. Phys. Chem. Lett. **8**, 1905 (2017)
18. P. Bampoulis, C. Castenmiller, D.J. Klaassen, J. van Mil, Y. Liu et al., Phys. Rev. Lett. **130**, 196401 (2023)
19. A. Zhao, B. Wang, APL Mater. **8**, 030701 (2020)
20. L. Xian, A.P. Paz, E. Bianco, P.M. Ajayan, A. Rubio, 2D Materials **4**, 041003 (2017)
21. S. Khatun, A. Banerjee, A.J. Pal, Nanoscale **11**, 3591 (2019)
22. C.-H. Hsu, Z.-Q. Huang, F.-C, Chuang, C.-C. Kuo, Y.-T. Liu et al., New J. Phys. **17**, 025005 (2015)
23. M.-Y. Liu, Q.-Y. Chen, C. Cao, Y. He, Phys. Chem. Chem. Phys. **21**, 2899 (2019)
24. X. Wang, G. Bian, C. Xu, P. Wang, H. Hu et al., Nanotechnology **28**, 395706 (2017)
25. F.-C. Chuang, L.-Z. Yao, Z.-Q. Huang, Y.-T. Liu, C.-H. Hsu et al., Nano Lett. **14**, 2505 (2014)
26. S.-D. Guo, Y.-T. Zhu, J. Phys. Condens. Matter **34**, 235501 (2022)
27. B. Weber, M.S. Fuhrer, X.-L. Sheng, S.A. Yang, R. Thomale et al., J. Phys. Mater. **7**, 022501 (2024)
28. L. Yan, P.-F. Liu, H. Li, Y. Tang, J. He et al., NPJ Comput. Mater. **6**, 94 (2020)
29. D. Fan, C. Yang, S. Lu, X. Hu, Appl. Phys. Lett. **117**, 013103 (2020)
30. C. Heil, S. Poncé, H. Lambert, M. Schlipf, E.R. Margine et al., Phys. Rev. Lett. **119**, 047003 (2017)
31. S.C. Ganguli, V. Vaňo, S. Kezilebieke, J.L. Lado, P. Liljeroth, Nano Lett. **22**, 1845 (2022)
32. H. Wang, X. Huang, J. Lin, J. Cui, Y. Chen et al., Nat. Commun. **18**, 394 (2017)
33. D. Wines, K. ChoOudhary, A.J. Biacchi, K.F. Garrity, F. Tavazza, Nano Lett. **23**, 969 (2023)
34. R.M. Ur, H. Chenqiang, L. Yunhao, Chin. Phys. B **29**, 057304 (2020)
35. C. Liu, W. Wan, J. Ma, W. Guo, Y. Yao, Nanoscale **10**, 7984 (2018)
36. P. Chen, X.-J. Zhang, B.-G. Liu, A.C.S. Appl, Nano Mater. **2**, 58 (2019)
37. A. Priydarshi, Y.S. Chauhan, S. Bhowmick, A. Agarwal, J. Appl. Phys. **131**, 034101 (2022)
38. W. Wan, C. Liu, W. Xiao, Y. Yao, Appl. Phys. Lett. **111**, 132904 (2017)
39. B. Xu, J. Deng, X. Ding, J. Sun, J.Z. Liu, NPJ Comput. Mater. **8**, 47 (2022)
40. P.K. Sachdeva, S. Gupta, C. Bera, Nanoscale Adv. **3**, 3279 (2021)
41. Y. Zhang, T. Ouyang, C. He, J. Li, C. Tang, Nanoscale **15**, 6363 (2023)
42. R. Fei, W. Li, J. Li, L. Yang, Appl. Phys. Lett. **107**, 173104 (2015)
43. M.M. Alyörük, Y. Aierken, D. Çakir, F.M. Peeters, C. Sevik, J. Phys. Chem. C **119**, 23231 (2015)
44. A. Li, J. Yang, Y. He, J. Wen, X. Jiang, Nanoscale Horiz. **9**, 365 (2024)

45. A. Wu, Y. Zhang, Y. Xia, L. Peng, H. Zhu, arXiv: 2206.02188v1 (2022)
46. D. Zhang, S. Hu, Y. Sun, X. Liu, H. Wang et al., ES Energy Environ. **10**, 59 (2020)
47. P. Yan, G.-Y. Gao, G.-Q. Ding, D. Qin, RSC Adv. **9**, 12394 (2019)
48. A.K. Bhojani, H.L. Kagdada, D.K. Singh, J. Appl. Phys. **135**, 095106 (2024)
49. H.Y. Lv, W.J. Lu, D.F. Shao, Y. Liu, Y.P. Sun, Phys. Rev. B **92**, 214419 (2015)
50. Y. Liu, S. Kwon, G.J. de Coster, R.K. Lake, M.R. Neupane, Phys. Rev. Mater. **6**, 084004 (2022)
51. M. Bonilla, S. Kolekar, Y. Ma, H.C. Diaz, V. Kalappattil et al., Nat. Nanotech. **13**, 289 (2018)
52. P. Manchanda, V. Sharma, H. Yu, D.J. Sellmyer, R. Skomski, Appl. Phys. Lett. **107**, 032402 (2015)
53. X. Zhao, Y. Gao, H. Zhang, H. Wang, T. Wang et al., J. Magn. Magn. Mater. **479**, 192 (2019)
54. X.-Y. Gao, J.-M, Zhang, A. Ali, X.-M. Wei, Y.-H. Huang, Thin Solid Films **732**, 138790 (2021)
55. D.M. Hoat, V.V. On, P.V. Huan, J. Guerrero-Sanchez, RSC Adv. **14**, 7241 (2024)
56. W. He, L. Kong, W. Zhao, P. Yu, Coatings **12**, 122 (2022)
57. H. Pan, Nanoscale Res. Lett. **11**, 113 (2016)
58. L. Wang, W. Gao, X. Chen, Y. Liu, A.H. Qureshi et al., J. Phys. Chem. C **125**, 12004 (2021)
59. M.H. Dalsaniya, T.K. Gajaria, N.N. Som, P.K. Jha, Phys. Chem. Chem. Phys. **22**, 19823 (2020)
60. J. Shi, Y. Huan, M. Hong, R. Xu, P. Yang et al., ACS Nano **13**, 8442 (2019)
61. H. Su, J. Jiang, S. Song, B. An, N. Li et al., Chin. J. Catal. **44**, 7 (2023)
62. Z. Zhao, J. Park, C. Choi, S. Hong, X. Hui et al., Innovation **3**, 100190 (2022)
63. T.-N. Do, M. Idrees, B. Amin, N.N. Hieu, H.V. Phuc et al., RSC Adv. **10**, 32027 (2020)
64. Z. Haman, N. Khossossi, M. Kibbou, I. Bouziani, D. Singh et al., Appl. Surf. Sci. **589**, 152997 (2022)
65. J. Zaffran, M.C. Toroker, ChemElectroChem **4**, 2764 (2017)
66. A. Kakekhani, S. Ismail-Beigi, ACS Catal. **5**, 4537 (2015)
67. N. Todorova, N. Minev, V. Marinova, K. Buchkov, V. Videva et al., Appl. Surf. Sci. **611**, 155534 (2023)
68. F. Haque, T. Daeneke, K. Kalantar-Zadeh, J.Z. Ou, Nano-Micro Lett. **10**, 23 (2018)
69. G. Giuffredi, T. Asset, Y. Liu, P. Atanassov, F.D. Fonzo, ACS Mater. Au **1**, 6 (2021)
70. M. Xie, S. Zhang, B. Cai, Y. Gu, X. Liu et al., Nano Energy **38**, 561 (2017)
71. M.O. Sauer, A. Taghizadeh, U. Petralanda, M. Ovesen, K. Sommer et al., NPJ Comp. Mater. **9**, 35 (2023)
72. J. Ren, C. Zhang, C. He, T. Ouyang, J. Li et al., J. Appl. Phys. **128**, 043103 (2020)
73. Q. Li, X. Tan, Y. Yang, X. Xiong, T. Zhang et al., Molecules **28**, 5390 (2023)
74. G. Qiu, A. Charnas, C. Niu, Y. Wang, W. Wu et al., NPJ 2D Mater. Appl. **6**, 17 (2022)
75. M.J. Mieczko, C. Zhang, H.R. Lee, H.-H. Kuo, B. Magyari-Köpe et al., Sci. Adv. **3**, e1700481 (2017)
76. S.T. Nguyen, T.T.T. Huong, N.X. Ca, C.Q. Nguyen, Nanoscale Adv. **6**, 1565 (2024)
77. A. Liu, X. Zhang, Z. Liu, Y. Li, X. Peng et al., Nano-Micro Lett. **16**, 119 (2024)
78. A. Samad, A. Shafique, Y.-H. Shin, Nanotechnology **28**, 175401 (2017)
79. W. Wang, Z. Sun, W. Zhang, Q. Fan, Q. Sun et al., RSC Adv. **6**, 54874 (2016)
80. Y. Cui, W. Fan, X. Liu, J. Ren, Y. Gao, Comput. Mater. Sci. **200**, 110767 (2021)
81. T. Liu, Z. Jin, D.-X. Liu, C. Du, L. Wang et al., Sci. Rep. **10**, 6897 (2020)
82. X. Zheng, S. Lin, D. Kong, Y. Wei, K. Pang et al., Adv. Theory Simul. **5**, 2100305 (2021)
83. K. Wang, Z. Guo, Y. Li, Y. Guo, H. Liu et al., ACS Appl. Nano Mater. **3**, 11363 (2020)
84. H. Wu, X. Zhou, J. Li, X. Li, B. Li et al., Small 1802276 (2018)
85. D. Chakraborty, P. Johari, A.C.S. Appl, Nano Mater. **3**, 5160 (2020)
86. J. Hu, S. Zhai, Q. Zhang, H. Cui, X. Jiang, Appl. Surf. Sci. **605**, 154572 (2022)
87. A. Kumar, A. Mirzaei, M.H. Lee, Z. Ghahremani, T.-U. Kim et al., J. Mater. Chem. A **12**, 3771 (2024)
88. J. Cao, Q. Chen, X. Wang, Q. Zhang, H.-D. Yu et al., Research **2021**, 9863038 (2021)

Index

P. Hess, *Bonding, Structure, and Performance of Two-Dimensional Materials*, NanoScience and Technology, https://doi.org/10.1007/978-3-031-74161-6

Q

R

S

W

X

Y

GPSR Compliance

The European Union's (EU) General Product Safety Regulation (GPSR) is a set of rules that requires consumer products to be safe and our obligations to ensure this.

If you have any concerns about our products, you can contact us on ProductSafety@springernature.com

In case Publisher is established outside the EU, the EU authorized representative is:

Springer Nature Customer Service Center GmbH
Europaplatz 3
69115 Heidelberg, Germany

Batch number: 10726056

Printed by Printforce, the Netherlands